Schmelzer (Ed.)
Glass

Also of Interest

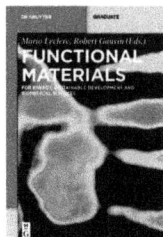

Functional Materials – For Energy,
Sustainable Development and Biomedical Sciences
Mario Leclerc, Robert Gauvin, 2014
ISBN 978-3-11-030781-8, e-ISBN 978-3-11-030782-5

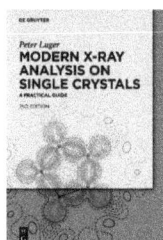

Modern X-Ray Analysis on Single Crystals – A practical Guide
Peter Luger, 2014
ISBN 978-3-11-030823-5, e-ISBN 978-3-11-030828-0,
Set-ISBN 978-3-11-030829-7

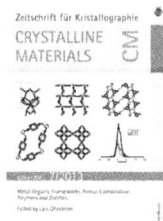

Crystalline Materials
Mathias Wickleder (Editor-in-Chief)
ISSN 2197-4578

Journal of the Mechanical Behavior of Materials
Elias C. Aifantis (Editor-in-Chief)
ISSN 2191-0243

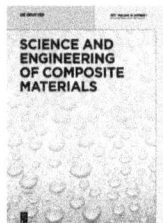

Science and Engineering of Composite Materials
Suong V. Hoa (Editor-in-Chief)
ISSN 2191-0359

Glass

Selected Properties and Crystallization

Edited by
Jürn W.P. Schmelzer

DE GRUYTER

Editor

Dr. rer. nat. habil. Jürn W.P. Schmelzer
University of Rostock
Institute of Physics
Wismarsche Str. 43–45
18057 Rostock
Germany

ISBN 978-3-11-055565-3
e-ISBN 978-3-11-029858-1
Set-ISBN 978-3-11-029859-8

Library of Congress Cataloging-in-Publication Data
A CIP catalog record for this book has been applied for at the Library of Congress.

Bibliographic information published by the Deutsche Nationalbibliothek
The Deutsche Nationalbibliothek lists this publication in the Deutsche Nationalbibliografie;
detailed bibliographic data are available in the Internet at http://dnb.dnb.de.

Cover image: The cover shows crystals of lithium meta-silicate on the surface of samples of

$80Li_2OSiO_2 \cdot 20CaOSiO_2$-glassforming melts formed during its cooling to a glass. The picture was
supplied by Prof. Vladimir M. Fokin (St. Petersburg, Russia & São Carlos, Brazil)
Typesetting: PTP-Berlin, Protago-TeX-Production GmbH, Berlin
Printing and binding: CPI buch bücher.de GmbH, Birkach
♾Printed on acid-free paper
Printed in Germany

www.degruyter.com

Foreword

The present monograph offers an overview by a team of renowned experts on glass science about established facts and open problems in the experimental and theoretical analysis of selected properties and crystallization of glasses.

The book is a continuation of two well-known monographs on glass science authored by the current editor, Jürn W.P. Schmelzer, in cooperation with Ivan S. Gutzow (I.S. Gutzow & J.W.P. Schmelzer, *The Vitreous State: Thermodynamics, Structure, Rheology, and Crystallization*, Springer, 1995; 2013 and J.W.P. Schmelzer & I.S. Gutzow, *Glasses and the Glass Transition*, Wiley-VCH, 2011). These two previous books describe general features of glasses, the glass transition and phase formation in glass-forming melts, their experimental and theoretical analysis. Additionally, in the second book, Oleg Mazurin and Alexander Priven present different theoretical approaches for the prediction of glass properties and data collection methods.

This successful cooperation among experienced glass experts, which started in the second of the mentioned books, is extended in this monograph to include a variety of problems in the analysis of glass properties and crystallization. In this new book, the general features of glasses described in the two previous monographs are exemplified for different types of glass-forming silicate, metallic and polymer systems. In addition, the wide field of phase formation processes and their effect on glass properties are illustrated by means of several theoretically and experimentally relevant examples. The book contains overviews on technologically important classes of glasses, their treatment to achieve desired properties, theoretical approaches for the description of structure-property relationships, and new concepts in the theoretical treatment of crystallization in glass-forming systems. The book contains overviews about the state of the art and about specific features for the analysis and application of important classes of glass-forming systems, and describes new developments in theoretical interpretation by well-known glass scientists. Thus, the book offers comprehensive and abundant information that is difficult to come by or has not yet been made public.

Six chapters are authored by excellent representatives of the Soviet/Russian school of glass science: Leko (crystallization kinetics of silica glass-forming systems); Pevzner & Tarakanov (origin of bubble formation in silica melts and methods of their removal); Vedishcheva & A.C. Wright, UK (structure property relationships and the chemical structure of oxide glasses); Polyakova (main phases of silica and the relation of some of their properties with the electronic structure of Si-Si and Si-O bonding); Fokin, Karamanov, Abyzov et al. (stress induced phase selection and bubble formation, their technological significance for the development of different types of glass-ceramics, and the understanding and control of the sinter crystallization process); and Baidakov (testing the classical nucleation theory by molecular dynamics simulations of crystallization of a Lennard-Jones model system). Most of these chapters are devoted to a comprehensive description of the authors' own experimental research and their theoretical interpretation of the main results, which have hardly any analogues in the literature of glass science. In a number of cases, these results have been published only in Russian, which considerably restricts access to them. This work is supplemented by a chapter written by representatives of another outstanding school of glass science that dates back to one of the founding groups (Stranski, Kaischew and coworkers) of the theory of first-order phases transitions, which has been headed for decades by Ivan Gutzow (I. Gutzow et al. (analysis of main physical properties of the crystalline and amorphous modifications of silica with main emphasis on their thermodynamic characterization, in particular the solubility of the different silica modifications in water and aqueous solutions)). This chapter offers an overview of the state of the art in the analysis of one of the most important and abundant glass systems, as well as new insights for a better understanding of sintering processes.

This spectrum of topics is further expanded by contributions on the theory of crystallization of glass-forming liquids. These contributions not only test the basic assumptions of the classical nucleation theory for a model system (Baidakov) but also cover aspects that go beyond the standard classical approaches (Johari & Schmelzer). They provide an overview of new perspectives opened up by ultrafast thermal processing and nano-calorimetry developed by C. Schick and coworkers (ultra-fast thermal processing and nano-calorimetry at heating and cooling rates up to 1 MK/s), and an overview of the initial stages of crystallization by G. Wilde (early stages of crystal formation in glass-forming metallic alloys). In addition to well-established results, a number of open problems are set forth that require further in-depth analysis (e.g., influence of thermal history on crystal nucleation and growth in polymers (Schick et al.), kinetic stability versus fragility as a measure of the glass-forming ability of metallic alloys, higher nucleation densities than those proposed by classical theory in some metallic systems, effect of rapid quenching and plastic deformation on crystal nucleation (Wilde), origin of the high sensitivity of silica glass crystallization on a variety of factors and the relation between the rate of crystallization and viscosity (Leko), confirmation of basic assumptions of classical nucleation theory for model Lennard-Jones systems (Baidakov), and the origin of deviations from classical theory in highly viscous glass-forming liquids (Johari & Schmelzer). A detailed overview of the content of the various chapters is given by the editor in the preface.

I congratulate Jürn W.P. Schmelzer for putting together this third monograph, which deals with several undeniably relevant aspects of the properties and crystallization of glass. I strongly recommend it as a must-read for teachers, glass scientists, graduate students and anyone interested and/or involved in the research and development of vitreous materials.

April 2014

Edgar Dutra Zanotto
Director of the Center for Research, Technology and Education in
Vitreous Materials (www.certev.ufscar.br)
Federal University of São Carlos, São Carlos, Brazil

Contents

Foreword —— v

Preface —— xv

List of contributing authors —— xxi

Christoph Schick, Evgeny Zhuravlev, René Androsch, Andreas Wurm,
and Jürn W.P. Schmelzer
1   Influence of Thermal Prehistory on Crystal Nucleation and Growth in
      Polymers —— 1
1.1   Introduction —— 1
1.2   State of the Art —— 2
1.2.1   Dependence of the Properties of Glass-forming Melts on Melt
          History —— 2
1.2.2   Polymer Crystallization —— 6
1.2.3   Differential Fast Scanning Calorimetry —— 9
1.3   Experimental —— 14
1.3.1   Samples —— 14
1.3.2   Suppression of Homogeneous Nucleation at Fast Cooling —— 16
1.3.3   Non-isothermal Ordering Kinetics —— 28
1.3.4   Isothermal Ordering Kinetics —— 36
1.3.5   Identification of Different Nuclei Populations —— 48
1.3.6   Enthalpy Relaxation and Crystal Nucleation in the Glassy State —— 52
1.3.7   Summary of Experimental Results and Conclusions —— 72
1.4   Illumination of the Nucleation and Growth Mechanism —— 74
1.4.1   Low-temperature Endotherms and Homogeneous Nucleation —— 74
1.4.2   Some Brief Theoretical Considerations —— 78
1.5   Conclusions and Outlook —— 80

Gerhard Wilde
2   Early Stages of Crystal Formation in Glass-forming Metallic Alloys —— 95
2.1   Introduction —— 95
2.2   Marginal Glass-formers —— 98
2.2.1   Nucleation versus Growth Control —— 98
2.2.2   Processing Pathway Modifications —— 101
2.2.3   Nucleation and Growth Kinetics —— 105
2.2.4   Characterization of the Amorphous Phase —— 109
2.2.5   Nanocrystal Formation at Temperatures Well Below T_g —— 115
2.3   Deformation-induced Nanocrystal Formation —— 124
2.4   Bulk Metallic Glasses —— 127
2.5   Conclusions and Hypotheses —— 131

Ivan Gutzow, Radost Pascova, Nikolai Jordanov, Stoyan Gutzov, Ivan Penkov,
Irena Markovska, Jürn W.P. Schmelzer, and Frank-Peter Ludwig

3   **Crystalline and Amorphous Modifications of Silica: Structure,**
        Thermodynamic Properties, Solubility, and Synthesis —— 137
3.1   Introduction —— 137
3.2   Properties of Silica Modifications: Literature Search —— 140
3.2.1   Classical SiO_2-literature —— 141
3.2.2   Original Literature Sources on the Different Silica Modifications —— 141
3.2.3   Internet Search —— 142
3.3   Phase Diagram of SiO_2 —— 142
3.3.1   Fenner's Classical Diagram —— 142
3.3.2   Flörke's Diagram —— 143
3.3.3   Contemporary ($p - T$)-phase Diagrams of SiO_2 —— 144
3.4   Modifications of SiO_2 and Their Synthesis —— 148
3.4.1   Mineralogical Characteristics of the SiO_2-modifications —— 148
3.4.2   Synthesis of Quartz —— 148
3.4.3   Synthesis and Stabilization of β-cristobalite —— 151
3.4.4   Synthesis of Keatite: Classical Aspects —— 159
3.4.5   Synthesis of Coesite —— 160
3.4.6   Stishovite: Synthesis and Thermal Stability —— 160
3.4.7   Synthesis of Amorphous Modifications of Silica —— 163
3.5   Structure and Thermodynamic Properties of the
            SiO_2-modifications —— 164
3.6   Solubility of the Different SiO_2-modifications —— 170
3.6.1   General Thermodynamic Dependencies —— 170
3.6.2   Solubility Diagram of SiO_2. Ostwald's Rule of Stages —— 175
3.6.3   Solubility of SiO_2: Size Effects —— 181
3.6.4   Different SiO_2-modifications at Hydrothermal Conditions:
            Technological Aspects —— 183
3.7   Resources of the Silica Modifications —— 186
3.7.1   Mineral Resources of Quartz —— 186
3.7.2   Plant Resources of Silica —— 187
3.7.3   Industrial Waste as Sources of Silica —— 188
3.7.4   Coesite and Stishovite as Impactite Remnants —— 188
3.8   Some Particularly Interesting Properties of Silica —— 189
3.9   General Discussion: Technical Perspectives —— 190

Irina G. Polyakova

4   **The Main Silica Phases and Some of Their Properties —— 197**
4.1   Introduction —— 197
4.2   Specific Properties of Silica Resulting from the Electronic Structure of
            Silicon —— 198

4.2.1   Specific Properties of Silica Compounds and Differences as Compared
           to Chemical Analogs: Silicon and Carbon — **198**
4.2.2   Electron Structure of the Silicon Atom and its Interaction with
           Oxygen — **201**
4.2.3   Consequences of π-Bonding in Silica — **202**
4.2.4   Increase in Silicon Coordination Number as a Result of
           s-p-d-hybridization — **203**
4.2.5   Implication of *s-p-d*-hybridization for Chemical Reactions and Physical
           Transformations of Silica — **205**
4.3   Phases of Silica and Their Properties — **207**
4.3.1   Dense Octahedral Silicas: High Pressure Phases — **209**
4.3.2   Clathrasils: Friable Silica Phases — **210**
4.3.3   Exception: Fibrous Silica — **211**
4.3.4   Proper Silicas — **211**
4.3.5   Main Crystalline Tetrahedral Silicas — **213**
4.3.6   Amorphous Silica — **223**
4.3.7   Polyamorphism — **225**
4.4   Quartz and Some of Its Properties — **228**
4.4.1   Enantiomorphism of Quartz — **228**
4.4.2   Twins (Zwillinge) in Quartz — **229**
4.4.3   Anisotropy of Quartz — **232**
4.4.4   Thermal Expansion of Quartz — **233**
4.4.5   High-Low or $(\alpha - \beta)$-Transformation in Quartz — **241**
4.4.6   Pressure-induced Amorphization of Crystalline Silica — **245**
4.5   Hydrothermal Synthesis of Quartz — **245**
4.5.1   Brief History — **246**
4.5.2   Temperature Drop Method — **247**
4.5.3   Main Problems of Hydrothermal Synthesis of Quartz — **250**
4.6   Concluding Remarks — **261**
4.7   Appendix: The Crystal Skulls — **261**

Natalia M. Vedishcheva and Adrian C. Wright
5   **Chemical Structure of Oxide Glasses: A Concept for Establishing
           Structure–Property Relationships — 269**
5.1   Introduction — **269**
5.2   Structural Models — **270**
5.3   Thermodynamic Approach — **274**
5.4   Concept of Chemical Structure — **277**
5.5   Short-range Order — **281**
5.5.1 $Na_2O–B_2O_3$ Glasses — **281**
5.5.2 $Li_2O–B_2O_3$ Glasses and Melts — **283**
5.5.3 $Na_2O–SiO_2$ Glasses — **287**
5.5.4 $Na_2O–B_2O_3–SiO_2$ Glasses — **289**

5.6   Intermediate-Range Order —— **289**

5.7   Structure–Property Relationships —— **293**

5.8   Summary and Conclusions —— **296**

Boris Z. Pevzner and Sergey V. Tarakanov

6   **Bubbles in Silica Melts: Formation, Evolution, and Methods of Removal —— 301**

Part I: Experimental Data and Basic Mechanisms —— **301**

6.1   Introduction —— **301**

6.2   Sources of Bubbles in Silica Melt and Glass —— **302**

6.2.1   Brief Account of the Technology of Silica Glass Production —— **302**

6.2.2   Raw Materials as a Source of Bubbles —— **303**

6.2.3   Furnace Atmosphere as a Source of Bubbles —— **305**

6.2.4   Interaction of Heaters and Form-shaping Equipment with the Melt as Source of Bubbles —— **308**

6.2.5   Concentrations of Impurities, Including Dissolved Gases, in Commercial Silica Glasses —— **308**

6.2.6   Experimental Study of Formation and Evolution of Bubbles in Silica Melts —— **309**

6.3   Physico-chemical Properties of Silica Melts Influencing the Formation and Evolution of Gas Bubbles —— **312**

6.3.1   Surface Tension —— **312**

6.3.2   Density —— **312**

6.3.3   Viscosity —— **313**

6.3.4   Solubility and Diffusion of Gases —— **315**

6.4   Summary to Part I —— **324**

Part II: Theoretical Analysis and Computer Simulation of the Process —— **325**

6.5   Introduction to Part II —— **325**

6.5.1   Main Stages of Fusion of Powdered Silica under Heating and Evolution of Bubble Structure —— **325**

6.5.2   Selection of Parameters for the Temperature Dependence Equations that describe the Properties of the Silica Melt Affecting the Kinetics of the Process —— **326**

6.6   Micro-rheological Model and Computer Simulation of the Process —— **327**

6.6.1   The Micro-rheological Model of Powder Sintering and Structuring of a Porous Body —— **328**

6.6.2   Influence of Some Technological Factors on Formation of Bubble Structure under Heating of Powdered Silica Glass: Computer Simulation of the Process —— **335**

6.7   Summary to Part II —— **343**

Part III: Mathematical Modeling and Computer Simulation of the Behavior of Gas-Filled Bubbles in Silica Melts —— **345**

6.8   Introduction —— **345**

6.9   Behavior of Isolated Bubbles —— **347**

6.10   Behavior of Solitary Gas-filled Bubbles under Mass Exchange with the Melt —— **348**

6.11   Two-phase Approach to the Description of Mono-disperse Ensembles of Bubbles —— **351**

6.12   Two-phase Approach to the Description of Poly-disperse Ensembles of Bubbles —— **356**

6.13   Diffusion of the Dissolved Gas in the Melt —— **360**

6.14   Relative Motion of Bubbles in the Melt: Modification of the Mathematical Model —— **364**

6.15   Flow of the Melt Governed by the Motion of the Bubbles: Complete System of Equations for Modeling of the Behavior of Gas-filled Bubble Ensembles in the Melt —— **369**

6.16   Summary to Part III —— **372**

Victor K. Leko

7   **Regularities and Peculiarities in the Crystallization Kinetics of Silica Glass —— 377**

7.1   Introduction —— **377**

7.2   Literature Review —— **381**

7.3   Development of Experimental Techniques —— **391**

7.4   Basic Phenomenological Features of the Crystallization Processes —— **394**

7.5   Influence of the Degree of Silica Reduction —— **398**

7.6   Influence of Concentration of "Structural Water" —— **402**

7.7   Influence of the Degree of Fusion Penetration of Quartz or Cristobalite Particles on Crystallization of Quartz Glasses —— **405**

7.8   Influence of Surface Contamination on Crystallization Kinetics —— **408**

7.9   Influence of the Composition of the Gas Medium on Crystallization of Quartz Glass —— **411**

7.9.1   Introductory Comments —— **411**

7.9.2   On Crystallization in Dry Gas Media —— **411**

7.9.3   Experiments on Crystallization in an Atmosphere Containing Water Vapor —— **413**

7.9.4   Crystallization of Quartz Glass in the Atmosphere of Gases in Equilibrium with the Melt —— **414**

7.10   Influence of the Drawing Process on the Crystallization Kinetics of Tubes of Quartz Glasses —— **417**

7.11   Summary of Results and Discussion —— **422**

7.11.1   Introductory Remarks —— **422**

7.11.2   Influence of Surface Reactions on Crystallization —— **423**
7.11.3   Relation Between Crystallization Rate and Viscosity —— **427**
7.12   Conclusions —— **435**

Vladimir M. Fokin, Alexander Karamanov, Alexander S. Abyzov, Jürn W.P. Schmelzer, and Edgar D. Zanotto

8   Stress-induced Pore Formation and Phase Selection in a Crystallizing Stretched Glass —— 441
8.1   Introduction —— **441**
8.2   Stress Induced Pore Formation and Phase Selection in a Crystallizing Stretched Glass of Regular Shape —— **443**
8.2.1   The Model —— **443**
8.2.2   Experiments —— **445**
8.2.3   Theoretical Interpretation: Classical Nucleation Theory —— **452**
8.2.4   Theoretical Interpretation: Generalized Gibbs Approach —— **460**
8.3   Sintered Diopside-albite Glass-ceramics Forming Crystallization-induced Porosity —— **467**
8.3.1   Introduction —— **467**
8.3.2   Experimental —— **468**
8.3.3   Results and Discussion —— **470**

Vladimir G. Baidakov

9   Crystallization of Undercooled Liquids: Results of Molecular Dynamics Simulations —— 481
9.1   Introduction —— **481**
9.2   Thermodynamics and Kinetics of Crystal Formation —— **484**
9.3   Description of the Systems under Investigation in the Present Study —— **487**
9.3.1   Models —— **487**
9.3.2   Phase Diagram —— **488**
9.4   Methods of Modeling of Spontaneous Crystallization —— **489**
9.4.1   Mean Life-time Method —— **489**
9.4.2   Mean First-passage Time Method —— **493**
9.4.3   Transition Interface Sampling —— **496**
9.5   Temperature Dependence of the Interfacial Free Energy Density Crystal-liquid for Planar Interfaces —— **498**
9.5.1   Triple Point —— **498**
9.5.2   Melting Line —— **499**
9.6   Kinetics of Crystallization in a cLJ-system —— **503**
9.6.1   Crystallization Parameters —— **503**
9.6.2   Nucleation Rate —— **507**
9.6.3   Comparison of Homogeneous Nucleation Theory with Computer Simulation —— **508**

9.6.4   Nucleation in the Region Below the Endpoint of the Melting
         Line — **509**
9.7   Kinetics of Crystallization in the mLJ-system and Free Energy of the
         Clusters of the Crystalline State — **512**
9.7.1   Pressure Dependence of the Nucleation Rate — **512**
9.7.2   Temperature Dependence of the Nucleation Rate — **513**
9.8   Discussion and Conclusions — **517**

Gyan P. Johari and Jürn W.P. Schmelzer

**10   Crystal Nucleation and Growth in Glass-forming Systems: Some New
         Results and Open Problems — 521**
10.1   Introduction — **522**
10.2   Consequences of Stochastic Structural Fluctuations in Ultraviscous
         Melts — **527**
10.2.1   Structure Fluctuations, Nucleation and Distribution of Relaxation
         Times — **527**
10.2.2   Structure Fluctuations and the Notion of Disordered Cluster
         Formation — **528**
10.3   A Case Study: Crystallization Kinetics of a Typical Metal Alloy
         Melt — **535**
10.3.1   General Considerations — **535**
10.3.2   One Experimental Example — **537**
10.3.3   Theoretical Interpretation in Terms of the KJMA-approach — **540**
10.3.4   Crystallization on Rate Heating — **543**
10.3.5   Differences Between Isothermal and Rate-heating
         Crystallization — **546**
10.3.6   Origin of the Second Peak for Crystallization on Rate-heating — **548**
10.4   Thermal Effects of Crystallization on Its Kinetics — **550**
10.4.1   General Remarks — **550**
10.4.2   Rayleigh–Bénard Convection Effects — **551**
10.4.3   Marangoni or Thermo-capillarity Convection Effect — **553**
10.5   Classical and Generalized Gibbs' Approaches to Cluster Formation and
         Growth — **554**
10.5.1   Basic Ideas — **554**
10.5.2   Application to Nucleation — **556**
10.5.3   Application to Cluster Growth Processes — **562**
10.5.4   Thermodynamics versus Kinetics: Ridge Crossing — **563**
10.6   Specific Interfacial Energy and the Skapski–Turnbull Relation — **568**
10.6.1   General Approach to the Determination of the Specific Interfacial
         Energy: Taylor Expansion — **568**
10.6.2   Stefan's Rule and Skapski–Turnbull Relation: Some Interpretation and
         Extension to Thermodynamic Non-equilibrium States — **570**

10.7   Dependence of Crystal Nucleation and Growth Processes on
           Pre-history —— 573
10.7.1   Introductory Comments —— 573
10.7.2   Kinetic Criteria for Glass-formation —— 574
10.7.3   On the Dependence of the State of the Melt on Cooling and Heating
           Rates and Its Relevance for Crystal Nucleus Formation and
           Growth —— 578
10.8   Conclusions —— 579

Index —— 587

Preface

The present book is devoted to problems of a very important state of condensed matter, the vitreous state. It presents overviews on experimental and theoretical investigations both on general properties of materials in the state of a glass and their specific realizations in different classes of glass-forming systems, such as silicate, metallic, and polymer glass. In addition, a detailed analysis of different aspects of phase formation processes in glass-forming systems and their effects on glass properties is given both from theoretical and experimental points of view. The chapters are written by renowned researches in the field giving an overview both on the general state of affairs in the respective field and own results. Below a brief overview on the content of the different chapters is conducted relying widely on the description as given by the authors.

In Chapter 1 (C. Schick et al.), a review is given on a new tool of experimental analysis of glass transition and phase formation in glass-forming systems, thin-film chip calorimetry with controllable cooling as well as heating rates up to 10^6 Ks^{-1}. This method has been already successfully employed for fast thermal processing and simultaneous calorimetric measurements of many polymer and metallic samples. It could be demonstrated, in particular, that the physical properties of the respective systems are generally strongly dependent on their thermal history. Besides, owing to the very small addenda heat capacity, the calorimeter is very sensitive to study samples of only several tenths of nanograms. With differential alternating current (AC) design, the sensitivity of the calorimeter could be increased to a few tenths of pico-Joules per Kelvin. In the chapter, following the discussion of the strategy to realize fast cooling, the static and dynamic thermal properties of the sensors are described used for the setup of the calorimeter. The method is applied to a comprehensive discussion of the effects of thermal history and crystal nucleation and growth processes.

In Chapter 2 (G. Wilde), metallic glasses as one of the presently most actively studied metallic materials are analyzed with special emphasis on the understanding of the early stages of crystal formation in metallic glass-forming melts. One of the classes of such systems, metallic alloy systems were thought of as easy describable molecular liquids. Yet, experimental experience has shown that metallic systems with sufficient stability against crystallization require a rather complex stoichiometry and considerable covalent contributions to the binding energy of the constituents. Chemically simple systems, e.g. binary glass-forming alloys, on the other hand, are kinetically unstable and do not allow experimental access to the undercooled liquid state. Two types of structure development in metallic systems are given special attention in the present analysis. As shown marginal glasses that are prone to nanocrystal formation provide the opportunities to study the early stage of the formation of a crystalline structure since crystallization proceeds sluggishly at temperatures very close to T_g. The second type of structure development analyzed in more detail is concerned with the temperature-dependent evolution of a dynamic structure in the undercooled liquid state of glass-forming material. Such structures are not necessarily related to the corresponding crystalline state of the material, but to associations of atoms within the fluctuating arrangement of the melt that have a higher degree of correlation. These correlated regions exist only temporarily in a given configuration, i.e. they do not form permanent clusters in the melt. However, the presence of an average fraction of more strongly correlated atoms in the liquid state might also affect the earliest stages of nanocrystal formation. Both aspects are considered in the discussion of the current knowledge on the earliest stages of crystal formation in metallic glasses.

In Chapters 3 (I. Gutzow et al.) and 4 (I.G. Polyakova) an overview on the current state of knowledge on a variety of aspects of the physics, structure, synthesis of the different modifications of SiO_2 is given and new theoretical results on their solubility are outlined. In detail, phase diagrams of SiO_2 are discussed, followed by an overview on realized or possible methods of synthesis of all the modifications of pure SiO_2, beginning with those stable at normal pressure, i.e. quartz, cristobalite, SiO_2-glass. Then the particular thermodynamic properties of both the "normal" and high pressure modifications of SiO_2 are reviewed. Particular significance is attributed in this analysis to those thermodynamic properties, which are of importance either for the synthesis or for the technological applications of a given SiO_2-structure or modification. As far as it turns out that all the modifications of SiO_2 can be synthesized hydrothermally, particular emphasis is given to the thermodynamics of SiO_2 aqueous solutions. Moreover a semi-empirical method is developed to calculate the solubility of all crystalline and amorphous modifications of SiO_2, using only one structurally significant property, their density. Particular attention is also devoted to the so-called polyamorphism of silica glass, thermal expansion and high-low transformation and accompanying them phenomena. An attempt is made to explain specifics of silica and its properties by the electronic structure of component atoms. The specific features of the electronic structure of the silicon atom enables them to form with oxygen multiple crystalline compounds. It also makes silica one of the strongest glass-formers.

Chapter 5 (N.M. Vedishcheva & A.C. Wright) is devoted to the chemical structure of oxide glasses as a concept for establishing structure-property relationships. The suggested by the authors concept of chemical structure of a glass enables the above problems to be successfully solved, since this approach takes into account the presence of network-modifying cations in glasses and considers structural changes in terms of the Gibbs free energy of a given system. As the approach allows both the structure and a variety of glass properties to be calculated, the structure-property relationship is established quantitatively. The approach also explains the relationship between two levels in the glass structure (the short- and intermediate-range order) and shows which of them determines properties of glasses. As compared to other existing structural models, the concept of the chemical structure of glasses/melts considered in the chapter has the following advantages: (i) It brings a rigorous physical meaning to the reasoning concerning the origin of various basic structural units, or super-structural units, in glasses (melts). In other words, this concept not only predicts what structural units are present in a given glass or melt but also why they form, in what quantities and what their most probable surroundings might be. (ii) It enables the difference between the structure of glasses (melts) and crystals to be quantitatively determined at the super-structural unit level. (iii) As compared to experimental studies, the concept of the chemical structure provides a more profound understanding of the temperature changes that occur in glasses/melts at the levels of structural and super-structural units.

Gases dissolved in the glass and bubbles formed out of them in transparent silica glasses are one of their main defects. They have frequently a negative effect on the appearance of the manufactured glass articles as well as on its mechanical and optical properties. Chapter 6 (B.Z. Pevzner & S.V. Tarakanov) is devoted to the resolution of one particularly important problem in the technology of glass production, the removal of bubbles from the melts. In the first part of the chapter, a brief overview is given on the main modes of production of silica glass as well as the resulting from this process concentration of impurities in silica glass, including dissolved gases, possible modes of formation of gas bubbles and the gas composition in the bubbles. Main sources of gases and bubbles are analyzed: raw materials, furnace atmosphere, and the interaction of heaters and form-shaping equipment with the melt. In the second part of the chapter a theoretical description and computer simulations of the course of evolution of isolated bubbles in the silica glass melt is given. The influence of the main physico-chemical and technological factors (size of silica particles, gas pressure, and temperature-time conditions in the furnace space) on formation and evolution of the bubbles are analyzed. In the third part, the results are extended to the description of the evolution of ensembles of bubbles in the melt and an overview on different possible technological strategies of bubble removal is outlined.

The analysis of the properties of silica and the chemical structure of oxide glasses, in general, is supplemented in Chapter 7 (V. Leko) by a comprehensive overview on existing data on the highly complex and sensitive to different factors crystallization behavior of the various industrial and laboratory silica glass samples studied. Here both the results of own extensive research are included as well as the results of other authors available in the literature. Considerable attention is hereby devoted to the outline of generally not well-known but very important papers published in Russian language containing partly results not having analogs in the existing English-language literature. Crystallization of silica glass samples is characterized by a variety of peculiarities. Partly this is due to the circumstances that crystallization proceeds here at temperatures exceeding 1200 °C. At such high temperatures, silica undergoes chemical reactions with the components of the gas phase surrounding the samples and the various impurities contained in them. Such chemical reactions may have a significant influence on the course of crystallization. In this respect, effects of technological preparation, concentration of impurities, of composition of the gas phase surrounding the samples on the crystallization kinetics are analyzed. In addition, samples of different shapes (quartz blocks and tubes) are investigated to explore the effects of sample shape on crystallization. A special analysis is devoted to the relation between crystallization kinetics and viscosity questioning the widespread believe that the crystallization kinetics of quartz glass is widely determined by viscosity. Possible alternative kinetic mechanisms for silica glass crystallization are reviewed.

In Chapter 8 (Fokin, Karamanov et al.) stress induced pore formation and phase selection in a crystallizing stretched glass is analyzed. In the first part, results of experimental and theoretical analysis are outlined of phase selection and nucleation of pores in small samples of undercooled diopside liquid when it is enclosed by a hard crystalline surface layer. The formation of the surface crystalline layer begins with nucleation and growth of highly dense diopside crystals. At the moment of impingement of these crystals on the sample surface, the crystallization pathway switches from diopside to a wollastonite-like (WL) phase. The WL-crystal produces less elastic stress energy in crystallization as compared to diopside due to its lower density, which is closer to the liquid density. The relative content of the two crystalline phases can be changed by varying the sample size. Due to the density misfit, the growth of the WL-crystalline layer leads to uniform stretching of the encapsulated liquid and finally to formation of one pore, which rapidly grows up to a size that almost eliminates the elastic stress and, therefore, dramatically reduces the driving force for pore nucleation. This nucleation process occurs in a very narrow range of negative pressures indicating that it proceeds via homogeneous nucleation. This result is corroborated by theoretical calculations of elastic stress fields and their effect on nucleation. Good qualitative and even quantitative agreement between experiment and theory is found. An overview on other systems with similar or related properties is included as well. The findings of this research are quite general because the densities of most glasses significantly differ from those of their iso-chemical crystals, and are thus of technological significance for glass-ceramics development and sinter-crystallization processes.

Chapter 9 (V.G. Baidakov) is devoted to molecular dynamics modeling of the kinetics of spontaneous crystal nucleation in under-cooled one-component Lennard-Jones liquids and a detailed comparison with the basic assumptions and results of classical nucleation theory (CNT). In the MD-computations the following spectrum of properties of the respective nucleating systems under consideration is determined: nucleation rate, diffusion coefficient of the crystal clusters in cluster size space, non-equilibrium Zeldovich factor, size of the critical crystal nucleus, and pressure inside the critical crystal nucleus. Based on these data, the interfacial energy density of the critical crystal nucleus is determined. Simultaneously, the interfacial energy density is computed by molecular dynamics methods for the planar interface liquid-crystal. It is found that for typical sizes of the critical nuclei in the range of 0.7–1.0 nm the value of the effective specific interfacial energy differs from that of the planar interface by less than 15 %. A comparison of the molecular dynamics results with classical nucleation theory shows that for the considered case of crystallization of one-component liquids

MD simulation results are in good agreement with the classical nucleation theory not only with respect to the final result, the nucleation frequency, but also with respect to the other specified above parameters. Consequently, the results of molecular dynamics simulations of crystallization in one-component liquids demonstrate the validity of the basic assumptions and the final results of CNT for this particular case of phase formation.

In Chapter 10 (G.P. Johari & J.W.P. Schmelzer), the question is posed, why in highly viscous glass-forming melts, at least, some of the assumptions employed in classical theory of nucleation and growth processes may not be adequate. In particular, the relation is critically analyzed between the typical sizes of supercritical nuclei *vis a vis* the sizes of co-operatively rearranging regions (CRR) of the configurational entropy theory and of the domains of heterogenous dynamics (DHD) envisaged in the structure of ultra-viscous melts. It is argued that stochastic structural fluctuations in a melt that produce supercritical nuclei are irreconcilable with those that produce the co-operatively rearranging regions and the dynamic heterogeneity domains – the nuclei have structural order and the other mentioned above regions do not. Since the nanometer size of such nuclei is the same as that of the disordered regions and domains, it seems improbable that nuclei would form inside these regions and domains. As a second aspect, the crystallization kinetics of a metal-alloy glass is described studied by calorimetry under, a) isothermal conditions and, b) heating at a fixed rate. As the third aspect, it is considered how buoyancy-induced mass transfer of the melt due to Rayleigh-Bénard convection and due to Marangoni-flow are expected to affect the crystallization process and thereby the diffusion-controlled crystallization kinetics and its activation energy. These effects can be resolved by studies performed in microgravity conditions or by subjecting a crystallizing melt to a centrifugal force and by studying the effects of pressure on crystallization kinetics. In the fourth aspect, new thermodynamic aspects in the derivation of the correct expression for the work of critical cluster formation are reviewed. Generalizing Gibbs' classical method of description of thermodynamically inhomogeneous systems, recently a thermodynamic description of non-equilibrium states consisting of clusters of arbitrary sizes and composition in the otherwise homogeneous ambient phase was developed by one of the authors. This approach leads not only to a sound foundation of the thermodynamic aspects of the theoretical description of cluster growth processes but also to a variety of principally new insights into the course of nucleation-growth or spinodal decomposition processes, in general. In particular, it leads to a different set of thermodynamic equilibrium conditions for the determination of the properties of the critical clusters as compared with Gibbs' classical treatment. It supplies us further with a tool allowing one to determine the dependence of the bulk and interfacial properties of the evolving clusters in dependence on their sizes. In addition, a generalization of the Skapski-Turnbull rule for the determination of the specific interfacial energy is developed for aggregates of the newly evolving phase not being in thermodynamic equilibrium with the ambient phase. The approach further allows one to give a novel answer to the question at what conditions the evolution of the cluster ensemble proceeds via a saddle point trajectory passing the critical cluster and under which conditions ridge crossing is preferred. These and some further of the most important consequences of this new approach in application both to nucleation and growth-dissolution processes are briefly reviewed. Finally, we complete the analysis with a discussion of prehistory effects in phase formation and possible ways to incorporate them into the theory of crystal phase formation. Summarizing the results of this final and the previous chapters, we may conclude: Despite an extensive work performed over more than ten decades and existing impressive results reflected partly in the present book, a broad spectrum of problems in the description of crystal nucleation and growth processes remain not finally resolved and requires intensive further studies.

Finally, it is a great pleasure to acknowledge the stimulating cooperation with all authors in preparing the different chapters of the present book. It is a particular pleasure to acknowledge also the valuable assistance of Prof. Alexander S. Abyzov (Kharkov, Ukraine) and Mrs. Anne A. Kovalchuk (Cherkassy, Ukraine) in the preparation of the present monograph for publication. The cover shows crystals of lithium meta-silicate on the surface of samples of $80Li_2OSiO_2 \cdot 20CaOSiO_2$-glass-forming melts formed during its cooling to a glass. The picture was supplied by Prof. Vladimir M. Fokin (St. Petersburg, Russia & São Carlos, Brazil), which is gratefully acknowledged as well.

April 2014

Jürn W.P. Schmelzer
Rostock, Germany and Dubna, Russia

List of contributing authors

Alexander S. Abyzov
National Science Center
Kharkov Institute of Physics and Technology
61108 Kharkov, Ukraine

René Androsch
Martin-Luther-University Halle-Wittenberg
Center of Engineering Sciences
06099 Halle/Saale, Germany

Vladimir G. Baidakov
Institute of Thermal Physics
Ural Branch of the Russian Academy of Sciences
Amundsen street 107a
620016 Ekaterinburg, Russia

Vladimir M. Fokin
Vavilov State Optical Institute
ul. Babushkina 36-1
193 171, St. Petersburg, Russia &
Federal University of São Carlos
UFSCar, 13565-905 São Carlos, SP, Brazil

Ivan S. Gutzow
Institute of Physical Chemistry
Bulgarian Academy of Sciences
1113 Sofia, Bulgaria

Stoyan Gutzov
Department of Physical Chemistry
Faculty of Chemistry and Pharmacy
University of Sofia "St. Kliment Ohridski"
1164 Sofia, Bulgaria

Gyan P. Johari
Department of Materials Science and
Engineering
McMaster University
Hamilton, Ontario L8S 4LS, Canada

Nikolai Jordanov
Institute of Physical Chemistry
Bulgarian Academy of Sciences
1113 Sofia, Bulgaria

Alexander Karamanov
Institute of Physical Chemistry
Bulgarian Academy of Sciences
1113 Sofia, Bulgaria

Victor K. Leko
OOO "GGA"
Naberezhnaya Chernoy Rechki 41, korp. 7
197342 St. Petersburg, Russia

Frank-Peter Ludwig
QSIL AG Quarzschmelze Ilmenau
Gewerbering 8
98704 Langewiesen, Germany

Irena Markovska
University "Prof. Dr. Assen Zlatarov"
8010 Bourgas, Bulgaria

Radost Pascova
Institute of Physical Chemistry
Bulgarian Academy of Sciences
Sofia 1113, Bulgaria

Ivan Penkov
Institute of Metal Science
Equipment and Technologies
Bulgarian Academy of Sciences
1574 Sofia, Bulgaria

Boris Z. Pevzner
Laboratory of Glass Properties
Varshavskaya uliza 5a
194086 St. Petersburg, Russia

Irina G. Polyakova
Grebenshchikov Institute of Silicate Chemistry
Russian Academy of Sciences, Nab. Makarova 2
St. Petersburg 199034, Russia

Christoph Schick
University of Rostock, Institute of Physics
Wismarsche Str. 43–45
18057 Rostock, Germany &
Faculty of Interdisciplinary Research
University of Rostock
Department "Life, Light and Matter"
18051 Rostock, Germany

Jürn W.P. Schmelzer
University of Rostock
Institute of Physics
Wismarsche Str. 43–45
18057 Rostock, Germany &
Joint Institute for Nuclear Research
Bogoliubov Laboratory of Theoretical Physics
141980 Dubna, Russia

Sergey V. Tarakanov
Laboratory of Glass Properties
Varshavskaya uliza 5a
194086 St. Petersburg, Russia

Natalia M. Vedishcheva
Grebenshchikov Institute of Silicate Chemistry
Russian Academy of Sciences
Nab. Makarova 2
St. Petersburg 199034, Russia

Gerhard Wilde
Institute of Materials Physics
Westfälische Wilhelms-Universität Münster
Wilhelm-Klemm Strasse 10
Münster 48149, Germany

Adrian C. Wright
J.J. Thomson Physical Laboratory
University of Reading, Whiteknights
Reading RG6 6AF, United Kingdom

Andreas Wurm
University of Rostock
Institute of Physics
Wismarsche Str. 43–45
18057 Rostock, Germany

Edgar D. Zanotto
Federal University of São Carlos
UFSCar
13565-905 São Carlos, SP, Brazil

Evgeny Zhuravlev
University of Rostock
Institute of Physics
Wismarsche Str. 43–45
18057 Rostock, Germany

Christoph Schick, Evgeny Zhuravlev, René Androsch,
Andreas Wurm, and Jürn W.P. Schmelzer

1 Influence of Thermal Prehistory on Crystal Nucleation and Growth in Polymers

Observations regarding the effect of thermal history of crystallizing polymer melts onto the outcome of crystal nucleation and growth processes are investigated experimentally. Some results can be at least on a qualitative basis explained by classical nucleation theory (CNT) while others are not easy to understand in the framework of CNT. The origin of the respective problems and possible extensions of CNT to overcome them are briefly discussed. We chose polymers as model systems because they allow one a separate investigation of nucleation and growth processes in a wide temperature and time range. Furthermore they are well suited to be analyzed experimentally by the recently developed fast scanning calorimetry. In particular, by applying fast scanning calorimetry we are able to investigate processes in bulk samples and do not need to use droplets to study homogeneous nucleation kinetics. Fast scanning calorimetry enables us to observe homogeneous nucleation in materials crystallizing relatively fast like most of the industrial relevant semicrystalline polymers. Quantitative analysis allowed us to judge the nucleation efficiency of additives in the whole range of temperatures, where polymers crystallize. Besides the observed nucleation and growth below the glass transition information about the nucleation activity of small crystals grown at different temperatures relative to the cold-crystallization temperature range were obtained. Finally we show how relaxation of the under-cooled melt (glass) influences homogeneous nucleation. These data may serve as input for a more general description of the interplay of nucleation-growth and relaxation processes within the framework of a structural order-parameter model.

1.1 Introduction

The basic theoretical concepts underlying the description of crystal nucleation and growth processes were developed 80–90 years ago. In a variety of cases, the classical methods of describing nucleation that result from these ideas, namely classical nucleation theory (CNT) and the classical theory of crystal growth, supply us with a satisfactory description of the respective processes. However, in a not much less, or possibly even larger number of cases severe deviations between the theoretical predictions and experiment are observed.

In this contribution we focus on some observations regarding the effect of thermal history of crystallizing polymer melts onto the outcome of crystal nucleation and growth processes. Some of them can be at least on a qualitative basis explained by CNT while others are not easy to understand in the framework of CNT. We chose polymers as model systems because they allow one a separate investigation of nucleation and growth processes in a wide temperature and time range. Furthermore they are well suited to be studied experimentally by the recently developed fast scanning calorimetry. In particular, by applying fast scanning calorimetry we are able to investigate processes in bulk samples and do not need to use droplets to study homogeneous nucleation kinetics.

The chapter is structured as follows: First we briefly discuss the state of art regarding theory in the framework of structural order-parameter descriptions. It is followed by the description of an experimental technique, differential fast scanning calorimetry, both providing a deeper insight into nucleation of crystals in dependence on melt history. In the main part of the contribution we then focus on experimental aspects applying the differential fast scanning calorimeter. We describe a way to generate samples with essentially no homogeneously formed nuclei at temperatures above and below the glass transition temperature. Since heterogeneities are never completely absent a strategy to minimize their impact on the observed crystallization is discussed. Making use of differential fast scanning calorimeters and combining both approaches finally allows us studying homogeneous nucleation kinetics under isothermal conditions. For experiments below the glass transition we show that nucleation starts only after significant volume and enthalpy relaxation towards the super-cooled liquid state. Here the influence of the state of the quenched melt regarding sub-T_g relaxation is important for nucleation even if the relaxation is not considered to generate order in the sample. Finally, we discuss the question how structures formed at annealing influence crystallization on heating. We show that crystals formed at temperatures above the cold-crystallization range do not act as nuclei there. But nuclei or crystals formed at so low temperatures that they melt on heating before the cold-crystallization range is reached are able to accelerate crystal growth.

1.2 State of the Art

1.2.1 Dependence of the Properties of Glass-forming Melts on Melt History

The proper account of the circle of problems sketched out above is a hard task. Fig. 1.1 shows the typical relationship between the glass transition temperature T_g (for conventional cooling rates) and the temperature T_{max} where the maximum of the steady-state nucleation rate is reached. It is evident that the maximum of the steady-state nucleation rate is found near to T_g. For this reason, one has to look carefully at the properties of the ambient glass-forming melt in order to determine correctly the thermodynamic driving force of the process of crystallization and the surface energy term.

The typical behavior of the density of glass-forming systems during vitrification is shown in Fig. 1.2. The density increases with decreasing temperature but its val-

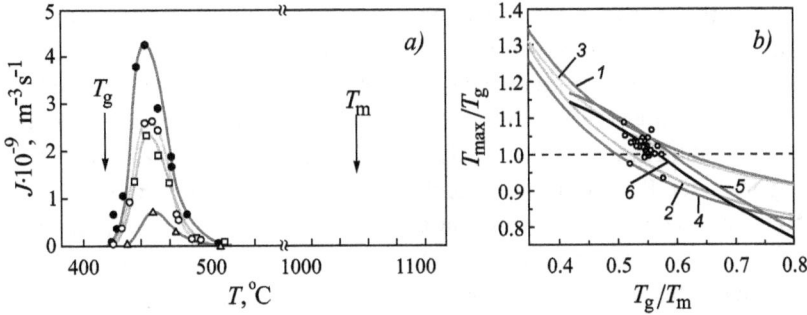

Fig. 1.1: (a) Dependence of the steady-state nucleation rates for α-$Li_2O_2SiO_2$ on temperature as obtained by different authors [1]. T_g is the glass transition temperature and T_m the melting temperature. (b) Relation between the temperature of the maximum nucleation rate and the glass transition temperature for a large class of glass-forming melts. Different systems are specified by the different numbers (for the details see [2]).

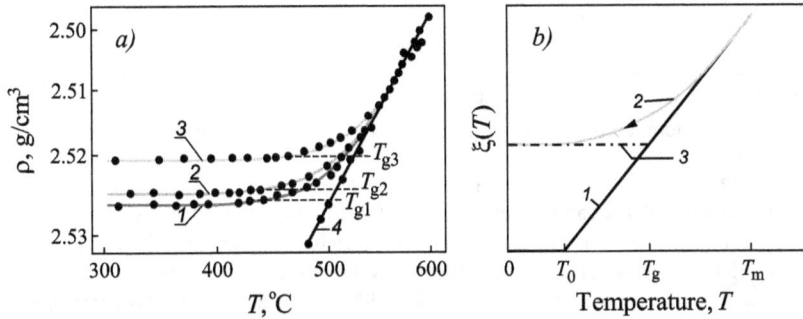

Fig. 1.2: (a) Typical dependence of the density of glass-forming melts on temperature during cooling (shown here for a borosilicate glass) for different cooling rates. With an increase of the cooling rate (from 1 → 2 → 3), the glass-transition temperature is shifted to higher values. (b) Qualitative interpretation of this behavior employing one structural order-parameter connected with the free volume of the system under consideration. Curve 3 refers to the equilibrium state of the melt, curve 2 describes cooling at some given rate and curve 3 describes the method of determination of the glass-transition temperature as employed in Fig. 1.2a [1, 3, 4, 5].

ues depend not only on the thermodynamic state parameters but also on cooling rate or, more generally on the melt history, i.e. the way how the glass-forming melt was brought into its current state. In order to describe the behavior in thermodynamic terms, one has to introduce, at least one additional structural order-parameter denoted here as ξ which we may associate, for example, with the free volume of the melt under consideration. The rate of change with time of this additional order-parameter can be described for isothermal and isobaric conditions by

$$\frac{d\xi}{dt} = -\frac{1}{\tau\left(p, T, \xi\right)}\left(\xi - \xi_e\right).$$ (1.1)

Here $\tau(p, T, \xi)$ is the characteristic relaxation time which depends on pressure, temperature and the structural order-parameter. It can be shown [5, 6] that such a relaxation equation can easily reproduce the often observed relaxation behavior of the form $\xi \sim t^{1/2}$ and can give a key to the theoretical understanding of the stretched exponential relaxation kinetics. In Eq. (1.1), ξ_e is the equilibrium value of the structural order-parameter. For given cooling and heating rates, $q = (dT/dt)$, Eq. (1.1) can be transformed into a relation describing the change of the structural order-parameter with temperature. The solution of this equation for constant cooling and heating rates results in the dependencies shown in Figs. 1.2 and 1.3.

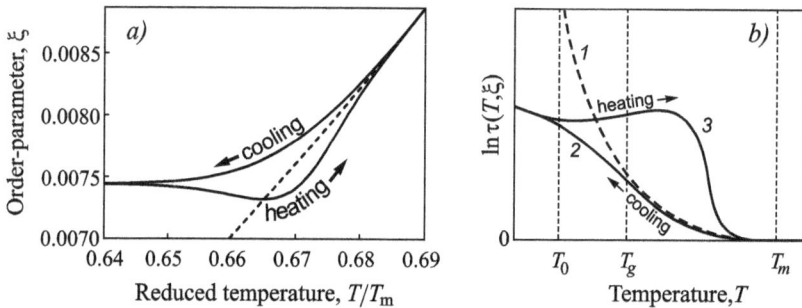

Fig. 1.3: (a) Dependence of the structural order-parameter on temperature for cooling and heating processes performed with the same absolute value of the rates of change of temperature. (b) Dependence of the characteristic relaxation time, $\tau(T, \xi)$, on temperature for cooling and heating processes [1, 5]. By a dashed curve (1), the equilibrium value of the relaxation time is shown. The values of the relaxation time in cooling (2) and heating (3) differ due to differences in the respective values of the structural order parameter as shown in Fig. 1.3a.

Since the structural order-parameter is a function of pressure and temperature and of the melt history (cooling and heating rates), the thermodynamic properties of the melt also depend on the same set of parameters. It follows as a consequence that the thermodynamic state parameters of the crystal cluster in the ambient phase are, as a rule, dependent on the melt history as well. Once the bulk properties depend on melt history, the surface properties also have to depend on it. Consequently, the kinetics of crystal nucleation and growth is affected, in general, by melt history and may proceed, in particular, in a different manner for cooling and heating processes. The degree to which such effects are of importance is determined by the ratio of the characteristic time scales for relaxation and critical crystal nucleus formation (cf. the discussion in Section 1.4.2). Further, since the thermodynamic state parameters are dependent on the structural order-parameter(s), the kinetic parameters have to depend, in general, on the structural order-parameter(s) as well. For the case discussed here, such possible dependence is demonstrated in Fig. 1.3b obtained by similar computations as the ones shown in Fig. 1.3a.

In the above considerations, the structural order-parameter is assumed to have the same value for the whole system i.e. it is assumed that the processes do not depend on the spatial coordinates. However, the intensity of fluctuations in the glass transition range is as a rule higher as compared to systems in thermodynamic equilibrium. In the glass transition range, local fluctuations are not damped out since the system is in a non-equilibrium state. A particular experimental realization of such peculiarity consists in the "fluctuation flashes" in heating observed and discussed in detail by Porai-Koshits et al. (see [7] for an overview). Consequently, in glasses also some spatial heterogeneity can evolve affecting the nucleation-growth kinetics.

Employing the structural order-parameter concept in this way for the description of glass-forming melts, the discussion of the dependence of the crystal nucleation and growth processes on the structure of glass-forming melts (see e.g. [8]) can be given a quantitative basis. In this treatment, structural properties are considered as additional parameters not determined uniquely by the conventional thermodynamic state variables such as number of moles, pressure and temperature but by additional parameters which can also be changed independently. Such approach is possible only if the system is out of equilibrium and is treated by introducing additional structural order-parameters. As a consequence, in such cases, in order to derive the nucleation and growth rates, one has to determine the bulk and surface properties as well as the dependence of kinetic parameters not only on numbers of moles, pressure and temperature but also on the (set of) structural order-parameters that depend, in general, on melt history (e.g. cooling and heating rates). A more extended analysis of the circle of problems sketched here is given in [6]. Consequently, in order to treat nucleation processes in glass-forming melts, as a rule, the melt history of the system under consideration has to be properly taken into account, in addition to the range of factors governing the crystallization process if the melt history can be neglected. An account of the melt history or the structure of the system can be made in the framework of the structural order-parameter approach as developed by De Donder. The application of this approach to the description of glass-forming systems is discussed in detail in [5] and the references cited therein.

Further on, the dynamics of phase formation at some given temperature depends also on the amount of the crystal phase already present in the system. This is another aspect of thermal history effects which has to be incorporated into the description appropriately as well. Both mentioned thermal history effects are considered highly important in the analysis in the whole range of slow to ultrafast nano-calorimetry at cooling and heating rates which can be varied by described here methods in between 10^{-4} Ks^{-1} up to 10^{6} Ks^{-1} [9, 10]. An experimental study of the possible effect of melt history on the kinetics of crystal nucleation and growth in polymers based on these methods is outlined in the present contribution (for additional supplementary information in this respect see also Chapters 2 and 10 in the present monograph).

1.2.2 Polymer Crystallization

The particular features of polymer crystallization have been of interest for decades since a large number of engineering polymers are semicrystalline [11]. Nevertheless, a general understanding of polymer crystallization has not yet been achieved [12–16]. In recent years new concepts describing polymer crystallization were developed. In particular, ideas put forward by Strobl [17, 18], Muthukumar [19], Olmsted, Ryan [20], Wunderlich et al. [21–23] and others have challenged the Hoffman-Lauritzen theory [24] (for a detailed discussion see The European Physical Journal E, Vol. 3, No. 2 (October 2000) pp. 165–200).

Polymer crystallization usually takes place far from equilibrium with kinetically controlled mechanisms [18]. Typical crystallization temperatures include the vitreous state going over into the glass transition region by further heating (cold crystallization) to temperatures rarely higher than 5 K below the melting transition (high-temperature crystallization). Molecular processes in such non-equilibrium conditions may only be rigorously followed by direct molecular level simulations, such as molecular dynamics (MD) or Monte Carlo (MC) methods ([25, 26] and Chapter 9 of the present monograph). However, due to its extremely slow dynamics and complexity, polymer crystallization has been so far out of reach of the conventional molecular simulations [19, 26]. In a recent attempt Luo and Sommer performed MD-simulations for crystallization and melting of a polymer on cooling and heating at rates of about $2 \cdot 10^7$ Ks^{-1} [15, 27–29]. These scanning rates, now available in computer simulations, are getting close to the 10^6 Ks^{-1} rates, which are possible in fast scanning calorimetry on heating [30] as well as on cooling [10].

For a direct comparison of experimental and computer simulation data, another point has to be considered. In most computer simulations, homogeneous nucleation is assumed to be the starting point of crystallization, while in experiments crystallization originates in most cases from heterogeneous nuclei. Fast scanning calorimetry, on the other hand, is one of the few techniques allowing one fast enough cooling in bulk samples to avoid heterogeneous nucleation on cooling [31] in contrast to the earlier observed homogeneous nucleation in droplets [32–35]. While in droplets the occurrence of one homogeneously formed nucleus is commonly observed only after complete crystallization of the whole droplet, in bulk samples differential fast scanning calorimetry (DFSC) is able to follow the growth until impingement. It is therefore an attractive task to study crystallization and homogeneous nucleation processes at short time and consequently length scales, approaching that of MD simulations. Even though the time scales do not overlap at the moment, there is a good chance to reach this goal in the near future due to progress in computation as well as in calorimetry.

Making use of an easy-to-operate DFSC [9, 36] we have studied isothermal nucleation and crystallization of fast crystallizing polymers, particularly poly(ε-caprolactone) (PCL), covering times from 10^{-4} s to 10^5 s and temperatures from about 25 K below the glass transition (T_g = 209 K at 10 K/min [37]) up to 330 K, which is close to the

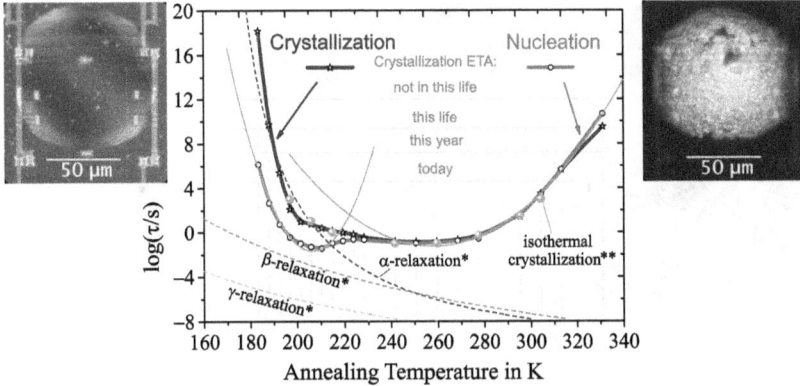

Fig. 1.4: Nucleation and crystallization half-times for PCL at temperatures from below the glass transition up to the melting temperature covering 20 orders of magnitude in time. The two processes are compared to dielectric relaxation data (dashed curves) [39] and isothermal experiments applying a single sensor device (green (in the color version) or filled (in the black-and-white version) spheres) [40].

equilibrium melting temperature (T_m = 342 K [37]). With the fast scanning calorimeter we were able to follow the development of crystals at one temperature over 9 orders of magnitude in time [38].

The final result of such an experiment is illustrated in Fig. 1.4, describing crystallization in the classical mode [1], assuming nuclei formation and subsequent growth. Half-times, τ, of crystallization (blue in the colored version) and of homogeneous nucleation (red in the colored version) are shown for PCL. Crystallization means the combined effect of nucleation and growth as measured e.g. by calorimetry (heat of crystallization) or X-ray scattering. The half-time is determined from an Avrami fit to the overall latent heat as described in Section 1.3.4.1. The half-time of nucleation corresponds to the increasing number of nuclei (clusters) able to grow at temperatures above the glass transition on heating (cold-crystallization). This half-time is the time when half of the maximum of cold-crystallization enthalpy is reached (see Section 1.3.4.1 for details).

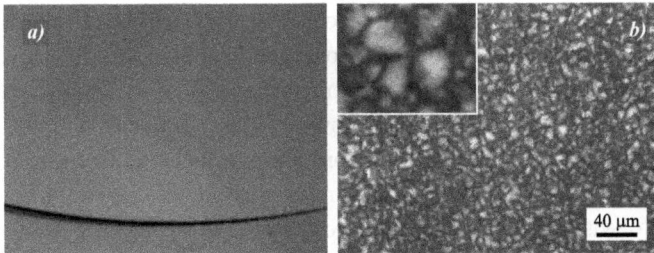

Fig. 1.5: (a) POM-images of PA 11 crystallized on heating an initially quenched sample and (b on slow cooling at 1 K min^{-1} [43].

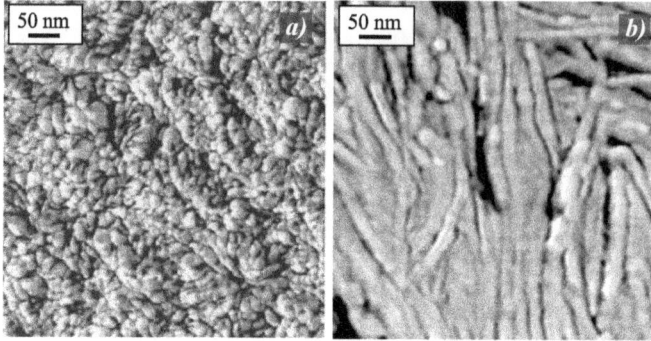

Fig. 1.6: (a) AFM phase-mode images of PA 11 crystallized on heating and annealing an initially quenched sample at 433 K and (b) on slow cooling at 10 K min^{-1} [43].

The rate of crystallization (both nucleation and growth) is zero at the equilibrium melting temperature, T_m, because the thermodynamic driving force for crystallization disappears at this point. Increasing the degree of under-cooling (decreasing temperature to lower values below T_m), the crystallization half-time τ decreases due to an increase of under-cooling. It reaches a minimum at the temperature ($T_{C\,min}$) and increases again at even lower temperatures due to slowing down of molecular diffusion. Nucleation is, compared to crystallization, a more local process and therefore remains faster at lower temperatures than crystallization and its half-time minimum occurs at a lower temperature. Such kind of behavior can be explained alternatively by the consideration of the dependencies of both nucleation and growth rate on temperature (see e.g. Fig. 6.8 in [1]). Both curves exhibit a maximum but the maximum of the growth rates is located as a rule at higher temperatures as compared to the maximum of the nucleation rate.

Figs. 1.5 and 1.6 show, as an example, optical and AFM microphotographs, respectively, for polyamide 11. The nodular, non-spherulitic and lamellar, spherulitic morphologies are seen in the photographs and are directly linked to crystallization starting with homogeneous (left side photographs) and heterogeneous (right side photographs) nucleation. The importance of the nucleation mechanism for the properties of the material is highlighted by the two microphotographs taken with crossed polarizers (POM). The right photo shows a bright image due to the birefringence of the spherulites formed at a few heterogeneous nuclei at low under-cooling of 25–50 K and grown to micro-meter size. In contrary, the left photo has been taken on a sample of polyamide, PA 11, which has been quenched to below the glass transition temperature and then crystallized on slow heating. Nuclei formation occurred in this case at an under-cooling of 160–180 K. The image is dark under crossed polarizers and the sample is transparent for visible light. This is due to crystallization starting from a large number of homogeneously formed nuclei and very limited crystal growth because of immediate space filling. The crystalline objects are of the order of 10 nm and therefore much smaller than the wave length of visible light (for a more detailed discussion of these observations and the relation to mechanical properties see e.g. [41–44]).

The direct measurement of nucleation is as a rule not possible by existing techniques due to the small size of the critical nuclei. Therefore indirect methods are usually employed [45]. In polarized optical microscopy the number of stabilized nuclei can be linked to the number of observed crystals or crystal aggregates such as spherulites, hedrites, or nodules ([23] (vol. 2) and [46]). In calorimetry, the cold-crystallization enthalpy on heating or at isothermal conditions was found to be dependent on the number of previously formed nuclei [47–53]. We follow these ideas but apply differential fast scanning calorimetry to extend the available scanning rate range of conventional DSC's (<10 Ks^{-1}) to cooling and heating rates as fast as 10^6 Ks^{-1}. The basic idea of such devices is described in the next section.

1.2.3 Differential Fast Scanning Calorimetry

For the investigation of polymer crystal nucleation and growth the conventional techniques, including DSC, cannot be employed since they do not allow one to cool fast enough to prevent nucleation on cooling for rapidly crystallizing polymers. An alternative is the single sensor fast scanning chip calorimeter [10, 54], which gives the possibility to perform heating and cooling of nanogram samples up to rates of 10^6 Ks^{-1}. But the single chip sensor calorimeter has several limitations regarding temperature control and the possibilities of quantitative measurements.

A differential scheme of two sensors with power compensation was therefore employed instead. The presence of an empty reference sensor reduces the influence of heat losses and addenda heat capacity on the obtained data. For a better scanning rate control, particularly in the transition regions, power compensation was introduced. Details of the device and data treatment are reported below and in [9, 36]. The differential fast scanning calorimeter (DFSC) is able to perform heat-flow measurements during controlled heating and cooling up to $5 \cdot 10^5$ Ks^{-1}. The power control circuit allows one to perform isothermal experiments for times shorter than 1 ms (with over- or undershoots < 1 K).

1.2.3.1 Fast Scanning Calorimetric Techniques

Conventional Differential Scanning Calorimetry (DSC) is one of the few techniques that have a relatively large dynamic range of scanning rates. It allows (quasi) isothermal measurements and scanning rates up to 10 Ks^{-1} for power compensated DSC's [55, 56]. Several approaches are known attempting to increase the scanning rate. Most of them are based on thin film techniques.

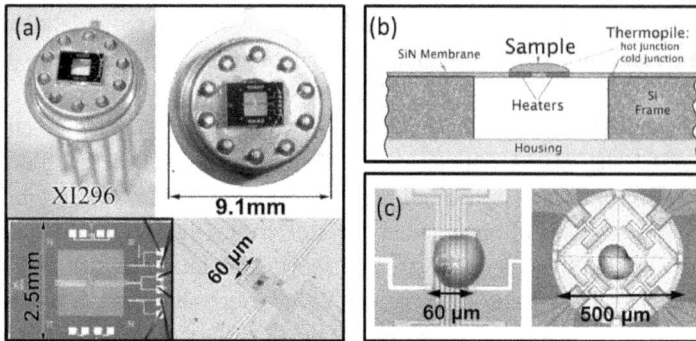

Fig. 1.7: Thin film chip sensor XI-396 based on a thin free standing SiN$_x$ film on a silicon frame and measuring area of 60 μm × 80 μm in the center of the film: (a) Different photographs of the sensor. (b) Schematic cross-section of the sample-loaded sensor (not to scale). (c) Photomicrograph of a sample loaded sensor XI-396 (left) and, for comparison, sensor UFS 1 (right) [9, 68].

Quasi-adiabatic scanning calorimetry at high heating rates, ca. 500 Ks^{-1}, was developed by Hager [57] and, even for rates up to 10^7 Ks^{-1}, by Allen and co-workers [30, 58, 59]. Similar approaches were used to study the behavior of metastable materials like vapor deposited films [60–65]. But, on the other hand, the investigation of metastable phase formation is possible only if the same high-controlled cooling rates are available too. Fastest cooling of a calorimetric cell is obtained if no heat is supplied to the cell (electrical power $P_0(T) = 0$ in the heat-balance equation Eq. (1.2)). The maximum possible cooling rate is therefore defined by the ratio between the heat-flow rate away from the measuring cell ($P_{loss}(T)$) and the heat capacities of the measuring cell (addenda: $C_0(T)$) and the sample ($C(T)$),

$$(C_0(T) + C(T)) \frac{dT}{dt} = P_0(T) - P_{loss}(T). \tag{1.2}$$

At a given temperature, $P_{loss}(T)$ is in a first approximation, assuming conductive and convection losses only, proportional to the temperature difference to the heat sink. Linear cooling is realized by keeping the right hand side of Eq. (1.2) at a constant negative value by controlling the heater power, $P_0(T)$. This idea was realized for fast scanning non-adiabatic nano-calorimeters based on thin film sensors (Fig. 1.7) [54] with extremely small addenda and sample heat capacity. It was shown that a gas is the optimum cooling agent to achieve largest cooling rates [10, 66, 67]. Furthermore, thermopiles are better suited as compared to resistive thermometers because they do not need an electrical current, which always generates some unwanted power. The technique described in [10, 54] is based on a single thin film chip sensor and is capable of applying both controlled heating and controlled cooling at rates up to 10^6 Ks^{-1}.

This single sensor ultra-fast scanning device [10] was successfully applied for the investigation of polymer melting and crystallization. The reorganization kinetics in PET and iPS was studied in combination with conventional DSC at scanning rates

covering 8 orders of magnitude [69–71]. Isothermal and non-isothermal crystallization and the formation of different crystal polymorphs were analyzed in a wide range
of temperatures and scanning rates applying DSC and the fast scanning setup in PE,
iPP and PVDF [54, 72–79]. The complex behavior in the temperature range between
glass transition and melting temperature was investigated in PBT [80]. The crystallization and cold-crystallization suppression in PA6 confined to droplets and in the
bulk was studied using such fast scanning calorimeter [31]. The dynamic range of
scanning rates of the device allowed us the investigation of superheating in linear
polymers like iPS, PET, PBT, and iPP [67, 81]. The examples given above have shown
that the effective range of controlled heating and cooling rates using different sensors
[68] is 100–10^6 Ks^{-1}. Unfortunately, at low rates the signal to noise ratio and therefore the sensitivity is reduced. Usually the device is limited to rates above 100 Ks^{-1} for
polymers. Nevertheless, the scanning rate range between conventional DSC and this
technique – the range $10\ldots100$ Ks^{-1} – is of high interest because several material
processing steps involve cooling rates just in this range [82, 83].

In order to improve fast scanning nano-calorimetry the very successfully applied
single sensor device as described in [10, 40, 54, 66, 84] was therefore first analyzed
and the weak points were identified in [9]. To overcome the identified problems a
differential scheme of two sensors with power compensation was constructed [9].
The presence of an empty reference sensor reduces the influence of heat losses and
addenda heat capacity on the obtained data dramatically. For a better sample temperature control, particularly in the transition regions, power compensation was introduced following in this way the work by Rodriguez-Viejo et al. [85, 86] and Merzlyakov
[87]. To improve signal to noise ratio and resolution of the device even under fast
scanning conditions an analog power compensation technique was implemented [9].
A differential power compensation scheme provides, under conditions of ideal symmetry of both sensors, directly the heat-flow rate into the sample, which simplifies
heat capacity calculation [36].

1.2.3.2 The Differential Fast Scanning Calorimeter (DFSC, Rostock)

The DFSC is intended to be employed to measure heat-flow rates into the sample as the
power difference between an empty and a sample-loaded sensor during fast temperature scans on heating and cooling at controlled rate. Finally, the obtained heat-flow
rate should be recalculated into heat capacity [36].

A very successful version of a power compensation differential scanning calorimeter was realized by PerkinElmer [88–91]. It is based on the measurement of the energy
difference required to keep both sides, sample and reference, at the same temperature throughout the analysis. The PerkinElmer scheme allows one for a relatively
simple determination of the heat-flow difference from the remaining temperature
difference between sample and reference cups [88, 90], not requiring measuring multiple signals or computing capabilities. But in this case both controllers must be fast

to avoid deviations from the programmed temperature even for the DSC scanning rates.

But as soon as we wanted to go to higher rates and sensitivity we were confronted with the problem of controller performance limitations. The origin of this problem is that the differential signal can contain fast events from the nanogram samples, requiring fast response of the controllers. Time resolution for the control of the average temperature could be much slower if the fast sample events would not be included. Output power range (dynamics) of the average controller is by orders of magnitude larger than needed for the compensation of the sample related effects. Therefore it may be beneficial to separate average and difference control totally avoiding any cross talk between both control loops.

Following this idea a power compensation scheme realizing such separate control loops as shown in Fig. 1.8 together with the changes in comparison to the PerkinElmer scheme was developed. First, we measure and control reference sensors' temperature alone. No average temperature is used. There is no influence found of any, even very strong, events occurring in the sample sensor on the reference temperature controller. This allows us to use a relatively slowly but precise PID controller for the reference temperature control. The differential controller detects any difference between reference and sample sensor temperatures and adds or subtracts its output voltage to the PID output voltage to the sample sensor alone. This way a total separation between both controllers is realized. It allows us to use a precisely but relatively slow working PID controller for the control of the reference temperature and a high sensitive and fast proportional controller for the difference controller. Compared to the PerkinElmer power compensation scheme this allows one a more precise control of the temperature

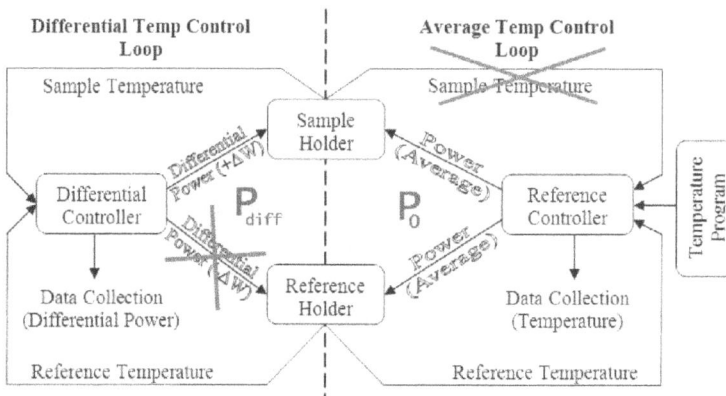

Fig. 1.8: Modified power compensation scheme for operation at high scanning rates: Separation of average temperature control (slow changes, large dynamic range) and differential control (fast events, small dynamic range) [9].

of both sensors at high rates. But the proportionality between the remaining temperature difference in the differential control loop and the differential heat-flow rate is lost. Therefore the new scheme requires the measurement of more than only one signal to allow one a recalculation of the power difference as it is described in [9, 36].

The separation of the two control loops makes recalculation of sample heat capacity more difficult in comparison to the symmetric power compensation scheme as it is used in the PerkinElmer power compensation DSC, but allows one going to higher rates with reliable average temperature control as shown in [9]. Deduction of heat capacity out of the measured signals consists of several steps and is not discussed here (for details see [36]). This device, and a commercial differential fast scanning calorimeter, the Flash-DSC 1 [92–96] from Mettler-Toledo, Switzerland, was finally used (i) to confirm the suppression of homogeneous nucleation at fast enough cooling and (ii) to study the successive formation of nuclei and crystals at annealing in a wide range of temperature and times after fast quenches.

1.2.3.3 Flash-DSC 1 (Mettler–Toledo)

A commercially available version of a chip based power-compensation fast scanning calorimeter is the Mettler–Toledo Flash-DSC 1 [93–95]. It was used in conjunction with a Huber intracooler TC90 reaching temperatures of –90 °C as starting temperature for the heating scans. Fig. 1.9 shows the device and the UFS 1 sensor.

The UFS 1 sensor was first conditioned and temperature-corrected according to the specification of the instrument provider, prior to positioning of the specimen in the center of the sample calorimeter. The furnace was permanently purged with dry nitrogen gas at a flow rate of 35 ml min^{-1}. The sample mass of a few hundred nanograms

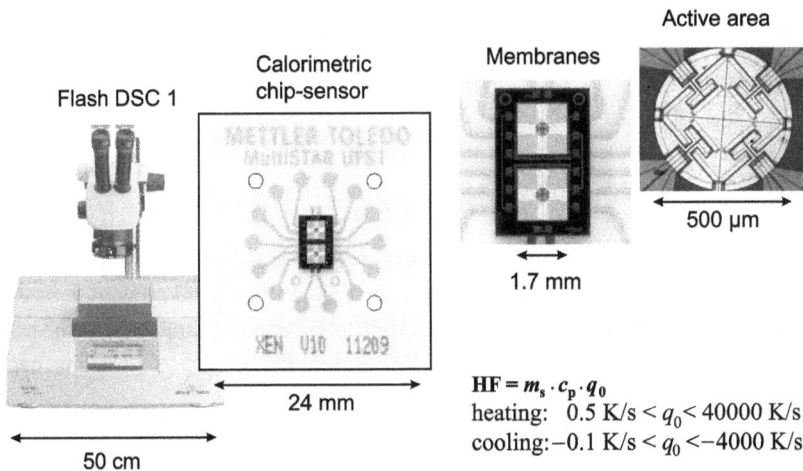

$$HF = m_s \cdot c_p \cdot q_0$$
heating: $0.5 \text{ K/s} < q_0 < 40000 \text{ K/s}$
cooling: $-0.1 \text{ K/s} < q_0 < -4000 \text{ K/s}$

Fig. 1.9: Mettler–Toledo Flash-DSC 1 and its sensor in different magnifications.

was estimated by comparing the measured absolute heat capacity of the liquid polymers with the expected specific heat capacity available in the ATHAS data base [37], and comparing the measured with the expected heat-capacity increment on heating a fully amorphous sample at the glass transition temperature. Specifications of the device and the sensor as well as calibration procedures are available elsewhere [93–97].

1.2.3.4 Sample Preparation

Specimens were prepared by cutting sections of ca. 10 µm thickness directly from a pellet or other shapes using a microtome or scalpel. The lateral size of the thin section was reduced to about 100 µm for the Flash-DSC 1 and about 30 µm for the DFSC, employing a scalpel under a stereo-microscope. Further details about sample preparation are reported in the corresponding papers referred to in the following sections. Fig. 1.10 shows different sensors with typical polymer samples on it.

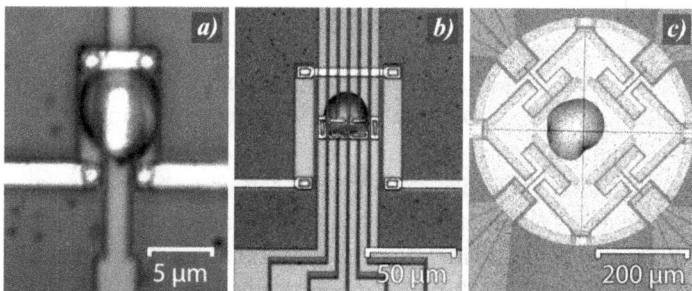

Fig. 1.10: Microphotographs of samples on different sensors: a) XEN 39394 one thermopile; b) XEN 39395; c) Mettler-Toledo UFS 1.

1.3 Experimental

1.3.1 Samples

1.3.1.1 Poly(ε-caprolactone) (PCL)

PCL was used as a model system because of its preferable properties for such investigations like fast crystallization kinetics, high thermal stability in the relevant temperature range due to the low melting temperature (equilibrium T_m = 342 K [37]), formation of geometrically stable samples on the DFSC sensor and not too low glass transition temperature (T_g = 209 K [37]). PCL, a linear aliphatic polylactone, is a commercial sample from Aldrich with a weight average molar mass of $2 \cdot 10^4$ g/mol (M_w/M_n = 1.73) and was used as received.

For comparison, some experiments were performed with nucleated PCL (by nucleating the identical PCL with multi-walled carbon nanotubes by melt mixing [98, 99]). In all cases the sample mass was approximately 20 ng. After a few heating-cooling cycles in dry nitrogen gas the samples established a good thermal contact to the sensor. Furthermore the about 5 μm thick samples were by then fully dried and the measured curves were highly reproducible. During the series of measurements the samples were several times cooled and reheated at 1000 Ks^{-1} to check their stability. Even after a few hundred heating cycles to 470 K (130 K above T_{m0}), where the samples are kept at temperatures above 370 K for 0.2 s each, the samples did not show any indication of degradation. The maximum temperature of 470 K was chosen to remove any order in the melt that could cause self-nucleation. This was checked by a series of measurements. Furthermore, a small piece of indium for in situ temperature calibration was placed on top of the PCL samples and requires heating above its melting temperature of 430 K (see [9] for details).

1.3.1.2 Isotactic Polypropylene (iPP)

We used an isotactic polypropylene with a mass-average molar mass of 373 kg/mol and a polydispersity of 6.2. The particular polymer is of commercial grade, that is, we expect that the material contains non-specified stabilizers or catalyst residues; however, presence of optical clarifiers or of β-nucleators as often found in industrial formulations of iPP is ruled out. The as-received pellets were compression-molded at 473 K to films of 100 micro μm thickness in a Collin Press, before further preparation of specimens for analyses by differential fast scanning chip calorimetry (DFSC).

Analysis of nucleation and growth of ordered structures has been performed using the DFSC, employing a sensor XI 395 from Xensor Integration (Netherlands). A small piece of iPP with a mass of about 3 ng was cut from the compression-molded film and placed on the chip, with the thermal contact between the sensor and the sample improved by silicon oil. It has been shown that larger sample mass did not allow one heating and cooling at rates faster about $1 \cdot 10^4$ Ks^{-1}. Experimentally observed heat-flow rate data were corrected for an instrumental baseline and converted into apparent heat capacities. A comparison of the overall heat capacity of liquid iPP with the corresponding specific heat capacity available in the ATHAS data base yielded the sample mass and ultimately allowed us calculation of specific apparent heat capacities. Note that a presence of impurities in the sample would not affect the estimation of the sample mass due to their extremely low amount in the sub-percentage range.

1.3.1.3 Isotactic Polybutene-1 (iPB-1)

We studied a commercial iPB-1 grade PB0110M from Basell Polyolefins synthesized using a Ziegler-Natta catalyst [100]. The mass-average molar mass and polydispersity are

711 kDa and 6.5, respectively, and the melt-flow rate (190°C, 2.16 kg) is 0.4 g (10 min)$^{-1}$ [101]. Samples are prepared as described in Section 1.2.3.4 for the Flash-DSC 1.

1.3.1.4 Polyamide 6 (PA6)

We employed a commercial PA 6 homopolymer grade Ultramid B27 from BASF, Germany, with a relative viscosity of 2.70. For Flash-DSC 1 analysis of the nucleation and crystallization kinetics, specimens were prepared as described in Section 1.2.3.4.

1.3.2 Suppression of Homogeneous Nucleation at Fast Cooling

1.3.2.1 Strategy

Studying homogeneous nucleation kinetics requires a starting situation without such nuclei. Mathot et al. [47] observed a reduction of the cold-crystallization peak on heating after cooling with increasing cooling rate. But even fast conventional DSC was not able to fully prevent nuclei formation on cooling in the slow crystallizing poly(L-lactic acid) (PLLA).

For low molecular mass compounds besides many others Angell et al. [51], Fokin et al. [52], Oguni et al. [48, 102–108] studied homogeneous nucleation based crystallization processes by DSC/DTA techniques at or below the common glass transition region (temperature defined by cooling at a few K/min). Vyazovkin et al. [49] studied nucleation below the common glass transition by crystallization on heating after annealing at different temperatures for different times. The latter two studies allowed one a comparison of the observed kinetics with known relaxation processes in the studied compounds. It was concluded that local relaxation processes and not the overall viscosity seem to control homogeneous nucleation. For polymers, on the other hand, nanoscale nodular structures are observed at crystallization near T_g, which do not grow to large scale crystals, as shown in [42, 43, 109–112], indicating homogeneous nucleation.

These results were taken as a basis for our isothermal crystallization investigation of poly(ε-caprolactone) (PCL) under isothermal conditions [38]. The possibility to avoid formation of homogeneous nuclei on cooling, as will be shown below, allowed us quenching the sample to any annealing temperature, T_a, to study nucleation and growth from a "truly" amorphous state. After a certain time annealing was interrupted and nuclei formation or crystal growth was studied by a subsequent heating scan. The cold-crystallization peak shows initially an increase in area which, based on the earlier work, might be interpreted as the presence of nuclei produced at the annealing temperature. As soon as the crystal growth starts already during annealing, the cold-crystallization peak decreases and the overall enthalpy change on heating, accounting for the latent heat of cold-crystallization <u>and</u> the overall heat of fusion, begins to increase, as described in more detail below. The identification of existing crystals and nuclei is performed by means of a heating scan after non-isothermal or isothermal heat treatments. The choice of the appropriate heating rate is discussed in the subsequent sections.

1.3.2.2 Analysis by Heating with 1000 Ks^{-1}

The temperature control system of the DFSC is capable to perform temperature scans of samples with a mass in the order of 10 ng from 1 Ks^{-1} up to 10^5 Ks^{-1} without significant temperature lag [9]. First, the influence of cooling rate on structure formation was studied at subsequent heating scans at a constant rate of 1000 Ks^{-1}. The scheme of the experiment is shown in Fig. 1.11.

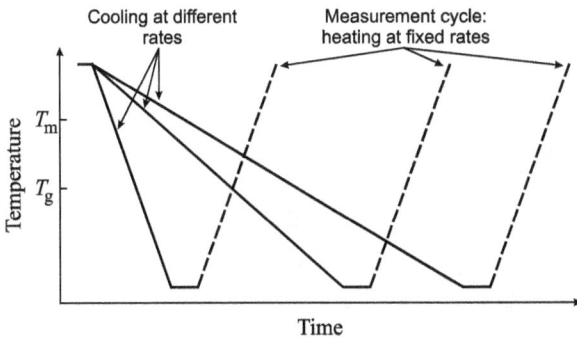

Fig. 1.11: Time-temperature profile of the experiment for investigation of non-isothermal nucleation and crystallization in PCL.

The sample was melted for 0.1 s at 470 K, ca. 130 K above the equilibrium melting temperature, to erase the thermal history, and then cooled at different rates between 50 and $5 \cdot 10^4$ Ks^{-1} down to 100 K. The glass transition of PCL is found at about 200 K. At 100 K changes in the sample during the fixed waiting time of 0.1 s are therefore unlikely to occur. The heating scan at 1000 Ks^{-1} was always performed after this waiting period at 100 K. A 50 Ks^{-1} cooling was slow enough to crystallize the sample on cooling and, therefore, slower cooling was not performed avoiding degradation at elevated temperatures. Due to fixed conditions of the measuring cycle (heating at 1000 Ks^{-1}) the observed changes in glass transition, cold crystallization and melting were attributed to the previous cooling.

Fig. 1.12 shows the analysis of PCL samples with different degrees of crystallinity, as developed on prior cooling with a variety of rates as detailed below. If the sample was crystallized by slow cooling with 50 Ks^{-1}, the thermal response on heating shows a reduced step in heat capacity at the glass transition and a heat of fusion (latent heat) as seen in Fig. 1.12a. The melting peak, ΔH_m, is in this case a measure of the crystallinity present in the sample before heating. In case one reaches a fully amor-

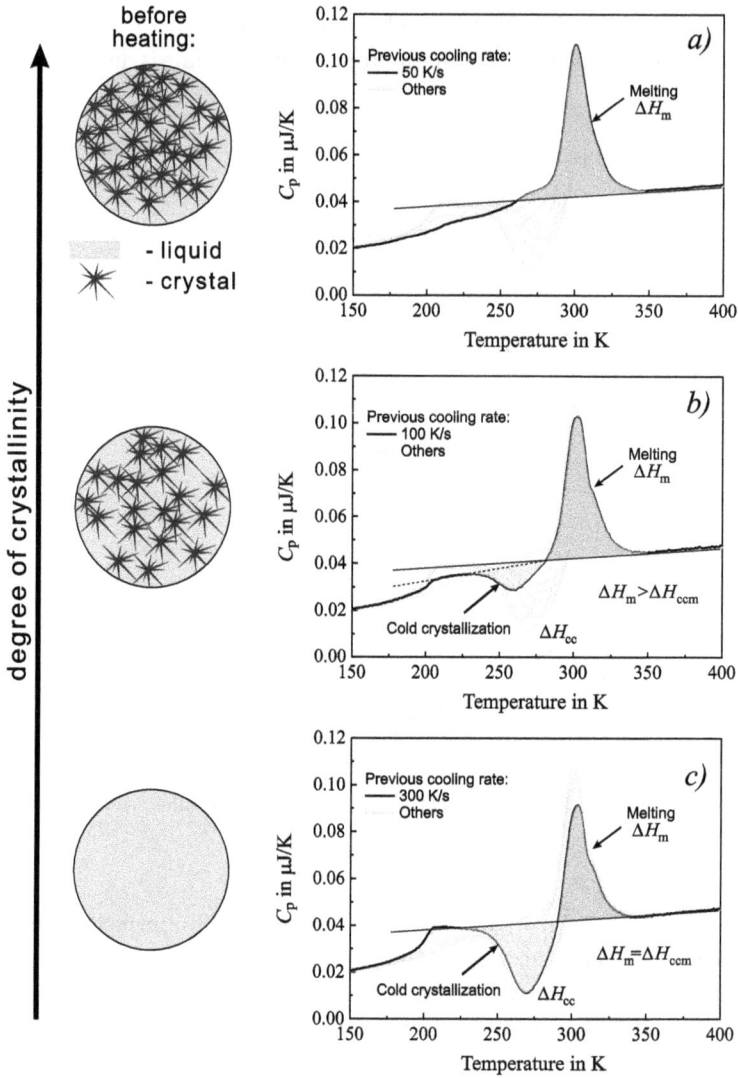

Fig. 1.12: Heating at a rate of 1000 Ks^{-1} of a sample with different degrees of crystallinity induced by cooling at a series of different cooling rates. Three cases are highlighted: (a) Crystallization occurred completely prior to heating, (b) only some crystallization prior to heating, and (c) no crystallization prior to heating but nucleation occurred before heating. The shaded areas show the quantitative evaluation of the corresponding melting and crystallization enthalpies on heating (see text for details).

phous state before heating (Fig. 1.12c), one sees a pronounced step in heat capacity at T_g, cold crystallization, ΔH_{cc}, and melting of the cold-crystallized phase, ΔH_{ccm}. In this case the overall (positive or endothermic) heat of fusion at high temperature plus the (negative or exothermic) cold-crystallization enthalpy equals zero. The term *cold crystallization* means crystallization on heating the sample, commonly from below T_g. It is assumed that cold crystallization reflects the growth of nuclei (clusters) at increasing temperatures where decreasing viscosity (increasing mobility) permits growth under the given heating conditions. Intermediate states of the material reveal a smaller step at the glass transition, reduced cold crystallization, and melting of the fewer crystals grown during cooling, added to the crystals formed on cold crystallization on heating, as shown in Fig. 1.12b. In the following we use pictograms to schematically highlight the state of the sample. The pictograms are explained in Table 1.1.

Table 1.1: Pictographs for different states of polymers during nucleation and crystallization at decreasing cooling rate or increasing time for isothermal treatment (from left to right)

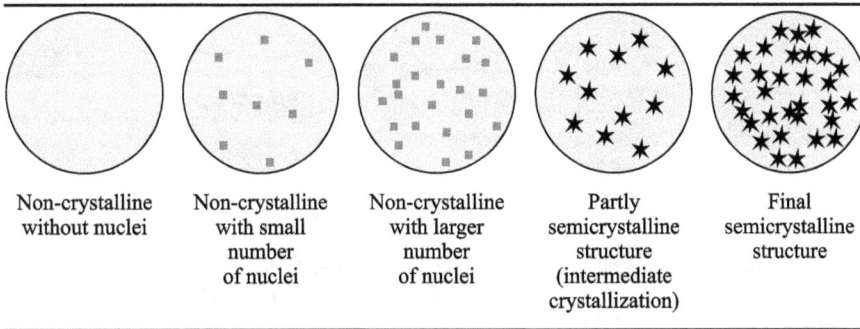

Non-crystalline without nuclei	Non-crystalline with small number of nuclei	Non-crystalline with larger number of nuclei	Partly semicrystalline structure (intermediate crystallization)	Final semicrystalline structure

The latent heats ($\Delta H_{\text{heating}} = \Delta H_m + \Delta H_{cc}$ shown in Fig. 1.12) are used as a measure of overall crystallization as a result of the temperature treatment preceding the heating scan. The baseline for peak integration is selected as described below, widely following Mathot's approach [113] (Fig. 1.12). The baseline heat capacity [113] is not known precisely for these samples because specific heat capacities can only be determined with an accuracy of about 10 % [36]. As a first approximation the melting peak is always integrated versus the liquid heat capacity of the sample (Fig. 1.12). The cold-crystallization peak area is integrated against liquid heat capacity in case of fully amorphous samples (Fig. 1.12c) or against some intermediate line between liquid and crystalline heat capacity for semi-crystalline samples (Fig. 1.12c). Compared to Mathot's approach [113], this yields an additional error, but does not change the overall picture significantly.

In similar calorimetric experiments, Oguni [107] and Mathot [47] have shown that in amorphous samples the cold-crystallization peak areas are dependent on the number of nuclei existing prior heating. Oguni et al. [107] demonstrated for the case of

amorphous sample
before heating

more nuclei

amorphous sample
before heating

fewer nuclei

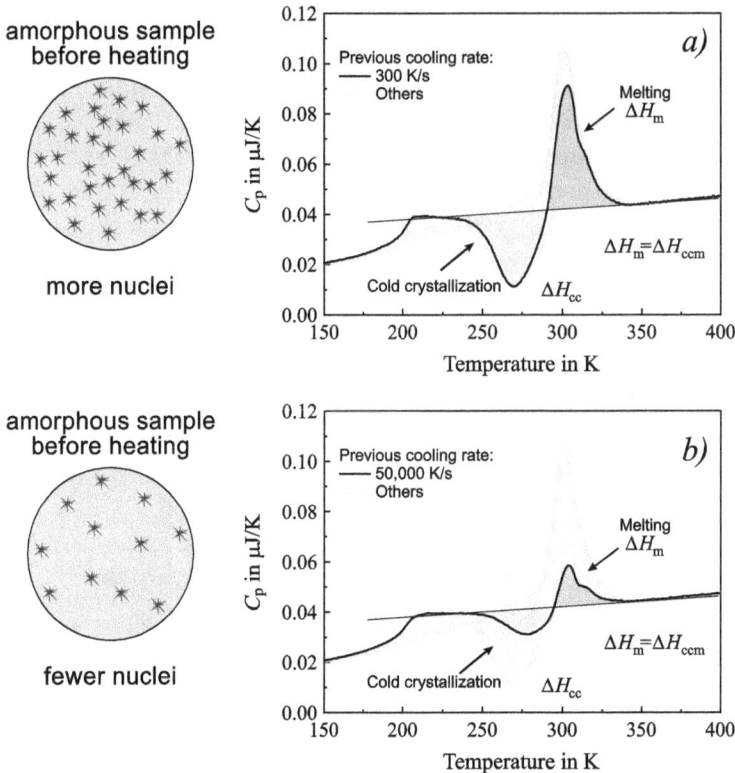

Fig. 1.13: Reduction of cold crystallization and subsequent melting in dependence on the number of nuclei in an amorphous sample before heating (note the different peak sizes when comparing Figs. 1.13a and b).

3,3'-dimethoxy-4,4'-bis(2,2-diphenylvinyl)biphenyl that cold crystallization vanished fully on heating with a rate of 10 K/min if no nuclei were formed by some prior heat treatment. Using PCL data, this is demonstrated by the schematic based on the extension of the highlighting of the experiments of Fig. 1.12 from cooling from 300 to $5 \cdot 10^4$ Ks^{-1} in Fig. 1.13.

Using this method the heating trace of the sample after the temperature treatment is usable for the detection of both processes: nucleation and overall crystallization. As long as the sample is always heated at the same rate, the influence of heterogeneous nucleation can also be established from the constant cold crystallization observed when homogeneous nucleation is completely avoided. Furthermore, the fixed heating rate allows one the application of the same evaluation procedure for all data, therefore minimizing errors, which can arise during evaluation of scans at different rates [36]. In addition, there is no need to evaluate small, usually noisy signals which occur during isothermal evaluations of latent heats at long crystallization times, like in [40]. Summarizing, based on these preliminary observations, both, the samples nucleated and crystallized with different cooling rates as well as those treated isothermally at constant annealing temperatures, T_a, were analyzed immediately thereafter with heating scans in the same device under the same experimental conditions as demonstrated in Figs. 1.12 and 1.13.

The formation of nuclei cannot be measured directly by calorimetry because the heat effects are below the detection limit. Therefore, an indirect method for quantitative separation of the nucleation and crystallization processes is needed. To probe crystallization on previous cooling and, if possible, to observe an influence on nucleation, the cold-crystallization and the overall latent heat on heating were evaluated, as shown in Figs. 1.12 and 1.13. The overall latent heat on heating, $\Delta H_{\text{heating}}$, was taken as the sum of the melting enthalpy (ΔH_m, positive) and the cold-crystallization enthalpy (ΔH_{cc}, negative). Fig. 1.14 shows the dependence of the overall latent heat (black squares) and the cold-crystallization enthalpies (triangles) versus previous cooling rate for neat PCL (solid symbols) and a heterogeneously nucleated PCL (open symbols). The error bars include the influence of baseline construction on the results of integration.

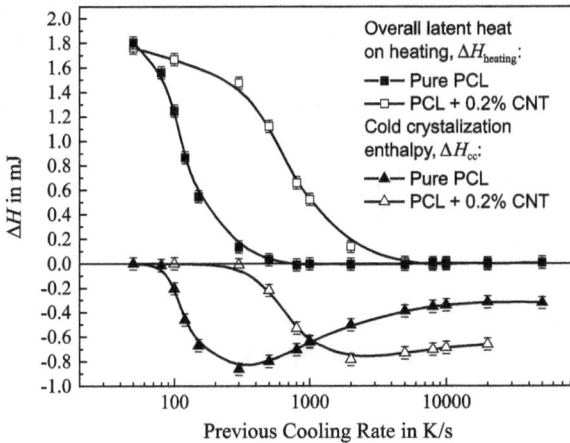

Fig. 1.14: Overall latent heat on heating, $\Delta H_{\text{heating}}$, and cold crystallization enthalpy, ΔH_{cc}, in dependence of previous cooling rate measured for neat PCL and PCL mixed with 0.2 wt.% MWCNT (multi-walled carbon nanotubes) functioning as added heterogeneous nuclei.

For the neat PCL the overall latent heat becomes equal to zero within error of measurement at previous cooling rates higher than 500 Ks^{-1}. This means that no crystallization occurs on cooling under this condition or at even higher rates. This finding coincides with results of previously performed non-isothermal experiments [40]. At cooling rates higher than 300 Ks^{-1} a reduction of cold-crystallization enthalpy is observed. This is interpreted as a reduction of nucleation on cooling in analogy to the experiments reported by Oguni et al. [107] and Mathot et al. [47] (Fig. 1.13). The reduction of the intensity of cold crystallization on heating without crystallization on previous cooling was explained by a reduction of homogeneous nuclei formation and the saturation was explained by the presence of unavoidable heterogeneous nuclei [38]. To justify the latter assumption, the same experiment was performed with the nucleated PCL. The nucleated PCL crystallizes much faster, as it is shown in Fig. 1.14. The overall latent heat at heating becomes zero only after cooling at rates faster than 5000 Ks^{-1}, i.e., up to this fast cooling rate some crystallization happens before heating. The reduction of the cold-crystallization peak after faster cooling is much less pronounced than in neat PCL. In parallel, the maximum cold-crystallization area for nucleated PCL is reached only on heating after cooling at 2000 Ks^{-1}.

The advantage of this indirect analysis (heating after preparation of the sample by cooling) compared to the direct isothermal measurement used in [40] is the possibility to obtain additional information about crystallization, nucleation and stability of the objects formed at controlled cooling. From changes at the glass transition temperature, T_g, and particularly from the heat-capacity increment at T_g, information about the interaction of the growing objects and the surrounding melt is available. Next, the cold-crystallization peak is linked to the available mobile and crystallizable material. Furthermore, in non-crystalline samples the cold-crystallization enthalpy is a relative measure of the number of available nuclei, and the temperature of the cold-crystallization range is a measure of the changes of the type of the nuclei present. Finally the melting temperature of the different species present prior to the analyses and created during heating, i.e., the sum of all exotherms and endotherms, can be linked to the stability of the objects and the total enthalpy change on heating provides the total crystallinity present in the sample before the heating scan. This strategy has been described in more detail in [38].

1.3.2.3 Analysis of a Low-molecular Mass Organic Compound (8OCB)

The above developed strategy works not only for polymers but is valid in general. Here we show that the molecular liquid crystal 4-cyano-4′-octyloxybiphenyl (8OCB) behaves in a very similar way. This liquid crystal molecule has an aliphatic tail attached by an oxygen link to the rigid biphenyl core and a polar cyano head group (M_w = 307.44 g/mol). 8OCB exhibits a weakly first-order isotropic to nematic transition at $T_{I-N} \cong 353$ K and a second-order nematic to smectic-A transition at $T_{N-A} \cong 340$ K. The strongly first-order crystal–Sm-A transition occurs reproducibly on heating at $T_{Cr-A} \cong 328$ K. Furthermore the crystal phase shows rich polymorphism, which makes the molecule interesting for DFSC studies [114].

8OCB undergoes several transitions even at high cooling rates as shown in Fig. 1.15. It is employed as a secondary standard for temperature calibration of DSC's particularly in the cooling mode [115–119]. Crystallization is prevented on cooling already at a cooling rate of 80 Ks^{-1} but the mesomorphic transitions, particularly the isotropic to nematic and the nematic to smectic-A transitions are not suppressed up

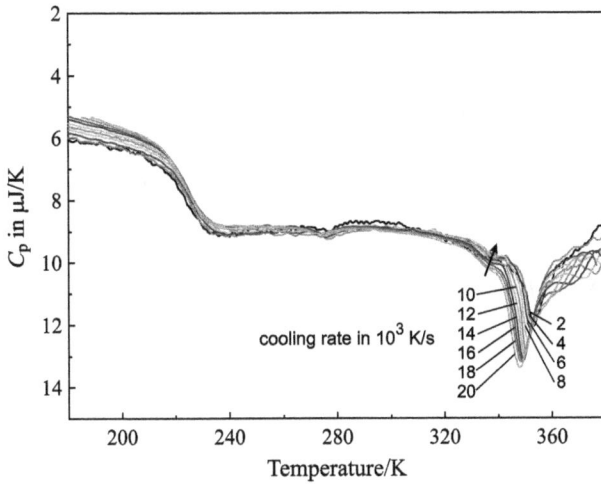

Fig. 1.15: Cooling curves of 8OCB from the melt by DFSC with different cooling rates.

to cooling rates of $2 \cdot 10^4$ Ks^{-1}. Consequently, the glasses of 8OCB created at cooling up to $2 \cdot 10^4$ Ks^{-1} are smectic-A phases and not amorphous. Fast cooling (quenching) is also desired, through which the transient or even evanescent polymorphs could be captured. Instruments with fast enough heating and cooling rates thus provide the possibility to elaborately control the samples thermal history. This way we are able to study fast evolving polymorphs, which was exemplified for the 8OCB SP form in [114]. Here we use the obtained data to show that the previously discussed scenario for preventing homogeneous nucleation does not only hold for polymers but also for the low-molecular mass organic liquid crystal 8OCB.

Fig. 1.15 shows the cooling curves of a 5 ng 8OCB sample with cooling rates from 2000 Ks^{-1} to $2 \cdot 10^4$ Ks^{-1}. Four transitions can be identified. They are the isotropic to nematic transition with peak temperatures ranging from 353 K to 348 K, the weak nematic to smectic transition (indicated by the arrow) with peak temperature from 339 K to 336 K, the weak exothermic peak (which may be due to surface induced heterogeneous nucleation and crystallization [120–123]) with fixed temperature at about 280 K and finally the glass transition with a T_g of about 232 K determined by the half-C_p step method. Although the transition temperatures of the four transitions show different cooling-rate dependencies related to their nature, the overall profile of the cooling traces is quite similar.

The small differences between the smectic glasses obtained by different cooling rates can be again investigated by subsequent heating measurements. A heating rate of $2 \cdot 10^4$ Ks^{-1} is used to avoid most of the recrystallization on heating. Fig. 1.16 shows differences in three regions depending on the previous cooling rate: the enthalpy overshoot at the glass transition because of the scanning rate asymmetry between heating

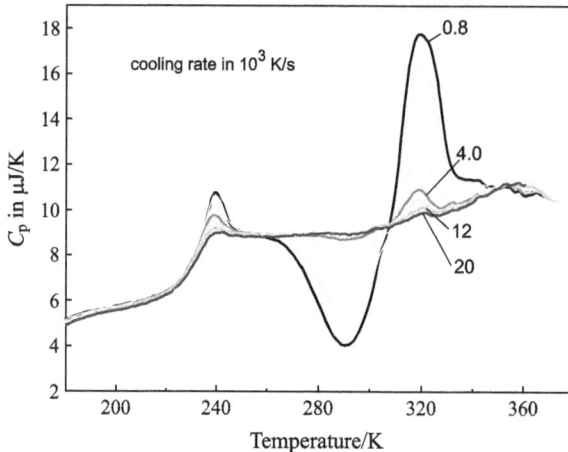

Fig. 1.16: Heating curves with rate of $2 \cdot 10^4$ Ks^{-1} of 8OCB quenched from the melt with different cooling rates (from 800 to $2 \cdot 10^4$ Ks^{-1}).

and cooling, cold crystallization just after the glass transition and melting of the crystals formed during cold crystallization.

Here, we focus on the relationship between cold crystallization and previous cooling history, which relates to the nucleation density [124]. At slow cooling (800 Ks^{-1}) a very strong cold crystallization is observed which gradually decreases with increasing cooling rate. At a cooling rate $2 \cdot 10^4$ Ks^{-1} and a heating rate $2 \cdot 10^4$ Ks^{-1} possibly no homogeneous nuclei are formed and the heterogeneous nucleation density is so low that no growth occurs at that fast heating. The important point is that at a cooling rate of $2 \cdot 10^4$ Ks^{-1}, we can obtain a smectic-A glass without homogeneous nuclei. The smectic-A glass obviously is ordered (structured) but this structure does not yield crystal nuclei. Comparing Fig. 1.16 with Fig. 1.13 above and Fig. 1.18 below shows how similar the general behavior is. For both materials, polymers and low-molar mass liquid crystals fast enough cooling prevents the formation of homogeneous nuclei at cooling.

1.3.2.4 Analysis by Heating at the Optimum Heating Rate

At heating with 1000 Ks^{-1} there is always cold crystallization seen for the neat PCL even after fastest cooling (Fig. 1.14). After fastest cooling no homogeneously formed nuclei should be present. The remaining cold crystallization is therefore due to heterogeneities, which are always present in polymers. Using cleaner samples reduces the problem but it is still present [125].

One way to avoid crystallization from unavoidable heterogeneities is using higher heating rates. At higher heating rates growth of crystals initiated from a few heterogeneities may be reduced and not result in measurable cold crystallization. The rate

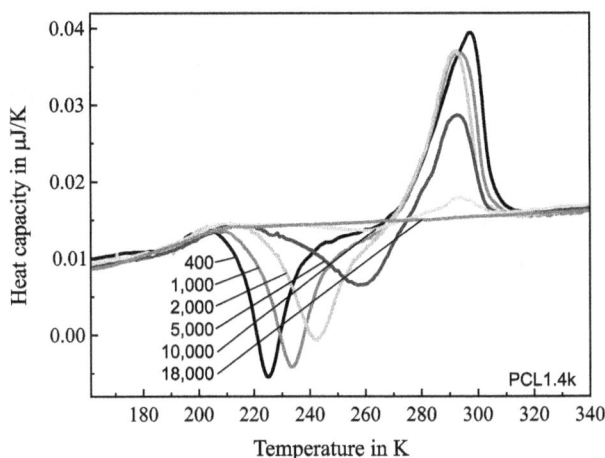

Fig. 1.17: Apparent heat capacity of about 10 ng PCL1.4k from DFSC on heating with different heating rates after cooling at 10^5 Ks^{-1}.

for the reheating scan (magenta (in color version) in Fig. 1.17, i.e., curve referring to $1.8 \cdot 10^4$ Ks^{-1}), where no cold-crystallization is seen for the untreated neat sample was determined to be $1.8 \cdot 10^4$ Ks^{-1} [125]. But it is required to check if at such high rate collection of information regarding the effects discussed above (e.g. cold crystallization originating from homogeneous nuclei, melting of crystals) is still possible.

The criteria that need to be fulfilled simultaneously are discussed here on hand of Fig. 1.17. The criteria can be summarized as follows: (i) the heating rate must be fast enough to prevent measurable growth of crystals from unavoidable heterogeneities in the neat sample. (ii) The heating rate must be fast enough to prevent formation of homogeneous nuclei on heating, i.e., a sample not containing a large number of heterogeneous or homogeneous nuclei at the beginning of the heating scan must not show cold crystallization. (iii) The heating rate must be slow enough allowing growth (cold crystallization) of a larger number of nuclei formed or added prior to heating. (iv) Melting of existing crystals or crystals grown on heating can be studied. These four conditions are well satisfied for PCL samples [125] at heating rates of $1.8 \cdot 10^4$ Ks^{-1}, as is verified next.

In the experiments displayed in Fig. 1.17, a low-molar mass sample (PCL1.4k) was initially cooled from the melt with 10^5 Ks^{-1}. At such high cooling rate homogeneous nucleation in PCL is prevented [38]. The heating rate needed to fulfill criteria (i) and (ii) was then determined by heating the initially rapidly cooled samples at different rates. Because the starting material is always amorphous, the glass transition of all samples is identical regarding the heat-capacity increment at T_g and the value of T_g within the experimental uncertainty, and the overall enthalpy change is always zero. The temperature and the enthalpy change of cold crystallization can be used for the analysis. The largest amount and lowest temperature of cold crystallization appears at low heating rates due to long enough time for nucleation near T_g and subsequent cold crystallization at higher temperatures. With increasing rate, the time for nucleation as well as for growth initiated by the few heterogeneous nuclei becomes reduced and cold crystallization finally vanishes at heating rates of $1.8 \cdot 10^4$ Ks^{-1}. Next, it was checked if for heating at $1.8 \cdot 10^4$ Ks^{-1} conditions (iii) and (iv) are simultaneously fulfilled too. Following the argumentation developed in Section 1.3.2.2 the influence of different cooling rates on nucleation and crystallization in the sample was measured using the scheme of Fig. 1.11. Fig. 1.18 shows the heat capacity curves on reheating with $1.8 \cdot 10^4$ Ks^{-1} measured immediately after cooling at rates between 50 Ks^{-1} and 10^5 Ks^{-1} for a low molar mass (PCL1.4k) sample.

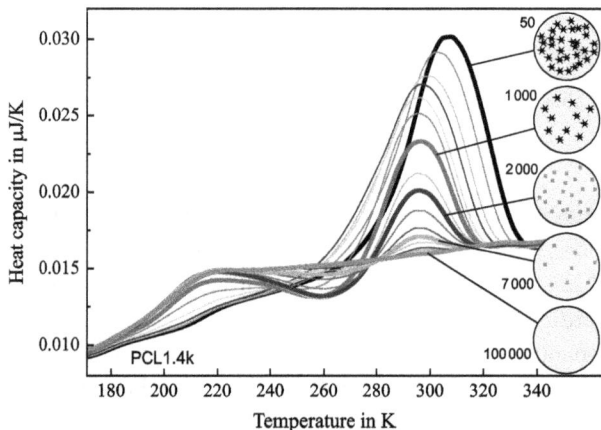

Fig. 1.18: Apparent heat capacity of about 10 ng PCL1.4k from DFSC on heating with $1.8 \cdot 10^4$ Ks^{-1} after cooling with rates between 50 and 10^5 Ks^{-1}.

The different situations highlighted in Table 1.1 are exemplified in Fig. 1.18 by marking the heating curves which are represented best by the chosen pictographs. For example, after cooling of PCL1.4k at 1000 Ks^{-1} (purple in colored version), the glass transition is clearly shifted to slightly higher temperatures and the step height is decreased in comparison to the non-crystalline samples as produced when cooling at 7000 Ks^{-1} or faster. Calculating the overall enthalpy change on heating after cooling with 1000 Ks^{-1} yields a value of about 15 J/g. It corresponds to a crystallinity of about 10 % at the beginning of the heating scan and can also be seen in the decreased step height of the glass transition. Some of the crystallinity after cooling at 1000 Ks^{-1} grows during heating, as can be seen from the presence of an exothermic cold-crystallization peak between 243 K and 273 K. The heating scan after cooling with 2000 Ks^{-1} (dark blue in colored version) shows no deviation from the glass transition of the amorphous PCL1.4k (after cooling at 10^5 Ks^{-1}). The overall enthalpy change was estimated to be 5 J/g, which is close to the error limit for the applied method. The interpretation is that there are no crystals in the sample prior to the analysis and all latent heat effects originate from the small amount of cold crystallization followed by its melting. More details are needed to understand these "small numbers of nuclei" of Fig. 1.18 .

The results of all experiments as exemplified in Fig. 1.18 are combined with those of samples of different molar masses in Fig. 1.19. The overall enthalpy change on heating as a measure of crystallinity (filled squares) and the cold-crystallization enthalpy as a relative measure of the number of active nuclei (triangles) are again plotted as a function of previous cooling rate. Clearly, several overlapping processes must be separated for interpretation and linked to their molar mass dependence. A main dividing line is found at about 2000 Ks^{-1}. At lower cooling rates crystallization occurs with an increasing intensity already during cooling and the glass transition decreases in magnitude and shifts to a higher temperature. This decrease is larger than the expected decrease from crystallization only. All of these observations indicate the formation of a rigid amorphous nanophase which ultimately stops the crystallization. The lowest molar-mass sample (PCL1.4k) shows most cold crystallization on heating (Fig. 1.17), while the higher molar-mass samples show less or no cold crystallization on heating. Not only the amount of cold crystallization on heating but also the cooling rate range after which cold crystallization occurs depends on molar mass.

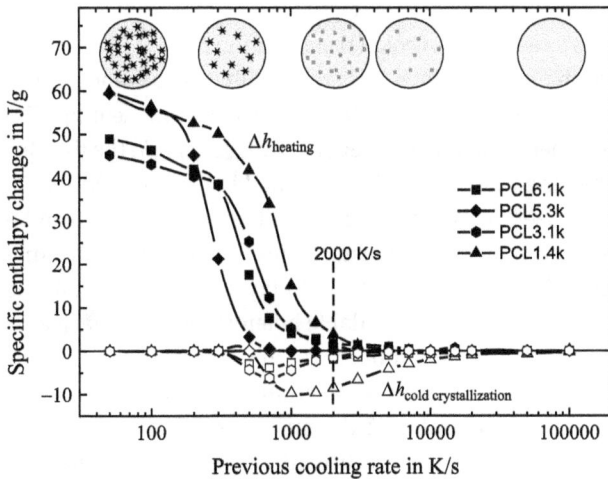

Fig. 1.19: Overall specific enthalpy change (squares) and specific enthalpy change on cold crystallization (triangles) of about 10 ng PCLs from DFSC on heating with $1.8 \cdot 10^4$ Ks^{-1} after cooling with rates between 50 and 10^5 Ks^{-1}. The pictograms are horizontally assigned for the PCL1.4k sample, but give also a general trend for all other samples.

From Fig. 1.19 the critical cooling rate for achieving a non-crystalline sample is estimated from the total enthalpy change (filled symbols). It reaches zero between 700 Ks^{-1} and 4000 Ks^{-1} depending on the sample under investigation. For the cold-crystallization enthalpy (open symbols) a strong dependence on cooling rate is observed too. The initial increase of cold-crystallization enthalpy with increasing cooling rate is caused by the reduction of crystallinity. When the sample is close to non-crystalline after cooling, the cold-crystallization enthalpy reaches a maximum. Later on, cold-crystallization enthalpy decreases with increasing cooling rate because the number of homogeneous nuclei is decreasing. For the sample used in [38] and discussed above the decrease was limited to a certain value due to heterogeneous nuclei which were present in the samples. The specially synthesized samples used for the analysis shown in Fig. 1.19 and a heating rate of $1.8 \cdot 10^4$ Ks^{-1} instead of 1000 Ks^{-1} as above, show a decrease of cold-crystallization enthalpy approaching zero, indicating negligible growth or activation of heterogeneous nuclei as expected. Figs. 1.18 and 1.19 show that for the PCL samples studied in [125] all mentioned conditions (i–iv) are fulfilled simultaneously for heating rates of $1.8 \cdot 10^4$ Ks^{-1}. Samples cooled at 10^5 Ks^{-1} therefore provide an excellent starting material for detailed studies of nucleation and crystallization kinetics.

1.3.3 Non-isothermal Ordering Kinetics

1.3.3.1 Neat PCL

Sufficiently fast cooling reveals a reduction of cold crystallization of neat PCL on subsequent heating. This reduction was explained to be caused by the absence of homogeneous nuclei after cooling, which could grow into a crystal during heating. The cold-crystallization enthalpy depends on the number of active nuclei and the time for growth (determined by the heating rate). Since the same heating rate of 1000 Ks^{-1}, respectively, $1.8 \cdot 10^4$ Ks^{-1} was used so far, the cold-crystallization enthalpy can be considered to be a relative measure of the number of nuclei present prior to the heating scan. The saturation of this effect before reaching zero cold-crystallization enthalpy is due to the presence of a constant number of unavoidable heterogeneous nuclei (causing nucleation on surfaces or impurities). Even though we cannot prove the reason for the remaining nucleation of cold-crystallization with certainty, the reduction of nucleation with increasing rate was much smaller in the nucleated PCL (Fig. 1.14) and Section 1.3.3.2. Such different behavior is expected from the hypothesis of heterogeneous nucleation. The cold-crystallization and overall latent heat on heating together are, thus, a measure of the number of heterogeneous and homogeneous nuclei grown on cooling and the crystallinity of the material prior to the heating scan.

To clarify what was detected on heating after preceding cooling from the melt at different rates, the results of the overall latent heat and cold-crystallization enthalpy on heating in Fig. 1.14 are summarized once more in Fig. 1.20 to quantify the scheme of Fig. 1.4. Depending on the prior cooling rates, the subsequent heating shows distinct regions which allow one a modeling of the different thermal histories:

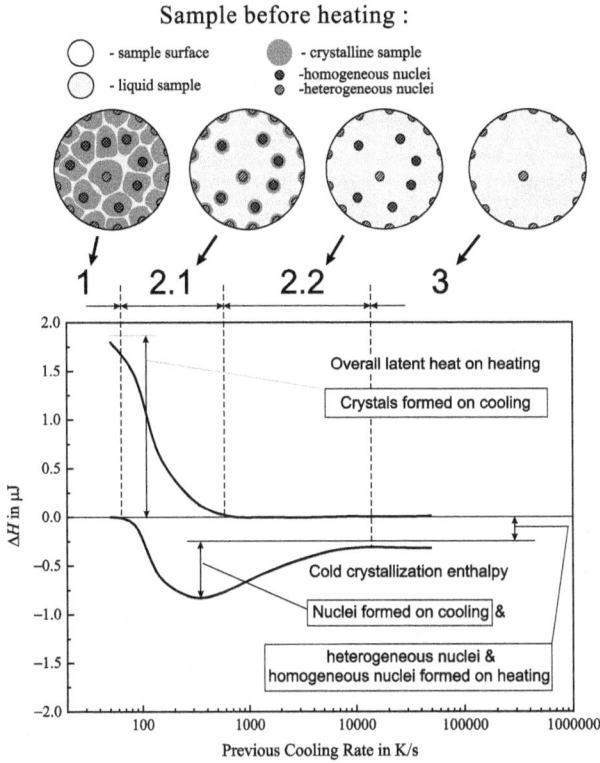

Fig. 1.20: Overall latent heat and cold crystallization enthalpy on heating at 1000 Ks^{-1} as a function of preceding cooling rate according to Fig. 1.14 (for details see text).

Region 1. The crystallization of the sample was completed during the slow cooling. The growth occurred mainly via heterogeneous nucleation, followed by crystal growth and annealing, to continue to the semicrystalline limit as set by the rigid amorphous fraction (RAF) [126]. The sum of all these processes is registered by the overall latent heat and expressed by the sketch above the curve of Fig. 1.20.

Region 2. The sample did not complete its crystallization during cooling. Initially (Region 2.1 in the sketch), homogeneous and heterogeneous nucleation still occurs but, compared to Region 1, fewer crystals grow and perfect. The remaining nuclei induce various amounts of cold crystallization. In Region 2.2 in Fig. 1.20, homogeneous and heterogeneous nuclei which do not reach measurable size during cooling, i.e., they do not contribute to the overall latent heat, can grow to higher-melting crystals in the cold-crystallization temperature range. The amount of the higher-melting crystals continues to decrease with shorter thermal pre-treatments, causing the minimum in the latent heat of cold crystallization before the beginning of Region 3.

In **Region 3**, there are no crystals produced by neither homogeneous nor hetero-geneous nucleation on cooling. All melting at high temperature is due to cold crystal-lization which is now caused exclusively by heterogeneous nucleation.

Two major results, which are important for studying nucleation and crystalliza-tion kinetics, were taken from these experiments: (i) The heating scan can be used as a relative measure of the number of previously formed homogeneous nuclei and sample crystallinity by evaluating cold-crystallization and overall latent heat, respec-tively. (ii) For PCL, DFSC allows one to perform temperature scans that avoid not only crystallization, but also homogeneous nucleation on cooling. The cold-crystallization enthalpy seen on analysis at a constant intermediate heating rate of 1000 Ks^{-1} [38] and at $1.8 \cdot 10^4$ Ks^{-1} [125] as well as in [127] was used as a measure for the number of prior formed nuclei and the overall latent heat was used as a measure of crystallinity formed on the preceding cooling or annealing. If a sample was amorphous, it shows zero overall latent heat, but the cold-crystallization enthalpy as well as its melting en-thalpy depends on the number of active nuclei in the sample. Because the same heat-ing rate was used for all experiments the rate-dependent errors are minimized and the changes due to the various thermal histories could be detected and compared.

Nevertheless, two assumptions were made in [38]: (i) the cold-crystallization seen after cooling at rates faster than 7000 Ks^{-1} is due to heterogeneous nuclei and not due to homogeneous nuclei formed during heating and (ii) the cold-crystallization enthalpy is a measure of the number of nuclei, in ideal case it is proportional to the number of nuclei. Assumption (i) could be proven if a heating rate could be found at which no cold crystallization is seen after very fast cooling but cold crys-tallization occurs if a certain number of nuclei is present. The second requirement is important because it is possible for most polymers to find a heating rate at which no crystallization occurs because the time for growth is too small even if a large number of nuclei is present (Section 1.3.2.4). In order to check if cold-crystallization enthalpy is proportional to the number of available nuclei, measurements on an-other series of multi-walled carbon nanotube (MWCNT) nucleated samples were per-formed.

1.3.3.2 Carbon Nanotube Nucleated PCL

A series of heating scans of a PCL sample nucleated by 5 wt% multi-walled carbon nanotubes (MWCNT) after cooling at different rates as sketched in Fig. 1.11 are shown in Fig. 1.21. They are all recalculated using the same procedure, which is described in [36]. Therefore the data can quantitatively be compared to each other. Additionally, the sample mass was estimated using ATHAS database data [37], as described above and in [36, 38], thus allowing comparison of crystallinity on an absolute scale for several samples.

The cold-crystallization enthalpy, Δh_{cc}, and the overall latent heat, $\Delta h_{cc} + \Delta h_m$, on successive heating after cooling at different rates, as shown in Fig. 1.22, provide in-

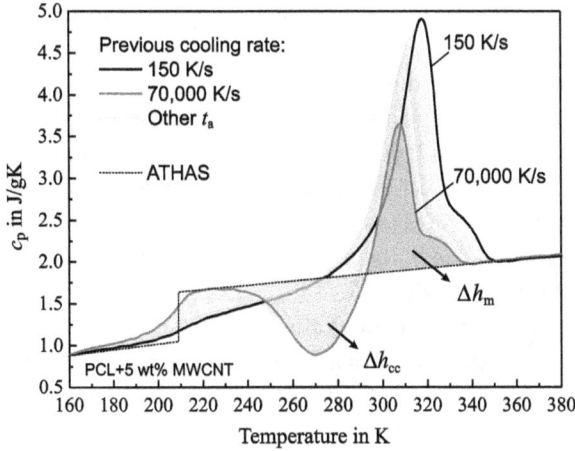

Fig. 1.21: Heating scans for PCL + 5 % MWCNT after cooling at different rates and an integration example.

formation on the number of active nuclei and the degree of crystallinity, respectively. Here we use specific enthalpies, h [J/g], while previously we discussed enthalpies, H [J]. This is needed for the comparison of the different samples. The overall latent

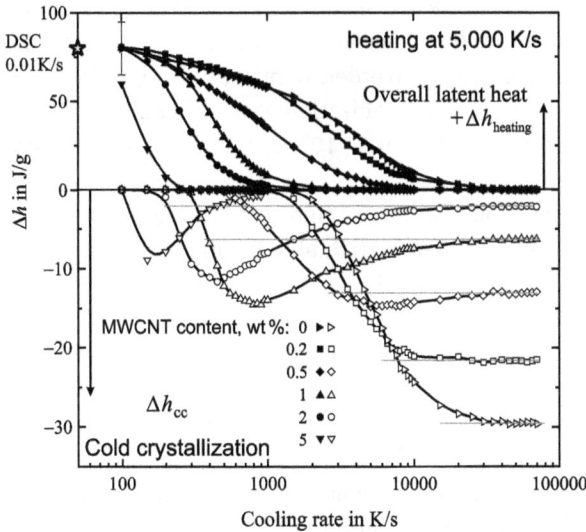

Fig. 1.22: Overall latent heat on heating, and cold-crystallization enthalpy (for clarity scales are different), as function of previous cooling rate measured for neat PCL and PCL with different concentrations of MWCNT. The star shows the heat of fusion after crystallization at 1 Kmin^{-1} (0.016 Ks^{-1}), measured in conventional DSC, which is within the determination error the same for all samples.

heat on heating was taken as the sum of the melting enthalpy (Δh_m, positive) and the cold-crystallization enthalpy (Δh_{cc}, negative). Fig. 1.22 shows the dependence of the overall latent heat ($\Delta h_{heating}$, upper part) and the cold-crystallization enthalpies (bottom) versus previous cooling rate for PCL with different concentrations of MWCNT. After crystallization at 1 K/min (0.016 Ks^{-1}) employing a conventional DSC (PerkinElmer Pyris 1 DSC), the melting enthalpy was almost the same for all samples and equals 80 ± 5 J/g (star on the ordinate in Fig. 1.22). This value was used for a final correction of the sample mass (neat polymer normalized), therefore all overall melting curves at slowest scanning rate saturate at the same value.

For the neat PCL, within the error of measurement, the overall latent heat at heating becomes zero at previous cooling rates higher than 300 Ks^{-1}. This means that no crystallization occurs on cooling under these conditions. At cooling rates higher than 200 Ks^{-1}, a reduction of the enthalpy of cold crystallization is observed. This reduction was explained to be caused by a reduced number of active nuclei after cooling, which could grow into a crystal during heating. The cold-crystallization enthalpy depends on the number of active nuclei and the time for growth (determined by the heating rate). Since the same heating rate of 5000 Ks^{-1} was used, the cold-crystallization enthalpy can be considered to be a relative measure of the number of nuclei present prior to the heating scan [38]. For cooling rates above 1000 Ks^{-1}, cold crystallization and corresponding melting peaks are almost invisible on successive heating for the neat PCL. The disappearance of cold crystallization indicates a dramatic reduction of the number of active nuclei in the sample. The nucleated samples show faster crystallization kinetics. PCL with an addition of 5 % MWCNT becomes amorphous only on cooling with rates higher than $5 \cdot 10^4$ Ks^{-1}, which is two orders of magnitude higher than for the neat polymer. The maximum of the cold-crystallization peak area is also shifted to higher cooling rates. It is reached after cooling at $3 \cdot 10^4$ Ks^{-1} and stays constant for faster cooling.

Sufficiently fast cooling causes disappearance of cold crystallization of neat PCL on subsequent heating. This was explained by a gradual reduction of the number of homogeneously formed nuclei during cooling with increasing cooling rate. For PCL with up to 1 wt% MWCNT a similar behavior is observed. But the cold-crystallization peak reduction in nucleated PCL saturates at a level, corresponding to the density of heterogeneous nuclei as seen in Figs. 1.22 and 1.23.

When studying the efficiency of nucleating agents (density of active nuclei), one needs to specify the temperature range of interest. Depending on crystallization temperature, homogeneous nucleation may compete with the added nuclei and overturn their influence [128]. Judging the nucleation efficiency is commonly done by comparing the crystallization peak temperature on cooling from the melt in differential scanning calorimeter (DSC) traces at cooling rates of the order of 10 K/min [129]. A higher crystallization temperature is considered to indicate higher nucleation efficiency. Fig. 1.23 shows data of pure and MWCNT nucleated PCL samples with different amounts of the nucleating agent. The increase of crystallization temperature

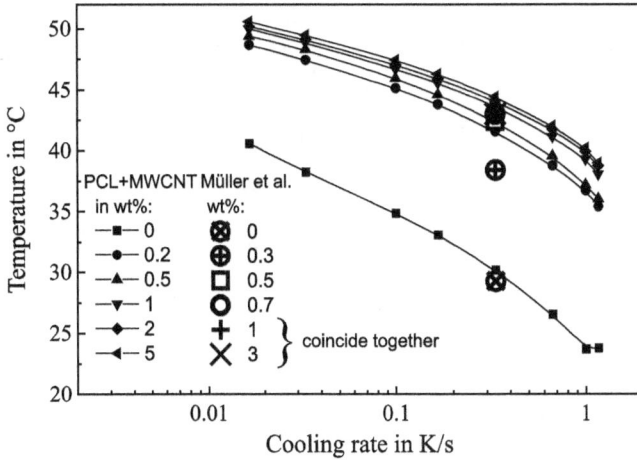

Fig. 1.23: Influence of multi-walled carbon nanotubes (MWCNT) on the crystallization peak temperature on cooling of poly(ε-caprolactone) (PCL), studied by conventional DSC (PerkinElmer Pyris 1 DSC). Crosses are data from [130].

by adding MWCNT to the PCL sample is well pronounced. From Fig. 1.23 one could conclude that already 0.2 % MWCNT yield a saturation of the nucleation effect. A similar study of the influence of MWCNT on nucleation of PCL was carried out by Müller et al. in [130]. They have found saturation of the nucleating effect at about 0.5 wt% MWCNT concentration as shown in Fig. 1.23.

Müller et al. determined the efficiency using the crystallization peak maximum temperature on slow cooling as proposed by Fillon et al. [129] and reached a value of 200 %. The isothermal crystallization of these samples was also followed in [130]. However, at that time the temperature range was very limited due to limited cooling abilities of conventional DSC's. The samples were crystallizing on cooling before reaching lower temperatures. For the neat PCL isothermal crystallization was measured in the temperature range from 318 K to 320 K, and for faster crystallizing PCL+MWCNT nano-composites the temperature range was limited to 322–328 K.

Our data on different PCL+MWCNT composites, Fig. 1.23, show the same trend and saturation of the nucleation effect is already observed at 0.2 wt% MWCNT. At such high crystallization temperatures a small number of active nuclei are sufficient to allow the sample to fully crystallize within the time defined by the slow DSC experiment. In order to clarify if nucleation by MWCNTs is indeed already saturated at 0.2 wt% MWCNT in PCL, we have performed the experiments described above. Going beyond the traditional methods of nucleation efficiency comparison, in [99] we suggest a comparison based on a quantitative measurement of isothermal crystallization rates. Furthermore, the influence of homogeneous nucleation on isothermal crystallization of pure and nucleated samples was investigated and compared with the nucleation efficiency of MWCNT as discussed below. The enthalpy data shown in Fig. 1.22 after

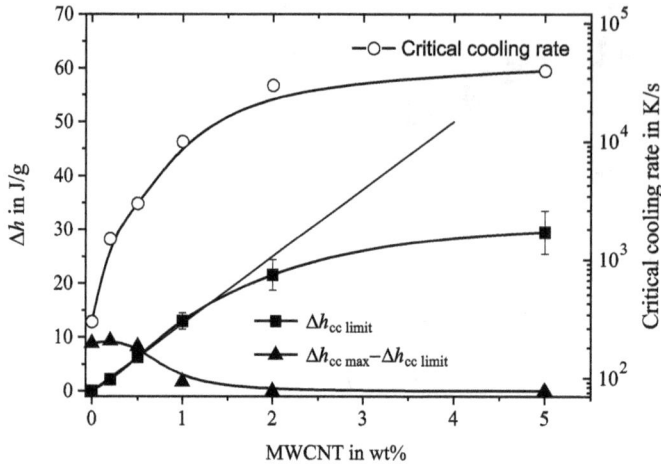

Fig. 1.24: Critical cooling rates to obtain non-crystalline samples and limiting cold-crystallization enthalpies, $\Delta h_{cc\,limit} = \Delta h_{cc}(7 \cdot 10^4\ \mathrm{Ks}^{-1})$ from Fig. 1.22, and the difference between the maximum value of cold-crystallization enthalpy, $\Delta h_{cc\,max}$ and $\Delta h_{cc\,limit}$ in Fig. 1.22 as a function of MWCNT loading. The lines are guides to the eye.

cooling at different rates obviously do not show saturation of the nucleating effect of MWCNTs at 0.2 wt% for the highest cooling rates employed. To quantify the observations, the critical cooling rates to obtain non-crystalline samples and the limiting cold-crystallization enthalpies, $\Delta h_{cc\,limit} = \Delta h_{cc}(7 \cdot 10^4\ \mathrm{Ks}^{-1})$, are presented in Fig. 1.24.

Fig. 1.24 shows a linear increase of the limiting cold-crystallization enthalpy up to about 2 wt% MWCNT. Only the sample with 5 wt% MWCNT is significantly off of the linear relation. This deviation is probably caused by agglomeration of the MWCNT at these high concentrations. A similar trend is observed for the critical cooling rate to make the samples non-crystalline. Up to 2 wt% MWCNT an increase by two orders of magnitude is present and for 5 wt% MWCNT only a small additional increase is seen. These two findings, the dramatic increase in critical cooling rate and the linear relation between $\Delta h_{cc\,limit}$ and the MWCNT loading up to 2 wt% MWCNT, contradict the apparent saturation of the nucleation efficiency at 0.2 to 0.5 wt% MWCNT as seen at low cooling rates in Fig. 1.23 and [98, 130]. The linear relationship for $\Delta h_{cc\,limit}$ indicates that the number of active nucleation sites increases proportional to the number of MWCNTs in the composite at least up to about 2 wt% MWCNT.

For intermediate cooling rates the cold-crystallization enthalpy in Fig. 1.22 shows a maximum. For the neat PCL this maximum is explained by the decreasing crystallinity with increasing cooling rate, allowing more pronounced cold crystallization, and at further increasing cooling rate a decreasing number of homogeneous nuclei eventually preventing cold crystallization at all. For the PCL+MWCNT nanocomposites a similar behavior is seen besides the different levels of saturation at high cooling rates depending on the MWCNT loading. The increasing limiting level of cold crystallization, $\Delta h_{cc\,limit}$, and the increasing critical cooling rate up to 2 wt% MWCNT shows that there is still a possibility of additional nucleation up to this level of MWCNT loading. For a given MWCNT loading additional nuclei can be formed by homogeneous nucleation on cooling. Particular cooling conditions, fast enough to prevent full crystallization on cooling but not as fast preventing homogeneous nucleation, yield additional nuclei which cause the peak in the cold-crystallization enthalpy.

As a relative measure of the number of homogeneous nuclei, in addition to the MWCNT heterogeneities, the difference between the peak value and the limiting value of cold-crystallization is shown in Fig. 1.24. The maximum number of homogeneous nuclei formed on cooling (~10 J/g) is about 25 % of the number of active nuclei provided by 5 wt% MWCNT (~40 J/g). This relation holds for non-isothermal crystallization, and in the following it will be compared to the results from isothermal experiments.

The nucleating effect of the MWCNTs is seen not only at high temperatures but also below the temperature of maximum crystallization rate (minimum in crystallization half-time). This temperature range is generally not accessible by non-isothermal experiments with linear cooling as schematically shown in Fig. 1.25. If the minimum in the curve is passed without significant crystallization (critical cooling rate or faster; lower (green in color version) curves for the nucleated sample) for the nucleated sample) crystallization is only possible under isothermal conditions after passing the minimum. For the neat PCL in Fig. 1.25 the critical cooling rate to prevent crystallization is low enough (400 Ks^{-1}) to allow homogeneous nucleation at lower temperatures

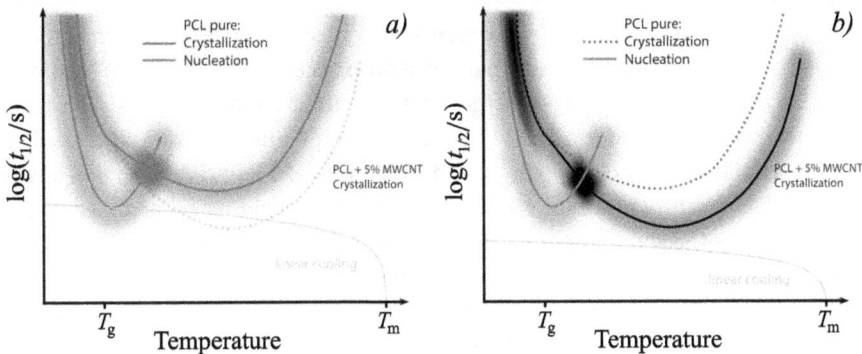

Fig. 1.25: Schematic representation of isothermal crystallization (blue curves (in color version), respectively, the right curve in the left figure surrounded by shadows and the dotted curve on the right figure) refer to PCL; black curves (in color version), respectively, the right curve in right figure surrounded by shadows and the dotted curve on the left figure) refer to PCL + 5 wt% MWCNT) and homogeneous nucleation kinetics (left curves (red in color version) in each of the figures surrounded by shadows). The shadows around the curves indicate the crystallization range and the green (lower) lines show a linear cooling from the melt. Latter curves describe: Left) Critical cooling rate to avoid crystallization for PCL (400 Ks^{-1}). Right) Critical cooling rate to avoid crystallization for PCL + 5 wt% MWCNT ($5 \cdot 10^4$ Ks^{-1}).

(crossing the red (left-side shadowed) curve). For the PCL with 5 wt% MWCNT homogeneous nucleation is not possible because critical cooling rate is so fast ($5 \cdot 10^4$ Ks^{-1}) that the curve for homogeneous nucleation (red (left-side shadowed)) is not crossed anymore. At such rates we therefore do not expect to see any additional homogeneous nucleation. For intermediate MWCNT concentrations the situation changes gradually and at rates around the critical cooling rates we first see the occurrence of homogeneously formed nuclei (the minimum in the bottom curves, Δh_{cc}, in Fig. 1.22) because the cooling intersects with the nucleation line (red (left-side shadowed)) in Fig. 1.25 and at higher cooling rates it does not intersect and the effect disappears. The data presented in Figs. 1.22 and 1.24 support this point of view. But the question remains if nucleation density due to the addition of 5 wt% MWCNT has already reached a limiting value. To check this we performed isothermal experiments discussed next.

1.3.4 Isothermal Ordering Kinetics

1.3.4.1 Neat PCL

To study nucleation as well as crystallization in PCL in detail, the ability to avoid crystal nucleation and growth on cooling is required and can be achieved by DFSC, as shown above. Next, another set of experiments was performed for which the sample was cooled so fast ($1 \cdot 10^4$ Ks^{-1} up to 10^5 Ks^{-1}, depending on the sample under investigation) that no homogeneous nucleation occurred. This method was used to study the kinetics of nucleation under isothermal conditions. The scheme of the experiments is shown in Fig. 1.26.

The base temperature for the experiments was chosen to be 100 K which is 100 K below the glass transition so that no changes should occur in the "frozen-in" sample. To equalize even unlikely changes of the sample at the base temperature, the sample was held there always for a fixed time of 0.1 s before annealing and the heating scan for analysis. The samples were, thus, heated up to 470 K and then cooled down to the base temperature at a rate $1 \cdot 10^4$ Ks^{-1} or faster, avoiding homogeneous nuclei formation. At the same rate the samples were heated to the annealing temperature, T_a, and held there for different times from 0.1 ms to 32 h. The final heating at 1000 Ks^{-1} was used as the measuring scan. The annealing was performed at selected temperatures for different times starting from 185 K ($T_g \cong 200$ K, half-step temperature measured on the used amorphous sample at 1000 Ks^{-1}) up to 330 K (equilibrium $T_m = 342$ K [37]).

Some of the resulting heating scans after isothermal annealing at various temperatures and for different times as described in Fig. 1.26 are displayed in Fig. 1.27. All of the approximately 500 DFSC runs to be discussed here are collected in the Supplementary Material of [38] at larger scale and in higher resolution compared to the examples shown in Fig. 1.27. The scan after 0.1 ms annealing equals a scan without any annealing. The 0.1 ms annealing was introduced to have in all experiments the same thermal

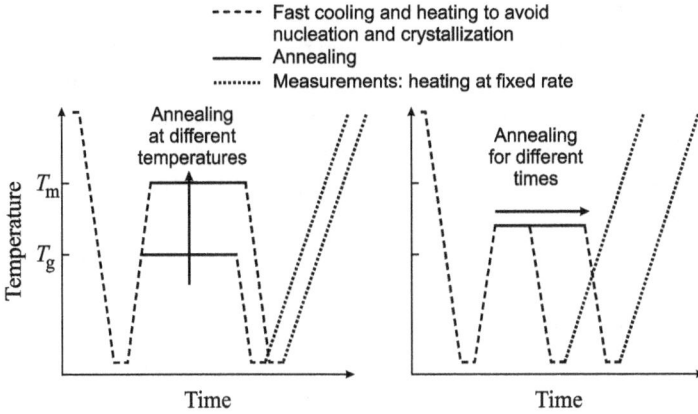

Fig. 1.26: Temperature-time profile for the annealing experiments at T_a. The sample was melted, quenched, and then annealed at different temperatures and times. All cooling and heating rates except the final measurement for analysis (dotted, 1000 Ks^{-1}) were performed at 10^4 Ks^{-1} to avoid formation of homogeneous nuclei and crystallization outside the annealing treatment.

history before the final measurement scan at 1000 Ks^{-1} (as illustrated in Fig. 1.26). The development of nuclei and crystallization during annealing was traced by changes in the glass transition, annealing peak, cold crystallization, and melting on heating in analogy to the experiments described in Fig. 1.14. The important difference is that then nuclei and crystal evolution can be studied at constant temperatures, and for specified times. The resulting overall latent heat of heating and latent heat of cold crystallization on heating for each annealing temperature and annealing time were calculated and are plotted versus annealing time in Fig. 1.28.

The overall latent heat displays the expected behavior. When the overall latent heat is zero, all crystals grew during the analysis by cold crystallization on heating and subsequently melted on further heating. In this case, the overall latent heat is zero i.e.

$$\Delta H_{\text{heating}} = \Delta H_{\text{cc}} + \Delta H_{\text{ccm}} = 0. \tag{1.3}$$

When there is any isothermal crystallization during annealing, $\Delta H_{\text{heating}}$ becomes positive due to the melting of the isothermally crystallized material at T_a with an enthalpy ΔH_{isom} and

$$\Delta H_{\text{heating}} = (\Delta H_{\text{cc}} + \Delta H_{\text{ccm}}) + \Delta H_{\text{isom}}. \tag{1.4}$$

Following the development of $\Delta H_{\text{heating}}$ with annealing time, therefore, allows one the observation of the isothermal crystallization kinetics at T_a. In most measured curves shown in Fig. 1.27 and the Supplementary Material of [38] the two contributions, one for the melting of isothermally formed crystals and the other for the cold crystallization or reorganization, are sufficiently separated for analysis.

The latent heat of cold-crystallization shown in the bottom graph of Fig. 1.28 has at short times a constant value due to unavoidable heterogeneous nucleation. This is followed later by additional crystal growth due to the formation of nuclei during annealing at T_a, and finally the cold crystallization decays due to crystallization occurring already at the annealing temperature out of these nuclei. Some of these crystals melt at low temperature, as discussed in Section 1.4. The presence of a 'nucleation' induced increase in cold crystallization is visible up to 290 K (black line with stars), being a measure of the time separation between nucleation and growth of crystals at T_a. For temperatures up to 230 K the nucleation effect of cold crystallization saturates at times before major crystallization at the annealing temperature is seen in the overall latent heat.

From the measured curves shown in Fig. 1.27 and the Supplementary Material of [38], the overall latent heat ($\Delta H_{\text{heating}}$) can be used as a measure of overall crystallization, and the latent heat of cold crystallization (ΔH_{cc}) is available as a qualitative measure of nucleation processes. Assuming that these two processes are independent permits a study of their kinetics. In the discussion in Section 1.4 an attempt is made

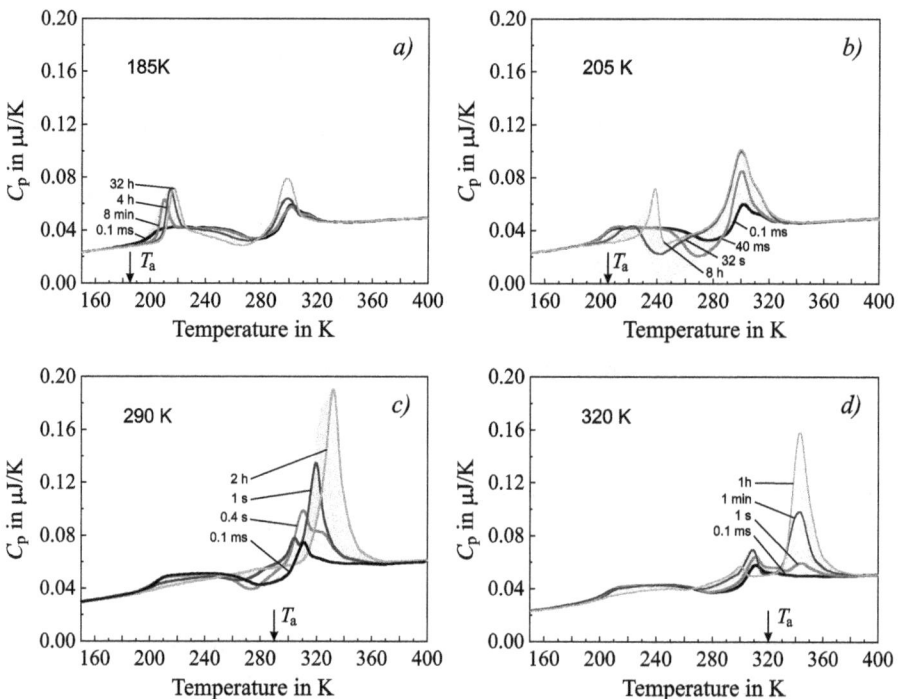

Fig. 1.27: Series of heating curves after annealing at the indicated temperatures for different times. Annealing close to T_g (a), near to T_m (d) and at intermediate temperatures ((b) and (c)). The complete set of measured curves can be found in the Supplementary Material of [38].

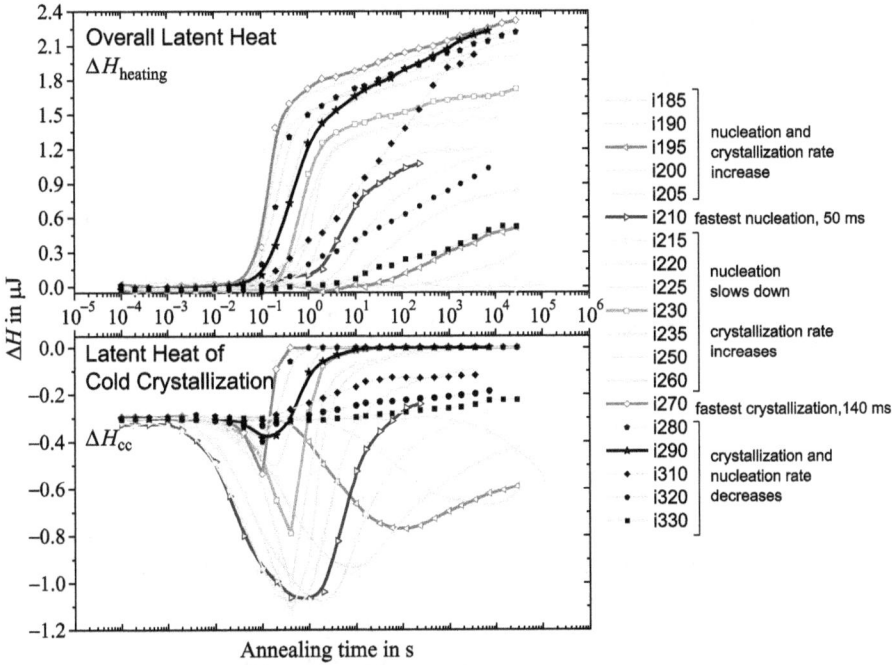

Fig. 1.28: Cold crystallization and overall latent heat for the annealing series. Each pair of curves corresponds to one annealing temperature (each point corresponds to a single measurement). The red (◇) curve shows the temperature with the highest crystallization rate (270 K), the blue (▷) curve that with the highest nucleation rate (210 K), the green (□) curve is an intermediate one (230 K), and the pink (◁) curve at $T_a < T_g$ (195 K) is discussed in more detail in Section 1.4. Curves with open symbols were taken at temperatures 270 K and below, filled black symbols, above 270 K.

first to rationalize the new data within the scope of the classical nucleation and growth model. When this proved unsuccessful, first suggestions for a revised model are advanced in Section 1.4. In this effort, use is made not only of the latent heats shown in Fig. 1.28, but also of the heat capacities in the glass transition regions for the different samples in terms of the mobile-amorphous and rigid-amorphous fractions (MAF and RAF, respectively) [126, 131]. Also, a link is attempted to be followed to consider several unresolved inconsistencies with the classical nucleation and growth model which already had appeared when reviewing the experiments known some 30 to 50 years ago (cf. [23, vol. 2]).

For the parametrization of the nucleation and crystallization results, an Avrami function [132] was used. The Avrami equation was chosen as a fitting equation to parameterize the measured curves, i.e., to obtain half-times for the two processes. We do not assume that the other parameters have any physical meaning in this particular case. The overall latent heat on heating, which equals the crystallization enthalpy during annealing, was therefore expressed as

$$\Delta H_{\text{heating}} = \Delta H_{c\infty} \left(1 - \exp\left(-\frac{t}{\tau_c} \ln 2\right)^{n_c} \right), \tag{1.5}$$

where $\Delta H_{c\infty}$ is the final enthalpy of primary crystallization at infinite time, t is the annealing time, τ_c is the crystallization half-time and n_c is the Avrami coefficient of crystallization. Taking into account a linear increase of enthalpy due to secondary crystallization [133] yields

$$\Delta H_{\text{heating}} = \Delta H_{c\infty} \left(1 - \exp\left(-\frac{t}{\tau_c} \ln 2\right)^{n_c} \right) \\ + A_2 \left(\ln(t - \tau_c)\right) \left(\frac{1}{2} \left(\frac{|t - \tau_c|}{t - \tau_c} + 1 \right) \right), \tag{1.6}$$

where A_2 is a secondary crystallization parameter.

The homogeneous nucleation process, investigated via the cold-crystallization enthalpy, ΔH_{cc}, can, for parametrization, as a first approximation also be described by an Avrami equation, i.e., by assessing the subsequent crystallization caused by the nuclei. In such approach, we arrive at

$$\Delta H_n = \Delta H_{n,\text{hom}}^{(\infty)} \left(1 - \exp\left(-\frac{t}{\tau_n} \ln 2\right)^2 \right) + \Delta H_{n,\text{het}}, \tag{1.7}$$

where τ_n is the nucleation half-time, $\Delta H_{n,\text{het}}$ is the enthalpy change due to cold crystallization at this heating rate on heterogeneities, and $\Delta H_{n,\text{hom}}^{(\infty)}$ is the limiting cold-crystallization enthalpy due to homogeneous nuclei. For nucleation, we use an Avrami parameter with a value equal to two ($n_c = 2$).

The development of the cold-crystallization peak is governed by the kinetics of nuclei formation (increase of the peak at short annealing times) and crystallization during annealing (decrease of the peak at larger annealing times). Therefore the peak area can be considered as the sum of two parts of opposite signs,

$$\Delta H_{cc} = -\Delta H_n + \Delta H_c. \tag{1.8}$$

The second term on the right side of this equation is the same as for the overall enthalpy change (crystallinity) and consequently ΔH_n can be obtained from a simultaneous fit of both curves. Such simultaneous fit of two measured curves with two functions is shown in Fig. 1.29. For simplicity the secondary crystallization term is not shown on the bottom part although it was used in the real fit.

Several assumptions are employed to improve fitting convergence. First n_c was restricted to values in the range between 2 and 4. The overall cold-crystallization enthalpy limit $\left(\Delta H_{n,\text{hom}}^{(\infty)} + \Delta H_{n,\text{ het}}\right)$ was manually restricted to $\Delta H_{c,\infty}$ at the first fitting iterations as well as A_2 was set equal to zero. After several iterations, when the parameters τ_n and τ_c were determined, the parameters of the fit were released and the fit was repeated until convergence. At an annealing temperature 185 K the algorithm described

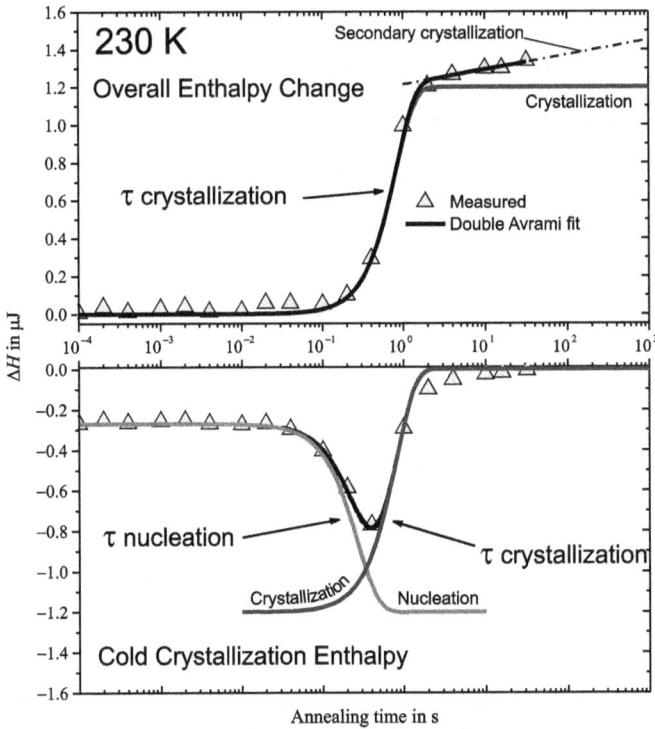

Fig. 1.29: Simultaneous fit of overall latent heat and the latent heat of cold crystallization on heating. The mathematical fit is detailed in the text.

above did not converge. Therefore the nucleation time constant was determined using a simple Avrami kinetics and $\left(\Delta H_{n,\text{hom}}^{(\infty)} + \Delta H_{n,\text{het}}\right) = 1.2\ \mu\text{J}$ as the final value of the crystallization enthalpy. The crystallization half-time was extrapolated using the achieved nucleation half-time and previous results for the crystallization half-time. The resulting large error is indicated in the final results below. The obtained half-times of crystal nucleation and crystallization are plotted in the activation diagram of Fig. 1.30.

At sufficiently low temperatures the nucleation half-time is ca. two orders of magnitude less than for crystallization and at higher temperature it assumes identical nucleation half-times as the crystallization process. Still, both processes change approximately parallel to the α-relaxation at low temperatures. The cold-crystallization peak growth during annealing was found to be sensitive to a time delay between nucleation and growth during annealing. A time separation can be followed up to 290 K, after that crystallization happens immediately after nucleation. The final results cover the full temperature range of crystallization from 185 K (T_g = 209 K [37]) to 340 K (T_m = 342 K [37]). The estimated time constants cover the range from $3 \cdot 10^{-2}$ s (nucleation at 215 K) to $3 \cdot 10^9$ s (crystallization at 185 K).

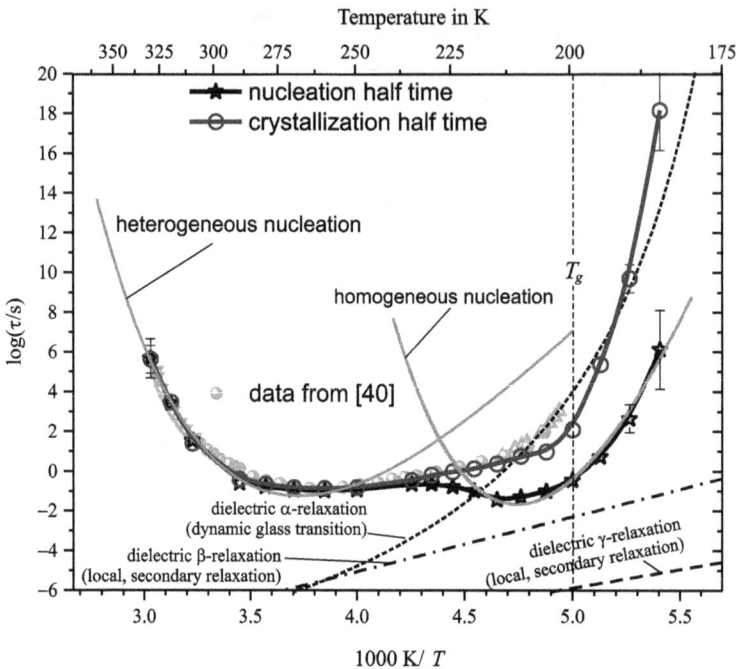

Fig. 1.30: Activation diagram for nucleation and crystallization at temperatures from below the glass transition up to the melting temperature covering 20 orders of magnitude in time. They are compared to dielectric relaxation data (dashed curves) [39] and isothermal experiments applying a single sensor device (green (in color version), respectively, partly filled spheres) [40].

The data of Fig. 1.27 and the Supplementary Material of [38] indicate that the newly developed DFSC opens the possibility to avoid crystallization and homogeneous nuclei formation on cooling to temperatures far below the glass transition. Being able to avoid formation of homogeneous nuclei at fast cooling allows us now to study isothermal nuclei formation and the subsequent crystallization. Both processes can be followed as they occur at the same temperature. Evaluating the heating scans, which were always collected at the same rate as described in Section 1.3.2, should provide information about nucleation and crystallization kinetics. The Avrami equation, commonly applied for crystallization processes, was used to fit the overall latent heat curves in order to determine a crystallization half-time for each annealing temperature. The resulting parameters were simultaneously used to fit the cold-crystallization curve, which includes both nucleation and overall crystallization kinetics (Fig. 1.29). Again a nucleation half-time was determined using the Avrami equation. The computation was based on the assumption that the transformation of the material occurs in a finite, shrinking volume at a constant nucleation rate. The obtained nucleation and crystallization half-times are shown in the activation diagram of Fig. 1.30.

Relative to Fig. 1.4, the crystallization kinetics as described in Fig. 1.30 shows a broadened curve at lower temperatures. Similar deviations relative to Fig. 1.4 were observed for PCL [40], polypropylene [75, 78, 134], polyamides [43, 135] and poly(butylene terephthalate) [80]. The reason for faster crystallization at low temperatures might be the dominating homogeneous nucleation at low temperatures (see the lower curve below 230 K (above about 4.4 in terms of 1000/T) in Fig. 1.30). The low-temperature part of the nucleation curve is connected to both homogeneous and heterogeneous nucleation whereas the high-temperature part corresponds to the activation time of heterogeneous nuclei. In the high-temperature region of the annealing pre-treatment the thermodynamic force for crystallization is low but mobility is high. Under such conditions the sample crystallizes as soon as nucleation occurs, but the rate of growth is slow. The separation between nucleation and crystallization is seen up to 290 K (Fig. 1.28), but becomes significant at low temperatures only. The validity of this argumentation was again verified by similar studies on MWCNT nucleated PCL samples (Section 1.3.4.3).

1.3.4.2 Isotactic Polybutene-1 (iPB-1)

Fig. 1.31 contains temperature-time profiles for the isothermal analysis of the kinetics of melt-crystallization (left) and cold crystallization (right). In the case of melt-crystallization, the sample has directly been cooled from 433 K to the analysis temperature at a rate of $1000 \, Ks^{-1}$. In case of cold crystallization, in contrast, the sample was first cooled to 213 K, kept there for a period of 0.1 s and then re-heated to the crystallization temperature. Crystallization rates were quantified by the half-time of crystallization, determined by interruption of the crystallization process at pre-defined time, and measurement of the corresponding enthalpy of melting on subsequent fast heating at $100 \, Ks^{-1}$. Due to the fast heating, cold ordering and reorganization are avoided and thus, the enthalpy of melting is equal to the enthalpy of prior isothermal crystallization. Plotting the enthalpy of crystallization as a function of the crystallization time yields conversion-time curves used for the determination of the half-time of crystallization.

The part of the temperature-time program for cold-crystallization experiments located below the temperature of 298 K in Fig. 1.31 indicates the expected temperature/time range for formation of additional crystal nuclei which then may accelerate the overall crystallization process. It is evident that the number of forming nuclei depends on the specific temperature-time profile used, that is, the rates of cooling and heating, the minimum temperature, and residence time at the minimum temperature. In the following, the conditions of cold crystallization summarized in Fig. 1.31 serve us as a reference for later discussion of the aging-controlled cold-crystallization behavior.

Fig. 1.32 shows half-times of crystallization of iPB-1 as a function of the crystallization temperature. The red (upper curves) and black (lower curve) colored symbols indicate melt- and cold crystallization, respectively, according to the temperature-time

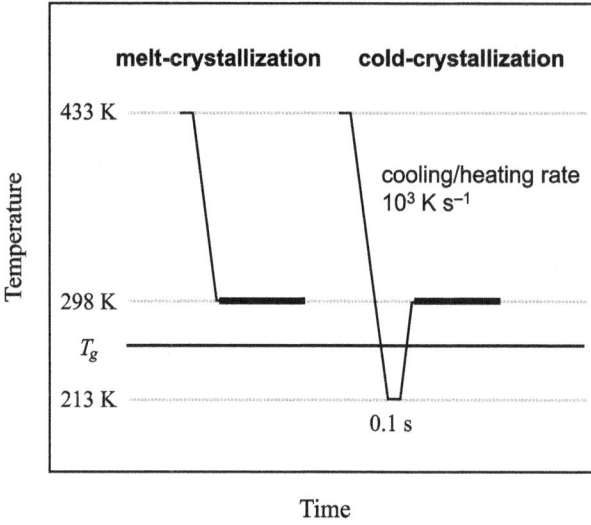

Fig. 1.31: Temperature-time program of FSC-experiments applied for the analysis of the kinetics of melt-crystallization (left) and cold-crystallization (right) of iPB-1.

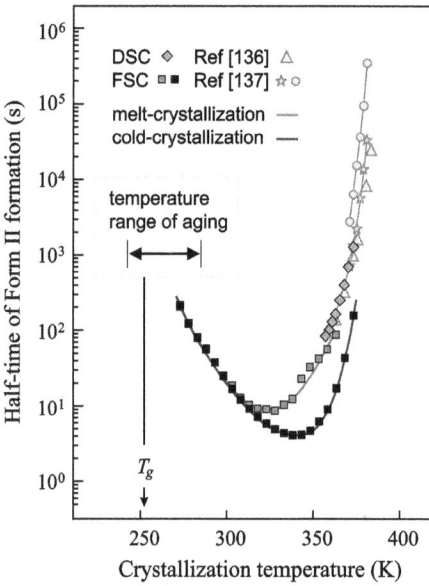

Fig. 1.32: Half-times of crystallization of iPB-1 as a function temperature. The red (upper curves) and black (lower curve) symbols (specified by diamonds and squares) indicate melt- and cold-crystallization, respectively, according to the temperature-time profiles shown in Fig. 1.31. Gray-filled symbols (triangles, stars and circles) refer tot data were adapted from the literature [136, 137].

profiles shown in Fig. 1.31. Both data sets exhibit a minimum which is expected from the theory of polymer crystallization [138, 139]. Melt-crystallization (red symbols, upper curves) is fastest at about 330 K while the maximum of the cold-crystallization rate is observed at slightly higher temperature. Data obtained by FSC (squares) are in agreement with data obtained by DSC (diamond symbols) as well as data available in the literature [136, 137]. Important in the context of the present work, it is pointed out that cold crystallization at the chosen reference conditions, shown in Fig. 1.31, is distinctly faster than melt crystallization. This is attributed to the additional residence time of the sample at temperatures lower than the transformation temperature. However, this observation is only true for crystallization temperatures higher than 315 K; at lower temperature, melt- and cold-crystallization rates are identical. Apparently, melt- and cold-crystallization processes at temperatures lower than 315 K, at the specified conditions shown in Fig. 1.31, rely on the instantaneous formation of nuclei, independent on the path the temperature is approached, pointing to a homogeneous nucleation mechanism at the crystallization temperature.

1.3.4.3 Carbon Nanotube Nucleated PCL

Nucleation and crystallization of the MWCNT nucleated PCL samples (see Section 1.3.3.2) were further studied by isothermal experiments after cooling at $7 \cdot 10^4 \, \text{Ks}^{-1}$ applying the temperature-time profile of Fig. 1.26. Measured curves for different samples after annealing at 195 K, that is, at a temperature close to the beginning of the glass transition, are shown in Fig. 1.33 as an example.

After annealing for 2 s at 195 K which is 5 K below T_g (red curves, 2 s) the glass transition (1) on heating is superimposed by an enthalpy recovery peak (1*), related to the non-equilibrium glassy state after quenching. The development of the enthalpy recovery peak and its transformation into melting of tiny isothermally formed crystals (peak 2 in the green curves, 200 s) at temperatures around the glass transition is discussed in more detail in Section 1.3.6. The leftover of the tiny crystals after melting initiates cold crystallization immediately after or already during melting. With this observation, a link to the well-known self-nucleation has been established.

Self-nucleation is known to occur when heating a semi-crystalline sample some 5 to 30 K above the melting or dissolution temperature before attempting crystallization at lower temperature [140]. This self-nuclei induced cold crystallization has a different peak maximum temperature (3a in the orange curves, 4000 s) compared to cold-crystallization of untreated heterogeneously nucleated samples (3b in black (0 s), red (2 s) and green (200 s) curves). As expected, the difference in crystallization temperatures results in different melting peak temperatures too (4a and 4b) for the nucleated samples. However, several melting events are overlapping since heating at 5000 Ks^{-1} is not fast enough to prevent reorganization on heating totally. To quantify the influence of the nucleating agent on PCL crystallization, the integration of the heat capacity curves on heating as discussed above (Fig. 1.21), was performed for all measurements

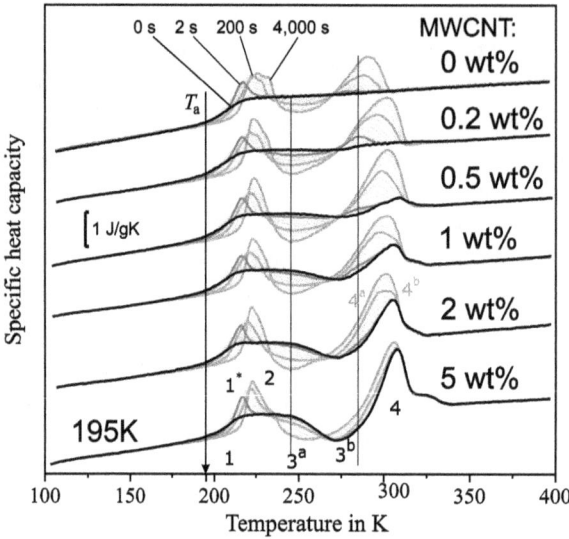

Fig. 1.33: Heating curves after annealing at 195 K for different times (from 0 to 4000 s) for MWCNT concentrations from 0 to 5 wt%. During annealing, homogeneous nuclei are formed, which is seen by the growing cold crystallization peak on the subsequent heating. Even for the highly nucleated sample (5 % MWCNT) a shift of the cold crystallization peak maximum is clearly seen (> 20 K), although the area of the peak does not significantly change with increasing annealing time.

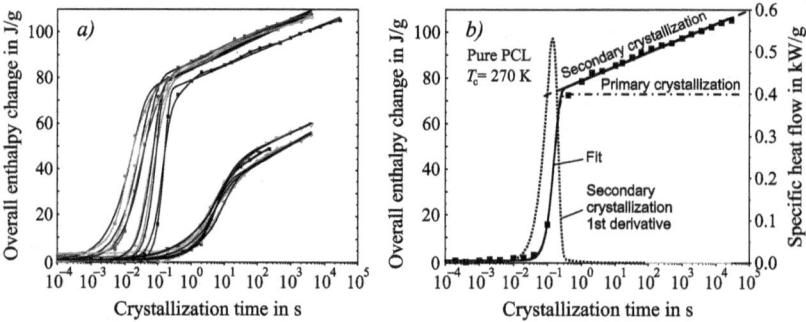

Fig. 1.34: Overall enthalpy change during isothermal crystallization at high temperatures (T_c = 270 K) and in the vicinity of the glass transition (T_g = 210 K).

and the latent heats were obtained. Two sets of results are shown in Fig. 1.34, where crystallization at two temperatures is compared for all samples.

The crystallization behavior at T_a = 210 K is, within the uncertainty of mass determination, the same for all samples, while at T_a = 270 K the MWCNT nanocomposites crystallize significantly faster than neat PCL. For parametrization of the crystallization results, the Kolmogorov-Johnson-Mehl-Avrami formalism [132] was used to fit the data as detailed above in Section 1.3.4.1. The resulting fit curve and its temperature deriva-

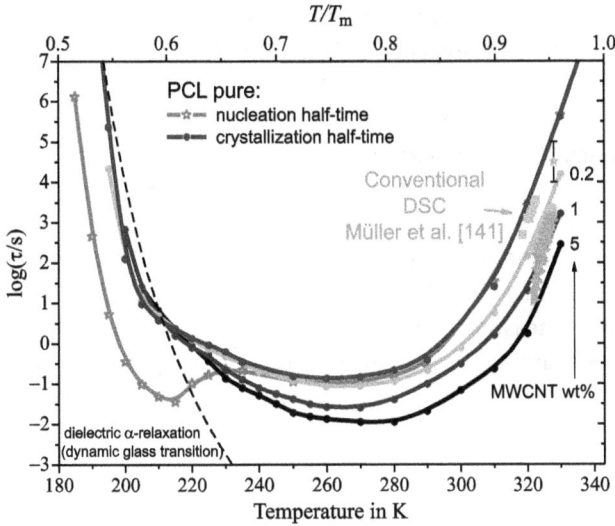

Fig. 1.35: Crystallization half-times of neat and heterogeneously nucleated PCL. Data from [38] and [141] are shown in addition. The star is another independent measurement [98] of the 0.2 wt% MWCNT sample at 328 K showing the error-bar in addition.

tive are shown in Fig. 1.34b. The results of fitting can be seen as black thin lines in Fig. 1.34a. The obtained half-times of crystallization were plotted in Fig. 1.35 together with the data from Fig. 1.30. The results cover the full temperature range of crystallization from 180 K ($T_g(10$ K/min$) = 209$ K [37]) to 340 K ($T_m = 342$ K [37]). The estimated time constants cover the range from 10^{-2} s (crystallization of PCL with 5 wt% MWCNT at 280 K) to $3 \cdot 10^9$ s (crystallization at 185 K).

In [38] it was shown that DFSC allows one to extend the temperature range for isothermal experiments for PCL from below the glass transition up to the melting temperature (185–340 K). The indirect measurement, evaluating heating scans at constant heating rate after the thermal treatment, allows one not only to investigate very slow crystallization processes, both at high and low temperatures, but the time resolution of the device makes it possible to follow also very fast (millisecond) ordering processes. Fig. 1.35 shows the obtained half-times for homogeneous nucleation for neat PCL [38] and the half-time of crystallization for neat PCL and the MWCNT composites. For temperatures down to about 220 K the nucleating effect of the MWCNTs is seen. At 270 K, the temperature of maximum crystallization rate for all composites, crystallization half-time is reduced by more than one order of magnitude. Here the MWCNTs are most efficient as nucleating agent. For temperatures below 220 K homogeneous nucleation becomes the dominant nucleation mechanism and the added heterogeneities (MWCNTs) do not speed up crystallization. The very good agreement for the temperature range of the minimum of the half-time of homogeneous nucleation (red (marked by open stars) curve in Fig. 1.35) and the coincidence of the crystallization kinetics data for the neat and the nucleated PCL very much support the dominance of the homogeneously formed nuclei over the heterogeneities below 220 K.

1.3.5 Identification of Different Nuclei Populations

Cold crystallization is seen in the measured heating curves as an exotherm that changes in size and location depending on the nucleation density and the activity of the nuclei present (Fig. 1.33). At a fixed heating rate the cold-crystallization enthalpy can be used as a relative measure of nucleation density. Additionally, a lower cold-crystallization temperature indicates more efficient nucleation and allows one to distinguish between different nuclei populations, but again only on a qualitative level.

Fig. 1.36a shows selected heating scans from Fig. 1.33 of PCL with 2 wt% MWCNT after annealing at 195 K for times ranging from zero to $4 \cdot 10^3$ s. Heating was per-

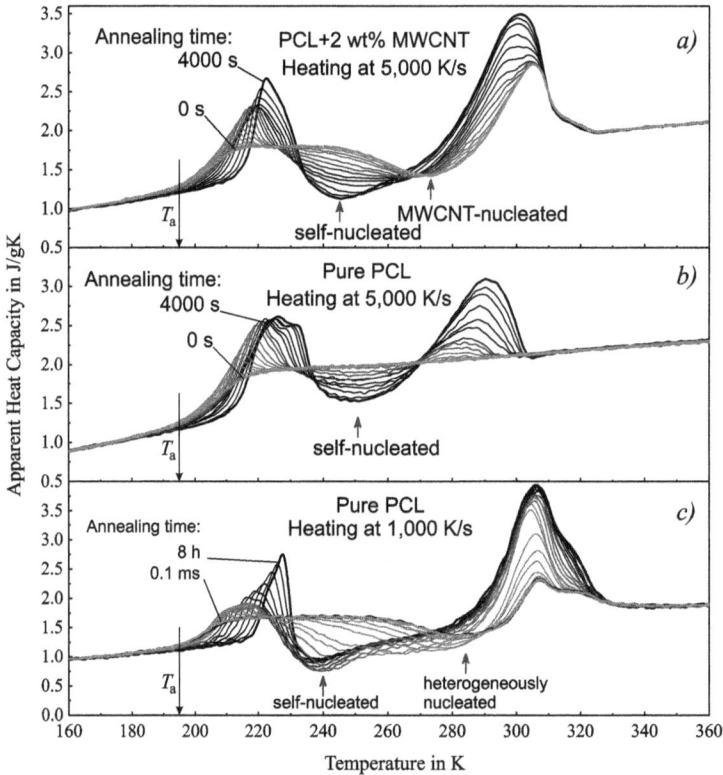

Fig. 1.36: Selected heating curves from Fig. 1.33 after annealing at 195 K for different times of PCL with 2 wt% CNT (a) and neat PCL (b and c). The curves shown in a) and b) were measured at 5000 Ks^{-1} and c) at 1000 Ks^{-1}. The annealing time changes nearly equidistantly on a logarithmic scale: 0 s (red); 1 ms; 2 ms; 4 ms; ...; 2000 s; 4000 s (black).

formed at 5000 Ks^{-1}. This rate is high enough to avoid crystallization in the neat PCL (Fig. 1.36b), but it is not too fast to prevent growth of crystals originating from the MWCNTs. Therefore even without annealing (red curve, 0 s) we see a cold-crystallization peak at about 275 K in Fig. 1.36a. With increasing annealing time, cold crystallization shifts to lower temperatures for about 30 K indicating the appearance of more active nuclei with longer annealing.

Two distinct cold-crystallization peaks are seen in the nucleated PCL (Fig. 1.36a). The high-temperature peak around 275 K, present from the very beginning, is caused by the added heterogeneities. The low-temperature peak develops due to nuclei homogeneously formed during annealing at 195 K. With increasing annealing time the high temperature peak decreases and eventually disappears after $4 \cdot 10^3$ s annealing. Then all cold-crystallization is initiated by the homogeneously formed nuclei at low temperature around 245 K. The cold-crystallization peak maximum for the neat PCL (b) is located close to this temperature (250 K). At a heating rate of 5000 Ks^{-1}, the few heterogeneities which are always present in a sample, cannot initiate measurable crystallization because the time interval for crystal growth is too short at this relatively high heating rate. At a slightly lower heating rate of 1000 Ks^{-1} this is changing. In Fig. 1.36c, without annealing, cold crystallization is already seen at heating the neat PCL at 1000 Ks^{-1}. With increasing annealing time homogeneously formed nuclei increasingly initiate crystallization and take over cold-crystallization enthalpy from the heterogeneously nucleated cold crystallization.

The general picture (Fig. 1.36c) is very similar to the nucleated PCL (Fig. 1.36a). These curves and all other curves in Fig. 1.33 clearly show that crystallization initiated by homogeneous nucleation is more efficient than by the added MWCNTs even at 5 wt% MWCNT. By this reason, the addition of MWCNT does not influence the crystallization kinetics or the final value of crystallinity when homogeneous nucleation is dominating the crystallization process as shown in Fig. 1.34a (T_a = 210 K). In other polymer nanocomposites, e.g. the polyamide layered silicate nanocomposites studied in [142], a significant reduction of the crystallization kinetics and the final degree of crystallinity was observed. This effect was explained by an immobilized polymer layer in the vicinity of the layered silicates (Rigid Amorphous Fraction, RAF). For the PCL+MWCNT composite the interaction between the polymer and the MWCNTs seems not to result in the formation of a significant RAF. It is worth mentioning that all this information is not available from non-isothermal crystallization experiments but only from isothermal annealing carried out in a wide range of temperatures.

Having at hand the possibility to perform annealing experiments in the whole temperature range of interest allows us to study the influence of annealing temperature on cold crystallization. Particularly we wanted to know how nuclei or small crystals influence cold crystallization at a given heating rate [125]. Therefore we compared heating experiments with the critical heating rate of $1.8 \cdot 10^4$ Ks^{-1} for PCL1.4k after cooling with 10^5 Ks^{-1} (see Section 1.3.2.4) at three selected temperatures. The first temperature, 202 K, is just above the glass transition, as shown in Fig. 1.37a. For short annealing times (0.0001 s up to 0.01 s) the sample remains non-crystalline on heating. After 0.1 s annealing, cold-crystallization on heating is seen and followed by its corresponding melting at higher temperature. The state of the sample at the beginning of the heating experiment can be evaluated (i) by comparing glass transition with the glass transition of the amorphous sample, (ii) by calculating the overall enthalpy change above glass transition, and (iii) by comparing the cold-crystallization enthalpies. Identical glass transitions and an overall enthalpy change of zero demonstrate the non-crystalline state even after 0.1 s annealing at 202 K. For longer annealing times both criteria indicate initial crystallinity – the heat capacity step at glass transition becomes smaller and is shifted to higher temperatures and the overall enthalpy change is endothermic.

The heating curves after annealing at 272 K in Fig. 1.37c show a much different behavior. Whereas for the short times, e.g. 0.01 s, the sample also remains amorphous, no cold crystallization can be detected after longer annealing times either. Only the melting endotherms just above the annealing temperature are present, related to the melting of isothermally formed crystals increasingly annealed at longer annealing

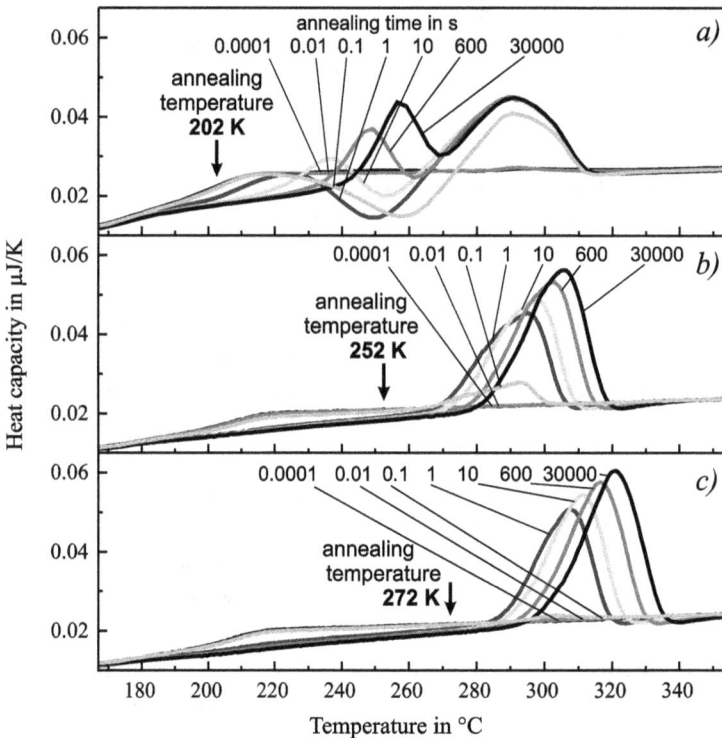

Fig. 1.37: Apparent heat capacity of PCL1.4k from DFSC on heating with $1.8 \cdot 10^4$ Ks^{-1} after annealing for different times at 202 K (a), 252 K (b), and 272 K (c).

times (commonly called "annealing peaks"). An intermediate annealing temperature of 252 K was chosen in the temperature range where the exothermic effect of cold crystallization or recrystallization on heating after low-temperature annealing is maximal. The heating curves are shown in Fig. 1.37b. Here the situation is similar as for high-temperature annealing. No cold crystallization and only melting can be observed. The shape of the melting curve supplies us now with an indication of two processes of crystallization, the initially grown crystals melt at the low-temperature side of the melting peak and the more prominent high-temperature portion of the melting peak is due to melting followed by reorganization processes. A similar behavior was found for several polymers and was also attributed to a melting and recrystallization processes on heating [47, 69, 71, 143].

In Fig. 1.37c, the melting peaks of the sample annealed at 272 K shift about 15 K towards higher temperatures with increasing annealing time. The higher melting temperature indicates higher stability of the crystals and can be associated with an internal stabilization of the crystals, observed for most polymers on isothermal crystallization from the melt [144] and was studied for PCL with X-ray diffraction by Strobl [145]. The existence of such a stabilization process during isothermal crystallization was confirmed by DMA [146] and X-ray scattering experiments [147]. A gradual, less complete shift of melting to higher temperatures due to stabilization with increasing annealing time is also observed after crystallization at 252 K in Fig. 1.37b. In contrast, low-temperature annealing at 202 K causes a constant final termination of melting at 313 K, independent on annealing time (see Fig. 1.37a). At 202 K the initially formed crystals melt already during a low-temperature endotherm at temperatures below 273 K. The shift of the melting peak from 223 K after annealing for 1 s to 255 K after annealing for $3 \cdot 10^4$ s is again the consequence of the internal stabilization process of the isothermally formed crystals. But on further heating, after total low-temperature melting, recrystallization followed by melting of the recrystallized species is observed. Heating at the same rate provides the same recrystallization conditions, resulting in an identical final melting temperature as it was also observed for PET [69], iPS [71], and iPP [74, 148].

Fig. 1.37 and above discussion show how nucleation and crystallization influence the glass transition. On the other hand Fig. 1.36 provides further interesting details about nucleation in PCL annealed below the glass transition temperature of the amorphous sample. Meanwhile it was shown for several polymers that ordering (nucleation and crystallization) in polymers can take place already in the glassy state [38, 43, 125, 127, 149, 150]. In the next section, we discuss the interplay between enthalpy relaxation in the glassy state and crystal nucleation in more detail.

1.3.6 Enthalpy Relaxation and Crystal Nucleation in the Glassy State

1.3.6.1 Neat and Nucleated PCL

The enthalpy relaxation, disordering and cold-crystallization peaks after annealing at temperatures below and slightly above the glass transition temperature, were analyzed separately in order to investigate the disordering (melting of crystals and nuclei) and its influence on crystallization on further heating in more detail. The sample containing 5 % MWCTN was used as an example because it also will demonstrate the high efficiency of homogeneous nucleation in this highly heterogeneously nucleated sample [99].

A series of heating scans after annealing for different times at 205 K is shown in Fig. 1.38. After 10 s annealing (blue curve in color version) the typical enthalpy recovery peak appears. It is located at the high temperature end of the glass transition region, with the glass transition temperature only marginally shifted. For longer annealing times, a sharp peak at an up to 20 K higher temperature develops. Annealing at 205 K causes an approach to the enthalpy of the supercooled liquid within a second (Fig. 1.39b), and at longer times saturation of all annealing effects is expected. The large shift of the peak and the long times where changes are observed do not allow one to assign this peak to enthalpy relaxation alone [151, 152]. That the two processes are overlapping in the temperature range just above the glass transition is seen for the neat PCL in Fig. 1.33. Two distinct peaks appear and the second was already assigned to the melting of a large number of crystals which immobilize the remaining amorphous matrix [38].

Careful integration was performed in order to confirm ordering/nucleation of polymers below the glass transition and its relation to enthalpy relaxation in analogy to studies discussed below [38, 43, 125, 127, 149, 150]. Fig. 1.39a shows integration ex-

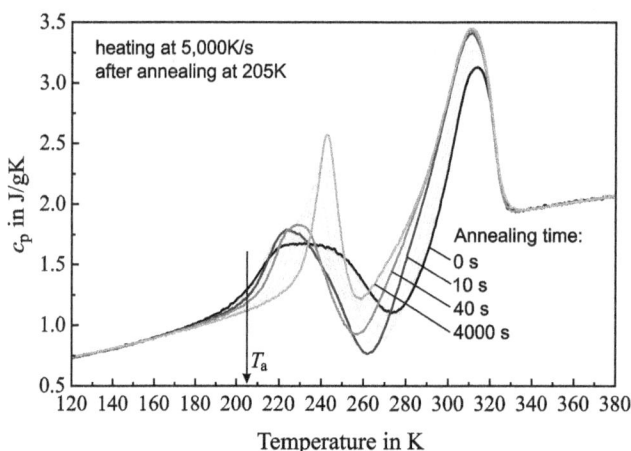

Fig. 1.38: PCL+5 wt% MWCNT, annealed at 205 K for different times and heated at 5000 Ks^{-1}.

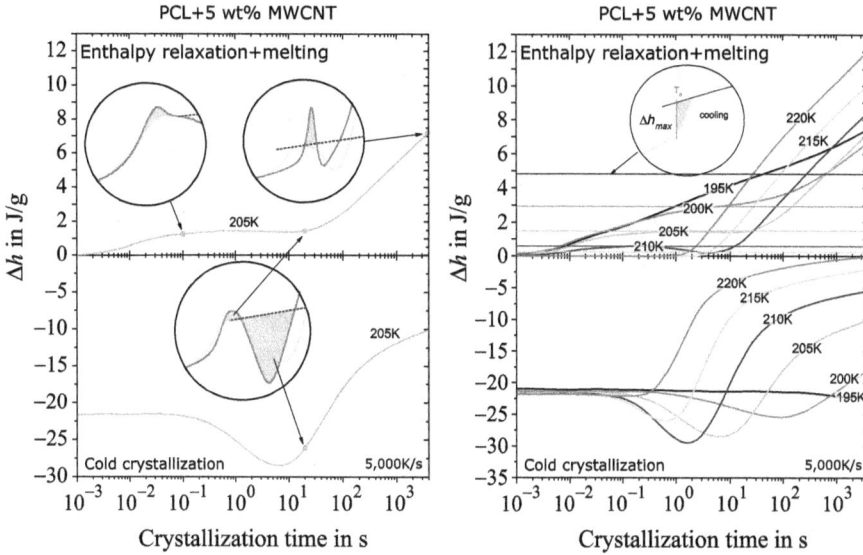

Fig. 1.39: (a) Analysis of the enthalpy of relaxation and melting peak development at different temperatures. (b) Comparison with the cold crystallization peak area (for details see text).

amples for curves from Fig. 1.38. The peaks originating from superimposed enthalpy relaxation and disordering are further superimposed by the apparent shift of the glass transition due to the occurrence of a large rigid amorphous fraction (RAF) [126]. In order not to falsify the peak areas by this effect, we used for all integrations the liquid heat capacity as baseline (dotted lines in Fig. 1.39a, top insertions) and count only the areas above this line.

Fig. 1.39b shows the evolution of the peak areas (enthalpy change) as a function of annealing time for different annealing temperatures. The upper panel presents the enthalpy change related to the first endothermic peak. In a first step, within less than 10 s, the enthalpy approaches the enthalpy of the supercooled liquid (enthalpy relaxation). The horizontal dashed lines are the limiting values estimated from the step of the heat capacity at T_g and the difference between T_g and T_a (see insert). The further increase of the enthalpy change cannot be assigned to enthalpy relaxation but to melting of crystals formed during annealing [38]. The very low melting temperature of these crystals (240 K) corresponds according to the Gibbs-Thomson equation to crystal sizes of about 2–3 nm (see [125] for details).

Next, we compare the development of enthalpy relaxation and melting of the crystals with the development of cold-crystallization enthalpy as a measure of nucleation density (Fig. 1.39b, lower panel). The starting level of cold-crystallization enthalpy of −22 J/g is due to the added 5 wt% MWCNT. Nevertheless, after annealing for 0.1 s at 205 K, isothermally formed nuclei increase the cold-crystallization enthalpy further (see discussion in Section 1.3.4). Comparing the corresponding curves in the upper and the lower panel of Fig. 1.39b shows that the increase in cold-crystallization enthalpy for all annealing temperatures coincides with the approach of the first step (enthalpy relaxation) to its limiting value (horizontal dashed lines). A similar behavior was found for neat PCL in [38] and other polymers as discussed below and in [43, 127, 149, 150].

The approach of the enthalpy change to its limiting value corresponds to a densification of the polymer. This densification presumably proceeds by cooperative rearrangements on a length scale of a few nanometers [153]. On the other hand an estimate from the Gibbs-Thomson equation yields a crystal size of about 2–3 nm for the crystals melting at 240 K [125]. Nucleation around 200 K is therefore expected to occur on an even smaller length scale. It seems plausible that the large scale cooperative rearrangements causing densification and enthalpy relaxation prevent the formation of overcritical nuclei as long as the process is active. Only after equilibrating the amorphous polymer with respect to density and enthalpy according to the corresponding liquid values nucleation can occur at the annealing temperature. This observation provides evidence for the influence of the state of the polymer (pre-history) on nucleation and crystallization, discussed in Section 1.4.

1.3.6.2 Isotactic Polypropylene (iPP)

Fig. 1.40 shows DFSC heating curves of initially glassy iPP after annealing for different periods of time at 220 K. The time of annealing was varied covering seven orders of magnitude from 10^{-3} to 10^4 s, including immediate reheating after the minimum temperature of the experiment of 100 K has been reached. The heating curve of the non-annealed sample (0 s annealing time, blue curve in color version) shows the glass transition at about 260 K, being superimposed by a small endothermic peak. We in-

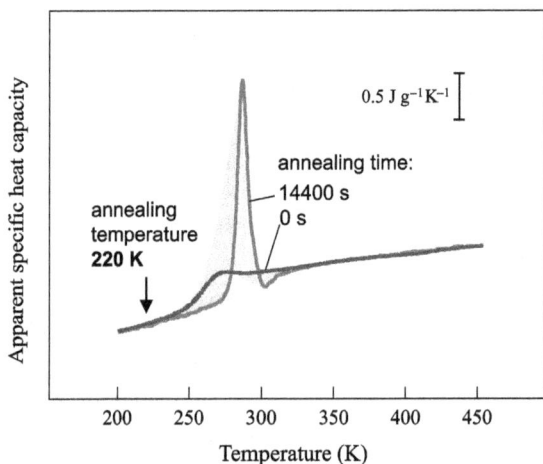

Fig. 1.40: Apparent specific heat capacity of iPP as a function of temperature, measured on heating at a rate of $3 \cdot 10^4$ Ks^{-1} after annealing initially glassy samples for different periods of time at 220 K.

terpret this peak as a classical hysteresis peak due to different rates of cooling and heating [152, 154]. Furthermore the data of Fig. 1.40 indicate that the glass transition apparently shifts to higher temperature and that the area and temperature of the endothermic peak increase with annealing time. Note that the small exothermic deviation at the high-temperature side of the endothermic peak is caused by the control loop of the calorimeter.

The annealing induced observations shown in Fig. 1.40 may be interpreted as being related either to the formation of nuclei and small ordered domains with latent heat or, more likely, to classical enthalpy relaxation [152]. Fast cooling (vitrification) of the liquid is connected with the formation of a non-equilibrium glass of high enthalpy, which is expected to relax during aging to lower values as a function of time, ultimately approaching the enthalpy of the liquid at the particular temperature of 220 K. On reheating, devitrification of the relaxed glass would then show a hysteresis and continuously be followed by an endothermic enthalpy-recovery peak. The maximum possible enthalpy of relaxation ($\Delta h_{\text{relax, max}}$) on annealing at 220 K is about 17 J/g [152],

$$\Delta c_p \left(T_g - T_a \right) = 0.46 \text{ J/gK} \cdot (257.5 - 220)\text{K} \cong 17 \text{ J/g,}$$

with T_g, T_a, and Δc_p being the glass transition temperature, the annealing temperature and the heat capacity difference of solid and liquid iPP at T_g, respectively. Estimation of the decrease of the enthalpy of the glass formed on cooling at $1 \cdot 10^4 \text{ Ks}^{-1}$ during annealing for $1.44 \cdot 10^4$ s at 220 K yields a value of only 7–8 J/g, which indicates that the changes of the DFSC heating curves due to annealing indeed can be caused by enthalpy relaxation only. The relaxation process, however, is not complete; extrapolation of the evolution of the decrease of the enthalpy during annealing with time suggests that equilibrium may only be reached after annealing for at least 10^{10} s (>300 years), impossible to adjust with the present experiments. In other words, at present we do not fully exclude ordering of chain segments during annealing at 220 K, although a quantitative analysis of the change of enthalpy with time, as discussed below, favors enthalpy relaxation.

In Fig. 1.41, DFSC heating curves of samples of iPP annealed at 230 K for different periods of time are shown. Analysis of the heating curves of the sample annealed at 220 K suggests classical enthalpy relaxation only and no indication of formation of homogeneous nuclei or even ordered domains at the annealing temperature; the estimate of the maximum possible decrease/relaxation of enthalpy has not been reached during annealing for $1.44 \cdot 10^4$ s. Heating of samples annealed at 230 K, in contrast, allows one not only detection of an endothermic peak at the glass transition but also exothermic cold-ordering between about 300 and 330 K, followed by disordering, reorganization and melting between 330 K and temperatures close to 400 K. Estimation of the maximum possible enthalpy of relaxation of the glass at 230 K yields a value of about 12 J/g,

$$\Delta c_p \left(T_g - T_a \right) = 0.46 \text{ J/gK} \cdot (257.5 - 230) \text{ K} \cong 12 \text{ J/g,}$$

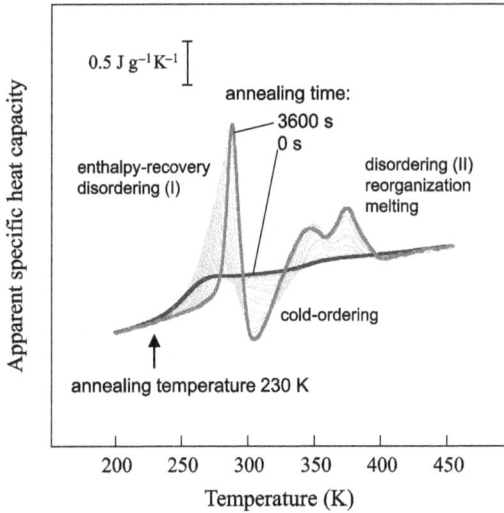

Fig. 1.41: Apparent specific heat capacity of iPP as a function of temperature in the glass transition range, measured on heating at a rate of $3 \cdot 10^4$ Ks^{-1} after annealing initially glassy samples for different periods of time at 230 K.

which, at least in part, causes the endothermic peak following the glass transition. While annealing up to few seconds seems to be connected to enthalpy relaxation only, continuation of annealing leads to additional formation of small ordered domains causing cold ordering on heating, and an increase of the enthalpy reduction during annealing to values larger 12 J/g after about 1000 s.

Fig. 1.42 is a plot of the overall change of the enthalpy as a function of the time of annealing at 230 K (dark gray squares). The curves of Fig. 1.41 have been integrated between 200 and 450 K, and the value of the non-annealed sample has subsequently been subtracted to gain information about the sole effect of annealing. As such, the data represent enthalpy recovery on heating, initially gained during isothermal annealing. The data of Fig. 1.42 suggest that annealing for periods of time less than a few seconds leads to enthalpy relaxation only without reaching the limit of 12 J/g. Extension of annealing for periods of time longer than about 10 s is connected with additional exchange of latent heat, attributed to the formation of ordered domains, leading to an increase of the slope of the data in Fig. 1.42, and observation of cold-ordering on heating (Fig. 1.41). Evidence for ordering processes during extended annealing at 230 K ultimately is gained by experimental observation of a value of the enthalpy below that of the equilibrium liquid; annealing longer than 1000 s leads to a decrease of the enthalpy larger than 12 J/g. For comparison, in Fig. 1.42 are also shown data calculated from the DFSC heating curves of Fig. 1.40, which is, of samples annealed at 220 K. In this case, cold-ordering has not been observed on heating, and the maximum value of enthalpy relaxation of about 17 J/g has not been reached within the time

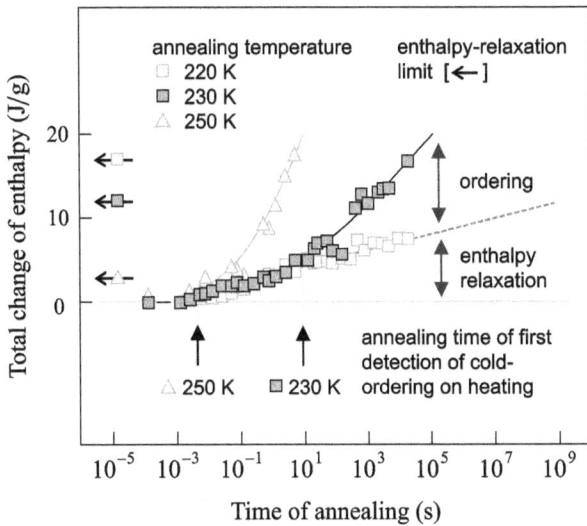

Fig. 1.42: Overall change of enthalpy of iPP as a function of the time of annealing at 230 K, calculated by integration the DFSC heating curves of Fig. 1.41, and subtraction of the enthalpy change of non-annealed amorphous iPP. For comparison, the plot contains data calculated from the DFSC heating curves collected after annealing at 220 K (gray squares) and 250 K (gray triangles). The horizontal arrows at the left-hand side indicate the maximum possible enthalpy relaxation (enthalpy relaxation limit) at the various annealing temperatures; changes of enthalpy beyond these limits unambiguously prove ordering at the annealing temperature connected with change of latent heat.

frame of the experiment as in case of annealing at 230 K. The enthalpy change during annealing obeys the expected exponential law without indication of additional formation of an ordered phase, as could be concluded either by exceeding the enthalpy relaxation limit of 17 J/g, or an increase of the slope of the enthalpy-change with time. Note again, extrapolation of the dashed line in Fig. 1.42 suggests that the enthalpy of the liquid would be reached only after annealing times larger than 10^{10} s.

For further discussion of the DFSC curves of Fig. 1.41 it is important to note that the areas of the cold-ordering peak and subsequent disordering (II), reorganization and melting peaks are approximately identical, at least if the annealing time is less than 120 s. In other words, ordered domains formed on annealing at 230 K disorder in conjunction with the enthalpy recovery peak on devitrification of the amorphous phase. However, for the sake of complete description of data it is reported that annealing longer about 120 s leads to superposition of the disordering peak (I) with subsequently observed exothermic cold ordering; as a consequence, quantitative analysis of the enthalpy of cold ordering is then complicated.

Notwithstanding, it has been suggested for PCL in Section 1.3.6.1 [38] that cold ordering on continuation of heating is initiated by self-nuclei left from incomplete randomization of formerly ordered domains during disordering (I). In case of the heating experiments of iPP, performed at a rate of $3 \cdot 10^4$ Ks^{-1}, growth of self-nuclei is detected at temperatures between 300 and 330 K, that is, if the rate of mesophase formation is maximal [143]. The cold ordering is then contiguously followed by disordering and reorganization of the cold-ordered mesophase into crystals which ultimately melt around 375 K; the particular reorganization behavior of semi-mesomorphic iPP has been discussed in detail elsewhere [143, 155]. In conclusion, cold ordering is considered as a strong indication for changes of structure during annealing. Note that heating of iPP, annealed at 220 K, did not result in cold ordering despite heating was performed at identical rate as in the case of samples annealed at 230 K.

In Fig. 1.43, DFSC heating curves obtained on samples annealed at 250 K for periods of time between 0 and 2500 s are shown. For the sake of clarity, the obtained heating curves are divided into two sets. The top set of curves represents samples which were annealed for periods of time between 0 and 0.1 s while the bottom set of curves represents samples annealed between 0.6 and 2500 s. Annealing at 250 K leads to negligible enthalpy relaxation compared to the annealing experiments at 220 and 230 K; the maximum possible enthalpy relaxation is here about 3 J/g, i.e.,

$$\Delta c_p \cdot \left(T_g - T_a\right) = 0.46 \text{ J/gK} \cdot (257.5 - 250) \text{ K} \cong 3 \text{ J/g}.$$

The low endothermic peak at the glass transition, after annealing for 0.1 s, reveals an enthalpy recovery less than 3 J/g suggesting that the major portion of this peak, indeed, could be caused by enthalpy relaxation. Notwithstanding, simultaneous

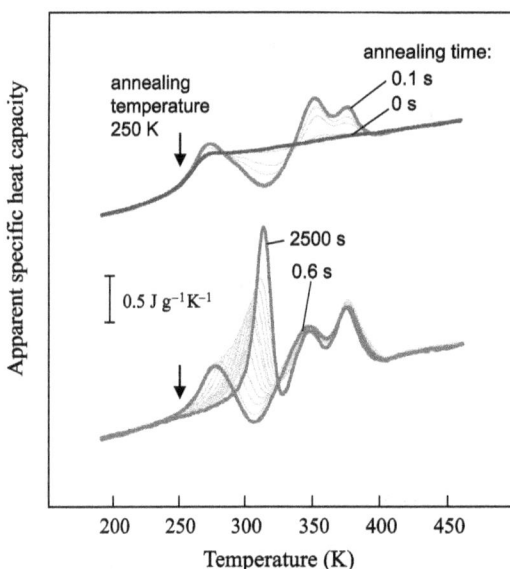

Fig. 1.43: Apparent specific heat capacity of iPP as a function of temperature, measured on heating at a rate of $3 \cdot 10^4$ Ks^{-1} after annealing initially glassy samples for different periods of time at 250 K.

observation of cold ordering on continued heating indicates that there must have been formed already small ordered domains/homogeneous nuclei during annealing. Though the enthalpy of disordering of these domains/nuclei is small and perhaps below the detection limit, we assume that they get destroyed on heating and leave self-nuclei for cold-ordering. As in case of annealing at 230 K, the cold-ordering peak is immediately followed by endothermic disordering. The enthalpies of cold ordering and disordering, reorganization and melting are approximately identical, suggesting that annealing at 250 K for 0.1 s, or shorter time, did not result in measurable formation of ordered domains with latent heat release.

Further increase of the time of annealing to values larger than about 0.5 s leads to a qualitative change of the heating curves, as is demonstrated with the bottom set of data. The cold-ordering peak apparently disappears and is replaced by an endothermic peak at approximately identical temperature, similar in appearance as in case of samples annealed at 230 K. We assume that cold ordering on heating still occurs, which, however, is just superimposed by the endothermic peak which we relate to minor amount of enthalpy recovery as a result of classical enthalpy relaxation of the unstable glass, and disordering of mesophase domains formed at 250 K. Again, the maximum possible enthalpy relaxation on annealing at 250 K is only 3 J/g. The major portion of the endothermic peak at the glass transition must therefore be caused by disordering of small mesophase particles. Worthwhile to be noted, ordering of molecule segments in the glassy state is connected with distinct immobilization of the amorphous phase, as is concluded from the increase of the glass transition temperature by more than 20 K [156].

A quantitative evidence for the formation of ordered domains during annealing at 250 K is provided by analysis of the change/decrease of the enthalpy with annealing time, shown in Fig. 1.42 with the gray triangles. The enthalpy decreases below the value of the equilibrium liquid after annealing for only about 10 s, that is, if the maximum possible value of about 3 J/g of enthalpy relaxation is exceeded, being in agreement with the observation of deviation of data from the enthalpy relaxation line.

1.3.6.3 Isotactic Polybutene-1 (iPB-1)

The analysis of the effect of low-temperature aging on non-isothermal cold crystallization is further demonstrated with Figs. 1.44–1.48. Fig. 1.44 shows the apparent specific heat capacity data of iPB-1 of different thermal history recorded on heating at a rate of 100 Ks^{-1}. The various curves were obtained after cooling the iPB-1 melt to 213 K and subsequent aging at 243 K, that is, below T_g, for different periods of time between 0 (black curve, 0 s) and $1 \cdot 10^4$ s (different other (blue in color version) curves). Regardless of the aging conditions, before the beginning of the heating experiment the samples were completely amorphous, that is, aging at 243 K did not result in crystallization within the time frame of the aging experiment, as detailed below. The curve of non-aged iPB-1 shows with the heat-capacity increment at around 255 K the glass tran-

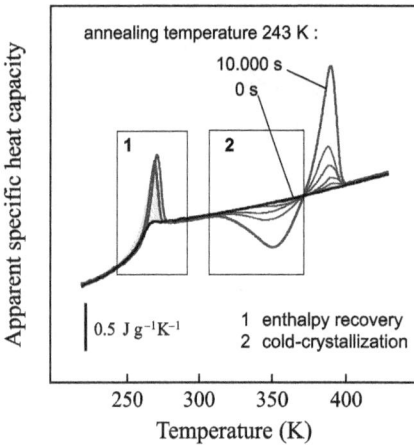

Fig. 1.44: Apparent specific heat capacity of iPB-1 of different thermal history as a function of temperature, obtained on heating at a rate of 100 Ks^{-1}. All samples were initially rapidly cooled at 1000 Ks^{-1} to 213 K, and then annealed at 243 K for different periods of time between 0 and 10^4 s, as is indicated in the legend. Annealing below T_g leads to enthalpy relaxation and nuclei formation which, on heating, is detected with the enthalpy-recovery peak (1), and cold-crystallization peak (2), respectively.

sition, however, on continued heating neither exothermic cold crystallization nor endothermic melting is observed. As such, heating at 100 Ks^{-1} is sufficiently fast to avoid cold crystallization in the absence of previously formed nuclei, being a pre-requisite for analysis of the nucleation kinetics at low temperature [38, 125].

With increasing aging time up to 100 s, we observed first relaxation of the enthalpy which leads on heating to an endothermic enthalpy-recovery peak in conjunction with the glass transition. In Fig. 1.44, the enthalpy-recovery peak is indicated with the box labeled '1'. Aging for periods of time larger than 100 s is connected with the formation of crystal nuclei which causes on heating exothermic cold crystallization, labeled '2' in Fig. 1.44. On continuation of heating, crystals formed during cold crystallization melt around 380 K. Since the areas of the cold-crystallization peak and melting peak for a sample of given annealing history are identical, it is concluded that crystals were absent at the beginning of the heating experiment.

Quantitative data about the enthalpy recovery at the glass transition and the enthalpy of cold-crystallization is provided with Fig. 1.45. It shows the areas of the enthalpy-recovery peak (top part) and the cold-crystallization peak (bottom part) as a function of the time of aging at 243 K. The area of the enthalpy-recovery peak, determined as the area between the curves for the non-annealed sample (0 s, black) and the sample under consideration in the temperature range '1', increases with increasing aging time to reach a plateau value after about 100 s. The obtained maximum enthalpy change is 5 J/g which is in accord with the maximum expected enthalpy of relaxation

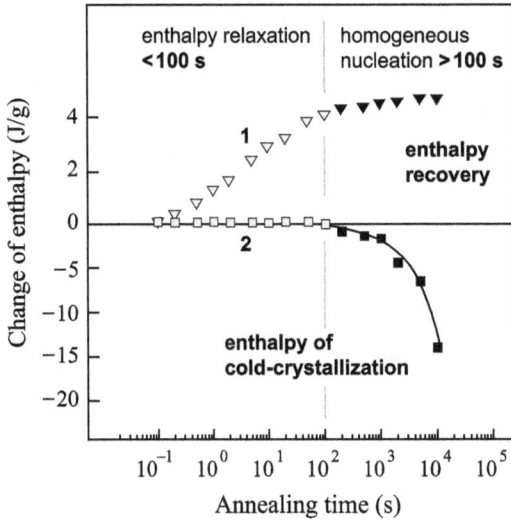

Fig. 1.45: Areas/enthalpies of the enthalpy-recovery peak (top part, curve 1) and cold-crystallization peak (bottom part, curve 2) of the FSC-curves obtained on iPB-1 of different thermal history shown in Fig. 1.44, with the data plotted as a function of the time of annealing at 243 K.

of the iPB-1 glass at 243 K. The latter amounts to about 4 J/g, and is determined as [152]

$$\Delta c_p \left(T_g - T_a \right) = 0.41 \text{ J/gK} \cdot (252 - 243) \text{ K} \cong 4 \text{ J/g},$$

with T_a and Δc_p being the annealing temperature and the heat capacity difference of solid and liquid iPB-1 at T_g, respectively. Fast vitrification of the supercooled liquid is connected with the formation of a glass of high enthalpy, which relaxes during aging, connected with a lowering of the enthalpy, ultimately approaching the enthalpy of the liquid at the particular temperature of 243 K. On subsequent heating, devitrification of the relaxed glass is contiguously followed by an endothermic enthalpy-recovery peak.

Cold crystallization in Fig. 1.45 is quantified as the area between the curves for the non-annealed sample (0 s, black) and the sample under consideration in the temperature range '2'. The data of Fig. 1.45 reveal that cold crystallization is only observed if the area of the enthalpy-recovery peak is nearly constant and has reached its expected maximum value. In other words, formation of crystal nuclei at 243 K only occurs if the enthalpy relaxation of the glass is completed. During isothermal annealing of the glass cooperative segmental mobility is accelerated by the driving force for relaxation, the distance from the equilibrium liquid state. This enhanced cooperative segmental mobility vanishes if the enthalpy reached the value of the hypothetic liquid phase at identical temperature.

We observed that nuclei formation only occurs if the densification of the glass is finished, which implies that the formation of overcritical clusters of parallel aligned molecule segments, that is, of crystal nuclei is not possible as long as the enhanced co-operative segmental mobility drives the sample towards the supercooled liquid state. This interpretation of the link between the enthalpy relaxation of the glass and the formation of crystal nuclei presumes a homogeneous nucleation mechanism, similar as it has been suggested in a study of the nucleation and crystallization behavior of PCL above [38].

The effect of aging of an initially fully amorphous iPB-1 at temperatures slightly above T_g is demonstrated with Figs. 1.46 and 1.47. Fig. 1.46 shows apparent specific heat capacity data of initially amorphous iPB-1 which has been aged at 273 K for different periods of time, as is indicated in the legend. The black curves were obtained on samples which have been aged at 273 K for periods of time up to 2 s. The data reveal – with the step-like increase of the heat capacity – the glass transition at about 255 K. However, on increasing the temperature no further thermal events are seen, leading to the conclusion that both nucleation and crystal growth were not initiated. If the annealing time is between 2 and 200 s (blue curves in the colored version) then, besides the glass transition at 255 K, exothermic cold-crystallization followed by endothermic melting is detected. The heat-capacity increment on devitrification of the glass and equal areas of the cold-crystallization and melting peaks prove that before the beginning of the FSC heating scan, samples were non-crystalline. Cold crystallization is caused by the formation of crystal nuclei during aging, with the area of the cold-crystallization peak being proportional to the number of nuclei formed. Further increase of the time of aging beyond 200 s leads to crystallization. The heat capacity increment at T_g is reduced to about 2/3 of the value expected for a fully amorphous sample, and the glass transition range becomes wider and is shifted to higher tem-

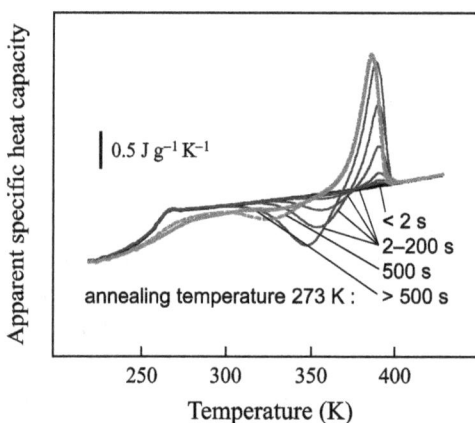

Fig. 1.46: Apparent specific heat capacity of iPB-1 of different thermal history as a function of temperature, obtained on heating at a rate of 100 Ks^{-1}. All samples were initially rapidly cooled at 1000 Ks^{-1} to 213 K, and then annealed at 273 K for different periods of time, as is indicated in the legend.

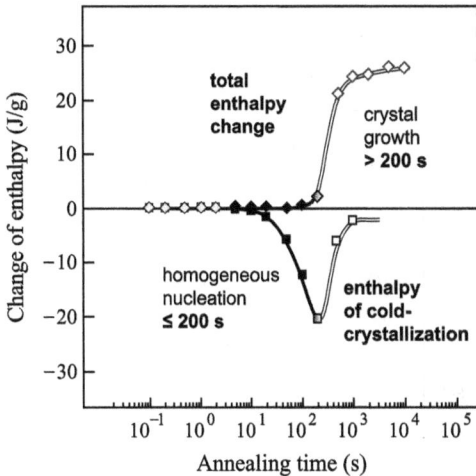

Fig. 1.47: Enthalpy of cold crystallization (squares) on heating iPB-1 of different thermal history at a rate of 100 Ks^{-1} (Fig. 1.46), plotted as a function of the time of prior annealing at 273 K (bottom part). The top part shows the overall change of the enthalpy during annealing at 273 K (diamond symbols). Black (central part) and white (right side) coloring of symbols indicates the time ranges of annealing at which formation of homogeneous crystal nuclei and crystal growth is observed, respectively.

peratures after aging for more than 500 s at 273 K. Moreover, cold crystallization is almost completely absent, and the observed overall change in enthalpy is related to crystals which were formed during annealing. The small endotherm around 315 K may be linked to partial melting associated to mobilization of rigid amorphous segments, which was proven to be completed in this temperature range [157].

A quantitative analysis of the FSC-curves of Fig. 1.46 is provided with Fig. 1.47 that visualizes the area of the cold-crystallization peak (bottom part) and overall change of the enthalpy during aging at 273 K as a function of the time of aging (top part). The overall enthalpy change on heating is determined in the temperature range from 280 to 400 K as the integral of the difference between the measured curve and a linear baseline connecting the integration limits. The gray-colored symbols at aging times less than 2 s indicate that neither cold-crystallization/nuclei formation during aging nor crystallization occurred. Aging between 2 and 200 s, indicated by black coloring of symbols, leads to cold-crystallization/formation of crystal nuclei, however, crystal growth during aging is not detected since the overall enthalpy change remains equal to zero. Only if the aging time exceeds 100 s then crystallization is observed, being in accordance with the half-times of crystallization shown in Fig. 1.32. It is worthwhile noting that the area of the cold-crystallization peak decreases as crystallization proceeds during annealing for times longer than 200 s.

In analogy to the experiments and data evaluation explained with Figs. 1.44–1.47, in Fig. 1.48 are shown the overall change of the enthalpy (top part) and the enthalpy of

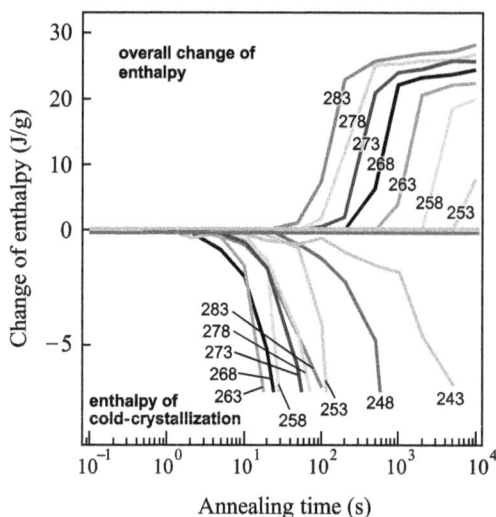

Fig. 1.48: Enthalpy of cold crystallization on heating iPB-1 of different thermal history at a rate of 100 Ks^{-1}, plotted as a function of the time of prior annealing at different temperatures as is indicated at each curve (bottom part). The top part shows the overall change of the enthalpy during annealing at the various temperatures as a function of the annealing time.

cold crystallization (bottom part) of iPB-1 aged at different temperatures as a function of the time of annealing. The top set of data shows that with decreasing aging temperature, within the frame of the analyzed temperature range from 243 to 283 K, the curves are shifted to larger times, that is, the crystallization rate decreases. This result is expected from the cold-crystallization experiments shown in Fig. 1.32 which revealed a maximum cold-crystallization rate at about 340 K. Analysis of the cold-crystallization behavior as a function of time and temperature of aging, provided by the bottom set of curves in Fig. 1.48, reveals that the nucleation rate is maximum at temperatures distinctly lower than 283 K. Aging at 263 and 268 K leads to cold crystallization on heating as a result of nuclei formation already after aging for 2 to 3 s. In contrast, if aging is performed at both either lower temperatures and or higher temperatures then the aging time to observe cold-crystallization/nuclei formation is increased.

Quantitative information about the kinetics of formation of crystal nuclei was collected by analysis of FSC heating scans obtained on iPB-1 of different aging history. The FSC scans of samples aged below T_g were evaluated regarding the enthalpy recovery peak caused by prior enthalpy relaxation/densification of the glass and regarding the cold-crystallization peak related to prior formation of crystal nuclei. It has been found that the formation of crystal nuclei in glassy iPB-1 only occurs after completion of the enthalpy relaxation of the glass, similar as has been observed for PCL (see Section 1.3.6.1 and [38]). The link between the enthalpy relaxation of the glass and crystal nucleation is interpreted as pointing to a homogeneous nucleation process. Analysis of the aging-time dependence of the enthalpy of cold crystallization on heating of samples aged at different temperatures leads to the conclusion that the rate of nucleation is maximal at around 265 K. In contrast, the maximum rate of form II crystallization is observed at distinctly higher temperatures in the range of 330–340 K.

1.3.6.4 Polyamid 6 (PA 6)

Fig. 1.49 shows schematically the temperature-time profile of the FSC nucleation/crystallization experiments performed for PA 6. The specimen was melted by heating to 523 K and then cooled to different annealing temperatures between 313 and 383 K after a dwelling time of 0.5 s. The annealing time at the respective final annealing temperature was varied from 0.1 to $1 \cdot 10^4$ s. The cooling rate of 1000 Ks^{-1} to approach the annealing temperature was selected to ensure absence of formation of both nuclei and crystals before the beginning of annealing (Section 1.3.2).

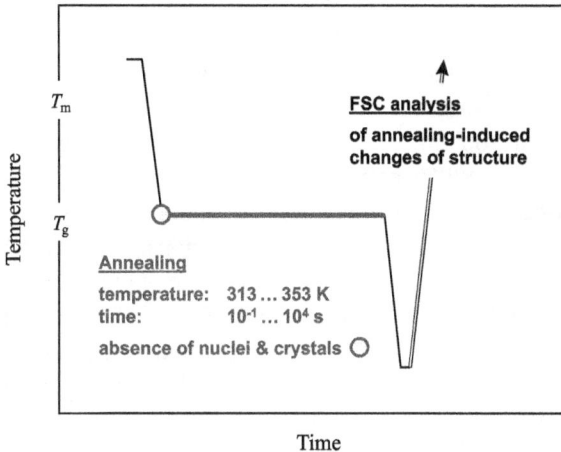

Fig. 1.49: Temperature-time profile for analysis of changes of structure of initially glassy or liquid PA 6 during isothermal annealing.

It has been shown in prior work that cooling the melt of PA 6 faster than 200 Ks^{-1} suppresses crystallization [95, 150]. The glass transition at the selected cooling rate of 1000 Ks^{-1} occurs around 330 K, which implies that annealing experiments were performed on both initially glassy and liquid samples. The changes of structure during annealing – including densification of the glass, formation of crystal nuclei and crystal growth, depending on temperature – were again analyzed on subsequent heating. The kinetics of the densification of the glass is analyzed by the area of the enthalpy-recovery peak detected on devitrification of the glass while the kinetics of nuclei formation is followed by analysis of the cold-crystallization peak. While the enthalpy of the enthalpy recovery peak related to prior enthalpy relaxation of the glass is nearly independent on the heating rate applied, the cold-crystallization behavior is strongly affected by the heating rate. In order to relate the enthalpy of cold crystallization to the formation of crystal nuclei during prior isothermal annealing, it is required that the heating rate is higher than a lower bound critical heating rate to avoid nuclei formation on heating. In addition, it is required that the heating rate is lower than a higher-bound critical heating rate to permit crystal growth as discussed in Section 1.3.2.

The lower-bound critical heating rate to avoid formation of crystal nuclei and cold crystallization on heating is about 500 Ks^{-1}. It is worthwhile noting that suppression of cold crystallization on heating requires a higher minimum rate of temperature change than is needed to suppress crystallization on cooling. For the particular homopolymer used in the present work the critical cooling rate to avoid melt-crystallization is only 100 Ks^{-1}, while suppression of cold crystallization requires heating faster than 500 Ks^{-1}. A major reason for this observation is that nuclei formed at rather low temperature, regardless whether on cooling at rates between 100 and 500 Ks^{-1}, or on immediate subsequent heating at rates lower than 500 Ks^{-1} are allowed to grow at relatively high temperature on heating. Note that the temperature of maximum (highest homogeneous) nucleation rate typically is distinctly lower than the temperature of maximum growth rate. Here non-isothermal nuclei formation is suppressed by cooling and heating the sample at 1000 Ks^{-1}. However, crystal growth on heating PA 6 at 1000 Ks^{-1} which contains nuclei still is permitted as is shown below.

In the following, selected annealing experiments on initially, that is, before begin of annealing fully amorphous PA 6 will be explained, for demonstration of the effect of the annealing temperature on the occurrence and sequence of enthalpy relaxation, crystal nucleation and crystal growth. In each example, Flash-DSC 1 heating curves in conjunction with quantitative evaluation of transition enthalpies, obtained on samples annealed for different time will be shown and discussed. The process of enthalpy relaxation is analyzed by the enthalpy-recovery peak, the formation of crystal nuclei is probed by analysis of the cold-crystallization peak area, and crystal growth during annealing is detected by the mismatch between the enthalpies of melting and cold crystallization on heating. Flash-DSC 1 curves and transition enthalpies are gray color-coded. Flash-DSC 1 curves and enthalpies of transition shown in black indicate that annealing led to enthalpy relaxation only. In case of formation of crystal nuclei, data are shown in dark gray, and if there has been observed crystal growth during annealing, the corresponding curves and enthalpy data are shown in light gray.

Fig. 1.50 shows heat-flow rate data of PA 6 as a function of temperature, obtained on cooling the melt at 1000 Ks^{-1}. The sample begins to vitrify around 335 K without prior crystallization at higher temperature. The vertical arrows at 313, 323, 333, and 348 K indicate the temperatures of the annealing experiments which are explained in detail below. The colors of the arrows indicate the changes of structure obtained during annealing for $1 \cdot 10^4$ s. Annealing amorphous PA 6 at 313 K for $1 \cdot 10^4$ s leads to enthalpy relaxation only, while annealing at 323 and 333 K for $1 \cdot 10^4$ s additionally

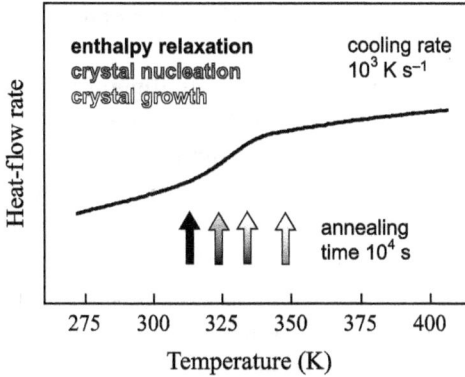

Fig. 1.50: Heat-flow rate of PA 6 as a function of temperature, obtained on cooling the melt at 1000 Ks^{-1}.

cause formation of crystal nuclei and eventually growth of crystals, respectively. In case of annealing at 348 K, enthalpy relaxation is absent, and only nuclei formation and crystal growth is detected.

Fig. 1.51a shows apparent specific heat capacity data of PA 6 as a function of temperature, obtained on heating at a rate of 1000 Ks^{-1}. The various curves represent samples which have been annealed for different periods of time between 0 (bold gray curve) and $1 \cdot 10^4$ s (bold black curve) at a temperature of 313 K, that is, about 20 K below T_g. Before the beginning of annealing, as a result of prior fast cooling of the melt at 1000 Ks^{-1}, the samples were amorphous. The Flash-DSC 1 curve of the non-annealed sample shows with the heat-capacity increment the glass transition at about 345 K. With increasing time of annealing, the glass transition gets superimposed by

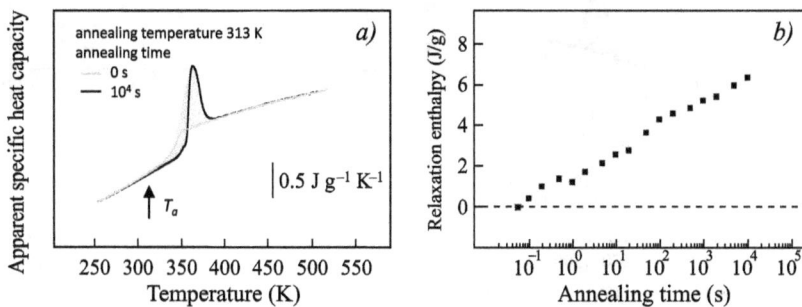

Fig. 1.51: Apparent specific heat capacity of PA 6 of different thermal history as a function of temperature, obtained on heating at 1000 Ks^{-1}. All samples were initially rapidly cooled at 1000 Ks^{-1} to 313 K, and then annealed for different periods of time between 0 and 10000 s, leading to enthalpy relaxation. The vertical arrow indicates the temperature of annealing (T_a). b) Enthalpy of relaxation the glass of PA 6 as a function of the time of annealing at 313 K. Prior to annealing, the melt was vitrified on fast cooling at 1000 Ks^{-1}.

an enthalpy-recovery peak, indicating prior relaxation of the glass at the annealing temperature.

The enthalpy of relaxation of the glass is shown as a function of the time of annealing in Fig. 1.51b. The data have been calculated by integration of the apparent specific heat capacities shown in Fig. 1.51a between 300 and 390 K, and subtraction of the enthalpy of the non-annealed sample. There is observed an increase of the enthalpy of relaxation of the glass with time, however, without the observation of an equilibrium value on extended annealing. Obviously, annealing the glass of PA 6 about 20 K below T_g for $1 \cdot 10^4$ s is insufficient to complete the relaxation process. The maximum expected enthalpy of relaxation, $\Delta h_{relax, max}$, amounts to about 8 J/g, as it can be estimated from the heat capacity difference Δc_p between solid and liquid PA 6 at T_g, and the temperature difference between T_g and the annealing temperature T_a, i.e. [152],

$$\Delta h_{relax, max} \cong \Delta c_p \left(T_g - T_a \right) = 0.44 \text{ J/gK} \cdot \left(T_g - 313 \text{ K} \right) \cong 8 \text{ J/g}.$$

Note, we consider the calculation of $\Delta H_{relax, max}$ as an estimate only, due to uncertainties in the definition of T_g and additional relaxation processes occurring during cooling and heating the glass subsequent to the isothermal annealing step. Annealing below T_g leads to enthalpy relaxation and nuclei formation which, on heating, is detected with the enthalpy-recovery peak (1), and cold-crystallization peak (2), respectively. The vertical arrow indicates the temperature of annealing (T_a).

Fig. 1.52a shows apparent specific heat capacity data of PA 6 of different thermal history as a function of temperature, obtained on heating at 1000 Ks^{-1}. Prior to the

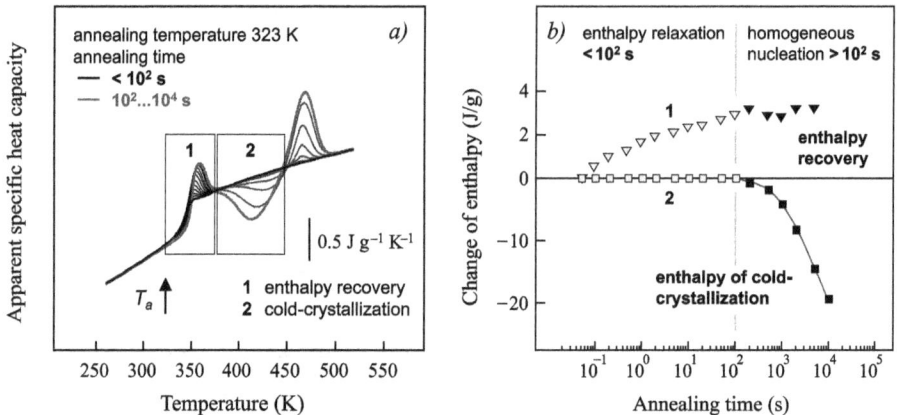

Fig. 1.52: (a) Apparent specific heat capacity of PA 6 of different thermal history as a function of temperature, obtained on heating at 1000 Ks^{-1}. All samples were initially cooled at 1000 Ks^{-1} to 323 K, and then annealed for different periods of time between 0 and 10^4 s. (b) Areas/enthalpies of the enthalpy-recovery peak (top part, curve 1) and cold crystallization peak (bottom part, curve 2) of the Flash-DSC 1 curves obtained on PA 6 of different thermal history shown in Fig. 1.52a, with the data plotted as a function of the time of annealing at 323 K.

heating experiment, samples were vitrified by fast cooling at 1000 Ks^{-1} to 323 K, that is, to a temperature about 10 K below the onset of the glass transition during cooling, and then annealed for different periods of time between 0 and $1 \cdot 10^4$ s. The Flash-DSC 1 heating curve of non-annealed PA 6 (black bold curve) shows with the heat-capacity increment at about 345 K the glass transition and then, on continued heating, no further events connected with exchange of latent heat. With increasing annealing time, up to 100 s (black curves), there is detected an enthalpy recovery peak, denoted '1' in Fig. 1.52a. If the annealing time exceeds 100 s, then the area of the enthalpy-recovery peak remains constant, and cold crystallization is observed in the temperature range between 375 and 450 K (blue curves in color version), followed by melting of crystals formed during cold crystallization. The areas of the cold-crystallization and the melting peaks are equal which proves absence of crystals before begin of the heating experiment. Fig. 1.52b shows the areas of enthalpy-recovery peak (top part, curve 1) and cold-crystallization peak (bottom part, curve 2) of the Flash-DSC 1 curves of Fig. 1.52a as a function of the time of annealing at 323 K. In contrast to the annealing experiment at 313 K, shown with Fig. 1.51, the maximum enthalpy of relaxation $\Delta H_{\text{relax, max}}$ of the glass of PA 6 on annealing at 323 K is decreased to about 3 J/g and the stabilization of the glass completes after about 100 s (see vertical line in Fig. 1.52b). Simultaneously, there is not observed formation of crystal nuclei during annealing for time periods less than 100 s, as it is concluded from the absence of the cold-crystallization peak on heating. The data of Fig. 1.52 suggest again that cold crystallization on heating, which is related to prior formation of crystal nuclei during isothermal annealing, occurs only if the process of enthalpy relaxation of the glass, quantified with the enthalpy-recovery peak, is completed.

Fig. 1.53a shows Flash-DSC 1 heating curves of initially rapidly cooled PA 6 which prior to the heating experiment has been annealed at 333 K, that is, at a temperature slightly lower than the onset of the glass transition during cooling. Annealing for periods of time less than 2 s leads to minor enthalpy relaxation as indicated with the small peak on completion of the glass transition on heating at 350–360 K (black (fully dark smooth) curves). If the annealing time is between 2 and 500 s then the enthalpy-recovery peak at the glass transition remains constant in area, and we observe cold crystallization followed by melting (blue curves in the color version, left figure bottom part). The enthalpies of cold crystallization and melting are equal, that is, annealing up to 500 s does not lead to crystallization at the annealing temperature. Only if the annealing time is longer than 500 s, then the enthalpy of melting is larger than the enthalpy of cold crystallization, indicating crystallization during isothermal annealing at 333 K (red curves in the color version, left figure upper set of curves). Crystallization at 333 K is connected with a shift of the glass transition temperature towards higher temperature, and the appearance of a melting peak at temperatures lower than 400 K, related to the beginning of melting of crystals formed at the annealing temperature. Simultaneously, the cold-crystallization enthalpy decreases due to the lower number of active nuclei, caused by their prior growth to crystals. The high-

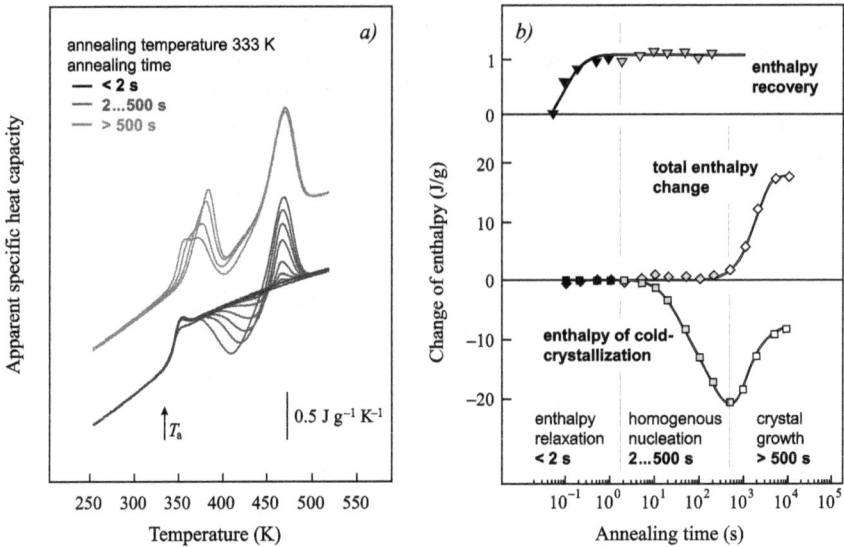

Fig. 1.53: (a) Apparent specific heat capacity of PA 6 of different thermal history as a function of temperature, obtained on heating at 1000 Ks^{-1}. All samples were initially cooled at 1000 Ks^{-1} to 333 K, and then annealed for different periods of time between 0 and 104 s. (b) Areas/enthalpies of the enthalpy-recovery peak (top part) and cold crystallization peak (bottom part) of the Flash-DSC 1 curves obtained on PA 6 of different thermal history shown in Fig. 1.53a, plotted as a function of the time of annealing at 333 K. The overall change of the enthalpy during annealing (central part) was calculated by integration of the Flash-DSC 1 curves covering the temperature range of the endothermic melting peaks and the exothermic cold-crystallization peak, and subtraction of the value of the non-annealed sample. Central and right hand side part symbols indicate time ranges of annealing at which formation of homogeneous crystal nuclei and crystal growth is observed, respectively. Annealing leads to enthalpy relaxation, nuclei formation and crystal growth as is seen for the other polymers too and described in the text. The vertical arrow indicates the temperature of annealing (T_a).

temperature melting peak around 470 K is interpreted as being related to melting of crystals formed by reorganization of crystals developed during annealing, as well as melting of crystals formed from the amorphous phase by cold-crystallization. For the sake of clarity in presentation, the set of curves obtained on samples annealed longer than 500 s (upper set of curves) is shifted relative to the curves obtained on samples annealed for shorter times.

In Fig. 1.53b are shown the areas of the enthalpy-recovery peak (top part), of the cold-crystallization peak (bottom part), and the overall change of enthalpy as a function of the time of annealing at 333 K, calculated from the Flash-DSC 1 curves of Fig. 1.53a. The overall change of the enthalpy during annealing (center part) was calculated by integration of the Flash-DSC 1 curves covering the temperature range of the endothermic melting peaks and the exothermic cold-crystallization peak, and subtraction of the value of the non-annealed sample. Despite the low maximum enthalpy relaxation during annealing of only about 1 J/g, the link between the enthalpy relaxation and nuclei formation, proposed with the annealing experiments performed at 323 K, is confirmed. Cold crystallization on heating is only obtained if the area of the enthalpy-recovery peaks remains constant, that is, if the enthalpy relaxation of the glass is complete. The decrease of the cold-crystallization enthalpy in samples annealed longer than 500 s is explained by growth of homogeneous nuclei to crystals during isothermal annealing. These nuclei are then not available to grow to crystals on heating, leading to reduced cold crystallization.

In analogy to Figs. 1.51–1.53, which showed Flash-DSC 1 heating scans and enthalpies of transitions of PA 6 samples annealed at 313, 323, and 333 K, respectively, in Fig. 1.54a the sequence of nuclei formation and crystal growth of initially, that is, before begin of annealing at 348 K, fully amorphous PA 6 is demonstrated. As is shown with the blue (color version), respectively, thick black (black-white version) curves (annealing time 2 . . . 10 s) in Fig. 1.54a, annealing for periods of time between 2 and 10 s leads only to formation of nuclei as is indicated with the observed cold-crystallization peak, and the equivalency of the enthalpies of cold-crystallization and subsequent melting. If the annealing time is larger than 10 s then the area of the endothermic peaks in Fig. 1.54a exceeds the area of the exothermic peak, proving formation of crystals during isothermal annealing. Crystallization at 348 K is con-

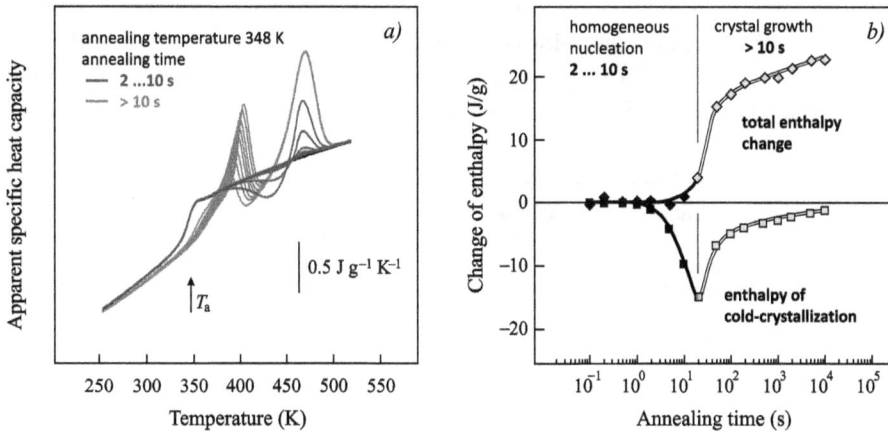

Fig. 1.54: (a) Apparent specific heat capacity of PA 6 of different thermal history as a function of temperature, obtained on heating at 1000 Ks^{-1}. All samples were initially cooled at 1000 Ks^{-1} to 348 K, and then annealed for different periods of time between 0 and 10^4 s. Annealing leads to nuclei formation and crystal growth as is described in the text. The vertical arrow indicates the temperature of annealing (T_a). (b) Enthalpy of the cold-crystallization peak of the Flash-DSC 1 curves obtained on PA 6 of different thermal history shown in Fig. 1.54a, plotted as a function of the time of annealing at 348 K. The top part shows the overall change of the enthalpy during annealing. Black and gray coloring of symbols indicates time ranges of annealing at which formation of homogeneous crystal nuclei and crystal growth is observed, respectively.

nected with a distinct increase of the glass transition temperature by about 10 K, that is, there is observed an immobilization of the amorphous phase due to the formation of crystals. Melting of crystals formed at 348 K occurs in conjunction with the devitrification of the amorphous phase, beginning at temperatures lower than about 400 K. The high temperature melting peak at about 470 K is related to the melting of both reorganized crystals which initially were grown during isothermal annealing at 348 K, and crystals formed during cold-crystallization. As discussed with Fig. 1.53b, the growth of nuclei to crystals during isothermal annealing reduces the number of nuclei which then would be available for cold crystallization. Consequently, the reduction of the enthalpy of cold crystallization parallels the increase of crystallinity during isothermal annealing.

1.3.7 Summary of Experimental Results and Conclusions

Fig. 1.55 shows in the top part a series of curves, each representing the change of the enthalpy of PA 6 due to crystallization as a function of the annealing time at a specific annealing temperature. The red (upper part) dotted curve was obtained on a sample which has been annealed at 328 K, and provides the information that formation of crystals began after 1000 s. With increasing temperature of annealing of the initially fully amorphous PA 6, the onset time of crystallization decreases. In case of annealing PA 6 at 333 K (red (in the color version) dashed-dotted curve in the upper part), crystallization begins after 100 s while on annealing at 338 K (solid (red in the color version) curve in the upper part) the onset time of crystallization is further reduced to

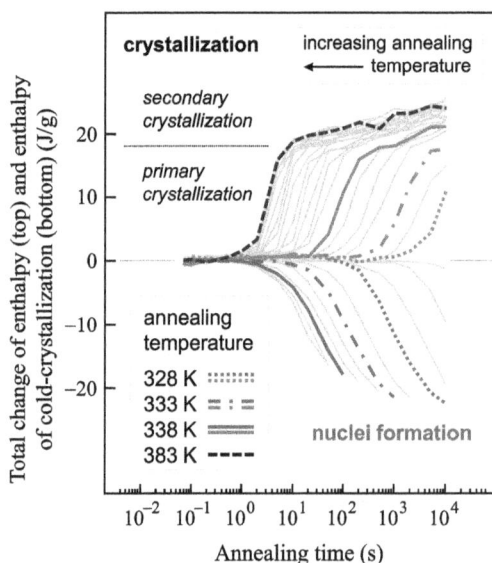

Fig. 1.55: Enthalpy of cold crystallization on heating PA 6 of different thermal history at a rate of 1000 Ks^{-1}, plotted as a function of the time of prior annealing at different temperatures, as is indicated in the legend (bottom part). The top part shows the overall change of the enthalpy during annealing at the various temperatures as a function of the annealing time, indicating crystallization.

about 10 s. In other words, with increasing annealing temperature between 328 K (dotted (red in the color version) curve in the upper part) and 383 K (dashed curve in the upper part) the crystallization rate increases, as it is expected from former research about the temperature dependence of the crystallization rate [150, 158]. It is known that the crystallization rate of PA 6 has its maximum around 410 K, with the decrease of the crystallization rate on lowering the temperature related to the loss of cooperative mobility of amorphous molecule segments. Though not in focus of the present study, the data obtained provide also information about the transition from primary to secondary crystallization, as it is deduced from the pronounced decrease of the slope of the curves when the conversion exceeds a critical value.

The bottom set of curves in Fig. 1.55 represents the enthalpy of cold crystallization as a function of the time and temperature of annealing, ultimately providing information about the temperature dependence of the rate of formation of nuclei. For easy comparison with the curves shown in the top part, the dotted, dashed-dotted, and solid blue (in the color version) curves were obtained on samples annealed at 328, 333, and 338 K, respectively. As in case of crystallization, an increase of the annealing temperature is connected with a shift of the curves to shorter time, that is, in the analyzed temperature range the nucleation rate decreases with increasing temperature. Comparison of the top and bottom sets of curves of Fig. 1.55 demonstrates that both the kinetics of nucleation and crystallization can advantageously be analyzed separately by the experimental approach applied here, and the companion studies on PCL, iPB-1, or iPP [38, 100, 125, 127]. For example, formation of crystal nuclei begins after about 10 s on annealing amorphous PA 6 at 333 K, while crystal growth connected with exchange of latent heat is only observed after 1000 s (see upper and lower dashed-dotted curves).

The data of Fig. 1.55 were further reduced by evaluation of extrapolated onset times of formation of crystal nuclei and of crystals, to obtain quantitative information about the temperature dependence of crystal nucleation and crystal growth. Fig. 1.56 shows such extrapolated onset times of crystallization (gray squares) and nuclei formation (dark gray (blue in the color version) squares) as a function of the annealing temperature, focusing on the temperature range close to the glass transition temperature; the temperature range of the glass transition on cooling at 1000 Ks^{-1} applied in this work is indicated in Fig. 1.56 with the gray shaded area. In addition, Fig. 1.56 includes with the gray circles information about the time required to complete the process of enthalpy relaxation.

It is concluded from visual inspection of the various data sets of Fig. 1.56 that formation of crystal nuclei in glassy PA 6 only occurs if the process of enthalpy relaxation of the glass is completed. The experimentally assessable decrease of the enthalpy of the glass to the value of the liquid state at identical temperature corresponds to a densification of the glass, involving cooperative rearrangement of molecule segments on a length scale of few nanometers [99, 153]. The data of Fig. 1.56 suggest that densification of the glass and disappearance of the relatively large scale motions connected to

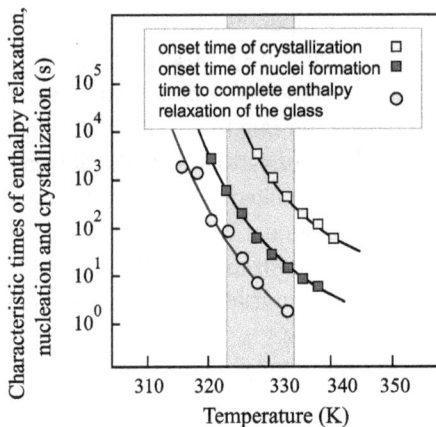

Fig. 1.56: Time to complete enthalpy relaxation, onset time of formation of crystal nuclei and onset time of crystallization as a function of the annealing temperature. The gray shaded area indicates the temperature range of the glass transition on cooling at a rate of 1000 Ks^{-1}.

enthalpy relaxation is completed prior to formation of crystal nuclei. Theoretical estimates show that for homogeneous nucleation frequently the characteristic relaxation times are considerably smaller than the typical times for formation of crystal nuclei of critical size (cf. [1] and Section 1.4). For heterogeneous nucleation, the respective difference between these time scales becomes smaller and nucleation may proceed even faster than relaxation. Thus, the experimentally in this work observed sequence of densification of the glass and nuclei formation is considered as evidence for a homogeneous nucleation mechanism.

The here presented data for PCL, iPP, iPB-1 and PA 6 show a very similar behavior regarding enthalpy relaxation and homogeneous nucleation during annealing below or slightly above the glass transition temperature. Since the studied polymers exhibit very different crystallization kinetics (PCL and iPP very fast and iPB-1 and PA 6 significantly slower) the observed behavior seems to be universal. In all cases homogeneous nucleation only starts after finalizing enthalpy relaxation. Obviously there is a strong coupling between the processes responsible for enthalpy relaxation and homogeneous nucleation, respectively. In other words, the thermodynamic state of the supercooled melt (measured by its enthalpy) plays an important role for the homogeneous nucleation. We try to summarize and generalize this aspect in the next section.

1.4 Illumination of the Nucleation and Growth Mechanism

1.4.1 Low-temperature Endotherms and Homogeneous Nucleation

The annealing treatment below the common glass transition of the polymers should allow us to gain information about details of homogeneous nucleation. In the classical model of nucleation and growth of crystals, one expects it to proceed from a critical nucleus to a supercritical nucleus (both unstable with a positive free enthalpy of formation) and continue on to an initial, tiny, stable crystal with a free enthalpy of formation less than zero, and finally grow to macroscopic sizes (see, for example [1, 23]).

In Fig. 1.57a, the results of thermal analysis are shown when heating a PCL sample after annealing at 195 K for different times. As for the other polymers under investigation, a structured, small endotherm appears close to the increase in heat capacity due to the glass transition. Such endotherms are normally single, small peaks, referred to as enthalpy relaxation of the amorphous phase. The development of the endotherm with annealing time is enlarged in Fig. 1.57b. The peak clearly separates into two events while the corresponding glass transition broadens and shifts to higher temperature. What is seen is a superposition of a typical enthalpy relaxation of the mainly amorphous sample with a small melting or disordering peak which indicates the presence of tiny crystals losing their thermodynamic stability somewhat above T_a, but before cold crystallization takes over. An early mention of similar observations can be found for selenium in its linear polymer allotrope and poly(ethylene terephthalate) (Figs. 2 and 1 on pages 617–618, respectively, in [159]).

A quantitative evaluation of the peak area was performed by integrating the measured curves vs. temperature over a baseline obtained from a sample heated without annealing (black curve in Fig. 1.57b). The result of integration versus annealing time is shown in Fig. 1.57c. For comparison, the maximum value of enthalpy relaxation for amorphous PCL at the annealing temperature $T_a = 195$ K

$$\Delta h_{relax, max} \cong \Delta c_p \left(T_g - T_a\right) = 0.012 \text{ J/gK} \cdot (200 \text{ K} - 195 \text{ K}) \cong 0.06 \text{ } \mu\text{J}$$

is given by the horizontal line in Fig. 1.57c. The value equals the difference between the enthalpy of the glass and that of the under-cooled liquid at T_a (as marked by the vertical (red in the color version) line in Fig. 1.57d). After 5 min annealing, the area of the peak becomes larger than possible due to enthalpy relaxation alone, therefore, the second peak in the (blue in the colored version, upper arrow) curve (1 min) of Fig. 1.57b must be considered to be due to melting. May be there is already before 0.4 s (red curve in the color version, lower arrow) a small shoulder due to melting at the high temperature flank of the enthalpy relaxation peak. The presence of a relatively small melting peak between the glass transition and the beginning of cold crystallization (accounted for by less than 10 % crystallinity after 8 h annealing) depresses the glass transition in a way as it is otherwise only observed for the highly crystalline material marked by the orange in the color version, (8 h 235 K)-curve in Fig. 1.57b and reached after 8 h at 235 K. This confirms the creation of a large amount of rigid amorphous material even around the tiny crystals.

To discuss the melting of the small crystals at T_{msc}, close to T_g, we will try to apply classical nucleation theory, as described, for example, in [1, 23]. The schematic enthalpy diagram of Fig. 1.57d follows the analysis path. The first solid vertical (red in the color version) segment marks the densification of the initial glass which, on analysis by heating, yields the hysteresis peak at the upper end of the glass transition.

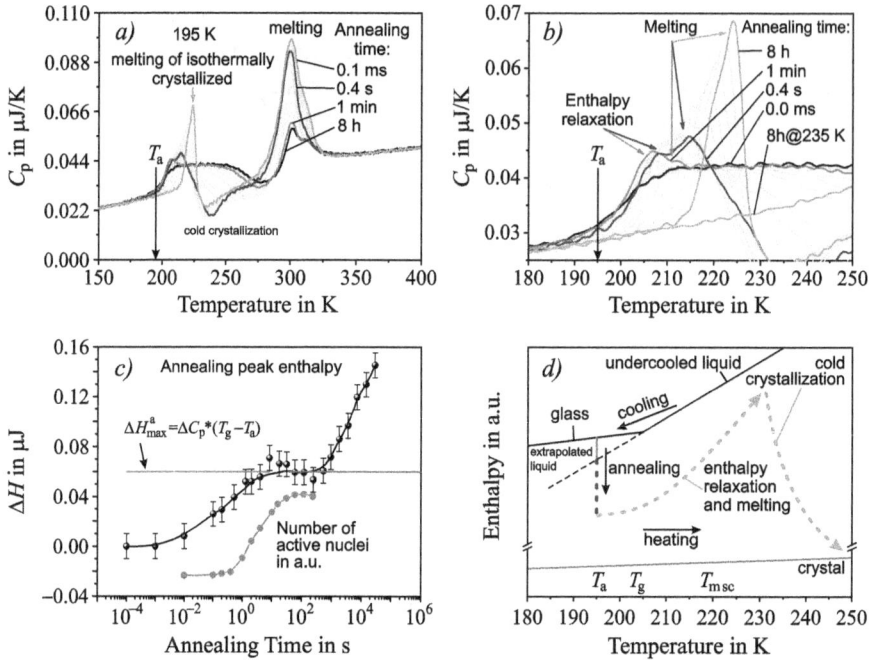

Fig. 1.57: The heating after annealing at 195 K. (a) Evolution of annealing (melting) peaks with annealing time (for comparison the curve of material crystallized for 8 h at 235 K is given). (b) The annealing peak enthalpy (integrated against heating of fully amorphous material without annealing, black curve in (b). (c) Schematic representation in an enthalpy-temperature diagram. Note that an axis break was introduced in the ordinate because the enthalpy of the crystal is considerably smaller than that of the glass (d).

Simultaneously, however, as the glass approaches the dashed line of the extrapolated liquid, ordering takes place, moving the system into the bold dotted vertical (blue in the color version) segment beyond the enthalpy of the extrapolated liquid. The small crystals, grown at the temperature T_a, must by then have passed over the critical nucleus barrier (the saddle point in the free enthalpy landscape), and, as supercritical nuclei, grown to reach the small crystal sizes with a negative free enthalpy of formation from the amorphous phase (glass). This process of growing crystals slows down and stops as soon as the rigid-amorphous fraction, RAF, influences all remaining amorphous material, as seen from the glass transition in Fig. 1.57b and is discussed in more detail in [38]. On subsequent heating, part of the RAF may go through its glass transition, increasing the slope of the grey (green in the color version) dots in the schematic from that of the glass, ultimately to that of the under-cooled liquid, corrected by the crystallinity. At T_{msc} melting or disordering begins. This scheme lets one note that the critical nucleus size at T_{msc} is larger than at T_a and, using the calculations in [23] for the free enthalpy for the case of polyethylene as a measure, many of the supercritical nuclei at T_a are expected to revert back into embryos on heating to $T_{m\,sc}$ and disappear without measurable latent heat if they are not able to stabilize themselves on heating. The small crystals grown at T_a remain stable up to their specific melting temperature, $T_{m\,sc}$, which depends on T_a as well as time for growth and perfection. Fig. 1.57b mirrors the distribution in perfection of these tiny crystals, and Fig. 1.28 illustrates the kinetics of the crystals grown at T_a.

Fig. 1.57c further depicts the relation between enthalpy relaxation and homogeneous nucleation followed by crystallization. A separate melting peak is first seen after 10 s annealing. This is the time needed to reach the enthalpy of the supercooled liquid (horizontal line in Fig. 1.57c, lower end of the full (red in the color version) vertical line in Fig. 1.57d). Nevertheless, an increase of the number of active nuclei is observed at similar times. Particularly for the studies of iPP, iPB-1 and PA 6 it is seen that nucleation only occurs after completion of enthalpy relaxation. This is in accordance with the data for the half-time of nucleation and crystallization shown in Fig. 1.30. Crystallization requires mobility on the time (length) scale of the dynamic glass transition (α-relaxation) while nucleation is much faster (shorter length scale) and may occur at time scales closer to local relaxations (β-relaxation). While the dielectric α-relaxation has been linked to the dynamic glass transition (cooperative, segmental relaxation), the β-relaxation has been linked to more local processes (secondary relaxation) [160]. The crystallization at low temperatures follows the dielectric α-relaxation in the activation diagram, confirming that under such conditions crystallization is long-range diffusion controlled. The nucleation at low temperatures is assumed to have a homogeneous origin and it becomes two orders of magnitude faster than crystallization, positioned in-between the dielectric α- and β-relaxation, in agreement with Oguni's observations [161], but, ultimately, it also approaches the slow-down of the α-process. One may speculate that the initial steps of nucleation involve both local and long range motion, something expected if in this stage the observation of molar mass segregation (molecular nucleation [21, 22]) is also taking place. Therefore it was an interesting task to study the influence of molecular mass and molecular mass distributions on the nucleation and crystallization kinetics as it was reported in [125]. Unfortunately, no clear answer was obtained regarding molar mass segregation (molecular nucleation).

The classical polymer crystallization theory does not give insight in these actual molecular motion processes taking place during the early stages of crystallization. One newer model on this subject was proposed by Muthukumar [19]. Based on molecular dynamics simulation, he showed that the organization of the nucleus occurs in several steps and requires significant time before crystal growth actually starts. This scenario seems to be in agreement with our data. The crystallization shows a significant delay after nucleation occurs. It is seen from Figs. 1.28 and 1.57c that as soon as large scale crystallization begins, additional nucleation seems to stop. In this way, the newly developed DFSC allows one a check of the existing theories for polymer nucleation and crystallization and may, by generating more quantitative information, contribute to a better understanding of polymer crystallization and nucleation. Based on the present observation, the classical nucleation and growth theory needs to be modified by coupling the thermodynamic landscape of the free energy with a mobility landscape (viscosity), which is also size (and perfection) dependent. As in the classical theory which identifies only one critical nucleus, it may be possible to identify a limited number of significant maxima, respectively, saddle points, for the multidimensional landscape, instead of identifying the complete path of the process.

1.4.2 Some Brief Theoretical Considerations

In the present section, we briefly summarize and discuss some theoretical results of the classical theory of nucleation and growth processes to have a "baseline" allowing one to classify which experimental results can be explained by classical theory and which not.

Considering nucleation at isothermal conditions after the melt was transferred into the respective state sufficiently fast to avoid nucleation and growth in cooling, in order to estimate the effect of possible deviations of the properties of the ambient liquid from equilibrium on crystallization, we have here to account for two different time scales, (i) the characteristic time of relaxation to the respective equilibrium state, τ_R, already discussed in Section 1.2.1 (and denoted there as τ), and (ii) the characteristic time-scale of nucleation, τ_{nucl}. The characteristic time-scale of nucleation, we can identify with the time-lag in nucleation, τ, being equal – in one of its interpretations – to the time of establishment of steady-state conditions in nucleation. In other words, this is the time required in order to allow the first supercritical nuclei to be formed by homogeneous nucleation. According to Gutzow and Schmelzer [1, Eqs. (6.193)–(6.196)], the time-lag in homogeneous nucleation can be written frequently in the form

$$\tau_{nucl}^{(hom)} \cong \text{constant } \tau_R j_c^{2/3}, \tag{1.9}$$

where j_c is the number of particles in the critical cluster and the constant is frequently estimated for glass-forming melts as having values in the range of 10^2–10^3 (cf. [1, 162] for details).

In the derivation of the respective equations it is assumed that the Stokes-Einstein relation is fulfilled, i.e., that the effective diffusion coefficients governing crystal cluster formation and growth can be replaced by Newtonian viscosity. This is frequently assumed but not always the case. In addition, a linear relationship between viscosity and relaxation time is supposed to exist. At such assumptions, it follows that the characteristic relaxation time is much smaller than the characteristic time of nucleus formation. By this reason, first – at the mentioned conditions – relaxation to the respective metastable state occurs and afterwards (homogeneous) nucleation is followed. Densification (or, more generally, stabilization) of a glass or glass-forming melt has under such conditions as a rule to be expected to proceed prior to nucleation, at least, as far as the process proceeds via homogeneous nucleation. These considerations allow us – as it has been already done here earlier – to give a straightforward interpretation of some of the experimental results obtained. Under mentioned conditions, enthalpy relaxation has to proceed prior to nucleation (cf. Section 1.3.6).

For heterogeneous nucleation, one can arrive at the estimate (cf. [1], Eq. (7.18))

$$\tau_{nucl}^{(het)} = \tau_{nucl}^{(hom)} \Phi^{1/3} \tag{1.10}$$

for the time-lag in heterogeneous nucleation. Here Φ is the activity of the heterogeneous nucleation core, defined via the factor Φ,

$$W_{het} = W_{hom} \Phi, \tag{1.11}$$

by which the work of critical cluster formation in heterogeneous nucleation, W_{het}, is reduced as compared with the case of homogeneous nucleation, W_{hom}. Consequently, indeed, if Φ is small, then heterogeneous nucleation may occur at the respective sites prior to relaxation. For heterogeneous nucleation, nucleation may proceed consequently under certain conditions prior to relaxation to the respective metastable equilibrium.

In connection with elastic stresses in silicate and other classes of glass-forming systems, similar problems, i.e., the relation between relaxation and nucleation, are discussed in detail in [162]. There it is shown by similar considerations that for crystallization processes elastic stresses can be of importance only if diffusion and viscosity (or relaxation) are independent parameters and not coupled via the Stokes-Einstein relation. Earlier, a lot of work has been done by some us [1] in describing the effect of elastic stresses for segregation in multi-component solutions [1, 163, 164]. In such segregation processes, different diffusion coefficients of segregating particles and ambient phase units are involved. In case, the latter are much smaller than the first, elastic stresses may occur growing more rapidly than linear with the cluster volume. In an alternative case, realized, for example, in gas segregation in polymers [165], growth of bubbles may lead to a decrease of the viscosity of the polymer-gas solution due to a decrease of its concentration in it and, as a consequence, to a shift of the glass transition temperature respectively a termination of bubble growth (this is the basic mechanism employed in polymeric foam formation at BASF). In such cases, the growth of the clusters may be also terminated by elastic stresses. There exists some deep as it seems analogy of these phenomena to processes of formation of rigid amorphous structures (RAF) playing an important role in polymer crystallization.

Elastic stresses do play an important role also in crystallization processes of different classes of glass-forming melts as discussed in detail in [1, 162]. Such behavior can be explained in classical terms only assuming that the characteristic relaxation times have to be again of similar order of magnitude or larger as compared to the characteristic times of nucleation. In order that elastic stresses may affect nucleation, critical clusters need to be formed in the presence of elastic stresses, i.e., critical clusters formation has to proceed at time-scales smaller than the characteristic relaxation time of stresses. Under such conditions also the dependence of the properties of the ambient melt on prehistory has to be expected to be of significant importance for crystal nucleation and growth. In addition, also for low-molecular systems cluster properties may deviate from the properties of the newly evolving macroscopic phases and the scenario of phase evolution may differ from the classical picture [1] showing some analogy to the features of polymer crystallization as discussed in the present chapter. A further detailed study of these problems is considered, therefore, to be of considerable interest (cf. also Chapter 10).

1.5 Conclusions and Outlook

Extensive research on the crystallization and nucleation of PCL, iPP, iPB-1 and PA 6 has demonstrated the enormous capabilities of chip-based fast scanning calorimetry as a new calorimetric technique. The starting amorphous melt and all intermediate stages of ordering to mesophases or crystals could be frozen for detailed subsequent analysis. It was shown that the classical model of crystal nucleation and growth developed on the basis of small motifs and a single viscosity term correcting for the slowing of molecular motion at lower temperatures must be modified to include the local as well as global mobility of polymer molecules. Based on this first insight, many additional experiments must be designed to collect specific, quantitative information on the time and temperature dependence of nucleation, ordering, molecular segregation, organization of the various levels of rigid-amorphous structures surrounding the ordered phase, annealing, and, finally, isotropization. Structures down to nanophase size can now be studied by assessing the changes in the heat capacity in the glass transition region which can test local molecular motion on a nanometer scale, as well as through the latent heat analysis on order/disorder transitions.

The crystallization of the studied polymers could be connected by the following working hypothesis through the two basic stages.

Stage 1: Homogeneous nucleation at a given temperature, T_a, is considered to be initiated as soon as a pertinent critical free-energy barrier can be overcome. The path of homogeneous nucleation may be clarified within the classical nomenclature. It starts with *embryos* which need increasing free energy for further growth in size and perfection. The embryo must move by fluctuation across the barrier of the *critical nucleus*, described as the saddle point in the size-and-perfection landscape of the free-enthalpy of formation. This leads to the *supercritical nucleus* which can grow with a thermodynamically permitted decrease in free enthalpy of formation. The actual rates of progress are additionally determined by *kinetic factors* which also are size and perfection dependent, creating a much more intricate, multidimensional free enthalpy landscape. In the past, only the rate of crossing the critical nucleus barrier was considered. It was described using the theory of absolute reaction rates which assumes the kinetics can be fitted with a single term similar in temperature dependence to the bulk viscosity. The molecular dynamics simulations should be able in the near future to supply more details about the complete path to a stable nucleus or *small crystal* (species with a zero or negative free enthalpy of formation).

The analysis of Fig. 1.30 indicates that crystal growth and nucleation cannot be fitted with a single viscosity-related term which slows the process in parallel to the bulk glass-transition kinetics. Particularly the nucleation needs much faster, local viscosity terms as are known from dielectric and heat capacity measurements for large-amplitude molecular motion. On the other hand it is surprising that homogeneous nucleation does not start immediately after reaching an annealing temperature below T_g. For all studied polymers it was observed that first enthalpy relaxation (densification) towards the supercooled liquid state finalizes and then homogeneous nucleation occurs. One possible explanation for this observation may be the following: Enthalpy relaxation is considered to proceed via cooperative rearrangements on length scales typical for the α-relaxation. These length scales are of the order of nanometer, according to Donth 2–5 nm [153]. From the melting temperature of the small crystals a size of about 2–3 nm was estimated, Section 1.3.6.1 and [125]. For the critical nuclei the size is even smaller and they exist only temporarily. Therefore it seems possible that the cooperative rearrangements responsible for the enthalpy relaxation below T_g and occurring on a length scale of about 2–5 nm overturn the formation of over-critical nuclei which result from sporadic fluctuations. Only after reaching the local equilibrium for the cooperative rearrangements by approaching the supercooled liquid state the driving force for the rearrangements vanishes and overcritical nuclei may survive. Obviously this is not the case for annealing temperatures above the glass transition temperature. There nucleation starts without any enthalpy relaxation as shown in Fig. 1.54.

Following nucleation, a growth of small to large crystals, depending on temperature, occurs as **Stage 2:** In the past, a new barrier to growth in the form of *secondary or molecular nucleation* was assumed [23]. The molecular nucleation was thought to introduce the molecular segregation on crystallization [21, 22]. The new data give a hint that the segregation may already occur before or in Stage 1. Further growth of the initial crystals at T_a is hindered by the slow, long-range diffusion, leading ultimately to a sufficient amount of RAF with a higher glass transition temperature to stop all further growth. The melting of crystals retains nuclei that can enhance cold crystallization for low melting temperatures. For high melting temperatures such nucleus retention has in the past been called *self-nucleation* and the nuclei were found to survive at heating above the equilibrium melting temperature. To study the activity of these self-nuclei cooling was necessary to temperatures that permit crystallization. This research has, thus, opened a new direction of inquiry to gather quantitative information about nucleation and growth of polymer crystals as it affects the amorphous and crystalline phases and their interrelationship.

The observation of missing cold crystallization for samples annealed at higher temperatures sheds some light on the two different stages of nucleation in PCL recently discovered [14]. Looking at Fig. 1.37, one notes that the three crystallization conditions chosen in the display start with the same amorphous glass, characterized at the shortest annealing times by the same glass transition (temperature and width). Next, (at least) two largely different ordering processes are obvious. At low annealing temperature (202 K) poorly ordered species grow first (nuclei and small crystals), both being able to lead to cold crystallization during the analysis by heating with $1.8 \cdot 10^4$ Ks^{-1}. At higher annealing temperatures (252 and 272 K) no cold crystallization is detected during the analysis. The devitrification of the partially ordered samples is also different after the annealing at various temperatures. In all cases the ordering is connected with an increase in glass transition temperature and its width. While at 202 K, the increase in glass transition temperature at longer annealing times goes parallel with the initial ordering (judged by the low-temperature endotherm) and beyond 1.0 s exceeds the annealing temperature, at 252 and 272 K the melting endotherms of the 1.0 s curves begin at the upper end of the glass transition (the T_g of the RAF close to the annealing temperature). Increasing the annealing times increases the T_g of the RAF, but less than at the 202 K annealing temperature. Finally, for all three annealing temperatures, the 0.1 s annealing time curve represents a unique separation between the (at least) two processes.

Comparing the the curves (green in the color version) after 0.1 s annealing in Fig. 1.37 yields for all annealing temperatures different observable effects. At 202 K, the lowest annealing temperature, the crystallinity is close to zero as seen from the zero overall enthalpy change on heating and the heat capacity throughout the glass transition. Nevertheless, a large number of nuclei is active in the temperature range between 230 K and 270 K, causing a strong cold-crystallization peak (exotherm). After annealing for 0.1 s at 272 K, the highest annealing temperature, only a small endothermic enthalpy change is observed, indicating a few percent crystallinity grown before heating of the sample, but there is practically no corresponding cold-crystallization seen before the final melting. The nuclei and crystals formed at longer times at 272 K are not able to initiate cold-crystallization.

One may argue that this is due to a mismatch between the thickness of the crystalline lamellae or the size of the nuclei formed at 272 K and the size of the crystals growing around 250 K. To check this, annealing was performed at 252 K, the temperature close to the maximum cold-crystallization rate seen for the sample annealed at 202 K. Crystals grown under this annealing condition should meet the growth condition, particularly with respect to the lamellar thickness, for further cold-crystallization in the same temperature range. But as seen from the pronounced melting peak in Fig. 1.37b a significant amount of crystals was present after annealing for 0.1 s at 252 K but by direct crystallization (followed by possible crystal perfection during the heating for analysis) without any major prior cold crystallization. This result is surprising because classical crystallization theories would predict a further growth of supercritical nuclei and metastable crystals present in the quite mobile polymer melt as follows from the pronounced step in the heat capacity at the glass transition. Summarizing, there are nuclei and crystals with the right size for further growth present in the melt and only a minor RAF, if any, is present. Obviously another mechanism than a size mismatch must prevent further growth. This result is somehow similar to the observation that stable seed crystals in a saturated polymer solution do not act as nuclei for further growth as shown already by Wunderlich [166].

Looking carefully at the curves for 0.1 s and 0.01 s annealing times (green and red, respectively, in the color version) in Fig. 1.37a and comparing them to the curves for the longer annealed samples, one can identify tiny endothermic peaks at approximately 220 K. The crystalline objects formed at 202 K seem to cause even after melting a remaining not fully relaxed melt which begins immediately thereafter with cold crystallization, although the exotherm of the curve representing 0.01 s annealing time (red in the color version) barely exceeds the error limit. The situation after annealing at higher temperatures is different. After annealing for 1 s or longer at 252 K a significant amount of crystals is formed. These crystals are not molten before passing through the cold-crystallization range and do not cause further crystal growth. In the amorphous part therefore no measurable amounts of poorly ordered, low-melting structures are present and the high level of the heat capacity shows that practically all amorphous material is mobile and able to crystallize. Only annealing below the cold-crystallization temperature range leads to an acceleration of cold crystallization and indicates enhanced nucleation. Annealing at higher temperatures yields more perfect ordered structures which do not act as nuclei on heating at $1.8 \cdot 10^4$ Ks^{-1}. The needed molecular nucleation seems to be supported only by mobile melts, as are generated after the low-temperature melting endotherm, seen in Fig. 1.37a. Comparing the heat capacities in the glass transition regions in Fig. 1.37a–c, indicates that, both, the growth of the poorly ordered, low melting structures in Fig. 1.37a as well as the better ordered, higher melting structures in Fig. 1.37b and c, shift and broaden the glass transitions up to the beginning of melting, i.e., generate a rigid amorphous fraction, RAF, that keeps the melt rigid [38, 167]. Note the terms mobile and rigid melts must be considered in reference to the chosen cooling and heating rates in the experiments in addition to the annealing temperatures and times. With the disordering of the small ordered structures, the RAF disappears and does not retard cold crystallization, as it was observed in other cases [142].

The annealing experiments and subsequently collected DFSC heating curves lead to the unexpected though clearly evidenced conclusion that annealing of fully amorphous polymer samples below the glass transition temperature allows formation of homogeneous nuclei and ordered structures with latent heat, often mesophases in polyolefines [127, 143, 168–170] but also in polyamides [42, 43, 171] and other polymers [172]. Disordering of the mesophase formed below the glass transition temperature occurs on devitrification of the glass as has been detected by an endothermic peak, denoted as disordering. Subsequent to the disordering of the ordered domains, cold crystallization has been observed which, in amount, is proportional to the fraction of previously disordered structures. This leads to the conclusion that self-nuclei after disordering were left, permitting/accelerating cold ordering on continued heating.

The nucleation and growth of small mesophase particles of perhaps sub-nanometer size in the glass of most polymers suggests that cooperative, large-amplitude mobility of unrestraint molecule segments is not a requirement for partial ordering of macromolecules, as is commonly assumed; non-cooperative local mobility of chain segments is sufficient to form ordered structures.

Modeling of the rate of nucleation and crystallization has in the past been performed using the well-known Turnbull-Fisher equation. It includes an estimation of the free enthalpy of activation of transport processes/diffusion of molecular segments at the boundary between ordered and non-ordered phases by the WLF equation [139, 173]. Since the WLF equation typically is applied for quantification of the viscosity of liquids only, obviously, prediction of ordering processes below the glass transition temperature is impossible. For this reason, the data of the present work strongly suggest that the classical nucleation and crystallization theories for polymers need modification from point-of-view of correction the restriction of nucleation and growth of ordered phases to temperatures between the equilibrium melting temperature and the glass transition temperature. This important conclusion is in accord with a number of recently published studies of homogeneous nucleation and crystallization of polymers in the glassy state, pointing to universality/generality of this observation.

Quantitative analysis allowed us to judge the nucleation efficiency of carbon nanotubes in PCL in the whole range of temperatures, where PCL crystallizes. Though highly nucleated samples show two orders of magnitude faster crystallization in the region of heterogeneous nucleation (small under-cooling, high temperatures), the homogeneous nucleation dominates at low temperatures, making crystallization rate at this region independent of the nucleating agent content. Even for the highest MWCNT loadings (up to 5 wt%) homogeneous nucleation is significantly more effective than the heterogeneities as seen from the shift in the cold-crystallization peak towards lower temperatures. This information is not available from non-isothermal experiments but requires isothermal treatments after very fast cooling where nucleation on cooling was prevented.

Differential fast scanning calorimetry (DFSC), a new powerful tool for the investigation of nucleation and crystallization in polymers enables us to study homogeneous nucleation in materials crystallizing relatively fast. Beside the observed nucleation and growth below the glass transition in polymers information about the nucleation activity of small crystals grown at different temperatures relative to the cold-crystallization temperature range were obtained. As a next step stability of the nuclei at further heating will be studied (Tammann's development method [45]). Fast scanning calorimetry allows us to vary the heating rate going from the nucleation temperature to the development temperature in a wide range of heating rates covering at least 7 orders of magnitude. The advantage of the new technique is the wide temperature-time range covering the kinetics of industrially important semi-crystalline polymers.

The ultra-fast cooling ability of the DFSC allows one an investigation of crystal nucleation and growth on time scales starting from 1 millisecond. Thus, previously unavailable temperatures of homogeneous and heterogeneously enhanced crystallization of fast crystallizing polymers became accessible and more interesting results are expected in near future.

Acknowledgement: A. W. and R. A. acknowledge support from the German Science Foundation (DFG SCHI331/21-1 and DFG AN212-8, -9, -12, respectively) and E. Z. from a European Union funded Marie Curie EST fellowship (ADVATEC).

Bibliography

[1]   I. Gutzow and J.W.P. Schmelzer, *The Vitreous State: Thermodynamics, Structure, Rheology, and Crystallization* (First Edition, Springer, Berlin, 1995; Second enlarged edition, Springer, Heidelberg, 2013).

[2]   V.M. Fokin, E.D. Zanotto, and J.W.P. Schmelzer, *Homogeneous nucleation versus glass transition temperature of silicate glasses*, Journal Non-Crystalline Solids **321**, 52 (2003).

[3]   H.N. Ritland, *Density Phenomena in the Transformation Range of a Borosilicate Crown Glass*, J. American Ceramic Society **37**, 370 (1954).

[4]   G.M. Bartenev, *On the Relation between the Glass Transition Temperature of Silicate Glass and Rate of Cooling or Heating*, Doklady Akademii Nauk SSSR **76**, 227 (1951).

[5]   J.W.P. Schmelzer and I.S. Gutzow, *Glasses and the Glass Transition* (WILEY, Berlin-Weinheim, 2011).

[6]   J. Schmelzer, W.P., and C. Schick, *Dependence of crystallization processes of glass-forming melts on melt history: a theoretical approach to a quantitative treatment*, Physics and Chemistry of Glasses – European Journal of Glass Science and Technology Part B **53**, 99 (2012).

[7]   I. Gutzow, J.W.P. Schmelzer, and S. Todorova, *Frozen-in fluctuations, immiscibility and crystallisation in oxidee melts and the structural and thermodynamic nature of glasses*, Physics and Chemistry of Glasses – European Journal of Glass Science and Technology Part B **49**, 136 (2008).

[8]   J. Deubener, *Structural aspects of volume nucleation in silicate glasses*, J. Non-Crystalline Solids **351**, 1500 (2005).

[9]   E. Zhuravlev, and C. Schick, *Fast scanning power compensated differential scanning nanocalorimeter: 1. The device*, Thermochim. Acta **505**, 1 (2010).

[10] A.A. Minakov, and C. Schick, *Ultrafast thermal processing and nanocalorimetry at heating and cooling rates up to 1 MKs^{-1}*, Rev. Sci. Instr. **78**, 073902 (2007).

[11] D.A. Ivanov, in *Polymer Science: A Comprehensive Reference*, Editors-in-Chief: M. Krzysztof and M. Martin (Elsevier, Amsterdam, 2012, pp. 227).

[12] C.K. Ober, S.Z.D. Cheng, P.T. Hammond, M. Muthukumar, E. Reichmanis, K.L. Wooley, and T.P. Lodge, *Research in Macromolecular Science: Challenges and Opportunities for the Next Decade*, Macromolecules **42**, 465 (2009).

[13] M. Muthukumar, in *Progress in Understanding of Polymer Crystallization*, edited by G. Reiter, and G. Strobl (Springer Berlin, Heidelberg, 2007, pp. 1).

[14] H. Janeschitz-Kriegl, *Some remarks on flow induced crystallization in polymer melts*, Journal of Rheology **57**, 1057 (2013).

[15]   C. Luo and J.-U. Sommer, *Disentanglement of Linear Polymer Chains Toward Unentangled Crystals*, ACS Macro Letters **2**, 31 (2012).

[16]   E. Piorkowska and G.C.R. (Eds.), in *Handbook of Polymer Crystallization* (John Wiley & Sons, Inc., 2013, pp. 1).

[17]   G. Strobl, *Colloquium: Laws controlling crystallization and melting in bulk polymers*, Rev. Mod. Phys. **81**, 1287 (2009).

[18]   G. Strobl, *Crystallization and melting of bulk polymers: New observations, conclusions and a thermodynamic scheme*, Progress in Polymer Science **31**, 398 (2006).

[19]   M. Muthukumar, *Modeling Polymer Crystallization*, Advances in Polymer Science **191**, 241 (2005).

[20]   P.D. Olmsted, W.C.K. Poon, T.C.B. McLeish, N.J. Terrill, and A.J. Ryan, *Spinodal-Assisted Crystallization in Polymer Melts*, Phys. Rev. Lett. **81**, 373 (1998).

[21]   A. Mehta and B. Wunderlich, *A study of molecular fractionation during the crystallization polymers*, Colloid & Polymer Science **253**, 193 (1975).

[22]   B. Wunderlich and A. Mehta, *Macromolecular nucleation*, J. Polymer Science. B: Polymer Physics Edition **12**, 255 (1974).

[23]   B. Wunderlich, *Macromolecular Physics, Vols. 1–3* (Academic Press, New York, 1973, 1976, and 1980, resp.), Vol. 1, Crystal structure, Morphology, Defects.

[24]   H.J.D. Lauritzen Jr., *Theory of formation of polymer crystals with folded chains in dilute solution*, Journal of Research of the National Bureau of Standards **64a**, 73 (1960).

[25]   T. Yamamoto, *Computer Modeling of Polymer Crystallization-Toward Computer-Assisted Materials Design*, Polymer **50**, 1975 (2009).

[26]   W. Hu and D. Frenkel, *Polymer Crystallization Driven by Anisotropic Interactions*, Advances in Polymer Science **191**, 1 (2005).

[27]   C. Luo and J.-U. Sommer, *Coexistence of Melting and Growth during Heating of a Semicrystalline Polymer*, Phys. Rev. Lett. **102**, 147801 (2009).

[28]   J.U. Sommer and C. Luo, *Molecular dynamics simulations of semicrystalline polymers: Crystallization, melting, and reorganization*, Journal of Polymer Science Part B: Polymer Physics **48**, 2222 (2010).

[29]   C. Luo and J.-U. Sommer, *Growth Pathway and Precursor States in Single Lamellar Crystallization: MD Simulations*, Macromolecules **44**, 1523 (2011).

[30]   L.H. Allen, G. Ramanath, S.L. Lai, Z. Ma, S. Lee, D.D.J. Allman, and K.P. Fuchs, *1000 000 C/S thin film electrical heater: In situ resistivity measurements of Al and Ti/Si thin films during ultra rapid thermal annealing*, Applied Physics Letters **64**, 417 (1994).

[31]   R.T. Tol, A.A. Minakov, S.A. Adamovsky, V.B.F. Mathot, and C. Schick, *Metastability of polymer crystallites formed at low temperature studied by Ultra fast calorimetry: Polyamide 6 confined in sub-micrometer droplets vs bulk PA 6*, Polymer **47**, 2172 (2006).

[32]   J. Ibarretxe, G. Groeninckx, L. Bremer, and V.B.F. Mathot, *Quantitative evaluation of fractionated and homogeneous nucleation of polydisperse distributions of water-dispersed maleic anhydride-grafted-polypropylene micro- and nano-sized droplets*, Polymer **50**, 4584 (2009).

[33]   D. Turnbull, *Correlation of Liquid-Solid Interfacial Energies Calculated from Supercooling of Small Droplets*, J. Chemical Physics **18**, 769 (1950).

[34]   L. Kailas, C. Vasilev, J.N. Audinot, H.N. Migeon, and J.K. Hobbs, *A Real-Time Study of Homogeneous Nucleation, Growth, and Phase Transformations in Nanodroplets of Low Molecular Weight Isotactic Polypropylene Using AFM*, Macromolecules **40**, 7223 (2007).

[35]   J. Carvalho and K. Dalnoki-Veress, *Surface nucleation in the crystallisation of polyethylene droplets*, The European Physical Journal E: Soft Matter and Biological Physics **34**, 1 (2011).

[36]   E. Zhuravlev and C. Schick, *Fast scanning power compensated differential scanning nanocalorimeter: 2. Heat capacity analysis*, Thermochim. Acta **505**, 14 (2010).

[37]   B. Wunderlich, *The ATHAS database on heat capacities of polymers see on WWW URL: http://www.springermaterials.com/docs/athas.html*, Pure and Appl. Chem. **67**, 1019 (1995).

[38]   E. Zhuravlev, J.W.P. Schmelzer, B. Wunderlich, and C. Schick, *Kinetics of nucleation and crystallization in poly(ε-caprolactone) (PCL)* Polymer **52**, 1983 (2011).

[39]   A. Wurm, R. Soliman, and C. Schick, *Early Stages of Polymer Crystallization – A Dielectric Stud*, Polymer **44**, 7467 (2003).

[40]   S. Adamovsky and C. Schick, *Ultra-fast isothermal calorimetry using thin film sensors*, Thermochim. Acta **415**, 1 (2004).

[41]   Q. Zia, H.-J. Radusch, and R. Androsch, *Deformation behavior of isotactic polypropylene crystallized via a mesophase*, Polym. Bull. **63**, 755 (2009).

[42]   D. Mileva, R. Androsch, E. Zhuravlev, and C. Schick, *Morphology of mesophase and crystals of polyamide 6 prepared in a fast scanning chip calorimeter*, Polymer **53**, 3994 (2012).

[43]   A. Mollova, R. Androsch, D. Mileva, C. Schick, and A. Benhamida, *Effect of Supercooling on Crystallization of Polyamide 11*, Macromolecules **46**, 828 (2013).

[44]   D. Mileva, R. Androsch, and H.J. Radusch, *Effect of structure on light transmission in isotactic polypropylene and random propylene-1-butene copolymers*, Polymer Bulletin **62**, 561 (2009).

[45]   G. Tammann, *Number of nuclei in supercooled liquids*, Zeitschrift für Physikalische Chemie **25**, 441 (1898).

[46]   B. Wunderlich and A. Mehta, *The Nucleation of Polyethylene Dendrites from Solution*, Journal of Materials Science **5**, 248 (1970).

[47]   M. SalmeronSanchez, V.B.F. Mathot, G. VandenPoel, and J.L. GomezRibelles, *Effect of the Cooling Rate on the Nucleation Kinetics of Poly(L-Lactic Acid) and Its Influence on Morphology*, Macromolecules **40**, 7989 (2007).

[48]   T. Hikima, Y. Adachi, M. Hanaya, and M. Oguni, *Determination of potentially homogeneous nucleation based crystallization in o-terphenyl and an interpretation of the nucleation enhancement mechanism*, Phys. Rev. B **52**, 3900 (1995).

[49]   S. Vyazovkin and I. Dranca, *Effect of physical aging on nucleation of amorphous indomethacin*, Journal of Physical Chemistry B **111**, 7283 (2007).

[50]   C.A. Angell, *Structural instability and relaxation in liquid and glassy phases near the fragile liquid limit*, J. Non-Crystalline Solids **102**, 205 (1988).

[51]   R.K. Kadiyala and C.A. Angell, *Separation of nucleation from crystallization kinetics by two step calorimetry experiments*, Colloids and Surfaces **11**, 341 (1984).

[52]   V.M. Fokin, A.A. Cabral, R.M.C.V. Reis, M.L.F. Nascimento, and E.D. Zanotto, *Critical assessment of DTA-DSC methods for the study of nucleation kinetics in glasses*, J. Non-Crystalline Solids **356**, 358 (2010).

[53]   A. Marotta, S. Saiello, F. Branda, and A. Buri, *Nucleation and Crystal Growth in $NA_2O \cdot 2CAO \cdot 3SIO_2$ Glass – A DTA Study*, Thermochim. Acta **46**, 123 (1981).

[54]   S.A. Adamovsky, A.A. Minakov, and C. Schick, *Scanning microcalorimetry at high cooling rate*, Thermochim. Acta **403**, 55 (2003).

[55]   M.F.J. Pijpers, V.B.F. Mathot, B. Goderis, R. Scherrenberg, and E. van der Vegte, *High-speed calorimetry for the analysis of kinetics of vitrification, crystallization and melting of macromolecule*, Macromolecules, **35**, 3601 (2002).

[56]   PerkinElmer_Inc., *Technical Specifications Thermal Analysis for the DSC 8000/8500 Differential Scanning Calorimeters* http://las.perkinelmer.com/content/RelatedMaterials/Specification Sheets/SPC_8000and8500.pdf, (2009).

[57]   N.E. Hager, *Thin heater calorimeter*, Rev. Sci. Instrum. **35**, 618 (1964).

[58]   M.Y. Efremov, E.A. Olson, M. Zhang, F. Schiettekatte, Z. Zhang, and L.H. Allen, *Ultrasensitive, fast, thin-film differential scanning calorimeter*, Rev. Sci. Instrum. **75**, 179 (2004).

[59]   M.Y. Efremov, E.A. Olson, M. Zhang, Z. Zhang, and L.H. Allen, *Probing Glass Transition of Ultrathin Polymer Films at a Time Scale of Seconds Using Fast Differential Scanning Calorimetry*, Macromolecules **37**, 4607 (2004).

[60]   M. Chonde, M. Brindza, and V. Sadtchenkoa, *Glass transition in pure and doped amorphous solid water: An ultrafast microcalorimetry study*, Journal of Chemical Physics **125**, 09501 (2006).

[61]   E. León-Gutierrez, G. Garcia, M.T. Clavaguera-Mora, and J. Rodríguez-Viejo, *Glass transition in vapor deposited thin films of toluene*, Thermochim. Acta **492**, 51 (2009).

[62]   S.A. McCartney and V. Sadtchenko, *Fast scanning calorimetry studies of the glass transition in doped amorphous solid water: Evidence for the existence of a unique vicinal phase*, J. Chemical Physics **138**, 084501 (2013).

[63]   A. Sepulveda, E. Leon-Gutierrez, M. Gonzalez-Silveira, C. Rodriguez-Tinoco, M.T. Clavaguera-Mora, and J. Rodriguez-Viejo, *Glass transition in ultrathin films of amorphous solid water*, The Journal of Chemical Physics **137**, 244506 (2012).

[64]   D. Bhattacharya, C.N. Payne, and V. Sadtchenko, *Bulk and Interfacial Glass Transitions of Water*, J. Physical Chemistry A **115**, 5965 (2011).

[65]   A. Sepúlveda, E. Leon-Gutierrez, M. Gonzalez-Silveira, M.T. Clavaguera-Mora, and J. Rodríguez-Viejo, *Anomalous Transformation of Vapor-Deposited Highly Stable Glasses of Toluene into Mixed Glassy States by Annealing Above Tg*, J. Physical Chemistry Letters **3**, 919 (2012).

[66]   A.A. Minakov, S.A. Adamovsky, and C. Schick, *Non adiabatic thin-film (chip) nanocalorimetry*, Thermochim. Acta **432**, 177 (2005).

[67]   A.A. Minakov, A.W. van Herwaarden, W. Wien, A. Wurm, and C. Schick, *Advanced non-adiabatic ultrafast nanocalorimetry and superheating phenomenon in linear polymers*, Thermochim. Acta **461**, 96 (2007).

[68]   A.W. van Herwaarden, *Overview of calorimeter chips for various applications*, Thermochim. Acta **432**, 192 (2005).

[69]   A.A. Minakov, D.A. Mordvintsev, and C. Schick, *Melting and Reorganization of Poly(ethylene Terephthalate) on Fast Heating (1000 Ks^{-1})*, Polymer **45**, 3755 (2004).

[70]   A.A. Minakov, D.A. Mordvintsev, and C. Schick, *Isothermal reorganization of poly(ethylene terephthalate) revealed by fast calorimetry (1000 Ks^{-1}; 5 ms)*, Faraday Discuss. **128**, 261 (2005).

[71]   A.A. Minakov, D.A. Mordvintsev, R. Tol, and C. Schick, *Melting and reorganization of the crystalline fraction and relaxation of the rigid amorphous fraction of isotactic polystyrene on fast heating (3 · 10^4 K/min)*, Thermochim. Acta **442**, 25 (2006).

[72]   A. Gradys, P. Sajkiewicz, A.A. Minakov, S. Adamovsky, C. Schick, T. Hashimoto, and K. Saijo, *Crystallization of polypropylene at various cooling rates*, Mater. Sci. Eng. A **413–414**, 442 (2005).

[73]   A. Gradys, P. Sajkiewicz, S. Adamovsky, A. Minakov, and C. Schick, *Crystallization of poly(vinylidene fluoride) during ultra-fast cooling*, Thermochim. Acta **461**, 153 (2007).

[74]   F. De Santis, S. Adamovsky, G. Titomanlio, and C. Schick, *Scanning nanocalorimetry at high cooling rate of isotactic polypropylene*, Macromolecules **39**, 2562 (2006).

[75]   F. De Santis, S. Adamovsky, G. Titomanlio, and C. Schick, *Isothermal nanocalorimetry of isotactic polypropylene*, Macromolecules **40**, 9026 (2007).

[76]   V.V. Ray, A.K. Banthia, and C. Schick, *Fast isothermal calorimetry of modified polypropylene clay nanocomposites*, Polymer **48**, 2404 (2007).

[77]   A. Krumme, A. Lehtinen, S. Adamovsky, C. Schick, J. Roots, and A. Viikna, *Crystallization behavior of some unimodal and bimodal linear low-density polyethylenes at moderate and high supercooling*, Journal of Polymer Science Part B: Polymer Physics **46**, 1577 (2008).

[78] C. Silvestre, S. Cimmino, D. Duraccio, and C. Schick, *Isothermal Crystallization of Isotactic Poly(propylene) Studied by Superfast Calorimetry*, Macromolecular Rapid Communications **28**, 875 (2007).

[79] C. Schick, *Differential scanning calorimetry (DSC) of semicrystalline polymers*, Analytical and Bioanalytical Chemistry, 1 (2009).

[80] M. Pyda, E. Nowak-Pyda, J. Heeg, H. Huth, A.A. Minakov, M.L.Di Lorenzo, C. Schick, and B. Wunderlich, *Melting and Crystallization of Poly(butylene Terephthalate) by Temperature-modulated and Superfast Calorimetry*, Journal of Polymer Science. B: Polymer Physics **44**, 1364 (2006).

[81] A. Minakov, A. Wurm, and C. Schick, *Superheating in linear polymers studied by ultrafast nanocalorimetry*, Eur. Phys. J. E Soft Matter **23**, 43 (2007).

[82] V. Brucato, S. Piccarolo, and V. La Carrubba, *An experimental methodology to study polymer crystallization under processing conditions. The influence of high cooling rates*, Chem. Eng. Sci. **57**, 4129 (2002).

[83] H. Janeschitz-Kriegl, *Crystallization Modalities in Polymer Melt Processing, Fundamental Aspects of Structure Formation* (Springer, Wien, New York, 2010).

[84] W. Chen, D. Zhou, G. Xue, and C. Schick, *Chip calorimetry for fast cooling and thin films: a review*, Front. Chem. China **4**, 229 (2009).

[85] A.F. Lopeandia, J. Valenzuela, and J. Rodríguez-Viejo, *Power compensated thin film calorimetry at fast heating rates*, Sensors and Actuators A: Physical **143**, 256 (2008).

[86] T. Naiki, H. Sugiyama, R. Tashiro, and T. Karino, *Flow-dependent concentration polarization of plasma proteins at the luminal surface of a cultured endothelial cell monolayer*, Biorheology **36**, 225 (1999).

[87] M. Merzlyakov, *Method of Rapid (10^5 Ks^{-1}) Controlled Cooling and Heating of Thin Samples*, Thermochim. Acta **442**, 52 (2006).

[88] M.J. O'Neill, *The analysis of a temperature-controlled scanning calorimeter*, Anal. Chem. **36**, 1238 (1964).

[89] M.J. O'Neill, *Measurement of Specific Heat Functions by Differential Scanning Calorimetry*, Anal. Chem. **38**, 1331 (1966).

[90] E.S. Watson, M.O. O'Neill, J. Justin, and N. Brenner, *A differential scanning calorimeter for quantitative differential thermal analysis*, Anal. Chem. **36**, 1233 (1964).

[91] E.S. Watson, and M.J. O'Neill, (The Perkin-Elmer Corporation, Norwalk, Conn., 1966).

[92] S. van Herwaardena, *Micro-sensors for Analysis Equipment: Research and Innovation*, Procedia Engineering **5**, 464 (2010).

[93] G. Poel, D. Istrate, A. Magon, and V. Mathot, *Performance and calibration of the Flash DSC 1, a new, MEMS-based fast scanning calorimeter*, Journal of Thermal Analysis and Calorimetry **110**, 1533 (2012).

[94] S. van Herwaarden, E. Iervolino, F. van Herwaarden, T. Wijffels, A. Leenaers, and V. Mathot, *Design, performance and analysis of thermal lag of the UFS1 twin-calorimeter chip for fast scanning calorimetry using the Mettler-Toledo Flash DSC 1*, Thermochim. Acta **522**, 46 (2011).

[95] V. Mathot, M. Pyda, T. Pijpers, G. Vanden Poel, E. van de Kerkhof, S. van Herwaarden, F. van Herwaarden, and A. Leenaers, *The Flash DSC 1, a power compensation twin-type, chip-based fast scanning calorimeter (FSC): First findings on polymers*, Thermochim. Acta **522**, 36 (2011).

[96] E. Iervolino, A.W. van Herwaarden, F.G. van Herwaarden, E. van de Kerkhof, P.P.W. van Grinsven, A.C.H.I. Leenaers, V.B.F. Mathot, and P.M. Sarro, *Temperature calibration and electrical characterization of the differential scanning calorimeter chip UFS1 for the Mettler-Toledo Flash DSC 1*, Thermochim. Acta **522**, 53 (2011).

[97]   G.V. Poel, A. Sargsyan, V. Mathot, G.V. Assche, A. Wurm, C. Schick, A. Krumme, and D. Zhou, *Recommendation for Temperature Calibration of Fast Scanning Calorimeters (FsCs) for Sample Mass and Scan Rate* (Beuth Verlag GmbH, Berlin, 2011).

[98]   A. Wurm, D. Lellinger, A.A. Minakov, T. Skipa, P. Pötschke, I. Alig, and C. Schick, *Crystallization of poly(ε-caprolactone)/MWCNT composites: A combined SAXS/WAXS, electrical and thermal conductivity study*, Polymer, submitted for publication.

[99]   E. Zhuravlev, P. Pötschke, R. Androsch, J.W.P. Schmelzer, and C. Schick, *Kinetics of nucleation and crystallization of poly(ε-caprolactone) – multiwalled carbon nanotube composites*, European Polymer Journal, submitted for publication.

[100]   I. Stolte, R. Androsch, M.L. Di Lorenzo, and C. Schick, *Effect of Aging the Glass of Isotactic Polybutene-1 on Form II Nucleation and Cold Crystallization*, The Journal of Physical Chemistry B **117**, 15196 (2013).

[101]   Product information available at https://polymers.lyondellbasell.com.

[102]   F. Paladi, and M. Oguni, *Generation and extinction of crystal nuclei in extremely non-equilibrium glassy state of salol*, Journal of Physics: Condensed Matter **15**, 3909 (2003).

[103]   F. Paladi, and M. Oguni, *Anomalous generation and extinction of crystal nuclei in nonequilibrium supercooled liquid o-benzylphenol*, Phys. Rev. B **65**, 144202 (2002).

[104]   T. Hikima, M. Hanaya, and M. Oguni, *Microscopic observation of a peculiar crystallization in the glass transition region and beta-process as potentially controlling the growth rate in triphenylethylene*, Journal of Molecular Structure **479**, 245 (1999).

[105]   T. Hikima, N. Okamoto, M. Hanaya, and M. Oguni, *Calorimetric study of triphenylethene: observation of homogeneous-nucleation-based crystallization*, Journal of Chemical Thermodynamics **30**, 509 (1998).

[106]   N. Okamoto, M. Oguni, and Y. Sagawa, *Generation and extinction of a crystal nucleus below the glass transition temperature*, Journal of Physics: Condensed Matter **9**, 9187 (1997).

[107]   N. Okamoto and M. Oguni, *Discovery of crystal nucleation proceeding much below the glass transition temperature in a supercooled liquid*, Solid State Communications **99**, 53 (1996).

[108]   T. Hikima, M. Hanaya, and M. Oguni, *Discovery of a potentially homogeneous-nucleation-based crystallization around the glass transition temperature in salol*, Solid State Communications **93**, 713 (1995).

[109]   G.S.Y. Yeh and P.H. Geil, *Crystallization of polyethylene terephthalate from the glassy amorphous state*, J. Macromol. Sci. (Phys.) **B1**, 235 (1967).

[110]   W. Frank, H. Goddar, and H.A. Stuart, *Electron microscopic investigations on amorphous polycarbonate*, Journal of Polymer Science Part B: Polymer Letters **5**, 711 (1967).

[111]   D.M. Gezovich, and P.H. Geil, *Morphology of quenched polypropylene*, Polymer Engineering & Science **8**, 202 (1968).

[112]   Q. Zia, E. Ingolič, and R. Androsch, *Surface and bulk morphology of cold-crystallized poly(ethylene terephthalate)*, Colloid & Polymer Science **288**, 819 (2010).

[113]   V.B.F. Mathot, *Calorimetry and Thermal Analysis of Polymers* (Hanser Publishers, München, 1994).

[114]   J. Jiang, E. Zhuravlev, Z. Huang, L. Wei, Q. Xu, M. Shan, G. Xue, D. Zhou, C. Schick, and W. Jiang, *A transient polymorph transition of 4-cyano-4′-octyloxybiphenyl (8OCB) revealed by ultrafast differential scanning calorimetry (UFDSC)*, Soft Matter **9**, 1488 (2013).

[115]   M. Chen, M. Du, J. Jiang, D. Li, W. Jiang, E. Zhuravlev, D. Zhou, C. Schick, and G. Xue, *Verifying the symmetry of ultra-fast scanning calorimeters using liquid crystal secondary temperature standards*, Thermochim. Acta **526**, 58 (2011).

[116]   S.M. Sarge, G.W.H. Hohne, H.K. Cammenga, W. Eysel, and E. Gmelin, *Temperature, heat and heat-flow rate calibration of scanning calorimeters in the cooling mode*, Thermochim. Acta **361**, 1 (2000).

[117] S. Neuenfeld, and C. Schick, *Verifying the symmetry of differential scanning calorimeters concerning heating and cooling using liquid crystal secondary temperature standards*, Thermochim. Acta **446**, 55 (2006).

[118] C. Schick, U. Jonsson, T. Vassilev, A. Minakov, J. Schawe, R. Scherrenberg, and D. Lörinczy, *Applicability of 8OCB for temperature calibration of temperature modulated calorimeters*, Thermochim. Acta **347**, 53 (2000).

[119] G.W.H. Hohne, J. Schawe, and C. Schick, *Temperature calibration on cooling using liquid crystal phase transitions*, Thermochim. Acta **221**, 129 (1993).

[120] O.D. Lavrentovich, M. Kleman, and V.M. Pergamenshchik, *Nucleation of Focal Conic Domains in Smectic-A Liquid Crystals*, J. Phys. II **4**, 377 (1994).

[121] S.H. Lee, Y.W. Jung, and R. Agarwal, *Size-Dependent Surface-Induced Heterogeneous Nucleation Driven Phase-Change in Ge2Sb2Te5 Nanowires*, Nano Lett. **8**, 3303 (2008).

[122] Y. Diao, A.S. Myerson, T.A. Hatton, and B.L. Trout, *Surface Design for Controlled Crystallization: The Role of Surface Chemistry and Nanoscale Pores in Heterogeneous Nucleation*, Langmuir **27**, 5324 (2011).

[123] J. Russo, and H. Tanaka, *Selection mechanism of polymorphs in the crystal nucleation of the Gaussian core model*, Soft Matter **8**, 4206 (2012).

[124] K. Konnecke, *Crystallization of Poly(Aryl Ether Ketones), 1. Crystallization Kinetics*, J. Macromol. Sci.-Phys. **B33**, 37 (1994).

[125] A. Wurm, E. Zhuravlev, K. Eckstein, D. Jehnichen, D. Pospiech, R. Androsch, B. Wunderlich, and C. Schick, *Crystallization and Homogeneous Nucleation Kinetics of Poly(ε-caprolactone) (PCL) with Different Molar Masses*, Macromolecules **45**, 3816 (2012).

[126] B. Wunderlich, *Reversible crystallization and the rigid-amorphous phase in semicrystalline macromolecules*, Progress in Polymer Science **28**, 383 (2003).

[127] M. Mileva, R. Androsch, E. Zhuravlev, C. Schick, and B. Wunderlich, *Homogeneous nucleation and mesophase formation in glassy isotactic polypropylene*, Polymer **53**, 277 (2012).

[128] J.W.P. Schmelzer and O. Hellmuth, *Nucleation Theory and Applications* (Joint Institute for Nuclear Research, Dubna, 2013).

[129] B. Fillon, A. Thierry, B. Lotz, and J.C. Wittmann, *Efficiency scale for polymer nucleating agents*, J. Therm. Anal. **42**, 721 (1994).

[130] M. Trujillo, M.L. Arnal, A.J. Müller, M.A. Mujica, C. Urbina de Navarro, B. Ruelle, and P. Dubois, *Supernucleation and crystallization regime change provoked by MWNT addition to poly(ε-caprolactone)*, Polymer **53**, 832 (2012).

[131] H. Suzuki, J. Grebowicz, and B. Wunderlich, *Glass transition of poly(oxymethylene)*, British Polymer Journal **17**, 1 (1985).

[132] M. Avrami, *Transformation-Time Relaxations for Random Distribution of Nuclei*, Journal of Chemical Physics **8**, 212 (1940).

[133] F. Rybnikar, *Mechanism of secondary crystallization in polymers*, Journal of Polymer Science Part A: General Papers **1**, 2031 (1963).

[134] P. Supaphol, and J.E. Spruiell, *Isothermal melt- and cold-crystallization kinetics and subsequent melting behavior in syndiotactic polypropylene: a differential scanning calorimetry study*, Polymer **42**, 699 (2001).

[135] I. Kolesov, R. Androsch, D. Mileva, W. Lebek, A. Benhamida, M. Kaci, and W. Focke, *Crystallization of a polyamide 11/organo-modified montmorillonite nanocomposite at rapid cooling*, Colloid Polym Sci **291**, 2541 (2013).

[136] J. Boor, and J.C. Mitchell, *Kinetics of crystallization and a crystal-crystal transition in poly-1-butene*, Journal of Polymer Science Part A: General Papers **1**, 59 (1963).

[137] F. Danusso, and G. Gianotti, *Isotactic polybutene-1: Formation and transformation of modification 2*, Die Makromolekulare Chemie **88**, 149 (1965).

[138] B. Wunderlich, *Macromolecular Physics* (Academic Press, New York, 1976), Vol. 2, Crystal Nucleation, Growth. Annealing.

[139] J.D. Hoffman, G.T. Davis, and J.I. Lauritzen, *Crystalline and noncrystalline solids* (Plenum Press, 1976), Vol. 3, In Treatise on Solid State Chemistry.

[140] B. Fillon, J.C. Wittmann, B. Lotz, and A. Thierry, *Self-Nucleation and Recrystallization of Isotactic Polypropylene (a Phase) Investigated by Differential Scanning Calorimetry*, J. Polym. Sci. B: Polym. Phys. **31**, 1383 (1993).

[141] A. Müller, Z. Hernández, M. Arnal, and J. Sánchez, *Successive self-nucleation/annealing (SSA): A novel technique to study molecular segregation during crystallization*, Polymer Bulletin **39**, 465 (1997).

[142] A. Wurm, M. Ismail, B. Kretzschmar, D. Pospiech, and C. Schick, *Retarded Crystallization in Polyamide/Layered Silicates Nanocomposites caused by an Immobilized Interphase*, Macromolecules **43**, 1480 (2010).

[143] D. Mileva, R. Androsch, E. Zhuravlev, C. Schick, and B. Wunderlich, *Formation and reorganization of the mesophase of random copolymers of propylene and 1-butene*, Polymer **52**, 1107 (2011).

[144] B. Wunderlich, *Crystal melting* (Academic Press, New York, 1980), Vol. 3, Macromolecular Physics.

[145] G. Strobl, *From the melt via mesomorphic and granular crystalline layers to lamellar crystallites:A major route followed in polymer crystallization?* Eur. Phys. J. E **3**, 165 (2000).

[146] A. Wurm, and C. Schick, *Development of thermal stability of polymer crystals during isothermal crystallisation*, e-Polymers **24**, 1 (2002).

[147] P. Kohn, and G. Strobl, *Continuous Changes in the Inner Order of the Crystalline Lamellae during Isothermal Crystallization of Poly(ε-caprolactone)*, Macromolecules **37**, 6823 (2004).

[148] D. Mileva, R. Androsch, E. Zhuravlev, and C. Schick, *Temperature of Melting of the Mesophase of Isotactic Polypropylene*, Macromolecules **42**, 7275 (2009).

[149] R. Androsch, and M.L. Di Lorenzo, *Crystal Nucleation in Glassy Poly(l-lactic acid)*, Macromolecules **46**, 6048 (2013).

[150] I. Kolesov, D. Mileva, R. Androsch, and C. Schick, *Structure formation of polyamide 6 from the glassy state by fast scanning chip calorimetry*, Polymer **52**, 5156 (2011).

[151] J.M. Hutchinson, S. Smith, B. Horne, and G.M. Gourlay, *Physical aging of polycarbonate: Enthalpy relaxation, creep response, and yielding behavior*, Macromolecules **32**, 5046 (1999).

[152] I.M. Hodge, *Enthalpy relaxation and recovery in amorphous materials [Review]*, J. Non-Crystalline Solids **169**, 211 (1994).

[153] E. Donth, *Glass Transition: Thermal Glass Transition – Glass temperature – Partial Freezing* (Springer, Berlin, 2001).

[154] B. Wunderlich, *Thermal Analysis of Polymeric Materials* (Springer, Berlin, Heidelberg, New York, 2005).

[155] D. Mileva, R. Androsch, E. Zhuravlev, C. Schick, and B. Wunderlich, *Isotropization, perfection and reorganization of the mesophase of isotactic polypropylene*, Thermochim. Acta **522**, 100 (2011).

[156] Q. Zia, D. Mileva, and R. Androsch, *Rigid Amorphous Fraction in Isotactic Polypropylene*, Macromolecules **41**, 8095 (2008).

[157] M.L. Di Lorenzo, M.C. Righetti, and B. Wunderlich, *Influence of Crystal Polymorphism on the Three-Phase Structure and on the Thermal Properties of Isotactic Poly(1-butene)*, Macromolecules **42**, 9312 (2009).

[158] W. Kozlowski, *Kinetics of crystallization of polyamide 6 from the glassy state*, Journal of Polymer Science Part C: Polymer Symposia **38**, 47 (1972).

[159] B. Wunderlich, *Beweglichkeit in kristallinen und amorphen Hochpolymeren aus kalorischen Messungen (Discussion with G. Hentze, figure on page 618)*, Kolloid-Zeitschrift & Zeitschrift für Polymere **231**, 605 (1969).

[160] F. Kremer, and A. Schönhals, *Broadband Dielectric Spectroscopy* (Springer, Heidelberg, 2002).

[161] M. Oguni, *Intra-Cluster Rearrangement Model for the Alpha-Process in Supercooled Liquids, as Opposed to Cooperative Rearrangement of Whole Molecules Within a Cluster*, J. Non-Crystal. Solids **210**, 171 (1997).

[162] J.W.P. Schmelzer, R. Müller, J. Möller, and I.S. Gutzow, *Theory of Nucleation in Viscoelastic Media: Application to Phase Formation in Glassforming Melts*, J. Non-Crystalline Solids **315**, 144 (2003).

[163] J. Schmelzer, R. Pascova, and I. Gutzow, *Cluster Growth and Ostwald Ripening in Viscoelastic Media*, physica status solidi **a 117**, 363 (1990).

[164] J. Schmelzer, I. Gutzow, and R. Pascova, *Kinetics of Phase Segregation in Elastic and Viscoelastic Media*, J. Crystal Growth **104**, 505 (1990).

[165] J. Schmelzer (under participation of V.V. Slezov, I.S. Gutzow, S. Todorova, J. Schmelzer Jr.), *Bubble Formation and Growth in Viscous Liquids* (Project Report for BASF Ludwigshafen, Rostock – Ludwigshafen, 1998, 203 pgs. (unpublished)).

[166] B. Wunderlich, and C.M. Cormier, *Seeding of supercooled polyethylene with extended chain crystals*, J. Phys. Chem. **70**, 1844 (1966).

[167] B. Wunderlich, *The Thermal Properties of Complex, Nanophase-Separated Macromolecules as Revealed by Temperature-Modulated Calorimetry*, Thermochim. Acta **403**, 1 (2003).

[168] D. Mileva, R. Androsch, D. Cavallo, and G.C. Alfonso, *Structure formation of random isotactic copolymers of propylene and 1-hexene or 1-octene at rapid cooling*, European Polymer Journal **48**, 1082 (2012).

[169] D. Cavallo, L. Gardella, G.C. Alfonso, D. Mileva, and R. Androsch, *Effect of comonomer partitioning on the kinetics of mesophase formation in random copolymers of propene and higher α-olefins*, Polymer **53**, 4429 (2012).

[170] R. Androsch, M.L.D. Lorenzo, C. Schick, and B. Wunderlich, *Mesophases in polyethylene, polypropylene, and poly(1-butene)*, Polymer **51**, 4639 (2010).

[171] D. Cavallo, L. Gardella, G. Alfonso, G. Portale, L. Balzano, and R. Androsch, *Effect of cooling rate on the crystal/mesophase polymorphism of polyamide 6*, Colloid & Polymer Science **289**, 1073 (2011).

[172] D. Cavallo, D. Mileva, G. Portale, L. Zhang, L. Balzano, G.C. Alfonso, and R. Androsch, *Mesophase-Mediated Crystallization of Poly(butylene-2,6-naphthalate): An Example of Ostwald's Rule of Stages*, ACS Macro Letters **1**, 1051 (2012).

[173] M.L. Williams, R.F. Landel, and D.J. Ferry, *The temperature dependence of relaxation mechanisms in amorphous polymers and other glass-forming liquids*, J. Am. Ceramic Soc. **77**, 3701 (1955).

Gerhard Wilde

2 Early Stages of Crystal Formation in Glass-forming Metallic Alloys

With metallic systems, kinetic stability rather than fragility is the parameter that provides a characteristic distinction between different glass-forming alloys. This distinction is related to either nucleation or growth control that governs the process of vitrification, i.e. the avoidance of crystallization. At the same time, the respective kinetic stability gives access to different regimes and measurement techniques, which in turn allow one analyzing the kinetics of phase formation under different perspectives. With marginally glass-forming alloy systems such as Al-based alloys with rare earth and transition metal additions, the pronounced nanocrystallization that occurs at low temperatures during reheating allows one studying the early stages of nucleation and growth of crystalline phases in detail. For $Al_{92}Sm_8$ and $Al_{88}Y_7Fe_5$ as model representatives of this class of kinetic stability, the origin of nanocrystal development and the nucleation and growth kinetics of the early stages of the evolution of crystalline structures in noncrystalline metallic alloys have been investigated. Additional assessment of the thermodynamics of these systems indicate the presence of precursor structures in the undercooled melt that effectively modify the cluster distributions such that copious nucleation is likely to occur. A rather new approach concerning the analysis of nucleation reactions in (metallic) glasses is based on plastic deformation of glasses at low homologous temperatures that creates localization of deformation in narrow regions of the glass that are called "shear bands". Within these mesoscopic defects, crystallization transformations are observed at drastically reduced temperatures. The kinetics of this crystal formation process is unexpectedly fast at temperatures well below the glass transition of the initial glass, yet the kinetics is still significantly slower than at temperatures near or above the glass transition, where the transformation normally occurs in an undeformed material of the same composition. Thus, the crystal formation process that occurs in shear bands allows one an unprecedented view into the early stages of crystal structure formation from amorphous precursors.

2.1 Introduction

With the discovery of glass formation in metallic alloy systems by Duwez and coworkers in 1960 [1], research on phase formation kinetics in undercooled melts and on the nature of the liquid state including the glass transition experienced increased attention since these systems were thought of as easy describable molecular liquids. Yet, experimental experience has shown that metallic systems with sufficient stability against crystallization that allow for bulk-glass formation require rather complex stoichiometries and considerable covalent contributions to the binding energy of the constituents [2]. As a result, strong tendencies towards short-range ordering into complex associations develop that are discussed with respect to either quasi-crystalline structures or crystalline aggregates with low symmetry that evolve and dissolve in a temperature-dependent dynamic equilibrium [3–6]. More recent attempts for structural descriptions even postulate the existence of non-crystalline aggregates with fixed structures that are forming the basic structural units in glass-forming metallic melts [7]. Chemically simple systems, e.g. binary glass-forming alloys, on the other hand, are kinetically unstable and do not allow experimental access to the undercooled liquid state. Consequently, no glass-forming alloy exists (except in computer simulations [8, 9]) that provides the opportunity to study the entire spectrum of questions that are concerned with the glass transition, the structure of the liquid state and the early stages of crystalline phase formation. Either metallic systems are simple and show a marginal glass-formation tendency with high transformation rates towards the crystalline state or the systems offer a high stability of the undercooled liquid state against crystallization but are chemically complex. However, metallic systems can be utilized as model systems if the variations in kinetic stability are used as a tool to select appropriate alloy systems for a limited range of problems to be investigated.

According to the classification scheme by Angell [10] (which is based on the earlier work by Oldekop [11], Laughlin and Uhlmann [12]), glass-forming systems are categorized with respect to the slope of the equilibrium viscosity of the undercooled liquid, η, in the temperature range of the glass transition. The steepness parameter that is called fragility-parameter, m, is defined as

$$m = \left.\frac{d\log_{10}(\eta/\mathrm{Pa}\cdot\mathrm{s})}{d(T_g/T)}\right|_{T=T_g} \tag{2.1}$$

and spans the range from $m = 20$ for strong glasses such as SiO_2 to about $m = 200$ for fragile systems such as ortho-terphenyl. Yet, glass-forming metallic alloys occur only in a rather limited range of intermediate fragility values between about $m = 35$ to $m = 55$. Thus, the viscosity vs. temperature curves for the metallic systems seem to be rather identical in the Arrhenius representation, shown in Fig. 2.1, if the fragility spectrum includes non-metallic systems. Estimated fragility values for many metallic alloys that have not yet been studied with respect to η render similar results [13]. Thus, in order to distinguish different metallic alloys according to their transformation behavior and with respect to the properties that are accessible in equilibrium measurements, fragility does not provide a suitable criterion for metallic systems.

Despite the mentioned similarity in the temperature dependence of the viscosity of the metallic glass-forming melts (which are distinguished following the mentioned classification into "strong" and "fragile" glass forming-liquids), a remarkable diversity

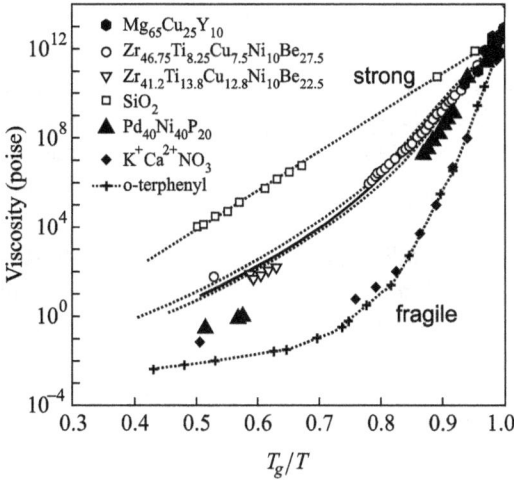

Fig. 2.1: "Angell"-plot of the viscosity of different glass-forming systems.

of kinetic stability and phase evolution trajectories have been observed for different metallic systems that range from marginally glass-forming alloys with a pronounced tendency for nanocrystallization to easy glass-forming bulk metallic glasses. The spectrum of kinetic stability that has been observed is displayed in Fig. 2.2 as a plot of the experimentally determined cooling rates sufficient for vitrification in dependence of the reduced glass transition temperature, $T_{rg} = (T_g/T_f)$, for glass-forming metallic systems. Here, T_f denotes the melting temperature, which, for glass-forming alloys, is often taken as the eutectic temperature of the system. It should be noted that the interval of sufficient cooling rates extents over almost eight orders of magnitude with only slight variations in fragility. According to Fig. 2.2, metallic glasses can roughly be

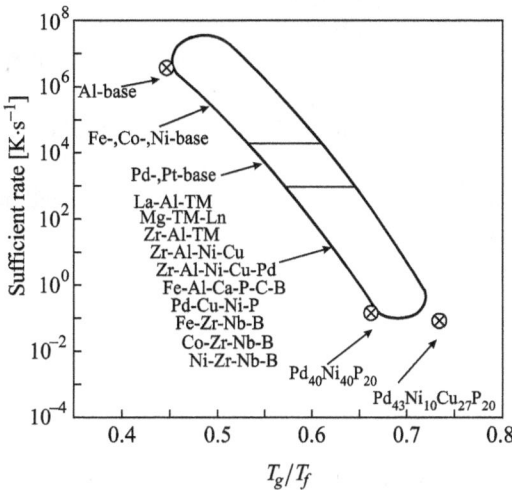

Fig. 2.2: Sufficient cooling rates for vitrification in dependence on the reduced glass temperature for metallic alloys. Note that the usually given "critical"-cooling rate applies rather to the respective quenching conditions. Thus, these rates are sufficient – the critical cooling rates might be lower.

subdivided into two classes according to the ease of glass-formation as indicated by the ability to form bulk amorphous samples during melt-quenching. Crossed circles in Fig. 2.2 mark two alloy classes that are representatives of the extreme cases of metallic glass formation. This distinction, although related to the kinetics of crystallization rather than to the properties of the undercooled melt, has an important impact on the properties that are measurable and consequently, on the aspects of structure development that can be assessed.

Structure formation in glass-forming metallic alloys has often been investigated with the viewpoint on crystallization of stable and metastable crystalline phases from the undercooled melt. With exception of the slower kinetics of nucleation and growth of the crystalline phases, which present the possibility to deeply undercool these materials, rather similar results for the basic mechanisms of crystalline phase formation are observed as in conventional solidification studies. Moreover, detailed investigations of nucleation and growth processes on bulk-glass forming alloys are additionally convoluted by the complex chemistry of these alloys.

Rather recently, a new type of structure development in metallic systems has attracted increased attention that is related to the occurrence of nano-scaled crystalline phases at low temperatures in several so-called marginal metallic glasses [14–18]. According to the resulting microstructure, this transformation is called nanocrystallization. From an application related perspective, these materials offer new and interesting properties: nanocrystallized Fe-based alloys, for example, show a high saturation magnetization and extremely low coercivity values since the pure Fe-nanocrystal precipitates are too small to pin magnetic domain-wall motion [19]. Al-based alloys that contain pure Al-nanocrystals have superior mechanical properties; in fact, the latest advances based on Al-nanocrystal-glass microstructures literally push the envelope of the domain of strength-density performance for all metallic structural materials [20]. However, although favorable with regard to improved materials properties, the origin of the nanocrystal dispersion is still a matter of controversy. Several of the Al-based marginal glass formers are found in binary or ternary systems without any metalloid or semi-metal additions. Therefore two alloys, $Al_{92}Sm_8$ and $Al_{88}Y_7Fe_5$, were chosen as representatives of the class of marginal glass formers to study the nanocrystallization process in detail. At the same time, these materials provide the opportunities to study the early stage of the formation of a crystalline structure since crystallization proceeds sluggishly at temperatures very close to T_g.

2.2 Marginal Glass-formers

2.2.1 Nucleation versus Growth Control

The observation of high levels of melt undercooling that is essential for access to different levels of metastability is closely related to the capability to circumvent the effect of active nucleation sites during cooling. In general, there are two main strategies for accomplishing this objective, which are based upon applying either nucleation or growth control conditions during processing. With nucleation control, catalysis is commonly avoided by the physical isolation of the nucleation site such as during the atomization of liquids [21]. In this case the liquid volumes that are not affected by the catalysts will exhibit a large undercooling. With growth control strategies, the nucleation site catalysis is bypassed by an effective temporal isolation. In this case, due to solidification growth kinetics limitations, the solidification front developing from a given nucleation site lags behind the thermal front, and, correspondingly, a zone of undercooled liquid develops that advances ahead of the solidification interface. Within this zone, alternate solidification products can form including metallic glass. The conditions for growth control require high rates of heat extraction that can be provided e.g. by melt spinning or surface melting methods. A uniform and fully amorphous state should be theoretically attainable even for growth-controlled systems, e.g. by applying higher cooling rates, but this may not be possible in practice.

Concerning the differences in kinetic stability of metallic glasses, nucleation controlled and growth controlled vitrification could be identified with bulk and marginal glass-forming behavior, respectively. In the case of bulk glasses, physical isolation of potent nucleation sites as required for nucleation controlled vitrification is attained by an effective removal of crystalline impurities, e.g. by flux treatments, combined with a low nucleation rate of the corresponding crystalline phases. For selected alloy compositions, the undercooling that can then be achieved during cooling bulk volumes bypasses the nucleation reaction and most importantly the cluster size distribution, $C(n)$ that may be retained by the quench does not overlap with the critical size, n^*, at T_x. As a result, there is no precursor reaction to influence the evolution of crystalline clusters during subsequent thermal treatment. In this way a clear separation in temperature between the T_g and T_x signals can be observed during reheating. During isothermal annealing at T_x, the heat evolution rate Q exhibits a clear delay before the onset of the nucleation reaction and a peak maximum associated with the completion of nucleation and continued growth. Actually, early experiments on the $Pd_{40}Ni_{40}P_{20}$-alloy by Kui and Turnbull [22] have shown that fluxed samples could be undercooled continuously at low rates to temperatures below T_g and subsequently reheated at a somewhat higher rate to T_f without crystallization.

The fact that the sample passes two times through the temperature region of maximum transformation rates without crystallizing strongly indicates that nucleation was completely avoided during the initial cooling process. Under growth control conditions some small fraction of crystallites may form initially during rapid quenching, but the rapidly rising viscosity with continued cooling near T_g prevents their development or a cluster distribution that is retained overlaps in size with the critical nucleation size at T_g. In either case, as indicated in Fig. 2.3, upon reheating a sample with pre-existing crystallites (i.e. quenched-in nuclei), rapid crystallization ensues at T_g

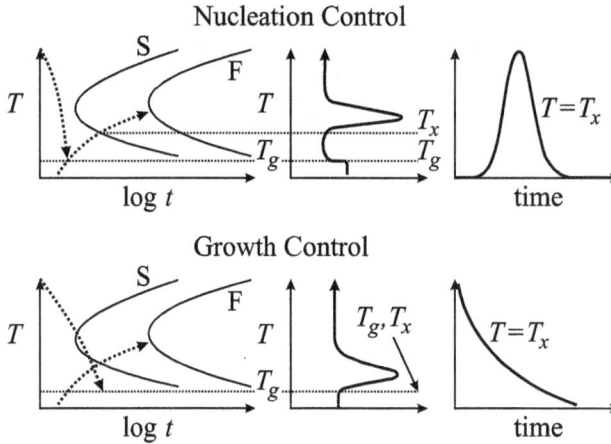

Fig. 2.3: Schematics showing the kinetics of metallic glass formation: nucleation control vs. growth control. Quenching and reheating paths are shown on the CCT-diagrams (S-start; F-finish) and the corresponding thermograms (dQ/dt: heat evolution rate). It should be noted that the relative positions (i.e. temperatures) of the maximum transformation rate reflect the maximum rates of the nucleation and growth processes. In general, the maximum rates will not occur at the same temperature and the relative positions will be sensitive to the transport characteristics of the undercooled melt (i.e. strong or fragile liquids).

which will essentially coincide with T_x. During isothermal annealing at T_x the growth of the pre-existing crystallites yields a continuously decreasing exothermic signal.

Actually, marginally glass-forming alloys often do not exhibit a clear glass transition signal upon reheating after rapid melt quenching, but instead exothermic maxima develop that indicate multiple stages of crystallization including the formation of a high density of nanocrystals within an amorphous matrix [23, 24]. However, homogeneous nucleation that occurs intensively at high levels of undercooling could also account for the high density of nanocrystals that develop during the primary crystallization stage of these alloys. Yet, although similar concerning the microstructural appearance, these modes of structural development are fundamentally different with respect to the kinetics of structure evolution since the activation barrier needs to be overcome at low mobilities, i.e. low attempt frequencies, in the case of homogeneous nucleation. Moreover, homogeneous nucleation is governed by temperature-dependent density fluctuations only. Thus, the formation of a static crystalline structure would occur inevitably in this temperature range without regard of the processing pathway, which would render the formation of a uniform amorphous structure impossible. Therefore, investigations of the structure evolution after controlled processing pathway variation can give further insight concerning the mechanism of early structure formation.

2.2.2 Processing Pathway Modifications

During solidification or other phase transformations, gradients of temperature and composition must exist within the phases. With conventional treatments the overall kinetics may be described by using diffusion equations to treat temperature and composition within the phases and by using the phase diagram to give the possible temperature and compositions for boundaries between the phases including corrections for interface curvature. This is called local interfacial equilibrium and is based on the concept that interfaces will equilibrate much more quickly than bulk phases.

As the rate of reaction becomes rapid, kinetic constraints that can arise from nucleation and growth limitations associated with an equilibrium product phase formation can develop and can expose alloy metastability. For the suppression of the equilibrium phase or the formation of a kinetically favored metastable phase, it is still possible to analyze reactions in terms of a metastable equilibrium that is used locally at interfaces. Under extreme conditions, significant loss of interfacial equilibrium for either a stable or metastable phase can develop when even interfacial relaxation becomes too slow. With the loss of interfacial equilibrium, thermodynamics can still be used to restrict the possible range of compositions that can exist at an interface at various temperatures since the selection must yield a net reduction in the free energy of the system [25]. One way to represent the thermodynamic restrictions is based upon the application of T_0-curves, which represent the limiting condition for partitionless transformation [26]. With isomorphous systems the T_0-curve is continuous with composition while for different structures each crystal phase has its own T_0-curve. Above the T_0-curves solute partitioning is required for solidification. Because of the diffusional constraint due to partitioning, crystallization can be inhibited by rapid quenching to promote glass formation [27].

Solid-state glass formation as an alternative vitrification route is often viewed as a non-equilibrium process resulting from the destabilization of crystalline phases when the maximum metastable solubility is exceeded. This perspective is also consistent with the general rules based upon a large negative heat of mixing and a large atomic size difference for the components that have been effective in identifying bulk glass forming compositions [28]. Within this framework, amorphization is depicted by a generalized phase diagram for partitionless transformations [29] as also indicated by Fig. 2.4, i.e. the T_0-curves of a terminal solid solution phase of an eutectic system and the composition-dependent glass temperature. According to this perspective, equivalent glassy states are obtained, which should follow identical transformation routes towards thermodynamic equilibrium during heating. However, amorphization by strain-induced mixing proceeds at low temperatures (i.e. low mobilities) while melt quenching from high temperatures involves rather fast diffusing atomic species in a homogeneous liquid phase that have to bypass the kinetic competition between vitrification and crystallization in order to form a glass. Therefore, in the case of nanocrystallization in marginally glass-forming alloys, the variation of the processing pathway

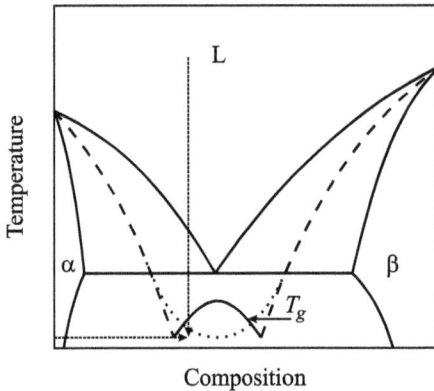

Composition

Fig. 2.4: The extensions of the T_0-curves to temperatures below a (metastable) eutectic are indicated as dashed curves. If the primary phases have different crystal structures and low mutual solubility, then the T_0-curves might not intersect. Such a situation favors glass formation in the composition range where the T_g-curve lies above the T_0-curves. The dashed arrow represents an isothermal solid-state amorphization process.

can lead to drastic modifications of the structure formation sequence if vitrification by rapid quenching is growth controlled [30].

Experimentally, both types of amorphization routes were applied on $Al_{92}Sm_8$, one of the chemically simplest glass-forming metallic alloys. Melt spinning was used as the rapid-quenching technique. Mechanical intermixing at ambient temperatures was achieved by cold-rolling stacked arrays of elemental foils of Al and Sm. The rolling procedure was used to avoid the detrimental temperature spikes that occur in high strain-rate processes such as ball milling. Energy dispersive X-ray analysis (EDX) on the cross-section of different samples confirmed that the composition of extensively cold-rolled samples corresponds to the nominal starting composition within the accuracy of the EDX-measurement (+0.5 at.%). TEM and XRD studies have shown that $Al_{92}Sm_8$ melt-spun ribbon (MSR) samples reveal a completely amorphous structure (Fig. 2.5a). Similar investigations on cold-rolled samples have yielded the same results: amorphous regions are free from crystalline fractions (Fig. 2.5b) [31]. Moreover, elemental analyses by EDX and by quantitative chemical analysis have revealed that the composition of the amorphous phases corresponds to the nominal composition for both sample types. However, on heating in a DSC, the samples prepared by melt spinning always showed a non-symmetrical exothermic signal in the temperature range between 170 °C and 220 °C, i.e. well below the crystallization of the eutectic $Al-Al_3Sm$ phases at 260 °C, T_x (Fig. 2.5c). The glass transition of the melt-spun sample could not be observed. In fact, even after annealing at a temperature of only 150 °C for 10 min a high number density of $3.5 \cdot 10^{22}$ m^{-3} Al-nanocrystals was observed within the amorphous matrix as indicated by Fig. 2.5e.

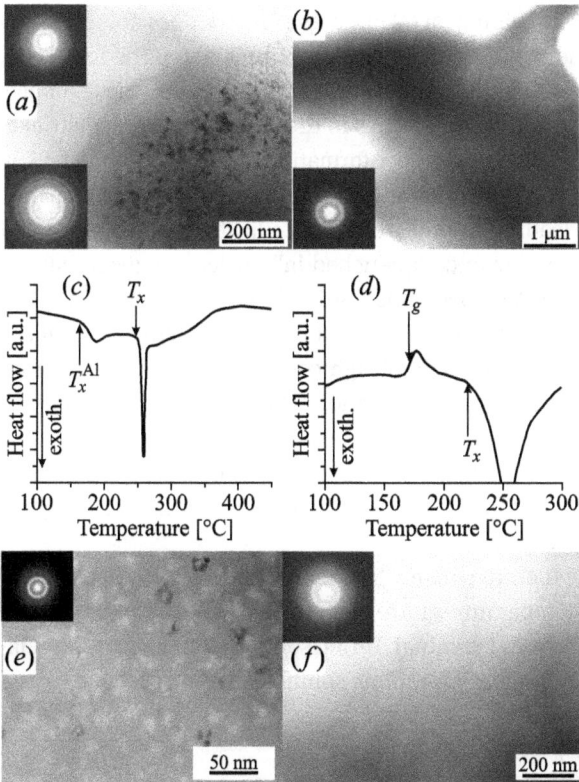

Fig. 2.5: Results of TEM and SAED on MSR (a) and on cold-rolled samples of $Al_{92}Sm_8$ (b) indicate similar amorphous regions. DSC-traces obtained from heating at a constant rate of 20 K/min show two exothermic events for the MSR (c) but the endothermic signal of the glass transition for the cold-rolled sample (d). After annealing at 150 °C for 10 min a large number density of nanocrystals developed in the MSR (e) while the cold-rolled sample is free from any nanocrystalline precipitates even after annealing at 150 °C for 60 min (f).

The DSC-signal on heating an $Al_{92}Sm_8$-sample at 20 K/min that was partially vitrified by cold rolling is given in Fig. 2.5d. The sample was annealed at 150 °C before the actual measurement in order to allow for relaxation of the vitreous state and to promote the development of any Al-nanocrystals. Two important characteristics can be observed on the heating curve in Fig. 2.5d. First, the calorimetric glass transition signal is clearly revealed at about $T_g = 172$ °C. This signal and the characteristic hysteresis, which is associated with the glass transition, could be observed reversibly on carefully heating and cooling through the transition region. Secondly, no exothermic signal occurs before the onset of the solid state reaction at about 230 °C at which the eutectic phases are formed. Moreover, TEM-experiments on cold rolled and annealed samples confirmed that Al-nanocrystals do not develop within the amorphous phase at temperatures near to or below the glass transition. The TEM-micrograph (Fig. 2.5f) taken on a sample directly after annealing for 60 minutes at 150 °C does not reveal any signs of crystallinity within the amorphous regions even after tilting the foil. Thus, calorimetry and microstructural analysis confirm that amorphous $Al_{92}Sm_8$-samples which have not been exposed to high temperatures in the liquid state before vitrification exhibit a clear T_g-signal and do not show the formation of a high number density of Al-nanocrystals. This result indicates that annealing at temperatures at or below T_g does not cause nucleation leading to premature crystallization, but that instead the nanocrystals develop by the growth of "quenched-in" nuclei [32] given that the amorphous phases are equivalent. Moreover, this result shows that the model of glass formation considering partitionless reactions, as visualized by the generalized phase diagram in Fig. 2.4, is only valid in the limit of high cooling rates R. The condition $R > R_{crit}$, with R_{crit} being the critical cooling rate for complete amorphization, has to be fulfilled in order to obtain similar amorphous states [31]. Yet, the result that a partially amorphous structure with remaining crystalline regions is kinetically more stable than a dispersed mixture of Al-nanocrystals and metallic glass is unexpected and requires additional attention.

Thus, in order to verify the thermodynamic boundary conditions as displayed in Fig. 2.4, calculations of the T_0-temperatures have been carried out. As an upper limit, T_0-curves have been calculated for the terminal Al-rich solid solution applying a relation derived by Boettinger and Coriell [33] for dilute solutions, i.e.

$$T_0 = T_f + m_l C_0 \ln\left(\frac{k}{k-1}\right) \tag{2.2}$$

with the melting point T_f, the liquidus slope m_l, the equilibrium distribution coefficient k and the alloy composition C_0. Equilibrium data on the solid solubility of Sm in Al are not available; phase diagram calculations indicate a negligible solubility. Thus, in order to define k, values of $2 \cdot 10^{-3}$ to $1 \cdot 10^{-1}$ at.% Sm were taken as the solubility limit at the eutectic temperature. The latter value was given by [34] as an upper bound. Additionally, a continuous T_0-curve of the metastable Al-Sm solid solution was also calculated by the use of tabulated data for liquid $Al_x Sm_{1-x}$ and for pure Al. The required lattice stability of the metastable fcc-Sm phase was estimated using the Clausius-Clapeyron equation on experimental high pressure data. For both methods of assessing the T_0-curve, the composition studied is well outside the range for partitionless crystallization for all temperatures at or above room temperature. Thus, thermodynamics does not account for the different stability against crystallization of the amorphous phases synthesized by different routes [32]. Yet, considering sluggish diffusion-limited growth, the different amount of interface area present in the samples that can easily reach four orders of magnitude, can explain the observed phase evolution differences. In that case, growth of the crystalline phases would involve only minor volume fractions in the cold-rolled samples that do not render a detectable DSC-signal. In contrast, the high amount of interface area in samples with a high density of nanocrystals would lead to a much higher fraction of transformed volume.

However, in order to account for the formation of crystal-like structures in the undercooled melt without subsequent fast growth that is thermodynamically favored, mechanisms related to growth rate limitations need to be considered. For marginally glass forming Al-based alloys, growth restrictions based on the low diffusivity of the rare earth component and the associated solute pile-up ahead of the interface of the evolved Al-clusters (so-called diffusion-field impingement) have been suggested and experimentally verified [35–37]. In order to obtain experimental evidence for diffusion-limited growth and solute pile-up at the interface, the studies on the nucleation and growth kinetics at lower temperatures, i.e. in the amorphous state, can also be utilized.

2.2.3 Nucleation and Growth Kinetics

Since many of the metallic glass-forming systems have been discovered based upon the empirical rule of adding solutes with a large difference in atomic size [38], a disparity in the solute diffusivities may be expected. The growth kinetics of nanocrystals during primary crystallization of these glass materials will be strongly affected by unequal solute diffusivities. Hence alloying strategies can be developed that exploit the effect of multi-component diffusion on the crystallization pathway and growth behavior. These strategies apply to any glass-forming material that has unequal diffusion coefficients and a region of glass stability, including the Al- and Fe-based materials.

In Al-TM-RE glasses, composition profile measurements [39] indicate that the rare earth element (RE) diffuses much more slowly that the transition metal (TM). The Coates-model accounts for diffusion limited growth of precipitates in ternary systems with unequal component diffusion coefficients [40, 41]. For a ternary system ABC, where B and C are solutes in A, the simplest case is given for $D_B = D_C$. Here the tie-line gives the interface compositions of the precipitate and the matrix (i.e. local equilibrium). If $D_B \neq D_C$, the interface compositions depart from the tie-line values. For differing component diffusivities, Coates has used the concept of an interface contour (IC), which includes all bulk alloy compositions in the two-phase field that yield a given set of precipitate and matrix compositions at the interface during growth before diffusion field impingement. The disparity in solute diffusivities produces different kinetic regimes that depend on the bulk alloy composition. The Coates model indicates that along the section of the IC that is essentially constant in B content, which includes the alloy composition at O, as shown in Fig. 2.6, the slow diffusing C-element limits the kinetics. Along the part of the IC that is essentially constant in yttrium which is restricted to near pure Al for the current conditions, the fast diffusing element (iron) governs the kinetics. Hence the delineation of the IC's in a multi-component alloy system is essential to gauging the kinetic response during growth.

The glass transition temperature in metallic glass-forming systems often develops a maximum within the ternary diagram [42], producing a region of glass surrounded by liquid in composition space. This tendency suggests two different strategies that

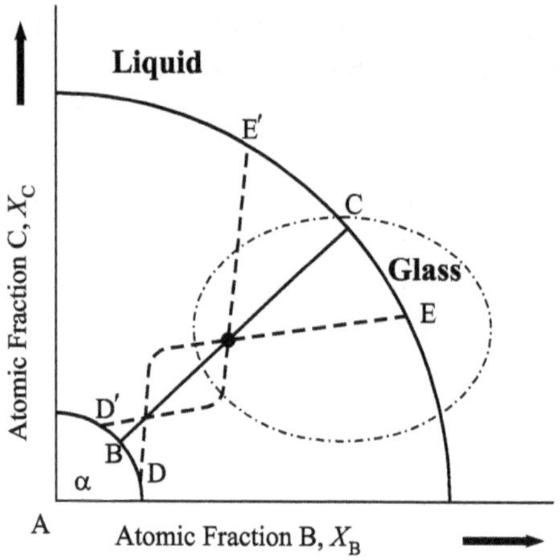

Fig. 2.6: Schematic isothermal ternary section illustrating the effects of multi-component diffusion on crystallization kinetics. The tie-line and interface contours for differing component diffusivities are shown. The dashed curve delineates the glass-forming region.

can be applied with a consideration of the multi-component diffusion effect. With $D_B < D_C$, the IC in Fig. 2.6 given by the curve DOE develops. Note that the interface composition E is now well within the glass region. Since the diffusivity of the glass is much smaller than that of the liquid, the growth rate of the α-phase is substantially reduced. This phenomenon allows for higher relative amounts of component C compared to component B, while retaining the interface composition of the amorphous matrix in the glass region rather than the liquid region. Thus, growth of the primary nanocrystals is limited by diffusion in the glass rather than in the liquid, significantly decreasing the growth rate. Note, though, that due to mass conservation new IC's will develop, and composition E will track along the liquidus boundary in the direction of C and eventually surpass it. The capability of increasing the amount of component C while retaining good kinetic stability of the material would be useful when additions of C yield improvement of desired material properties. This situation would allow extended elevated-temperature capability for a nanocrystal/amorphous matrix composite material for the alloy composition O.

However, if $D_B > D_C$, the matrix composition E' of the IC given by D'OE' will lie on the liquidus boundary to the left of composition C rather than to the right, as is shown in Fig. 2.6, and the kinetics would be based on growth into the liquid rather than the glass. The metastable Al-Y-Fe phase diagram including the fcc-liquid equilibria and the IC's has been calculated [43]. Here, iron was assumed to diffuse rapidly and to adjust its composition in the matrix as the composition gradient of yttrium evolves. This expectation has been confirmed experimentally for the similar alloy Al-Ni-Ce by field ion microscopy results [44]. Thus, during growth the matrix composition will move along the phase boundary to compositions higher in iron and lower in yttrium and establish new ICs. This process will continue until the reaction reaches completion. In general, a complete description of the kinetics requires that the trajectory of the ICs be modeled. But due to the restricted range of matrix compositions at the interface during the evolution of the ICs in the case of Al-Y-Fe, the primary effect is due to the diffusion coefficient and the number density of nanocrystals. Hence, a constant interface composition given by the tie-line can be used for describing the growth kinetics of this alloy.

At sufficiently high temperature, i.e. at temperatures where the cluster size distribution overlaps with the critical radius for nucleation, the Al-nanocrystals start to grow into the amorphous matrix, thereby rejecting the rare earth and the transition metal elements. Eventually, the diffusion fields for the TM and the RE ahead of the solidification front of different Al-nanocrystals will overlap causing the so-called diffusion-field impingement that limits nanocrystal growth. In this case, the kinetics of diffusion-limited precipitate growth will deviate from parabolic growth behavior (i.e. $r \propto \sqrt{Dt}$, with r, particle radius, D, diffusion constant and t, time), described by the well-known Johnson-Mehl-Avrami-Kolmogorov equation. For a more complete analysis, Ham has developed an expression for time-dependent particle growth under the condition of diffusion-field impingement [45]. The model considers a cubic array of identical particles growing under diffusion control with a composition-independent diffusivity and treats the composition profile in the matrix as an average quantity. The application of this analysis to the growth of Al-nanocrystals that has been discussed in detail elsewhere [43], can be used to model the shape of the DSC-signal that corresponds to the primary nanocrystallization. To further illustrate the importance of impingement, the diffusion fields of two adjacent particles at 270 °C were each calculated by assuming growth into an infinite matrix (Fig. 2.7).

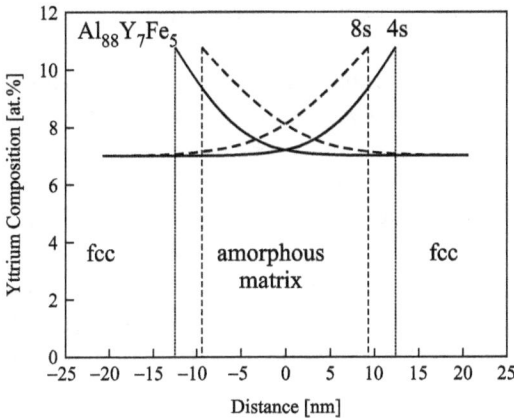

Fig. 2.7: Calculated diffusion fields for yttrium for particles 40 nm apart with midpoint between nanocrystals at zero after 4s (solid lines) and after 8s (dashed lines) at 270 °C. Vertical lines represent the interfaces between Al and the amorphous matrix [43].

It should be noted that with conditions similar to those found to fit the nanocrystallization exotherm, diffusion field impingement becomes significant at about eight seconds. The tie-line through $Al_{88}Y_7Fe_5$ gives the fraction of the fcc-phase as 35 %, thus with uniform particles and a spacing of 40 nm, each particle will grow to a radius of 14 nm. The Ham-analysis predicts a rapid deviation from parabolic growth highlighting the need to consider diffusion-field impingement in a growth kinetics analysis involving high particle densities. Moreover, the result indicates that for increasing the temperature stability and the overall performance of the material, an increase of the initial nucleant density is the key.

Based upon the experimental results discussed above, it is apparent that the formation of the large nanocrystal number densities is inherently coupled to the presence of a high nucleation site density in the undercooled melt [46]. In fact, the nucleation site density required to be in accordance with experimental microstructure observations is by several orders of magnitude too high to be explained by classical nucleation theory [35]. Yet, the nature of these nucleation catalysts is still a matter of hypothesis. Impurity-based catalysis can be ruled out regarding the impurity level required. Phase separation would provide another rationale for the formation of high number densities, which has been advanced in several publications [47, 48]. However, the topology of the phase diagrams, especially the location of the glass-forming composition near pure Al, renders this hypothesis less likely on first sight. At the same time there are other structural arrangements in the melt that may act to serve as a high density of viable catalysts at high undercooling. Some of the possibilities have been suggested by Turnbull [49] as either heterophase or homophase catalysts where minute levels of solute either serve as discrete cores for a catalyst or promote aggregations that act as catalysts. These catalyst effects can be active at low solute levels in the ppm-range, but their impact may be detected by careful kinetics measurements. Actually, there are some results from X-ray [50] and neutron [51] scattering measurements that tend to support this type of catalyst effect. However, more detailed models of catalysts effects have to await more complete crystallization kinetics analyses that cover the undercooled liquid and the glassy state.

In former investigations, the development of nanocrystals at low temperatures (i.e. below T_g) has been observed only by TEM-analyses, the heat-flow associated with the nanocrystal formation was below the detection limit of the respective calorimeter devices [52]. In order to monitor the early stages of nanocrystal development in the vitreous state at temperatures below T_g directly, isothermal micro-calorimetry measurements on long time scales have been performed. With the micro-calorimeter used here (Thermometric, 2277 TAM), a baseline stability better than 0.025 µW/h is obtained, and the temperature of the samples is maintained with an accuracy better than $2 \cdot 10^{-4}$ °C. For the measurements, $Al_{92}Sm_8$-samples of about 40 mg in weight were cut from the as-spun ribbon, filled into stainless steel capsules and inserted into the system after relaxing the samples at 110 °C for one hour. The calorimeter signal was calibrated electrically by measuring a defined (electrical) power pulse. These measurements utilize the slow kinetics of the nanocrystallization reaction in the Al-rich alloys, which allows one observing the overall kinetics without disturbance of the main heat-flow signal by the initial transient of the apparatus that takes about 20 min to become negligible [53]. For comparison, a baseline was measured under identical conditions using an empty crucible as a reference.

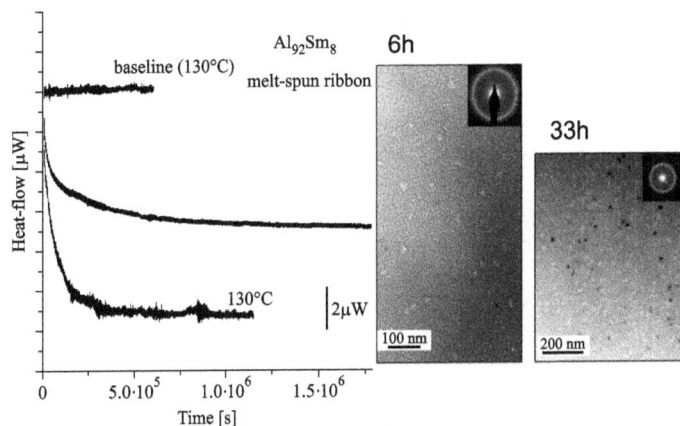

Fig. 2.8: The heat-flow that evolved during isothermally annealing $Al_{92}Sm_8$ samples at 120 °C and 130 °C, respectively. The monotonic decrease of the heat-flow signal that was observed for the Al-Sm samples indicates that nanocrystallization proceeds via diffusional growth at these temperatures. The baseline that was measured under identical conditions for about seven days using an empty capsule indicates the long-time stability of the apparatus. The respective curves have been shifted vertically for clarity. The TEM-images on the right hand side verify the correlation of the observed signal with a growing fraction of the crystalline (fcc-Al)-phase.

The results of the measurements at 120 °C and 130 °C are displayed in Fig. 2.8. The baseline measurement clearly indicates the absence of any instrumental effects that exceed the scatter of the data during the entire measurement duration of about seven days. In contrast, the sample measurements show a monotonic decrease of the heat-flow signal that extends for about 85 h at 130 °C (about 195 h at 120 °C). The absence of any maximum in the heat-flow signal as well as the monotonically decreasing signal indicates that diffusional growth of pre-existing nuclei instead of nucleation and growth of Al-nanocrystals occurs in the samples.

2.2.4 Characterization of the Amorphous Phase

In Section 2.2.2 it was shown that the alteration of the processing route for a glass-forming (Al–Sm)-alloy yields a higher kinetic stability for the material that has been synthesized via solid-state deformation mixing. This unexpected behavior can be explained in terms of the kinetic competition between crystallization and vitrification that is avoided by solid-state processing routes in conjunction with the low diffusivity of the rare earth component that effectively limits crystal growth. However, the modification of the processing route makes a separate examination of the melt-quenched samples necessary to ensure that an identical amorphous state has been attained in both cases. For this reason, amorphous melt-spun ribbon samples of Al-base alloys such as $Al_{92}Sm_8$ and $Al_{88}Y_7Fe_5$ were investigated by temperature-modulated differential scanning calorimetry (TMDSC) [54]. In this mode, a periodically varying temperature oscillation is superimposed on a constant heating or cooling rate, β. The time dependence of the sample temperature is given by

$$T = T_0 + R_0 t + T_a \sin(w_0 t) \tag{2.3}$$

with the initial temperature, T_0, the amplitude of the temperature modulation, T_a, the scanning (cooling) rate, R_0, and the angular frequency of the temperature modulation, w_0.

Under the condition that the system is close to a local equilibrium, i.e. for sufficiently small temperature modulations, the oscillating heat-flow signal can be analyzed within the framework of linear response theory [55]. Within this framework, the response function to an oscillatory attenuation consists of two contributions that are in and out of phase with respect to the input signal. Using complex notation for the vector sum of the Fourier-transformed contributions to the dynamic specific heat signal gives

$$C^*(w) = C'(w) - iC''(w), \tag{2.4}$$

where $C^*(w)$ is the frequency-dependent complex specific heat; $C'(w)$ is the real part or storage specific heat; and $C''(w)$ is the imaginary part or loss specific heat [55]. While dissipative processes such as the relaxation of the glass that are related to entropy production contribute to C'', the reversible molecular rearrangement that occurs during a glass transition contributes to C'. In equilibrium, C' thus equals the thermodynamic equilibrium specific heat, C_p and correspondingly C'' vanishes. In the case of a time-dependent specific heat there is dispersion in C' and therefore C'' becomes finite (and positive) as required by the Kramers-Kronig relations and the second law of thermodynamics.

With DSC-measurements the heat-flow rate into the sample, $\bar{\varphi}$, is the measured quantity. In the linear response regime, $\bar{\varphi}$ can be separated into a contribution from the underlying heat flow due to the constant rate R_0 and an oscillating heat flow as [55]

$$\bar{\varphi}(T(t)) = C_R(T) R_0 + w_0 T_a |C^*(T, w_0)| \cos(w_0 t - \phi), \tag{2.5}$$

$$|C^*(T, w_0)| = \sqrt{C'^2 + C''^2}$$

with $C_R(T)$ being the apparent specific heat in the presence of time-dependent phenomena that would be measured at a constant rate in conventional DSC-experiments and ϕ, the phase shift between the cooling rate oscillation and the temperature oscillation of the sample. This formalism, concerning the complex specific heat, C^*, will be of special importance in direct measurements of the cooperative relaxation modes of deeply undercooled metallic liquids that are discussed at the end of this chapter.

The results in this respect obtained on $Al_{92}Sm_8$ and $Al_{88}Y_7Fe_5$ MSR are shown in Fig. 2.9 in comparison to the DSC-traces obtained from the underlying static heat flow. The exothermic signal that corresponds to the non-reversible primary crystallization

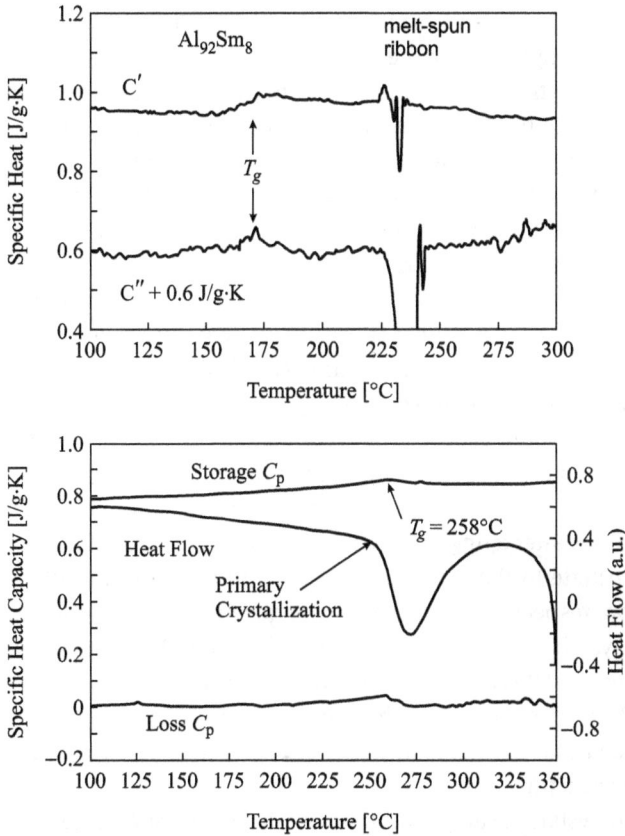

Fig. 2.9: The results on the real and imaginary parts of the complex specific heat capacity function obtained by TMDSC-measurements on (top) $Al_{92}Sm_8$ and (bottom) $Al_{88}Y_7Fe_5$-glasses that had been synthesized by rapid melt quenching via melt spinning. Additionally, in the lower figure the underlying heat flow of the nanocrystal formation reaction is also indicated.

process is excluded by the time series analysis of the modulated heat-flow signal, and the endothermic signal due to the glass transition is observed in the storage specific heat curves at about 172 °C for Al-Sm and at about 258 °C for Al–Y–Fe. The TMDSC-results confirm that a truly vitreous state has actually been achieved during the melt-spinning process [56]. The observation of the glass transition signal in calorimetric measurements provides evidence that the glass state had been attained by mechanical intermixing as well as by rapid melt-quenching. The equivalence of the amorphous states obtained for $Al_{92}Sm_8$ by rapid solidification and deformation mixing (except for secondary processes such as structural relaxation) indicates further that nucleation at temperatures at or below T_g does not cause the nanocrystallization, but that instead the nanocrystals of the observed sizes between 15–25 nm in diameter (i.e. corresponding to the order of 10^5 unit cells of Al) develop by the growth of "quenched-in nuclei", which are of the order of the critical size with a radius of about 0.6 nm at the respective high undercooling (i.e. corresponding to the order of ten unit cells of Al). At this point it might be helpful to note that these estimations are based on pure Al as the crystallizing entity, which has also been observed in the experiments for all nanocrystal sizes that allowed analyzing the composition. In the case of melt-spun $Al_{92}Sm_8$, a high nucleation density is initiated during the quenching process, but nuclei growth is limited by the reduced temperature during quenching and by solute rejection.

Besides the experimental assessment of the glass transition, the TMDSC-measurements yield thermodynamic information that can be used in conjunction with a Kirchhoff cycle and additional measurements of the melting parameters to extract an estimate for the isentropic Kauzmann temperature, T_K, of the respective alloy. For the analysis of thermodynamic properties of the glassy phase such mechanically alloyed samples are not favored because the resulting microstructure consists of a non-uniform mixture of amorphous and crystalline regions. Melt-spun ribbon samples, however, show a uniform composition after quenching and a uniform distribution of Al-nanocrystallites after aging at temperatures below T_x. Therefore, the sample fractions that contribute to the calorimetric signals of the nanocrystallization and the glass transition, respectively, can be determined with a higher accuracy. After completion of nanocrystallization at a temperature of about 593 K during slow heating, the impinged diffusion fields arrest grain growth. Annealing at that temperature for several minutes does not change the subsequently observed microstructure. Therefore, it can be assumed that the microstructure observed after fast cooling is representative for the microstructure present at 593 K.

The fraction of Al-nanocrystals of samples at that arrested state amounts to 29 % of the sample volume (25.4 % of the sample mass), as shown in Fig. 2.10a. Thus, the specific enthalpy released by the growing Al-nanocrystals, $\Delta H_x(Al_{nano})$, is obtained from the integrated heat-flow difference, ΔH_{exp}, of an as-quenched sample as $\Delta H_x(Al_{nano}) = \Delta H_{exp} \cdot (100/25.4) = 214$ J/g. Assuming a constant specific heat of the undercooled liquid Al equal to the specific heat of the stable liquid state, C_p^l, the crystallization enthalpy of pure Al, $\Delta H_x(Al)$, in a (hypothetical) deeply undercooled state was estimated. At 573 K, i.e. in the temperature range where the enthalpy release of the nanocrystallization process has a maximum, a value of $\Delta H_x(Al) = 370$ J/g is obtained. The magnitude of the difference between $\Delta H_x(Al)$ and $\Delta H_x(Al_{nano})$ cannot be explained by a variation of C_p^l. Yet, for the initial stages of nanocrystallization, finite-size effects have to be considered.

Recently, it was shown that besides the melting temperature, the melting enthalpy is also a strong function of the system size if the crystals are of the order of 50 nm or less in diameter. Actually, for pure Pb-nanocrystals that are embedded in a pure Al-matrix, the decrease of the melting enthalpy with decreasing crystal size was much higher than the respective decrease of the melting temperature [57]. At a mean diameter of about 10 nm, ΔH_f is reduced to merely one half of the respective bulk value. Similar results have also been reported for tin particles [58]. Thus, a decreased value of

$\Delta H_x(\text{Al}_{\text{nano}}) = 0.58\Delta H_x(\text{Al})$ for Al-nanocrystals in the observed size range of 10–20 nm in diameter can be accounted for by the size dependence of the chemical potential without considering further energetic contributions, e.g. due to volume strain that is induced by the solute pile-up.

However, the result that about 70 % of the sample volume is still amorphous at a temperature of 573 K agrees well with the DSC-signal that corresponds to the formation of the intermetallic phases. Both, the rapidity of the heat release and the magnitude of the thermal effect indicate that the major part of the sample contributes to the transformation and that the effective diffusivity involved is rather large. TEM-investigations in conjunction with subsequent image analysis indicate that under the conditions present during the TMDSC-measurement, shown in Fig. 2.9b, about 92 % of the sample mass contributed to the glass transition signal. The average composition of the remaining amorphous phase at that temperature then amounts to $\text{Al}_{86.8}\text{Y}_{7.7}\text{Fe}_{5.5}$. Therefore, the glass-transition temperature that was determined by TMDSC-measurements corresponds to an alloy of this modified composition. However, thermal analysis measurements of the liquidus temperatures, T_l, of Al-rich Al–Y–Fe alloys indicate that T_l does not vary strongly with composition in the range between 4–6 at.% Fe and 6–8 at.% Y [59]. Moreover, the glass temperatures of alloys of a given system do not depend strongly on the composition within a limited so-called glass-forming range and the residual variation scales with the composition dependence of T_l [60]. Therefore, the value for the glass transition temperature measured by TMDSC can be regarded as representative for the glass temperature of a completely amorphous $\text{Al}_{88}\text{Y}_7\text{Fe}_5$ alloy.

According to Kirchhoff's rule, a complete thermodynamic cycle of melting and crystallization of the $\text{Al}_{88}\text{Y}_7\text{Fe}_5$ alloy is given by

$$\Delta H_f = \Delta H_x(\text{Al}_{\text{nano}}) + \int_{T_1}^{T_x} \left[\left(C_p^{\text{Al}_{\text{nano}}} + C_p^\alpha\right) - C_p^{x_{equ}}\right]dT \tag{2.6}$$

$$+ (\Delta H_x + \Delta H_u) + \int_{T_u}^{T_f} \left(C_p^l - C_p^{x_{equ}}\right)dT \,,$$

with the melting enthalpy, $\Delta H_f = (460 \pm 15)$ J/g, the crystallization enthalpy of the intermetallic phase, ΔH_x, the enthalpy released upon formation of the equilibrium phases from the primary crystallization products, ΔH_u, with $\Delta H_x + \Delta H_u = (200\pm5)$ J/g. $C_p^{\text{Al}_{\text{nano}}}$ denotes the specific heat of the nanocrystalline Al-precipitates, $C_p^{x_{equ}}$ is the specific heat of the equilibrium crystalline phases, C_p^α denotes the specific heat of the residual amorphous (undercooled liquid) phase, T_f is the melting temperature, T_u denotes the offset temperature of the signal that corresponds to the formation of the equilibrium phases and T_1 denotes the maximum temperature for the pre-crystallization. Except C_p^α and C_p^l, all quantities in Eq. (2.6) have been determined experimentally. Without pre-crystallization, C_p^α and C_p^l would be identical for the same temperature interval. Thus, Eq. (2.6) allows one to estimate the average specific heat difference between the undercooled liquid and the equilibrium crystalline phases, which yields $\Delta C_p = 0.12$ J/gK.

As expected for an alloy with a rather low glass-forming tendency, the average specific heat of the undercooled liquid ($C_p^l(740$ K$) = 0.97$ J/gK) does not deviate strongly from the (weighted average) specific heat of the ideal liquid Al-Y-Fe solution that amounts to 1.01 J/gK. However, this result indicates that the value for the specific heat difference at T_g, obtained directly from the TMDSC-measurements ($\Delta C_p(T_g) = 0.07$ J/gK), is deteriorated by the simultaneously proceeding crystallization that accelerates at temperatures in the immediate vicinity of T_g (Fig. 2.10a). Following above considerations, the differences in entropy, ΔS, and Gibbs free energy, ΔG, between the undercooled liquid and the equilibrium crystalline phases can be calculated as a function of temperature. The Kauzmann-temperature, T_K, defines the low-temperature limit of the liquid state. The melting entropy of the $Al_{88}Y_7Fe_5$-alloy was obtained by dividing the melting range in temperature intervals of 5 K width and summation of the partial melting entropies of each interval. Fig. 2.10b shows the estimated results for ΔG and ΔS between the undercooled liquid and the equilibrium crystalline phases.

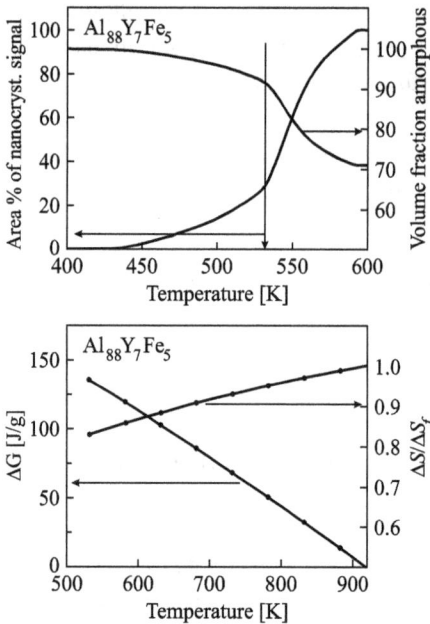

Fig. 2.10: (a) Area fraction of the nanocrystallization signal and the remaining amorphous volume fraction of the sample. (b) Calculated differences of the Gibbs free energy and the entropy between the undercooled liquid and the equilibrium crystalline phases.

The results indicate an unreasonably low value for the isentropic Kauzmann-temperature [61]. T_K, defined by $\Delta S(T_K) = 0$, yields $T_K \leq 100$ K. However, even if the specific heat difference increases rapidly at temperatures below T_g, still T_K is expected to be lower than the ambient temperature. The difference $T_g - T_K$ is related to the fragility of the material: a wide temperature interval between T_g and T_K, as observed here for the Al-Y-Fe alloy, indicates a small slope of the equilibrium viscosity in the glass-transition range or, equivalently, a small dependence of the residual entropy on temperature, which is characteristic for strong glass-formers. In analogy to the results of tracer diffusion experiments on Zr-rich alloys that also show strong behavior, this result can be interpreted in terms of a "rigid-cage"-model [62]. It is suggested that the majority species, i.e. Al, shows a pronounced short-range order that effectively limits the dynamics of the (undercooled) liquid resulting in the observed strong glass-forming behavior. This strong glass-forming behavior seems to be in contradiction to the observed kinetic stability, since strong glass-formers are expected to show bulk glass formation. However, neutron diffraction studies indicate the presence of enhanced short-range order correlations of about 15 Å in Al–Y–Ni [63] and strong $(s - d)$-hybridization in Al–Fe–Ce [64]. It was proposed by Egami [65] that this preferential bonding is also effective in the liquid state, leading to the presence of aggregates or clusters in the glass and the undercooled melt.

Concerning the classification of these glass-formers according to the fragility concept, it is suggested that the short-range order that is present in the undercooled liquid state effectively limits the dynamics of the melt resulting in strong glass-forming behavior. With respect to nucleation during fast cooling, this short-range order could effectively shift the inherent cluster distribution towards the critical size thus favoring copious nucleation. With the restricted mobility of the solute atoms leading to a kinetic stabilization of the residual undercooled melt against complete crystallization, this scheme can explain the observed nucleation and growth behavior as well as the microstructures obtained after the respective processing pathway modifications.

2.2.5 Nanocrystal Formation at Temperatures Well Below T_g

In order to analyze the hypothesis extracted from the experimental observations described above, further analyses on the earliest stages of crystal formation have been conducted. Detailed analyses of the evolution of the number density as well as of the size distribution based on calorimetry, microcalorimetry and quantitative transmission electron microscopy [66, 67] have revealed that the evolution of the nanocrystalline fraction, specifically at the earliest stages, is not in accordance with theoretical descriptions based on a simple nucleation and growth process. Especially, a significant bimodality of the crystallite size was found, with one fraction of crystallites (that form the fraction of larger average size) appearing at extremely short times. Additionally, the evolution of the transformed volume did not follow a parabolic growth law at short times as expected for diffusion-controlled transformations. These observations are unexplained so far (except in one theoretical approach discussed below). However, they clearly point to the presence of processes that act in addition to conventional nucleation and growth. Moreover, this also becomes apparent when comparing the diffusion coefficients, D, estimated from the microstructure evolution observed in this work by assuming a parabolic growth [68–70]:

$$R(t) = (2DSt)^{1/2}. \tag{2.7}$$

Here $R(t)$ is the particle radius after annealing at the time t. With a super-saturation $S = 1.5$, the estimation yields a diffusion coefficient of $D = 9 \cdot 10^{-20}$ m^2/s at 140 °C and $D = 2 \cdot 10^{-21}$ m^2/s at 210 °C. These diffusion coefficients are not in accordance with the expected Arrhenius behavior.

Specifically concerning the occurrence of bimodal nanocrystal (or cluster) size distributions as well as concerning the observation of nanocrystal densities that are too high with respect to the expectations from classical nucleation theory, it has been shown theoretically in a series of papers by Schmelzer et al. [71, 72, 73, 74] that these effects are a distinct result of changes of the bulk and surface properties of nanocrystals in dependence on their sizes. These effects can can be interpreted in a description based on a generalization of Gibbs' approach. In fact, due to a distinct dependence of state variables such as the composition of the crystalline aggregates on the size of the growing cluster or nanocrystal, these effects are expected to occur in the early stages of the evolution of large ensembles of clusters within a highly viscous amorphous matrix. These theoretical analyses and their results are described in more detail in Chapter 10 of the present volume and the cited there additional references.

In order to further analyze these problems, experiments investigating the relationship between the glass transition and the primary crystallization reaction have been performed. The glass transition has been explored by MDSC, high heating rates DSC and conventional DSC. The impact of annealing the samples in the temperature region in which primary crystals are generated has been investigated and compared to the results obtained after annealing the samples at temperatures where primary crystallization does not occur. Microcalorimetry studies complemented by XRD and microstructure analyses by TEM were performed to resolve the onset of the low temperature primary crystallization and to analyze in detail the processes that govern the very early stages of crystal formation in a model Al-rich glass forming alloy. Isothermal microcalorimetry measurements were performed on as-spun samples at 60 °C, 100 °C, 120 °C, and 140 °C, respectively. The results are shown in Fig. 2.11. XRD-measurements performed after the isothermal measurements show crystallization only in the case of the sample annealed at 140 °C (Fig. 2.13). However, the sample appears fully amorphous in TEM-micrographs (both, bright field and dark field images, not shown here). In conventional DSC measurements, the as-spun glassy ribbons do not show a glass transition (Fig. 2.14a). However, a glass transition was observed by using modulated DSC (MDSC) [75], higher heating rates [76–78] or an increased yttrium content, which stabilizes the amorphous matrix against crystallization [79]. In the presented case of Al$_{89}$Y$_6$Fe$_5$, the use of a high heating rate does not immediately reveal a glass transition. However, the slow decrease of the signal from point 3 (see Fig. 2.14a) in mea-

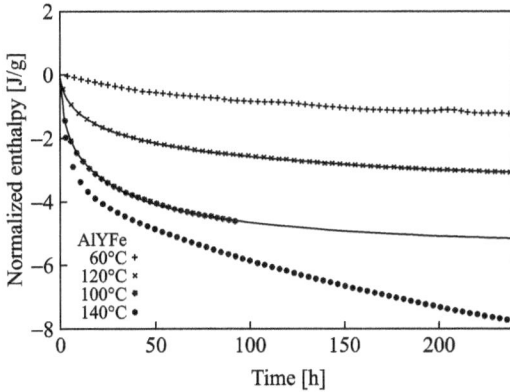

Fig. 2.11: Isothermal microcalorimetry measurements at various temperatures and KWW-function fits (black lines). At 60 °C, the signal is too small for a meaningful analysis. The good fitting obtained for isothermally treated samples at 100 °C and 120 °C is attributed to a pure relaxation process. At 140 °C, the signal is no longer caused by pure relaxation but is due to a superposition of relaxation and crystallization contributions.

Fig. 2.12: TEM-micrograph of $Al_{89}Y_6Fe_5$ annealed for three hours at 210 °C. The inset shows the corresponding selected area electron diffraction pattern. Apart from the diffuse halos corresponding to the amorphous matrix, Bragg reflexes corresponding to [111] (within the brightest halo) and [200] Al-lattice planes (on the outer rim) are visible.

surements performed at 20 °C/min splits into an endothermic event (point 3a) and an exothermic event (point 3b) (see Fig. 2.14b). If the sample is subjected to a first DSC-measurement up to 250 °C at a heating rate of 300 °C/min, the subsequent measurement (Fig. 2.14b, dashed line) shows a distinct endothermic event prior to the primary crystallization reaction, which is interpreted as the signal corresponding to the glass transition. Another possibility to obtain the apparent glass transition signal is presented in Fig. 2.14c. The DSC-run was performed after annealing the sample for three hours at 210 °C during which 10^{20} m^{-3} aluminium nanocrystals with an average diameter of 16 nm form.

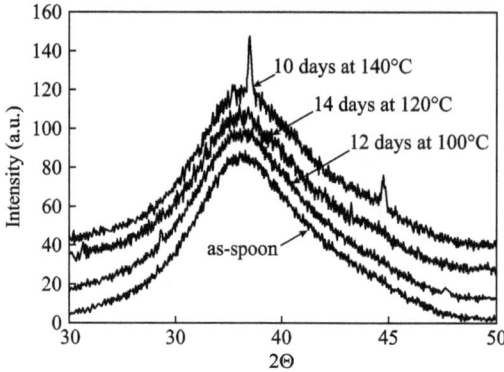

Fig. 2.13: XRD-measurements performed after isothermal microcalorimetry measurements presented in Fig. 2.11. The sharp crystal peaks that appear after annealing at 140 °C confirm that crystallization occurs at this temperature.

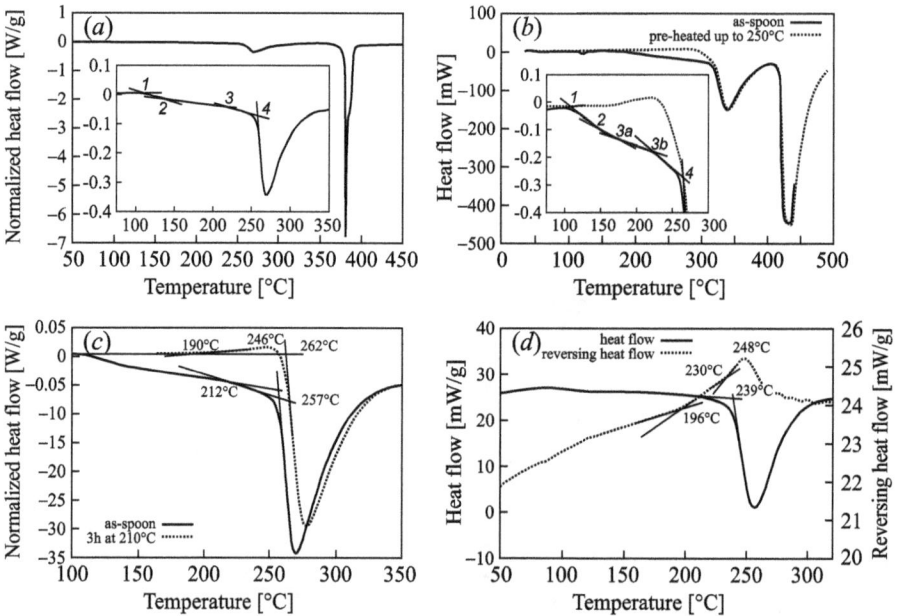

Fig. 2.14: (a) DSC run at 20 °C/min of an as-spun sample showing four points of interest: (1) and (2) would account for the onset of relaxation; (4) is the onset of the primary crystallization peak; (b) DSC-runs at 300 °C/min of an as-spun sample and of a sample preheated to 250 °C. The exothermic tendency at point (3) splits up into two points at a heating rate of 300 °C/min. Point (3a) is interpreted as the onset of the glass transition and (3b) as a contribution of the initial stage primary crystallization; (c) DSC-run at 20 °C/min of an as-spun sample and of a sample annealed for three hours at 210 °C. Unlike the as-spun sample, the annealed sample shows a glass transition starting at 190 °C; (d) glass transition as observed by MDSC.

The corresponding microstructure is shown in Fig. 2.12. The remaining amorphous matrix has a higher yttrium content and thus is more stable against the primary crystallization reaction [79]. As a consequence, the DSC-run shown in Fig. 2.11c exhibits a clear endothermic event starting at about 190 °C and a primary crystallization temperature that is 5 °C higher than for the as-spun sample. Fig. 2.14d shows the heat-flow signal and the reversing heat-flow signal obtained in a single MDSC measurement on an as-spun sample. The measurement confirms the results of Wu et al. [75] and also those of the measurements conducted at high heating rates and after prolonged annealing treatments, since a positive change of the reversing heat flow in the temperature range of the onset of primary crystallization (that is observed as an exothermic signal on the total heat flow) clearly indicates the occurrence of a glass transition.

MDSC-measurements were performed on a series of $Al_{89}Y_6Fe_5$ samples that were previously subjected to different annealing treatments (Fig. 2.15). A systematic shift of the glass transition onset was observed. While the glass transition of the as-spun sample (Fig. 2.14d) starts at 196 °C, an annealing period at temperatures below this value shifts the onset temperature towards the annealing temperature. However, if the temperature is sufficiently high for crystallization to occur, an extended annealing time will shift the onset temperature to higher values. Thus, the lowest onset temperature observed was 137 °C for an annealing of five days at 100 °C. For an annealing at 140 °C, the onset temperature was lowered to 167 °C after three hours and raised to 193 °C after annealing for four weeks. A sample annealed for three hours at 210 °C shows a glass transition onset of 221 °C. Since the glass transition is only a shallow peak, the error for the onset values is ±5 °C. Table 2.1 summarizes the obtained results.

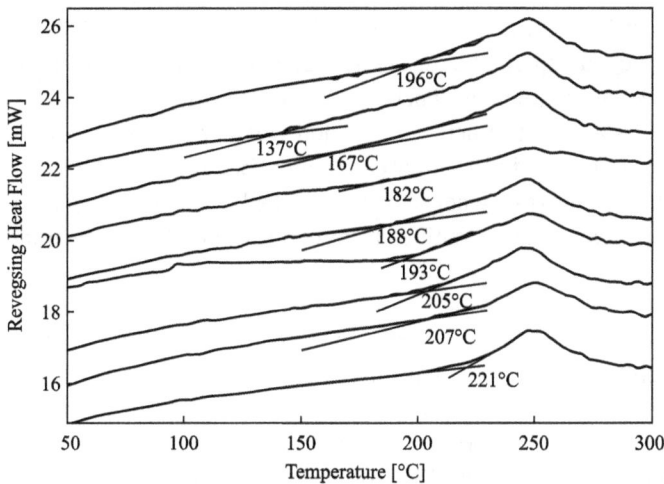

Fig. 2.15: MDSC-measurements of $Al_{89}Y_6Fe_5$ samples, previously subjected to different heat treatments. From top to bottom the pre-treatment conditions were as follows: as-spun, five days at 100 °C, three hours at 140 °C, two weeks at 120 °C, pre-run up to 200 °C, four weeks at 140 °C, ten minutes at 210 °C followed by three hours at 140 °C, ten minutes at 230 °C and three hours at 210 °C.

Table 2.1: Results of MDSC-measurements performed after different annealing treatments

Temperature	Time	T_{onset}
as-spun		196 °C
100 °C	5 days	137 °C
140 °C	3 hours	167 °C
120 °C	2 weeks	182 °C
200 °C	pre-run to 200 °C	188 °C
140 °C	4 weeks	193 °C
210 °C	10 minutes	
+140 °C	+3 hours	205 °C
230 °C	10 minutes	207 °C
210 °C	3 hours	221 °C

The measurements presented here were performed near the temperature region in which the DSC-measurements show the first exothermic signal (compare Fig. 2.14a, point 1 and 2). In order to discern whether the signal is solely caused by structural relaxation or not, the experimental data were fitted to a Kohlrausch–Williams–Watts-function (KWW)

$$H(t) = \Delta H \exp\left(\left(-\frac{t}{\tau} \right)^{\beta} \right)$$
(2.8)

with the enthalpy difference, ΔH, between the initial state and the fully relaxed state, the time, t, relaxation time-constant, τ, and the stretching exponent, β. The KWW-function describes a stretched exponential decay which serves as an envelope function for the individual relaxation functions that contribute to the structural relaxation of a glass [80]. So far, it is a matter of active discussion whether the individual relaxation processes follow a stretched or a simple exponential behaviour, respectively. However, the relaxation processes as a whole can be described by a KWW-function in either case. Therefore, a signal can be identified as pure relaxation if it can be fitted by a KWW-function.

Although in the case of the measurement at 60 °C a fit to a KWW-function is possible, it is not meaningful to do so because of the small magnitude of the signal. As shown in Fig. 2.11, a significant heat release during isothermal treatments at 100 °C and 120 °C allows a good fit by a KWW-function. This is however not the case for the measurement at 140 °C. In XRD-measurements (Fig. 2.13) no Bragg peaks corresponding to aluminium nanocrystals are detected in samples treated at 100 °C. At 120 °C only weak crystallization signals are visible. This indicates that the applied heat treatments do not induce the development of a significant crystalline volume fraction. In contrast, after an annealing treatment of 240 hours at 140 °C, sharp Bragg peaks corresponding to fcc-aluminium are visible. Thus, the good agreement between the KWW-fit and the experimental measurements performed at 100 °C and 120 °C indicates that the exothermic heat-flow signal is related to almost pure structural relaxation. In the case of the isothermal measurement at 140 °C, the underlying crystallization event makes it impossible to fit the measured data with a KWW-function. Thus, the present measurements supply us with direct evidence for crystal formation at temperatures that are up to about 50 °C lower than the glass transition onset of the as-quenched material (as determined by MDSC). On a relative scale, with T_g^{onset} as the reference temperature that marks the onset of the liquid state including liquid-like mobility behaviour, the first crystallization reaction evolves at $T_{rg} = \left(T_{cryst}[K]/T_g^{onset}[K] \right) = 0.88$. The observation of crystallization at such low temperatures is quite unusual since the viscosity of the glass attains high values at such low reduced temperatures that are conventionally related to solid state behaviour. Since crystal formation in metallic glasses (such as the present Al-Y-Fe alloy) necessarily involves long-range diffusion, it is expected that crystal formation proceeds only at higher temperatures that are in the range of the glass transition. Therefore, observing clear evidence for the evolution of new crystallites at such low temperatures directly shows that the underlying processes that govern the structure formation cannot be explained solely on the basis of conventional nucleation and growth in a homogeneous amorphous material. It should be noted that this finding is not in contradiction to the results obtained by alternative processing pathways [81] indicating the importance of quenched-in structures, since the present study focuses on the (smaller) number fraction of nanocrystals that evolve during the very early stages of devitrification.

An analysis of the [111] Bragg peak using the Debye-Scherrer formula yields an average crystal size of 730 nm. However, this value has to be treated with care since the Debye-Scherrer formula can only be accurately applied for average crystal sizes of up to 100 nm [82]. With the measured value of the crystallization enthalpy of $-2.8 \cdot 10^8$ J/m^3 [83], contributing the total heat release of 7.7 J/g to crystallization leads to an upper limit for the transformed volume fraction of 9 %. The volume which can be inspected in an average TEM specimen is about 10^{-18} m^3 (ten micrographs of the dimension 1 μm · 1 μm · 100 nm) and consequently, the particle number density must be greater than 10^{18} m^{-3} for the particles to be readily identifiable. Thus, the particle number density of samples annealed at 140 °C must be lower than 10^{18} m^3 in order to appear completely amorphous in TEM and, at the same time to appear partially crystalline in XRD. Together with a transformed volume fraction of 9 % this upper limit for the particle density results in a lower boundary of 300 nm for the mean particle size, which is in agreement with the Debye-Scherrer analysis, if proper confidence ranges are taken into account.

As shown in Fig. 2.14, the glass transition observed in MDSC is in agreement with the glass transition obtained by advanced conventional DSC-techniques. However, the overall shape of the glass transitions observed in MDSC measurements does not exhibit the expected features. From fundamental works on the glass transition of metallic glasses [61, 75] one expects a step in the heat capacity marked by a small peak. The glass transition observed in MDSC does not show a considerable step in heat capacity. Instead, it consists of a single peak with slow rising flank and a kink just before the peak is reached. In addition to the unusual signal shape, a strong variation of the onset of the glass transition has been observed.

A possible explanation for the shift of the glass transition onset to lower temperatures is given by a phase-separation reaction that occurs during the annealing treatment. In this case, the glass transition would no longer occur homogeneously throughout the sample. Instead, depending on the composition and size of the phase-separated regions [84], a spectrum of overlapping glass transitions would occur and a superposition of these "partial" glass transition signals would lead to the "total" signal observed in MDSC-measurements. Apart from accounting for the rather atypical shape of the glass transition, the phase separation scenario can provide an explanation for the nanocrystallization reaction observed. We suggest that the primary crystallization occurs in phase separated areas that have already undergone a glass transition or that are undergoing the glass transition at the same time. This scenario is illustrated schematically in Fig. 2.16 and also bears resemblance to the "structurally ordered clusters aggregation" model, proposed by Oguni and co-workers [85–87].

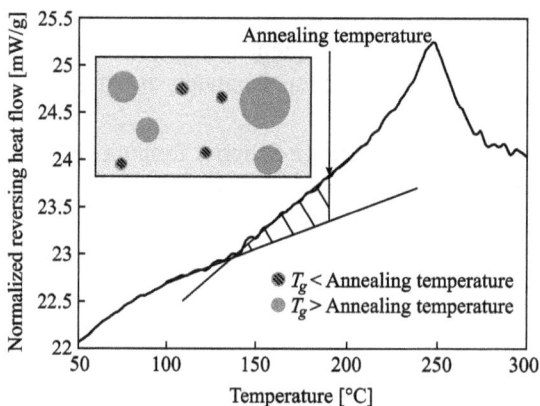

Fig. 2.16: Principle of the proposed crystallization reaction. During the annealing treatment the areas contributing to the glass transition signal in the shaded area crystallize.

In their model, Oguni et al. assumed that the crystallization observed at temperatures below the "standard" glass transition was due to the more mobile molecules between the structurally ordered regions. Yet, it is probably more appropriate to think in terms of two at least partially percolating clusters with different mobilities and not in terms of regions of higher (lower) mobility that are completely surrounded by a matrix of lower (higher) mobility. For the metallic systems studied here, where crystallization below T_g is still at temperatures of the order of half the melting temperature, sufficient mobility for crystallization to occur would also be expected to be available in the structurally ordered regions, and since they require only smaller rearrangements for forming a critical nucleus, the preferred crystallization of these clusters would be anticipated. Yet, this scenario might be reversed for large organic molecules and for situations, where crystallization below T_g occurs at low absolute temperatures that do not allow for molecular rearrangements within the time scale of the experiments if the structure is ordered.

An experimental evidence for this mechanism is given by the good agreement of the rather sharp onset of low temperature primary crystallization as observed by XRD and by microcalorimetry at about 140 °C on the one hand, with the lowest observed glass transition onset temperature observed in MDSC (137 °C) on the other hand. The shift of the glass transition onset to higher temperatures at higher annealing temperatures is also explained by this mechanism. If the annealing temperature exceeds the glass transition temperature of a particular phase separated region, this region crystallizes and no longer contributes to the observed glass transition. In a subsequent MDSC-heating experiment the glass transition onset temperature shifts to higher temperatures, as observed experimentally. Furthermore, this offers an explanation for the wide span of crystal number densities reported throughout the literature: at any given annealing temperature, nucleation of fcc-aluminium crystals only takes place in regions near or above their respective glass transition temperatures.

The idea of decomposition as a precursor to devitrification is not new to the field of metallic glasses [88, 89]. Up to now, the experimental evidences for decomposition in metallic glasses of the Al–TM–RE-group have been ambivalent. Reports on phase separation observed in TEM ($Al_{88}Gd_6La_2Ni_4$) [90] have been shown to be preparation artefacts [91]. 3D-atom probe measurements on $Al_{89}Ni_6La_5$ [92] reveal a large number of structural heterogeneities in as-spun samples in the form of Al-rich and Ni-rich regions that indicate a slight decomposition tendency during rapid quenching. At a first glance and from a thermodynamic point of view, the Al–TM–RE-systems are not the candidates for phase separation. Because of the high aluminum content coupled with negative enthalpies of mixing, the liquid alloys should be in a region of solute solubility. However, the association model for compound-forming metallic melts [93] permits the occurrence of miscibility gaps at high liquid undercooling in alloys with a high negative enthalpy of mixing. Thus, the existence of a miscibility gap in Al–TM–RE systems cannot be discarded up front. In fact, recent XAFS-measurements [94] support this scenario. It is reported that in Al–RE–TM metallic glasses the rare earth atoms are on average surrounded by 16 aluminum atoms. This may be interpreted as the formation of associates of identical atoms in the melt.

Apart from phase separation caused by a miscibility gap, decomposition by compound formation must also be considered. In this case, the structural units found by XAFS [94] can be interpreted as non-crystalline compounds with an average stoichiometry corresponding to $Al_{16}Y$. This scenario does not have the difficulty of explaining its existence at such high aluminum concentrations. Furthermore, the wide span of glass transition temperatures is more readily explainable for short-range ordered regions with different structure, as highlighted by the strong shift of the glass transition temperature of binary Al–RE-glasses such as $Al_{92}Sm_8$, where the glass transition occurs at 172 °C, as reported in the literature [75]. This scenario can explain the observed discrepancies concerning the crystallization kinetics during the early stages of nanocrystal formation in Al-rich metallic glasses and, due to the high undercooling levels normally involved in the crystallization of marginal glasses, might be of wider applicability for a range of different amorphous alloys.

2.3 Deformation-induced Nanocrystal Formation

One new opportunity for enhancing the number density of nanocrystals is presented by severe plastic deformation treatments of amorphous quenching products [95, 96]. In addition to nanostructure formation, the deformation treatment can also serve as a consolidation step, which is important for producing bulk shapes. This new advancement of deformation-based methods of nanostructure formation has been initiated by observations of nanocrystal formation during nano-indentation experiments on metallic glasses [97].

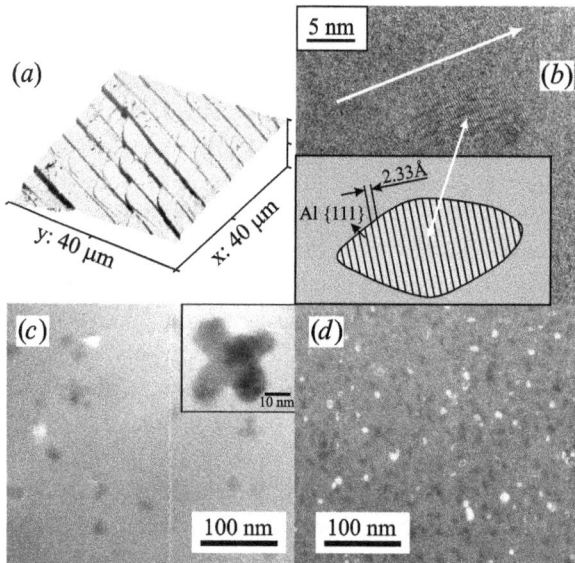

Fig. 2.17: Microstructures of metallic glasses that developed during plastic deformation: (a) shear band formation after cold rolling, imaged by atomic force microscopy. The pattern represents the displacements resulting from slip along the individual shear bands; (b) high resolution transmission electron microscopy image of an $Al_{88}Y_7Fe_5$-alloy that was deformed by cold rolling. The white arrow indicates the in-plane direction of a shear band. The fringe contrast inside the shear band region is due to a nanocrystal that is schematically depicted in the inset; (c) bright-field TEM-image after annealing an as-quenched $Al_{88}Y_7Fe_5$ alloy at 245 °C for 30 min. The nanocrystal number density is drastically lower compared to the material after severe plastic deformation by high-pressure torsion straining, as shown in (d). The dark-field transmission electron microscopy image shows a high number density of Al-nanocrystals with a size of about 12 nm that developed during the deformation within the amorphous matrix.

It is well known that amorphous materials do not show homogeneous plastic deformation upon applying external strain. Instead, the deformation is localized in narrow regions, so-called shear bands that carry the entire response of the material to external stresses [98], as also shown in Fig. 2.17a. While recent experiments indicate that, to a certain extent, the amorphous material might also be affected in its volume by externally applied stresses [99], the major part of the deformation is certainly localized in the shear band regions. These regions are distinctly different from shear bands in plastically deformed nanocrystalline materials: while the latter are the microstructural response to strain localization due to negligible strain hardening capacity and low strain rate sensitivity of the flow stress [100], shear bands in amorphous materials are to be treated as an effective second phase. These shear bands are characterized by a modified and often decreased mass density that, for example, changes the contrast in transmission electron microscopy analyses and that provides for faster atomic transport [101]. Recent measurements on the mass density in shear bands by combining different analytical TEM-methods indicate that the mass density varies from shear band to shear band and even within a shear band from region to region but that the mean density change amounts to a reduction between 1–2% [102]. This value is in excellent agreement with results obtained by acoustic emission spectroscopy [103]. In fact, measurements by radiotracer methods of the diffusion rate within shear bands after the stress has ceased [104] have shown a diffusion rate enhancement by about six orders of magnitude in comparison to the diffusion rate in the undeformed glass at identical temperatures! Since the most excited states will relax most quickly, it is expected that the atomic mobility inside the shear bands during the shear event is even larger.

In a simplified picture, the increase of the diffusion rate and a decrease of the mass density are directly coupled through an increased amount of volume that can be freely distributed in a glass, so-called *free volume* [105, 106], and that allows for atomic rearrangements under applied stress. Thus, the formation of shear bands leads to an effective work-softening, which gives rise to plastic instability in tension and to inadequate fatigue test performance of bulk metallic glasses. In this context it is still actively debated whether the shear softening and related microstructural features, such as the vein pattern on fracture surfaces of amorphous materials, are related to a local temperature rise due to adiabatic heating. Estimates between several tenths of a Kelvin and several thousand Kelvin for the temperature rise of the material within a shear band have been published [107, 108]. While recent studies that utilized a clever and simple local melting probe indicate that the temperature rise during shear band formation is significant and affects a rather large volume of several 100 nm in the direction perpendicular to the shear plane [109], it is not clear how the magnitude of the slip along the shear band affects the heat balance. Independent studies on shear band generation by bending (with significantly smaller shear displacement) find that the microstructural response of the material adjacent to a shear band is different for the compressive and the tensile side [110]. Such a finding is at least unexpected if one would assume a massive heat release during shear band formation. In fact, in-situ straining and annealing experiments within a transmission electron microscope have revealed that, in an Al-base marginal glass-former, shear band initiation at small total shear did not initiate nanocrystal formation. However, after additional thermal activation, the nanocrystallization transformation occurred at significantly reduced temperatures as compared to the regions of the sample outside the shear bands [111]. Thus, at present it appears that both underlying processes – an increase of the amount of free volume and a localized increase of the temperature – contribute to the observed nanocrystal formation at ambient conditions within shear bands.

Interestingly, recent investigations on the deformation of amorphous Al-based alloys indicate that the nanocrystals within shear bands induced by cold rolling have an oblong morphology with the long axis oriented in parallel to the direction of the shear bands (Fig. 2.17b) [99]. The elongated morphology of the deformation-induced nanocrystals, along with the common orientation of the (111) lattice planes relative to the longer axis of the nanocrystals, indicates the importance of the details concerning the redistribution of free volume. In addition to generating nanocrystals by plastic deformation, for partially devitrified Al-rich material it was also shown that plastic deformation caused the disruption of the nanocrystals that exceeded a critical size of about 15 nm [112]. The resulting smaller nanocrystals were found to be dispersed in the matrix. Dislocations that were observed in few remaining larger nanocrystals suggest a shear-induced fragmentation mechanism. These results indicate that plastic deformation can yield nanocrystals within the amorphous matrix and that it can also limit the size of these nanocrystals [112]. Thus, nanocrystalline-amorphous composites with a rather narrow grain size distribution of small nanocrystals and with extremely high nanocrystal number densities might be obtainable if a deformation treatment is applied that affects the volume of the material. A recent investigation on an Al-rich amorphous alloy that was plastically deformed by high-pressure torsion verifies this expectation [95]. Fig. 2.17 also displays representative examples of partially nanocrystallized $Al_{88}Y_7Fe_5$ samples after (c) thermally induced and (d) deformation-induced nanocrystallization. The comparison clearly indicates the enhanced nanocrystal number density that can be obtained by combining rapid quenching and severe plastic deformation pathways sequentially. Additionally, the increase of the nanocrystal number density with the applied strain demonstrates the possibility to produce homogeneous ultrafine nanostructures in bulk shape that yet wait to be explored.

Concerning the early stages of nanocrystal formation, the results obtained on nanocrystallization inside shear bands, particularly the observation that massive growth can occur even a few hundred degrees below T_g, which cannot be explained by diffusion-controlled nucleation and growth in an ideal random structure, substantiates that local structures exist in these materials already after quenching that can be activated thermally or by mechanical energy input to transform into the crystalline state.

2.4 Bulk Metallic Glasses

In order to analyze the scenario for nanocrystal formation in marginal glasses as described above, it might be of interest to compare the behavior to that of bulk metallic glasses concerning the average size range of the dynamic structures that are present in the undercooled liquid in the framework of dynamic heterogeneity. With glass-forming systems of high kinetic stability, the thermodynamic properties of the deeply undercooled liquid state near to the glass transition can be accessed if the characteristic relaxation time and the time to perform the measurement are shorter than the onset time for nucleation. Thus, for some metallic glasses, such as $Pd_{40}Ni_{40}P_{20}$, it is possible to measure e.g. the specific heat in the entire undercooling interval during slow cooling if special precautions to avoid nucleation, such as fluxing, are applied [113].

Above, the early stages of the development of a crystalline structure in glass-forming metallic melts have been discussed with Al-based alloys as representatives of marginal glass-formers. One consequence of these investigations is the indication of precursor structures in the undercooled melt that should be in the size range of the first or second coordination shell. Neutron scattering and X-ray diffraction measurements on similar alloy systems support this conclusion obtained from the kinetic analyses. For glass-forming liquids at high temperatures, i.e. high mobilities, MCT was applied successfully to predict and describe the different modes of molecular relaxation, i.e. α and β processes, respectively. Within this theory, the characteristic non-exponential relaxation is explained by the strongly cooperative motion of molecules in the liquid state that results in a "caging"-effect at higher density or lower temperatures, respectively. For liquid dynamics on a pico-second scale, these dynamic structures have been experimentally observed for a metallic melt by inelastic neutron scattering. For temperatures near the glass transition, i.e. in the deeply undercooled melt, MCT describes the dynamics of the material as glass-like. Yet, thermodynamically, the system is still in the liquid state and macroscopic flow, i.e. liquid-like viscous behavior, governs the response to mechanical shear stress. Within this regime, the dynamics of the liquid slows down dramatically, finally leading to motional freezing that leads to the formation of a noncrystalline solid.

In the framework of so-called dynamic heterogeneity [114], the undercooled liquid state of glass-forming materials is described as a spatially heterogeneous dynamic structure that consists of regions of low and high mobility, respectively. Besides the obvious importance for a deeper understanding of the glass transition, such dynamic structures, and especially the presence of regions with a low mobility in the undercooled melt, might also be important with respect to the development of crystalline structures since the nucleation probability depends sensitively on the time-averaged cluster distribution and the local atomic mobilities. However, for metallic systems, the methods that involve the direct observation of the dynamics in localized spatial regions as well as some well-known spectroscopic methods (e.g. dielectric loss spectroscopy) that probe the intrinsic dynamics directly [115] are not applicable.

Yet, besides mechanical spectroscopy (i.e. frequency-dependent rheology) [116], heat capacity spectroscopy is a powerful tool concerning the monitoring of the intrinsic relaxation times of the undercooled melt since the dynamic specific heat couples to all modes of molecular movement in contrast to e.g. dielectric loss spectra that probe the response of polarization-sensitive modes only. An important drawback of heat capacity spectroscopy is its restriction in applicable frequencies due to the thermal momentum of the measurement devices and due to the limited heat transfer kinetics.

After alloying and flux-cleaning in dehydrated B_2O_3, the Pd-based alloys allow the formation of bulk amorphous rods of 4 mm in diameter and 30 mm in length by injection casting into a copper mold. The vitrification of much larger volumes is possible, however not necessary for the present investigations. Small disks of about 100 mg in mass were cut from these rods for the TMDSC-experiments. The absence of detectable crystalline fractions within the samples was verified by X-ray diffraction using (Mo-K_α)-radiation. Temperature-modulated heat capacity measurements were performed with a DSC-device (Perkin Elmer, Pyris 1) operating in the modulation mode. The specific heat measurements as well as the temperature scale for the temperature-modulated measurements were calibrated by measuring the TMDSC-response of a pure (99.999 % purity) Ni-sample in the range of its Curie temperature. Time periods for one heating rate oscillation ranged from 30 s to 1000 s with amplitudes of the corresponding temperature oscillations always less than 1 K. The TMDSC-measurements were performed during cooling the metastable melt from T = 623 K, i.e. well above the kinetic glass transition, at low average cooling rates that ranged from 0.05 to 0.5 K/min. Thus, it was ensured that the kinetic glass transition occurred always at temperatures that are below the temperature range where the frequency-dependent dynamic glass transition was observed. Subsequent XRD-measurements confirmed the absence of detectable crystalline fractions in the samples.

Fig. 2.18 shows the experimentally obtained data for C' and C'' together with the smoothed signal of the underlying heat flow rate, $\bar{\varphi}(T)$. It is apparent that the kinetic glass transition that is identified on the $\bar{\varphi}(T)$-curve occurs at temperatures well below the interval where the dynamic glass transition is observed on the C' and C'' signals [117]. The curves that are drawn together with the C' and C'' data represent the respective real and the imaginary parts of a complex Cole-Davidson *stretched exponential* function in the temperature domain [118]:

$$C_\omega(T) = C_\infty(T) + [C_s(T) - C_\infty(T)]\left(1 + i\tau(T)\, 2\pi t_p^{-1}\right)^{-\beta},\qquad(2.9)$$

$$\tau(T) = \tau_0 \exp\left(-\frac{B}{T}\right)$$

with C_∞, the contribution of quickly equilibrating degrees of freedom, t_p, the duration of one oscillation period, β, the stretching exponent, $\tau_0 = 1.5 \cdot 10^{35}$ s, a scaling factor and (B = 50,000 K) an activation term. The entire set of data can be fitted by the real and imaginary parts of this function with one set of parameters, as shown in Fig. 2.19, except for β that varies somewhat between 0.65 and 0.74 due to the variation of the

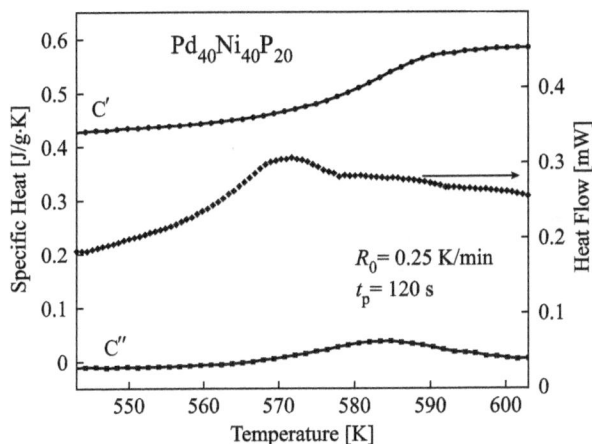

Fig. 2.18: C' and C'' of $Pd_{40}Ni_{40}P_{20}$ obtained during cooling. The solid curves that are drawn with the C' and C''-data represent the real and the imaginary part of a Cole-Davidson stretched exponential function, respectively. The smoothed heat-flow signal (closed diamonds) was obtained during heating at the same underlying rate and applying identical modulation parameters. The corresponding curve, obtained during cooling, would be shifted to even lower temperatures but would appear less characteristic.

width of the curves at different frequencies. The results shown in Fig. 2.19 clearly indicate that the experimentally obtained data obeys the Kramers-Kronig relations, as required from irreversible thermodynamics.

Furthermore, it is evident that a Debye relaxation law (with $\beta = 1$) does not fit the data – the peaks in C'' are wider than those obtained from a relaxation process with a single characteristic relaxation time indicating a distribution of relaxation times, a result that is consistent with the *dynamic heterogeneity* of glasses [114]. However, in order to fit Eq. (2.9) to the experimental data, a distinct function describing the temperature dependence of the average relaxation time needs to be assumed. Besides the fact that the values of the parameters B and τ_0 that are found to fit the data do not have any physical relevance, it is the temperature dependence of the average relaxation time and the shape of $C''(\omega)$ that are the characteristic properties concerning liquid dynamics. Therefore, the data that are obtained in dependence on temperature at a constant temperature-oscillation frequency, v, have been transformed into frequency-dependent C''-values at constant temperature. In the frequency domain, the respective DC-functions

$$C^*(v) = C_\infty + [C_s(T) - C_\infty(T)] \left[1 + i\left(\frac{v}{v_p}\right)\right]^{-\beta} \tag{2.10}$$

give the average relaxation time, $\tau_a = 1/v_p$, directly. The results for $C''(v)$ are shown in Fig. 2.19 together with the calculated imaginary parts of the respective DC-functions. The resulting peaks are about 1.5 decades wide, i.e. they are significantly broader than

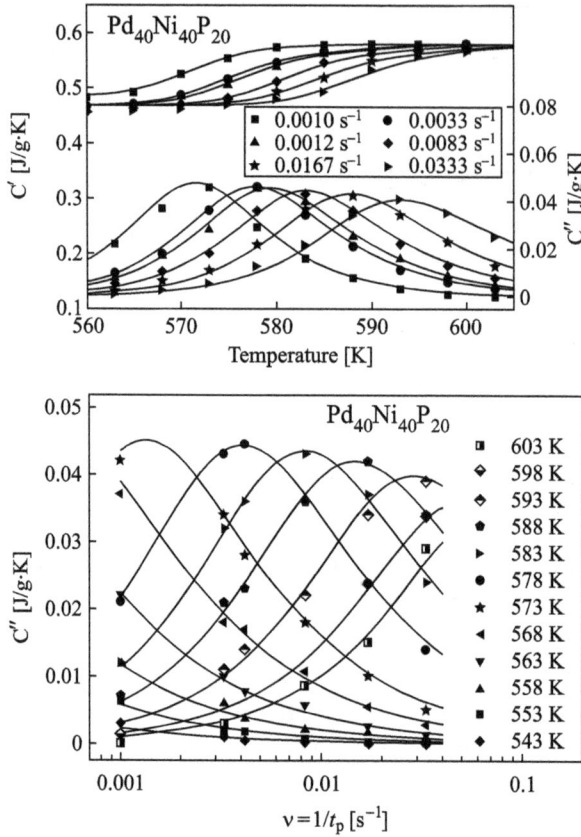

Fig. 2.19: Top: Reduced set of experimental data (symbols) and the real and imaginary parts of the respective DC-fits (solid lines) displaying the TMDSC-results in the entire range of temperature-modulation frequencies applied during cooling. Some data at intermediate frequencies have been omitted for clarity. Bottom: Isothermal $C''(v)$-curves obtained by transforming the experimental data shown above.

a Debye function. Similar widths have been reported for C''-curves obtained by HCS on non-metallic glass formers [115]. Thus, the $(T - \tau_p)$-data field can be extended (at a lower accuracy) by fitting the low- and high-frequency tails of incomplete dispersion data, as indicated in Fig. 2.19, e.g. for the curve corresponding to $T = 563$ K [119].

Adopting the fluctuation concept developed by Donth [120], a characteristic length (i.e. a correlation length), ξ_a, can be obtained from the width of the $C''(T)$-curves, δT. According to this framework, the characteristic correlation length for the size of molecular cooperativity can be obtained as

$$\xi_a^3 = k_B T^2 \left(\frac{\Delta C_p}{\bar{C}_p^2} \right) \rho \delta T^2 \qquad (2.11)$$

with the density, ρ, the average specific heat of undercooled liquid and glass, $\bar{C}_p$, the step height of C_p at T_g, ΔC_p and the Boltzmann-constant, k_B. Since all parameters are available from the TMDSC-experiments and the measurements of the specific volume [121], ξ_a could be determined. For different temperatures, i.e. TMDSC-curves at different frequencies, values for ξ_a were found in the range between 1 to 2 nm. These values are smaller than the respective values obtained on polymers, which seems reasonable accounting for the different nature of molecular binding forces in the respective materials.

2.5 Conclusions and Hypotheses

Detailed investigations by structural and thermodynamic methods have shown that true glass-formation is achieved for marginally glass-forming systems by melt-quenching and mechanical alloying and that the large number densities of nanocrystals originate from quenched-in nuclei. The thermodynamic assessment as well as results from scattering experiments indicate that inherent dynamic structures in the undercooled melt modify the cluster distribution such that copious nucleation is favored. This assessment is also supported by the observation of nanocrystal formation in shear bands at low homologous temperatures. The preservation of the nanocrystal dispersion that is embedded in the residual amorphous matrix is facilitated by the sluggish diffusion-controlled growth. These results that should hold in general for marginally glass-forming systems indicate that a high initial number density of nuclei is the key for synthesizing structures with even higher nanocrystal densities.

Bulk glass-forming alloys, on the other hand, provide the opportunity to investigate the thermodynamic properties of deeply undercooled metallic liquids as well as their relaxation behavior in detail. The strong increase of the viscosity near the glass transition is one indication of the presence of inherent dynamic structures in the melt that have an increasing correlation length with decreasing temperature. The characteristics of these regions of cooperative molecular motion, as well as an estimate of the respective correlation length, were obtained from TMDSC-measurements, indicating a characteristic correlation length of the order of one nanometer. With bulk metallic glasses, these regions of cooperative dynamics lead to vitrification and the characteristic glassy dynamics of the deeply undercooled liquid state. However, with marginally glass-forming systems the existence of more strongly correlated molecular assemblies of the order of one nanometer can modify the delicate kinetic balance towards copious nucleation. Thus, both the formation of crystalline structures and the formation of the glassy state might depend on the specific molecular interactions that are characteristic for glass-forming liquids and that are summarized in the framework of dynamic heterogeneity.

Acknowledgement: The author gratefully acknowledges the stimulating discussions with Prof. J. H. Perepezko and the fruitful work with Drs. Nancy Boucharat and Harald Rösner. The support of large parts of this work by the Deutsche Forschungsgemeinschaft is most gratefully acknowledged.

Bibliography

[1]   W. Klement, R.H. Willens, and P. Duwez, Nature **187**, 869 (1960).
[2]   A. Inoue, X.M. Wang, and W. Zhang, Reviews Advanced Materials Science **18**, 1 (2008).
[3]   S. Sachdev and D.R. Nelson, Phys. Rev. **B 32**, 4592 (1985).
[4]   E. Donth, Acta Polym. **30**, 481 (1979).
[5]   C. Dasgupta, A.V. Indrani, S. Ramaswamy, and M.K. Phani, Europhys. Lett. **15**, 307 (1991).
[6]   F. Sommer, J. Non-Crystalline Solids **117–118**, 505 (1990).
[7]   Y.Q. Cheng, E. Ma, and H.W. Sheng, Phys. Rev. Lett. **102**, 245501 (2009).
[8]   H. Teichler, Phys. Rev. Lett. **76**, 62 (1996).
[9]   W. Kob and H.C. Andersen, Phys. Rev. E **51**, 4626 (1995).
[10]   C.A. Angell, J. Phys. Chem. Solids **49**, 863 (1988).
[11]   W. von Oldekop, Glastechnische Berichte **30**, 8 (1957).
[12]   W.T. Laughlin and D.R. Uhlmann, J. Phys. Chem. **76**, 2317 (1972).
[13]   D.N. Perera, J. Phys.: Condens. Matter **11**, 3807 (1999).
[14]   Y. He, S.J. Poon, and G.J. Shiflet, Science **241**, 1640 (1988).
[15]   J.H. Perepezko and G. Wilde, J. Non-Crystalline Solids **274**, 271 (2000).
[16]   G. Wilde, N. Boucharat, R.J. Hebert, H. Rösner, S. Tong, and J.H. Perepezko, Adv. Engr. Mater. **5**, 125 (2003).
[17]   R. Sahu, A.K. Gangopadhyay, K.F. Kelton, S. Chatterjee, and K.L. Sahoo, Scripta Mater. **61**, 588 (2009).
[18]   J. Bokeloh, N. Boucharat, H. Rösner, and G. Wilde, Acta Mater. **58**, 3919 (2010).
[19]   K. Suzuki, N. Kataoka, A. Inoue, and T. Masumoto, Mater. Trans. JIM **32**, 93 (1991).
[20]   A.L. Greer, Science **267**, 1947 (1995).
[21]   J.H. Perepezko, in: R.D. Shull, J. Joshi (Eds.), *Thermal Analysis in Metallurgy* (TMS, Warrendale, PA, 1992, p. 121).
[22]   H.W. Kui and D. Turnbull, Appl. Phys. Lett. **47**, 796 (1985).
[23]   L. Battezzati, M. Baricco, P. Schumacher, W.C. Shih, and A.L. Greer, Mater. Sci. Eng. **A 179–180**, 600 (1994).
[24]   Y. He, S.J. Poon, and G.J. Shiflet, Science **241**, 1640 (1988).
[25]   J.H. Perepezko and G. Wilde, Ber. Bunsenges. Phys. Chem. **102**, 1974 (1998).
[26]   J.W. Cahn, Bull. Alloy Phase Diag. **1**, 27 (1980).
[27]   W.J. Boettinger, in: B.H. Kear, B.C. Giessen, M. Cohen (Eds.), *Rapidly Solidified Amorphous and Crystalline Alloys* (North-Holland, Amsterdam, 1982, p. 15).
[28]   A. Inoue, Mat. Sci. Forum **179–181**, 691 (1995).
[29]   W.L. Johnson, Prog. Mater. Sci. **30**, 81 (1986).
[30]   P. Rizzi, C. Antonione, M. Baricco, L. Battezzati, L. Armelao, E. Tondello, M. Fabrizo, and S. Daolio, Nanostruct. Mater. **10**, 767 (1998).
[31]   G. Wilde, H. Sieber, and J.H. Perepezko, J. Non-Crystalline Solids **250–252**, 621 (1999).
[32]   G. Wilde, H. Sieber, and J.H. Perepezko, Scr. Mater. **40**, 779 (1999).
[33]   W.J. Boettinger and S.R. Coriell, in: *Science and Technology of the Undercooled Melt* (NATO ASI Series, Dordrecht, E-No. 114, 1986, p. 81).

[34] L.F. Mondolfo, in: *Aluminum Alloys: Structure and Properties* (Butterworth's, London, 1976, p. 376).

[35] D.R. Allen, J.C. Foley, and J.H. Perepezko, Acta Mat. **46**, 431 (1998).

[36] K. Hono, Y. Zhang, A.P. Tsai, A. Inoue, and T. Sakurai, Scripta Met. **32**, 191 (1995).

[37] J.H. Perepezko and G. Wilde, J. Non-Crystalline Solids **274**, 271 (2000).

[38] A. Inoue, T. Zhang, and T. Masumoto, J. Non-Crystalline Solids **156–158**, 866 (1993).

[39] K. Hono, Y. Zhang, A. Inoue, and T. Sakurai, Mater. Trans. JIM **36**, 909 (1995).

[40] D.E. Coates, Metall. Trans. **A3**, 1203 (1972).

[41] D.E. Coates, Metall. Trans. **A4**, 1077 (1973).

[42] A. Inoue, Mater. Trans. JIM **36**, 866 (1995).

[43] D.R. Allen, J.C. Foley, and J.H. Perepezko, Acta Mat. **46**, 431 (1998).

[44] K. Hono, Y. Zhang, A.P. Tsai, A. Inoue, and T. Sakurai, Scripta Met. **32**, 191 (1995).

[45] F.S. Ham, J. Phys. Chem. Solids **6**, 335 (1958).

[46] S.K. Das, J.H. Perepezko, R.I. Wu, and G. Wilde, Mat. Sci. Eng. **A 304**, 159 (2001).

[47] P. Rizzi, C. Antonione, M. Baricco, L. Battezzati, L. Armelao, E. Tondello, M. Fabrizio, and S. Daolio, Nano-Struct. Mater. **10**, 767 (1998).

[48] K.F. Kelton, T.K. Croat, A.K. Gangopadhyay, L.Q. Xing, A.L. Greer, M. Weyland, X. Li, and K.J. Rajan, J. Non-Crystalline Solids **317**, 71 (2003).

[49] D. Turnbull, in: Prog. Mat. Sci. – Chalmers Ann. Vol. 1981. J.W. Christian et al. (Eds.), Pergamon, 1981, p. 269.

[50] A.P. Tsai, T. Kamiyama, Y. Kawamura, A. Inoue, and T. Masumoto, Acta Mater. **45**, 1477 (1997).

[51] H.Y. Hsieh, B.H. Toby, T. Egami, Y. He, S.J. Poon, and G.J. Shiflet, J. Mater. Res. **5**, 2807 (1990).

[52] J.C. Foley, D.R. Allen, and J.H. Perepezko, Scripta Mater. **35**, 655 (1996).

[53] J.H. Perepezko, R.J. Hebert, W.S. Tong, J. Hamann, H. Rösner, and G. Wilde, Materials Transactions JIM **44**, 1982 (2003).

[54] M. Reading, D. Elliot, and V.H. Hill, J. Therm. Analysis **40**, 941 (1993).

[55] J.E.K. Schawe, Thermochim. Acta **260**, 1 (1995).

[56] G. Wilde, R.I. Wu, and J.H. Perepezko, Advances Solid State Phys. **40**, 391 (2000).

[57] H. Ehrhardt, J. Weissmüller, and G. Wilde, Mat. Res. Soc. Symp. Proc. (2013), in print.

[58] S.L. Lai, J.Y. Guo, V. Petrova, G. Ramanath, and L.H. Allen, Phys. Rev. Lett. **77**, 99 (1996).

[59] J.C. Foley, PhD-Thesis, Madison, (1997).

[60] H.S. Chen, Mat. Sci. Eng. **23**, 151 (1976).

[61] W. Kauzmann, Chem. Rev. **43**, 2191 (1948).

[62] R. Busch, JOM **7**, 39 (2000).

[63] Z. Altounian, S. Saini, J. Mainville, and R. Bellissent, Physica **B 241–243**, 915 (1998).

[64] T. Egami, J. Non-Crystalline Solids **205–207**, 575 (1996).

[65] T. Egami, Mat. Sci. Eng. **A 226–228**, 261 (1997).

[66] M. Kusy, P. Riello, and L. Battezatti, Acta mater. **52**, 5031 (2004).

[67] B. Radiguet, D. Blavette, N. Wanderka, J. Banhart, and K.L. Sahoo, Appl. Phys. Lett. **92**, 103126 (2008).

[68] F.C. Frank, Proc. Royal Society **A 201**, 586 (1950).

[69] F.S. Ham,. J. Phys. Chem. Solids **6**, 335 (1958).

[70] D.R. Allen, J.C. Foley, and J.H. Perepezko, Acta mater. **46**, 431 (1998).

[71] J. Bartels, U. Lembke, R. Pascova, J.W.P. Schmelzer, and I. Gutzow, J. Non-Crystalline Solids **136**, 181 (1991).

[72] J.W.P. Schmelzer, A.R. Gokhman, and V.M. Fokin, J. Colloid and Interface Science **272**, 109 (2004).

[73] J.W.P. Schmelzer, G.S. Boltachev, and V.G. Baidakov, J. Chem. Phys. **124**, 194503 (2006).

[74]   J.W.P. Schmelzer, V.M. Fokin, A.S. Abyzov, E.D. Zanotto, and I. Gutzow, International Journal of Applied Glass Science 1, 16 (2010).

[75]   R.I. Wu, G. Wilde, and J.H. Perepezko, Mater. Sci. Eng. A 301, 12 (2001).

[76]   M. Kusy, P. Riello, and L. Battezatti, Acta mater. 52, 5031 (2004).

[77]   S. Saini, A. Zaluska, and Z. Altounian, J. Non-Crystalline Solids 250, 714 (1999).

[78]   X.Y. Jiang, Z.C. Zhong, and A.L. Greer, Phil. Magazine B 76, 419 (1997).

[79]   H. Nitsche, F. Sommer, and E.J. Mittemeijer, J. Non-Crystalline Solids 351, 3760 (2005).

[80]   M.D. Ediger, C.A. Angell, and S.R. Nagel, J. Chem. Phys. 100, 13200 (1996).

[81]   G. Wilde, H. Sieber, and J.H. Perepezko, Scripta Mater. 40, 779 (1999).

[82]   C.E. Krill and R. Birringer, Phil. Magazine 3, 621 (1998).

[83]   F.C. Foley, E.R. Allen, and J.H. Perepezko, Scripta Mater. 35, 655 (1996).

[84]   G. Barut, P. Pissis, R. Pelster, and G. Nimtz, Physical Review Letters 80, 3543 (1998).

[85]   H. Fujimori and M. Oguni, Solid State Commun. 94, 153 (1995).

[86]   M. Oguni, J. Non-Crystalline Solids 210, 171 (1997).

[87]   S. Tomitaka, M. Mizukami, F. Paladi, and M. Oguni, J. Thermal Analysis and Calorimetry 81, 637 (2005).

[88]   J.F. Löffler and W.L. Johnson, Scripta mater. 44, 1251 (2001).

[89]   S.C. Glade, J.F. Löffler, S. Bossuyt, W.L. Johnson, and M.K. Miller, J. Appl. Phys. 89, 1573 (2001).

[90]   A.K. Gangopadhyay, T.K. Croat, and K.F. Kelton, Acta mater. 48, 4035 (2000).

[91]   B.B. Sun, Y.B. Wang, J. Wen, H. Yang, M.L. Sui, J.Q. Wang, and E. Ma, Scripta mater. 53, 805 (2005).

[92]   B. Radiguet, D. Blavette, N. Wanderka, J. Banhart, and K.L. Sahoo, Appl. Phys. Lett. 92, 103126 (2008).

[93]   F. Sommer, Z. Metallkunde 73, 77 (1982).

[94]   W. Zalewski, J. Antonowicz, R. Bacewicz, and J. Latuch, J. Alloy. Comp. 468, 40 (2009).

[95]   N. Boucharat, R.J. Hebert, H. Rösner, R.Z. Valiev, and G. Wilde, Scripta Mater. 53, 823 (2005).

[96]   N. Boucharat, H. Rösner, R. Valiev, and G. Wilde, Arch. Mater. Sci. 25, 357 (2005).

[97]   H. Chen, Y. He, G.J. Shiflet, and S.J. Poon, Nature 367, 541 (1994).

[98]   A.S. Argon, Acta Metall. 27, 47 (1979).

[99]   R.J. Hebert, J.H. Perepezko, H. Rösner, and G. Wilde, Scripta Mater. 54, 25 (2006).

[100] Q. Wei, D. Jia, K.T. Ramesh, and E. Ma, Appl. Phys. Lett. 81, 1240 (2002).

[101] Q.Y. Zhu, A.S. Argon, and R.E. Cohen, Polymer 42, 613 (2001).

[102] H. Rösner, M. Peterlechner, C. Kübel, and G. Wilde, submitted to Ultramicroscopy.

[103] D. Klaumünzer, A. Lazarev, R. Maaß, F.H. DallaTorre, J.F. Löffler, and A. Vinogradov, Physical Review Letters 107, 185502 (2011).

[104] J. Bokeloh, G. Reglitz, S. Divinski, and G. Wilde, Physical Review Letters 107, 235503/1-5 (2011).

[105] M.H. Cohen and D. Turnbull, J. Chem. Phys. 31, 1164 (1959).

[106] M.H. Cohen and G.S. Grest, Phys. Rev. B 20, 1077 (1979).

[107] K.M. Flores and R.H. Dauskardt, J. Mater. Res. 14, 638 (1999).

[108] H.A. Bruck, A.J. Rosakis, and W.L. Johnson, J. Mater. Res. 11, 503 (1996).

[109] J.J. Lewandowski and A.L. Greer, Nature Mater. 5, 15 (2006).

[110] W.H. Jiang and M. Atzmon, Acta Mater. 51, 4095 (2003).

[111] G. Wilde and Rösner, Applied Physics Letters 98, 251904 (2011).

[112] G. Wilde, N. Boucharat, R. Hebert, H. Rösner, and R. Valiev, Mater. Sci. Eng. A 449–451, 825 (2007).

[113] G. Wilde, G.P. Görler, R. Willnecker, and G. Dietz, Appl. Phys. Lett. 65, 397 (1994).

[114] J. Jäckle, Phys. Blätter 52, 351 (1996).

[115] N.O. Birge, Phys. Rev. **B 34**, 1631 (1986).

[116] K. Schröter, G. Wilde, R. Willnecker, M. Weiss, K. Samwer, and E. Donth, Eur. Phys. J. **5**, 1 (1998).

[117] G. Wilde, Appl. Phys. Lett. **79**, 1986 (2001).

[118] O. Bustin, M. Descamps, J. Chem. Phys. **110**, 10982 (1999)

[119] G. Wilde, J. Non-Crystalline Solids **312–314**, 49 (2002).

[120] M. Beiner, J. Korus, H. Lockwenz, K. Schröter, and E. Donth, Macromolecules **29**, 5183 (1996).

[121] G. Wilde, G.P. Görler, R. Willnecker, and H.J. Fecht, J. Appl. Phys. **87**, 1141 (2000).

Ivan Gutzow, Radost Pascova, Nikolai Jordanov, Stoyan Gutzov,
Ivan Penkov, Irena Markovska, Jürn W.P. Schmelzer, and
Frank-Peter Ludwig

3 Crystalline and Amorphous Modifications of Silica: Structure, Thermodynamic Properties, Solubility, and Synthesis

An analysis of the main physical properties of the crystalline and amorphous modifications of silica is given. Hereby the main emphasis is directed to their thermodynamic characterization, determining their location in the phase diagram, their synthesis and possible applications. An approximative method is developed allowing one to estimate the standard entropy and enthalpy of all the modifications of silica, based on the knowledge of their respective density data, only. In this way the thermodynamic properties even of those silica modifications can be computed with sufficient accuracy, for which at present the corresponding measurements are lacking or have given controversial results. Particular attention is also devoted to possibilities of determining the solubility of the silica modifications in water and aqueous solutions at hydrothermal conditions. Thus, new possible variants of hydrothermal synthesis are anticipated. In addition, various methods of synthesis of one of the crystalline forms of silica – of cristobalite – are analyzed in detail. This interest is caused, at part, by the particular properties both its modifications have: the negative value of the Poisson coefficient of its α-modification and the high temperature resistivity of its β-form. A detailed analysis of pre-activated sinter-crystallization, of sol-gel methods of its formation is given together with estimates of the utility of classical hydrothermal synthesis routes of this modification. The problems of silica deposits, their mining and possible extraction of SiO_2 from industrial or natural waste products, even of plant origin, are also discussed at the final part of the present review.

3.1 Introduction

In the contemporary international scientific literature claims can be found on the supposed existence of more than forty different modifications (or better *structural varieties*) of silicon dioxide, SiO_2, traditionally called *silica*. More or less well-defined and recognized modifications of SiO_2 at normal pressure are, however, only the three well-known crystalline modifications (quartz, tridymite, cristobalite) and quartz glass.

Both quartz and cristobalite have low temperature (α) and high temperature (β) modifications. Moreover, for tridymite several, usually three modifications (α, β, and γ) are under discussion. However, in present day literature the opinion prevails that tridymite has not to be considered as a "full constituent of the SiO_2-family" and of the crystalline silica modifications: its existence and stability are due to the influence of different chemical stabilizers, in most cases alkaline oxides. This is the reason why in the present contribution only relatively little place is given to the description of this SiO_2 modification (or better structural variety) of silica. Nevertheless, tridymite has an enormous significance in several industrial applications, mainly connected with the production of refractory materials for glass industry and metallurgy (the so-called "tridymitization" of DINAS-refractories).

In the mineralogical literature, the structure and the properties of a great variety of SiO_2-forms (crystalline, amorphous or semi-crystalline) are discussed. Some of these varieties are of significance as gemstones (mostly different vitro-crystalline variants of quartz-like silica, sometimes with a relatively high water content) like chalcedony, opals etc. or as the SiO_2-constituent of magmatic rocks. In the present review, we consider, however, mainly (or even only) those pure SiO_2-modifications (crystalline or amorphous) having more or less distinct positions in the thermodynamically more or less precisely founded phase diagrams at both normal and especially at increased pressure, p.

With such restriction, from an academic point of view the two high-pressure, high-density crystalline modifications of SiO_2, coesite and stishovite, are of particular significance. They possess a well-defined structure, significantly differing from that of the other modifications of SiO_2 and thus they also display specific thermodynamic properties. This statement applies especially to stishovite. Many investigations have been devoted to this modification in the last twenty years. A third high-pressure modification, keatite, was artificially synthesized at the end of the 1960s. It is assumed widely that it (similarly to tridymite) only exists in the presence of stabilizing dopants. Its chemically stabilized structure is intermediate between those of quartz and coesite. In the present review, also relatively little attention is devoted to it.

Of particular geophysical and even cosmological significance are the newly described in the literature, the so-called post-stishovite, silica modifications. They are at present only artificially synthesized at colossal pressures (80–100 GPa). Most probably, they are of importance for the structure of the Earth's mantle and especially for the propagation of earthquakes etc. It is assumed in geology that stishovite is synthesized in nature in the depths of the Earth at 300 km below the surface of our planet. The structure of these new high-pressure phases of SiO_2 and especially of stishovite can be compared with that of some of the structural forms of elemental carbon – with diamond, in particular. In some respects only carbon has so numerous structural varieties as shown by SiO_2: graphite, diamond, fullerenes, nanotubes, vitreous carbon, and the still experimentally not realized hypothetical structure of liquid carbon. The newly synthesized post-stishovite modifications of SiO_2 are (as well as stishovite itself) the hardest known oxides, matching in hardness only with diamond and with cubic boron nitride. In eliminating the mentioned mineralogical varieties of quartz and the amorphous modifications of SiO_2 as well as structures like tridymite and keatite (existing only in the presence of stabilizing dopants), the crystalline modifications of SiO_2 are reduced to only eight properly defined modifications. In geophysics and in Earth's geology, they seemingly play a role similar to that of carbon and its organic compounds in living organisms.

In discussing the structure of the SiO_2-modifications, usually its analogy with the structure of water and of the known seven or eight crystalline modifications of ice is underlined. Water, similarly to SiO_2 and carbon, plays an essential role in geophysics, cosmophysics and life. In this sense, it is also interesting to mention here the great number of investigations connecting the structure of the crystalline modifications of SiO_2 with those of water. The interplay between SiO_2 and H_2O is of great significance for both the synthesis of the different modifications of SiO_2 and for the existence of life on our planet. Particular interest in Earth science literature of recent years is also given to the mechanism of penetration of water through the most densely structured SiO_2-modifications: stishovite and post-stishovites. The distinguishing feature of stishovite and of its higher pressure analogues consists in the fact that in them the highest possible density of structural packing of oxygen is realized.

The present review is organized in the following way: First, we outline a brief description of how we performed our literature search (Section 3.2). As the next step, the general phase diagram of SiO_2 is discussed (Section 3.3). It is followed by the analysis of existing or possible methods of synthesis of all modifications of pure SiO_2, beginning with those stable at normal pressure: quartz, cristobalite and of silica glass (Section 3.4). Then the particular thermodynamic properties of both the "normal" and high pressure modifications of SiO_2 are discussed (Section 3.5). In this analysis, special attention is devoted to those thermodynamic properties (Section 3.5), which are of importance either for the synthesis or for industrial applications of the existing SiO_2-modifications. As far as all the crystal forms of SiO_2 can be synthesized hydrothermally, particular emphasis is given to the thermodynamics of the SiO_2 aqueous solutions. Special attention is devoted to the possibilities of cristobalite synthesis and of the stabilization of its high-temperature β-modification. Both modifications of cristobalite display exceptional properties and in this respect both "ceramic" and sol-gel methods of its synthesis, known from other fields of silicate technology, have been of particular interest. Traditionally the methods of quartz synthesis are usually discussed in more details in reports, connected with the different modifications of silica. Knowing the high importance of quartz single crystals in present day optics and instrument-building, we have also included a brief historical survey on the hydrothermal synthesis of quartz.

Hydrothermal synthesis in its different variants allows one to synthesize practically all eight above mentioned crystalline modifications of silica. A thorough knowledge of the thermodynamic properties of all the crystalline and amorphous modifications of silica is a necessary requirement for realization of any of the possible ways of their synthesis. However, the analysis and description of thermodynamics of silica in general and of any of its modifications in particular is a difficult task: high temperatures and extreme pressures are required. The determination of solubility data of the silica modifications at hydrothermal conditions turns out to be an even more difficult task. It involves high temperatures (around and above 1000 °C) and pressures in the GPa-region (as in the case of stishovite synthesis). This is the reason why, in the framework of the present analysis, we developed a particular semi-empirical method to calculate the solubility of all the crystalline and amorphous modifications of SiO_2. In this approach, we employ only one structurally significant property, the density, ρ. This method, based on the generic application of the thermodynamics of known SiO_2-modifications is developed in details in Section 3.6. It is performed in order to calculate the solubility and to predict a way for the hydrothermal synthesis even of the less investigated modifications, for which only their density, ρ, is known with sufficient accuracy.

Particular attention in this generalized analysis of the hydrothermal synthesis of the crystalline silica modifications is given to the solubility of silica glass and to the possible synthesis of quartz, cristobalite, keatite, coesite from aqueous solutions of silica glass. The possibilities for production of stishovite (and may be even of post-stishovite) in an industrial scale via the hydrothermal way are also analyzed. Such synthesis could be eventually of practical importance taking into consideration the extraordinarily interesting mechanical properties of substances like the low-temperature α-cristobalite or of stishovite and post-stishovite.

The main source of silica are the geochemically well investigated quartz deposits of both hydrothermal and magmatic origin. In industrial quantities, they are known to exist in Brazil and in the Ural mountains in Russia. The world's industrial production of the respective pure silica modifications is depending mainly on these deposits. In the last section of the present review (Section 3.7), also some information is summarized on another possibility to obtain silica in industrial scales: out of natural plant products, containing a relatively high percentage of SiO_2 and of natural waste products, like rice husk, as possible raw materials for high purity silica production. A brief overview on some particularly interesting properties of the different silica modifications (Section 3.8) and the discussion of the results (Section 3.9) completes the chapter.

As far as the chemical nature of silicon (Si-Si) and silica (Si-O) bonding and of the structural and chemical similarities of carbon compounds and of silicates are given in greater details in the review of Dr. Irina G. Polyakova (Chapter 4 in the present monograph), we do not include here a discussion of these problems.

3.2 Properties of Silica Modifications: Literature Search

The present review is based on a thorough literature search in which we used three different channels to obtain the information about the properties, the synthesis, the thermodynamics and the general physicochemical properties of the various modifications of SiO_2. First a search was made through the classical reference literature, connected with SiO_2 and silicates.

3.2.1 Classical SiO_2-literature

The physical chemistry of SiO_2-modifications, silicates and glasses, of natural and synthetic silicates are refereed in Eitel's book, "The Physical Chemistry of Silicates", in its two editions: from 1954 and from 1976 [1, 2]. Further on comes the book of Iler [3] on the chemistry of silica and Toropov's reference books on the phase diagrams of silicate systems [4, 5]. Also we took into account Levy's well known classical reference book on phase diagrams of silicates.

The data on the thermodynamic properties of the SiO_2-modifications were mainly taken from Landolt-Boernstein [6, 7] and from the Russian series of reference books on the thermodynamics of inorganic substances, published under the editorship of Glushko [8] and compared with Barin and Knacke's thermodynamic reference data book [9]. Several of the classical books on the properties of silica [10, 11], published in Russian literature, summarizing the results obtained in the Leningrad (St. Petersburg) Institute of Silicate Chemistry, were also helpful. Of general use were also the two volumes of Hinz's book [12], "Silikate". The books "The Thermodynamics of Silicates" by Babushkin, Matveev and Mchedlov – Petrosyan [13], the book by Gutzow and Schmelzer on the properties and thermodynamics of the vitreous state [14] and Morey's "Properties of Glass" [15] were also of significance in generalizing the necessary thermodynamic data.

With respect to structural problems, the monographs "Structural Inorganic Chemistry" of Wells [16] and "Structural Chemistry" of Evans [17] were mainly employed. Several of the well-known German encyclopedic reference book series and especially the newest editions of "Roempp's Chemistry Lexicon" [18] and "Ullmann's Encyclopedia of Industrial Chemistry" [19] were utilized as well. The mineralogical side of the SiO_2-modifications was treated referring to several reference books out of which we would like to cite here the optical mineralogy by Winchell [20], Betechtin's course of mineralogy [21] and the "Crystal Habits of Minerals" by Kostov and Kostov [22]. Two books on the properties of silicate glasses written by Vogel [23] and by Scholze [24] were also employed in the present review.

3.2.2 Original Literature Sources on the Different Silica Modifications

Particular emphasis was directed to the original literature on synthesis, properties, thermodynamics, and transformations between different phases of SiO_2 as it is predominantly given in mineralogical, geochemical and cosmo-chemical publications. The respective literature is cited in detail in the subsequent outline. Prior to that, here we would like to mention only several general review articles which are of significance in analyzing thermodynamics and the phase status of the silica modifications (Swami and Saxena et al. [25], Holm et al. [26], Dorogokupets et al. [27] and Richet, Botinga et al. [28]). We also consulted several papers in which attempts are developed for the calculation of thermodynamic properties of SiO_2 out of first principles by using more or less known molecular-statistics models. A typical example in this respect is the paper of Keskar et al. [29]. However, it turns out that this and similar papers are based mainly on Lennard – Jones potentials and the mentioned first principle calculations are far from giving useful information for the task of the present review.

3.2.3 Internet Search

General information was also provided by an Internet search. We performed this search in a fan-like manner by browsing the database for keywords in the abstracts for a period starting from 1999 up to present day publications. This approach resulted in approximately 950 summaries on publications in international scientific literature, devoted to general properties of SiO_2 and its main modifications (quartz, tridymite, cristobalite, keatite, coesite and stishovite have been used as our key words). In addition, standard search engines were employed to obtain an additional update mainly on products of technical use. This gave us an additional output of about fifty literature sources, mainly connected with natural plant products containing SiO_2.

3.3 Phase Diagram of SiO_2

In the literature, one can find several versions of phase diagrams of SiO_2. Some of them are only of historic interest and mostly illustrative, in many details even misleading, but from most of the present day phase diagrams one can really derive distinct conclusions concerning the properties of SiO_2.

3.3.1 Fenner's Classical Diagram

The analysis and discussion of the different crystalline and amorphous modifications of SiO_2 is commonly started with the classical phase diagram of Fenner from 1913 (see Fig. 3.1). However, this diagram gives only a qualitative picture of the relative stability of the SiO_2-modifications using the virtually constructed temperature dependence of the vapor pressure of the considered amorphous and crystalline forms of SiO_2 as a criterion of phase stability. These vapor pressures of the silica modifications are in fact extremely low, they cannot be measured experimentally and they can only be calculated from the respective thermodynamic potential differences. Latter quantities

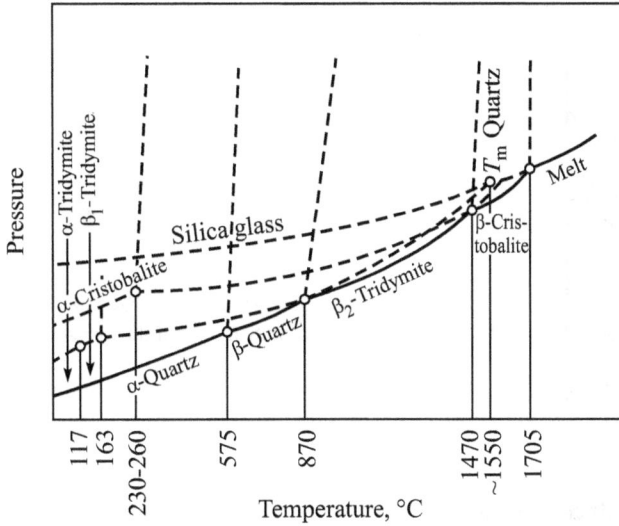

Fig. 3.1: Fenner's phase diagram of SiO$_2$ in the representation given by Eitel [1]. Bold lines give the expected course of the vapor pressure of the phases, stable in the respective temperature range. Their broken extrapolations denote metastable states. Equilibrium temperatures indicated at the abscissa are given by circles. The nearly vertical dashed lines present the expected change of the equilibrium temperatures with increasing pressure.

were, however, not known at Fenner's times. Thus, Fenner's diagram is only a qualitative illustration of the thermodynamic potential diagram, which could be constructed only from the present-day knowledge of the thermodynamics of SiO$_2$. Moreover, this diagram is criticized for the inclusion of tridymite as an independent stable phase of SiO$_2$. As mentioned already above, more recent investigations have shown, however, that the three modifications of tridymite can be formed and exist only in the presence of a measurable amount of dopants, mostly alkali oxides and especially K$_2$O. In Fenner's diagram, the high pressure crystalline modifications of SiO$_2$, synthesized many years after Fenner had published his phase diagram, are not included: neither coesite, nor keatite and stishovite appear on this picture.

3.3.2 Flörke's Diagram

In addition to Fenner's proposal, another diagram often referred to in literature has been established by Flörke [30] in 1956 (see Fig. 3.2). Here, the stability regions of the three modifications (quartz, cristobalite and liquid SiO$_2$ known at the time when this diagram was constructed) are again schematically indicated, however, without tridymite. Flörke was in fact the first author to prove that tridymite is not formed in the absence of alkali oxides.

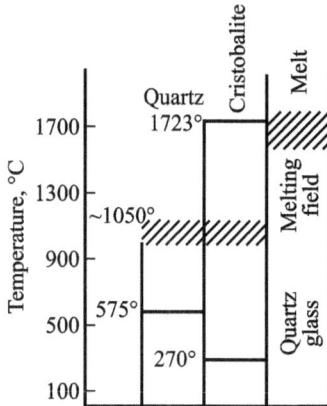

Fig. 3.2: Flörke's phase diagram [4, 30] of SiO_2. The vertical bold lines give the stability ranges of the respective modifications at normal pressure. Shaded regions indicate the corresponding metastable states.

Another schematic diagram for the description of the low-pressure phases of SiO_2 (see Fig. 3.3) is employed since the end of the 1970s in one of the best known reference books of Russian silicate literature [4]. Here, however, (already) coesite was introduced. The most interesting point in this diagram is that the quartz-coesite coexistence curve was constructed employing thermodynamic measurements. These computations were supported by evidence concerning the transition kinetics in between the different phases. In this diagram, tridymite is schematically considered again as a natural member of the silica family, while stishovite, just synthesized at that time, is still absent in it. The synthesis of stishovite was carried out under pressures exceeding 80 kbar. Further on, coesite was also synthesized and demonstrated to be stable at pressures exceeding 10 kbar and at temperatures higher than 1400 °C as evident from Fig. 3.3.

3.3.3 Contemporary ($p - T$)-phase Diagrams of SiO_2

Quantitative phase diagrams of SiO_2 have been constructed employing experimentally determined thermodynamic properties of the SiO_2-modifications. However, calorimetry of the SiO_2-modifications even at normal pressure is a very difficult task: the specific heats have to be measured up to 2000 K and the melting enthalpy has to be determined at approximately 2000 K. The transformation heats between the stable SiO_2-phases at normal pressure are traditionally determined by solution experiments in hydrofluoric acid, performed in platinum calorimeters. However measurements of the specific heats in the indicated temperature limits are a tedious task and such investigations were performed only in several cases (see [26–28]). Direct determinations of the melting enthalpy of cristobalite are lacking. For this quantity, in reference literature usually only the difference between the dissolution heats of SiO_2-*glass* (not as necessary: *of the melt*) and of *stabilized* (i.e chemically doped) high-temperature β-cristobalite is given. Moreover, in such measurements (necessarily performed at

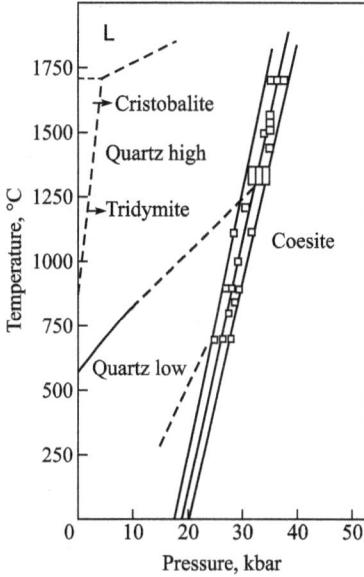

Fig. 3.3: Thermodynamically calculated (p, T)-phase diagram of SiO₂ in its preliminary form (still without stishovite). The (p, T)-lines of the "normal" phases are more or less qualitatively drawn, the quartz – cristobalite – coesite line is thermodynamically calculated accounting for the respective volume changes. The open squares are kinetic experimental results confirming the general thermodynamic expectation (according to [4]).

room temperatures) often even low temperature α-cristobalite is employed and considered as β-cristobalite (and vice versa). In this way only estimates of the melting enthalpy ΔH_m of cristobalite are in fact given in the literature. The transformation enthalpies between the high-pressure modifications (coesite, stishovite) are also usually determined only at normal pressure and extrapolated using Hess' law. Nevertheless, in this way quantitative data of eight of the crystalline modifications of SiO₂ are given in the reference literature [8, 9].

In our present calculations of the solubility of quartz, cristobalite, coesite, and stishovite, the following dependence

$$\Delta H_m \cong 2\Delta H_g^{298}(T), \quad \Delta H_g^{298}(T) = \left(H_{SiO_2\text{-glass}}^{298} - H_{\alpha\text{-quartz}}^{298} \right) \tag{3.1}$$

is used as an estimate for ΔH_m. Here $\Delta H_g^{298}(T)$ is the difference between the enthalpy of quartz glass and of α-quartz at room temperature. This estimate followed from an empirical dependence, first mentioned sixty years ago by G. Tammann. It is widely discussed in [14] (see also literature cited there). Here it is employed in the series of calculations, summarized in Section 3.6. At present, the validity of this approximation is confirmed for more than 100 substances (various silicate, borate, phosphate, and organic glasses) for which it is known that in fact $\Delta H_g \sim (1/2)\Delta H_m$ [14]. That is why we preferred to use this estimate in our calculations in Section 3.6 as a better, empirically founded approximation, instead of taking the difference between the enthalpy of glass and the respective stable crystal modifications as a measure for ΔH_m.

Stishov made a further significant step forward as compared to the results as given on Fig. 3.3. He proposed, using thermodynamic data and the respective pressure dependencies as functions of temperature, the approximation (see [4])

$$p(T) = (97500 \pm 5000) + 20.33T \tag{3.2}$$

for the line dividing the regions of stable existence of coesite (given on Fig. 3.3) and stishovite in the (p, T)-diagram. This border line (not introduced in Fig. 3.3) is essential for the whole physics and structure of SiO_2 in its different modifications. It divides the regions of the two main structural forms in the crystal-chemistry of this substance: quartz or coesite-like from stishovite-like structures. In some respect it is similar in its significance to the Simon-Leipunskii-line in the physics of carbon, dividing graphite-like from diamond-like structures (see [31]).

The further developments of the thermodynamics of the SiO_2-modifications were mainly directed to the account of the high-temperature modifications of SiO_2. This interest is connected mainly with their particular importance in Earth science and tectonics. It was assumed, and thermodynamic calculus verified this hypothesis, that in the depth of the Earth's mantilla the predominant modifications of SiO_2 are coesite and stishovite, the latter being stable at depths below 300 km. Later on coesite and stishovite were found in meteoritic impact craters on the Earths surface (the most significant example being the Arizona crater). The thermodynamics of stishovite and coesite gained also crucial importance in cosmology and cosmogony.

From a series of measurements performed from the late 1960s till the end of the 1980s mainly in geological laboratories the specific heats and enthalpies of transitions of quartz, cristobalite and SiO_2 glass, coesite and stishovite have been determined at least for their stable modifications [25–29]. Using the expected and calculated change of the thermodynamic properties with pressure (i.e. at the quartz-coesite and the coesite-stishovite $p(T)$-phase boundary lines) the real phase diagrams of SiO_2 (including stishovite) were constructed as they are presented on Figs. 3.4, 3.5 and 3.6. On Figs. 3.4 and 3.5 one can see the experimentally determined kinetic transition points these diagrams are also based upon. Fig. 3.6 is given in a more convenient representation in terms of pressure, p, vs. temperature, T. A comparison of these three phase diagrams shows the way the final commonly employed now diagram as presented in Fig. 3.6 is constructed. It is seen that Figs. 3.4, 3.5, and 3.6 are comparable in a qualitative way: the differences between them are due mainly to the lack of exact experimental thermodynamic data on the molten SiO_2-modifications. Nevertheless, the most significant features of the SiO_2 phase diagram are clearly seen on these figures. They illustrate the stability regions of the high pressure phases (stishovite and coesite) and their relation to the phases of silica (quartz and cristobalite), which are stable at normal pressure.

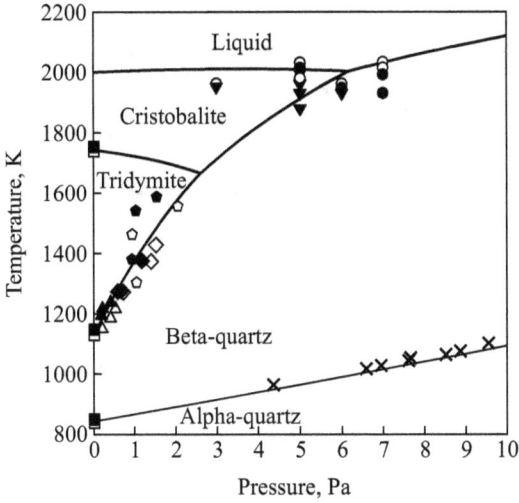

Fig. 3.4: Thermodynamically constructed phase diagram of SiO$_2$ with kinetic data from different authors summarized by Swami, Saxena et al. [25]. This diagram is, in fact, an extension of the diagram shown in Fig. 3.3 and the next stage in the construction of the SiO$_2$ phase diagrams. This generalization was introduced by Swami, Saxena et al. [25].

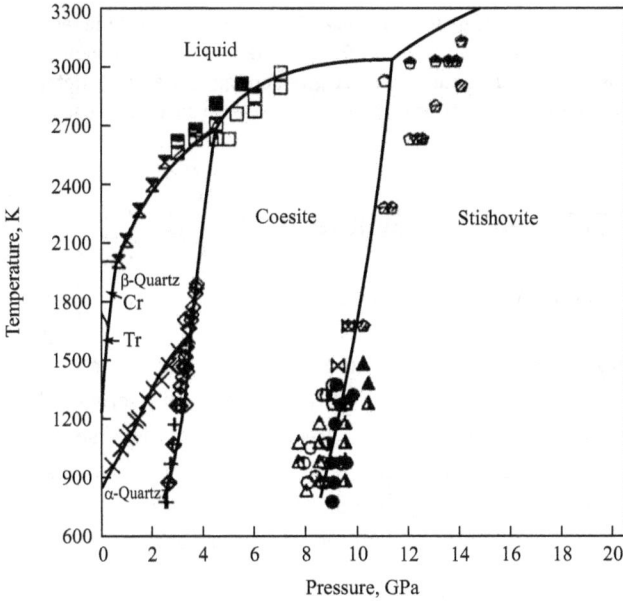

Fig. 3.5: Extended SiO$_2$-phase diagram generalized by Swami, Saxena et al. [25] in its full pressure and temperature ranges. Note that Fig. 3.4 is only a part of Fig. 3.5.

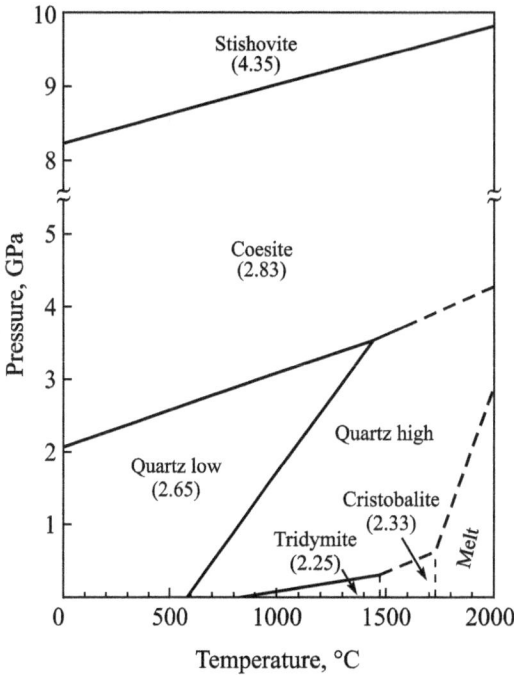

Fig. 3.6: Present day representation of the SiO$_2$ phase diagram as it is recommended in the reference literature [18] and used throughout the present review. The densities of the corresponding phases are given in brackets. It is based on diagrams like those given on Figs. 3.4 and 3.5. Note, however, that here the more comfortable presentation with temperature as an abscissa is used.

3.4 Modifications of SiO$_2$ and Their Synthesis

3.4.1 Mineralogical Characteristics of the SiO$_2$-modifications

At present the following crystalline and amorphous SiO$_2$-modifications are considered in literature and are discussed here: α- and β-quartz, α- and β-cristobalite, keatite, coesite, stishovite, SiO$_2$-glass, amorphous precipitates (or disperse amorphous), and molten SiO$_2$. Their space groups and temperature stability ranges are given in Table 3.1, while the main structural types of SiO$_2$ are seen on Fig. 3.7. Besides, Table 3.1 also contains data for several "non-systematic" modifications.

3.4.2 Synthesis of Quartz

The formation of quartz in nature follows both a hydrothermal way [1, 2] or proceeds via crystallization of magmatic rocks. The first artificial synthesis of quartz was per-

Table 3.1: Silica polymorphs

Name	Crystal system	Space group	Stability range
Low quartz (α)	trigonal	$C3_12$ and $C3_22$	Up to 573 °C
High quartz (β)	hexagonal	$C6_22$ and $C6_43$	573–870 °C
Low tridymite (α)	orthorhombic	–	
Middle tridymite ($\beta1$ or β)	hexagonal (trigonal)	–	870–1470 °C
High tridymite ($\beta2$ or γ)	hexagonal	$C\frac{6}{m}mc$	
Low cristobalite (α)	tetragonal (trigonal)	$P4_12_1$	280 °C
High cristobalite (β)	isometric (cubic)	$P2_13$	1470–1713 °C
Keatite (silica K)	tetragonal	$P4_12_1$ and $P4_32_1$	–
Coesite (silica C)	monoclinic	$C\frac{2}{c}$ or Cc	1300 °C
Stishovite	tetragonal (rutile type)	–	800 °C
Post-stishovite 1	Tetragonal (CaCl$_2$ type)	–	?
Post-stishovite 2	tetragonal (α-PbO$_2$ type)	–	?
Lechatelierite (silica glass)	amorphous	–	1100 °C
Amorphous silica	amorphous	–	–
Chalcedone	trigonal	–	–

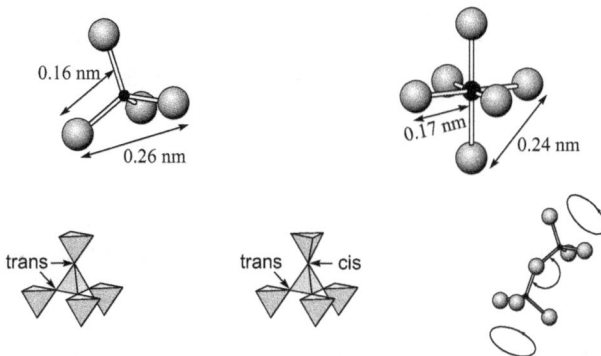

Fig. 3.7: The SiO$_4$/2 tetrahedron and the SiO$_6$/3 octahedron (top) together with the corresponding distances. Bottom row gives the cis-trans arrangement of tetrahedra in quartz; right-hand scheme: the possible rotations of the two tetrahedra (from [19]).

formed hydrothermally using the different solubility of crystalline quartz and of quartz glass [32–34].

3.4.2.1 Isothermal Methods: Wooster and Nacken

As it will be discussed in more details further on, the increased thermodynamic potential of quartz glass as a frozen-in system determines its higher solubility. This was expected already both by Wooster [33, 35] and Nacken [34] in England and Germany respectively during the years of World War II. In both cases, the relatively high super-saturations at temperatures and pressures reached in the autoclaves (see Section 3.6 and our calculations there) yielded very poor results only: no growth of large quartz crystals (which was the task of both authors) could be realized with these isothermal methods. The details of the isothermal synthesis, in which the route SiO_2 (glass) → SiO_2 (quartz) was exploited, are given critically and in great details in Smakula's book [32] and may be followed in Wooster's publications [33, 35] and Nacken's report [34] (see also [36]). In the already mentioned survey of I.G. Polyakova in the present book some additional details of this way of synthesis are summarized.

3.4.2.2 Gradient Methods

The failure of the isothermal SiO_2-glass → α-quartz route is the reason why different temperature gradient methods were soon developed for single crystal quartz synthesis (see [32, 36] and Barrer [37]). In this way of synthesis, silicate glasses, quartz glass, quartz crystals were suspended in autoclaves, in which supercritical aqueous solutions of NaOH and of Na_2SiO_3 were formed and an externally applied temperature gradient determined a relatively low supersaturation between the SiO_2-precursor and the growing crystal. In this way, large quartz single crystals were grown. They could be directly used in the piezo-technique. In both World War I and II the hydrophones used in the anti-submarine warfare were constructed exploiting the singular piezo-electric properties of quartz single crystals.

3.4.2.3 Isothermal Synthesis of Quartz from Cristobalite

The isothermal synthesis of quartz (and especially of fine grained quartz) was also attempted using α- or β-cristobalite as a precursor in the autoclaves. These two SiO_2-modifications being metastable at the temperatures used in hydrothermal synthesis have an elevated thermodynamic potential and thus higher solubility than quartz (see [38] and references cited there). In Section 3.6 a thorough discussion of this topic is given in terms of the solubility curves we construct there.

3.4.2.4 Inclusion of Impurity Droplets into Quartz

Usually, in the literature the synthesis of quartz is discussed assuming processes of formation and growth of nearly perfect large quartz single crystals. For the synthesis of ultra pure glass out of quartz, the synthesis of fine grained quartz is also of interest. In such case, the crystals are so small that the inclusion of bubbles, containing residual alkaline traces and filled with the precursor aqueous solutions, is minimal (see Fig. 3.8). This problem is discussed in detail in the technological literature [19].

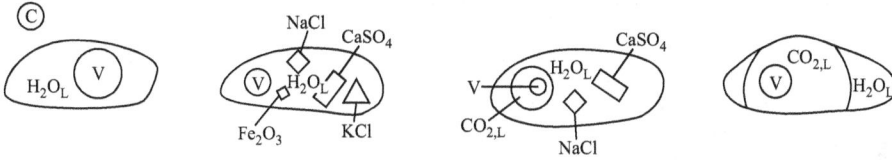

Fig. 3.8: Impurity distribution in quartz single crystals (from Ullmann's Encyclopedia [19]): types of fluid inclusions. V – vapor, L – liquid in the system H_2O–CO_2–SiO_2. Squares, rectangles and triangles schematically represent different crystals.

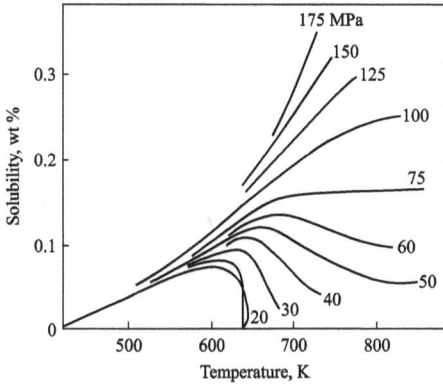

Fig. 3.9: Solubility diagram of amorphous SiO_2 in pure water at different pressures given as a parameter to each curve. Mineralizers (NaOH, Na_2CO_3) additionally increase the indicated SiO_2-solubility (according to [19]).

Of particular significance in performing hydrothermal synthesis of quartz and of other modifications of SiO_2 are the solubility diagrams of SiO_2 as given on Fig. 3.9 (as an amorphous SiO_2-phase in pure H_2O). The pressure resulting in an autoclave with a given degree of initial filling with an aqueous solution is seen on Fig. 3.10. In [19, 32], detailed information concerning the construction and the maintenance of different autoclave models both for industrial and laboratory applications for scientific investigations are given.

3.4.3 Synthesis and Stabilization of β-cristobalite

Cristobalite is one of the most interesting and promising modifications of SiO_2. Its low temperature or α-modification possesses a negative (auxetic) Poisson ratio, i.e. it becomes wider when stretched and thinner when compressed [40, 41]. The high temperature or β-form of cristobalite is stable up to the melting point of SiO_2 (1725 °C) and it is characterized by a very low coefficient of thermal expansion (almost equal to zero) [42–46]. However, there is a general disadvantage in using β-cristobalite as a refractory material. At about 250 °C cubic β-cristobalite transforms by a first order displacive phase transition (as it is commonly denoted) into its low temperature polymorph, i.e., into α-cristobalite possessing a tetragonal structure. This transition is connected with a considerable volume dilatation (of about 5 %) resulting in crack formation or even

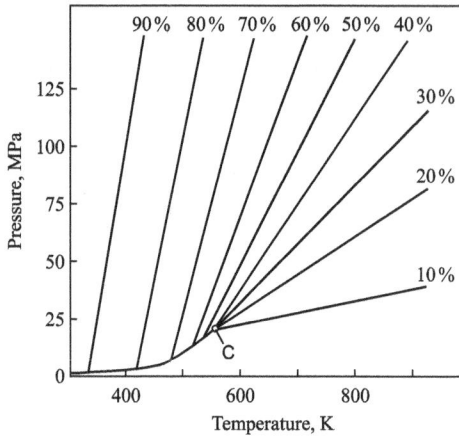

Fig. 3.10: $(p - T)$-diagram of water: Isochores for various percentages of autoclave filling (given as a parameter to each line) are shown here. The critical point of water (22.04 MPa, 574.4 K) is indicated with C (after [19]).

in the destruction of pure β-cristobalite samples in the course of cooling to room temperatures. That is why for many years intensive investigations have been performed to stabilize the high temperature β-modification of cristobalite in such a way as to avoid or at least to suppress its $\beta \rightarrow \alpha$ transition [44, 47–58].

One way to inhibit the $\beta \rightarrow \alpha$ transformation of cristobalite is the employment of the so-called "stuffing" mechanism. This mechanism is similar to methods used for the stabilization e.g. of the high temperature modifications of quartz or of zirconium (ZrO_2) [59–61]. The suppression of the $\beta \rightarrow \alpha$ transition by this mechanism is due, as first stated by Buerger, to the incorporation of appropriate (alkaline or alkaline-earth) "stuffing" cations into the large voids of the cristobalite structure built up by six-membered rings of $SiO_{4/2}$ tetrahedra [48]. The charge of the dopant "stuffing" cations is compensated by replacing a part of the Si^{4+}-ions in the SiO_2-network by Al^{3+} or by other three valence cations. This way of stabilization is also known as chemical stabilization, it is caused by a change of chemical composition, which is determined by the dopants introduced.

As it seems, Li [50] was the first to synthesize alkali-free glass-ceramic materials containing stabilized β-cristobalite-like solid solutions as the main crystalline phase. For this purpose Li used a conventional glass-ceramic technology: glasses with the desired dopant concentration (up to several molar percent, typically 3–6% CaO, 3–6% Al_2O_3) were synthesized just above the melting point T_m of silica and then sinter-crystallized at lower temperatures. A serious disadvantage of this technology is, however, the extremely high temperatures (nearly 1800 °C) and the prolonged melting times (of about 100 h) needed to produce the initial silica reach precursor glasses with a homogenous dopant distribution. The uniform dopant distribution as shown by a number of investigations is an obligatory prerequisite to prepare glass-ceramics with a high content of stabilized β-cristobalite (or more correctly a β-cristobalite-like solid solution). Very long heat-treatment times (also up to 100 h) are also necessary for the second stage of the synthesis: for the formation of the stabilized β-cristobalite phase by sinter-crystallization of the powders, obtained after milling of the pre-melted glasses.

Another possible route for the synthesis of stabilized β-cristobalite is connected with the application of different sol-gel techniques [52–58]. Their most important advantage is that relatively low temperatures are needed to achieve homogeneous distribution of stabilizing cations of the above mentioned concentration in the SiO$_2$-network of the precursor glass. Shortcomings of this way of synthesis are, however, the high prices of the reagents usually employed in the SiO$_2$-sol preparation and the complex and time consuming sol-gel transition procedures. That is why alternative intermediate approaches, also involving sol-gel routes, are of particular interest, e.g. the so called "incipient-wetness" techniques [62]. Other methods for the synthesis of β-cristobalite glass ceramic materials were also developed by the authors of the present review. One of them, which we called the activated reaction sinter-crystallization approach [63], and modified sol-gel techniques [64] are described in the next sections.

3.4.3.1 Activated Reaction Sinter-crystallization Approach in β-cristobalite Glass Ceramics Synthesis

The main advantage of such an approach in the synthesis of stabilized β-cristobalite glass-ceramics, which could be also useful in other cases, is that the high temperature glass pre-melting stage typical for the conventional glass-ceramic technology is avoided. It is based on an activated reaction sinter-crystallization process of precursor mixtures containing all necessary components in a chemically, mechanically or thermally pre-activated state. The stuffing Ca^{2+}-cations are introduced as freshly de-carbonated highly reactive calcium oxide (CaO), instead of CaCO$_3$ conventionally used. Finely milled alumina (Al$_2$O$_3$) and quartz glass (both with particle sizes smaller than 40 μm) are employed as pre-activated sources of the Al^{3+}-cations and of SiO$_2$ itself. In this way, it becomes possible to *directly* synthesize glass-ceramic materials with a high percentage of the stabilized β-cristobalite-like phase in the course of a single chemically activated sinter-crystallization process at relatively low temperatures (not higher than 1450 °C) and short heat-treatment times (of about 20 h), omitting the usually employed high-temperature pre-melting procedure. Two main problems have to be resolved in applying this technique for the synthesis of stabilized β-cristobalite glass-ceramics, one has to know: (i) The influence of the dopant concentration and sinter-crystallization heat-treatment regime on the stability and on the content of the β-cristobalite-like phase formed; (ii) the influence of the concentration of the stabilizing dopants on the nature of the ($\alpha \rightarrow \beta$)-cristobalite transition. Figs. 3.11a and b illustrate how the concentration of the stabilizing additives and the sinter-crystallization heat-treatment regime influence the content of the β-cristobalite-like phase.

The nature of the ($\alpha \rightarrow \beta$)-cristobalite transformation was studied in detail in [63]. The main results of these investigations are shown in Figs. 3.12a–c. As seen in these

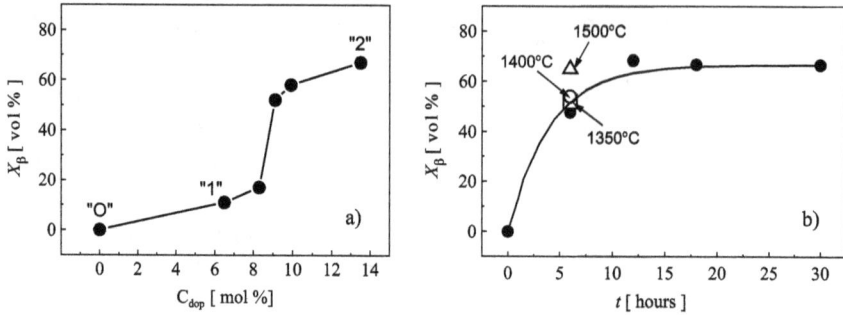

Fig. 3.11: Content X_β of the β_x-cristobalite-like phase in glass-ceramic samples obtained by the activated reactive sinter-crystallization method. (a) Dependence of X_β on of the dopant concentrations C_{dop}. All samples in Fig. 3.11a are heat-treated employing a two-stage heat-treatment regime: 1 h at 1350 °C and 18 h at 1500 °C. Here "O", "1" and "2" denote the pure cristobalite sample and samples containing 6.5 and 13.5 mol% dopants (CaO and Al_2O_3 in equimolar ratios), respectively. (b) Kinetics of formation of a X_β-cristobalite-like phase in glass-ceramics samples containing 13.5 mol% dopants at 1450 °C. The values of X_β are also given for three other temperatures: 1350 °C (hollow square), 1400 °C (hollow circle) and 1500 °C (hollow triangle) each one after 6 h exposure.

figures the increase of the dopant concentration C_{dop} lowers the thermal effect and diminishes the step-like first-order type of volume changes. Moreover, the temperature of this transition is shifted to lower values with the increase of C_{dop}. Such a lowering of the $(\alpha \to \beta)$-transition point was observed with respect to the α- and β-modifications of quartz [65].

The analysis of Figs. 3.12a and b and the results of the high-temperature X-ray investigations, shown in Fig. 3.12c, indicate that by increasing the dopant concentration the transformation (having originally the character of a first-order displacive transition) can be transformed into a second-order transition between the α- and β-modifications of cristobalite. The shift to lower temperatures of the $(\alpha \to \beta)$-transition point immobilizes the building units of the system (both in cristobalite and in quartz) and converts the modification change from its originally displacive mechanism to a more or less pronounced process of tilting of the $SiO_{4/2}$-tetrahedra. In increasing the dopant percentage the formation of α- and β-cristobalite-like solid solutions with nearly equal lattice parameters becomes possible. Thus, the volume change at the $(\alpha \to \beta)$-transition is minimized or even nullified as in the case of the sol-gel derived sample "T" (see Figs. 3.12a and b). This conversion of the first-order displacive $(\alpha \to \beta)$-transition into a second-order phase change caused by the incorporation of an appropriate concentration of stabilizing cations into the cristobalite structure is most probably of high technological importance.

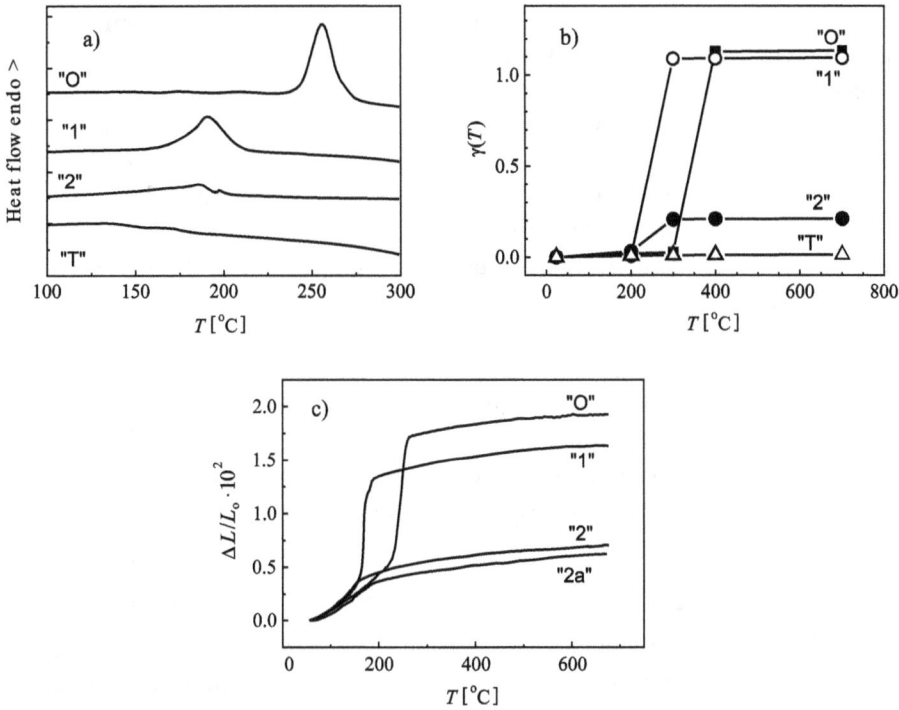

Fig. 3.12: Influence of the dopant content on the thermal (a), structural (b) and dilatational (c) behavior of the stabilized β-cristobalite glass-ceramics investigated. Here "T" denotes the sol-gel derived sample doped with 3.25 mol% CaO and 3.25 mol% Al$_2$O$_3$ and synthesized according to methods described in Sections 3.3.1 and 3.3.3. Samples "1" and "2" have the same dopant percentage as the respective samples in Fig. 3.11a. Sample "T" is heat-treated for 4 hours at 1350 °C, two stage heat-treatment regime is employed to samples "0", "1" and "2": 1 hour at 1350 °C and 18 hours at 1450 °C. Sample "2a" possesses the same composition as sample "2", however, with respect to sample "2a" a longer second heat-treatment stage of 30 hours instead of 18 hours is employed. (a) The heat flow curves proportional to the specific heat $C_p(T)$-dependencies of the samples. (b) The relative change $\gamma(T) = (V - V_o)/V_o$ in the volume V of the elementary cell of the pure cristobalite phase crystallized in the reference sample "0" and of the cristobalite-like solid solutions formed in samples "1", "2" and "T". (c) The relative elongation $\Delta L(T)/L_o$ of the samples as a function of temperature. Here, $\Delta L(T) = L(T) - L_o$, where $L(T)$ and L_o are the lengths of the samples at temperature T and at room temperature, respectively. Note the shift of the peak positions in the heat flow curves and of the 'inflection' points in the $\gamma(T)$- and $\Delta L(T)/L_o$-curves to lower temperatures with the increase of the dopant concentration from sample "0" to sample "2". Note also the absence of a thermal effect and of a stepwise change in the $\gamma(T)$- dependence of sample "T" indicating for the full inhibition of the α/β-cristobalite transition in this sample.

3.4.3.2 Hydrothermal Synthesis of Cristobalite

The hydrothermal synthesis of cristobalite is up to now described in several publications. In these references, mainly the production of micro- or even nano-sized cristobalite powders out of alkaline aqueous solutions at hydrothermal conditions is described both for a realization in small experimental devices but also in relatively large autoclaves [66, 67]. In this way of cristobalite synthesis, only alkaline solutions have been employed up to now. This feature makes it difficult or even excludes the use of hydrothermally prepared cristobalite as a constituent of alkaline-free materials. Here, the possibility of introducing Ca^{2+}- and Al^{3+}-ions into the aqueous solution should be investigated at conditions described in more details in Section 3.6.4.

It is interesting to note that according to [66, 67] hydrothermally synthesized high-temperature β-cristobalite is relatively stable at room temperature because of the high content of alkaline dopants included into it from the host solution. However, the thermal stability of β-cristobalite and the mechanism of the $(\alpha \rightarrow \beta)$-transition between hydrothermally grown α- and β-cristobalite modifications are according to our experience not sufficiently understood up to now. Usually, the low-temperature α-cristobalite is obtained at lower autoclave temperatures. Its use as an auxetic material (i.e. as crystals or micro-crystals with a negative Poisson ratio) is also discussed in Section 3.6.4 and at the end of the present chapter.

3.4.3.3 Sol-Gel Methods of Synthesis

Three different sol-gel techniques were developed for reproducible synthesis of glass-ceramic materials with a high content of the stabilized β-cristobalite like phase. The first one is analogous to those used in previous publications of one of the present authors for the preparation of optical materials, especially of silica doped with Ho^{3+} or Tb^{3+}-ions [68, 69]. The second and the third methods are derivatives of the so called "wet impregnation methods" also used in the preparation of optical materials, in which the precursor oxide powders are impregnated with a doping solution containing a soluble nitrate or chloride dopant. Below, these three methods for stabilization of cristobalite are discussed in details.

Pure sol-gel method: As a first step in this way of synthesis a homogeneous, transparent sol with a composition $n_{TEOS} : n_{EtOH} : n_{H_2O} = 1 : 1 : 4$ is prepared at room temperature by mixing of the necessary amounts of tetraetoxysilane (TEOS) used as a SiO_2-source, ethyl alcohol (EtOH) and 0.55M aqueous solutions of $Ca(NO_3)_2$ and $Al(NO_3)_3$ for the introduction of the stabilizing cations. It is followed by acid hydrolysis at pH = 2 (2 hours under stirring), catalyzed by HCl, and successive gelation of the sol (48 hours at 50 °C in closed boxes). Typical starting amounts of TEOS are 5 ml. As a result transparent gels were obtained, revealing by their transparency that a molecular distribution of the dopand in the gel has taken place.

It is well known from sol-gel chemistry, that gelation times can be decreased by increasing the pH-value at the gelation temperature. In such sol-gel procedures, there always exists the possibility for an unwanted phase separation of the doping oxides. The next steps were drying (120 hours at 50 °C), grinding, second drying (1 hour at 90 °C) and sintering (4 hours at 700 °C). The amorphous powders thus obtained were milled again and pressed using aqueous solution of polyvinyl alcohol (PVA) as a binder. The compact pellets were heat-treated at different temperatures in the range from 1350 up to 1500 °C. It turned out that the highest content of the stabilized β-cristobalite like phase (of about 95–98 vol%) was achieved for the sol-gel synthesized samples containing 3.25 mol% CaO and 3.25 mol% Al$_2$O$_3$ and heated 4 hours at 1350 °C in air.

The disadvantage of this sol-gel technique is the relatively high amount of residual water (about 25 %) remaining in the samples causing difficulties for their further sintering. To overcome this problem two ways were tested. The first one was to modify the described sol-gel technique by increasing the drying time (from 1 to 4 hours) and the temperature and the duration of the sintering (from 4 hours at 700 °C to 5 hours at 800 °C). Further enhancement of sample sintering was achieved by replacing 2% of SiO$_2$ by TiO$_2$. The X-ray diffractogram of a sample prepared by this second improved sol-gel method is given in Fig. 3.13. The samples synthesized in this way are characterized by a full suppression of the (α/β)-cristobalite transition and are used as reference samples in the investigation of the glass-ceramics obtained by the activated reaction sinter-crystallization approach (see Figs. 3.13). Typically, the sintered powders prepared using a "pure" sol-gel method do not contain mullite traces, they contain only cristobalite.

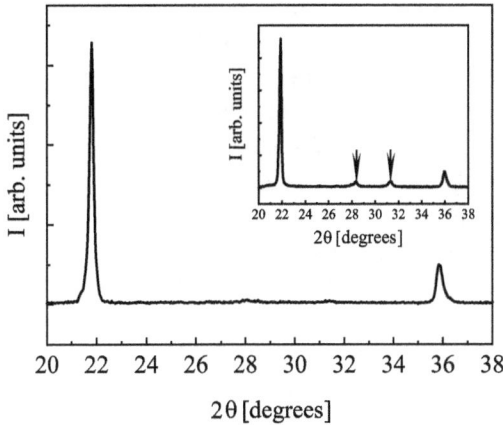

Fig. 3.13: *X*-ray diffraction patterns of the sol-gel derived reference sample "T" (see Figs. 3.12a and 3.12b). The sample is doped with 3.25 mol% CaO and 3.25 mol% Al$_2$O$_3$ and heat-treated 4 hours at 1350 °C. In the insert the X-ray diffractogram of a pure cristobalite sample obtained by sinter-crystallization of quartz glass powder is also shown. Note that in the *X*-ray diffractogram of the reference sample "T" the two characteristic peaks of α-cristobalite (denoted by arrows in the insert) are missing which indicates for the full suppression of the α/β-cristobalite transition in this sample.

Mixed sol-gel method: Another main aim of the present investigations was to find cheaper substitutes for the expensive sol-gel reagents (e.g. TEOS) usually used as a SiO_2-source in the sol-gel techniques. For this purpose a "mixed" sol-gel method was developed in which the main part of silica (up to 90 wt%) is introduced using a micrometer grained quartz-glass powder suspension under stirring, containing 0.174 mol Si/l. The rest (10 wt%) of SiO_2 was added in the form of an acid silica sol with a SiO_2-concentration of about 2.75 mol/l. The idea of this preparation way was to check the use of commercial sol products like LEVASIL® (BAYER AG, Leverkusen) for the preparation of larger amounts of β-cristobalite. The stabilizing dopants were introduced using 0.55 M aqueous solutions of $Ca(NO_3)_2$ and $Al(NO_3)_3$. This suspension was stirred (24 hours at room temperature) followed by solvent evaporation at heating and drying of the obtained powders. After grinding the samples were sintered for 4 hours at 700 °C and than sintered for 4 hours at 1350 °C to obtain β-cristobalite.

Wet impregnation method: In addition, a third preparation method, widely similar to above described procedure but without using of a colloidal silica sol, was also employed in the described sol-gel techniques in order to obtain β-cristobalite. This method can be characterized as a "wet impregnation method". As a silica source, only a micrometer grained quartz glass powder suspension under stirring containing 0.174 mol Si/l is used. Impregnation, stirring, evaporation, drying and heating conditions are the same as described in the "mixed sol-gel method".

The proposed here sol-gel techniques and, in particular, the "wet impregnation method" could be used for a relatively low cost production of small cristobalite articles (e.g. for details with high electric resistance at elevated temperatures, e.g. for application in electronic industry). Employing this "wet impregnation method" glass-ceramics samples were obtained which contained up to 60–70 vol% stabilized β-cristobalite. It seems that the β-cristobalite content is not substantially affected (only about a 10 % increase) by the use of colloidal silica suspensions. On the other hand, a content of 70 % stabilized β-cristobalite phase can be obtained by improving the homogenization conditions of the "wet" impregnation method. It is important to note, that powders obtained by the "wet impregnation method" and by the "mixed sol-gel method" contain mullite as a second phase after heating 4 hours at 1350 °C. The "pure sol-gel method", however, leads to a high content of β-cristobalite (about 95 %) and α-cristobalite traces only.

Another important feature of the sol-gel synthesis of cristobalite and cristobalite–rich materials is that a high percentage of the stabilized β-cristobalite phase is achieved at temperatures about 100 °C lower and at heat-treatment regimes times substantially shorter than those, needed in the production of the glass-ceramic materials, synthesized by the reaction sinter-crystallization approach. This effect in the sol-gel approach is again due to the formation of chemically very active nano-sized dopants: here of silica nano-particles, facilitating the low temperature homogeneous distribution of the stabilizing dopants and their incorporation into the SiO_2-matrix.

However, up to now both the analysis of international literature and evidence from experimental work, obtained in the framework of the work on the present review, indicate that it is impossible to form via the sol-gel route articles of dimensions and with the technological properties, obtained via direct ceramic technologies indicated above.

In comparison with the also discussed hydrothermal methods of synthesis of cristobalite the sol-gel route also offers some advantages: the expensive (and requiring very particular attention, e.g. anti-explosion measures and provision) hydrothermal autoclave equipment is not required. On the other side present-day sol-gel equipment cannot secure production and yields in industrial scales, possible for both hydrothermal methods and in the also discussed sinter-crystallization glass-ceramic ways of production of cristobalite or any other of the silica modifications. Maybe the also mentioned small scale production of particularly important cristobalite wares or of cristobalite in sand-pile state could be the perimeter for both hydrothermal and sol-gel technical synthesis.

3.4.4 Synthesis of Keatite: Classical Aspects

The SiO$_2$-modification, named by its initial synthesizer P.P. Keat, is the first artificially synthesized SiO$_2$-modification. It was obtained under hydrothermal conditions in 1954. The conditions employed are described in great details by Keat himself in the respective publication [70]. He determined the density of keatite and drew attention to the fact that the refractive indices of the SiO$_2$-modifications change linearly from vitreous silica via tridymite, cristobalite, keatite and quartz (from 1.460 to 1.540) with their densities (increasing from 2.20 g/cm^3 for vitreous silica to 2.60 g/cm^3 for α-quartz).

Later on, keatite turned out to exist only in the form of solid solutions and to possess a structure intermediate between those of quartz and coesite. Nevertheless, the success of P.P. Keat stimulated the search for other possible high pressure crystalline modifications of silica. They were soon found by L. Coes, Jr., S.M. Stishov and S.V. Popova [18, 19] in form of two modifications bearing the names of two of the mentioned scientists, who were the first to synthesize them. These two modifications, obtained at high pressure, showed the same linear dependence of the refractive index, n, on density, ρ, as presented on Fig. 3.14. An important result of this development was the establishment of the fact that stishovite has a structure differing from all other crystalline silica modifications known till that times: its structural units are not the classical SiO$_{4/2}$-tetrahedra, but SiO$_{6/2}$-octahedra (see Fig. 3.7).

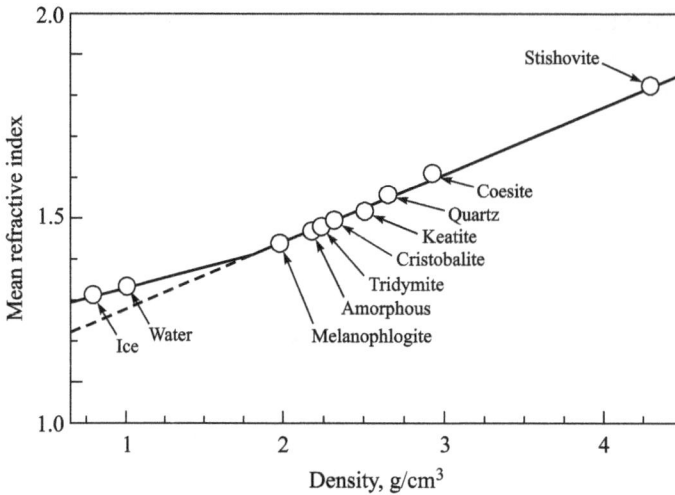

Fig. 3.14: Dependence of the coefficient of optical refraction, n, on density, ρ [in g/cm^3], for the different phases of silica and water (according to Iler [3]).

3.4.5 Synthesis of Coesite

The structure of coesite is constituted (as that of cristobalite and quartz) of $SiO_{4/2}$-tetrahedra, however, in a particular, denser arrangement, than in quartz and cristobalite. Coesite is usually synthesized at 500–800 °C at 35 kbar and is characterized by a relatively high density ($\rho = 2.93$g/cm^3). In nature it is found in meteorite impact craters on sandstone soils and in the kimberlite tubes. In both cases these locations are rocks with a prehistory of enormous pressures and high temperatures (the South African kimberlites are a well known deposit of natural diamonds). According to our estimate of the solubility of coesite (made by using thermodynamic data on density (see Section 3.6)), this silica modification could be also synthesized hydrothermally. At normal pressure coesite decomposes at 1300 °C giving quartz. Up to now no technological process is developed for its industrial production: only microscopic crystals are obtained of this SiO$_2$-polymorph.

3.4.6 Stishovite: Synthesis and Thermal Stability

For the synthesis of stishovite, two general methods have been employed up to now, hydrothermal nucleation and growth at high temperatures and pressures higher than 80 kbar [71] in belt, anvil and hammer cameras similar to those employed for the synthesis of diamond (cf. Figs. 3.15 and 3.16). However, up to now only stishovite microcrystals have been synthesized and their price is even higher than that of diamond micro-crystals of good quality.

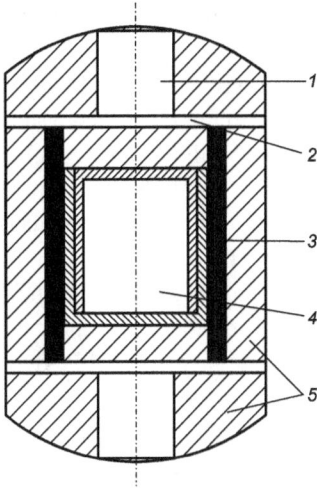

Fig. 3.15: The small ultra-high pressure autoclave cell of Lityagina, Dyuzheva et al. [71] used to synthesize stishovite crystals at 9.5 GPa and 1170–1770 K: (1) current feed-through; (2) metal disk; (3) graphite heater; (4) inner autoclave ampoule; (5) outer part of autoclave.

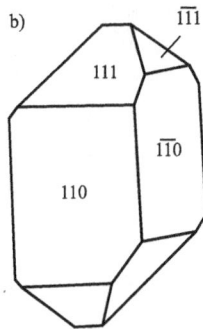

Fig. 3.16: Stishovite crystal synthesized by the authors of [71]: (a) real crystal and (b) face symbols.

The initial synthesis of stishovite was performed by Stishov under pressures of approximately 100 kbar. During the last ten years, several investigations have been published, in which the synthesis of the so-called post-stishovite modifications of SiO_2 is described. These crystalline silica modifications have been obtained and are stable as it seems at even higher pressures than stishovite (up to 800 kbar!).

The interest in stishovite arises from both experimental findings and theoretical predictions [72, 73] indicating that stishovite and post-stishovite structures are most probably the hardest oxides ever known and ever synthesized. It is claimed [72, 73] that their mechanical properties are surpassed only by diamond and cubic boron nitride. However, the thermal stability of stishovite is even less than that of coesite: it decomposes at normal pressures even below 800 °C into SiO_2-glass, then into cristobalite and, at the end of the process, into quartz (see Fig. 3.15). Stishovite (like coesite) has been also found in the meteorite impact crater of Arizona.

3.4.6.1 Hydrothermal Methods, Anvil and Hammer, and Belt Cameras. Direct High Pressure Synthesis

In a recent paper of Litiyagina et al. [71] the hydrothermal synthesis and growth of stishovite is described in detail. The synthesis is performed in the H_2O/SiO_2-system in a high pressure cell (in fact a little autoclave with a high filling percentage) at pressures from 90 to 95 kbar (see Fig. 3.15). This pressure is developed by heating the autoclave to 1200 K. In this way, stishovite crystals up to a size of 1 mm have been obtained and thoroughly investigated [71]. This investigation, performed in the Institute of High Pressure Physics of the Russian Academy of Sciences in Troitzk, indicates the technical developments which could be used in order to obtain even larger stishovite crystals. A schematic picture of the cell employed in [71] is given on Fig. 3.15. The shape of the stishovite crystals, grown in such way, is illustrated in Fig. 3.16.

3.4.6.2 Post-stishovite Modifications of SiO_2 with $CaCl_2$- and PbO_2-structure

The anvil and hammer, respectively, belt high pressure cameras for direct synthesis of stishovite are described by Dubrovinskaya and Dubrovinsky in [72, 73], where the ways of synthesis of stishovite-like structures with (α-PbO_2) and $CaCl_2$-structure are outlined. As mentioned, stishovite has been observed in meteorite impact material on the Earth as discussed in several publications [74, 75].

3.4.6.3 Amorphization of Stishovite under Normal Pressure and at Elevated Temperatures

It was pointed out above that stishovite, when heated up to 600–1000 °C, disintegrates into amorphous silica with a structure analogous to the one of the usual silica-glass obtained e.g. by melting of quartz. This process has been thoroughly investigated in [75–77], mostly by employing DSC- and DTA-techniques. Further heating of this amorphous SiO_2-material leads to the formation of quartz crystals (see Fig. 3.17). Here, it is to be especially mentioned that this is the interesting phenomenon of the phase transition of a crystal (stishovite) stable only under very high pressures to another crystal (cristobalite, quartz) through the intermediate structure of an amorphous solid: the silica-glass. It seems that the direct transformation of crystalline stishovite into crystalline coesite is impossible or at least very difficult to realize because of the enormous structural change connected with this process. The transformation of stishovite into amorphous SiO_2 was investigated from a thermodynamic point of view in [25, 28].

From the technical point of view, the disintegration of stishovite into amorphous SiO_2 and quartz strongly limits its possible application at normal pressures as an ultra-hard oxide to temperatures only below 600 or 700 °C. The mentioned stishovite transition is, however, very significant from the geochemical point of view, because it is a process taking place in the Earth's mantle. It is also of interest for all the possible trans-

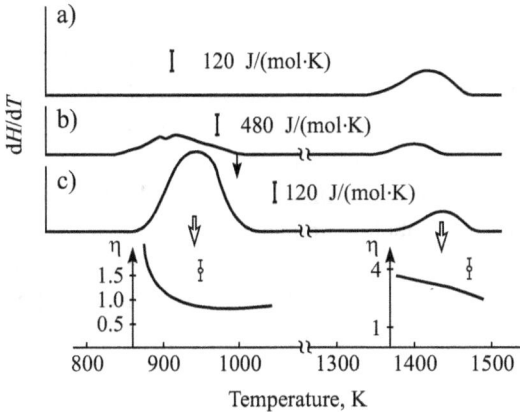

Fig. 3.17: Change of coesite (a) and stishovite ((b) and (c)) during heat-treatment at normal pressure. The heat released during the process is indicated as a parameter to the respective curve: (a, b) heating runs obtained at 5 K/min and (c) at 20 K/min. The first peak on (b, c) represents the transition of stishovite into amorphous SiO$_2$ (glass); the second peak describes the subsequent crystallization of cristobalite (according to [38]).

formations stishovite → coesite → quartz which may take place in volcanic eruptions and in catastrophic meteorite impacts onto the Earth's surface. These problems are analyzed in many publications in the geological literature.

3.4.7 Synthesis of Amorphous Modifications of Silica

The synthesis of SiO$_2$-glass via melting of quartz is a well-known process and needs no particular description here. At present, of higher technical significance are chemical methods of synthesis of super-pure SiO$_2$ employing the following gas reaction,

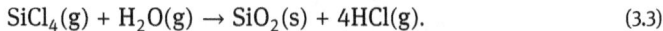

$$SiCl_4(g) + H_2O(g) \rightarrow SiO_2(s) + 4HCl(g). \tag{3.3}$$

It results in a condensate, called Suprasil, and similar products, which are obtained by above reaction in a voluminous cotton-like form. These products are, however, relatively expensive and their technical application is restricted only to special cases.

The method of production of micron sized SiO$_2$-glass spheres (ballotines) is also well-known. In this process SiO$_2$-glass powder of the desired fraction is heated up above the melting point of silica on carbon soot powders. The synthesis of SiO$_2$ in amorphous form via sol-gel reaction is described in Section 3.4.2, where the sol-gel synthesis of cristobalite is described. In this way super-clean SiO$_2$-glass can be obtained: however, again at a very high cost.

3.4.7.1 The Problem of Existence of Stishovite Glasses

It is not clear as yet, whether stishovite, melted e.g. at 100 kbar, could produce a 6-fold coordinated stishovite glass. In some respects this possibility is analogous to the attempts to synthesize (besides "normal" graphite–like sp^2-structured vitreous carbons) also carbon glasses with a diamond-like sp^3-structure.

The usual amorphous SiO_2-modifications, synthesized and existing in stable form at normal pressure, are built (as quartz, cristobalite, and coesite) out of four coordinated SiO_2, i.e. out of $SiO_{4/2}$-tetrahedra. It has been suggested in the literature that, similarly to water and liquid carbon, a "poly-amorphism" of vitreous silica could be also expected if melting of this substance proceeds at extremely high pressures (in the stability region of stishovite (see Fig. 3.6)). In this sense, stishovite and maybe coesite-like liquids and glasses could be expected in the development of further synthesis of the amorphous modifications of SiO_2 (coesite glass: with a denser stochastic arrangement than the "normal" quartz glass). Up to now, no direct synthesis of these possible structural polymorphs of amorphous SiO_2 has been performed or reported in literature.

3.5 Structure and Thermodynamic Properties of the SiO₂-modifications

As mentioned in the introduction, SiO_2 is the major constituent of rock forming minerals in Earth's magmatic and metamorphic rocks. It is also an important component of sediments and soils. Bound in the form of silicates, SiO_2 accounts for approximately 75 wt% of the Earth's crust [19]. Free silica predominantly occurs in nature as quartz, which makes 12–14 wt% of the lithosphere. Quartz is the thermodynamically stable modification of silica at ambient pressure (see Figs. 3.1–3.6), occurring as one of the main products of slowly cooled silica rich magmas, granites, granodiorites and related rocks. Despite of its chemical simplicity, SiO_2 displays a remarkable diversity of crystal structures. Apart from the ultra-high pressure modification, stishovite with octahedral coordinated silicon (see Fig. 3.7) and some artificial silica products, where chains of edge sharing tetrahedral occur, all other crystal structures of silica are formed from a three dimensional framework of corner sharing tetrahedra (see Fig. 3.7 and [16, 17]).

The non-crystalline silica phases existing and stable at normal pressure also consist of continuous random three-dimensional networks of corner sharing tetrahedra [14]. The Si–O–Si bond has a mixed character: about 50 % ionic and 50 % covalent. The bonding results in inter-tetrahedral Si–O–Si bonds which are bent in the range from 120 to 180° with a mean value of about 147° (see again Fig. 3.7). The strong bonds and the three dimensional connectivity of the $SiO_{4/2}$ tetrahedrons are the reason for the following properties: high hardness of SiO_2-modifications (quartz has a Mohs hardness of 7, stishovite, however, approaches 10), lack of good cleavage, high elasticity, high melting point (approximately 2000 K of cristobalite), high activation temperature of the quartz-cristobalite transformation (ca. 1300 K), and high glass transition temperature, T_g, of silica glass (1473 K). In contrast to these properties, the resistivity of silica to irradiation damage is relatively low.

Silica in all its forms is an insulator. This property is utilized in the fabrication of silicon based micro-electronic devices and in the semiconductor technique. The SiO$_2$-tetrahedra of the crystal modifications, stable at normal conditions, can be considered as rigid structural units. They remain almost unchanged upon thermal expansion or compression by high pressures and can only rotate or tilt. The variability of the inter-tetrahedra Si-O-Si angles and the unrestricted torsion angle of connected tetrahedra account for the topological diversity of crystal structure of silica and of its high tendency for glass formation.

Quartz is the stable SiO$_2$-form under normal on the Earth conditions. The thermal expansion coefficient of α-quartz is relatively high, i.e., $\alpha_{11} = 13.3 \cdot 10^{-6}$ K^{-1}, $\alpha_{33} = 7.1 \cdot 10^{-6}$ K^{-1} and drops to even slightly negative values at the displacive phase transition to the high-temperature β-quartz modification at 573 °C (Fig. 3.18). The respective volume change ($\Delta V/V$) is shown on Fig. 3.18.

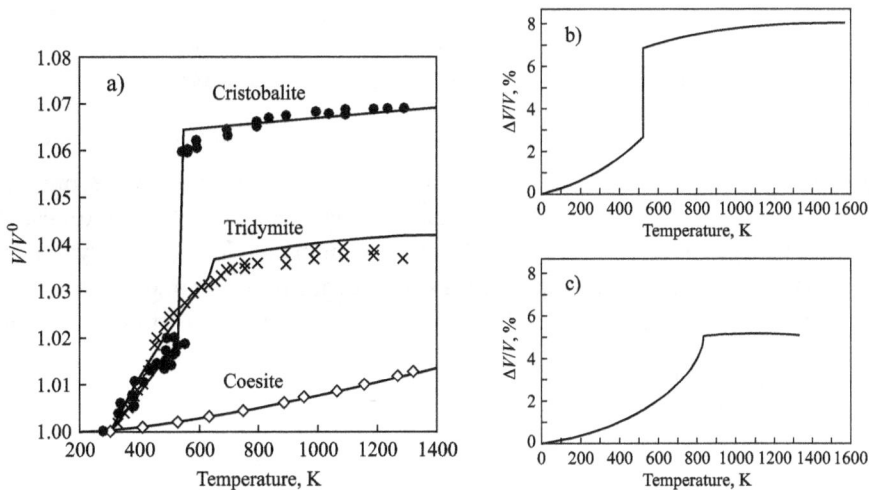

Fig. 3.18: Volume changes at phase transformations of SiO$_2$-modifications upon heating at normal pressure: (a) cristobalite, tridymite, coesite after [25]; (b) cristobalite, and (c) quartz (after Ullmann [19]).

Cristobalite is the low-pressure high-temperature modification of silica. It persists as a metastable phase (α-cristobalite) at low temperatures. A displacive phase transition from the tetragonal low-temperature α-cristobalite form to the cubic high-temperature form (β-cristobalite) takes place at nearly 540 K. This transition is usually accepted to be of the first order (as witnessed by the ΔV change in Fig. 3.18a and b) with a hysteresis of more than 20 K (at heating! – which is quite unusual) and a volume discontinuity of approximately 5 %. However, opinions are also expressed in literature that this $(\alpha - \beta)$-transition is of mixed first and second order. Our own experimental results, described here in Section 3.4.3, strongly support this second interpretation. The crystal structure of β-cristobalite is a derivative structure of the diamond structural type, in which carbon is replaced by silicon, and oxygen is located midway between neighboring Si-atoms. This network can be described as a stacking of parallel layers of six-membered rings of tetrahedra, alternately directed upwards and downwards.

A very particular property of α-cristobalite is that the low temperature form of SiO_2 exhibits a negative Poisson coefficient, i.e. at elongation the crystal is not thinned, but swelled. The well-expressed volume discontinuity at the $(\alpha - \beta)$-cristobalite transition leads to the generation of micro-cracks. Up to now the displacive cubic to tetragonal $(\beta \rightarrow \alpha)$-transition could not be suppressed by quenching, however, the structure of β-cristobalite can be stabilized in the way of chemical stuffing and toughening as this was already discussed in the previous sections. Other compounds with β-cristobalite structure type are the high temperature forms of $AlPO_4$ and $GaPO_4$ [57]. This is also used as an idea for possible toughening of the structure of β-cristobalite with the help of these compounds. At high temperatures β-cristobalite has a coefficient of thermal expansion nearly equal to zero. For more details in this direction we refer to the respective sections in the present review, as well as to [44, 45] entirely devoted to this problem.

Tridymite is formed in the range of 1200–1800 K at ambient pressure, however, only in the presence of foreign ions (K^+ or Na^+). Its crystal structure is a derivative of the hexagonal wurtzite type. With respect to the structural bonding requirements tridymite is much less balanced than cristobalite. As a consequence tridymite undergoes a cascade of displacive phase transitions (α, β, γ, see Fig. 3.1). In the present analysis, as far as we are interested only in modifications of pure SiO_2, no further discussion of tridymite and these modifications is given.

Coesite is a high pressure modification of silica with the densest framework of $SiO_{4/2}$-tetrahedra. It is composed out of four-membered rings of tetrahedra which are linked in the form of chains of rings. Coesite exists as a metastable crystalline silica polymorph at normal ambient pressure and at temperatures up to 1300 K, then it is transformed to quartz (see Figs. 3.3 and 3.6).

Stishovite crystallizes in the rutile structural type with silicon atoms in the mentioned six-fold oxygen coordination. The $SiO_{6/3}$-octahedra form chains by sharing opposite edges. These arrangements result in a very close packing of oxygen even though the Si-O-bonds are longer than in SiO_4-tetrahedra (see Fig. 3.7). Stishovite is about 43 % denser than coesite. In the mentioned post-stishovite modifications with $CaCl_2$

or α-PbO$_2$ structure the highest density of packing of oxygen atoms, theoretically possible, seems to be achieved. The change to the even denser CaCl$_2$-structure type is expected to involve a peculiar tilting of the octahedra. Moreover, in mineralogical crystallographic literature several additional modifications of SiO$_2$ are discussed. They are essential for the understanding of the structure of chalcedones, chalcites and opals. Accounting for the purpose of the present analysis, we will not go into detail with respect to such varieties of crystalline and amorphous SiO$_2$ like moganite, melanophlogopite, etc.

Keatite has a tetragonal framework of corner sharing SiO$_{4/2}$-tetrahedra similar to the structure of β-spodumene (LiAlSi$_2$O$_6$). Another variety of SiO$_2$ is also known. It is formed at temperatures above 1500 K via the reaction

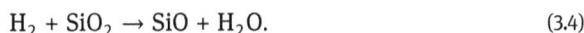

$$H_2 + SiO_2 \rightarrow SiO + H_2O. \tag{3.4}$$

This is in fact a silicon monoxide (SiO) rich modification.

In the mineralogical literature, glassy silica of natural occurrence is called lechatelierite. It originates from molten silica formed by lightning strikes (forming the so called fulgurites) or by meteoritic impacts on quartz sand or silica rocks. The volumetric properties of the SiO$_2$-modifications are summarized in Table 3.2. The thermodynamic properties of the SiO$_2$-modifications, as they are given in the literature [8, 9, 25–27], are reproduced here in Tables 3.3, 3.4 and 3.5. The heats and entropies of transition between the different phases, stable or metastable at normal pressure, are best illustrated according to [13] on Fig. 3.19, where the respective Gibbs free energies of transition are given with respect to cristobalite.

Table 3.2: Bulk properties of SiO$_2$-phases (* at 623 K; ** at 1673 K, however, in fact as a glass)

Phase	Density, g · cm^{-3}	Molar volume, cm^3 · mol^{-1}
Low quartz (α)	2.65	22.69
High quartz (β)	2.60	22.86
Low tridymite (α)	2.30	26.53
Middle tridymite ($\beta1$)	2.30	–
High tridymite ($\beta2$)	2.27	27.51
Low cristobalite (α)	2.32	25.74
High cristobalite (β)	2.21	27.40*
Glass	2.20	27.31
Keatite	2.50	24.04
Coesite	3.01	20.64
Stishovite	4.35	14.01
Amorphous	–	30.04
Liquid **	?	27.20 (?)

Table 3.3: Thermodynamic properties of the silica phases at 298 K and normal pressure. Note that the authors of [25], in fact, treat the SiO_2-glass as a liquid

Phase Literature source	Enthalpy of formation kJ mol^{-1} [25]	Entropy of formation J mol^{-1} K^{-1} [25]	Enthalpy of formation kJ mol^{-1} [8]	Entropy of formation J mol^{-1} K^{-1} [8]	Enthalpy of formation kJ mol^{-1} [103]	Entropy of formation J mol^{-1} K^{-1} [103]
Low quartz (α)	−910.70	41.46	−910.57	41.80	−870.70	42.22
High quartz (β)	−910.50	41.70	–	–	–	–
Low tridymite (α)	–	–	–	–	–	–
Middle tridymite ($\beta 1$)	−906.91	45.12	–	–	–	–
High tridymite ($\beta 2$)	–	–	–	–	–	–
Low cristobalite (α)	−906.03	46.06	−904.55	43.47	−859.41	43.26
High cristobalite (β)	–	–	−907.39	42.64	–	–
Glass	–	–	−900.70	46.80 [13]	–	46.82
Keatite	–	–	–	–	–	–
Coesite	−906.90	40.50	−905.00	40.34	–	–
Stishovite	−864.00	29.50	−860.70	27.76	–	–
Liquid	−901.00 [25]	49.00	–	–	–	–
Amorphous (disperse)	–	–	−895.98	–	–	–

Table 3.4: Specific heats of the SiO$_2$-phases at 298 K and normal pressure

Phase Literature sources	Specific heats, J mol^{-1} K^{-1} [25]	Specific heats, J mol^{-1} K^{-1} [8]
Low quartz (α)	44.59	44.39
High quartz (β)	44.59	–
Low tridymite (α)		–
Middle tridymite (β_1)	44.25	–
High tridymite (β_2)		–
Low cristobalite (α)		44.14
High cristobalite (β)	44.30	44.56
Glass	–	–
Keatite	–	–
Coesite	42.79	45.35
Stishovite	42.16	42.93
Liquid	44.22	–
Amorphous (disperse)	–	43.47

Table 3.5: Molar enthalpies and entropies of transition between the different phases of SiO$_2$

Transition	T, [8] K	ΔH, [8] kJ mol^{-1}	ΔS, [8] J mol^{-1}K^{-1}	ΔH, [26] kJ mol^{-1}	ΔS, [26] J mol^{-1}K^{-1}	ΔH, [103] kJ mol^{-1}	ΔS, [103] J mol^{-1}K^{-1}
α-quartz → β-quartz	846	0.63	0.75	–	–	0.75	–
β-quartz → tridymite	1140	0.50	0.46	–	–	–	–
α-quartz → melt	1883	8.53	4.51	–	–	14.21	–
α-cristob. → β-cristobalite	515	1.30	2.51	–	–	–	–
β-cristob. → melt	2001	7.70	3.85	–	–	8.78	–
α-quartz → cristobalite	–	–	–	2.68	–	–	–
α-quartz → coesite	–	–	–	5.06	–	–	–
α-quartz → silica glass	–	–	–	8.99	–	12.54	–
quartz → stishovite	–	–	–	49.32	–	–	–
cristobalite → silica glass	–	–	–	6.31	–	–	–

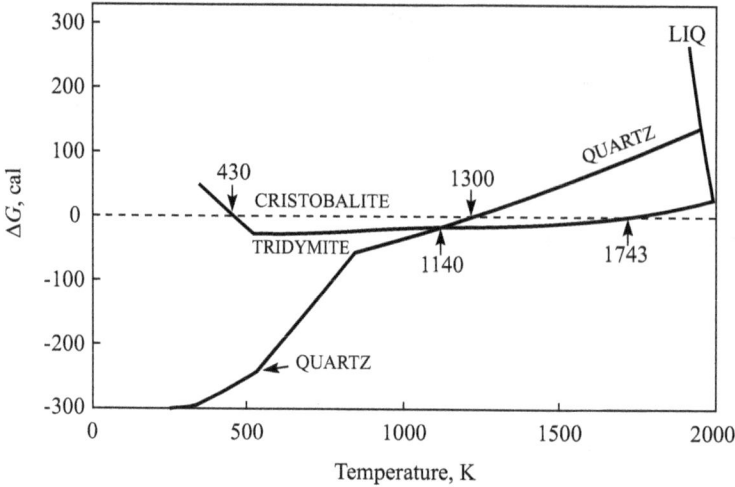

Fig. 3.19: Difference ΔG [in cal/mol] in the thermodynamic potential between the SiO_2-phases exist-ing in stable or metastable equilibrium at normal pressure. This difference is expressed with respect to G^{298} of β-cristobalite (given as a zero ordinate). The numbers above or below the arrows indicate the temperatures for transition to or from β-cristobalite (data [13]).

3.6 Solubility of the Different SiO_2-modifications

3.6.1 General Thermodynamic Dependencies

The solubility of the different modifications of SiO_2 is determined in the present sec-tion. Hereby we use available thermodynamic data, results of our previous investi-gations summarized in [14] and the classical approach, introduced by I.F. Schroeder into the physical chemistry of solutions at the end of the 19-th century (see [78] and literature cited there).

In calculating the solubility of the different modifications of silica we assume that the solubility of SiO_2 under hydrothermal conditions at temperature T can be described with sufficient accuracy by the dependence

$$\log C_x \cong -\frac{\Delta H_x^{298}}{2.3RT} + \frac{\Delta S_x^{298}}{2.3R}. \tag{3.5}$$

Here ΔH_x^{298} and ΔS_x^{298} are molar enthalpy and molar entropy of dissolution of the re-spective x–modification in the hydrothermal solution, respectively, and R is the uni-versal gas constant. As known, hydrothermal SiO_2-solutions are highly diluted. In pure water C_x corresponds to concentrations in the order of several ppm at temper-atures of 300–500 K. In this case, the classical dissolution model of Schroeder can be employed, according to which in Eq. (3.5) ΔH_x^{298} and ΔS_x^{298} can be substituted as follows:

$$\Delta H_x^{298} \cong \Delta H_{mx}^{298} \tag{3.6}$$

and

$$\Delta S_x^{298} \approx \Delta S_{mx}^{298}. \tag{3.7}$$

Here ΔH_{mx} and ΔS_{mx} denote the molar enthalpy and entropy of melting of the considered SiO$_2$-modification, respectively. Thus, in correspondence with Schroeder's model of dissolution of very slightly soluble crystals [78], we assume that $\Delta H_m^{mix} \cong 0$ and $\Delta S_m^{mix} \cong 0$, where ΔH_m^{mix} and ΔS_m^{mix} are molar enthalpy and molar entropy of mixing corresponding to the dissolved quantity of the respective substance, in our case the considered silica polymorph. Because of the low solubility of SiO$_2$ both quantities (ΔH_m^{mix} and ΔS_m^{mix}) can be neglected, as indicated here. Accounting for the mentioned low concentrations of the SiO$_2$ modifications at hydrothermal conditions (in molar terms: $C = 10^{-4}$ to 10^{-6} mol l^{-1}) this is quite an acceptable approximation.

From a more general kinetic and thermodynamic point of view, Eq. (3.5) with constant ΔH and ΔS-values is a typical case of the so-called first approximation of Ulich [70] which is widely accepted as a most direct way in the thermodynamical calculus of chemical reaction kinetics and solubility. This second approximation (the assumption of $\Delta H_m = \Delta H(T) =$ const. and $\Delta S_m = \Delta S(T) =$ const.) is quite satisfactory in our case at temperatures $T \ll T_m(\text{SiO}_2)$. It is shown experimentally that for temperatures lower than the critical temperature of H$_2$O, the log C vs. $1/T$ dependencies give in fact straight lines corresponding to Ulich's approximation. This is seen from the three cases of solutions of glass-forming systems, illustrated in Fig. 3.20. Schroeder's model of dissolution of crystals assumes that the process determining the dissolution

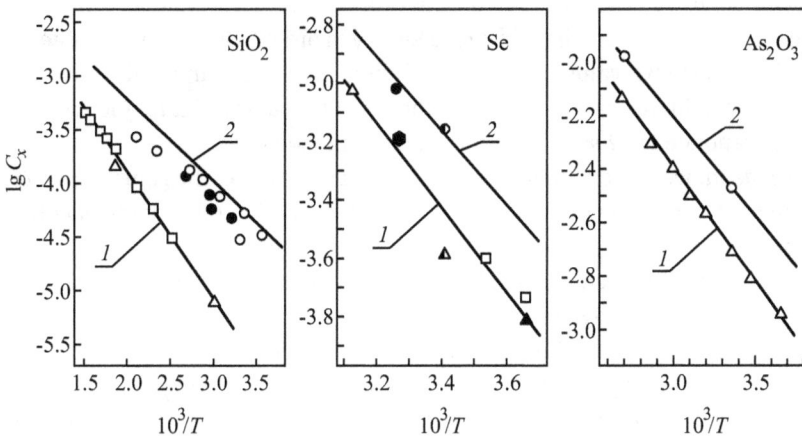

Fig. 3.20: Experimental data on the solubility C_x of simple inorganic glass-forming systems in coordinates log C_x vs. $(1/T)$ according to Eq. (3.5). As a measure of solubility the respective molar fraction C is used. Solubility of SiO$_2$ in water (from hydrothermal autoclave experiments); Solubility of Se in liquid CS$_2$; Solubility of As$_2$O$_3$ in water. In all cases, the respective crystalline phase (quartz for SiO$_2$, monoclinic Se and cubic As$_2$O$_3$) is specified by the number 1 and with 2 the respective vitreous phase (from [14]).

of a slightly soluble crystal is the melting of its crystalline structure in order to reach the liquid state of the considered solution.

In order to determine enthalpy and entropy of melting (cf. Eqs. (3.5)–(3.7)) we use, employing Ulich's approximation, the following estimate

$$\Delta H_{mx}^{298} \cong H_x^{298}(\text{liquid}) - H_x^{298}(\text{solid}), \tag{3.8}$$

i.e. we assume that $\Delta H_{mx}(T_m) \sim \Delta H_x^{298}$ holds. Under "solid" in the above approximations we also include the enthalpy of dissolution of the respective vitreous modifications of SiO_2 or of any other solute. However, as a rule we do not know the respective melting enthalpies and entropies of the SiO_2 modifications. This is so, because of the extremely high melting temperatures of cristobalite (approaching 2000 K and even the calculated melting temperature of quartz $\sim$ 1880 K). Usually one knows only the differences between the enthalpies ΔH_g and the entropies ΔS_g between the SiO_2-glass and the respective crystal modification because these quantities can be easily determined in any dissolution calorimeter operating with hydrofluoric acid as a solvent. This is the reason why we decided to use two empirical dependencies known in glass science [14] according to which

$$\Delta H_g \cong \frac{1}{2}\Delta H_m, \tag{3.9}$$

$$\Delta S_g \cong \frac{1}{3}\Delta S_m. \tag{3.10}$$

These two dependencies, proposed years ago by Tammann (1932) and Gutzow (1971), are satisfactorily fulfilled for more than 100 cases of glass-forming systems. This result gives us the foundation to use them in a reverse way in order to estimate the values of ΔH_m and ΔS_m from known values of ΔH_g and ΔS_g, which can be found in the literature [8, 9, 13]. However, in the case of SiO_2 even the standard values of ΔH_m and ΔS_m (i.e. the values extrapolated to 298 K) are not known for all modifications.

By this reason, we decided to follow an easier way to predict the solubility of the different modifications of SiO_2 via Eqs. (3.5)–(3.10) by exploiting the simple fact that many properties of the SiO_2-modifications can be related to their density, ρ. From a more general standpoint such a dependence of ΔH_m and ΔS_m on density could be expected as far as both thermodynamic quantities are essentially configurational in their nature: they have to depend on the relative free volume of the respective crystal, determining its configurational properties. In particular, for the melting entropy we expect (see [14]) $\Delta S_m \sim R\ln(\Delta V_o)$ where ΔV_o is the difference in between the relative free volumes V_o of the melt and the respective crystal. Expanding the logarithm in above simple approximative dependence for $V_o < 1$, we arrive at

$$\Delta S_x^{298} = a_o + b\rho, \tag{3.11}$$

where a_o and b_o are constants. As seen from Fig. 3.21, such a dependence between the known ΔS_x^{298}-values and ρ gives an acceptable straight line for all the modifications, with $a_o = 61.9$ and $b_o = -7.2$, when ΔS is expressed in J mol^{-1} K^{-1}. It is seen that

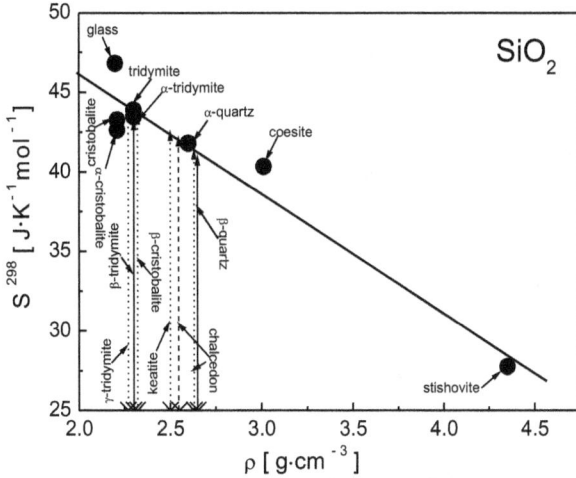

Fig. 3.21: Molar entropy of formation S^{298} at 298 K of the SiO$_2$-modifications in dependence on their density, ρ. Only those cases are presented (as black points) for which the thermodynamic data are available (cf. Tables 3.2 and 3.3) in the reference literature [8, 9]. The straight line corresponds to Eq. (3.11) with the respective constants given in the text.

this dependence in some respects corresponds to the already cited linear relation (see Fig. 3.14) between the refractive index, n, of the crystalline and of the amorphous modifications of SiO$_2$ and their density, ρ.

The enthalpy difference ΔH should be in general a more complicated function of density: in fact the enthalpy according even to very simple mean field models described in [14] is a quadratic function of the free volume (e.g. $\Delta H \sim$ const.$(1 - \rho)\rho$ as indicated in [14]). In fact as seen from Fig. 3.22

$$\Delta H^{298}_{mx} \cong c_0 + g\rho + h\rho^2 \tag{3.12}$$

where the constants are equal to $c_0 = -746$, $g = -119.3$, $h = 21.6$, ΔH is expressed in kJ mol^{-1} and ρ is given in g cm^{-3}. Thus, we can rewrite Eq. (3.5) in the form

$$\log C_x = -\frac{\Delta H^{298}_{mx}}{2.3RT} + \frac{\Delta S^{298}_{mx}}{2.3R} \tag{3.13}$$

$$= \frac{1}{2.3R}\left[-\left(\frac{H^{298}(\text{liquid}) - H^{298}(\text{cryst})}{T}\right) + \left(S^{298}(\text{liquid}) - S^{298}(\text{cryst})\right)\right].$$

The respective enthalpy and entropy values can be taken, when known, from reference literature. In our case, using Eqs. (3.11) and (3.12), from Eq. (3.13) we obtain

$$\log C_x = \frac{1}{2.3R}\left[-(A_0 - g\rho - h\rho^2)\frac{1}{T} + B_0 - b\rho\right]. \tag{3.14}$$

Here A_0 and B_0 are constants depending on the units in which C_x is expressed.

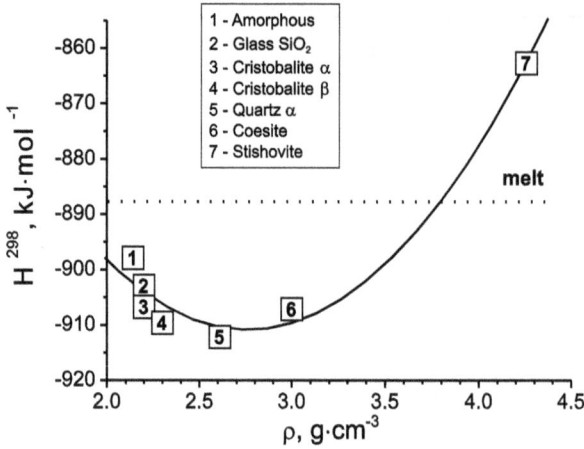

Fig. 3.22: Molar enthalpy of formation H^{298} in dependence on density, ρ, of the SiO_2-modifications for which data are available in [8, 9]. The quadratic dependence (Eq. (3.13), solid line) is drawn with the values of the constants given there. The horizontal dotted line presents the enthalpy H^{298} of the SiO_2-melt, using Eq. (3.9) with $\Delta H^{298} = H^{298}_{glass} - H^{298}_{cryst}$ in accordance with points 2 and 5.

When we know the solubility and the respective $\log C$ vs. $(1/T)$-dependence of one of the modifications (e.g. for $x = 1$), the solubility ratio with another substance ($x = 2$) can be expressed via the following relation

$$\log\left(\frac{C_1}{C_2}\right) = \frac{1}{2.3R}\left[-\frac{g\left(\rho_1 - \rho_2\right) + \left(\rho_1 - \rho_2\right)^2 h}{T} + b\left(\rho_1 - \rho_2\right)\right]. \tag{3.15}$$

In a more condensed form the above expression can be written as

$$\log\left(\frac{C_1}{C_2}\right) = -\frac{\left(\rho_1 - \rho_2\right)}{2.3R}\left[\frac{g + \left(\rho_1 - \rho_2\right)h}{T} - b\right]. \tag{3.16}$$

In this way, using above described simple formalism by knowing the solubility curve of one of the modifications of SiO_2 we can calculate the solubility curve of any other SiO_2-modification with known density, ρ.

In the next section we construct the solubility curves of all the SiO_2-modifications using the van Lier dependence (see Iler's book [3]). According to this relation, in the whole range from room temperature to the critical point of water the solubility of quartz can be expressed via

$$\log C_{quartz} = 0.151 - \frac{1162}{T}. \tag{3.17}$$

In this way using Eq. (3.17) for C_{quartz} we constructed Fig. 3.23 presented in the next section for all other modifications of SiO_2. By accepting Eq. (3.17) as a sufficiently good approximation for the solubility data of quartz at hydrothermal conditions, we change

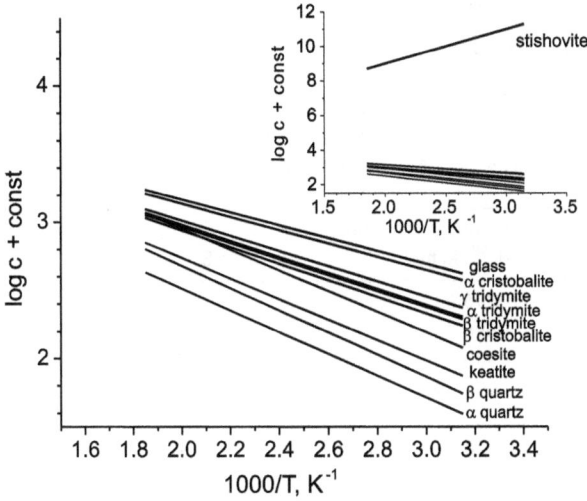

Fig. 3.23: The calculated solubility of all SiO$_2$-modifications (including tridymite) in water using the data from Figs. 3.21 and 3.22 and Eqs. (3.13) and (3.15) with the constants given there. In order to allow us a qualitative comparison with experimental data given in Fig. 3.24, the solubility of quartz is determined employing van Lier's relation (Eq. (3.17)) and introduced as $C_1(1/T)$ into Eq. (3.15). Note that Eqs. (3.13) and (3.15) refer only to aqueous solutions below the critical point of water. In the insert the solubility of stishovite is presented. It is computed employing the assumption that it is not transformed into SiO$_2$-glass at the hydrothermal experiment. Note the positive slope of the log C vs. $(1/T)$-dependence for stishovite.

in fact the value of the right-hand side additive term in Eqs. (3.5), (3.13) and (3.16) to $[(\Delta S_x^{298}/2.3R) + \text{constant}]$. Here the value of the constant depends on the dimensions used in expressing C_x.

3.6.2 Solubility Diagram of SiO$_2$. Ostwald's Rule of Stages

We constructed the solubility diagram of SiO$_2$ shown in Fig. 3.23 employing the S^{298} vs. ρ and H^{298} vs. ρ dependencies given in Figs. 3.21 and 3.22. In order to proceed, we first introduced into both Figs. 3.21 and 3.22 the values of S^{298}(liquid) and H^{298}(liquid) in the form

$$S^{298}(\text{liquid}) = S^{298}(\text{quartz}) + 3\Delta S_g^{298} \tag{3.18}$$

and

$$H^{298}(\text{liquid}) = H^{298}(\text{quartz}) + 2\Delta H_g \tag{3.19}$$

utilizing Eqs. (3.9) and (3.10). In this way, taking

$$S^{298}(\text{quartz}) = 41.8\,\text{Jmol}^{-1}\text{K}^{-1}, \quad H^{298}(\text{quartz}) = -910.57\,\text{kJmol}^{-1} \tag{3.20}$$

Table 3.6: Peer values of the thermodynamic properties of the phases of silica at 298 K and normal pressure. These values are used in the calculation of the solubility at hydrothermal conditions: [a] Data according to Glushko et al. [8] compared and corrected employing the data of Swami and Saxena [25]; [b] H^{298} (liquid) calculated as H^{298} (quartz) + ΔH_m^{298} (quartz) and S^{298} (liquid) calculated as S^{298} (quartz) + ΔS_m^{298} (quartz); [c] ΔH_m and ΔS_m were calculated according to the known empirical dependencies like $\Delta H_g = (1/2)\Delta H_m$ and $\Delta S_g = (1/3)\Delta S_m$; [d,e] estimates given by Glushko et al. [8]

Phases	Enthalpy of formation [a] H^{298}, kJ mol^{-1}	Entropy of formation [a] S^{298}, J mol^{-1} K^{-1}
Liquid [b]	−890.83	56.80
Amorphous (disperse)	−896.00	22.74
Glass	−900.70	46.80
High cristobalite (β)	−907.39	42.64
Low cristobalite (α)	−904.55	43.47
Low quartz (α)	−910.57	41.80
High quartz (β)	−910.50	–
Coesite	−905.00	41.70
Stishovite	−860.70	27.76
Phase transition	**Enthalpy of change** [d] ΔH, kJ mol^{-1}	**Entropy of change** [e] ΔS, J mol^{-1} K^{-1}
Melting of low quartz (α) [c]	19.74	15.00
Glass crystallization → low quartz (α)	9.87	5.00
Melting of high cristobalite (β)	(7.69)	(3.85)

and

$$\Delta S_g = 5 \, \mathrm{Jmol}^{-1}\mathrm{K}^{-1}, \quad \Delta H_g = 9.87 \, \mathrm{kJ \, mol}^{-1} \tag{3.21}$$

(cf. Table 3.6), we obtain

$$H^{298}(\text{liquid}) = -890.8 \, \mathrm{kJ \, mol}^{-1}, \quad S^{298}(\text{liquid}) = 56.8 \, \mathrm{Jmol}^{-1}\mathrm{K}^{-1}. \tag{3.22}$$

In order to compare our thus calculated diagram with existing experimental data we employed the equation of van Lier, Eq. (3.17), for the solubility of quartz. In this relation, the solubility C is expressed in ppm. On Fig. 3.24, known data for the solubility of several different modifications of SiO_2, which are stable at normal pressure, are given as they are summarized in the book of Iler [3]. In this figure, results not only for temperatures below the critical point of water (647 K) are shown, but also those corresponding to the solubility above the critical point (see the solubility data on the figure below the left-hand side maximum). As it is evident, our schematic construction on Fig. 3.23 is in satisfactory agreement with existing experimental data given on Fig. 3.24 as compiled by Iler [3] for the solubility of quartz, cristobalite and vitreous SiO_2. For each of the SiO_2 modifications we constructed the respective log C vs.

Temperature, °C

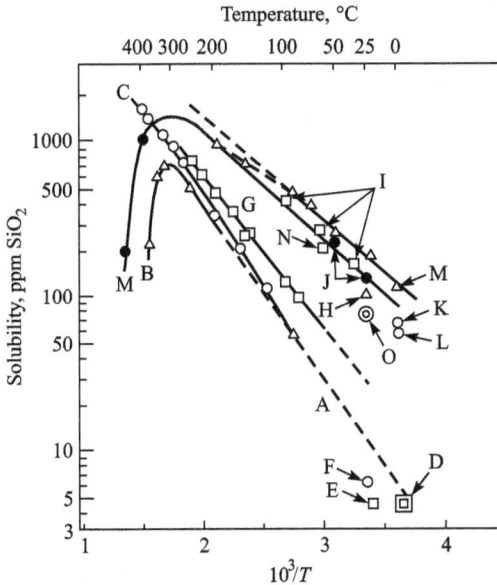

Fig. 3.24: Hydrothermal solubility of the SiO$_2$-modifications according to data compiled by Iler [3]. Above the critical point of water (574.4 K) the log C vs. (1/T) dependence changes its slope. With different letters are indicated Quartz: A) van Lier equation (dashed line); B) Morey; C) Morey – 1000 bars; D) Willey – in sea water; E) Mackenzie and Gees; F) Morey, Fournier and Rowe. Cristobalite: G) Fournier and Rowe. Amorphous: H) Stöber; I) Elmer and Nordberg; J) Lagerström, in 0.5 M NaClO$_4$; K) Willey; L) Jones and Pytkowicz; M) Goto; N) Okkerse; O) Jorgensen, in 1.0 M NaClO$_4$.

1/T straight line in order to determine the values of $\Delta H_x^{298} = H^{298}(\text{liquid}) - H_x^{298}$ and $\Delta S_x^{298} = S^{298}(\text{liquid}) - S_x^{298}$ in Eq. (3.5). It is seen that from vitreous SiO$_2$ to coesite the logC vs. 1/T solid lines follow the expected course, using the H and S values given in Figs. 3.21 and 3.22.

With respect to its solubility, stishovite takes a very particular position. If this SiO$_2$-modification could exist at pressures of the order of 300–400 bars and temperatures up to 800 K at hydrothermal conditions, its solubility would be by many times higher than that of glassy SiO$_2$. Accounting for the particular six-fold symmetry SiO$_{6/3}$-structure of stishovite and the extremely low values of its entropy and high value of its enthalpy at 298 K, the solubility of stishovite in coordinates of Fig. 3.23 would be represented by a straight line as it is given in the insert of this figure. However, as already mentioned stishovite rapidly decomposes at elevated temperatures and low pressures (i.e. outside its field of stable existence). It forms SiO$_2$-glass (see [75–77] and Fig. 3.17). In this way it seems impossible to measure at "normal" pressure conditions the real solubility of stishovite which is obtained in hydrothermal synthesis in "normal" autoclaves (i.e. not at ultra-high pressures).

A similar situation is also found with respect to coesite. The latter also decomposes outside its stability region (see the phase diagram on Fig. 3.6). Under the conditions of "normal" autoclave synthesis it could be expected that coesite should grow in the presence of amorphous or glassy SiO_2 as all other modifications. However, in considering further applications of coesite it should be remembered that it decomposes rapidly at normal pressure when heated above 1300 K.

The SiO_2-diagram we constructed and showed on Fig. 3.23 illustrates that glassy or any other amorphous form of SiO_2 could be used as a precursor phase in the isothermal hydrothermal synthesis of all other modifications of SiO_2 except stishovite. Accounting for the already cited results on the synthesis of post-stishovite modifications of SiO_2 (of the α-PbO_2 and $CaCl_2$ type), they should have even higher values of H^{298} and lower S^{298}-values when compared with the respective thermodynamic properties of stishovite. Hence, it could be expected that log C vs. $1/T$-curves for these modifications should lie higher than that of stishovite and should also have a positive slope (see the insert on Fig. 3.23). The above summarized findings can be of distinct significance for industrial production of any SiO_2 crystal modification for which relatively cheap amorphous (i. e. glassy) SiO_2 is obtained as a technological residual product with sufficient purity.

In considering solubility and relative stability of the different modifications of SiO_2 it is interesting to compare them with another system exhibiting many modifications, the carbon system. In such comparison it is of importance to note that the metastable form of carbon (i.e., diamond), which is thermodynamically stable only at high pressures (exceeding 50 kbar), begins to graphitize at normal pressures at temperatures exceeding 1600 K. In this way diamond can grow either under metastable conditions (i.e. at elevated temperatures and normal pressure) or in solution (e.g. in metallic solutions like Ni-alloys). However, at temperatures lower than 1600 K a sufficiently high attachment rate for diamond growth is to be expected. The same can be anticipated as it is experimentally verified in [31] at temperatures from 1000 to 1300 K also at gas-exchange reactions, in which diamond grows using vitreous carbon as a precursor of C-saturation (see evidence and literature, given in [31]).

The low stability of stishovite at normal pressures and elevated temperatures seems to prevent any possibility either to grow or synthesize this SiO_2-modification at metastable conditions (i.e. at normal pressure) either hydrothermally or using gaseous exchange reactions, as done with diamond. The foregoing results in international literature and our experience summarized in Fig. 3.23 shows, however, that all other "normal" or high-pressure modifications of SiO_2 could be hydrothermally synthesized or grown in a way analogous to the growth of diamond either by gaseous reactions or from solution using vitreous SiO_2 as precursor. The diagram on Fig. 3.23 also gives a direct possibility to determine the thermodynamic driving force $\Delta \mu_{1,2}$ and the relative supersaturation $\gamma_{1,2}$ for the synthesis of any of the SiO_2-modifications from any other crystal form of SiO_2 employing the well known formalism of the theory of crystal nucleation and growth [14, 79–81].

According to this formalism, in our case we have

$$\Delta\mu_{1,2} = RT\ln\left(\frac{C_1}{C_2}\right), \tag{3.23}$$

where R is again the universal gas constant, T is the considered temperature at which the phase 1 with solubility C_1 is the precursor for the growth of the phase 2 with solubility C_2. The relative supersaturation, playing a major role in determining nucleation and growth conditions, is given as

$$\gamma_{1,2} = \frac{\Delta\mu_{1,2}}{RT}. \tag{3.24}$$

Thus, in our case we have

$$\gamma_{1,2} = \ln\left(\frac{C_1}{C_2}\right) = 2.3\log\left(\frac{C_1}{C_2}\right) \approx 2.3\left[\frac{C_1 - C_2}{C_2}\right]. \tag{3.25}$$

Now we can see that in the already cited experiments of Wooster [33] and Nacken [34] the relative supersaturation for quartz growing out of amorphous SiO_2 as the precursor substance, relative supersaturations ($\gamma_{1,2} > 1$) are obtained leading to nucleation. On the other hand, if we imagine a process of quartz synthesis at hydrothermal conditions and using cristobalite and not directly SiO_2-glass as a precursor substance, the value of $\gamma_{1,2}$ is considerably lower, guaranteeing the smooth growth of quartz at $\gamma_{1,2} < 1$. On contrary, the synthesis of high dispersity quartz requires a relative supersaturation $\gamma_{1,2} > 1$ approaching even $\gamma_{1,2} = 2 - 4$.

In this way, the synthesis of ideal single quartz crystals should be realized under different conditions compared to other interesting cases of formation of micro- or even nano-sized quartz populations e.g. in using pre-crystallization in order to purify SiO_2 [38]. This is true also for the crystallization of cristobalite. The sequence of straight lines, plotted on Fig. 3.23, gives, in fact, the sequence of supersaturation or relative supersaturations using Eqs. (3.23), (3.25) and the respective $\ln(C_1/C_2)$-data for a chosen temperature. However, the real sequence of formation of the different SiO_2-modifications is not governed directly by the thermodynamics of nucleation (i.e. by the respective $\gamma_{1,2}$-values) but by the kinetics of these processes, i.e. by kinetic restrictions and recommendations following from Ostwald's Rule of Stages [14, 81–84]. According to this rule, at a given supersaturation $\gamma_{1,2}$, not the formation of the thermodynamically most stable phase as the first precipitate (as indicated by Fig. 3.23) is to be expected. Instead, the process proceeds in another sequence as it is predicted by the contemporary formulation of Ostwald's Rule of Stages. This rule states that first those phases occur the formation of which is kinetically favored. Thus, in most cases of condensation from the vapor phase the initially formed phase is not the most stable one. Instead an intermediate metastable liquid phase is formed first, which afterwards transforms into the thermodynamically more stable crystalline condensate.

The kinetic interpretation of Ostwald's Rule of Stages was initially proposed by Stranski and Totomanov [83] and then further developed in a series of publications by

Gutzow and Avramov [84] and Gutzow, Schmelzer, and Möller [14, 84–86]. Remaining in the framework of classical theory of nucleation, it was shown in cited papers [14, 84] that the formation first of the metastable phase f from an initial supersaturated phase i instead of the formation of the stable phase c from i (in accordance with Ostwald's Rule of Stages) occurs, when the inequality

$$\left(\frac{\varphi_{i,c}}{\varphi_{i,f}}\right)\left(\frac{\sigma_{i,c}}{\sigma_{i,f}}\right)^3\left(\frac{V_c}{V_f}\right)^2 > \left(\frac{\Delta\mu_{(i\to c)}}{\Delta\mu_{(i\to f)}}\right)^2 > 1 \tag{3.26}$$

is fulfilled. Here $\mu_{(i\to c)}$ and $\mu_{(i\to f)}$ are the thermodynamic driving forces or the supersaturation for the transformation of the initial phase i into either the metastable phase f or the stable phase c, respectively. With σ_{if}, $\sigma_{i,c}$ and σ_{if} are denoted the interfacial energies at the i-phase/c-phase and i-phase/f-phase interfaces, respectively. The symbols $\varphi_{i,c}$ and $\varphi_{i,f}$ indicate the nucleation activities of eventually introduced active nucleation cores, favoring the formation either of the metastable phase f or of the stable phase c.

The inequality Eq. (3.26) is easily derived. Indeed, according to classical nucleation theory, the nucleation rate, J, for a given phase is determined via

$$J = \mathrm{const}_1 N_i Z_i \cdot \exp\left[-\frac{\Delta G_{i\to x}}{RT}\right], \tag{3.27}$$

where Z_i is the impingement rate of ambient phase molecules to the critical clusters of the newly formed phase x, $\Delta G_{i\to x}$ is the nucleation barrier for critical cluster formation, and N_i is the number of molecules per unit volume of the ambient phase. Treating the critical cluster as having a spherical shape, the thermodynamic barrier $\Delta G_{i\to x}$ (or the work of critical cluster formation as it is denoted in nucleation theory) can be written as

$$\Delta G = \frac{4\pi\sigma_{i,x}^3 V_x^2}{3\Delta\mu_{i,x}}\Phi_{i,x}. \tag{3.28}$$

In all above given equations V_x (or V_f, and V_c) are the molar volumes of the respective phases.

Employing the kinetic interpretation of Ostwald's Rule of Stages, we have thus to compare the nucleation rates $J_{i,f}$ and $J_{i,c}$. We assume hereby an approximate equality of the impingement rates Z_i in both cases of phase formation. From Eqs. (3.27) and (3.28) we thus arrive directly at Eq. (3.26). In considering the applicability of Eq. (3.26) we have to know σ_{ix} for both phases. This is not so easy to be done and here in developments given by Gutzow and Schmelzer [14] the formula of Stefan, Skapski, Turnbull is used, according to which the specific interfacial energy at the ix interface can be estimated via

$$\sigma_{i,x} = \alpha_0\frac{\Delta H_{i,x}}{N_A^{1/3} V_x^{2/3}}, \tag{3.29}$$

where ΔH_{ix} is the respective change of enthalpy when going from the i to the x phase, N_A is Avogadro's number and V_x is the already introduced value of the molar volume of the newly formed phase. Thus instead of the interfacial energy, the respective values of the molar enthalpy of change $\Delta H_{i,x}$ upon the corresponding phase formation or the respective entropies $\Delta S_{i,x}$ ($\Delta H_{i,x} = \Delta S_{i,x} T_x$) can be introduced into Eq. (3.26). In this way, the inequality Eq. (3.26) can be transformed (as done by Gutzow, Dobreva and Schmelzer [85]) into dependencies, stating that in some ambient phase that phase will be formed which resembles in its structure the initial phase. Employing such argumentation, Fig. 3.23 can be transformed from a thermodynamic picture into a kinetic scheme giving the way, the probability and the paths of preferred formation of the desired phases in the ambient phase.

We also have to mention here that, according to the investigations performed in [87], the formation of cristobalite is observed in the presence of Cu or Au nucleation cores of micrometer dimensions. In this sense, Eq. (3.26) also supplies us with possibilities to induce the formation of different phases by introducing active nucleation cores favoring the formation either of the c or of the f-phase. The nucleation activity of different substances and especially of noble metals is analyzed in detail in [88].

3.6.3 Solubility of SiO$_2$: Size Effects

According to the well-known formula of the classical thermodynamic theory of nucleation, the Gibbs-Thomson equation, the radius r of the critical cluster can be expressed as

$$r = \frac{2\sigma_{i,x} V_m}{\Delta\mu}.$$
(3.30)

Thus, the solubility C_r of a phase x of radius, r, when compared with its solubility C_∞ of sufficiently large crystallites can be given by $\Delta\mu_r$, where

$$\Delta\mu_r = RT \ln\left(\frac{C_r}{C_\infty}\right).$$
(3.31)

From Eq. (3.30) it follows that at $T = $ const. the solubility C changes with r, as

$$\ln C_r = \frac{2\sigma_{i,x} V_x}{RT} \frac{1}{r} + \ln C_\infty.$$
(3.32)

By adopting again the equation of Stefan-Skapski-Turnbull, Eq. (3.29) and accounting for the simple relation

$$\left(\frac{V_x}{N_A}\right)^{1/3} \cong d_{o,x},$$
(3.33)

where d_{ox} is the mean intermolecular distance in the phase x, we can write

$$\ln\left(\frac{C}{C_0}\right) \cong 2\alpha_0 \frac{\Delta S_{m,x}}{R} \frac{T_{m,x}}{T} d_{ox} \frac{1}{r}.$$
(3.34)

Consequently, from the known values of the entropy of dissolution, $\Delta S_{m,x}$, of the melting point, $T_{m,x}$, and of the intermolecular distance, d_{ox}, of the phase x, we can determine α_0 from latter relation or directly via Eq. (3.29).

In Iler's book [3] the size dependence of the solubility of amorphous (or glassy) SiO_2 in hydrothermal solutions at 25 °C at pH-values ~ 8 is given. In coordinates $\log C_r$ vs. reciprocal particle diameter the experimental data result in a straight line with a slope of $1.5 \cdot 10^{-7}$ cm^{-1} (see Figs. 3.25 and 3.26). Together with Eq. (3.34) at $\alpha_0 = 0.5$ a value of 0.4 for $\Delta S_g/R$ is obtained. This result is a quite satisfactory estimate when compared with the value of ΔS_g for quartz glass given in Table 3.6, according to which $\Delta S_g/R = 0.6$. It is evident that a change of α_0 by 20 % would give the expected value of $(\Delta S_g/R)$ obtained experimentally. Expressing ΔS_m through the well known dependence $\Delta H_m = \Delta S_m T_m$ we can introduce $\Delta H_{m,x}$ in Eq. (3.34) provided this quantity is known from direct experimental observations. From the slope of the log C vs. 1/r dependence

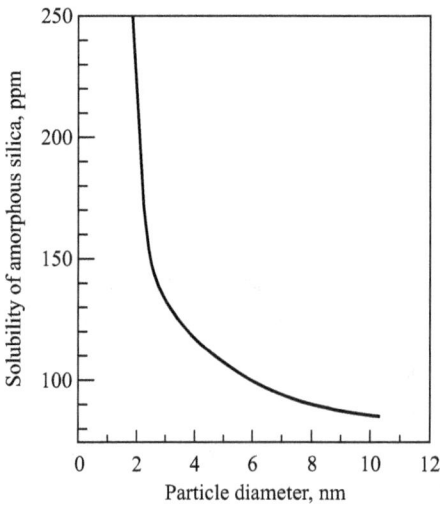

Fig. 3.25: Solubility of amorphous nano-grained SiO_2 at 25 °C in dependence on mean particle diameter.

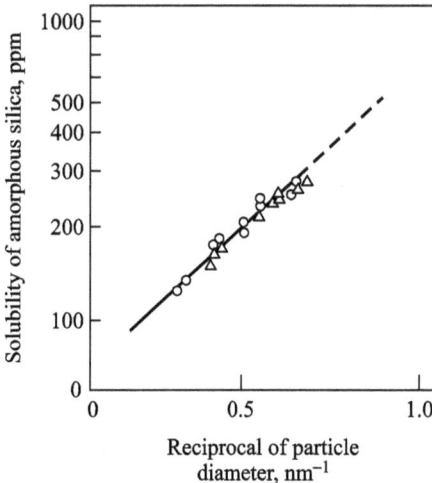

Fig. 3.26: Experimental data on the solubility of amorphous SiO_2 vs. the reciprocal particle diameter according to Eq. (3.34). Data in Figs. 3.25 and 3.26 are taken from Iler [3].

in Fig. 3.26 a value of $155 \cdot 10^{-3}$ J m^{-2} is obtained for the specific surface energy $\sigma_{i,x}$ at the SiO$_2$ glass-aqueous solution interface. Comparing this value with the data from solubility experiments with various substances it can be claimed that the classical theory of nucleation, as in many other cases, gives nearly satisfactory results when the processes of nucleation are treated in the framework of its capillary approximation. In this way, the results on the size-dependence of the solubility of amorphous SiO$_2$ indicate that the treatment of the precipitation kinetics in terms of Ostwald's Rule of Stages, introduced in the framework of the classical theory of nucleation, could also give satisfactory results, at least, appropriate for initial technical use.

3.6.4 Different SiO$_2$-modifications at Hydrothermal Conditions: Technological Aspects

In the preceding section (Section 3.6.3) we discussed the possibilities to synthesize different modifications of SiO$_2$ by a process of nucleation and growth, using quartz glass as a precursor phase in the autoclave. However, utilizing cristobalite as a precursor, e.g. in the growth of quartz crystals, could be a more convenient way in order to achieve the same aim. According to general consequences of classical nucleation theory, growth could be expected to occur in the cases when the relative supersaturation $\gamma_{1,2}$ is restricted by an upper value of 1.5 or even to values as low as 0.1–0.2. Under such conditions, especially at $\gamma_{1,2} < 1$, smooth growth of single crystals could be expected. Our solubility diagram (Fig. 3.23) shows that such a possibility exists, firstly, for the growth of α-cristobalite single crystals out of SiO$_2$ glass. Growing directly β-cristobalite by using SiO$_2$ glass as a precursor could be an even more interesting possibility. According to Fig. 3.23 it is obvious that the introduced β-cristobalite micro-crystals could grow smoothly at relatively low temperatures.

There are indications in the literature [38] that by changing the temperature the hydrothermal synthesis of the different modifications of cristobalite is possible. Our diagram gives the possibilities to discuss and initiate experiments in which at relatively low temperatures a hydrothermal growth of large cubic β-cristobalite crystals could be realized if appropriate cations are added to the solution stabilizing the high temperature modification of cristobalite. It has been also proposed in the literature to stabilize hydrothermally grown β-cristobalite by introducing sodium and aluminium cations into the hydrothermal solution. Cristobalite hydrothermal synthesis by using, instead of NaOH-solutions, Ca(OH)$_2$-solutions could be another interesting possibility when the presence of alkali cations has to be excluded. The realization of such approach would require a study of the solubility and precipitation of Ca(OH)$_2$ under hydrothermal conditions. It could give a new way of toughening of β-cristobalite with Ca^{2+} and Al^{3+} in hydrothermal solutions. Another interesting possibility follows from our results summarized on Fig. 3.23, it consists in the high-temperature nucleation and further growth of large coesite crystals at hydrothermal conditions.

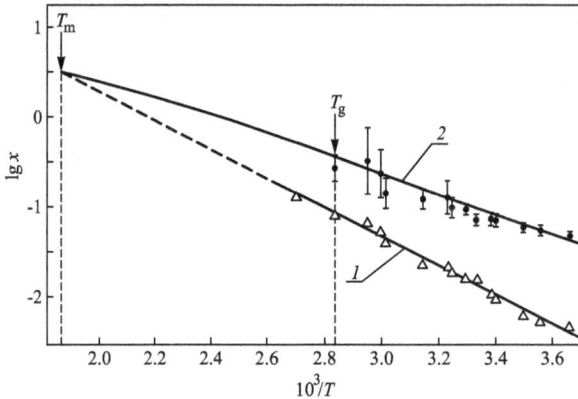

Fig. 3.27: Solubility of phenolphtaleine in water as a crystal (1) and as a glass (2) according to measurements of Grantscharova and Gutzow [14, 89]. T_m and T_g are the melting point of phenolphtaleine and its glass transition temperature.

In model experiments, performed by Grantscharova and Gutzow [89, 90], the growth of a model glass-forming system, phenolphthalein, from aqueous solutions was studied. On Fig. 3.27 the solubility of phenolphthalein is given as a glass and as a crystal in the temperature range from 10 to approximately 100 °C. The respective solubilities were determined by analyzing the kinetics of dissolution of both crystalline and glassy phenolphthalein as shown on Fig. 3.28. Similar experiments, carried out with different modifications of SiO_2, could give more precise solubility values than those calculated or given in the literature and leading to pictures like Figs. 3.20 or 3.24. Usually, in existing measurements the solubility is determined under static conditions and not by kinetic experiments as presented in Fig. 3.28. We performed (see [89, 90]) growth experiments with the phenolphthalein-water system under isothermal conditions. In the

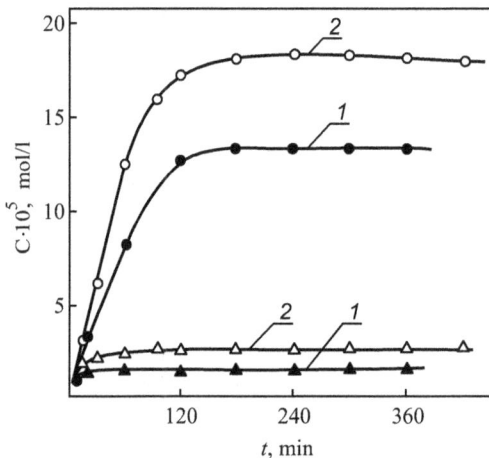

Fig. 3.28: Dissolution curves of phenolphtaleine in water at two temperatures: at 22 °C (the upper two curves) and at 13 °C (the bottom two curves) for crystalline (1) and for vitreous (2) phenolphtaleine.

experiments, we could achieve a smooth growth of phenolphthalein single crystals using phenolphthalein glass as a precursor. Wooster and Nacken [33, 34] employed in their experiments also quartz glass as a precursor. However, as seen on Fig. 3.23, relatively high supersaturations $\gamma_{1,2}$ are to be expected in the temperature interval (300–400 K), used by these authors. Under these conditions, as already mentioned, a large number of small quartz crystallites has been formed and a smooth growth of large quartz single crystals could not be achieved.

The hydrothermal growth of β- and α-cristobalite crystals to relatively large dimensions could be of particular scientific and technological interest. In a number of present day investigations [91, 92] it is demonstrated, that α-cristobalite exhibits the unusual property of becoming wider when stretched and thinner when compressed: thus it is a material with negative Poisson ratio. Such auxetic properties of larger crystals could be of exceptional significance in instrument making industry. The results, shown in Fig. 3.23, guaranty very low supersaturations and smooth growth even at high temperatures. They also open thus the possibility of growth of coesite at hydrothermal conditions using β-cristobalite as a precursor.

In our foregoing experimental investigations on the solubility of vitreous phenolphthalein and the crystalline modification of the same substance, we also constructed a simple apparatus, shown on Fig. 3.29b. Fig. 3.29a supplies us with the principal scheme of growth of diamond single crystals at isothermal conditions using vitreous carbon as a precursor and the gas reactions as the carrier agent. In both cases (phenolphthalein and diamond) glass was used as a precursor for the growth of single crystals either in solution or at chemical transport reaction. In the case of diamond, these transport reactions involved CCl$_4$, CS$_2$ or CH$_4$ and the respective chemical equilibria [31].

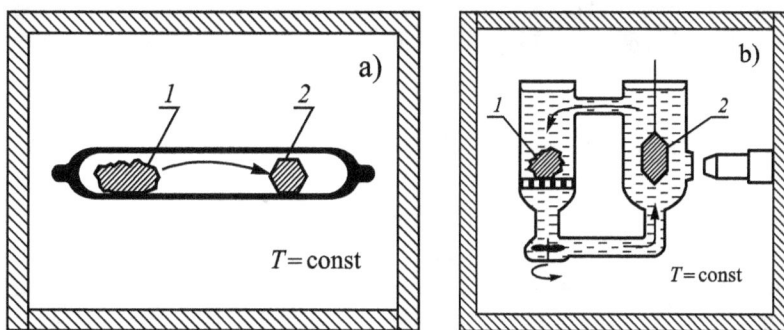

Fig. 3.29: Two possibilities to use glass as a precursor for crystal growth at isothermal conditions: (a) ampoule growth of diamond (2) from vitreous carbon (1) using gas transport reaction at 1050 °C (here: C + S$_2$ → CS$_2$; see Gutzow et al. [31]). (b) The growth of crystalline phenolphtaleine (2) from glassy phenolphtaleine (1) in aqueous solution (Grantscharova and Gutzow [14, 63]).

The hydrothermal growth of SiO_2 in closed volumes supplies us with an exceptional possibility to regulate the pressure resulting in the autoclave by the relative filling of its volume. The pressure, obtained in an autoclave by its filling to different degrees and heating the system to different temperatures, is presented in Fig. 3.10. On Fig. 3.9 this dependence is given in an extended form including known solubilities of SiO_2 in the range from room temperatures up to temperatures higher than the critical point of water. As evident, hydrothermal synthesis gives extraordinary possibilities of regulating temperature, pressure and solubility in the autoclave in a simple way within wide limits.

In a recent publication, the construction of a simple camera autoclave (Fig. 3.15) with relatively small volume was reported. In this device, growth of stishovite crystals up to dimensions of several millimeters was realized (see Fig. 3.16). The same device could be also used for coesite growth. Stishovite according to a number of literature sources is one of the hardest materials ever known and the hardest oxide ever synthesized. In this respect the synthesis of stishovite single crystals could result in materials with exceptional properties, however, for prices nearly equal or even surpassing the price of diamond and cubic boron nitride. In comparing the properties of stishovite and coesite as single crystals and the possibilities for their metastable growth (i.e. at pressures outside the stability pressure region of both crystals) it should be remembered that stishovite and coesite decompose to SiO_2-glass at temperatures approaching $600-800\,°C$ and $1300\,°C$, respectively.

3.7 Resources of the Silica Modifications

Silicon dioxide is the major constituent of rock forming minerals and an important component of sediments and soils. In the form of silicates it accounts for 75 wt% of the Earth's crust. It is also a constituent of many plants in a measurable quantity.

3.7.1 Mineral Resources of Quartz

Quartz in its low temperature α-modification forms 12 wt% of the Earth's lithosphere and is after feldspars the most widespread mineral on the Earth. At hydrothermal conditions in the Earth crust single crystals are formed, reaching up to 1.5–2 m in size. The SiO_2-tetrahedra in quartz are arranged along its trigonal c-axis. Two types can be distinguished in this arrangement: left- and right-handed. This property results in the optical activity of quartz with a considerable polarization power for visible light. The right- and left-handed crystals are mirror images of each other. The two-fold symmetry axis, perpendicular to the screw axis, is polar and mechanical stress along the axis produces a direct piezo-electric effect. The reverse effect is utilized in oscillatory devices. Mainly due to this property, high quantities of quartz crystals are produced most often hydrothermally. The thermal expansion of α-quartz is high ($13.3 \cdot 10^{-6}$ K^{-1}) and drops to even slightly negative values after the displacive transition into its high temperature modification, β-quartz.

There exist many special, trade and trivial, names for silica rocks, often with various meanings and imprecise definitions. Different classifications are accordingly given in literature. Silica rocks typically contain up to more than 90 wt% SiO$_2$. Ordinary sandstones contain only about 65 wt% silica, mainly as quartz grains. The rest is matrix and cement (feldspars, CaCO$_3$, mica, clay, etc.). Different transformation processes occur in silica rocks with increasing the burial depths in the Earth's crust which results in a reduction of porosity and water content. Thus, quartz can be solidified to quartz sandstone and grades by crystallization into quartzites under pressure at elevated temperatures. Similarly, under compression non-crystalline biogenic silica remnants of diatoms, radiolarian etc. are transformed into porcelanite and finally to chert. When the biogenic origin is known these rocks are denoted as diatomite, radiolarite or spiculite. Lidite is a rock formed from the remnants of radiolarites from the Paleozoic era. A great number of different semi-crystalline or quasi-amorphous minerals are known and described in mineralogy and geology.

Of technical importance for all silicate industries, and in particular for the glass, ceramic and porcelain industries are high quality (especially: free of iron oxides) quartz-sand deposits like the quartz-sand fields in Hoehenbocka or Weferlingen in Germany. Of global significance for the production of high-purity quartz single crystals are the huge quartz deposits on Madagascar, in Brazil and in the Ural Mountains: in most cases they are the result of growth processes under natural hydrothermal conditions. In several places (also in Europe) venous quartz formations of magmatic origin are also in technical utilization.

3.7.2 Plant Resources of Silica

Of great significance in future developments may also be the SiO$_2$-content in organic residuals and especially in rice husks, bamboo, cocoa shells, oats, etc. There are several different methods to obtain and use the so called white ash of rice husk (i.e. the amorphous SiO$_2$ resulting from the oxidation heat treatment of rice husk). Micro- and nano-cristobalite crystalline phases are easily obtained at relatively low temperatures (of 850–950 °C) upon oxidation heat-treatment or directed pyrolysis. Taking into account that rice is a basic food of millions of people in the Asiatic regions of our world, rice husk can be considered as a possible natural source of SiO$_2$ for industrial applications.

In this respect, two factors are of importance: First, a relatively high percentage of alkali oxides (either Na$_2$O or K$_2$O) are usually present in rice husks and any other natural product. Such impurities which sometimes take three or five percent of the mentioned white ash can be, however, easily washed out with mineral acids. Second, heavy metals are found only in relatively low concentrations in plant silica: the plants themselves prevent to absorb oxides, toxic for their growth. The third essential point in possible plant silica production and purification is the circumstance that SiO_2 is contained in the plants structure in an amorphous nano- or micro-sized form and is obtained in the "white ashes" both as amorphous silica, or (in dependence of heat-treatment history and of dopants present) as α- or β-cristobalite. Plant silica is thus chemically very active and can be easily dissolved by mineral acids. In this way, it can be subjected to various purification procedures.

The present day status of the both scientifically, ecologically and technically very important problems of the state of SiO_2 in the plants is interrelated with the processes of plant pyrolysis [93–96]. The possibility of formation and directed synthesis of silicon carbides, of silica and of active carbons (all of them with possible industrial applications!) is treated in great detail in international literature. An extract of them is given here with the mentioned papers and especially with [94]. It is also to be noted that the results obtained by one of the present authors with plant silica [97] and with the analysis of the respective processes of pyrolysis [98, 99] and of active silica crystallization initiated some of the developments, connected with the polymorphous transitions of cristobalite and amorphous SiO_2, with quartz etc., described in more or less detail in the present review. Biological activities of the animal world (of worm populations [100]) and of enzymes [101] have been also proposed as possible methods of degradation of rice husk waste products and their following transformation into more or less useful forms of silica and silicon compounds.

3.7.3 Industrial Waste as Sources of Silica

Fly ashes from chemical and metallurgical industry can also supply SiO_2 in a highly active form. However, here, contamination with toxic oxides and heavy metals can be a severe problem. It has only to be mentioned that the active amorphous phase formed out of plant SiO_2 is directly transformed into high temperature β-quartz, which, however, gives again α-cristobalite at low temperatures.

3.7.4 Coesite and Stishovite as Impactite Remnants

Coesite and stishovite which initially were artificially synthesized are not only found in impactite minerals at meteoritic craters like the great Arizona meteorite crater. Of greater significance is the experimentally verified geological fact, that in the depths of the Earth's crust quartz minerals are transformed into coesite and stishovite and thus these minerals are of exceptional significance in following, predicting and analyzing earthquakes and other processes in the Earth's crust.

3.8 Some Particularly Interesting Properties of Silica

Finally, we would like to emphasize here several specific properties of silica both in its crystalline and amorphous modifications. Usually one draws the attention only to the properties of quartz and quartz sand. As the result, several of the already mentioned unusual properties of SiO_2 are disregarded. They can be, however, of significance in technology, instrument production and also of commercial interest. By this reason, we return here to the discussion of several of such properties briefly already mentioned in the previous sections.

In order to reconsider the possible significance of SiO_2-materials in a new light, beginning with the most dense and most unusual modifications of SiO_2, it should be repeated that stishovite and the two post-stishovite modifications are the hardest oxides synthesized up to now. Their hardness is surpassed only by cubic boron nitride and diamond. When comparing stishovite and post-stishovites with cubic boron nitride and diamond it should be, however, taken into account that stishovite has a lower thermal stability than diamond under normal pressure: diamond can be used up to 1100–1200 °C while stishovite transforms into SiO_2-glass at temperatures beginning from 650–700 °C. Nevertheless, it could be a very useful material for the production tools when only silicon and oxygen should be in contact with the material under treatment. The already mentioned hydrothermal synthesis of stishovite at super-high pressures could be eventually developed to an industrial scale.

Cristobalite in its high temperature β-form is the SiO_2 crystalline phase with the highest refractoriness: it could be used at temperatures above 1500 °C. Moreover, its coefficient of thermal expansion approaches zero at high temperatures. This property opens additional perspectives for the utilization of β-cristobalite as a major constituent in both glass-ceramic materials and – in the form of little appropriately preformed substrates etc. – in high-temperature electronic technique. Of particular significance is, on the other hand, the fact that α-cristobalite has a negative Poisson ratio. This property α-cristobalite shares with only four or five inorganic substances. Materials, possessing such a property are extraordinary resistant against the impact of projectiles entering its structure even with a very high speed (e.g. bullets in a modern chain-plated armor). Many other applications of such auxetic materials are discussed in present day literature, partially mentioned above (see [91, 92]). It is also interesting to note that only the low temperature α-modification of cristobalite possesses this property: in the $(\alpha - \beta)$-transition the high temperature β-cristobalite emerges without the auxetic properties of the α-modification. On the other hand, it is also to be noted that with quartz the reverse case is observed: here the high temperature β-quartz is slightly auxetic, while the low-temperature α-modification has a normal (positive) Poisson ratio.

The chemical resistivity of cristobalite is another important property. It makes this material a promising and competitive substance for the construction of high temperature refractory chemically resistant devices for chemical applications. The same applies also to β-cristobalite chemically stabilized with CaO and Al_2O_3 in the already described way.

Silica in its amorphous form as a glass has exceptional applications because of its chemical inertness, temperature resistance (however only up to 1000–1050 °C! Then, crystallization begins) and its extraordinary low coefficient of thermal expansion ($0.3 \cdot 10^{-6}$ K^{-1}). The properties of amorphous SiO_2 in disperse form, as it results e.g. from the oxidation heat treatment of natural plant products, are also well known and used in applications connected with absorption, etc.

3.9 General Discussion: Technical Perspectives

The foregoing sections give the possibility for several conclusions:

1. The most promising material (with exception of quartz glass) out of all possible modifications and forms of existence of SiO_2 is stabilized β-cristobalite. Its particular properties, concerning the coefficient of thermal expansion, thermal stability, chemical stability, makes this modification a very promising material with many applications. The results described in literature concerning its synthesis show that it can be synthesized in several quite different ways, any of them applicable for different purposes:

    (a) The activated reaction sinter-crystallization approach of synthesis with the introduction of chemical stabilizers (CaO + Al_2O_3), developed recently by the authors of the present review, shows that this method could lead (eventually combined with hot pressing) to the production of different technical devices or parts of it.

    (b) Results obtained with a combined sol-gel – ceramic way of synthesis (the so-called "wet impregnation" method) could also be a promising method of developing objects even on a larger scale. This method could give additional perspectives because it requires lower temperatures of sinter-crystallization.

    (c) As a method of direct formation of small-size objects useful in high temperature electronics could be the direct sol-gel method of synthesis of chemically stabilized β-cristobalite, described in one of the previous sections.

    (d) Cristobalite can also be synthesized by hydrothermal methods as this is demonstrated in literature. It could be produced in such a way as to give micro- or even nano-sized powders of β-cristobalite, stabilized, however, with alkaline oxides.

2. The hydrothermal method of synthesis can be used to produce not only quartz (as known for more than 50 years) but also cristobalite and coesite in commercial quantity and in different forms and sizes, using either SiO_2-glass (in the synthesis of cristobalite, coesite and quartz) or cristobalite for the synthesis of quartz. However, it has to be accounted for that in most applications the autoclave hy-

drothermal method works efficiently only in alkaline solutions: in this way only crystals with a measurable content of alkaline pollutants can be formed.

3. It should be of interest to develop methods of hydrothermal synthesis based either on pure aqueous solutions of SiO_2 (without any alkalizing additives) or on the development of hydrothermal methods of synthesis in which solutions with high pH-values could be formed by exploiting the basic character of CaO or other alkaline-earth oxides. Here, however, up to now no experimental evidence or theoretical considerations are known to the authors of this review.

4. Hydrothermal synthesis could be also considered as a method of purification of both quartz crystals and quartz micro- and nano-crystals, e.g. in synthesizing micro-quartz crystallites from either glassy SiO_2 or cristobalite as a precursor. Due to the small sizes of the micro-crystals, the number of included bubbles, containing initial solvent could be brought to a minimum. However, the alkaline solutions, with pH ~ 8–9, employed in present day hydrothermal synthesis could bring additional problems. Vitreous or any other form of amorphous SiO_2 can be more easily purified.

5. The synthesis of cristobalite as micro-disperse powders could give other possibilities for the method of production of quartz glass macro-tubes employing the plasma technique. Here, it could be of particular significance that the density of β-cristobalite is nearly equal to the density of quartz glass. This, however, should require chemically stabilized β-cristobalite to be used. This would bring CaO and Al_2O_3 into the quartz glass.

6. SiO_2 is a substance which, like carbon, exists in a great variety of modifications. Up to now only two of these forms – quartz glass and quartz macro-crystals – are implemented in industry. We hope that the theoretical predictions concerning the water solubility of SiO_2-modifications reveal new perspectives for the synthesis and future applications not only of the "common" ambient pressure forms of SiO_2, but also of the their "exotic" high pressure forms. The method of activated reaction sinter-crystallization synthesis of stabilized β-cristobalite glass-ceramics, developed by the authors of this review, opens the possibility for their economically profitable industrial production and many fields for their application.

In the last time, several attempts to use different "unusual" sources of SiO_2 for the synthesis of stabilized β-cristobalite glass-ceramics e.g. purified diatomite [102] have been made. Here, the most important problem is to find effective, simple and affordable methods for the purification of similar biogenic resources of SiO_2. In addition, new possibilities have been realized in this respect, namely in the purification of rice husk ashes, considered, as mentioned, as a promising plant source of active silica. Since in these ashes SiO_2 is in amorphous nano- and micro-sized form, it can be expected that the preparation of stabilized β-cristobalite glass-ceramics, as well as of other SiO_2-containing materials, could be realized at substantially lower temperatures and shorter heat-treatment times.

Bibliography

[1] W. Eitel, *The Physical Chemistry of the Silicates* (Chicago University Press, Chicago, 1954).

[2] W. Eitel, *Silicate Science*, 2: *Glasses*, 7: *Glass Science* (Academic Press, New York, London 1965 (Second Enlarged Edition, 1976)).

[3] R.K. Iler, *The Chemistry of Silica (Solubility, Polymerization, Colloid and Surface Properties, and Biochemistry)* (John Wiley & Sons, New York, 1979).

[4] N.A. Toropov, V.P. Barzakovskii, V.V. Lapin, N.N. Kurtseva, and A.I. Boikova, *Phase Diagrams of Silicate Systems* (Reference Data), 1: Binary Systems (Nauka Publishers, Moscow, Leningrad, 1965 (in Russian)).

[5] N.A. Toropov, V.P. Barzakovskii, V.V. Lapin, N.N. Kurtseva, and A.I. Boikova, *Phase Diagrams of Silicate Systems* (2nd edition), 2: Metal-Oxide Compounds of Silicate Systems (Nauka Publishers, Moscow, Leningrad, 1970 (in Russian)).

[6] A. Landolt-Boernstein, *Zahlenwerte und Funktionen aus Physik, Chemie, Astronomie, Geophysik und Technik; Eigenschaften der Materie in ihren Aggregatzuständen*, Bd. II, 2b, Kalorische Eigenschaften, 6. Auflage (Springer, Berlin, 1962).

[7] A. Landolt-Boernstein, *Numerical Data and Functional Relationships in Science and Technology, New Series: Group IV, Physical Chemistry*, 19A, Editor: W. Martienssen, Heidelberg, New York, *Thermodynamic Properties of Inorganic Materials*, compiled by SGTE (Springer, Berlin, 2001).

[8] V.P. Glushko (Editor in Chief) et al., *Thermal Constants of Matter*, 4 (C, Si, Ge, Sn, Pb) (Academy of Sciences of the USSR Publishers, Moscow, 1970).

[9] I. Barin, O. Knacke, and O. Kubaschewski, *Thermodynamical Properties of Inorganic Substances*, Vol. 1 and 2 (Springer, Berlin, 1991).

[10] N.A. Toropov and V.P. Barzakovskii, *High Temperature Chemistry of Silicate and Oxide Systems* (Academy of Sciences of the USSR Publishers, Moscow, Leningrad, 1963).

[11] E.K. Kazenas and D.M. Chizhikov, *Vapor Pressure of the Oxides of the Chemical Elements* (Nauka Publishers, Moscow, 1976).

[12] W. Hinz, *Silikate, Grundlagen der Silikatwissenschaft und Silikattechnik*, Band 2, *Die Silikatsysteme und die technischen Silikate* (Verlag für Bauwesen, Berlin, 1970).

[13] V.I. Babuskin, G.M. Matveev, and O.P. Mcedlov-Petrosyan, *Thermodynamik der Silikate* (Verlag für Bauwesen, Berlin, 1966); see also: 2nd Russian Edition, Stroyizdat Publishers, Moscow, 1986.

[14] I. Gutzow and J. Schmelzer, *The Vitreous State: Thermodynamics, Structure, Rheology and Crystallization* (Springer, Berlin, 1995; 2nd edition, Springer, Heidelberg, 2013).

[15] G.W. Morey, *The Properties of Glass*, 2nd edition (Reinhold Publishing Corporation, New York, 1954).

[16] A.F. Wells, *Structural Inorganic Chemistry*, 4th edition (Clarendon Press, Oxford, 1975).

[17] R.C. Evans, *Einführung in die Kristallchemie* (J.A. Barth, Leipzig, 1954).

[18] H. Römpp, J. Falbe, and M. Regitz, *Römpp Chemie Lexikon*, 10 Auflage, volumes 1–6 (Georg Thieme, Stuttgart, New York, 1997).

[19] *Ullman's Encyclopedia of Industrial Chemistry*, 5th edition **A23**, Wiley-VCH, 1993 (a); (b) *Ullman's Encyclopedia of Industrial Chemistry*, 6th edition **32**, Wiley-VCH, 2003.

[20] A.N. Winchell and H. Winchell, *Elements of Optical Mineralogy* (van Nostrand, New York, 1951).

[21] A.G. Betechtin, *Mineralogy*, 2nd edition (GNTI Publishers, Moscow, 1956 (in Russian)).

[22] I. Kostov and R.I. Kostov, *Crystal Habits of Minerals* (Academic Publishing House, Sofia, 1999).

[23] W. Vogel, *Chemistry of Glass*, 2nd edition (Springer, Berlin, New York, 1992).

[24] H. Scholze, *Glass: Nature, Structure and Properties* (Springer, New York, London, 1990).

[25] V. Swami, V. Saxena, K. Surendra, B. Sundman, and J. Zhang, *A Thermodynamic Assesment of the Silica Phase Diagram*, J. Geophys. Res.; **99**, 11787–11794 (1994).

[26] J.L. Holm, E. F. Westrum, Jr., and O.J. Kleppa, *Thermodynamics of Polymorphic Transformations in Silica*, Geochim. Cosmolog. Acta **31**, 2289–2307 (1967).

[27] P.I. Dobrokupets et al., *Thermodynamics of the Polymorphic Forms of Silica*, Proc. (Izv.) AN UdSSR, Geological Series **11**, 87–97 (1988) (in Russian).

[28] P. Richet, Y. Bottinga, L. Denielou, J.P. Petitet, and C. Tequi, *Thermodynamic Properties of Quartz, Cristobalite and Amorphous SiO_2: Drop Calorimetry Measurements between 1000 and 1800 K and a Review from 0 to 2000 K*, Geochim. Cosmochim. Acta **46**, 2639–2658 (1982).

[29] N.R. Keskar and J.R. Chelikowsky, *Calculated Thermodynamic Properties of Silica Polymorphs*, Phys. Chem. Minerals **22**, 233–240 (1995).

[30] O.W. Floerke, *Über das Einstoffsystem SiO_2*, Naturwissenschaften **43**, 419–423 (1956).

[31] I. Gutzow, S. Todorova, L. Kostadinov, E. Stoyanov, V. Guencheva, G. Völksch, H. Dunken, and C. Rüssel, *Diamonds by Transport Reactions with Vitreous Carbon and from the Plasma Torch*, In: *In Nucleation Theory and Applications*. Ed. J.W.P. Schmelzer (Wiley-VCH, Berlin-Weinheim, Chapter 8: 256–308, 2005).

[32] A. Smakula, *Einkristalle – Wachstum, Herstellung und Anwendung* (Springer, Berlin, Heidelberg, 1962).

[33] N. Wooster and W.A. Wooster, *Preparation of Synthetic Quartz*, Nature (London), No.3984, 297 (1946).

[34] R. Nacken, *Hydrothermale Synthese von Quarzkristallen aus Quarzglass*, FIAT *Rev. Ger. Sci. Report* **641**, 11 (1945); Chemiker Zeitung **74**, 745 (1950), see also [32], pp 158–159.

[35] L.A. Thomas, N. Wooster, and W.A. Wooster, *The Hydrothermal Synthesis of Quartz*, Disc. Faraday Society (London) **5**, 341–345 (1949).

[36] A.C. Swinnerton, G.E. Owen and J.F. Corwin, *Some Aspects of The Growth of Quartz Crystals*, Disc. Faraday Soc. (London) **5**, 172–180 (1949).

[37] R.M. Barrer, *Preparation of Synthetic Quartz*, Nature (London) **157**, 734–734 (1946).

[38] M. Hosaka and T. Miyata, *Hydrothermal Growth of α-Quartz Using High-Purity α-Cristobalite as Feed Material*, Mat. Res. Bulletin **28**, 1201–1208 (1993).

[39] M.H. Mueser and K. Binder, *Molecular Dynamics Study of the $\alpha-\beta$ Transition in Quartz: Elastic Properties, Finite Size Effects, and Hysteresis in the Local Structure*, Phys. Chem. Minerals **28**, 746–755 (2001).

[40] A. Yeganeh-Haeri, D.J. Weidner, and J.B. Parise, *Elasticity of α-Cristobalite: A Silicon Dioxide with a Negative Poisson's Ratio*, Science **257**, 650–652 (1992).

[41] N.R. Keskar and J.R. Chelikowsky, *Negative Poisson Ratios in Crystalline SiO_2 from First-Principles Calculations*, Nature **358**, 222–224 (1992).

[42] W. Buessem, M. Bluth, and G. Grochtmann, *Roentgenographische Ausdehnungsmessungen an Kristallinen Massen I.*, Ber. Dtsch. Keram. Ges.**16**, 381–392 (1935).

[43] A.F. Wright and A.J. Leadbetter, *The Structure of the β-Cristobalite Phases of SiO_2 and $AlPO_4$*, Philos. Mag. **31**, 1391–1401 (1975).

[44] F. Aumento, *Stability, Lattice Parameters, and Thermal Expansion of β-Cristobalite*, Am. Mineral. **51**, 1167–1176 (1966).

[45] I.P. Swainson and M.T. Dove, *On the Thermal Expansion of β-Cristobalite*, Phys. Chem. Minerals **22**, 61–65 (1995).

[46] D.R. Peacor, *High-Temperature Single-Crystal Study of the Cristobalite Inversion*, Z. Kristallographie **138**, 274–298 (1973).

[47] D.M. Roy and R. Roy, *Tridymite-Cristobalite Relations and Stable Solid Solutions*, Am. Mineral. **49**, 952–962 (1964).

[48]   M.J. Buerger, *The Stuffed Derivatives of the Silica Structure*, Am. Mineral. **39**, 600–614 (1954).

[49]   J.F. MacDowell, *Alpha- and Beta-Cristobalite Glass-Ceramics and Methods*, U.S. Patent 3445252, 1966.

[50]   C.-T. Li, *Glasses, Thermally Stable High (Beta)-Cristobalite Glass-Ceramics and Method*, U.S. Patent 4073655, 1978.

[51]   J.A. Kaduk, *Synthetic Cristobalite*, U.S. Patent 4395388, 1983.

[52]   A.J. Perrotta, D.K. Grubbs, E.S. Martin, N.R. Dando, H.A. McKinstry, and C.-Y. Huang, *Chemical Stabilization of Beta-Cristobalite*, J. Am. Ceram. Soc. **12**, 41–47 (1989).

[53]   J. Perrotta, D.K. Grubbs, and E.S. Martin, *Process for Preparing Stabilized High Cristobalite*, U.S. Patent 4818729, 1989.

[54]   S.L. Bors, S.C. Winchester, M.A. Saltzberg, and H.E. Bergna, *Process for Making Chemically Stabilized Cristobalite*, EP 0482534, 1992.

[55]   Y.H. Hu and M.A. Saltzberg, *Chemically Stabilized Cristobalite*, U.S. patent 5096857, 1992.

[56]   P.L. Gai-Boyes, M.A. Saltzberg, and A. Vega, *Structures and Stabilization Mechanism in Chemically Stabilized Ceramics*, J. Solid State Chem. **106**, 35–47 (1993).

[57]   E.S. Thomas, J.G. Thompson, R.L. Withers, M. Sterns, Y. Xiao, and R.J. Kirkpatrick, *Further Investigation of the Stabilization of Beta-Cristobalite*, J. Amer. Ceram. Soc. **77**, 49–56 (1994).

[58]   O. San and C. Özgür, *Investigation of a High Stable β-Cristobalite Ceramic Powder from* CaO–Al_2O_3–SiO_2 *System*, J. Eur. Ceram. Soc. **29**, 2945–2949 (2009).

[59]   G.H. Beall, *Glass-Ceramic Bodies and Method for Making Them*, U.S. Patent 3252811, 1966.

[60]   G.H. Beall, B.R. Karstetter, and H.L. Rittler, *Crystallization and Chemical Strengthening of Stuffed β-Quartz Glass-Ceramics*, J. Am. Ceram. Soc. **50**, 181–190 (1967).

[61]   A.G. Karaulov, E.I. Zoz, I.N. Rudyak, and T.E. Sudarkina, *Structure and Properties of Solid Solutions in the Systems* ZrO_2-MgO, ZrO_2-CaO, *and* ZrO_2-Y_2O_3, Refract. Ind. Ceram. **24**, 452–458 (1984).

[62]   M.D. Alcala, C. Real, and J.M. Criado, *A New Incipient-Wetness Method for the Synthesis of Chemically Stabilized β-Cristobalite*, J. Am. Ceram. Soc. **79**, 1681–1684 (1996).

[63]   R. Pascova, I. Penkov, G. Avdeev, F.-P. Ludwig, J.W.P. Schmelzer, and I. Gutzow, *Refractory Alkali-Free Cristobalite Glass-Ceramics: Activated Reaction Sinter-Crystallization Synthesis and Properties*, International Journal of Applied Glass Science **3**, 75–87 (2012).

[64]   N. Danchova, *Preparation and properties of doped sol-gel materials*, PhD Dissertation, University of Sofia (2012), Bulgaria.

[65]   Z. Strnad, *Glass-Ceramic Materials*, 81–101 (Elsevier, Amsterdam, 1986).

[66]   Y. Zhu, K. Yanagisawa, A. Onda, and K. Kajiyoshi, *The Preparation of Nano-Crystallized Cristobalite under Hydrothermal Conditions*, J. Mater. Sci. **40**, 3829–3831 (2005).

[67]   A.M. Bychkov, V.S. Rusakov, and G.A. Sukhadol'skii, *Crystallization of Quartz and Cristobalite in the Presence of Iron under Low-Temperature Hydrothermal Conditions*, Geokhimiya **10**, 1019–1023 (1996).

[68]   S. Gutzov, C. Berger, M. Bredol, and C.L. Lengauer, *Preparation and Optical Properties of Holmium Doped Silica Xerogels*, J. Mater. Sci. Letters **21**, 1105–1107 (2002).

[69]   S. Gutzov, G. Ahmed, N. Petkova, E. Füglein, and I. Petkov, *Preparation and optical properties of samarium doped sol-gel materials*, J Non-Crystalline Solids **354**, 3438–3442 (2008).

[70]   P.P. Keat, *A New Crystalline Silica*, Science **120**, 328–330 (1954).

[71]   L.M. Lityagina, T.I. Dyuzheva, A. Nikolaev, and N.A. Bendeliani, *Hydrothermal Crystal Growth of Stishovite* SiO_2, J. Crystal Growth **222**, 627–629 (2001).

[72]   N.A. Dubrovinskaya and L.S. Dubrovinsky, *High-Pressure Silica Polymorphs as Hardest Known Oxides*, Mater. Chem. Phys. **68**, 77–79 (2001).

[73]   L.S. Dubrovinsky, N.A. Dubrovinskaya, V. Prakapenka, F. Seifert, F. Langenhorst, V. Dmitriev, H.-P. Weber, and T. Le Bihan, *A Class of New High-Pressure Silica Polymorphs*, Phys. Earth Planet. Interiors **143–144**, 231–240 (2004).

[74]   A.E. Goresy, A.L. Dubrovinsky, T.G. Sharp, and M. Chen, *Stishovite and Post-Stishovite Polymorphs of Silica in The Shergotti Meteorite: Their Nature, Petrographic Settings Versus Theoretical Predictions and Relevance to Earth's Mantle*, J. Phys. Chem. Solids **65**, 1597–1608 (2004).

[75]   V.V. Brazhkin, R.N. Voloshin, and S.V. Popova, *The Kinetics of the Transition of the Metastable Phases of SiO_2, Stishovite and Coesite to the Amorphous State*, J. Non-Crystalline Solids **136**, 241–248 (1991).

[76]   X. Xue, J.F. Stebbins, and M. Kanzaki, *A ^{29}Si MAS NMR Study of Sub-T_g Amorphization of Stishovite at Ambient Pressure*, Phys. Chem. Minerals **19**, 480–485 (1993).

[77]   G. Serghiou, A. Zerr, and L. Boehler, *The Coesite – Stishovite Transition in a Laser-Heated Diamond Cell*, Geophys. Res. Letters **22**, 441–444 (1995).

[78]   I. Prigogine and R. Defay, *Chemical Thermodynamics* (Longmans Green, London, New York, 1954).

[79]   B. Mutaftschiev, *The Atomistic Nature of Crystal Growth* (Springer Series in Materials Science, Springer, Berlin, 2001).

[80]   J.W.P. Schmelzer, *Phases, Phase Transitions, and Nucleation Theory*, In: Encyclopedia of Surface and Colloid Science (M. Dekker Publ., New York, 4017–4029, 2002).

[81]   M. Volmer, *Kinetik der Phasenbildung* (Th. Steinkopff, Dresden, 1939).

[82]   W. Ostwald, *Studien ueber die Bildung und Umwandlung fester Körper*, Z. Phys. Chem. **22**, 289–315 (1897).

[83]   I.N. Stranski and D. Totomanov, *Keimbildungsgeschwindigkeit und Ostwaldsche Stufenregel*, Z. Phys. Chem. **A 163**, 399–408 (1933).

[84]   I. Gutzow and I. Avramov, *On the Mechanism of Formation of Amorphous Condensates from the Vapor Phase (I): General Theory*, J. Non-Crystalline Solids **16**, 128–142 (1974).

[85]   I. Gutzow, J. Schmelzer, and A. Dobreva, *Kinetics of Transient Nucleation in Glass-Forming Liquids: A Retrospective and Recent Results*, J. Non-Crystalline Solids **219**, 1–16 (1997).

[86]   J. Schmelzer, J. Möller, and I. Gutzow, *Ostwald's Rule of Stages: The Effect of Elastic Strain and External Pressure*, Z. Phys. Chem. **204**, 171–181 (1998).

[87]   V.G. Pol, A. Gedanken, and J. Calderon-Moreno, *Deposition of Gold Nanoparticles on Silica Spheres: A Sonochemical Approach*, Chem. Mater. **15**, 1111–1118 (2003).

[88]   V. Guencheva, E. Stoyanov, I. Gutzow, G. Carl, and C. Rüssel, *Induced Crystallization of Glass-Forming Melts, Part 1*, Glass Sci. & Technol. **77**, 217–228 (2004).

[89]   E. Grantscharova and I. Gutzow, *Vapor Pressure, Solubility and Affinity of Undercooled Melts and Glasses*, J. Non-Crystalline Solids **81**, 99–127 (1986).

[90]   E. Grantscharova, I. Avramov, and I. Gutzow, *Die thermodynamischen Parameter und die Löslichkeitskurve von glasartigen Substanzen*, Die Naturwissenschaften **73**, 95–96 (1986).

[91]   J.N. Grima, R. Gatt, A. Alderson, and K.E. Evans, *An Alternative Explanation for the Negative Poisson's Ratios in α-Cristobalite*, Mater. Sci. & Eng. **A 423**, 219–224 (2006).

[92]   H. Kimizuka, H. Kaburaki, and Y. Kogure, *Mechanism for Negative Poisson Ratios over the $\alpha - \beta$ Transition of Cristobalite – SiO_2: A Molecular-Dynamics Study*, Phys. Rev. Letters **84**, 5548–5551 (2000).

[93]   E. Natarajan, A. Nordin, and A.N. Rao, *Overview on combustion and gasification of rice husk in fluidized bed reactors*, Biomass and Bioenergy **14**, 533–546 (1998).

[94]   S. Chandrasekhar, K.G. Satyanarayana, P.N. Pramada, and P. Raghavan, *Processing, properties and applications of reactive silica from rice husk – an overview*, J. Mater. Science **38**, 3159–3168 (2003).

[95]   V.P. Della, I. Kuhn, and D. Hotza, *Rice husk ash as an alternate source for active silica production*, Mater. Lett. **57**, 818–821 (2002).

[96]   Y. Ma, X. Zhao, H. Zhang, and Z. Wang, *Comprehensive utilization of the hydrolyzed productions from rice hull*, Industrial Crops Prod. **33**, 403–408 (2011).

[97]   I.G. Markovska, *Investigation on the properties of waste rice husk and alumina and the possibilities for its utilization as ceramic matrix*, PhD Thesis, Prof. A. Zlatarov University, Bourgas 2005.

[98]   T. Vlaev, I.G. Markovska, and L.A. Lyubchev, *Non-isothermal kinetics of pyrolisis of rice husk*, Thermochimica Acta **406**, 1–7 (2003).

[99]   I.G. Markovska and L.A. Lyubchev, *A study on the thermal destruction of rice husk in air and nitrogen atmosphere*, J. Thermal Analysis Calorimetry, **89**, 809–814 (2007).

[100] O. San and C. Özgür, *Preparation of a Stabilized β -Cristobalite Ceramic from Diatomite*, J. Alloys Comp. **484**, 920–923 (2009).

[101] M. Estevez, S. Vargas, V.M. Castano, and R. Rodriguez, *Silica nano-particles produced by worms through a bio-digestion process of rice husk*, J. Non-Crystalline Solids **355**, 844–850 (2009).

[102] S. Wattanasiriwech, D. Wattanasiriwech, and J. Svasti, *Production of amorphous silica nano-particles from rice straw with microbial hydrolysis pretreatment* **356**, 1228–1232 (2010)

[103] E.W. Brizke and A.F. Kapustinski (Eds.), *Caloric Constants of Inorganic Substances* (Academy of Sciences Press, Moscow, Leningrad, 1949 (in Russian)).

Irina G. Polyakova

4 The Main Silica Phases and Some of Their Properties

The present chapter is devoted to a review of both historical and modern aspects of the investigation of the structure and properties of silica – one of the most abundant substances on Earth. Silicas are substances with the same chemical formula, SiO_2, but different structures. The variety of crystal forms of silicas seems to be unique; they vary from loose clathrasils to the built of tetrahedral units proper silicas such as quartz and cristobalite and to very dense high-pressure phases built of SiO_6-octahedrons. Specific features of the crystal structures, conditions of their formation and distribution in nature are discussed. Special attention is given to quartz and some of its properties such as the existence of right and left quartz, its anisotropy; uncommon thermal expansion and the mechanism underlying it; the phenomena accompanying the $(\alpha - \beta)$-conversion – opalescence and appearance of incommensurate phases in a very narrow temperature interval; pressure induced amorphization etc. Physicochemical problems of hydrothermal synthesis of quartz single crystals are also discussed. Some relatively new phenomena, pressure induced transformations in amorphous silica and/or polyamorphism, are briefly reviewed as well. The electronic structure of silica is analyzed; its specific features explain the broad variety of silicas as well as their unexpectedly high reactivity and the negative thermal expansion coefficients of high-temperature polymorphs of proper silicas built of SiO_4-networks.

4.1 Introduction

The concept of zero is said to be a great finding of ancient Indian mathematicians. Just in the 9th century they have comprehended that nothing is something. If we want to fill something, we should provide ourselves with an appropriate void. To have a void is to have something. Thus, zero in a number shows that we have no unities in a given number position, but it also shows us that the number contains this position.

A similar situation is found in chemistry. Atoms with empty outer orbitals have more possibilities for chemical bond formation than atoms with the same electronic configuration but lacking empty orbitals. The presence of empty d-orbitals in silicon atoms imparts a number of specific features to silicon dioxide which are absent for carbon dioxide although carbon is the nearest analogue of silicon. This "chemical void" can be partially filled and involved in orbital hybridization. As a result, we get such properties as an extraordinary glass forming ability of silica; a great number of polymorphs for this elementary glass forming substance; ability to form five-coordinated complexes contrary to pronounced tetravalence of silicon; ability to form easily activated complexes resulting in an unexpectedly high reactivity at very low temperatures. The latter property, together with ability to catalysis, leads to high-temperature phases (like quartz and cristobalite) growth from solutions at 300 °C or even at 100 °C. Finally, involving empty d-orbitals in hybridization in the long run leads to zero thermal expansion of SiO_4 tetrahedra.

As it seems to me, the reasons of the listed and some other features of silica often remain incomprehensible for specialists in glass science because these reasons have their origin in a quite far field. In the present review, I am going to discuss some spectrum of the interesting properties of silica and, whenever it is possible, to supply an explanation for them. In addition, some amazing investigations are described which have been performed already about fifty years ago, very prolonged and laborious and very precise ones. Apparently, they could not be executed in full detail in a similar way presently. In my presentation, I employ mainly the monographs [1] and [2] as a starting point. So, for information on some further details eventually of interest the reader is referred to these references.

4.2 Specific Properties of Silica Resulting from the Electronic Structure of Silicon

4.2.1 Specific Properties of Silica Compounds and Differences as Compared to Chemical Analogs: Silicon and Carbon

As mentioned by Robert B. Sosman [3]: "*The beginning chemist might predict that silicon, having the same external electron structure as carbon, will form two oxides – SiO and SiO_2, both gaseous; further, that SiO, like CO, will be stable at high temperatures but will disproportionate at lower temperatures into SiO_2 and Si (as does CO into CO_2 and graphite). Why is he so nearly right about SiO and so completely wrong about SiO_2?*"

A great diversity of silicon compounds can be found in nature. After carbon, silicon forms the largest number of compounds with other elements. On the one hand, this is a result of silicon position in the main sub-group of the fourth group of the periodic system of elements which is similar to carbon. Consequently, it is determined by the same factors that provide such a wide spectrum for carbon compounds. However, the latter statement is true only partly, and the author of the periodic system of elements, D.I. Mendeleev, was the first to emphasize the significant differences in the properties of CO_2 and SiO_2 [4]. The great number of carbon compounds also results from the closeness in C–C, C–O and C–H bonding energies (Table 4.1). As a result, these bonds occur with approximately equal probability.

In contrast, the energy for the bond Si–O considerably exceeds the Si–H-bonding energy and is larger by a factor of two as compared to the Si–Si-bonding energy (Table 4.1). Therefore the basis of silicon chemistry is made up not from –X–X–X-chains, common for carbon, but from –Si–O–Si–O–Si-chains. Only a relatively small

Table 4.1: Average bonding energies (kJ/mol) for some bonds of carbon and silicon [1]

Bond	X	
	C	Si
X–X	346	222
X–O	358	452
X–H	413	318

number of silicon compounds may be considered as analogs of organic compounds of carbon. All three types of the bonds (X–X, X–H and X–O) are nearly equal with respect to their energetic preference in carbon chemistry. In silicon chemistry there exists only one energetically preferential type of bonding, i.e. (X–O), therefore a less number of compounds is known for silicon as compared to carbon. At the same time, silicon compounds have specific features that make them unique among other classes of compounds. Silica is really unique by the abundance of its polymorphic modifications; the structural variety of silicates is large indeed and by far exceeds the variety of carbon compounds.

The difference between the chemical analogues, C and Si, is not only of quantitative but even of qualitative nature. The highest coordination number for carbon is four. In contrast, for silicon it can reach five or even six. The valence angle of oxygen in organic compounds of R–O–R-type (where R = CH_3, C_2H_5 and others) is equal to the tetrahedral angle $109\,°28'$ and it seems that the same should be valid for the similar Si–O–Si-angle (Fig. 4.1). So, the question arises whether the Si–O-bond can be considered as mainly ionic or mainly covalent. In fact, in silica, silicates and organo-silicon compounds this angle changes in wide limits in the range of $120\text{–}180\,°$.

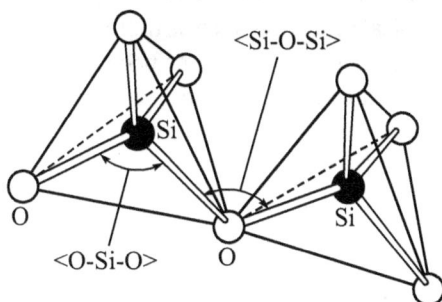

Fig. 4.1: Interior tetrahedral O–Si–O and exterior inter-tetrahedron Si–O–Si angles.

As it is seen from Fig. 4.2, constructed according to the most precise measurements of Si–O–Si-angles in silicates, the value of the valence angle of $139\,°$ is frequently observed. For pure silica polymorphs, the most frequent value of this angle is even higher, i.e., $147\,°$. Compounds with a tetrahedral Si–O–Si-angle, equal to $109\,°28'$, obviously do not exist in silicas and silicates! Moreover, the valence angles Si–O–Si

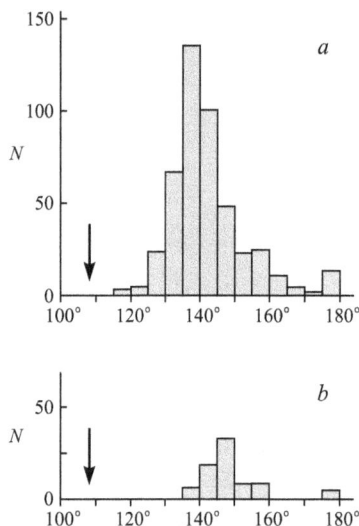

Fig. 4.2: Histograms for the Si–O–Si-angle distribution established for different compounds: (a) silicates (468 angle values); (b) polymorphs of silica (80 angle values) [1]. Arrows mark the tetrahedral angle.

in silicas and silicates not only vary from structure to structure, but for every given structure they can change with temperature or as a result of chemical substitution in solid solutions over rather wide limits (Fig. 4.3). This particular feature is not found in carbon-oxygen compounds; the valent C–O–C-angle is rigid and differs only slightly from the tetrahedral angle. Another specific feature of silicon compounds is the too large difference between the Si–O-bond length calculated from the covalent Si and O radii (1.83 Å) and the bonds established experimentally, in particular, the bond length (1.62 Å) as the average for silicas and silicates. The origin of the above mentioned and some other features of silica compounds such as the variety in mechanisms of chemical reactions, catalytic character of silica glass crystallization and chemical reaction with silica participation, specifics of some physical properties such as thermal expansion results from the electronic structure of the silicon atom and some peculiarities of its interaction with oxygen. These topics will be addressed in the next section.

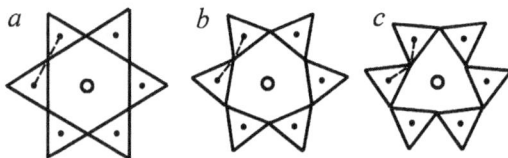

Fig. 4.3: The effect of temperature and cation size on the Si–O–Si angle of double-layered aluminasilicates: (a) Ba[Al$_2$Si$_2$O$_8$] above 570 °C; (b) Ba[Al$_2$Si$_2$O$_8$] below 570 °C; (c) Ca[Al$_2$Si$_2$O$_8$]. Open points refer to Ba or Ca cations, solid points to Si and Al atoms.

4.2.2 Electron Structure of the Silicon Atom and its Interaction with Oxygen

The configuration of the valence shells of carbon and silicon atoms seems at a first glance to be the same, s^2p^2. A transformation of the valence shells usually takes place in the course of formation of chemical compounds. In an excited state, the atom has one s and three p electrons at the outer shell (Fig. 4.4), which form four identical sp^3 hybrid orbitals. Silicon atoms in the ordinary four-valent state use just these tetrahedral directed bonds with an angle of nearly 109 °. The overlap of $3sp^3$ orbitals of silicon with $2p$ orbitals of oxygen forms a chemical bonding of σ-type which is the bond with maximal overlapping of electronic density on the line connecting Si and O atoms. These four hybrid sp^3 orbitals are quite stable and thus of primary importance for silicon chemistry. The major part of silicon compounds is constructed from SiO_4-tetrahedra; they are the main constructional elements of silicas, silicates and organosilicons.

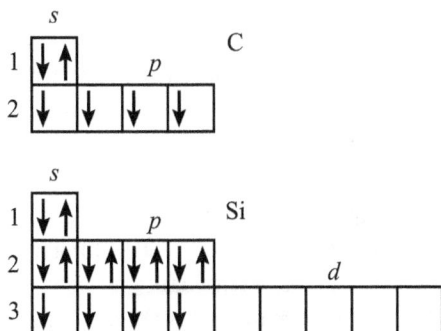

Fig. 4.4: Electron shells of carbon and silicon in an exited state.

However, there exists a very significant difference in between the valence shells of carbon and silicon: silicon has five vacant d-orbitals that are absent in the outer shell of carbon. Linus Carl Pauling was the first to understand that a vacant place differs from an absent one. In his famous book "The Nature of the Chemical Bond" he advanced the idea of participation of $3d$-orbitals in the chemical bonding of silicon with electronegative atoms (oxygen, for example) in order to explain the differences between silicon and carbon chemistry and numerous discrepancies in the theoretically expected and experimentally established properties of the silicon bonds. Electronegative substitute of silicon increases its positive effective charge and promotes contraction of diffusive $3d$-orbitals, which become with respect to their energy commensurable with $3s$ and $3p$-orbitals and thereby enhance valent capabilities of silicon. According to modern concepts based on manifold experimental evidences, $3d$-orbitals of silicon are involved in π-bond formation together with an unshared p-electron pair of electronegative substitute (oxygen), i.e. (p_π–d_π)-coupling. Furthermore, formation of additional σ-bonds is also possible by $3sp^3d$ and $3sp^3d^2$ hybridization. Since they are of high importance for silicon chemistry, we briefly discuss the consequences of the occurrence of these two additional types of bonding for the case of silicon-oxygen interaction. A detailed review of related topics and the analysis of available literature can be found in the monograph [1] for X-ray investigations and in [2] for the chemical approach.

4.2.3 Consequences of π-bonding in Silica

The occurrence of π-bonding in SiO_4-tetrahedra is clearly confirmed by the above-mentioned difference between the calculated and experimentally measured single Si–O-bond length. The bond length is 1.83 Å for the covalent model and contracts to 1.76 Å with due regards for the partially ionic character of the bond, but according to X-ray analysis the true length varies within the range of 1.59–1.63 Å for different forms of silica with tetrahedral coordination. Such large contraction of the bond indicates strong π-bonding inside the SiO_4-group and increasing in the bond order because of a shift of an unshared p-electron pair of oxygen atom to one of the empty d-orbitals of silicon. Besides the bond contraction, π-bonding leads to a reduction in the dipole moment of Si–O-bonds, the decrease of the effective positive charge of silicon to +2 and the negative charge of oxygen to –1. The availability of additional $(p_\pi - d_\pi)$-bonding can also explain the more acid character of silanols in comparison with carbinols, the higher Si–O-bond energy (419–494 kJ/mol) as compared to a more dipolar C–O-bond (358 kJ/mol) as well as the absence or significant weakening of donor-acceptor ability of siloxane (Si-O-Si) bond in siloxanes and polysiloxanes. Participation of both unshared electron pairs of oxygen of bridging Si-O-Si bond in $(p_\pi - d_\pi)$-bonding affects the valent Si–O–Si-angle and leads the latter to increase (Fig. 4.2). Valent Si–O–Si-angles are not rigid but have a considerable flexibility because of diffusivity and different orientation of the five $3d$-orbitals of silicon resulting in π-bond formation for just any spatial location of oxygen atoms. This feature appears to explain the large number of polymorphic modifications of silica with different types of packing of SiO_4-tetrahedra. Clear correlations between Si–O–Si-angles and Si–O-distances may be observed for a given crystal structure, but they are not always comparable for different modifications. The approaching of the valent Si–O–Si-angle to a value of 180° is not typical for silica modifications and occurs only in individual structural positions of high-temperature cristobalite and coesite.

Thus, π-bonding manifests itself quantitatively in Si–O-bond contraction and strengthening as a result of the increase in the bond order, the latter varying in silicas and organosilicon compounds in the range from 1.2 to 1.5. The most suitable orbitals for π-bonding are d_{xy}, d_{xz} and d_{yz}, which are directed at an angle of 45° to the (x, y, z)-axes. Maximal overlapping of the electron clouds takes place in this case away of the line connecting Si and O atoms. One interesting consequence of the latter fact is considered below in the paragraph devoted to thermal expansion of quartz. Another consequence is the very high tendency of silica melt to glass formation. SiO_2 is the most famous glass former, its analogue in the periodic system GeO_2 is less known but comparable in glass formation tendency in contrast to CO_2 which was transferred into an amorphous silica-like state only very recently under very high pressure [5]. The inter-tetrahedron Si–O–Si-angle in quartz glass varies from 120 to 180° with a most probable value of nearly 144°.

4.2.4 Increase in Silicon Coordination Number as a Result of *s-p-d*-hybridization

Five vacant *d*-orbitals of silicon may be used apart from the Si–O-strengthening by π-bonding for formation of additional directed σ-bonds with strongly electronegative atoms like fluorine, oxygen or nitrogen. In this case, silicon coordination number increases to five and six. Presently, a number of compounds, being quite stable at ambient conditions, is known, for which the five-coordinated state of silicon was established by X-ray single crystal investigations. For some other compounds, the groups [SiA$_5$] with five-coordinated silicon were established by spectroscopy and chemical methods. In these compounds, the first coordination sphere of silicon contains at least one strongly electronegative atom besides oxygen. These groups form via $sp^3 d_{z^2}$-hybridization, and the coordination polyhedron of silicon is usually represented by a slightly distorted trigonal bipyramid. For ideal $sp^3 d_{z^2}$-hybrid orbitals (and trigonal bipyramids), the valent X–Si–X-angle is equal to 120°.

Fig. 4.5 demonstrates the interrelation between average values for three X–Si–X-angles in 15 refined crystal structures involving five-coordinated silicon and average values of X-Si-Y and X-Si-Y' angles (where Si-Y is the shortest and Si-Y' is the largest distance between the Si-atom and apical ligands [6]). Point 1 in the figure refers to forsterite $Mg_2[SiO_4]$ with distorted tetra-coordination of silicon; point 2 refers to the ideal tetrahedral sp^3-hybridization in a hexagonal close-packed lattice with the angle of 109°28'; point 17 to the the ideal $sp^3 d_{z^2}$-hybridization. The experimental data for compounds with five-fold silicon are located on two lines passing through ideal

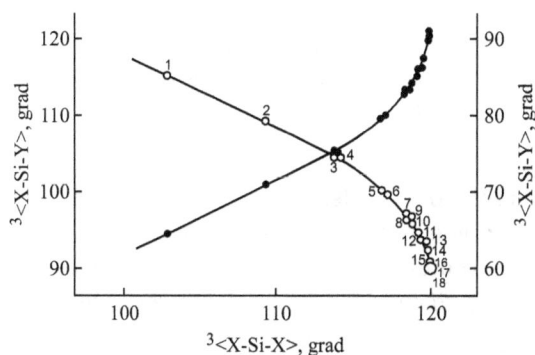

Fig. 4.5: Dependence of average values for three X–Si–X angles on X–Si–Y and X–Si–Y' angles.

tetrahedral and trigonal bipyramidal positions. Thus, there is a more or less continuous change from sp^3- to $sp^3 d_{z^2}$-hybridization and participation degree of $3d_{x^2-y^2}$- and $3d_{z^2}$-orbitals of silicon in its bonding with ligands increases in this row; correlations between interatomic distances d(Si–A) and corresponding angles confirm this conclusion [6].

Six-fold coordination of silicon appears due to $sp^3 d^2$-hybrid orbitals formation directed to octahedron vertices. It may be formed at ambient conditions in silicates with the general formula $M_r Si_s O_t$ containing Si–O–M bonds, provided the electronegativity of the M-atom is sufficiently high as it is the case for C, H, P or F. Two polymorphic modifications of silicon pyrophosphate SiP_2O_7 are examples of compounds with SiO_6-octahedrons which are stable at ambient conditions. Among the compounds with $Si(OH)_6$-groups there is the mineral thaumasite [7] which was also found in industrial concretes. However, the major part of compounds with six-coordinated silicon is stable only under high and super-high pressure; a list of these compounds together with structural information on them can be found in [1].

The most famous example in this respect is stishovite, one of high-pressure modifications of SiO_2 (Fig. 4.6). Involving six relatively large oxygen atoms in the coordination sphere of the relatively small silicon atom, its formation is aided by ultrahigh pressures (16–18 GPa) because of the necessity to overcome strong electrostatic repulsion of oxygen. The Si–O-bonds inside the octahedrons of stishovite are taken to be predominantly covalent, four shorter bonds (0.1716 nm at Fig. 4.6) being $(p - d)_\pi$-bonded to some extent and two longer bonds (0.1872 nm) being single.

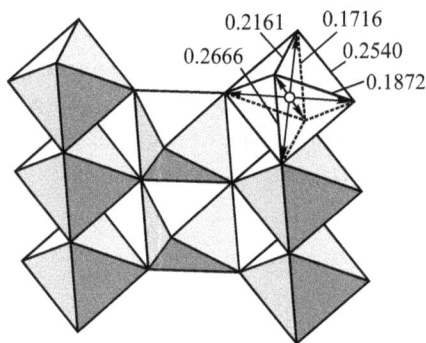

Fig. 4.6: Crystal structure of stishovite [8]. All silicon atoms are located inside oxygen octahedrons.

Polymerization of silicic acid in water solutions also leads to an increase in coordination number of silicon. For example, in a series of X-ray studies reviewed in [2] Mitzyuk and coauthors showed that the average coordination number of silicon with respect to oxygen in hydrogels of polysilicic acid changes from 4.5 to 5.8 in dependence on water content and degree of the gelskeleton ageing. A similar situation is found in methanol-replaced gels. The fact was explained by donor-acceptor complexes formation between water or methanol molecules and silicon atoms and confirmed by IR-spectroscopy data. Thus, high-coordinated silicon complexes can be easily formed at ambient conditions.

It is well known that the larger is the atom in size, the higher can be its coordination number because of deminution in Coulomb repulsion of ligand atoms. Germanium is larger than silicon and also has vacant d-orbitals, it easily forms GeO_6-octahedrons at atmospheric pressure. Stishovite can exist at atmospheric pressure and was found in small amounts in nature but only in meteorite craters where it was formed by percussion metamorphism produced at meteoric impact.

4.2.5 Implication of *s-p-d*-hybridization for Chemical Reactions and Physical Transformations of Silica

The possibility for a silicon atom to form (s-p-d)-hybrid orbitals in addition to the main four-valent sp^3-bonds is of decisive importance for chemical reactions with silicon participation and explains the extreme sensibility of silica phase transformations to the presence of minor additions of some substances. Comprehensive evidence exists demonstrating that chemical processes with participation of silica phases, such as dilution or growth from a solution, pass through formation of five- or six-coordinated activated complexes. A detailed literature review and further references can be found in the monograph [2], here we only briefly summarize the main specific features of these processes.

Any physical or chemical transformation of silica proceeds via a switch of (≡Si–O–Si≡)-bonds which are the main structural element for all types of silicas as well as silicates. The bond is especially inert, and its decomposition is a limiting stage for dilution or polymerization of silica in water; the hydrolysis mechanism is catalytic and depends on different agents present in water. In neutral and alkaline medium, the process is accelerated by OH^- and in an acid medium by H^+ and F^- ions. The presence of some salts in water solutions accelerates silica dilution at pH = 2–4 and higher because nucleophilic anions attack in the row $F^- > SO_4^{2-} > Cl^-$. The ability of siloxane bond Si–O–Si to decompose heterolytically at the presence of these agents was also repeatedly verified for organosilicon compounds, the process is always accompanied by the formation of intermediate five-fold complexes. The redistribution of electronic density in ≡Si–O–Si≡ groups favors the heterolytic decomposition of Si–O-bonds. Actually, the oxygen atom of the water molecule forms a donor-acceptor bonding with the attacked silicon atom, lowers its positive effective charge and weakens the $(p_\pi - d_\pi)$-coupling in the Si–O-bond. As the result, an electrophilic attack of a hydrogen atom of the H_2O molecule onto the bridging oxygen is facilitated and the siloxane bond breaks resulting in the formation of strongly hydroxylated products.

In the presence of OH⁻ ions in water solution, they apparently take part in initial stages of silica dilution with intermediate activated complex formation:

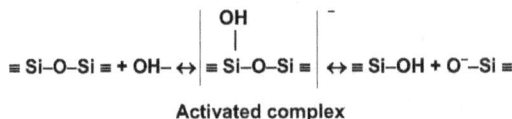

$$
\equiv Si\text{–}O\text{–}Si \equiv + \; OH\text{–} \;\leftrightarrow\; \left| \begin{array}{c} OH \\ | \\ \equiv Si\text{–}O\text{–}Si \equiv \end{array} \right|^{-} \;\leftrightarrow\; \equiv Si\text{–}OH + O^{-}\text{–}Si \equiv
$$

Activated complex

The intermediate activated complexes are not stable and easily decompose with ≡Si-O⁻-anions formation; the latter can participate in the reverse reaction. In dependence on pH-value, nature and state of the medium, they may be deactivated to a variable degree and thereby regulate the rate of dilution. In acid solutions the process of silica dilution apparently proceeds via the formation of intermediate activated complexes with closed chains. In the case of weakly dissociable hydrofluoric acid, bond decomposition takes place according to the following scheme:

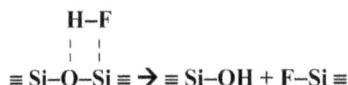

$$
\begin{array}{c} H\text{–}F \\ |\; \vdots \\ \equiv Si\text{–}O\text{–}Si \equiv \end{array} \;\rightarrow\; \equiv Si\text{–}OH + F\text{–}Si \equiv
$$

It is easy to see that one of the silicon atoms in the four-centered activated complex is five-coordinated. Strong acids like HCl and H_2SO_4 also form closed-chain complexes but with water molecules. The more detailed regularities of water solubility of quartz and the influence of different agents will be discussed in the section devoted to hydrothermal synthesis.

Hydrothermal crystallization of quartz is also a complex process depending on a large number of parameters. Acidity of the growth medium and presence of salt additions act on the rate of growth of quartz in addition to temperature and pressure. The ability of siloxane bonds to form a five-folded activated complex is the basic origin of this catalytic activity. In manufacturing environment of hydrothermal synthesis the temperature-pressure conditions may be chosen such to guaranty direct formation of quartz from the water solution. However, at lower values of these parameters the process is realized via a passage through intermediate phases. The type of the crystallizing phases including clathrasils, their stability and sequence of formation also depends on acidity or alkalinity of the medium and the nature of salt additions provided all other factors being the same.

As we have seen earlier, the five-fold activated complexes are different for different mediums; they stimulate formation of clathrasils with different structures, silica-X and silica-Y (see the next part for details). It is significant that silica-X changes with environment, in particular, with the nature and state of the medium. Different types of silica-X and their forming conditions as well as crystallization paths of silica in water solutions for different temperature/pressure conditions are described in detail in [2]. In particular, it was shown that temperature resistant silica phases like cristobalite and quartz can be formed in water solutions as a result of reconstructive transformations of other crystalline phases at 100–300 °C and even at room temperature. Such processes seems to be impossible taking into consideration the high strength of Si–O-bonding and the fact that it is the only type of bonding in silica. Nevertheless, the formation of activated complexes sharply reduces the energy barrier for the transformations and makes them possible.

Crystallization of silica and silicate glasses takes place along comparable pathways. According to own experience, cristobalite and quartz can be produced in sodium silicate glasses of appropriate compositions immediately above the glass transition region, at 540 °C. The alkali ions play here the role of a transformation catalyst. Hence, silicate glasses have intrinsic catalysts in their structure that provoke siloxane bond splitting followed by high-temperature phase (like quartz or cristobalite) crystallization at relatively low temperatures. In silica glass, this role is taken over by the impurities. An extremal sensitivity of silica glass crystallization to the mode of production and experiment conditions as well as non-repeatability of crystallization results (this is true not only for coefficients in the kinetic kinetic equations but also with respect to the qualitative behavior) has been systematically studied and explained by Leko & Komarova (see, for example [9, 10–14] and here Chapter 7) as a result of trifling fortuitous pollution. In superpure silica glass any trifling admixture or pollution are extraordinary active nucleation catalysts and play a decisive role in crystal nucleation and growth. Kinetic dependencies, which are really intrinsic to silica glass, can be obtained only under special conditions as described by above mentioned authors.

4.3 Phases of Silica and Their Properties

The word "silica" denotes a substance with chemical formula SiO_2. It is the most abundant substance on the Earth. Nearly 58 % of the lithosphere consists of bonded SiO_2, and 12 % of this silica is found in form of separate rocks as quartz, chalcedony, opal, etc. Silica is unique among natural and artificial compounds because of a wide variety of different modifications. The major part of them has for a long time been known as minerals, but the development of high-pressure techniques brought about the creation of a number of dense forms of silica, both crystalline and amorphous. Different authors give different lists of silica polymorphs, which exist at quite special conditions such as different gas or liquid environment, its acidity and pressure. The phase diagram for pure SiO_2 under normal conditions is a topic of debate and there is no finally established set of proper silica phases (cf. also Chapter 3).

We consider here some problems of compact silica. The dispersed forms of silica are a separate and very special subject, which we do not touch upon in this chapter. All compact crystalline as well as amorphous silicas (except for fibrous silica-W) have three-dimensionally connected networks, and all silica networks (except for the three densest ones) are built from SiO_4-tetrahedra connected by their vertices. Silica phases and compounds are amazing substances in many respects. Let us take a look at these phases, their structures, abundance in nature and some properties (cf. Tables 4.2 and 4.3).

Table 4.2: Symmetry, calculated crystallographic density and number Z of $SiO_{4/2}$-tetrahedra in the unit cell for the main silica crystal phases (according to [1, 2, 15, 16]). The latter reference is the largest on-line mineral database and mineralogy reference website on the internet. Data for the high-temperature modifications have been obtained at different temperatures; hence they should be compared with the low-temperature data with caution. In the table they are marked with grey color ((1) Meteorite low-tridymite, (2) natural low-tridymite, (3) volcanic low-tridymite)

Mineral name	Modification	Crystal system	Z	T, °C	Density, g/cm²	Typical structural elements of crystal lattice [2]
Zeolite H-ZSM5		orthorhombic	96	room	1.79	Large cavities
Melanophlogite 46[SiO₂]·6[N₂,CO₂]·2[CH₄,N₂]		cubic	46	room	1.95	Large cavities 0.5 and 0.65 nm
Fibrous silica-W (oxide_Si)		orthorhombic	4	room	1.97	Chains of SiO₄ tetrahedrons bonded by edges
Zeolite theta-1		orthorhombic	24	room	1.97	Large cavities
Lechatelierite (silica glass)		disorded	–	room	2.201	Continuous three-dimensional cristobalite-like network formed of tetrahedrons
Tridymite	high	hexagonal	4	>420	2.19	Layers of 6-membered rings in the (0001)-plane
		hexagonal	8	220	2.21	
	medium		24	105-180	2.23	
	Low⁽¹⁾	rhombic	48	room	2.25-2.27	Layers of 6-membered oval rings in the (001)-plane
	Low⁽²⁾	rhombic	64, 160	room		
	Low⁽³⁾	rhombic	320	room		
Cristobalite	high	cubic	8	248	2.19	Layers of 6-membered rings of tetrahedrons in the (111)-plane
	low	tetragonal	4	28	2.32	Layers of 6-membered rings of tetrahedrons in (101)-plane

Table 4.3: Continuation of Table 4.2

Mineral name	Modifi-cation	Crystal system	Z	T, °C	Density, g/cm^2	Typical structural elements of crystal lattice [2]
Keatite		tetragonal	12	room	2.50	Spiral 4-membered chains around quadric axis
Moganite	high	orthorhombic	12	1354	2.56	left- and right-handed threefold spirals
	low	monoclinic	12	room	2.62	left- and right-handed threefold spirals
Quartz	high	hexagonal	3	575	2.54	Spiral chains around hexagonal axis
	low	trigonal	3	22	2.655	Spiral chains around triple axis
Coesite		monoclinic	16	room	2.95	Chains of 4-membered rings of feldspar type
Stishovite		tetragonal	2	room	4.28	Rutile-like structure built from SiO_6 octahedrons bonded by edges
				1 atm		
Seifertite		orthorhombic	4	room	4.29-4.30	α-PbO_2-like structure built from SiO_6 octahedrons

4.3.1 Dense Octahedral Silicas: High Pressure Phases

The three densest phases of silica are constructed from SiO_6-octahedrons. One of these phases – the quite famous **stishovite** – has been described in the previous section (Fig. 4.6). The second phase, **seifertite**, is the so far densest and hardest polymorph of silica found in nature, it has a scrutinyite (α-PbO_2) type structure. The mineral was named after Friedrich A. Seifert (born 1941, founding Director of the Bayrisches Geoinstitut, Universität Bayreuth, Germany) for his seminal contributions to high-pressure geoscience. This phase was predicted in 2007 by the metadynamics method [17]. It was found in 2008 [18] as lamellae occurring together with dense silica glass lamellae in composite silica grains in the heavily shocked Martian meteorite Shergotty. It was inferred that seifertite was formed by shock-induced solid-state transformations of either tridymite or cristobalite on Mars at an estimated minimum equilibrium shock pressure in excess of 35 GPa. With respect to density the mineral corresponds to the density of the last solid layer which is situated above the liquid Earth core. The mineral was also intergrown in some grains with minor stishovite and a new (third) **unnamed monoclinic dense silica polymorph** constructed from SiO_6 with a ZrO_2-type structure.

4.3.2 Clathrasils: Friable Silica Phases

All silica phases built of tetrahedra may be divided into two groups: proper silicas and filled silicas or clathrasils [1]. The proper silicas like quartz, cristobalite, keatite and coesite can be formed from pure SiO_2 at appropriate temperature-pressure conditions. Clathrasils, in contrast, can be formed only at the presence of some organic or inorganic molecules. Low-molecular forms of silicic acid condense around the guest molecules, the latter being not cations but neutral compounds which organize the silica network around themselves and play the role of seeds in the crystallization process. The guest molecules appear to be locked inside polyhedral cavities (cages) of the network and the shape of the cage is determined by the size and the form of the guest molecule. In some cases, organic guest molecules may be removed from the system by burning. Then, a skeleton of pure SiO_2 remains which may keep its stability (zeolites). In other cases, the structure collapses after removal of the guest molecules (melanophlogite).

A common chemical formula of **melanophlogite** is $46SiO_2 \cdot 2M^{12} \cdot 6M^{14}$, where M^{12} and M^{14} are guest molecules in 12- and 14-hedral cages, respectively. Possible M^{12} guests are N_2, Kr, Xe, CH_4; possible M^{14} guests are N_2, N_2O, CO_2, Kr, Xe, CH_3NH_2 [1], as well as S and water [2]. The structure of the cubic silica framework of melanophlogite is presented in Fig. 4.7 (according to [2]). Melanophlogite is a rare mineral; it is found as a sublimation product at fumaroles near volcanoes. After prolonged air curing (during 30 years) in natural weathering conditions melanophlogite loses its structure stabilizing organic molecules and water and turns into cristobalite. Melanophlogite turns black under heating because of burning-out of organic guest molecules but keeps the cubic silica structure until 900 °C and then is gradually transformed into cristobalite above this temperature. Because of its open cage-like molecular structure, melanophlogite is sometimes considered as a relative of zeolites which we do not consider here because zeolites are a very special type of crystals. Nevertheless, since the Database MINCRYST [15] includes two of them in the list of silicas, we present some

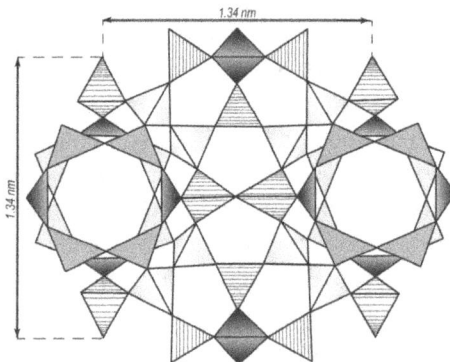

Fig. 4.7: Projection of structure of melanophlogite to the plane (001).

information on them in Tables 4.2 and 4.3 in one row with melanophlogite to demonstrate typical properties of clathrasils – their low density and friable unwieldy structure with large number of $SiO_{4/2}$-tetrahedra per unit cell (Z).

4.3.3 Exception: Fibrous Silica

The main crystalline silica phases are presented in Tables 4.2 and 4.3 in ascending order of density for the low-temperature modifications. Many of them occur in nature and all of them have their individual mineral names because of the great importance of silicas for people. The clathrasils begin the table and octahedral silicas finish it; the usual tetrahedral silicas are placed in between them. The only exception is a strange phase which is placed among clathrasils because of its low density. However, it does not belong to this family of silicas because of its low Z and relatively simple structure. This is fibrous **silica-W** with a structure formed by chains of SiO_4-tetrahedra bonded by the edges (Fig. 4.8). It was synthesized in laboratory [19] and is different with respect to its structure as compared to all other tetrahedral silicas which are built from tetrahedra bonded by corners. Silica-W is formed by oxidation of gaseous SiO, either directly by O_2 or through disproportionation, according to the equation $2SiO=SiO_2+Si$. The structure is isotypic with SiS_2 and $SiSe_2$, both of which are fibrous. The phase is unstable at atmospheric conditions because it absorbs water and transforms into amorphous hydrated silica [3].

Fig. 4.8: Two-element isolated chain in the structure of silica-W.

4.3.4 Proper Silicas

Among pure tetrahedral silicas there are five modifications with topologically different frameworks – coesite, keatite, quartz, moganite and cristobalite. Tridymite has the same topology of layers as cristobalite. As we will see later, it can be considered as filled silica. A brief characterization of these phases is given in the central part of Tables 4.2 and 4.3.

Three of the most familiar silicas (quartz, cristobalite, and tridymite) have high and low temperature polymorphs. In the scientific literature there is no general rule for their specification. In physical chemistry high-temperature modifications are denoted as α-phases because these phases can be established with high reliability whereas some probability always exists to find a new modification at low-temperatures. Physical chemists specify the low-temperature phases as β, γ, δ and so on. The $(\alpha - \beta)$-transformations of quartz, cristobalite and tridymite are completely reversible and it is practically impossible to quench their high-temperature phases at atmospheric conditions. On the contrary, geologists denote by α that modification they can take into their hands, i.e. the low-temperature phase. It means that there is no chance for a geologist to come in nature into direct contact with high-temperature modifications of these silicas. To avoid confusion with respect to the notations (α and β modification), in the modern literature it is accepted to specify silica as high and low polymorphs.

The dependence of the refractive indices on density for silicas is linear (Fig. 4.9, according to [2]) both for the natural and synthetic modifications. From a more general point of view, up to now, the specification of the thermodynamic parameters of the main silicas remains a topic of permanent and persistent discussion. Tridymite takes here a special place because it always contains some impurities and cannot be formed from quartz and cristobalite in 'dry' conditions. Employing the suggestion that tridymite does not belong to proper silicas, a phase diagram can be constructed as presented in Fig. 4.10 (according to [2]). The more traditional version (based on the famous Fenner's diagram and including tridymite) one can find for example in [20] and in Chapter 3 of the present book. To our opinion, not quartz but cristobalite is the only stable silica phase at atmospheric conditions and its field should be prolonged down along the abscissa. The phase equilibra of silicas, however, are not the main topic of this chapter, we will return briefly to this problem in the discussion of tridymite. Let us consider the main crystalline tetrahedral silicas.

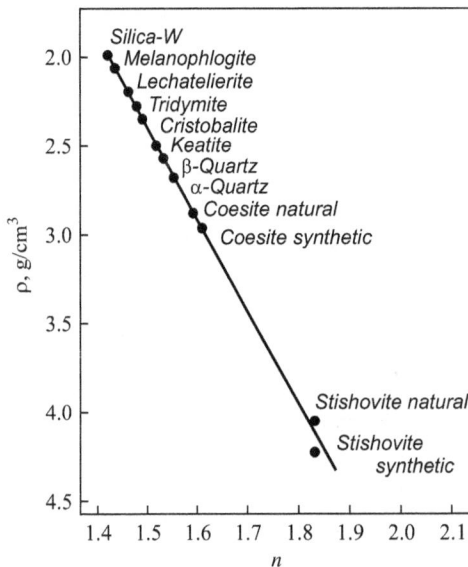

Fig. 4.9: Density dependence of refraction index for silica modifications.

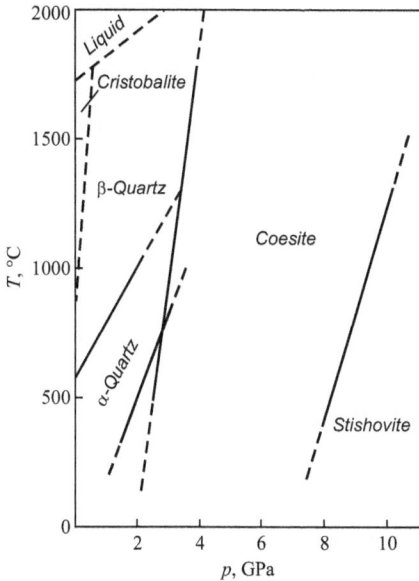

Fig. 4.10: Phase equilibrium diagram for SiO_2-polymorphs.

4.3.5 Main Crystalline Tetrahedral Silicas

4.3.5.1 Quartz

Already the ancient Greeks knew quartz and named it "cold similar ice", i.e. χρυσταλ-λος that sounds as 'crystallos'. Thus, quartz as the most ambient terrestrial mineral with its striking regular faceting and transparency gave its name for designation of crystalline solids, in general. Its contemporary name quartz gained from the Bo-hemian mines of the 14th century. The miners named gob and nonmetallic minerals as *Querz*. In this way, contrary to the approach followed in ancient Greece, quartz got its name from the collective denomination of a large class of minerals.

Quartz is the most widespread mineral in the Earth's crust after feldspars. Quartz is ubiquitous in nature; it is a substantial part of the composition of sedimentary and igneous rocks, as well as crystalline slates. It is also an important constituent of vein and mineral deposits of various origin. Because of its high hardness and low solubility quartz accumulates as sands and pebbles. It seems that quartz needs oxygen for its formation, and it is very rare on planets devoid of atmosphere like Moon and Mars or meteorites [2, 21]. According to modern data quartz does not form at normal pressure in the absence of impurities including water [2] which stabilizes its structure.

The crystal structure of quartz is nicely presented and explained on the remark-able web-site *The Quartz Page* [22]. By this reason, we do not give here an extended description but mention briefly only the main features of quartz structures, both high and low modifications (Fig. 4.11). In the next section, which is devoted especially to quartz properties, we present some specific features of it. The network structure of quartz (as well as of cristobalite and tridymite) with $SiO_{4/2}$-tetrahedra bonded by their

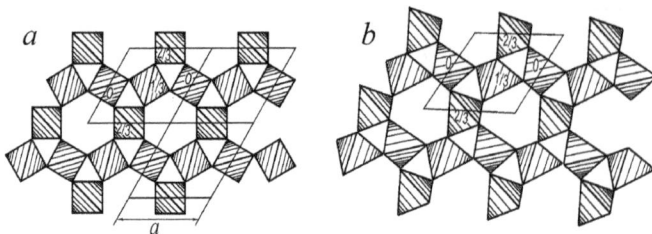

Fig. 4.11: Plane (0001)-projection of the structure of high (a) and low (b) modifications of quartz.

vertices is quite flexible and its high-low ($\beta - \alpha$)-transition occurs without structure reconstruction but only due to mutual turn and rotation of the tetrahedra. Thus, the structures of the both quartz polymorphs are very similar, but the symmetry of the high-temperature form is higher. Along the c-axis, which is perpendicular to the plane of the figure, the quartz structure is built of continuous helixes of tetrahedra. Fractional numbers on the tetrahedra in Fig. 4.11 show their heights inside the unit cell in fractions of the c-parameter. The helixes form relatively large channels (about 2 nm crosswise) penetrating the quartz structure. Small cations like H^+, Li^+ or Na^+ can enter the channels for compensation of negative charge and thereby enlarge the unit cell parameters of quartz. The entry of large foreign atoms into the quartz structure results in the formation of so-called defect-channels with cross-section of 0.02–0.05 micrometers [2]. The defect-channels go perpendicular or parallel to the optical axis of quartz and are important for its physical properties. Impurities inside the defect-channels begin to move under electric field and provide quartz conductivity [23, 24]. The inversion temperature of quartz (temperature of ($\alpha - \beta$) or low-high transition) is close to 573 °C; impurities in natural samples vary it slightly within ±2 °C, but for synthetic quartz a temperature deviation of the conversion point may reach 35 °C.

4.3.5.2 Moganite

This novel silica polymorph has first been identified in 1976 [25] in volcanic rocks of the Mogan formation on Gran Canaria islands, Spain. It later turned out to be identical to **lutecite**, a so-called length-slow chalcedony type that was commonly found in chalcedony. Moganite is always intergrown with cryptocrystalline quartz to form chalcedony [26]. For some time both names, moganite and lutecite, were used in crystallographic literature, but now moganite is the conventional name of the mineral.

A review of early investigations of the moganite crystal structure together with molecular dynamics simulations is given in [27]. The structure of moganite was repeatedly analyzed [28–30]. Moganite crystallizes in monoclinic crystal form in contrast to trigonal low-quartz. Despite that fact, their structures are very similar (Fig. 4.12; compare with Fig. 4.11b): distorted six-membered canals are found along one of the axes in both structures. Unlike quartz, moganite features alternating fragments of both

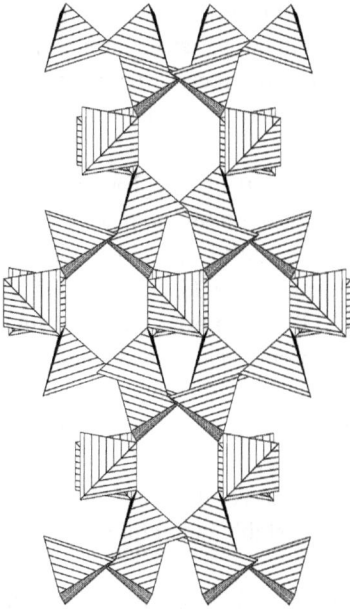

Fig. 4.12: Crystal structure of low moganite in the plane (011). The six-membered canals go along the *a*-axis.

left- and right-handed threefold spirals cut by (011) plains. As a result, closed four-membered rings are presented in moganite, and the unit cell dimension doubles along the corresponding axis of quartz structure [27].

The structural similarity of quartz and moganite results in their morphological and radiographic similarity which makes their diagnostics difficult. Thus, similar to quartz crystallites in chalcedony, moganite is intensely twinned according to the Brazil law and forms rocks morphologically similar to quartz [26]. A certain similarity in *X*-ray patterns of moganite and quartz is also evident (Fig. 4.13). Positions of the

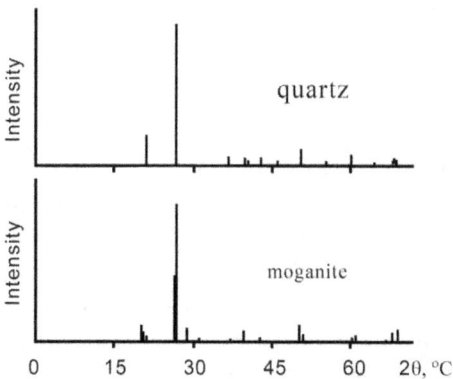

Fig. 4.13: *X*-ray patterns of quartz and moganite according to PCPDFWin database.

strongest peaks are very close, and there is some correspondence in the positions of other significant peaks, but they split or have satellites in the pattern of less symmetric moganite. Together with morphological similarity of the rocks, this difficulty of X-ray identification seems to be the reason why moganite has been identified as an individual silica polymorph only in the modern time.

A careful X-ray examination of more than 150 specimens of fine-grained quartz varieties from places all around the world has revealed that more than 10 % and as much as 80 % of the silica in many samples is actually moganite [31]. Large amounts of moganite (>30 %), found in charts from arid alkaline environments, may resurrect the idea of length-slow growth of these silica being an indicator of evaporative regimes, and the absence of moganite in weathered and hydrothermally altered silica samples may be a useful measure of fluid-rock interaction. The amount of moganite seems to decrease with time as it is slowly converted into chalcedony, and agates older than approximately 100–150 million years seem to be almost devoid of it [26]. On the other hand, recent experiments [32] established that, at a pressure of 100 MPa, moganite is a low temperature polymorph of low quartz and that it is stable rather for kinetic then for thermodynamic reasons.

Moganite has a high-temperature orthorhombic β-modification, its crystal structure determination has been performed at a temperature 1354 °C in [30]. Molecular dynamics (MD) simulations modeling high-pressure conditions [27] showed that moganite at 300 K has to exhibit two crystal-crystal phase transformations, at about 5 and 21 GPa, and these high-pressure forms appear to be the most stable phases among quartz family of silica in the temperature range from 100 to 1100 K and the pressure range from 0 to 30 GPa. Results of Raman spectroscopy studies of the anomalous behavior of moganite in the course of a pressure induced transformation are reported in [33]. Comparison of natural and MD simulated moganite structures permits us to suggest that natural moganite is a high-pressure phase existing in the range between 5 and 21 GPa and conserving as a metastable phase at normal conditions. The MD simulated orthorhombic phase which is stable at normal conditions and up to 5 GPa seems to be close to the high-temperature modification of natural moganite.

4.3.5.3 Tridymite

We have already discussed a serious objection against the point of view that tridymite is a pure silica. It is based on the impossibility to produce tridymite without stabilizing impurities or water. In this respect very impressive results were obtained by Flörke [34]. He tried the experiment of electrolyzing the foreign oxides out of a disk of tridymite with direct current at 1200 and 1350 °C. The final product at the anode proved to be pure silica but in the form of cristobalite while tridymite still persisted at the cathode, where the other oxides had accumulated. A similar experiment at 1050 °C yielded a mixture of quartz and cristobalite at the anode. Flörke concluded that tridymite has no place on the equilibrium diagram of pure silica.

Another fundamental argument against attempts to allocated tridymite to pure silica polymorphs was provided by methods of physical chemistry. Holmquist [35] investigated phase equilibria in the high-silica regions of Li_2O–SiO_2, Na_2O–SiO_2 and K_2O–SiO_2 systems. He found that tridymite is a binary incongruently melting phase containing between 0.5 and 1 % of a metal oxide which has its own separate field of existence on the phase diagrams separated from pure silica (cristobalite) with two-phase region. This conclusion is quite radical and does not give place for doubts of the extrinsic nature of tridymite – but only for those who knows the phase rule.

Our own experiments gave a similar result [36]; a reductive schema of phase equilibria in the sodium-silicate system according to these data is presented in Fig. 4.14. It was found that tridymite is the only equilibrium phase existing in a concentration range of 1.75–2.5 wt% Na_2O and over a temperature range of 900–1250 °C. In this region the X-ray pattern of tridymite continuously changes with concentration and reflects a continuous change in the structure of tridymite. Over a concentration range 0.6–1.75 Na_2O tridymite coexists with cristobalite. In this region the amount of tridymite changes from 100 % to negligible parts with reduction in sodium oxide content; the amount of cristobalite simultaneously increases. Existence of the two-phase region (Tr + Cr in the picture) with gradual transition from tridymite to cristobalite shows that tridymite cannot be a pure silica phase. Instead it is a binary (or foreign multi-component) phase and its structure changes with the type and content of impurities. It is necessary to emphasize that concentration limits in the diagrams in [35] and in Fig. 4.14 are given by synthesis without taking alkali volatilization into account. The real boundaries of the single-phase field of tridymite are located much closer to pure silica than they were presented in the diagrams discussed.

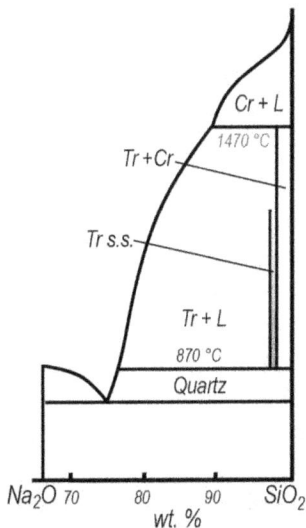

Fig. 4.14: Phase equilibria in the silica-rich part of Na_2O–SiO_2 system. Tr s.s. denotes the single-phase region of the tridymite solid solution.

An example of possible changes in the tridymite structure in dependence on the type and the amount of additional components in its composition is given in Tables 4.2 and 4.3. The tridymite species described there differ with respect to the parameter Z per unit cell. These differences are a consequence of the different origin and therefore of the deviations in the chemical composition. The other result of structural variety of tridymites is the strange fact that up to now nobody knows how many polymorphs it has [2, 3, 37]. In the structure of hexagonal high temperature tridymite there are planar layers parallel to the (0001) crystallographic plane (Fig. 4.15). In vertical direction the layers are bonded by tetrahedra vertices directed towards each other. The layers alternate as ABAB... (hexagonal packing). In the ideal structure there are two such layers per unit cell, but in some natural samples the unit cell can consist of ten and more distorted layers alternating as ABCBA, DBCBD, etc.

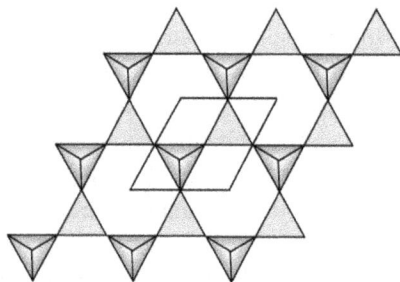

Fig. 4.15: Idealized layer in the structure of high-tridymite.

All above described properties of tridymite give convincing evidence that it is a binary compound with a very low content of cations (alkali or others) and it may be considered as cation-filled silica. The tridymite position in Tables 4.2 and 4.3 between the typical clathrasils and the proper silicas is in good correspondence with this attribution. Tridymite is quite rare as an individual mineral. It is commonly found as three twinned plates that intersect one another at an angle of 35°18' (see photo at *The Quartz Page* [22]); its name is related to this peculiarity.

4.3.5.4 Cristobalite

Cristobalite was found for the first time at the end of the 19th century during microscopic investigations of rocks from the *San Cristobal* deposit in Mexico. It exists in two polymorphic forms (presented in Tables 4.2 and 4.3) which can transform one into another without frame reconstruction but owing to rotation of tetrahedra SiO_4. The structure of the low-temperature tetragonal modification is presented in Fig. 4.16. The structure of the high-temperature cubic modification contains planar layers parallel to (111) and is constructed from six-membered rings of tetrahedra. These layers are exactly the same as the ones of high-tridymite in the crystallographic plane (001)

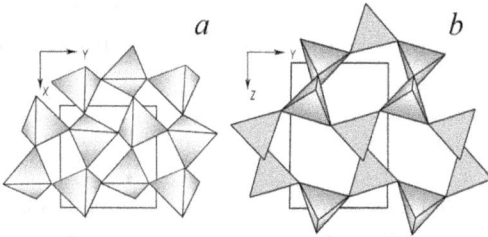

Fig. 4.16: Projection of the low-cristobalite structure onto the plane (001) (left) and (100) (right).

(Fig. 4.15). However, in cristobalite the layers alternate as three-step cubic close packing ABCABC... instead of double-step hexagonal close packing in tridymite. This specific feature enables intergrowing of cristobalite and tridymite structures. According to the layers alternation, the crystal can be mainly cristobalite or mainly tridymite; in some cases it may be even impossible to distinguish them [2]. Such hybrid forms are not rare in nature, they are denoted as cristobalite-tridymite opals (opal-CT, their mineral name is *lussatite* [20]).

The temperature of the $(\alpha - \beta)$-transformation of cristobalite is rather sensitive to any disordering in its structure. Well-ordered cristobalite undergoes an abrupt transition at 270 °C, a hysteresis being almost absent. Structure disorder reduces the transformation temperature of cristobalite to 130 °C and broadens its temperature hysteresis [38]. Flörke mentioned that for opals the transformation temperature can be reduced to 60–100 °C [39].

The low density of cristobalite is an indication (verified by close examination of its structure) that its framework is relatively friable in comparison with quartz. The most obvious illustrative example of this difference is provided by the amount and shape of structural interstices of quartz and cristobalite calculated by molecular dynamics methods [40] (Fig. 4.17). As we can see from the figure, the volume of structural interstices in cristobalite mediates between quartz and quartz glass. Quartz is a much denser substance than cristobalite. Nevertheless still it contains structural interstices along its *c*-axes. Thus, it was supposed that these substances, having flexible frames

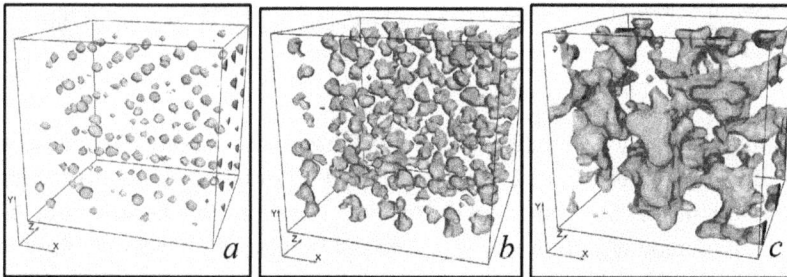

Fig. 4.17: Perspective projections of the structural interstices of quartz (a), cristobalite (b) and silica glass (c).

with sufficient structural interstices, can be converted into much denser tetrahedral modifications. Indeed, such a modification was really obtained by Coes [41] and afterwards named *coesite*.

4.3.5.5 Coesite

It was shown that the most suitable material for coesite synthesis is amorphous silica. At a temperature $T = 580\,°C$ and a pressure $P = 3GPa$, it can be completely transformed into coesite in one hour. In the field of its stability (Fig. 4.10) coesite can be produced from different forms of silica including quartz. It was also produced by shock compression of natural and artificial quartz-containing porous materials. According to the phase diagram (Fig. 4.10), coesite should be transformed into quartz after pressure relief, but it exists in a metastable state under ambient conditions. Under atmospheric pressure and at temperatures below $1100\,°C$ coesite is not transformed into quartz during a long time, but at $1700\,°C$ it is directly converted into cristobalite.

One main reason of coesite stability in the metastable region is its specific crystal structure with SiO_4-tetrahedra, associated not in six-member but in four-member rings (Fig. 4.18). The structure of coesite is a three-dimensional net with four-member rings of tetrahedra which are parallel to (010) and (001) planes. Infinite chains of the four-member rings do not incorporate into each other within their planes but connect together through equivalent chains in overlying and underlying layers. One of the four-member rings of overlying layer is presented in Fig. 4.18 (the filled tetrahedra) according to [3].

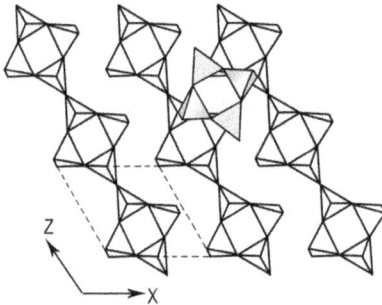

Fig. 4.18: Chains $[Si_4O_{11}]$ in plane (010) in the coesite structure.

After laboratory discovery of coesite this modification was also observed in nature, in the famous Barringer Meteor Crater, which was produced by a large meteorite impact in Arizona desert (history and geology of the crater are presented on the website of the Barringer Crater Company [42]). It was also detected in diamond kimberlite pipes of Yakutia and South Africa. Later coesite was found in many young meteorite craters. By this reason, now it is considered as an indicator of a recent (in geological time scales) meteoritic explosion. In young craters coesite may reach for about 40 % of the total silica content. In ancient craters its quantity does not exceed hundredth parts of percent.

4.3.5.6 Intermediate Phases

Hydrothermal synthesis of quartz from amorphous silica proceeds quickly till full completion at high temperatures (300–600 °C) and pressures (50–400 MPa), but at lower temperatures the process goes much slower and passes through formation of intermediate crystal phases. These phases may be of two types: pure silicas and cation-filled phases like the already discussed melanophlogite and tridymite or opal-CT. Some other cation-filled phases, forming during quartz synthesis, are denoted as *silica-X* and *silica-Y*. They have no constant composition because it changes with the nature and concentration of cations in the hydrothermal solution; nevertheless their X-ray patterns have clear specific features, which permit to identify each phase and confidently specity their subspecies like *silica-X1* and *silica-X2* [32].

A detailed investigation of physico-chemical properties of silica-X and silica-Y and the necessary conditions for their formation has been performed in [2]. Fig. 4.19 illustrates temperature changes in hydrothermal transformations of amorphous silica in alkaline medium at saturated vapor pressure. We see that two silica phases of very high thermodynamic stability, quartz and cristobalite, can form at relatively soft conditions, at comparatively low temperature and pressure. The path of quartz formation passes through origination and decay of intermediate phases, and metastable silica-X2 is much more stable here than cristobalite.

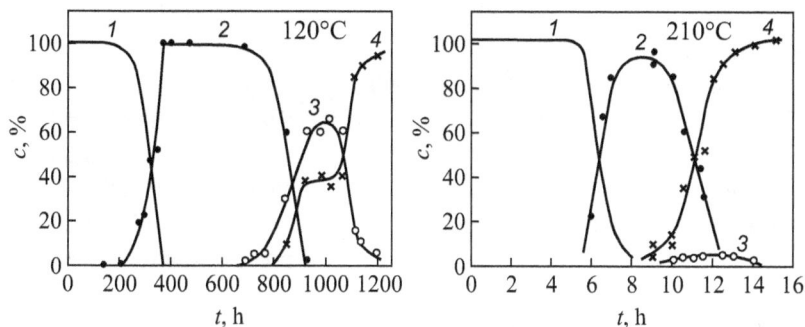

Fig. 4.19: Content of silica phases in dependence on time during their transformations in hydrothermal conditions [2]: 1) amorphous silica; 2) silica-X2; 3) cristobalite; 4) quartz.

Two bedded minerals close to *silica-Y*, *magadiite* ($Na_2Si_{14}O_{29}11H_2O$) and *kenyaite* ($Na_2Si_{22}O_{41}(OH)_8 6H_2O$), are described in the Mineralogy Database [43]. Here also the structure of magadiite is discussed. These minerals were named by the place of their origin – the soda Magadi lake in Kenya where the transformation of amorphous silica into quartz takes place in natural conditions. The above-described clathrasil

a   SiO$_2$-X1→SiO$_2$-X2→SiO$_2$-X3

amorphous ————→ SiO$_2$-Y ————————→ cristobalite→ quartz
silica
alkaline → magadiite
medium → melanophlogite

b

amorphous ————————————→ quartz
silica → keatite
p,T ↑ cristobalite

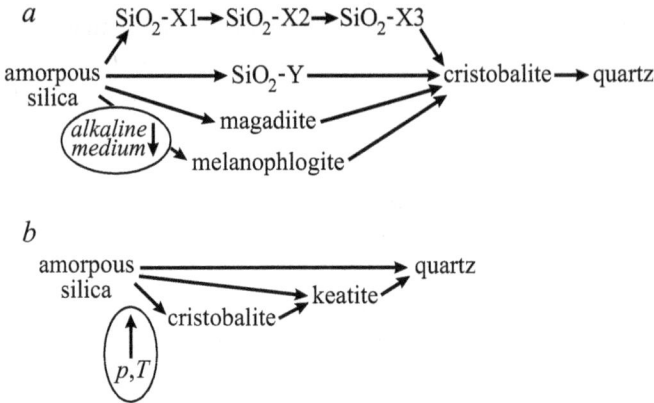

Fig. 4.20: Intermediate phases of hydrothermal transformation of amorphous silica into quartz in alkaline solutions [2] at moderate temperature and pressure (a) and in relatively pure water solutions at higher temperature and pressure (b). Arrows in ovals show the direction of the operating parameter increase.

melanophlogite can also form as an intermediate phase in natural conditions. Possible diagrams for transformation of amorphous silica into quartz in alkaline media and at moderate temperature/pressure conditions are presented in Fig. 4.20a. In relatively pure media and at higher temperatures and pressures the reaction proceeds via the formation of proper silica phases – cristobalite and the denser modification, *keatite* (Fig. 4.20b).

4.3.5.7 Keatite

Keatite is medium in its density having values in between cristobalite and quartz (see Tables 4.2 and 4.3). It was first produced in 1954 by P.P. Keat [44] during hydrothermal synthesis of quartz in alkaline solutions, but till now it is not found in nature (the *Quartz Page* suggests that it may be found in stratospheric dust particles). Keatite can be obtained as an intermediate product in a large number of hydrothermal reactions. It forms over the temperature range 380–585 °C and pressure from 30 to 120 MPa in the systems SiO$_2$–H$_2$O [45] and Al$_2$O$_3$–SiO$_2$–H$_2$O [46]. The structure of keatite is built from tetrahedra coupled by vertices similar to the main silica phases like quartz, cristobalite, tridymite or coesite. In contrast to them, the SiO$_4$-tetrahedra in keatite are not crystallographically equivalent.

Two types of tetrahedra can be found in the keatite structure (Fig. 4.21). Eight of 12 tetrahedra in the unit cell are in general positions and form di-ortho-groups, which make 4-fold helixes along the *c*-axis (the darker tetrahedra in the figure), as opposed to four other tetrahedra (the lighter ones) which are located in the particular positions at 2-fold axes and join the 4-fold helixes among themselves in endless wollastonite

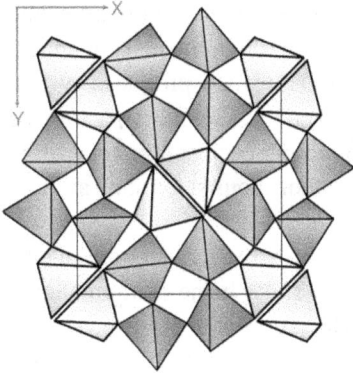

Fig. 4.21: The (001)-plane projection of keatite structure.

chains along the z-axis. Under heating up to 430 °C keatite has a negative thermal expansion due to its contraction along the z-axis. Keatite fully transforms into cristobalite when calcinated in air at 1620 °C during 3 hours [2].

4.3.6 Amorphous Silica

The natural silica glass has its own mineral name, *lechatelierite*. It forms by rapid cooling of molten silica and occurs as so-called *fulgurites* at places where a lightning has struck into quartz sand. High temperatures cause the quartz sand melt along the branched and irregular paths of the lightning through the sand. Simultaneously, the molten quartz is pushed away from the lightning because of the repelling forces between the charged particles. As a result, hollow tubes of silica glass form [47]. Lechatelierite can also be found at impact craters of meteorites. This mineral (Fig. 4.22a) seams to be radically different from commercial quartz glasses (Fig. 4.22b) but they are the same material with the density 2.20 g/cm^3 and similar structure. In Fig. 4.9, the point referring to lechatelierite falls exactly onto the common line of crystalline silicas and coincides with the point of artificial silica glass.

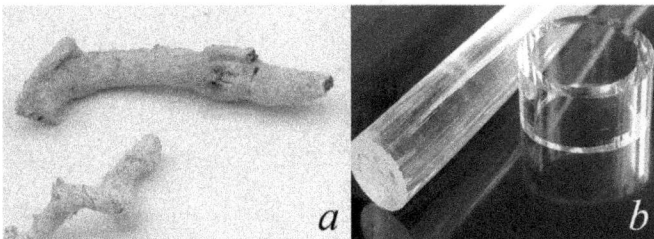

Fig. 4.22: (a) Fulgurites of lechatelierite with approximate size 150–200 mm (this rare photo has been taken from *The Quartz Page*) and (b) typical products from quartz glass.

The structure of artificial silica glass is now well-known owing to both experimental and simulation methods. Therefore, we do not consider it in the present analysis. It should be only emphasized that, although silica glass is usually produced by quartz melting and often designated as quartz glass, it has a cristobalite-like structure. Designations a-silica or a-SiO_2 without any interpretation or specification are currently found in the literature. By these notations, compact amorphous silica such as molten quartz glass or CT-opal are specified.

Note that there exists a certain interesting analogy between water and a-silica. They both have similar anomalies of their properties. The scheme of anomalies, according to [48] and [49], is presented in Fig. 4.23. The latter reference is a very interesting review of more then 70 anomalies of water; literature on anomalous properties of a-silica is presented in [50]. The anomalies appear as a hierarchy of effects with different bounds. The 'structural' bounds indicate where water and a-silica are more disordered when compressed; the 'diffusion' or dynamic bounds indicate where diffusion increases with density, and the 'density' or thermodynamic bounds show where there is a temperature of the maximum density. All above-listed phenomena were first discovered and investigated for water and only then were carried over to amorphous silica. However, there is one exception: silica is known to exist in the vitreous state in nature for long times. In contrast, vitreous water was observed in nature only in modern times in meteorites. Quite probably, the nuclei of comets are formed of vitreous water. In addition to the already mentioned anomalies both water and a-silica exhibit transformations inside the liquid state referred to as polyamorphism. This effect we will study in the next section.

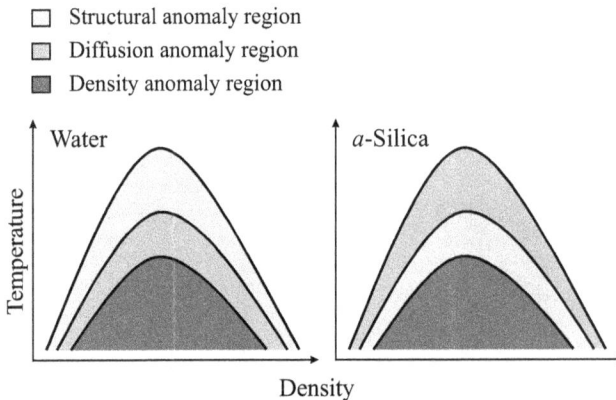

Fig. 4.23: Scheme of anomalies in water and a-silica.

4.3.7 Polyamorphism

Polyamorphism is a relatively new term in application to liquids denoting a phenomenon similar to polymorphism for crystalline solids. Being confronted with relevant experiments, we have to admit the fact of existence of liquids and amorphous solids with the same composition but with different densities and therefore with different structures. It was the discovery of a high-density modification of water [51] which introduced as common knowledge into scientific community the idea of polyamorphism [52–54].

Refractive indices for silicas built of tetrahedra are linearly proportional to their densities (Fig. 4.9). A significant change in the density or in the refractive index is a convincing indicator of the structure change. As pointed out above, a standard silica glass has cristobalite-like structure; its density is approximately 2.201 g/cm^3 and the refractive index is the smallest for glasses, $n_D = 1.459$. Quartz is denser than the glass (Tables 4.2 and 4.3) and its refractive index is higher, $n_D = 1.544$ for ordinary rays (quartz is birefringent).

Fig. 4.24 presents a partially melted piece of quartz. The sample was heated at 1700 °C for a few hours and then quenched. The refractive indices were measured for the residual quartz (right side of the sample) and for a vitreous band at its edge (left side). This simple experiment was conducted in the 1960s, it exhibited clearly that a denser form of silica glass exists. Really, the refractive index of the just melted glass immediately adjacent to the crystalline part of the sample is very close to that of quartz. Immediately after melting the structure of the melt is quartz-like and only after a few hours it is transformed into a usual cristobalite-like glass with the corresponding refractive index. The results of this experiment were not noticed by the scientific community. Only modern success in high-pressure investigations in connection with intensive and widely reviewed studies of dense forms of water stimulated specialists in silica glass to recognize its importance, its proof of polyamorphism.

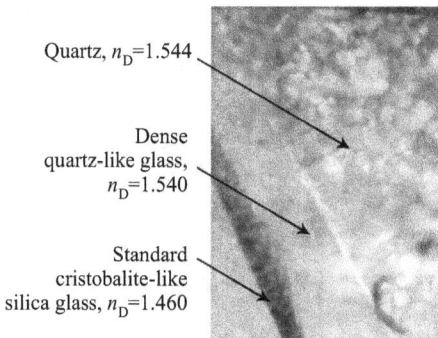

Quartz, n_D=1.544

Dense quartz-like glass, n_D=1.540

Standard cristobalite-like silica glass, n_D=1.460

Fig. 4.24: Changes in the refractive index in the process of crystalline quartz melting (according to [37]).

In the 1960s it was also established that fast neutron irradiation produces a significant change in the structure of silica glass and at a dose of the order of 10^{20} per cm^2 the initial glass completely transforms into the so-called metamict phase. The mentioned dose also leads to amorphization of crystalline quartz with formation of the metamict phase. A brief review of early investigations of the metamict phase in context of the discussion on existence of liquid-liquid transformations can be found in [55]. The results of modern Raman scattering investigations indicate that concentration of three- and four-member rings of tetrahedra in the irradiated glass increases with the dose leading to an enhancement of coesite-like features of the structure [56]. It is known that all glasses have a universal form of the low-frequency Raman boson peak [57] and with respect to this property the metamict phase is really a glass despite of the excess of free volume generated in the explosion-like interaction between neutrons and the substance. In [56] it was also demonstrated that the maximum of the low-frequency Raman spectrum (boson peak) shifts with an increase in the irradiation dose, and the size of the region with medium-range order decreases from 25Å for the initial glass to 19Å for the sample subjected to irradiation at a maximum dose.

A large list of references on amorphous-amorphous transitions and densification of silica glass is presented in [50]. It comprises both MD-simulation and experimental studies of dense modifications of a-SiO$_2$ [58]. Two mechanisms of a-silica densification are established presently. One of them takes place at room temperature under pressures in the range of 12–40 GPa. Investigations of the glass structure, Raman and Brillouin scattering and MD simulation show that this transformation proceeds via a change in silicon atom co-ordination from 4 to 6. It represents an analogue of the quartz-stishovite transformation. The second type of silica glass densification proceeds at room temperatures at lower pressures in the range 8–10 GPa. So far, however, difficulties exist in the description of its origin and regularities. Here we present a brief review of an *in situ* experimental study of the latter type of transformation according to [58] to give an indication of this relatively new phenomenon.

The volume of glassy a-SiO$_2$ was measured upon compression to 9 GPa at high temperatures up to 730 K and at both pressure increase and decrease (Fig. 4.25). It was established that the residual densification of a-SiO$_2$ after high-pressure treatment was due to an irreversible transformation accompanied by a small change in the volume directly under pressure. The bulk modulus of the new amorphous modification was appreciably higher (80 % more than its original value), giving rise to residual densification as high as 18 % under normal conditions for densification at 700 K (for densification temperature 545 K in Fig. 4.25 the residual densification is 12 %). It was shown that the transformation pressure shifted to a lower pressure of about 3–4 GPa with a rise in temperature up to the crystallization interval. Heating of the dense silica glass with the rate 20 K/min at normal pressure showed a reverse transformation into the low-density phase at 1000–1100 K. The above-discussed densification of a-SiO$_2$ at normal temperature and pressures higher than 9 GPa was demonstrated to be accompa-

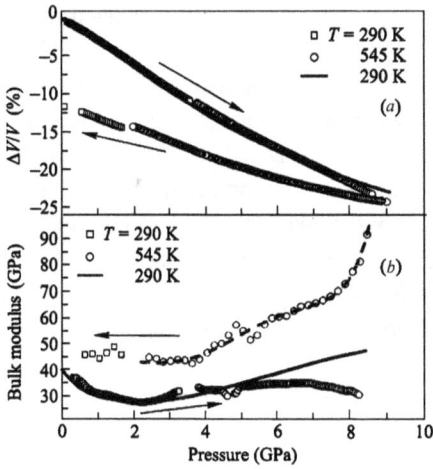

Fig. 4.25: Relative volume change (a) and bulk modulus (b) for silica glass under pressure increase and decrease (according to [58]). The arrows show the direction of pressure changes. The breaks in the curves correspond to heating from 290 K to 545 K and backwards. Solid lines specify the glass compression at 290 K.

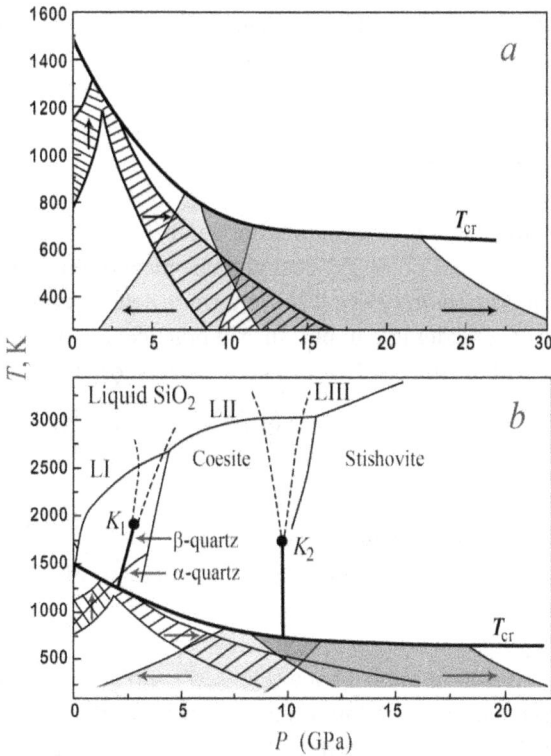

Fig. 4.26: Phase diagram for a-silica polyamorphism below crystallization temperatures (a) and the thermodynamic phase diagram and hypothetic diagram for liquid SiO_2 (b) (according to [58]).

nied by tetrahedral rearrangement being a manifestation of the same phase transformation as the quartz-coesite transition.

The authors of [58] made a conclusion about the existence of at least two pressure-induced phase transitions accompanied by structure rearrangement in a-SiO$_2$. They suggested a nonequilibrium phase diagram for a-silica (Fig. 4.26a). The hatched gaps correspond to the straight and forward transitions between the usual cristobalite-like silica glass (LI) and the densified tetrahedral coesite-like glass (LII). The blank gaps correspond to transitions between tetrahedral LII phase and octahedral stishovite-like phase (LIII). The solid lines limit the straight transitions, the dashed lines – the forward ones. Phase diagrams for SiO$_2$ transformations in the solid and liquid states are compared in Fig. 4.26b.

Similar pressure-induced transformations are known for many glasses, in particular for another famous glass-former oxide B$_2$O$_3$ [59]. The measurements of the relative volume change under compression together with structure investigations and computer simulations reveal the basic features of the phase transitions in B$_2$O$_3$ glass. Similar to a-silica, both direct and reverse transitions are smeared in pressure. Analogous results were also obtained for a-GeO$_2$ [60].

4.4 Quartz and Some of Its Properties

General information on quartz properties with links to other databases on quartz is presented in [61]; the full-length explanation of physical and chemical origin of the quartz properties one can find on *The Quartz Page* [22]. Here we review in more detail some of the properties being of considerable importance in practical application, however, little known especially with respect to their physico-chemical origin.

4.4.1 Enantiomorphism of Quartz

The absence of planes and a center of symmetry among symmetry elements of low-quartz gives it the possibility to form both right and left-hand structures which are mirror identical or enantiomorphic. L_3 axes of quartz are polar because the chains of tetrahedra form spirals along these axes with right or left screw. Crystals of right and left quartz are different in their crystallographic shape (Fig. 4.27) and also in some other physical properties. The plane of polarization rotates clockwise in right quartz and anticlockwise in the left one. Right and left quartzes differ in their etch figures, in percussion patterns on crystal face (11$\bar{2}$0), in patterns of Brazilian twin seams. However, they are indistinguishable with respect to their thermal, electrical and optical properties (except rotation of plane of polarization).

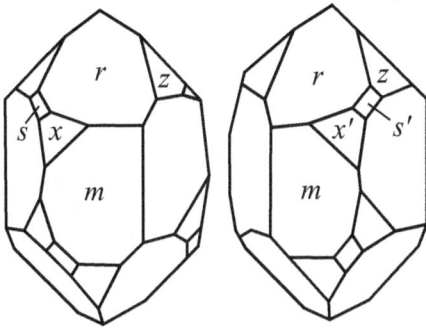

Fig. 4.27: Habits of left and right-hand quartz.

4.4.2 Twins (Zwillinge) in Quartz

The Great Encyclopedia of Cyril and Methody defines twins as regular joins of two similar crystals in which one crystal differs from the other in a mirror-reflection plane, with respect to rotation around a symmetry axis or with respect to reflection in an inversion center. A joint plane in the twin is not a phase boundary because the structure change or reflection takes place at this plane without bond breaking.

Twins may be different in nature. Growth twins form by coalescence or reciprocal intergrowth of crystals, but the mechanism of these processes remains incompletely understood so far. Transformation twins arise under structure transformation during polymorphic transitions. For example, when hexagonal high-quartz transforms into trigonal low-quartz under cooling, some parts of the structure may be turned through $180°$ relatively to each other around the L_3-axis and form so-called Dauphiné twins, which are very typical for quartz. Deformation twins arise under mechanical loading of a crystal during plastic deformation. Twins of this kind appear only in crystals with hindered sliding deformation, quartz and tiff being the most striking examples. An ordinary pressure exerted by a knife blade on an edge of the tiff rhombohedron shifts a part of the crystal into a twinning position.

Problems of quartz twinning were systematically studied in the course of thirty years by E.V. Tzinzerling and then described in a very interesting monograph [62].

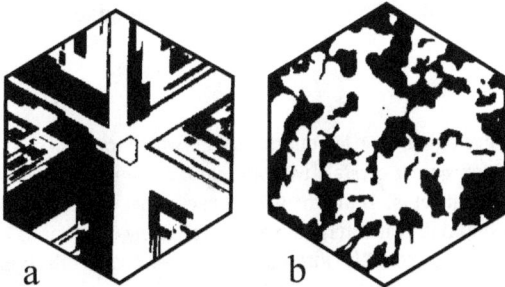

Fig. 4.28: Twinning boundaries revealed by etching a basal plane (0001) (perpendicular to L_3-axis): (a) according to the Brazilian law; (b) according to Dauphiné law.

She started her work under the guidance of the famous Russian crystallographer A.V. Shubnikov and realized his idea of artificial transformation of a twinned quartz crystal into single crystal and back. In the subsequent part of this section we briefly review her main results.

Twinning laws are especially multiform for quartz because of its enantiomorphism, which represents the capability to form irreducible right and left crystals. Besides the mentioned Dauphiné twins, Brazilian and Japanese[1] twinning laws are abundant in quartz (Fig. 4.28). From the optical point of view twinned quartz is a single crystal and may be used in optical industry but it is not convenient for electronics if the twins contained are Dauphiné or Brazilian. Electrical axes are antiparallel in components of these types of twins. If the twin contains 50% of one and 50% of the other twin component in area extent, the total piezoelectric effect of the crystal is zero. Conditioned single crystals free from twins are seldom found for quartz and therefore quite expensive, but just these crystals are needed for electronics. Artificial twinning and untwinning (transformation of a quartz crystal with twins into a single crystal) is possible only for Dauphiné twins, for other types with non-parallel twin axes untwinning is equivalent to destruction of the crystal. In the process of Dauphiné twinning the atoms rearrange in twin position inside the unit cell without any macroscopic displacement of crystal matter (Fig. 4.29).

Based on her laborious investigations Tzinzerling demonstrated clearly that twinning as a result of any kind of mechanical deformation as well as a result of only thermal action on the crystal or of voltage failure is a mechanical phenomenon by its nature. The difference consists merely in the origin of the shifting strain: an outside force or internal stresses in the crystal as a result of anisotropy of quartz thermal expansion. The latter is maximal just below the $(\alpha - \beta)$-inversion point, from 573 to 550 °C, and remains significant until 300 °C. Retwinning occurs readily in this temperature interval, and as it was pointed out, in quartz crystals even during some technological procedures. Investigators noted the random character of thermal retwinning patterns. Tzinserling showed that the system in the twinning pattern becomes evident only for pure crystals, free from impurities, inclusions and internal cracks. She discovered the method to free ill-conditioned quartz crystals from twins by double low-high-low temperature inversion with following slow cooling. As a result, the crystals with a twin affection of 50% in area extent and with zero piezoelectric effect converted into predominantly single crystals with only 12% of twins concentrated at the edges of the

---

1 Dauphiné twins consist of an aggregate of two right or two left quartz crystals with parallel L_3-axes. Brazilian twins represent joins of right and left quartz crystals with antiparallel L_3-axes. Japanese twins are twins of growth. Axes [0001] of separate crystals are inclined to each other at an angle of $84°34'$. The typical view of Japanese and Dauphiné twins in quartz as well as many other kinds of twins is accessible at the site "Twins in the world of crystals" (in Russian; http://files.school-collection. edu.ru/dlrstore/16774b92-93f0-80a6-2d02-aed3a1c1784e/48-51_01_2003.pdf)

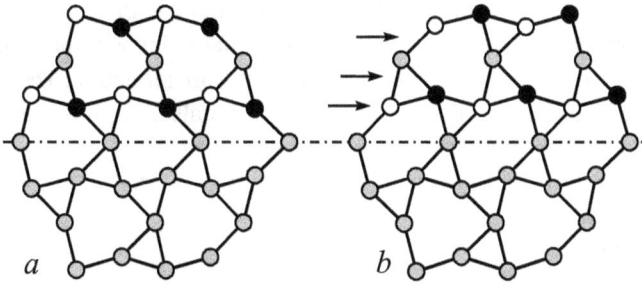

Fig. 4.29: Scheme of twinning of Dauphiné type under a shifting force according to A.V. Shubnikov: (a) low-quartz structure in projection on the basis plane (0001); (b) a Dauphiné twin in low-quartz. An arrow shows the direction of action of the force. "Grey" atoms keep their positions, "white" and "black" atoms move.

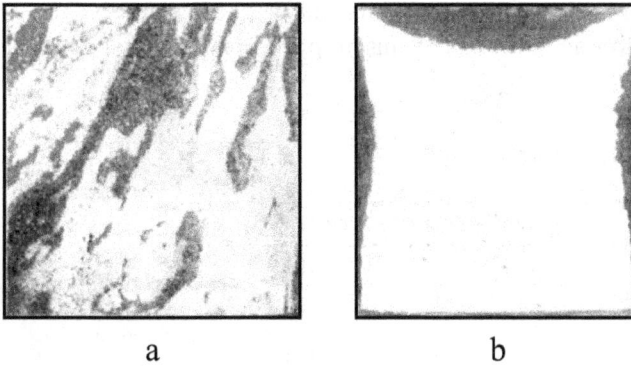

Fig. 4.30: Effect of the double low-high-low inversion on Dauphiné twins in quartz: (a) a pattern of the initial twins on the quartz plate; (b) a curtain pattern on the same plate after retwinning procedure. Size of the plate is 25x22 mm^2.

quartz plates. The typical form of twins after this retwinning procedure is presented in Fig. 4.30 and received the name "curtain". Making of "curtains" has some similarity with single crystal growing: a moderate temperature gradient combined with careful slow heating and cooling in the vicinity of the inversion point are required. After the retwinning procedure the edges of the "curtains" should be carefully deleted by polishing and, as the result, the piezoelectric properties of the quartz plate are completely recovered.

The first successful experiments of Dauphiné twins elimination from quartz by torsional deformation were performed in [63, 64] simultaneously with the author of [62]. The latter author showed that all types of Dauphiné twins, independently of their origin, defects and patterns, can be untwinned into single crystals by a multiple prolonged and enlarged loading followed by a subsequent heating. However, quartz containing impurities keeps the memory about its initial twins. After complete untwinning by torsional deformation quartz crystals restore the initial twins with faithful copy of former twin boundaries if they undergo low-high-low transformation. These works provided a possibility to create untwinning of quartz crystals in technological scales.

4.4.3 Anisotropy of Quartz

A variety of physical properties shows a sharp anisotropy with respect to the orientations of the quartz crystals predominantly in the directions which are parallel and perpendicular to the major axis L_3 [65]. The anisotropy shows up not only in structural but in all vectorial (such as thermal, mechanical, optical, electrical, etc.) properties. Durability of quartz is maximal and thermal expansion is minimal in L_3-direction because the structure is harder in this direction than in the perpendicular one. Table 4.4 (according to [62, 66]) gives an indication of anisotropy for some physical properties of quartz.

Table 4.4: Anisotropy of some properties of quartz

Physical property	in parallel with L_3	perpendicular to L_3
Thermal expansion at 40 °C, K^{-1}	$7.81 \cdot 10^{-6}$	$14.19 \cdot 10^{-6}$
Electroconductivity, $Ohm^{-1} m^{-1}$	2.50	0.16
Resistivity at 20 °C, Ohm·cm	$1 \cdot 10^{14}$	$2 \cdot 10^{16}$
Thermal conductivity at 0 °C, cal/(cm·s·K)	0.0325	1.0173
Refractive index, n_D	1.553	1.544
Breaking point, kg/cm^2		
compressive	28020	27380
tensile	1210	930
bending	1790	1180
Hardness, dyn/cm^2	$22.5 \cdot 10^9$	$30.2 \cdot 10^9$

Quartz possesses a very low positive birefringence. The optical axis in quartz corresponds to the L_3-axis of the unit cell, so there is no birefringence when light passes through the crystal from tip to tip. The maximum birefringence occurs when the light passes perpendicular to the optical axis. Light that passes through the crystal along the L_3-axis will not be split into two rays of opposite polarization either [67].

4.4.4 Thermal Expansion of Quartz

Many people do not distinguish in between quartz and quartz glass and allocate properties of the glass (which is better known to users) also to crystals. It is widely believed that quartz has a very low thermal expansion coefficient and can be easily heated to high temperatures and quenched from them without occurrence of mechanical stresses. However, such kind of behavior is true not for quartz but for silica glass, which has a thermal expansion coefficient of about 1/18 of that of ordinary glass. This uncommon property permits one to use silica glass for chemical glassware production.

4.4.4.1 Common Features of Thermal Expansion of Solids

Near room temperature quartz has a thermal expansion coefficient that is only slightly lower than that of ordinary glass but at higher temperatures the thermal expansion of quartz becomes very unusual. The reason is the following. For most of the solids their thermal expansion is determined by the skewness of the atomic potentials resulting in anharmonicity of lattice thermal vibrations (Fig. 4.31 [68]). A coefficient of anharmonicity γ is defined by the third-order derivative of the potential energy. The linear thermal expansion coefficient α is directly proportional to γ, and their signs coincide. Thermal expansion of crystalline solids is usually anisotropic; a typical example is presented in Fig. 4.32. Both parameters are given at the same scale. We see that magnesite becomes longer at heating predominantly along the c-axis and almost does not expand along the a-axis. If we cut out a ball of a single crystal of such a material and heat it, the ball will change its form and transform into an ellipsoid in correspondence with the thermal expansion coefficients along each of the crystal axes.

A parabolic law usually gives a good interpolation for thermal expansion of solids resulting in the constant α. In some cases the structure of solids is such that at heating

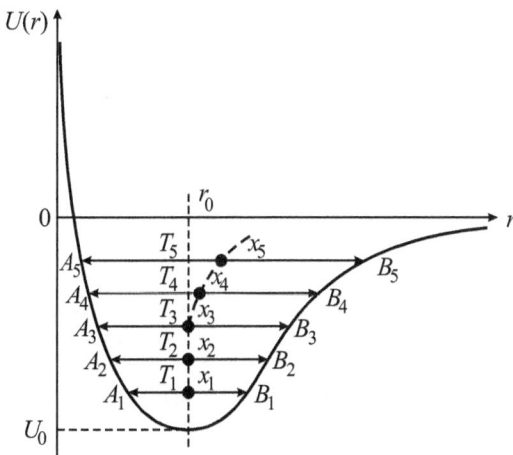

Fig. 4.31: Potential energy of interaction between two atoms for the case of anharmonic oscillations in dependence on the distance between the atoms. Here r_0 is the equilibrium atom position at zero temperature; $x_1, x_2, x_3, \ldots$ are the mean atom positions at temperatures $T_1 < T_2 < T_3 < \ldots$

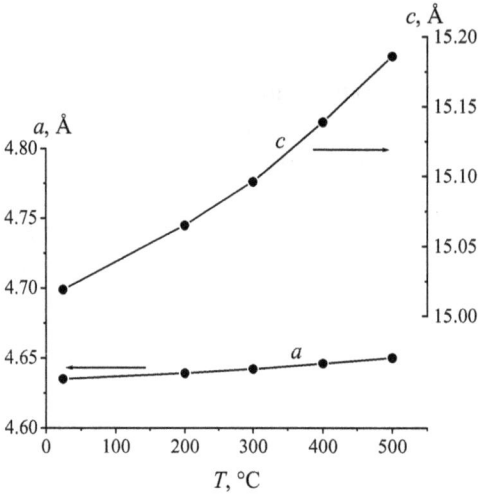

Fig. 4.32: Temperature changes of the unit-cell parameters of magnesite $MgCO_3$.

it contracts in one direction and expands in the others, similar to an elastic band, but in doing so its volume increases with temperature.

4.4.4.2 Specific Features of Thermal Expansion in Low Quartz

The above-sketched consideration was outlined in order to describe the usual thermal expansion and to appreciate the uncommon thermal behavior of quartz [69, 70, 71], which is presented in Fig. 4.33. As we can see, the thermal behavior of the low and

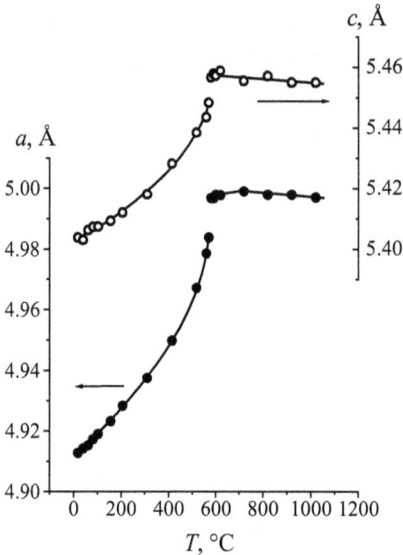

Fig. 4.33: The unit cell parameters of high purity standard polycrystalline quartz in dependence on temperature [69]. The break in the interpolated curves corresponds to the $(\alpha - \beta)$-conversion point.

high modifications of quartz are radically different and both are far from the standards described in the previous section. The parabolic law does not hold as a fit of the low-quartz thermal expansion.

It is known that, at a correct fitting, the experimental data points are located in a random manner on both sides of the approximating curve. An improper approximation leads to an alternation of groups of experimental points lying on one and the other side of the fit curve. The parabolic law results just in such kind of approximation. A proper approximation for the low-quartz expansion law is given by a power law with a fractional index and a vertex at the point of the high-low inversion, i.e.,

$$a_{low} = a_{high} + k_a(T_{cr} - T)^\eta, \quad c_{low} = c_{high} + k_c(T_{cr} - T)^\eta. \tag{4.1}$$

Here a_{low}, c_{low} and a_{high}, c_{high} are the unit cell parameters of the low and high phases, correspondingly; k_a and k_c are coefficients, T_{cr} is the inversion temperature and η is the fractional index which may be considered as giving an indication of the order of the transition: 1/2 for first-order and 1/3 for second-order. Both indices give comprehensible approximation, but require the introduction of some adjustable parameters. In the case of $\eta = 1/2$ an appropriate approximation can be achieved by correcting the inversion temperature, $T^*_{cr} = T_{cr} + \Delta$, with a value of the temperature lag Δ of about 7–9 °C. As we shall see later, this approach corresponds to a first-order transformation although the opinion to treat it as a second-order transformation is rather widespread.

The proportions of the unit cell of quartz change with temperature (Fig. 4.34) since the a-axis lengthens quicker than the c-axis. From room temperature up to 500 °C the c/a value changes linearly, but above 500 °C the slope of the plot increases. The latter fact shows the peculiarity of the thermal expansion process in low-quartz in a relatively wide vicinity of the transformation. In high-quartz the changes of the c/a proportions are very small but also quite well measurable.

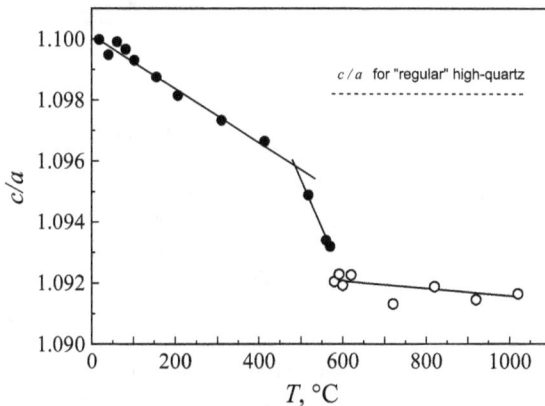

Fig. 4.34: Temperature dependence for the lattice parameters ratio, c/a, for low and high quartz according to [69]. Solid dots: low-quartz, open dots: high quartz.

4.4.4.3 Thermal Expansion of High Quartz

From Fig. 4.33 it is obvious that both parameters of high-quartz decrease at heating, with constant $\alpha_a = -3.8 \cdot 10^{-7}$ K^{-1} and $\alpha_c = -1.31 \cdot 10^{-6}$ K^{-1}, therefore, its volume decreases too. As a consequence, high-quartz is compressed at heating. Such a phenomenon occurs rather seldom and needs an explanation.

4.4.4.4 Linear Thermal Expansion Coefficient of Quartz

The temperature change of both considered linear thermal expansion coefficients is presented in Fig. 4.35 according to [69]. At negative temperatures both $\alpha(T)$-coefficients for low-quartz can be considered as linear. However, above zero in the Celsius temperature scale the rate of their change begins to increase. Above a temperature of about 450–500 °C both coefficients grow up critically and in approaching the high-low conversion temperature of 573 °C reach rather high values. At a change of temperature, the thermal expansion takes place almost instantaneously. At these conditions, such behavior of the thermal expansion coefficient near the conversion point results in a frequent fracture of the samples being the basis of the thermo-crushing method. In this method quartz samples are heated up to 900–1000 °C and then dropped into water. A violent contraction of quartz below the inversion point and the presence of the temperature gradient in depth of the samples result in their destruction.

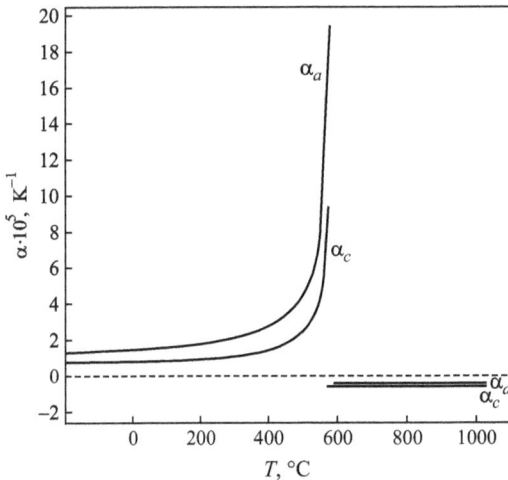

Fig. 4.35: Temperature dependencies of linear thermal expansion coefficients of quartz along the a-axis (α_a) and the c-axis (α_c). Above 600 °C, α_a virtually coincides with an abscissa axis on the scale of the figure.

Own X-ray experiments with thin single-crystal plates will be described in detail later; in these experiments the temperature was slowly varied around the conversion temperature by 3–5°. As a result, about every second the plate has burst because of the small temperature gradient along the plates. For comparison, a red hot silica glass rod can be dipped into water without cracking. It is also known that clays involving quartz sand crack at heating up to 575 °C but not far from this temperature. Quartz is not used for refractory production; cristobalite and tridymite have jumps of the volumes at high-low transformation, which are comparable with that of quartz, but their thermal expansion coefficients do not exhibit any critical behavior in the vicinity of the transformations. This is sufficient in order to guaranty that products from them are not destroyed in the transformation.

4.4.4.5 Nature of Thermal Expansion of High and Low Quartz

We know that the quartz structure is built of only one type of structural units, namely, tetrahedra SiO_4 connected by vertices. X-ray structural analysis allows one determining co-ordinates of each atom in the unit cell and permits us to follow temperature changes of a single tetrahedron. The high-temperature modification of quartz belongs to the highest crystal system, hexagonal, and tetrahedra in that structure are regular, with equal Si–O-bonds. Low-temperature quartz belongs to the lower trigonal system, and under the action of the crystal field skewness the tetrahedra are distorted. Two of four Si–O-bonds in each tetrahedron have one length and the others two have another one. Fig. 4.36 shows how tetrahedral Si–O-bonds change with temperature.

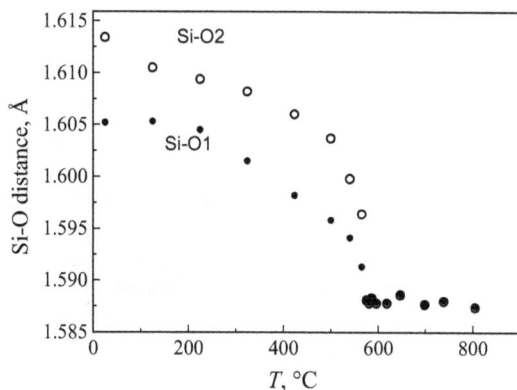

Fig. 4.36: Temperature dependence of the Si–O-bond lengths in quartz tetrahedra [72].

In low quartz both types of bonds, Si-O1 and Si-O2, behave in the same way – they contract with temperature. At heating from room temperature up to the high-low conversion at 573 °C the Si–O1-bond decreases by 1.1 % and the Si–O2-bond length by 1.6 %. At approach to the point of transformation, the rate of contraction of tetrahedra increases, whereas the structure (generated by these tetrahedra) steadily expands. It is obvious that the thermal expansion of low-quartz has its origin in a rearrangement of tetrahedra. At the conversion point the bond lengths Si-O1 and Si-O2 become equal. At further heating of high quartz we observe only a very slow contraction of regular tetrahedra with the thermal expansion coefficient of the order of $-1.2 \cdot 10^6$ K^{-1}. This result gives a suitable explanation of the thermal expansion of high quartz.

Let us consider a geometric model explaining the mechanism of thermal expansion of low-quartz (Fig. 4.37 [73]). We assume that the tetrahedra are regular as in the case of high-quartz. All lengths in the model are dimensionless and given in units of the a-axis in the high-phase. Then the lattice constants in the high temperature phase are:

$$a = 1, \quad c = 2\sqrt{3}R, \quad R = \frac{\sqrt{3}}{2}\frac{1}{1 + \sqrt{3}}, \tag{4.2}$$

R being the Si–O-distance. Vertical displacements of tetrahedra in Fig. 4.37 (numbers in the circles) are also given in the same units. For "regular" quartz (regular SiO$_4$-tetrahedra) there is no open parameter in the high-phase except for a scaling factor. As the authors of [73] wrote, "the conditions of rigid and connected tetrahedra leave only one way to produce the low-temperature phase from this structure. This is the simultaneous tilt of the tetrahedra around the three axes d_{1-3}... Because of tetrahedra being linked, this "tilt operation" shortens the dimensions (a, c) of the unit cell."

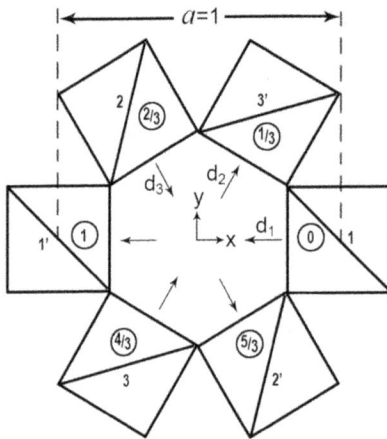

Fig. 4.37: Projection of right-hand high-quartz structure onto the (x, y)-plane perpendicular to the c-axis directed towards the observer. The figures in the circles show the relative height of the Si-atoms above the (x, y)-plane. The tilt axes d_1, d_2 and d_3 are indicated.

A simple geometrical consideration allows one to formulate the structure of "regular" low-quartz in terms of only one parameter – the tilt angle δ. In particular, for the unit-cell parameters of low-quartz the authors got the following expressions:

$$a = 1 - 2R(1 - \cos \delta), \quad c = 2\sqrt{3}R \cos \delta. \tag{4.3}$$

The unit-cell parameters of low-quartz do not depend on the sign of the tilt angle δ but the atomic co-ordinates do, and the δ-sign determines the kind of Dauphiné twin. A check of the validity of the tilt model for "regular" quartz by comparing these equations with the respective experimental data at room temperature ($\delta = 16.3°$) and

Table 4.5: Comparison of the tilt model of "regular" quartz with experiment. By convention, the a-parameter of high-quartz is taken as unity

| Variables | 25 °C | | 600 °C | |
	Exp.	Model	Exp.	Model
a	0.9831	0.9745	1.0 (fixed)	1.0 (fixed)
c	1.0813	1.0539	1.0921	1.0981

at 600 °C (δ = 0, above the phase transition) shows a surprisingly good agreement (Table 4.5).

The authors of [73] considered the tilt angle δ as the order parameter. They proposed the analytical form of δ (T) assuming a Landau-type free energy expansion to be valid. They suggested, in addition, that the high-low transition is of first order. As Dauphiné twins are energetically equivalent, only even powers enter the free energy expansion with respect to the tilt angle δ. Mentioned authors also obtained an equation for calculating the angle δ from X-ray structural data. Combining both results they concluded that at the transition temperature a jump of the tilt angle has to occur with a value of the order of 7.3 ° (Fig. 4.38). In the framework of this model, the temperature lag was estimated as 10 °C. This value is close to the lag derived approximating the unit cell parameter by a power law with a fractional index η = (1/2).

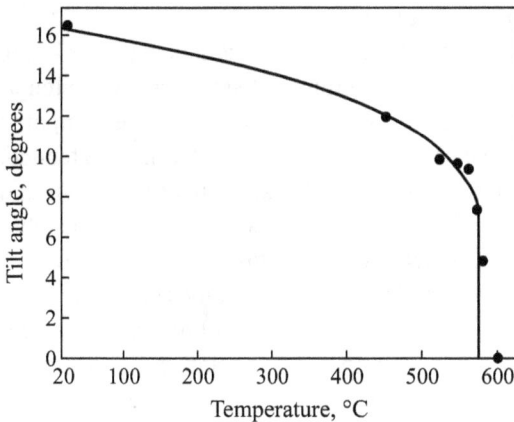

Fig. 4.38: Comparison of the experimentally determined tilt angle δ with the theoretically expected temperature dependence of δ on the basis of a Landau-type expansion for the free energy [73].

4.4.4.6 Origin of Thermal Contraction of the Tetrahedra

The nature of thermal contraction of the XO_4^{n-} tetrahedral ions is well explained in [74] with the significant title "Role of 3d-orbitals in π-bonds between (a) silicon, phosphorus, sulfur, or chlorine and (b) oxygen or nitrogen". As the author writes, "two strong π-bonding molecular orbitals are formed with the $3d_{x^2-y^2}$ and $3d_{z^2}$ orbitals of X and

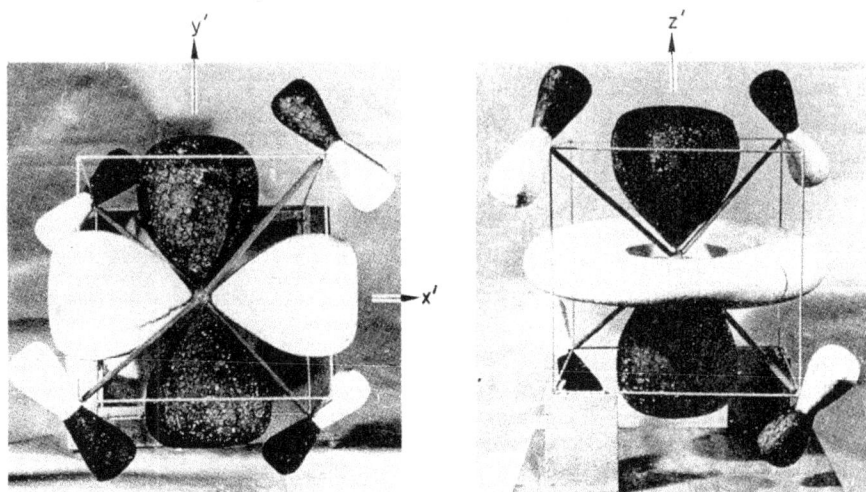

Fig. 4.39: Left: overlap of $X(d_{x^2-y^2})$ with oxygen $2p\pi$ in XO_4^{n-}. Right: overlap of $X(d_{z^2})$ with oxygen $2p\pi$ [74].

the appropriate $2p\pi$ and $2p\pi'$-orbitals". The author illustrates the electron structure of the tetrahedral ion with the help of handmade models (Fig. 4.39). Thermal expansion of these orbitals results in increasing their overlapping and growth of the bond order. The bonds become stronger and shorter at heating.

From [74] it becomes obvious that not only silicon but also its neighbors in the third group of the Periodical System can create tetrahedral structures with similar properties, although some tetrahedral ions are much more complicated as compared with the SiO_4-tetrahedron. Today the list of tetrahedral ions is by far more extended. In particular, a quartz-like structure of $AlPO_4$ expands at heating in the low phase and has an almost constant volume in the high phase [75]. Hence, aluminum and phosphorus should be added to the list (a) and (b) of [74], correspondingly. Specific features of high-low transitions in quartz and quartz-like $AlPO_4$ are similar, too [76].

Contraction of SiO_4-tetrahedra takes place not only in quartz but in all silicas, for example, in cristobalite (Fig. 4.40). We see again that, when relative displacements of tetrahedra are exhausted, the structure loses its ability to expand. The mechanism of thermal expansion of low-cristobalite is described in [77] together with a review of cristobalite-like structures constructed of tetrahedra and showing a similar type of thermal expansion behavior. Low-quartz and low-cristobalite have "corrugate" structures; straightening of them provides their thermal expansion at heating despite of the absence of the expansion of the "constructive bricks", the tetrahedra. The structure of silica glass is loose (Fig. 4.17c) with a minimal corrugation. As a result, relative tilts of the tetrahedra at heating are insignificant and the thermal expansion depends on the inherent thermal expansion of the tetrahedra or rather on its absence.

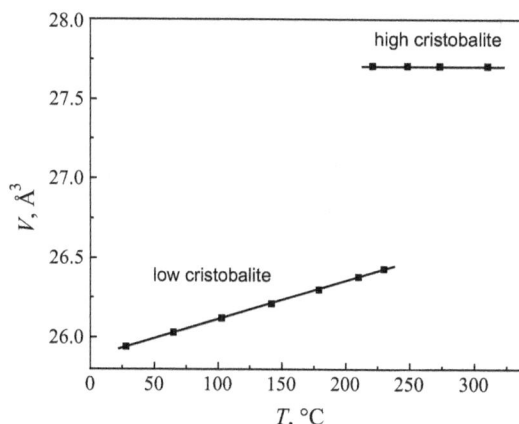

Fig. 4.40: Thermal expansion of low and high cristobalite [78].

4.4.5 High-Low or ($\alpha - \beta$)-transformation in Quartz

For the last sixty years and to less extent even considerably earlier [79, 80] much attention of researchers has been directed to the high-low transition in quartz. The reason of this interest is that this transformation is accompanied by very specific phenomena and anomalies [81–87]. The literature devoted to this question is numerous. It seems that it is the most investigated phase transformation. Nevertheless, new investigations lead to new questions. Here we make only a very brief review of some aspects of the transition.

4.4.5.1 Order of the High-low Transition in Quartz

A lambda-like behavior of the second-order derivatives of the thermodynamic potentials, e.g., of the thermal expansion coefficient, is typical for second-order phase transitions according to the classical Ehrenfest's classification. Other similar properties (such as the specific heat capacity [88], elastic modulus, and light scattering intensity (Fig. 4.41)) still in the 1950s gave rise to the opinion that high-low transformation in quartz is of second order. Nevertheless, not the lambda-like trend of the second-order derivatives is the main criterion of a second-order transition but primarily the absence of a jump of the first-order derivatives of the Gibbs free energy and, first of all, a jump in the specific volume.

In the 1960s, the results of the powder diffraction study [88] seemed to allow one to suggest that the unit cell parameters of quartz undergo a jump at the conversion point. However, this conclusion could be easily questioned because of the missing in the study accuracy of temperature measurement and insufficient number of measurements in the transformation region. To avoid these problems a special method, the so-called X-ray radiography with oscillating temperature, was developed [69, 91]. In this approach, the temperature of the sample varies continuously around the conversion

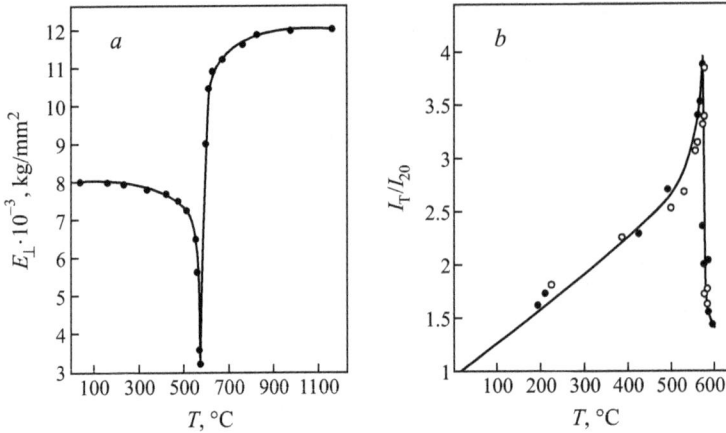

Fig. 4.41: Temperature dependencies of (a) elastic modulus in the direction perpendicular to the optical axis of quartz [89] and (b) scattered light intensity (solid dots: heating; circles: cooling [90]).

temperature within the limits ± 2 °C. If the crystal structure continuously changes with temperature, the measured 2θ-values of the selected reflection should be distributed normally in the narrow temperature interval because of random errors. If the function $2\theta(T)$ has a break inside the limits of the temperature variation, two nearly normal distributions of 2θ should be registered. The reflections 214 ($2\theta \approx 95$ ° at Cu_{K_a} radiation) and 231 ($2\theta \approx 104$ °) were measured for quartz powder approximately hundred times for both cases. The result is presented in Fig. 4.42. Two separate distributions of the 2θ-angle frequency are observed for both reflections. In both cases a forbidden band is found for 2θ, for the reflection 321 it reaches a value which is larger than the width of the 2θ-distributions.

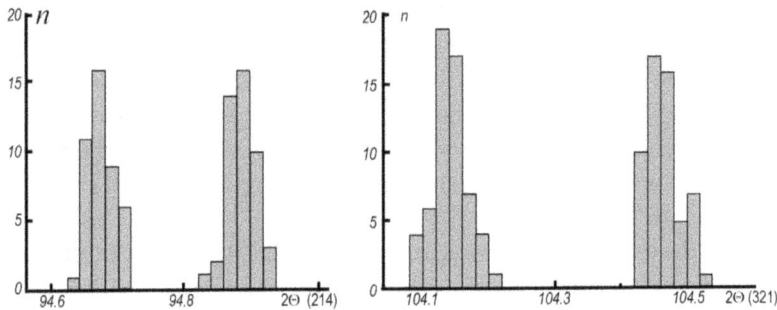

Fig. 4.42: Bar graphs for 2θ-values of quartz reflections 214 and 231 at the inversion point. The left distribution refers to high-quartz, the right one, to low-quartz.

A similar investigation was carried out for a single crystal of quartz. The reflection 060 ($2\theta \approx 150°$) was measured nearly fifty times with a similar result; for such remote reflection a distance between the peaks of the high- and low-quartz distributions reaches 0.8°. There do not exist experiments in which any reflection within the forbidden bands could be possibly observed. We register here a jump of the structural parameters of quartz during high-low inversion, thus this is a first-order transformation. The unit cell volume jump, calculated by using these data, is only 0.6%. This value is significantly smaller than the volume jump (2.5–5%) which can be found for the quartz conversion in the literature because the latter relates to a "destructive" part of thermal expansion. The real jump of the volume is small, but it does exist. Investigations performed by other sensitive methods also confirm the first-order character of the transformation [92, 93].

4.4.5.2 Transitional Opalescence

Molecular light scattering in transparent solids has been first discovered in 1928 by the Soviet physicists G.S. Landsberg and L.I. Mandelstam. First it was revealed and investigated just in crystalline quartz. At room temperature, molecular light scattering in quartz is of very low intensity, only about 10^{-7} of the incident light intensity. The intensity of molecular light scattering in quartz linearly increases with temperature up to 450 °C (Fig. 4.41b) and then begins to rise critically in approaching the high-low inversion temperature [90]. In the latter mentioned work, performed in 1956–1957, a high-temperature optical chamber was designed with a temperature gradient in the direction of the incident light beam from 0.03 to 0.1°/mm and with a vertical gradient of 0.01°/mm. At these conditions, the authors observed (as they supposed) a "critical" opalescence in quartz in a very narrow temperature interval of about 0.1 °C near to the high-low inversion point 573 °C. In the photos (shown in Fig. 4.43) low-temperature quartz is located on one side of the nebulous band and high-quartz, on the other side. Under continuous heating the opalescent band moves from the hot side of the chamber to the low one.

The described phenomenon is reversible and can be repeated as many times as one wants until the crystal is destroyed. The authors of [90], following an independent theoretical consideration [94], concluded that the presence of the narrow nebulous band between two polymorphous modifications of quartz does not leave any possibility to interpret the low-high inversion in quartz as a first-order transition. They supposed that the observed phenomenon is light scattering on optical inhomogeneities produced by thermal fluctuations.

Since that times, the transitional opalescence in quartz has been studied repeatedly [95–97]. The intensity of the transitional opalescence was shown to depend on lattice defects, the size and shape of which can be diagnosed relying on frequency and angle dependencies in the opalescence region [98]. A quality control method for piezoelectric quartz was suggested. It is based on the measurement of the intensity

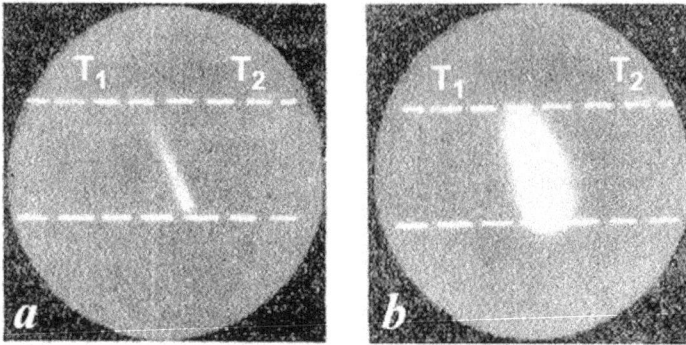

Fig. 4.43: Photographs of opalescence in quartz with horizontal temperature gradient of 1 (a) and 0.03 (b) degree/mm. White dashed lines show the direction of the incident light beam. The exposition time was one second; the width of the opalescence zone in the photo (a) is 0.5 mm, in (b) it equals 3 mm; light scattering by the nebulous band in the direction of observation is by a factor of $1.4 \cdot 10^4$ more intense than at room temperature or in high-quartz at 600 °C.

of iso-frequency light scattering at the $(\alpha - \beta)$ phase transition [99]. The idea that the opalescence in quartz is originated by thermal density fluctuations was not confirmed by later investigations. The measured opalescence turned out to be not a dynamic but a static phenomenon, the optical inhomogeneities are invariable in time and space [98, 100, 101]. They are generated by a so-called incommensurate phase [95, 102] (the latter work is a review with the indicative title: "The α–<incommensurate phase>–β transition of quartz: a century of research on displacive phase transitions"). The suggestion of irregular Dauphiné micro-twinning as the reason of the opalescence was advanced as well [62, 100].

4.4.5.3 Incommensurate Phases

Fig. 4.44 schematically presents the process of formation of an incommensurate or modulated phase (in the literature it may be designated as IC or inc) in a crystal [103]. In modulated phases the atoms are slightly displaced from their lattice positions following a periodic law. A superposition of two or three displacement waves can coexist in the crystal. In the case of a single wave the modulated phase is referred to as $1q$. If two or three waves coexist the modulated phases are denoted as $2q$ and $3q$, respectively. When the ratio of the lattice parameter of the modulated phase to the modulation wavelength is an irrational number, this phase is termed incommensurate. However, since the exact value of the actual ratio is unknown, it is usually represented in the form of an irreducible fraction such as 20/41 [104].

In dielectric crystals the modulated phase arises from the high-symmetry commensurate (normal) phase due to a phase transition, which "is caused by the disap-

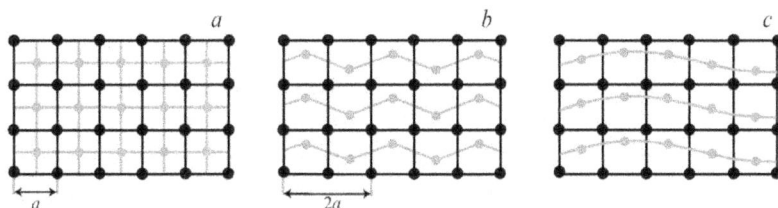

Fig. 4.44: Superstructure formation in the model two-dimensional crystal: (a) the normal phase with the unit cell parameter a; (b) the commensurate phase with the doubled parameter in one of the directions; (c) the incommensurate phase, a period of the displacement wave is incommensurate with the unit cell parameter.

pearance of the acoustic mode with a wave vector lying within the first Brillouin zone" [104]. In quartz upon cooling the following sequence of phase transformations is observed: normal phase (high-quartz) $\rightarrow 1q \rightarrow 3q \rightarrow$ commensurate phase (low-quartz) [104, 105]. It was shown that the incommensurate phase near the $(\alpha - \beta)$-transition point in quartz is improper ferroelastic and should be split into domains [106]. Electron microscope images of the domains are demonstrated, for example, in [101], a review of simulation methods for the sequence of transformations is presented in [104] as well as the author's simulation of the domain structures formation at the different stages of the inversion. Acoustic and light scattering anomalies observed in the IC phase are qualitatively consistent with its ferroelastic nature [106].

4.4.6 Pressure-induced Amorphization of Crystalline Silica

After the discovery of pressure-induced amorphization of ice the crystalline-to-amorphous transformations in the solid state became the subject of intensive studies (see the review [107]). It was established [108] that quartz and coesite as well as other materials with quartz-type structure like $Si_{0.56}P_{0.44}O_{1.56}N_{0.44}$ or $AlPO_4$-$GaPO_4$ may transform into amorphous solids at 25–35 GPa and 300 K [109]. The phenomenon was confirmed by molecular dynamics simulations [110] and a similar behavior was predicted for cristobalite [111]. It was also shown that melting is the physical phenomenon responsible for pressure-induced amorphization, and that the structure of a "pressure glass" is similar to that of a very rapidly (10^{13} to 10^{14} K/s) quenched thermal glass.

4.5 Hydrothermal Synthesis of Quartz

The basic principles of single crystal growth were repeatedly exposed. One can find them, for example, in the books of R.A. Laudise [113, 114]. In addition, a description of the hardware implementations is given there. The monographs [112–114] were employed in writing the present section.

Hydrothermal synthesis is a special case of growth in solutions. This method is used when the solubility of a pure crystal phase which is attempted to be grown is too low. The main ways to raise it are addition of an appropriate mineralizator and temperature and pressure increase. Many minerals were formed in such conditions in the interior of the Earth via very slow growth from water solutions at high temperature and pressure. The most abundant mineral formed in this way is quartz. That is the reason why geologists were the first to perform experiments on hydrothermal synthesis of quartz in order to understand how it could evolve on Earth.

4.5.1 Brief History

K. Schaufheutl and independently H. de Senarmont first grew microscopic crystals of artificial quartz from water solutions in the middle of the 19th century. These works as well as the subsequent ones had the task to explain natural quartz formation; therefore, the authors were quite satisfied with small sizes of the crystals grown by them. However, in 1880–1881 the brothers Pierre and Paul-Jacques Curie discovered the direct and inverse piezoelectric effects for quartz (see also [65] for details of the phenomenon) and some other non-central symmetrical crystals. This effect gave the possibility to use quartz for transformation of electric signals to sound and back. However, for this purpose quartz crystals should be large, perfect and free of twinnings. Thus, an interest arose in artificial crystals production since natural crystals with necessary properties are very rare. The first large quartz single crystals were grown on seeds only in 1905–1909 by G. Spezia. His single crystals were produced from sodium silicate solution with addition of NaCl and after five-month experiments had nearly 2 cm length along the c-axis.

Very soon, radio engineering expansion led to an insistent need in large single quartz crystals. Massive defect-free crystals were required in order to develop systems for frequency stabilization by means of piezoelectric quartz resonators. In the 1930s, systematic efforts were directed at developing a method suitable for commercial production of piezoelectric quartz. In Germany, R. Nacken succeeded in growing relatively large crystals (up to 5 g) in isothermal conditions (400 °C) with silica glass powder as a feed material. Solubility of silica glass is an order of magnitude higher than that of crystalline quartz; hence the solution was supersaturated with respect to quartz and growth was possible. It seems that the process of glass solution with subsequent transfer of the dissolved silica to a seed may proceed until the glass exhausts and large crystals may be produced in this way. Unfortunately, quartz grew not only on the seed but on the glass, too. Crystallization stopped dissolution of the glass, and cyclical renewal of the feed material lowered the crystal quality. Nevertheless, these first experiments showed that free of twinning quartz can grow relatively fast on high-quality single-crystal seeds. The largest crystals, synthesized by Nacken, reached 2 cm after 90 days. Hence, the task of artificial growth of piezoelectric quartz was solvable in principle.

World War II gave a further impulse to advance radio industry. After the war strong attempts to find a way for industrial synthesis of piezoelectric quartz were made in UK, USA, and in the USSR. At the end of the 1940s it became evident that Nacken's method has no perspective for several reasons. The most important of them is that the method should not be conducted isothermally. The English scientists W. Wooster and L.A. Thomas were the first to show the principal possibility to produce sufficiently large quartz crystals by continuous transfer of matter from a feed material to the seed under temperature drop conditions. This work had a profound impact on the further industrial development of hydrothermal synthesis of quartz.

In the USA the "Bell Telephone" company achieved the most significant success. In 1956 the general procedure for synthetic piezoelectric quartz was laid down and the industrial production was organized by "Western Electric". In 1958 they started up with a pilot plant in Massachusetts and soon satisfied to a considerable degree the needs of radio-electronic industry with respect to quartz. Independently, T. Sawyer made a major contribution into the development of the temperature drop method. In 1956 he organized his own company "Sawyer research production" and began to produce high-quality quartz crystals.

Japan has no own deposits of natural piezoelectric quartz. The electronics company "Toyocom" was one of the first in Japan interested in the creation of an independent source of quartz raw materials. In 1955 this company started experimental investigations of piezoelectric quartz synthesis and already in 1960 mass-production of large crystals was established.

In the USSR A.V. Shubnikov undertook in 1939 his first exploration work on quartz synthesis. After the war this work was resumed in the Institute of Crystallography of the Academy of Sciences of the USSR. At the beginning of the 1950s, a new Scientific Research Institute of piezoelectric raw materials ("VNIIP") was created together with a pilot plant in Alexandrov city not far from Moscow. Later the institute was renamed as "VNIISIMS", "All-Union Scientific Research Institute for Synthesis of Mineral Raw Materials". In 1957, the first trial production of piezoelectric quartz at a plant was realized. The industrial production of synthetic quartz was started in 1961–1962.

4.5.2 Temperature Drop Method

The method consists in creation of some temperature difference between dissolution and crystallization chambers and can be used if the solubility of the considered substance in a working medium changes with temperature. As a rule, the solubility of quartz in the used solvent has a positive temperature coefficient at working conditions.

In this method, a temperature T_1 is established in a region where the substance is dissolved and a lower temperature T_2 is set in a region of the crystal growth, the growth chamber being placing below the dilution chamber (Fig. 4.45). As a result, the solution in the hot chamber with temperature T_1 begins to float upwards and arrives into the colder upper chamber. As quartz solubility at T_1 is higher than at T_2, the solution is found to be supersaturated at T_2 and may feed the growing quartz seed. Then the liquid with T_2 moves downward and ascending and descending flows (due to the action of the buoyancy force) establish a closed loop of free convection. Convective mass transport is very important for the process realization because it provides continuous entry of the supersaturated solution from the hotter charge to the colder seed. In a rare instance of retrograde solubility, the location of the chambers for dilution and growth should be inversed.

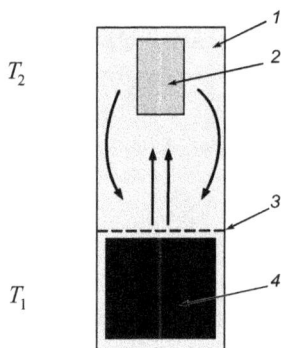

Fig. 4.45: The scheme of crystal growth employing the method of temperature drop (here the relation $T_1 > T_2$ holds): (1) pressure vessel with water solution; (2) growing quartz seed; (3) perforated diaphragm; (4) container with diluting quartz charge. Arrows show the convectional circulation of the solution.

A liquid growing medium has to assure a sufficiently large absolute value of solubility for crystal growth. Weakly concentrated soda and alkaline water solutions are usually used for quartz growth because of their higher (in comparison with pure water) quartz solubility; fluoride, acid and some other solutions are also used for special purposes. For the major part of manufacturing technologies a sufficient level of quartz solubility is reached at temperatures above 300 °C. Diluted water solutions may exist at these temperatures only under high pressure; the latter also stimulates the crystallization process and permits to produce crystals of higher quality. In practice, pressures of 70–200 MPa are commonly used. The choice of a particular level of pressure for a given technology depends on specific features of growth and the desired quality of the crystals. For the practical realization of hydrothermal synthesis it is important that the relation between the necessary levels of temperature and pressure allows one to realize the process in isochoric conditions without external sources of pressure. In this case, the necessary pressure at the given temperature may be created only due to temperature expansion of the working medium (the water solution) and depends on the relative value of the working space infill. The infill factor is the main pressure governor for hydrothermal synthesis; the necessary value of it may be found from the (p, V, T)-diagram of the working solutions, the latter being determined in separate investigations.

From the above-given considerations it appears that hydrothermal synthesis of quartz should be realized in upright autoclaves (autoclave is a high-pressure vessel without an external source of pressure) with controlled systems for heating and heat shield. Specially oriented quartz seeds are placed in the growth chamber at metal frames, the quartz charge is placed into the dilution chamber in metal containers. As a rule, natural crystalline materials of size ranging from 20 to 40 mm are used as a charge. Surface area of the charge should be five times larger than the total surface area of the seeds; in this case charge dilution does not restrict the seeds growth.

The volume of the autoclaves may vary within wide limits, from a few cubic centimeters for laboratory bombs to a few cubic meters for industrial equipments. It was shown that the overall growth rate is higher in smaller reactors than in larger ones, all other parameters being the same (temperature and temperature drop). Small-size vessels with an internal diameter of less than 300 mm are frequently employed because they are not heavy and are convenient in manufacturing and sealing during exploitation. However, the autoclaves of chief commercial importance have an internal diameter of above 600 mm. Large autoclaves provide a possibility to grow larger and more perfect crystals because the large mass and thermal inertia of the vessel allow one to retain the appropriate conditions for stable growth. They are also more economical in exploitation because they need less maintenance staff, working area and so on.

In addition, the geometrical shape of the working space plays a great role for the realization of a successful synthesis. The ratio of the height of the working space to its diameter has to be in limits of 8–15. Otherwise, instabilities of the thermal regime may arise in too much elongated reactors. Nevertheless, autoclaves with an elongation factor of 20–30 are utilized and working successfully in Japan. An additional constructional feature, which helped to solve the problem of stable and controlled mass transfer providing, is a perforated diaphragm that divides the reactor into two parts: a hotter chamber for quartz charge dilution in the bottom of the reactor and a colder chamber for crystallization above it. The structure of the diaphragm often undergoes various modifications to improve and optimize the mass transfer in the vessel. However, the structure complication leads to reliability degradation and does not compensate the expected improvement of the solution circulation. For this reason, relatively simple perforated diaphragms are usually employed in practice. Anyway, the proper choice of a flow section value is one of the most important problems of synthesis.

Just at the start of the process, the working space is partially filled with a working solution according to the chosen infill factor. Pressure starts to grow on heating but this growth is not uniform (Fig. 4.46). At an early stage of heating the working medium is in a two-phase state, the working solution and its saturated vapor being located in the vessel. At this stage the pressure depends only on temperature and does not depend on the infill factor. The increase of pressure at this stage is relatively slow having values in the range 0.01–0.05 MPa/degree. If heating proceeds, the system reaches the critical point (p_c, T_c) and converts into a single-phase state. The increase of pressure due to temperature increase becomes approximately linear (1–2 MPa/degree) in the homogeneous stage and uniquely depends on the infill factor.

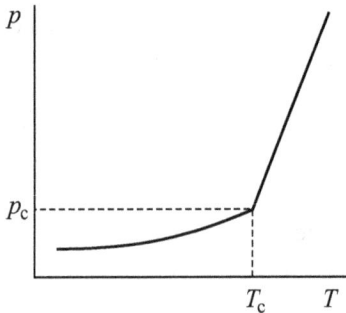

Fig. 4.46: Schematic temperature dependence of pressure for the working solution inside the heated autoclave.

Processes of charge dilution and crystal growth are energy-reciprocal, so a moderate energy supply is required to keep the mass transfer at a given temperature to the condition of appropriate thermal protection of the autoclave. The duration of quartz growth by hydrothermal methods can reach about 300–400 days, and the accuracy of temperature and pressure maintenance has not to exceed the limits of 0.5–2 °C and 0.5–1 MPa, respectively. Autoclaves must be safe and reliable; at the same time they should be serviceable for assembling and dismantling because of recurrence of the growth process and necessity to clean internal parts of the vessel from spontaneously formed crystals at the top and heavy fluid at the bottom.

4.5.3 Main Problems of Hydrothermal Synthesis of Quartz

4.5.3.1 Phase Equilibria in the System SiO_2–H_2O

Hydrothermal synthesis is a recrystallization process under relatively high temperature and pressure conditions. This process is realized by means of continuous mass transfer from a dissolving charge to a growing seed, the driving force of the process being the supersaturation. To create the supersaturation, which is necessary for the crystal growth in a preset dissolvent with a given rate, it is necessary to know the phase diagram of the system that is the temperature and pressure dependence of the solubility of the respective substance. A region with an appropriate value of the solubility coefficient may be then chosen according to this dependence to provide the mass transfer, but some additional kinetic requirements should be taken into account as well to provide the desired high quality of the grown crystals.

The first qualitative information on the solubility of different forms of silica was obtained in the middle of the 1930s. Two groups of scientists were involved in these studies, scientists interested in artificial crystal production and geologists. Geologists are faced with consequences of hydrothermal synthesis widely occurring in nature, in particular with abundance of crystalline quartz. A clarification of the mechanism and the conditions of formation of quartz single crystals was for geologists the way to comprehend the Earth's crust formation. This is the reason why so much information on this topic can be found in the geological literature.

The most comprehensive study of phase equilibrium in the quartz-water system over a wide range of temperature and pressure was performed in 1950 by G.C. Kennedy [115]. He thoroughly investigated the temperature range from 160 to 610 °C at pressures up to 170 MPa. These are just the hydrothermal conditions which are typical for quartz formation in nature. Quartz solubility was studied in isothermal conditions by the quenching method but not only quartz samples but the overall autoclaves were quenched. Quartz solubility was determined from the mass loss of the sample stated at regular intervals. The attainment of a steady state was established from termination of mass loss changes. Such type of behavior could be observed over a long period. The results of this extended laborious work are presented in Fig. 4.47.

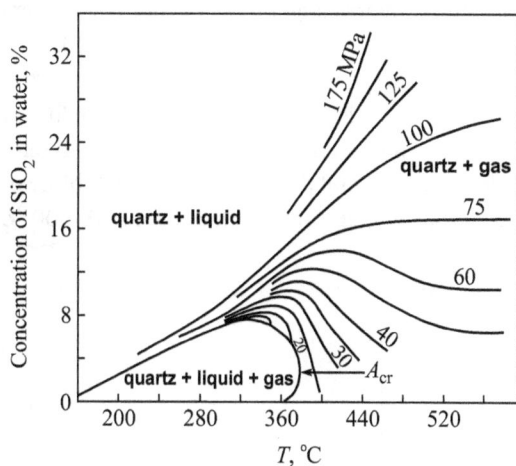

Fig. 4.47: Temperature dependence of quartz solubility in water at hydrothermal conditions according to [115].

Three different regions may be singled out for the system from the presented phase diagram: a three-phase region for heterogeneous equilibrium "quartz + water solution + vapor" and two regions for two-phase equilibria, "quartz + water solution" and "quartz + vapor". A dividing line between two-phase regions is conventional and may be considered as corresponding to the critical temperature for water, i.e. approximately 374 °C. Water solubility of quartz increases with temperature in the three-phase region as far as nearly 332 °C and reaches here its maximal value 0.075 %. On further heating up to the critical temperature (very close to 374.11 °C for pure water) the solubility begins to reduce to 0.023 %.

Solubility of quartz in the saturated vapor in equilibrium with liquid water and quartz (that is along the low boundary of the three-phase region) is extremely small up to $T = 360\,°C$ and $p \cong 21$ MPa. At moderate temperature and pressure, a percentage distribution of diluted silica between liquid and vapor phases is proportional to the ratio of densities for these phases; therefore, it is rather small. However, it begins to increase rapidly over 360–374 °C interval and reaches 0.023 % in the critical point. In the point A_{cr} (Fig. 4.47) at critical temperature and pressure, the liquid and its vapor become indistinguishable and form a so-called fluid phase or supercritical liquid. Since water solubility of quartz is rather low, the point A_{cr} is very close to the critical point of pure water. Above this point, the system is in a two-phase state: "quartz + fluid phase". The point A_{cr} is the lower critical point of the system SiO_2–H_2O as distinct from the upper critical point of the three-phase region "quartz + water solution + vapor" concerning the region of the melts. In the region of a homogeneous state of the solvent, i.e. above the curve of the three-phase equilibrium (Fig. 4.47), a retrograde water solubility of quartz is found at pressures below 70 MPa, the solubility being temperature independent at 70 MPa and sharply increasing at pressures above this value.

Kennedy's data were later supplemented by investigations of other researchers who studied water solubility of quartz up to 900 °C and 1.0 GPa [116] (Fig. 4.48). As it follows from the experimental data, the main factor, which is responsible for solubility of quartz in near-critical and supercritical water, is the state of the fluid and, first of all, its density. If the density is kept constant, the solubility of quartz grows with temperature. Thermodynamic analysis of the experimental data showed that solubility of quartz in dependence on density of water at temperatures in the range 200–600 °C

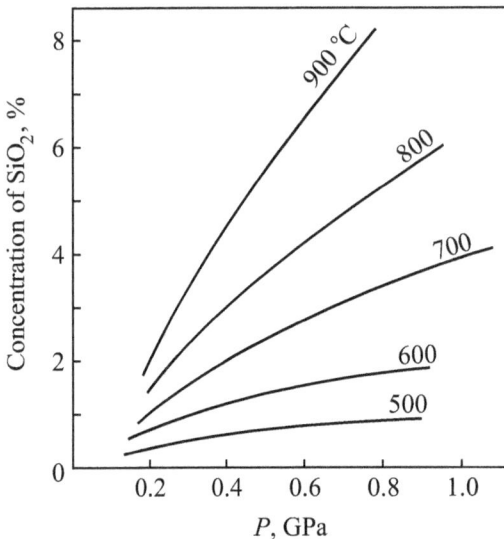

Fig. 4.48: Solubility of quartz in supercritical water according to [116].

and pressures below 200 MPa can be described by the following equation [117]:

$$c \cong \rho^2 \exp\left(\frac{Q}{RT} + h\right),$$

(4.4)

where c is the molar fraction of dissolved silica; ρ is the density of the water solution; Q is the differential heat of solution which is equal on average to 39.6 Joule/mol; $h = 0.362$ is the constant of integration.

Another approximation for Kennedy's data was suggested by F.G. Smith [118], but the extrapolation of the proposed by him equation to the region of high $(p - T)$-values resulted in a significant deviation from the experimental data of the work [116]. A similar extrapolation of the equation of the work [117] gives only a slightly better result. Since different forms of silica have different solubilities, the $(\alpha - \beta)$-transformation of quartz on heating and its further transformation into tridymite can be the possible reason of this failure. It is also possible that the deviations are related to particular features of structure and properties of supercritical water under high temperature-pressure conditions, especially its degree of dissociation.

The presence of two critical points on the phase diagram is typical for binary systems if one of the components has a very high volatility like H_2O and the other component has a low solubility like SiO_2. Already the first schematic phase diagram for the SiO_2-H_2O system [119] predicted two critical points and was in many respects true. The first quantitative experimental study of the upper critical point of the system was performed in [120] for $T = 1000-1300$ °C and $p = 120-200$ MPa but the final phase diagram was constructed again by G.C. Kennedy and coauthors [121]. They stated the position of the invariant curve up to 1080 °C and 970 MPa and constructed the full diagram involving the previous results (Fig. 4.49).

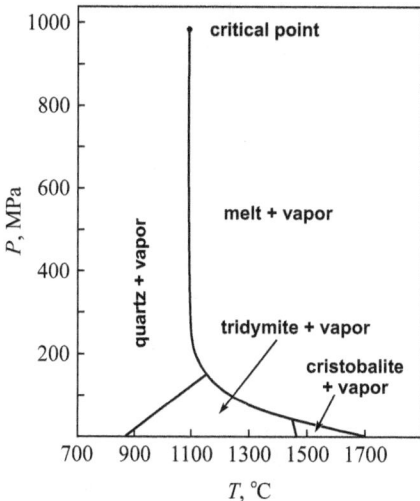

Fig. 4.49: The melting curve of silica in equilibrium with water [121].

Water reduces the melting temperature of silica even at a moderate pressure of the vapor. The cristobalite melting temperature decreases with the vapor presence as the pressure rises up to 40 MPa. At higher pressures, cristobalite becomes metastable but tridymite is stable up to 150 MPa. Quartz is still stable at larger pressures and, similar to other forms of silica, it gives the melt with significant water content. Hence, there are two quadric points in the system SiO_2–H_2O according to the presented diagram: the first point ($T = 1160\,°C, p = 150\,MPa$) corresponds to equilibrium between quartz, tridymite and two fluid phases (the melt and the vapor); the second point, which corresponds to equilibrium between tridymite, cristobalite, melt and vapor, is located at $T = 1470\,°C$ and $p = 40\,MPa$. It was also shown [122] that quartz can remain stable at temperatures considerably above its quadric points if the pressure is sufficiently large. Neither cristobalite nor tridymite can keep their stability under elevated pressure.

From the diagram in Fig. 4.49 it is evident that the melting temperature of silica decreases sharply with pressure in the fields of cristobalite and tridymite stability but remains almost the same in the field of quartz. Thus, change in pressure from 200 MPa to the critical value of 970 MPa reduces the melting temperature of quartz by 50 °C only. For comparison, in waterless conditions the melting temperature of quartz grows with pressure from 1870 °C in vacuum to 2150 °C at 700 MPa and 2300 °C at 1200 MPa [123].

The compositions of two fluid phases, coexisting along the upper three-phase boundary, were also established in [121] (Fig. 4.50). The left branch of the cupola refers to the water-rich vapor, which is in equilibrium with the silica-rich melt and the solid phase. The right branch relates to the melt, which is in equilibrium with the solid and the vapor. The solubility of quartz in water-rich fluid (vapor) changes with pressure from 5.7 % at 200 MPa to nearly 75 % at 970 MPa in the vicinity of the critical point. On the contrary, water content in the silica-rich melt rises slowly from 4.4 % at 200 MPa

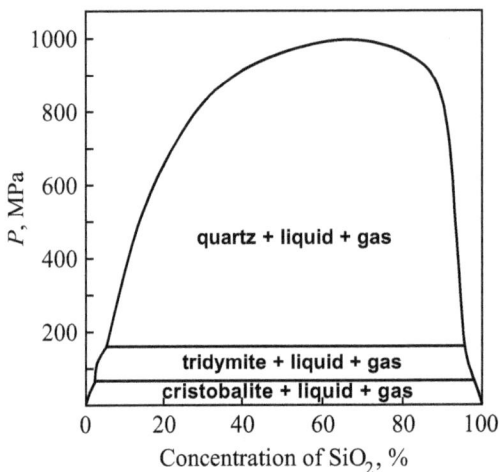

Fig. 4.50: Compositions of the fluid phases along the upper three-phase boundary for the system SiO_2–H_2O.

to nearly 6% at 600 MPa. At further elevation of pressure water content in the melt rises steeply up to 25% at the critical pressure. A comparison of Figs. 4.49 and 4.50 shows the difference of the upper and lower three-phase regions in the system SiO_2–H_2O. The compositions of both fluids vary over wide limits in the upper three-phase region and change moderately in the lower region, reaching the maximal value for silica solubility in water of 0.075%. Hydrothermal synthesis is usually performed in $(p - T)$-conditions close to the lower critical point.

The effect of water on silica melting is in accordance with the thermodynamic treatment [121]. However a theoretical melting curve calculated with regard to thermodynamic data is appreciably different from the experimental data for pressures 200–900 MPa, i.e. when the melting temperature of quartz changes very weakly but water solubility in the melt and silica solubility in the vapor rise regularly. The simulated curve is S-shaped but the experimental curve is nearly straight. This appears to be related to peculiarities of water solution in the melts and changes in the vapor state with pressure [2].

4.5.3.2 Silica Dissolution in Water Solutions of Salts and Liquid Phase Separation at Hydrothermal Conditions

For hydrothermal synthesis, it is important to increase the solubility of the charge. It was established that different phases of silica have different solubilities in water (Fig. 4.51), and the highest solubility is exhibited by amorphous silica. Unfortunately, as it was mentioned above, the crystallization ability of amorphous silica is also very high, and spontaneous crystallization stops soon after the dilution process. Additions of salts to water can significantly elevate water solubility of quartz and therefore were investigated frequently. Some of the typical results are presented in Figs. 4.52 and 4.53 (citation according to [112]). At the same time, as it is often the case in hydrothermal synthesis, the phenomenon appears to be much more complicated and its application is rather restricted.

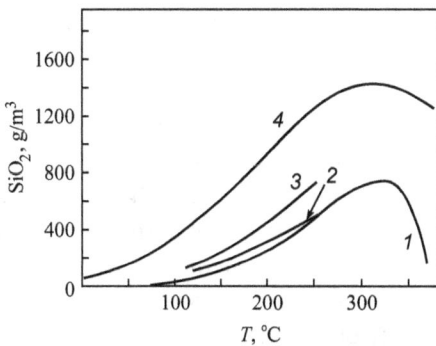

Fig. 4.51: Water solubility of quartz (1), chalcedony (2), cristobalite (3), and amorphous silica (4) according to [2].

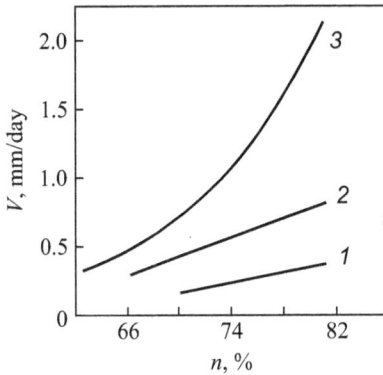

Fig. 4.52: Temperature dependence of solubility of quartz in water solutions of Na_2CO_3 for different content of soda. The infill factor is 70 % (according to [112]).

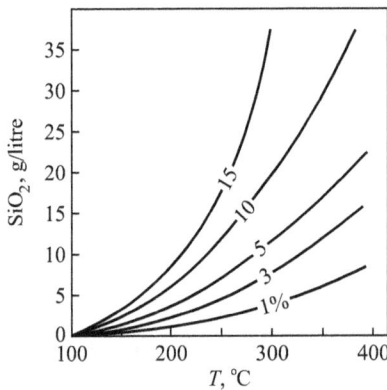

Fig. 4.53: Temperature dependence of solubility of quartz in water solutions of NaH_4F for different content of the solute (according to [112]).

Already one of the first systematic works on quartz solubility in the system H_2O–Na_2O–SiO_2 [124] revealed the main problems of this method although the later studies of the same and many other researchers significantly refined and supplemented the obtained information. The authors of [124] performed their hydrothermal experiments with NaOH solutions at 250, 300 and 350 °C (temperatures of 400 and 500 °C were studied later), the content of sodium hydroxide changing from 1.5 to 38.8 wt%. Autoclaves of minor capacity permit to measure only temperature but not pressure. This is the reason why the constructed phase diagrams were specified as polybaric; one of the diagrams is presented in Fig. 4.54. The investigated area is divided into two principal regions by the complicated curve H_2O–A–B–C–D–E–F. All compositions to the left of the curve are unsaturated at 250 °C, and all the mixtures to the right of the curve consist of two liquids or of one liquid and one or more crystalline phases. Let us consider this ternary diagram from the point of view of hydrothermal synthesis.

The lower axis of concentration triangles corresponds to the binary system H_2O–SiO_2. As it was discussed above, water solubility of silica is quite low. However, an addition of alkali increases it. The solubility increases when the point, presenting the

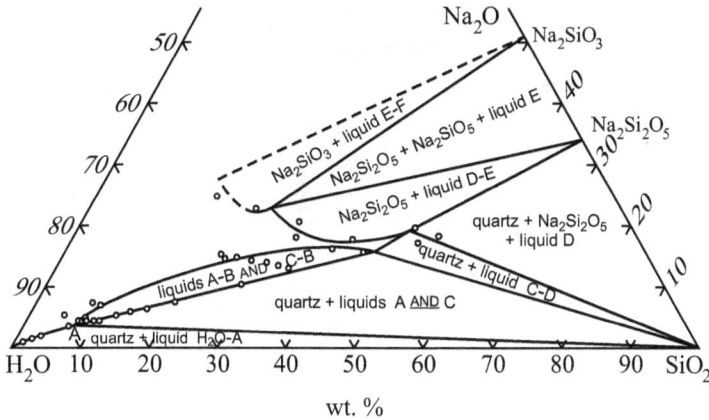

Fig. 4.54: Polybaric saturation relations at 250 °C in the system H_2O-Na_2O-SiO_2 [124].

composition of the system, moves upwards from the side of H_2O-SiO_2. After the first alkali oxide additions the point enters the narrow two-phase region where silica is in equilibrium with the liquid, which has a composition in between H_2O and A. A further addition of alkali leads to the further motion of the figurative point upwards and it comes into the spacious triangle A–C–SiO_2. This region is of great importance for hydrothermal synthesis because it includes working compositions. Quartz coexists here with two liquids. One of them is a "thin" water-rich liquid A, and another one is "a clear water-soluble glass" with composition represented by point C [124]. In the Tuttle and Friedman experiments, the glass always occupied the lowest point in the bomb and therefore was denoted as heavy phase. As the temperature increases, the point C shifts to the right, towards the Na_2O–SiO_2 system and the heavy phase enriches with silica. If after heat-treatment at 250 °C this quenched glass has a hardness of about 2.5 and is readily soluble in water, after treatment at 300 °C it is noticeably less soluble, and the 350 °C glass is very brittle. It has a hardness of approximately 5 and is weakly soluble in water. At the temperatures of the experiment the system was shown to be a liquid, it was then quenched during the bomb cooling.

Liquid phase separation of the working solution is a damaging factor for crystal growth by hydrothermal synthesis. In this case, the water-rich liquid plays the role of the working medium. The heavy phase localizes at the bottom of the autoclave and partially covers the batch preventing its dilution in the working medium. In addition, small drops of the heavy phase are carried away with convection upstream, incorporate into the growing crystal as a non-structural admixture and thereby worsen its quality. A decrease of the magnitude in the temperature drop decreases the convection rate and prevents the incorporation of the heavy phase admixture into the crystal, but at the same time, it decreases its growth rate.

A further increase in alkali content leads to a transfer of the figurative point into regions where sodium silicates have to be crystallized. Silicate formation is also detrimental for hydrothermal synthesis because the crystals precipitate on the walls of the working chamber, cover the batch and stop its further dilution. As a result, it becomes necessary to clean the autoclave periodically. The increase in temperature aggravates the situation because it leads to a broadening of silicates crystallization fields. From the above considerations, it follows that attempts to increase quartz solubility in water by increasing its alkali content may initiate some detrimental processes. Hence, the correct choice of the synthesis parameters, including type and amount of additions, is a delicate question. Its successful solution depends in many respects on the operator's experience.

The phase diagrams constructed in [124] give the principal understanding of the behavior. Unfortunately, they are quantitatively incorrect. Weight losses of quartz in the presence of two liquids do not define its solubility at the given temperature (as it correctly took place in the Kennedy's experiments but incorrectly assumed in Fig. 4.54). The reason is that silica diluted is divided in some proportion between the two liquids. The knowledge of the true quartz solubility in a "thin" water-rich phase is necessary for hydrothermal synthesis because it is precisely this phase that is the convective medium of the synthesis. The later investigations significantly refined quantitative data [124] on quartz solubility. In Fig. 4.52 the curve for 2 % Na_2CO_3 belongs to the solution without phase separation, but the curves for 5 and 11 % correspond to phase separated liquids.

4.5.3.3 Seeds

A specific feature of quartz growing on a seed consists of using plate seeds instead of usual point ones. At any physicochemical conditions of synthesis, quartz crystals are not necessarily built upon m-faces of the hexagonal prism. As a result, a crystal in the form of a needle extended along the L_3-axis grows from the point seed. Shape, size and crystallographic orientation of the seeds determine significantly the shape and size of the growing quartz crystals. To produce a large crystal of quartz it is necessary to set its size in the section perpendicular to the c-axis by a lengthy seed. Besides the plates with (0001) c orientation, ($11\bar{2}0$) ($\pm x$), ($10\bar{1}1$) R and ($01\bar{1}1$) r orientations are often used, the two latter are employed for jewelry quartz production. Other orientations, including irrational ones, may also be used for special purposes.

The capture of structural and non-structural impurities depends to a considerable degree on the crystallographic orientation of the seed. Impurity distribution over growth pyramids and zones leads to formation of sectorial and zonal structures together with parasitic growth pyramids (a secondary sectoriality) and twins in the growing individual crystals. As a result, real faces of crystals produced by hydrothermal method are not ideal planes but always have some specific (for the given crystal face) relief. The mechanism of formation of some types of crystal face reliefs, like

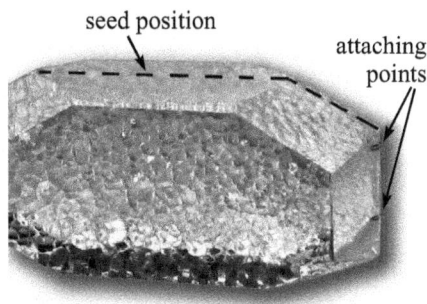

Fig. 4.55: The cellular relief on the faces of the *c*-crystal of quartz grown on a seed with (0001) *c*-orientation.

the cellular structure shown in Fig. 4.55 (named also a cobblestone road), is comprehended in a general way [113].

If the crystal grows not from its melt but from a solution, concentration gradients evolve near the growing faces. In some conditions it leads to so-called concentration supercooling in the close vicinity of the crystallization front (Fig. 4.56). If in the growing crystal the equilibrium concentration of the admixture is lower than in the feeding solution, the admixture is rejected by the crystal and accumulated in the near-boundary layer (Fig. 4.56a). It is well known that commonly an admixture reduces the melting temperature. In Fig. 4.56b the melting temperature of the crystal, T_m, decreases to T_{m-s}; if in the close vicinity of the boundary T_{m-s} is lower than the medium temperature, the crystal growth stops. Furthermore, if there is a sufficiently small temperature gradient near the growth surface, the temperature of the solution may occur lower than T_{m-s} (the region of concentration supercooling) in the near-boundary layer of the solution, and crystal growth is possible here. It means that any bump on the smooth crystal surface has a tendency to move in the solution until it reaches the point D where the temperature is equal to the melting temperature. Diffusion along sidewall of bumps makes this process easier. Size and form of the cells on the crystal face depend on several factors such as temperature gradient, concentration and diffusion rate of the admixture, surface free energy on the crystal-solution boundary for the different crystallographic planes. The morphology of the crystallization front changes with temperature gradient and cellular growth changes into dendritic one under very high supercooling. Frost flowers (patterns) on window glasses are the most prominent example of this case. Natural quartz crystals never have cellular relief [21] because the extreme slowness of the growth processes in nature results in concentration equalization in the solution. For industrial production of high-quality piezoelectric quartz, it is necessary to create and regularly supplement a bank of seeding material produced from defect-free single crystals of natural quartz.

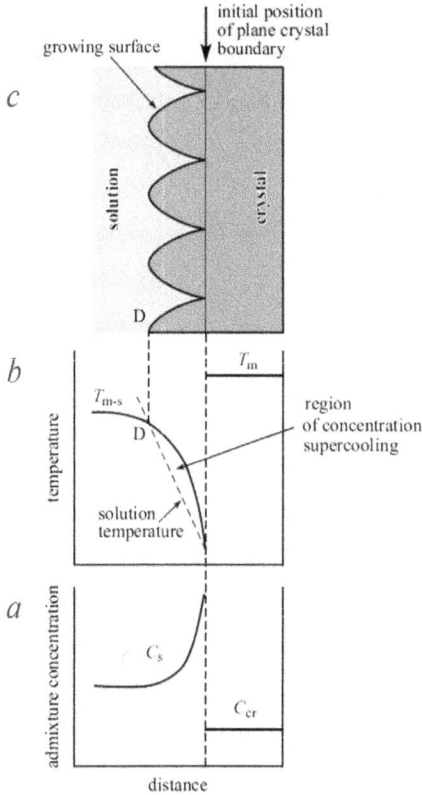

Fig. 4.56: Cellular growth, caused by concentration supercooling: (a) concentration of admixture in the crystal (c_{cr}) and solution (c_s); (b) temperature gradient near the growth surface (dashed line) and melting temperature of the crystal in dependence on the admixture content (T_m and T_{m-s}); (c) a form of the growth surface in the region of concentration supercooling.

The main factors, which act on the growth rate in the temperature drop method, are: (i) Value of the temperature drop, ΔT, between the dilution and growth chambers; (ii) surface area of the seed plate; (iii) area of the active surface of the diluting charge; (iv) intensity of the solution circulation (mass flow) between the chambers of dilution and growth; (v) thickness of the diffusion layer around the growing crystal faces; (vi) diluting surface of the quartz charge. The growth rate of the crystal faces is a linear function of ΔT, but there is an upper limit for legitimate value of ΔT, above which a spontaneous crystallization begins on the inner walls of the autoclave. The limiting value is equal to approximately 15–17 °C for soda solutions. It has a higher value in NaOH solutions and permits to reach higher crystallization rates (up to 4 mm/day for short-term experiments) in them. In long-term experiments, spontaneous crystallization occurs even at the growth rate of 3.5 mm/day. Because of the low temperature coefficient of quartz solubility in NaOH solutions, the specific growth rate is lower as compared with soda solutions. The latter phenomenon is of great practical importance since for NaOH solutions there is no need in such a fine temperature control as for soda solutions. If all other crystallization conditions are fixed, the increase of pressure leads to the increase of the crystal growth rate due to the increase in the solubility of quartz and sodium silicates, the latter may form simultaneously with quartz and disturb the process.

4.6 Concluding Remarks

Quartz is the usual sand under our feet but it is a very unusual substance. Intensive research is devoted to the understanding of its properties, but this analysis often leads to more questions then to give answers. I tried to describe its important but poorly known properties. I also tried to present an explanation of these properties, at least, of some of them. This explanation is connected with the electronic structure of the silicon atom and its interaction with oxygen. The same reason – presence of empty d-orbitals in silicon – leads to very different consequences: to the great variety of structural forms of silicas and silicates; to unusually high chemical reactivity that permits to form heat-resistant crystals at low temperatures; to thermal contraction of the SiO_4-tetrahedra. May be, it explains also the mysterious so far properties of quartz? Who knows?!

4.7 Appendix: The Crystal Skulls

A recent movie about the adventures of Indiana Jones in his search of the mysterious crystal skulls seems to be a fully artificially constructed story. However this is not the case because crystal skulls do exist in reality. Here, I do not have in mind present-day objects made of semiprecious stones and priced from tens of dollars to tens of thousands dollars. These crystal skulls (of all possible shapes and sizes) are usually fabricated by carvers in Brazil and China. However, there are other such objects: enigmatical artifacts that can be found on display in such famous museums as the British Museum, Musée de l'Homme (Trocadéro Museum) in Paris or the National Museum of the American Indians in New York. The presented here discussion on this topic is based on press and internet sources.

There are two types of ancient crystal skulls. Wikipedia attributes smaller bead-sized skulls to Mesoamerican beads or rosary beads of Mexican Catholics carved not so long ago (the first specimens appeared in the middle of the 19th century). However, there is evidence that smaller crystal skulls were rather widespread already in Italy and South America in the 15th and 16th centuries. Larger skulls are practically life-sized; some of them are quite primitive and carved from opaque quartz minerals (Fig. 4.57), while others are transparent, as the skull from the British Museum, and also have an amazing degree of perfection, as the famous Mitchell-Hedges Skull (Fig. 4.58). All of the photos were taken from [125]. Some other high quality photos can be found in [126].

Independent experts investigated these skulls, carefully and repeatedly, in attempts to establish the period of their preparation but the results were contradictory.

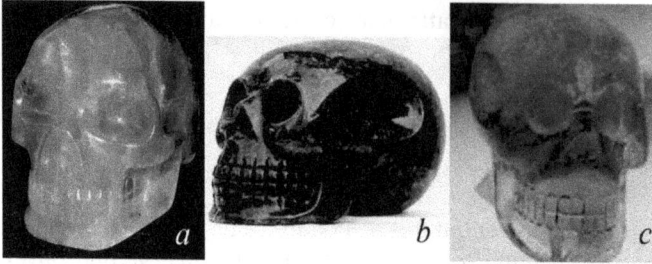

Fig. 4.57: The large skulls from opaque quartz minerals. The Rose Quartz skull was found in Mexico and called Rosie (a). The Amethyst skull made from unbroken pieces of Amethyst; it was discovered in the early 1900s in Guatemala and is now kept in Japan (b). The Mayan Crystal skull was discovered in the early 1900s in Mexico (c).

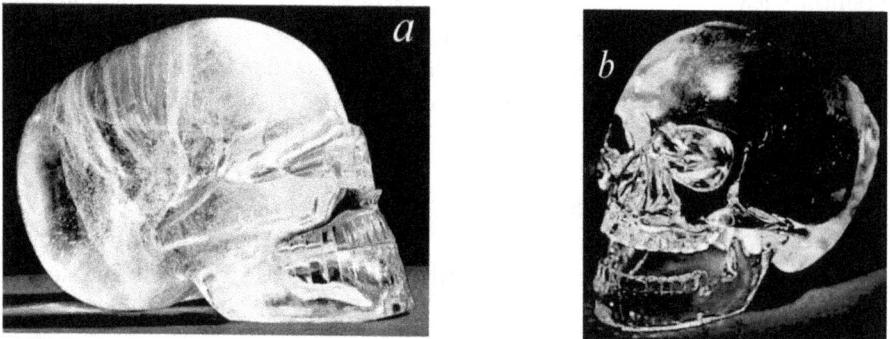

Fig. 4.58: Large transparent skulls produced from unbroken pieces of quartz: (a) British Museum skull: it was brought from Mexico by a Spanish officer before the French occupation (in 1863) and was sold to an English collector. The details can be found at the web site of the British Museum [127]. (b) Mitchell-Hedges crystal skull: It is the most famous and most studied skull. It was found in 1924 by F.A. Mitchell-Hedges in Lubaantun, Belize. A history of the skull and its famous owners can be found in [128].

Let us briefly consider the main problems that exist in connection with large skulls. There is no independent method for determining the age of quartz handicrafts. Quite effective geological methods allow one to determine the age of quartz with an accuracy of millions years. But when were the skulls carved from it? Contrary to a popular belief, no satisfactory scientific technique is available for an accurate determination of the period when a stony object was carved. There are two ways to answer the question: either to find a place of given artifacts in a known culture of some nation or to investigate the manufacturing technique and to relate it to history of technology.

All large crystal skulls were found in more or less modern time, and none of the skulls in museum collections came from documented excavations. All of them were cut from unitary blocks of quartz materials, and the transparent skulls were made from quartz single crystals. A detachable jaw of the Mitchell-Hedges crystal skull was produced from the same crystal of quartz as the cranium. The majority of the large skulls were investigated by scientists who faced an unusual problem: it was impossible to carve the skull using modern methods.

Crystalline quartz is a substance with the hardness of 7, which is lower than that of diamond, whose hardness rates 10. It requires either other quartz crystals or special diamond tools to attempt to carve a crystal skull. The hardest parts to duplicate in carving a crystal skull are the various angles and shapes of the bone, which comprise the skull and the jaw. Even utilizing the most sophisticated lasers of today, it would be an incredible challenge to precisely cut a crystal skull to exactly match our own human skull [129]. Art restorer Frank Dorland, who had studied the Mitchell-Hedges skull for six years, set up the hypothesis for making the skull, according to which, it was roughly hewn out with cutting by diamond tools, this procedure being followed by a thorough finishing treatment with the use of a mild mixture of silicon sand and water. The exhausting job – assuming it could, possibly, be done in this way – would have required man-hours adding up to 300 years to complete [130].

Several ancient crystal skulls were brought to Hewlett-Packard (HP) laboratories, located near to the San Francisco area in California. HP has long been known as a manufacturer of computer printers and systems, in addition to having one of the most extensive scientific and crystal research laboratories in the world. There the Mitchell-Hedges skull was lowered into a vat of benzyl alcohol, where it became nearly invisible. This result was the proof that the skull was actually quartz crystal (alcohol and quartz have the same refraction index). The HP researchers also determined that the skull was carved from a single piece of quartz [131]. It was reported that "the lab found that the skull had been carved against the natural axis of the crystal. Modern crystal sculptors always take into account the axis or orientation of the crystal's molecular symmetry, because if they carve "against the grain", the piece is bound to shatter – even with the use of lasers and other high-tech cutting methods". There was another strange feature: at Hewlett-Packard lab the researchers could not find any microscopic scratches on the crystal which would indicate it had been carved with metal instruments [130].

In addition to the Mitchell-Hedges skull, other skulls (shown in Fig. 4.57) were studied at HP as well. They were also found to be inexplicably cut against the axis of the crystals. Later investigations of some skulls (the British Museum skull and the Smithsonian Institution skull) performed by the British museum using high-power microscopy showed some traces of rotary lapidary tools on the surface of the skulls. The tool marks on the skulls are very different as compared to those on ancient Mexican rock crystal objects, which were carved by hand. Pre-Columbian MesoAmerican lapidary techniques never included rotary wheels, thus the skulls could not be carved in ancient times. A more detailed and critical description of investigations of the skulls, including SEM-images of the skull surfaces, with reference to original scientific papers is presented in [132].

According to [125], the enigma of the skulls, however, does not end with just their fabrication. The zygomatic arches of the Mitchell-Hedges skull (the bone arch extending along the sides and front of the cranium) are accurately separated from the skull piece, and act as light pipes, using principles similar to modern optics, to channel light from the base of the skull to the eye sockets. The eye sockets, in turn, are miniature concave lenses that also transfer light from a source below into the upper cranium. Finally, in the interior of the skull there is a ribbon prism and tiny light tunnels, by which objects, held beneath the skull, are magnified and brightened. It seems that the skull was designed to be placed over an upward shining beam. As a result, with the various light transfers and prismatic effects, they would illuminate the entire skull and cause the sockets to become glowing eyes.

In addition, some unusual phenomena occurred around the crystal skulls. Observers of the Mitchell-Hedges skull reported that it appears to influence all five human feelings. It changes color and light, it emits odours, it creates sound, it gives off sensations of heat and cold to those who touch it, even though the crystal always remained at a temperature of 70 degrees. Observers noticed strange scenes, reflected in the eye sockets, buildings and other objects, even though the skull is at rest and put against a black background [125, 129, 133].

A considerable amount of information on crystal skulls is available in internet, books and documentary films. Returning to Indiana Jones, it is necessary to say that, in connection with ancient skulls, also some espionage interest was indeed involved and theft took place but not from the direction of KGB. The secret mystical organization of Nazis, Ahnenerbe, tried to purloin the crystal skulls of 'the Goddess of Death' from museums. Several agents of Ahnenerbe were detained in 1943 in Brazil and gave evidence that they had fulfilled the Abwehr task to get hold of the Brazilian skulls. It is possible that some of non-caught agents succeeded in their mission. Indeed, some of the larger crystal skulls were really stolen. The so-called Rose quartz (a skull comparable in its perfection with the Mitchell-Hedges skull) has enigmatically disappeared in Honduras.

The enigma of the large crystal skulls is not something unique. For example, a large quartz dish in the form of a three-leaved flower is kept in the Historical Museum of Cairo. Similar to the skulls, the dish is carved from a quartz single crystal against its natural axes. In St. Petersburg, two large Egyptian sphinxes have been looking, for many years, towards each other on an embankment of the river Neva, opposite to St. Isaac's Cathedral. The surface of the sphinxes is prepared with surprisingly high quality and radically differs from the machined surface of the great columns of the St. Isaac's Cathedral. As it seems, at present, the riddles of ancient technologies are not fully understood, and the time for their solution has not come yet.

Bibliography

[1] Liebau F., *Structural chemistry of silicates* (Springer-Verlag, Berlin – Heidelberg – New York – Tokyo; Russian Edition, Mir, Moscow, 1988).

[2] Mitzyuk B.M., Gorogotzkaya L.I., *Fiziko-khimicheskie prevrashcheniya kremnezema v usloviyah metamorfizma (Physicochemical transformation of silica under metamorphism condition*, in Russian, Naukova Dumka, Kiev, 1980).

[3] Sosman R.B., *The Phases of Silica* (Rutgers University Press, New Brunswick, New Jersey, 1965).

[4] Mendeleev D.I., *Osnovy khimii (Chemistry principles)*, (in Russian, 13th edition, **2**, Goskhimizdat, Moscow-Leningrad, 1947).

[5] Santoro M., Gorelli F.A., Bini R., Ruocco G., Scandolo S., Crichton W.A., Nature **441**, 857 (2006).

[6] Liebau F., Inorg. Chim. Acta **89**, 1 (1984).

[7] Duffy J.A., Macphee D.E., J. Phys. Chem. B **111**, 8740 (2007).

[8] Stishov S.M., Belov N.V., Dokl. AN SSSR **143**, 951 (1962) (in Russian).

[9] Leko V.K., Komarova L.A., Neorgan. Mater. **7**, 2240 (1971).

[10] Leko V.K., Komarova L.A., Neorgan. Mater. **8**, 1125 (1972).

[11] Leko V.K., Komarova L.A., Neorgan. Mater. **10**, 1872 (1974).

[12] Leko V.K., Komarova L.A., Neorgan. Mater. **11**, 1115 (1975).

[13] Leko V.K., Komarova L.A., Neorgan. Mater. **11**, 2046 (1975).

[14] Leko V.K., Komarova L.A., Neorgan. Mater. **11**, 2106 (1975).

[15] Crystallographic and Crystallochemical Database for Minerals and their Structural Analogues WWW-MINCRYST, http://database.iem.ac.ru/mincryst/.

[16] http://www.mindat.org.

[17] Martonak R., Donadio D., Oganov A.R., Parrinello M.. Phys. Rev. B **76**, 014120 (2007).

[18] El Goresy A., Dera P., Sharp T.G., Prewitt C.T., Chen M., Dubrovinsky L., Wopenka B., Boctor N.Z., Hemley R.J., Eur. J. Mineral. **20**, 523 (2008).

[19] Weiss A., Weiss A., Naturwissenschaften **41**, 12 (1954); Z. Anorg. Chem. **276**, 95 (1954).

[20] http://www.quartzpage.de/gen_mod.html#pdsp.

[21] O'Donoghue M., *Quartz* (Butterworths, London, 1987).

[22] http://www.quartzpage.de.

[23] Mortly W.S., Nature **221**, 359 (1969).

[24] Frenninger H.H., Laves F., Naturwissenschaften **48**, 23 (1961).

[25] Flörke O.W., Mielke H.G., Weichert J.Z., Kristallogr. **143**, 156 (1976).

[26] http://www.quartzpage.de/gen_mod.html#moganite.

[27] Murashov V.V., Svishchev I.M., Phys. Rev. **B 57**, 5639 (1998).

[28] Miehe G., Graetsch H. Eur. J. Mineral. **4**, 693 (1992).

[29] Miehe G., Graetsch H., Flörke O.W., Fuess H.Z., Kristallogr. **182**, 183 (1988).

[30] Heaney P.J., Post J.E., Amer. Mineral. **86**, 1358 (2001).

[31] Heaney P.J., Post J.E., Science **255**, 441 (1992).

[32] Schäf O., Ghobarkar H., Garnier A., Vagner C., Lindner J.K.N., Hanss J., Reller A., Solid State Science **8**, 625 (2006).

[33] Heaney P.J., McKeown D.A., Post J.E., Amer. Mineral. **92**, 631 (2007).

[34] Flörke, O.W. Bull. Amer. Ceram. Soc. **36**, 142 (1957).

[35] Holmquist S.B., J. Amer. Ceram. Soc. **44**, 82 (1961).

[36] Polyakova I.G., Proc. *19th Int. Congress on Glass*, Edinburgh, U.K. Invited lecture, 272 (2001)

[37] Pryanishnikov V.P., *Sistema kremnezema (The system of silica, in Russian)* (Izd. Liter. po Stroitel'stvu, Leningrad, 1971).

[38]   Berezhnoi A.S., *Mnogokomponentnye sistemy okislov (Multicomponent oxide systems, in Russian)* (Naukova Dumka, Kiev, 1970).

[39]   Flörke, O.W., Silicates Ind. **26**, 415 (1961).

[40]   Boiko G.G., Berezhnoi G.V., Glastechnische Berichte, Glass. Sci. Techn. **77**, 346 (2004).

[41]   Coes L., Science **118**, 131 (1953).

[42]   http://www.barringercrater.com/science/.

[43]   http://webmineral.com/data/Magadiite.shtml, http://webmineral.com/data/Kenyaite.shtml.

[44]   Keat P.P., Science **120**, 328 (1954).

[45]   Carr R.M., Fyfe W.S., Amer. Mineral. **43**, 908 (1958).

[46]   Pesty L., Tomschey O., Acta Geo. Acad. Sci. Hung. **17**, 121 (1973).

[47]   http://www.quartzpage.de/gen_mod.html#T.

[48]   Ryzhov V.N., Gribova N.V., Tareeva E.E., Fomin Yu. D., Tziok E.N., *Presentation at the XI Conf. of Young Scientists of the Institute for High Pressure Physics "Problems of Solid State and High Pressure Physics"*, http://www.hppi.troitsk.ru/meetings/school/XI-2010/V.N.Ryzhov.pdf.

[49]   http://www.lsbu.ac.uk/water/anmlies.html.

[50]   Huang L., Duffrène L., Kieffer J., J. Non-Cryst. Solids **349**, 1 (2004).

[51]   Mishima O., Stanley H.E., Nature **396**, 329 (1998).

[52]   Mishima O., Calvert L., Whalley E., Nature **314**, 76 (1985).

[53]   Poole P.H., Sciortino F., Essmann U., Stanley H.E., Nature **360**, 324 (1992).

[54]   Mukherjee G.D., Vaidya S.N., Sugandhi V., Phys. Rev. Lett. **87**, 195501 (2001).

[55]   Landa L., Landa K., Thomsen S., *Uncommon description of common glasses. Vol. I: Fundamentals of the united theory of glass formation and glass transition* (Yanus Publishing House, St. Petersburg, 2004).

[56]   Malinovsky V.K., Novikov V.N., Surovtsev N.V., Shebanin A.P., Phys. Solid State **42**, 65 (2000).

[57]   Malinovsky V.K., Sokolov A.P., Solid State Commun. **37**, 751 (1986).

[58]   El'kin F.S., Brazhkin V.V., Khvostantsev L.G., Tsiok O.B., Lyapin A.G., JETP Letters **75**, 342 (2002).

[59]   Brazhkin V.V., Tsiok O.B., Katayama Y., JETP Letters **89**, 285 (2009).

[60]   Tsiok O.B., Brazhkin V.V., Lyapin A.G., Khvostantsev L.G., Phys. Rev. Lett. **80**, 999 (1998).

[61]   http://www.webmineral.com/data/Quartz.shtml.

[62]   Tzinzerling E.V., *Iskusstvennoe dvoynikovanie kvartza (Artificial twining of quartz, in Russian)* (Academy Sciences Publishers, USSR, Moscow, 1961).

[63]   Wooster W.A., Wooster N, Ricroft J.L., Thomas L.A., J. Inst. Electr. Engin. (III A) **94**, 927 (1947).

[64]   Perez J.-P., C.R. **230**, 849 (1950).

[65]   http://www.quartzpage.de/gen_phys.html.

[66]   Larikov L.I., *Teplovye svoystva metallov i splavov (Thermal properties of metals and alloys, in Russian)* (Naukova Dumka, Kiev, 1985)

[67]   http://www.quartzpage.de/gen_phys.html#i02.

[68]   http://dssp.petrsu.ru /files/tutorial/ftt/Part6/part6_4.htm.

[69]   Gaikovoi A.G., *Master's thesis: Study on thermal deformations and the nature of α − β inversion in quartz.* (Leningrad, 1981).

[70]   Ackermann, R.J., Sorrell, C.A., J. Appl Cryst. **7**, 461 (1974).

[71]   Jay A.H., Proc. Royal Society **A 142**, 237 (1933).

[72]   Kuniaki Kihara, Eur. J. Mineral. **2**, 63 (1990).

[73]   Grimm H., Dorner B., J. Phys. Chem. Solids **36**, 407 (1975).

[74]   Cruickshank D.W.J., J. Chem. Soc. **1077**, 5486 (1961).

[75]   Achary S.N., Jayakumar O.D., Tyagi A.K., Kulshreshtha S.K., J. Solid State Chem. **176**, 37 (2003).

[76]   van Tendeloo G., van Landuyt J., Amelinckx S., Phys. Status Solidi (A) Appl. Res. **33**, 723 (1976).

[77]   Thompson J.G., Withers R.L., Palethorpe S.R., Melnitchenko A., J. Solid State Chem. **141**, 29 (1998).

[78]   Peacor D.R., Z. Kristallogr. **138**, 274 (1973).

[79]   Steinwehr, H.E., Z. Kristallogr. **99**, 292 (1932).

[80]   Raman C.V., Nedungadi T.M.R ., Nature **145**, 45 (1940).

[81]   Bachheimer, J.P., J. de phys. Lett. **41**, 559 (1980).

[82]   Spearing D.R., Farnan I., Stebbins J.F., Phys. Chem. Minerals **19**, 307 (1992).

[83]   Nikitin A.N., Markova G.V., Balagurov A.M., Vasin R.N., Alekseeva O.V., Crystallography Reports **52**, 428 (2007).

[84]   Heaney P.J., Veblen D.R., Amer. Mineralogist **76**, 1018 (1991).

[85]   Heaney P.J., McKeown D.A., Post J.E., Amer. Mineralogist **92**, 631 (2007).

[86]   Kimizuka H., Kaburaki H., Phys. Stat. Sol. (B) Basic Research **242**, 607 (2005).

[87]   Carpenter M.A., Salje E.K.H., Graeme-Barber A., Wruck B., Dove M.T., Knight K.S., Amer. Mineralogist **83**, 2 (1998).

[88]   Sinel'nikov N.N., Dokl. Akad. Nauk SSSR, Ser. Fizich. Khimiya **92**, 369 (1953) (in Russian).

[89]   Perrier A., Mandrot R., Comp. Rend. **175**, 622 (1922).

[90]   Yakovkev I.A., Velichkina T.S., Uspekhi fizicheskikh nauk **63**, 411 (1957) (in Russian).

[91]   Filatov S.K. Polyakova I.G., Gaikovoi A.G., Kamentzev I.E., Kristallografiya **27**, 624 (1982) (in Russian) .

[92]   Banda E.J.K.B., Craven R.A., Parks R.D., Horn P.M., Blume M., Solid State Communications **17**, 11 (1975).

[93]   Höchli U.T., Solid State Communications **8**, 1487 (1970).

[94]   Ginzburg V.L., DAN SSSR **105**, 240 (1955) (in Russian).

[95]   Shapiro S.M., Cummins H.Z., Phys. Rev. Lett. **21**, 178 (1968).

[96]   Ginzburg V.L., Levanyuk A.P., Phys. Lett. A **47**, 345 (1974).

[97]   Bartis F.J., J. Phys. C: Solid State Phys. **6**, L295 (1973).

[98]   Koziev K.S., Ph.D. thesis: *Investigation on structural phase transitions in solid states with defects* (Hudzhand, Tadjikistan, 2004).

[99]   Umarov B.S. DPhil thesis: *Raman scattering spectroscopy in noncentro-symmetrical crystals in the presence of external actions and impurities* (Dushanbe, Tadjikistan, 1984).

[100]   Dolino G, Bastie P., J. Phys.: Condens. Matter **13**, 11485 (2001).

[101]   Yamamoto N., Tsuda K., Yagi K., J. Phys. Soc. Japan **57**, 1352 (1988).

[102]   Dolino G., Phase Trans. **21**, 59 (1990).

[103]   Gridnev S.D.: *Surprises of incommensurate phase in ferroelectrics*, grant N 01-02-16097 of the Russian Foundation for Basic Research.

[104]   Dmitriev S.V., Phys. Sol. State **45**, 352 (2003).

[105]   Abe K., Kawasaki K., Kowada K, Shigenari T., J. Phys. Soc. Japan **60**, 404 (1991).

[106]   Aslanyan T.A., Shigenari T., Abe K., Ukigaya N., J. Korean Phys. Soc. **32**, S914 (1998).

[107]   Richet P., Gillet P., Europ. J. Mineralogy **9**, 907 (1997).

[108]   Hemley R.J., Jephcoat A.P., Mao H.K., Ming L.C., Manghnani M.H., Nature **334**, 6177 (1988).

[109]   Haines J., Cambon O., Le Parc R., Levelut C., Phase Transitions **80**, 1039 (2007).

[110]   Badro, J, Gillet, P., Barrat, J.-L., Europhysics Letters **42**, 643 (1998).

[111]   Huang L., Kieffer J., J. Chem. Phys. **118**, 1487 (2003).

[112]   Khadzhi V.E., Tzinnober L.I., Shterenlikt L.M. and others: *Sintez mineralov* (*Synthesis of minerals*, in Russian) **1** (Nedra, Moscow, 1987).

[113]   Laudise R.A., *The growth of single crystals* (Prentice-Hall, Inc. Englewood Cliffs, New Jersey, 1970).

[114] Laudise R.A., in the book: *Crystal Growth*, ed. H.S. Peiser, Pergamon, New York, 1967, p. 3.

[115] Kennedy G.C., Econ. Geol. **45**, 629 (1950).

[116] Anderson G.M., Burnham C.W., Amer. J. Sci. **263**, 494 (1965).

[117] Mosebach R.J., Geol. **65**, 347 (1957).

[118] Smith F.G., Can. Mineral. **6**, 210 (1958).

[119] Smiths A., Rec. Trav. Chim. **49**, 962 (1930).

[120] Tuttle O.F., England J.L., Bull. Geol. Soc. Amer. **66**, 149 (1955).

[121] Kennedy G.C., Waserburg G.J., Heard H. C ., Newton R.C., Amer. J. Sci. **260**, 501 (1962).

[122] Ostrovsky I.A., Geol. J. **5**, 127 (1966).

[123] Boganov A.G., Popov S.A., Rudenko V.S., Dokl. AN SSSR **202**, 1099 (1971) (in Russian).

[124] Tuttle O.F., Friedman I.I., J. Amer. Chem. Soc. **70**, 919 (1948).

[125] http://www.crystalinks.com/crystalskulls.html.

[126] http://www.crystalskullexplorers.com/csarticle-p2.htm

[127] http://www.britishmuseum.org/explore/highlights/highlight_objects/aoa/r/rock_crystal_skull.aspx.

[128] http://www.mitchell-hedges.com/.

[129] http://www.v-j-enterprises.com/csartaus.html.

[130] http://www.skepdic.com/crystalskull.html.

[131] http://science.howstuffworks.com/crystal-skull.htm.

[132] http://www.badarchaeology.net/data/ooparts/crystal.php.

[133] http://www.paranormalnews.com/article.asp?ArticleID=927.

Natalia M. Vedishcheva and Adrian C. Wright

5 Chemical Structure of Oxide Glasses: A Concept for Establishing Structure–Property Relationships

In a series of papers by the present authors, it has been shown that the concept of the chemical structure of glasses allows the calculation of a wide variety of glass properties with an accuracy that is comparable to that of reliable experimental studies. The aim of the present chapter is to give an overview of the aforementioned investigations, and to generalize them in order to demonstrate how this concept can be used to predict the species and quantities of the (super)structural units expected to be present in glasses from different systems. It will also be demonstrated that this approach permits the establishment of the relationship between the different levels in glass structure, and the quantitative determination of structure-property relationships.

5.1 Introduction

For fundamental studies, it is more essential not to discover (establish) new phenomena or even regularities but rather to get insight into their nature. In other words, an investigator should not restrict himself to the question "how" but should try to get to know "why it is so". Surely, the answer to the second question will immediately entail new questions and this gradual approach to the ultimate truth is really infinite.

Evgenii Alexandrovich Porai-Koshits (1988) [1]

Glass is one of the oldest man-made materials. Although it has been produced in different parts of the world for at least 4000 years, its systematic study started only about 300 years ago. The investigation of glass properties began in the middle of the 18th century and was initiated by the works on glass coloring performed by the outstanding Russian scientist Mikhail V. Lomonosov. Studies of glass structure became possible only in the late 1920s, following the development of the X-ray diffraction technique. Subsequently, other methods (NMR, Raman and IR spectroscopy, neutron scattering, SAXS, etc.) were developed that allowed information to be obtained not only on the short-range order but also about the intermediate-range order and the longer-range fluctuations of density and composition present in glasses.

However, despite the great body of excellent experimental data on the structure of glasses accumulated in the last few decades, researchers are still unable to explain specific features they observe in various glass-forming systems. Thus, in case of the short-range order in the structure of binary borate glasses, the following long-established facts remain inexplicable in terms of the structural models known to the present authors: (i) the fraction of 4-fold coordinated boron atoms, $BØ_4^-$, becomes larger when the content of a modifying oxide increases to ~25–30 mol%; (ii) at a higher modifier content, this fraction decreases, $BØ_4^-$ units being replaced by boron-oxygen triangles with non-bridging oxygen atoms; (iii) the maximum fraction of $BØ_4^-$ units in glasses never reaches 0.5, which is the maximum value for the ambient pressure crystalline polymorphs; (iv) the fraction of boron-oxygen triangles with all bridging oxygen atoms does not depend on temperature, while those of $BØ_4^-$ tetrahedra and $BØ_2O^-$ triangles decrease and increase, respectively, with increasing temperature.

As to the next level in the structure of borate glasses (the intermediate-range order), which is characterized by superstructural units, structural studies only enable qualitative data to be obtained concerning the types of the units present and the way in which their content varies (increase or decrease) with a variation in the glass composition or temperature. The experimentally observed changes in the structure of silica-containing glasses are not explained, either. This concerns not only the data available in the literature for such complex systems as borosilicate glasses but also for more simple ones, e.g. alkali silicate glasses.

The above facts reveal that, despite the development of more and more sophisticated techniques, which allow increasingly more accurate data to be obtained on the short- and intermediate-range structure, there is still no understanding as to *why* specific basic structural units form in a given glass or melt and transform into each other as the composition or temperature vary or *how* various levels in the glass (melt) structure are related. The available experimental results also do not allow the relationship between two levels in the glass structure to be established. All this clearly points to the necessity of developing new approaches towards the interpretation of structural data, as emphasized by the quotation at the head of this section, taken from Evgenii Alexandrovich's presentation at the XV. International Congress on Glass held in Leningrad in 1989. In the present chapter, the factors responsible for the above limitations are analyzed and an approach that enables all of the above questions to be answered is proposed. It will also be shown that this approach allows the relationship between the short- and intermediate-range order as well as structure-property relationships to be determined quantitatively.

5.2 Structural Models

To explain the origin of the various basic structural units in glasses and their transformation into each other as the glass composition and temperature vary, it is necessary to look for a general approach that enables structural changes occurring in various glass-forming systems to be analyzed on some unified basis. The majority of models put forward before the early 2000s have been developed for individual systems. Therefore, they have a limited applicability and cannot predict the character of the structural changes as new components are introduced into a glass or synthesis conditions (the ambient atmosphere, temperature, the quench rate, etc.) change. For the same reason, none of the individual structural models allows an equally reliable description of properties in the systems incorporating different glass-forming oxides (e.g. sodium borate and sodium silicate glasses) or different modifying oxides (lithium to caesium borate glasses) or in a series of systems with different numbers of components (boron oxide – alkali borate glasses – alkali borosilicate glasses). These models were developed for calculating properties, mainly the densities, in individual binary systems on the basis of structural information. Rather frequently, simplified approaches were used, e.g. when an apparent similarity in the character of changes in glass properties and in structural characteristics is considered as a sign of an unambiguous relationship between them. However, the linking of such similarities may lead to erroneous conclusions concerning the influence of glass structure on properties. Although in recent years more elaborate structural approaches have been put forward for calculating a large variety of properties of multi-component (borosilicate) glasses [2, 3], it is necessary to analyze the factors responsible for the limited applicability of these previous individual models, since these factors are equally important when more general models are developed.

As an example, the densities of lithium borate glasses (Fig. 5.1) vary, as a function of composition, in the same manner as the fraction of 4-fold coordinated boron atoms, $BØ_4^-$, does in binary alkali borate systems (Fig. 5.2). Namely, the density gradually increases and reaches its maximum as the Li_2O content goes up to approximately 30 mol% and then, over the higher alkali region, the density decreases. With this similarity, it is easy to attribute the above changes in the density to an increase in the number of additional B–O bonds due to the formation of boron-oxygen tetrahedra, $BØ_4^-$, in the region with 0–30 mol% Li_2O, followed by the decrease in the number of these bonds at a higher content of lithium oxide, when the fraction of $BØ_4^-$ tetrahedra goes down. However, Fig. 5.3 shows that the composition dependence of the molar volume of lithium borate glasses has no common features with the known dependence for the units $BØ_4^-$, and this is true for the molar volumes, V_{gl}, of all other alkali borate glasses as well. This points to the fact that the molar volume of alkali borate glasses

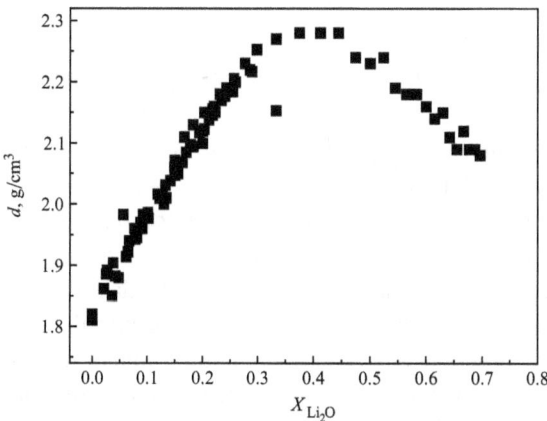

Fig. 5.1: Experimental densities of lithium borate glasses [4].

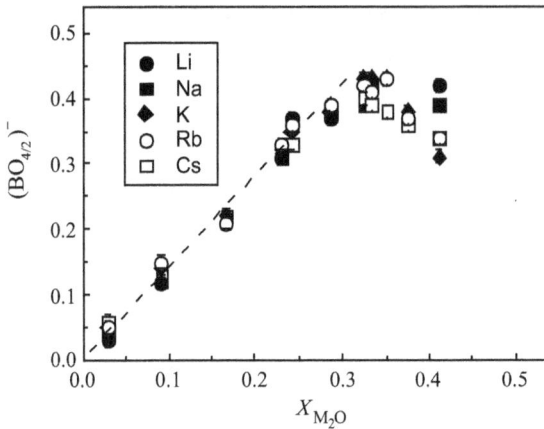

Fig. 5.2: Experimental fractions of 4-fold-coordinated boron atoms [5] in alkali borate glasses.

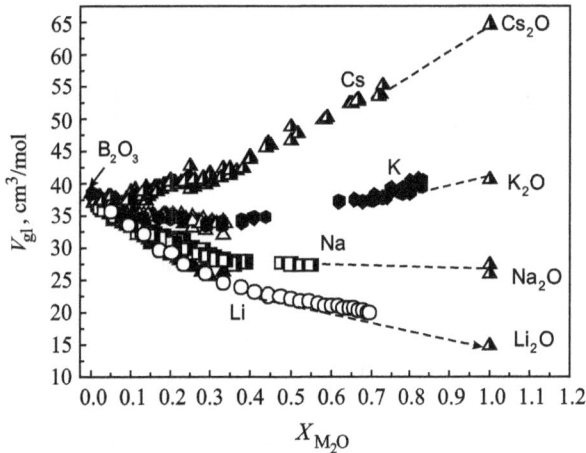

Fig. 5.3: Experimental molar volumes of alkali borate glasses [4].

cannot be explained exclusively in terms of their content of 4-fold coordinated boron atoms.

According to Fig. 5.2, over the composition region up to 30 mol% M_2O (M = Li, Na, K and Cs), the short-range order of alkali borate glasses is identical, since their content of the $BØ_4^-$ units is practically the same, irrespective of the chemical nature of the modifying oxide. Since it is commonly known that the structure of glasses determines their properties, one could expect a similarity in the molar volumes of glasses in different systems as well. However, Fig. 5.3 shows that this property noticeably depends on the type of alkali, which unambiguously points to the important role of modifying oxides for understanding the behavior of glass properties. An underestimation of this evidence is the key factor responsible for the limited success of the known structural models. Their limitations arise from the fact that they are mainly based on experimental results obtained for vitreous networks by NMR, Raman spectroscopy or neutron scattering. Since these methods frequently yield only limited information on metal-oxygen polyhedra, MO_n, the models mainly consider only boron-oxygen, silicon-oxygen, germanium-oxygen or other similar basic structural units. However, as follows from high temperature neutron diffraction [6] and Raman spectroscopy [7] studies as well as thermodynamic calculations [8], the transformations of borate structural units into each other,

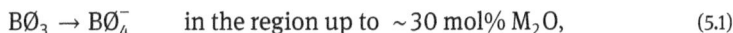

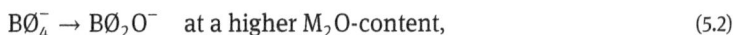

$$B\varnothing_3 \rightarrow B\varnothing_4^- \qquad \text{in the region up to } \sim 30 \text{ mol\% } M_2O, \qquad (5.1)$$

$$B\varnothing_4^- \rightarrow B\varnothing_2O^- \quad \text{at a higher } M_2O\text{-content}, \qquad (5.2)$$

proceed with endothermic effects, not exceeding approximately +50 kJ/mol. Hence, these processes cannot account for the experimentally observed exothermic effect of glass formation in the alkali borate systems [8]. In other words, if the reactions described by Eqs. (5.1) and (5.2) were the only processes taking place when the oxides M_2O and B_2O_3 interact, there would be no gain in energy in the systems considered and borate glasses (crystals, melts) would not form. In reality, the above reactions cannot be considered alone, since their progress is associated with another process, viz. the formation of additional metal-oxygen bonds in glasses as compared to metal oxides, via

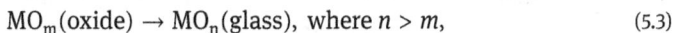

$$MO_m(\text{oxide}) \rightarrow MO_n(\text{glass}), \text{ where } n > m, \qquad (5.3)$$

which is a highly exothermic process.

As is indicated in [8], in the case of alkali metals, the enthalpy of a single M-O bond varies, depending on the nature of the metal, in the range $-(50\text{-}150)$ kJ/mol, and the number of additional M-O bonds increases in the series from Li- to Cs-borate glasses. The total exothermic effect of formation of additional bonds noticeably exceeds the endothermic contribution from the reactions Eqs. (5.1) and (5.2), the degree of this excess increasing in the series from lithium borate to caesium borate glasses. This tendency explains the experimental observation [8], according to which the enthalpy of formation of alkali borate glasses from the oxides, M_2O and B_2O_3, becomes more and more negative in the same series, Li $\rightarrow$ Cs.

From the above, it follows that structural models that do not take into account the presence of modifying oxides in glasses cannot be used either for a *rigorous* quantitative description of the experimentally observed structural changes occurring with variations in the glass (melt) composition and temperature or for quantitatively establishing structure-property relationships in glasses. This is due to the fact that these models are based on incomplete information about the glass (melt) structure, which concerns *only the atoms forming the network* of a given system (e.g. boron-oxygen units, solely, in borate glasses and melts), whilst the presence of metal-oxygen polyhedra, which are also basic structural units, is not taken into account. It should be particularly emphasized that, in terms of the energy gain, structural changes that take place in the boron–

oxygen network are of no significance whatsoever. Thus, in the structural models, the requirements of the mass and charge balance and that of the minimum Gibbs free energy of a given system are not observed. However, the violation of these laws remains unnoticed because this inadequacy is compensated for by use of adjustable parameters, either explicit or implicit, which not always have a physical meaning. Since the models are usually developed for a particular system, a description of glasses (melts) in any other system requires an elaboration of new individual models and their corresponding adjustable parameters. As a result, the structural models do not have such an important predictive ability as that discussed in the next section.

5.3 Thermodynamic Approach

All of the above flaws are successfully overcome in an approach based on the model of associated solutions, since this takes into account the presence of network-modifying cations in glasses and considers structural changes in terms of the minimum Gibbs free energy of a given system. The approach was developed in the late 1970s by Boris Shakhmatkin, initially for calculating the entire set of thermodynamic potentials of alkali borate melts [9, 10] and sodium silicate melts [11]. In [10], the first attempt was made to calculate the fraction of 4-fold coordinated boron atoms, $B\varnothing_4^-$, in lithium borate and sodium borate melts. This approach, which became internationally known in 1994 [12], uses no adjustable parameters and enables the calculation of a large variety of glass properties and structural characteristics in different systems and the solution of the problem of the structure-property relationship [13–15]. It considers glasses formed from components with different chemical nature as solutions, whose constituents are the unreacted oxides and the products of their interaction. It is assumed that (i) these products form chemical groupings similar in stoichiometry to those of the crystalline compounds existing in the phase diagram of the system in question, (ii) a structural similarity between the groupings and crystals also exists, at least, in terms of the ratio of the basic structural units, and (iii) the groupings and the unreacted oxides form an ideal solution.

The assumption about the stoichiometric similarity of the groupings and crystals is based on the known fact that the chemical interaction between substances proceeds in a similar way, i.e. with the formation of the same products, regardless of the state of aggregation (solid, liquid or gaseous) of the substances involved. In addition, as follows from the authors previous calculations made for various binary and ternary glasses and melts, it is the set of stoichiometries (similar to those of the crystalline compounds) that yields the minimum Gibbs free energy of a given system, which is a requirement of the model formalism [12, 13].

The assumption concerning the structural similarity between the groupings and the corresponding crystals also agrees with the requirement for the minimum Gibbs free energy. The experimental thermodynamic data on vitreous and crystalline al-

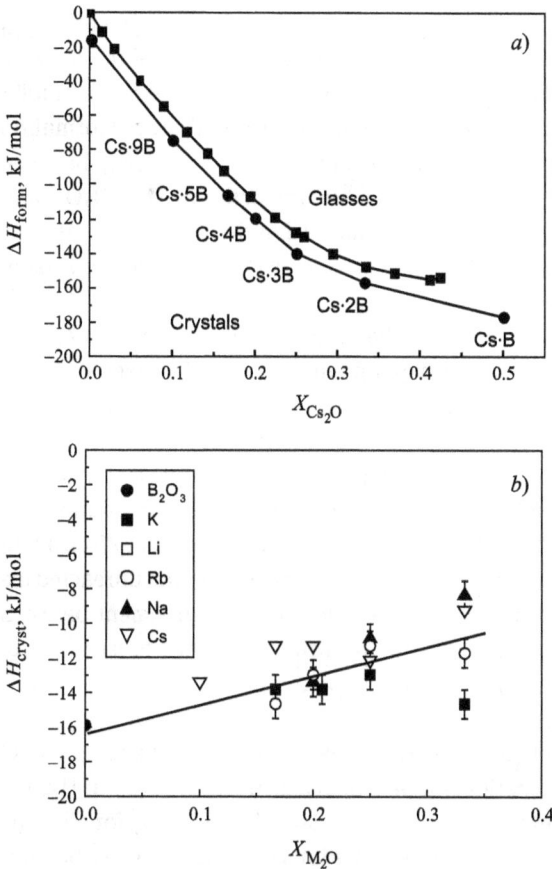

Fig. 5.4: (a) Experimental enthalpies of formation of caesium borate glasses and crystals from oxides, and (b) the enthalpies of crystallization of alkali borate glasses [16].

kali borates reported in [16] also confirm this assumption. Fig. 5.4a shows that the enthalpies of formation of caesium borate glasses and crystals, ΔH_{form}, from the respective oxides are rather close over the whole composition region considered. The difference between the enthalpies ΔH_{form} for crystals and glasses is equal to the enthalpy of crystallization for the glasses, ΔH_{cryst}. Fig. 5.4b shows these enthalpies for five systems, $M_2O-B_2O_3$ (M = Li, Na, K, Rb and Cs), the values of ΔH_{cryst} being within the range of $-(12 \pm 2)$ kJ/mol. A comparison of this value with the enthalpies of formation of glasses reveals that, over the region 15–35 mol% Cs_2O, where the thermodynamic potentials, ΔH_{form}, vary from approximately -90 kJ/mol to -145 kJ/mol, the value of ΔH_{cryst} comprises no more than 10–15 % of the enthalpy of formation of the glasses, which points to similar internal energies for glasses and crystals. Since the energy of any system unambiguously determines its structure, the above comparison provides strong support for the assumption concerning the structural similarity between glasses (considered as a solution of various chemical groupings) and the relevant crystals that form in the composition interval 15–35 mol% M_2O. In the low-alkali region, when the content of alkali oxide is below 10 mol% M_2O, the value of ΔH_{cryst} gradually becomes comparable to that of the enthalpy of formation of the glasses, comprising up to 60–80 % of the ΔH_{form} value. For glasses containing less than 2 mol% M_2O, the enthalpy of crystallization exceeds (in magnitude) the enthalpy of formation of the glasses. This undoubtedly points to the lack of structural similarity between low-alkali glasses, whose prevailing constituent is the unreacted vitreous B_2O_3, and crystalline boron oxide (B_2O_3–I). It is commonly known that the structure of vitreous B_2O_3 is characterized by the presence of boroxol groups, B_3O_6, which do not exist in the structure of crystalline B_2O_3–I.

Finally, the structural similarity between the groupings and the corresponding crystals is likewise confirmed by numerous experimental studies of glasses and crystals by X-ray and neutron scattering, NMR, EPR, IR and Raman spectroscopy (see [6–21] quoted by Shultz et al. [8] and [17, 18]). It should be noted that, taking into account the above assumptions, the set of groupings expected to be present in a given glass-forming system is chosen on the basis of the set of crystalline compounds that form in the system and can be found in its phase diagram. The third assumption, about the ideality of the solution formed from the chemical groupings and the unreacted oxides, was made *a priori*. However, numerous examples of good agreement between the calculated and experimental properties and the structural characteristics, accumulated meanwhile for a large variety of systems, strongly support the validity of this assumption.

The thermodynamic approach considered here is based on the rigorous model of associated solutions. Its mathematical formalism is described in detail in [12, 13]. Briefly, it consists in solving the set of equations for the law of mass action for all of the reactions proceeding in a given system, and the equations of the law of mass balance of the components. If sodium borate glasses (melts) are considered as an example, it is necessary to take into account the formation of the following chemical groupings, the crystalline analogues of which can be found in the phase diagram of the Na_2O–B_2O_3 system [19]: $Na_2O\cdot 9B_2O_3$, $Na_2O\cdot 5B_2O_3$, $Na_2O\cdot 4B_2O_3$, $Na_2O\cdot 3B_2O_3$, $3Na_2O\cdot 7B_2O_3$, $6Na_2O\cdot 13B_2O_3$, $Na_2O\cdot 2B_2O_3$, $Na_2O\cdot B_2O_3$, $2Na_2O\cdot B_2O_3$ as well as $3Na_2O\cdot B_2O_3$. The groupings form from oxides, according to the reaction

$$mNa_2O + nB_2O_3 = mNa_2O \cdot nB_2O_3, \tag{5.4}$$

where m is equal to 1, 2, 3 and 6, and n is equal to 1 to 5, 7 and 9, and 13, respectively. The set of equations to be solved comprises the following relationships:

$$K_i = \frac{X_{mNa_2O\cdot nB_2O_3}}{X_{Na_2O}^m \cdot X_{B_2O_3}^n}, \tag{5.5}$$

$$X^*_{Na_2O} = n_{Na_2O} + n_{Na_2O\cdot 9B_2O_3} + n_{Na_2O\cdot 5B_2O_3} + n_{Na_2O\cdot 4B_2O_3} \tag{5.6}$$

$$+ \, n_{Na_2O\cdot 3B_2O_3} + 3n_{3Na_2O\cdot 7B_2O_3} + 6n_{6Na_2O\cdot 13B_2O_3} + n_{Na_2O\cdot 2B_2O_3}$$

$$+ \, n_{Na_2O\cdot B_2O_3} + 2n_{2Na_2O\cdot B_2O_3} + 3n_{3Na_2O\cdot B_2O_3},$$

$$X^*_{B_2O_3} = n_{B_2O_3} + 9n_{Na_2O\cdot 9B_2O_3} + 5n_{Na_2O\cdot 5B_2O_3} + 4n_{Na_2O\cdot 4B_2O_3} \tag{5.7}$$

$$+ \, 3n_{Na_2O\cdot 3B_2O_3} + 7n_{3Na_2O\cdot 7B_2O_3} + 13n_{6Na_2O\cdot 13B_2O_3} + 2n_{Na_2O\cdot 2B_2O_3}$$

$$+ \, n_{Na_2O\cdot B_2O_3} + n_{2Na_2O\cdot B_2O_3} + n_{3Na_2O\cdot B_2O_3},$$

$$K = \exp\left(-\frac{\Delta G^0_f}{RT}\right), \tag{5.8}$$

$$X_i = \frac{n_i}{\Sigma n}. \tag{5.9}$$

Equation (5.5) is the law of mass action written in an ideal form for reaction Eq. (5.4). The symbol K_i represents the equilibrium constant of reaction Eq. (5.4), $X_{mNa_2O\cdot nB_2O_3}$, $X^m_{Na_2O}$ and $X^n_{B_2O_3}$ are the equilibrium concentrations of the above chemical groupings and the unreacted oxides.

The relationships Eqs. (5.6) and (5.7) are the equations of the law of mass balance, where the symbols $n_{mNa_2O\cdot nB_2O_3}$, n_{Na_2O} and $n_{B_2O_3}$ depict, respectively, the numbers of moles of the chemical groupings and the unreacted oxides in a given glass (melt), and $X^*_{Na_2O}$ and $X^*_{B_2O_3}$ represent its analytical composition in mole fractions. Equation (5.8) is used for calculating the equilibrium constants, K_i, on the basis of the standard Gibbs free energies of formation from oxides, ΔG^0_f, of the compounds existing in the system. These potentials are obtained using the data from reference books, hence avoiding the use of adjustable parameters. Equation (5.9) describes the relationship between a mole fraction, X_i, and the numbers of moles, n_i, of the species i in a given glass or melt. The index i refers both to the chemical groupings and the unreacted oxides. The symbol Σn denotes the total number of their moles. The solution of the set of equations Eqs. (5.5)–(5.8) yields information on the numbers of moles of the chemical groupings and the unreacted oxides, which enables the chemical structure of glasses (melts) to be determined. Note that, as it is shown in [20], at constant temperature, pressure and numbers of moles of the initial substances, the above set of nonlinear equations has only one solution.

5.4 Concept of Chemical Structure

The concept of chemical structure is the corner-stone of the approach analyzed here and hence deserves a more detailed consideration. The notion of the chemical structure of glasses formed from components with different chemical nature implies the content of the various chemical groupings that are the products of the chemical interaction between the constituent oxides, together with the unreacted oxides themselves.

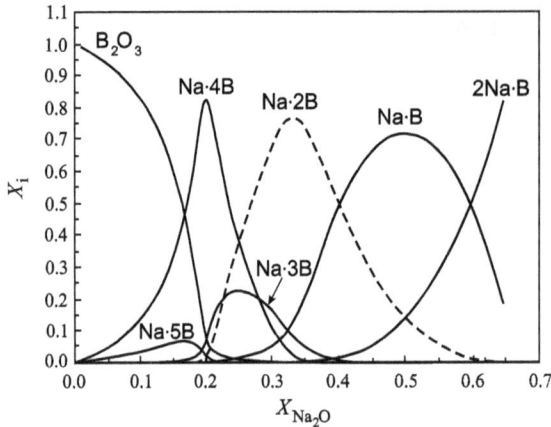

Fig. 5.5: Chemical structure of sodium borate glasses at 800 K. The ratio of the oxides, Na_2O and B_2O_3, in the chemical groupings is denoted by mNa·nB.

The content of all of the species can be presented as absolute values, i.e. in terms of the numbers of moles, n_i, or as relative values, the molar fractions, X_i, determined by Eq. (5.9).

As an example, we consider the chemical structure of sodium borate glasses at 800 K shown as relative values, X_i, in Fig. 5.5. Note that, although the formation of the groupings $Na_2O\cdot9B_2O_3$, $3Na_2O\cdot7B_2O_3$, $6Na_2O\cdot13B_2O_3$ and $3Na_2O\cdot B_2O_3$ is taken into consideration by introducing the relevant equations into Eqs. (5.5)–(5.9), their content in glasses is so low (much less than 1 %) that these species are not shown in the graph. It is also seen that, over the low-alkali region of 0–10 mol% Na_2O, the unreacted B_2O_3, whose content (fraction) varies from 1.0 to 0.8, dominates the chemical structure. At a higher Na_2O content, e.g. 20 mol%, the fraction of B_2O_3 becomes as low as ~0.08, and $Na_2O\cdot4B_2O_3$ becomes the dominating grouping ($X_{Na_2O\cdot4B_2O_3}$ ~0.8). Its structural similarity to the relevant crystal is confirmed by the value of the ratio $(\Delta H_{cryst}/\Delta H_{form}) \approx 0.12$ between the enthalpy of crystallization and the enthalpy of formation of the glass [16]. The above clearly explains the poorly pronounced crystallization ability of low-alkali glasses. In view of a significant difference between the structures of vitreous and crystalline B_2O_3–I (no boroxol rings in the latter), the process of crystallization requires a substantial re-arrangement of the structural units in glasses with a low alkali content. A larger degree of structural similarity between glasses and relevant crystals favors crystallization. For illustration, in [21], the bulk crystallization of vitreous $Cs_2O\cdot9B_2O_3$ required as long as 550 h, whilst in the case of the glasses $Cs_2O\cdot4B_2O_3$ and $Na_2O\cdot4B_2O_3$, this process took about 100 h.

Note that the chemical groupings considered in this approach are formed from the basic structural units, the ratio of these units being similar to that in the crystalline compounds with the same stoichiometry. The observance of the principle of the minimum Gibbs free energy of a system in the vitreous state requires that in the groupings the basic structural units are combined together in the same manner as those in the relevant crystalline compounds. In other words, it can be expected that the chemical groupings present in glasses comprise associates similar to the superstructural units present in the crystals that form in the system under consideration. Hence, the concept of the chemical structure enables both the short-range order and the intermediate-range order in the glass structure to be described, at the levels of the basic structural units and superstructural units, respectively. Table 5.1 shows which (super)structural units and in what quantities are included in the chemical groupings present in glasses of the systems $Li_2O-B_2O_3$, $Na_2O-B_2O_3$, $Na_2O-B_2O_3-SiO_2$. This information is based on the data reported in [17, 22, 23]. The basic structural units present in glasses of the system Na_2O-SiO_2 are also shown in Table 5.1 since the short-range order in the structure of these glasses can similarly be calculated using the same approach.

Table 5.1: The chemical groupings present in glasses of the systems $Li_2O-B_2O_3$, $Na_2O-B_2O_3$, $Na_2O-B_2O_3-SiO_2$, and Na_2O-SiO_2, and their relation to the short-range and intermediate-range order in the glass structure

Chemical groupings	Types and numbers of the basic structural units and superstructural units introduced into the glasses by 1 mole of each chemical grouping	
	Basic structural units	Superstructural units
Lithium borate, sodium borate and sodium boroslicate glasses		
B_2O_3	$2BØ_3$	Boroxol ring (½), $BØ_3$ (½)
$Na_2O \cdot 9B_2O_3$*)	$2[BØ_4]^-$, $16BØ_3$	Boroxol rings (4), Triborate rings (2)
$Na_2O \cdot 5B_2O_3$	$2[BØ_4]^-$, $8BØ_3$	Pentaborate rings (2)
$Na_2O \cdot 4B_2O_3$	$2[BØ_4]^-$, $6BØ_3$	Pentaborate ring (1), Triborate ring (1)
$Li_2O \cdot 3B_2O_3$ & $Na_2O \cdot 3B_2O_3$	$2[BØ_4]^-$, $4BØ_3$	Triborate rings (2)
$Li_2O \cdot 2B_2O_3$ & $Na_2O \cdot 2B_2O_3$	$2[BØ_4]^-$, $2BØ_3$	Diborate ring (1)
$Li_2O \cdot B_2O_3$ $Na_2O \cdot B_2O_3$	$2BØ_2O^-$	Chains of triangles Cyclic metaborate ring (²/₃)
$2Li_2O \cdot B_2O_3$ & $2Na_2O \cdot B_2O_3$	$2BØO_2{}^{2-}$	None
$Na_2O \cdot B_2O_3 \cdot 2SiO_2$	$2[BØ_4]^-$, $2Q^4$	Danburite ring(1)
$Na_2O \cdot B_2O_3 \cdot 6SiO_2$	$2[BØ_4]^-$, $6Q^4$	Reedmergnerite rings (2)
Sodium silicate glasses		
SiO_2	Q^4	
$3Na_2O \cdot 8\ SiO_2$	$2Q^4$, $6Q^3$	
$Na_2O \cdot 2SiO_2$	$2Q^3$	None
$Na_2O \cdot SiO_2$	Q^2	
$3Na_2O \cdot 2SiO_2$	$2Q^1$	
$2Na_2O \cdot SiO_2$	Q^0	

(*) The presence of the grouping $Na_2O \cdot 9B_2O_3$ becomes noticeable at temperatures below 800 K. As an example, at 700 K, its maximum content is 2%.

The concept of the chemical structure explains the difference between the structures of glasses and crystals in terms of the model of associated solutions. The model considers glasses as a superposition of various chemical groupings, whose stoichiometries correspond to those of the crystalline compounds forming in the given system. All of these groupings are present in any glass of the system but in different quantities that depend on the glass composition. As a result, in glasses of stoichiometric compositions, the characteristic grouping is always present together with other species. Fig. 5.5 shows that in sodium diborate glass the content of the grouping $Na_2O \cdot 2B_2O_3$ is 76 %, the other species being $Na_2O \cdot 3B_2O_3$ (9 %), $Na_2O \cdot B_2O_3$ (12 %) and $Na_2O \cdot 4B_2O_3$ (3 %). In some cases, the characteristic grouping does not dominate the chemical structure; e.g. in sodium triborate glass the content of the grouping $Na_2O \cdot 3B_2O_3$ is only 22 %, whilst that of the groupings $Na_2O \cdot 4B_2O_3$ and $Na_2O \cdot 2B_2O_3$ is noticeably higher (38 % each). It can be assumed that, being an incongruently melting compound, $Na_2O \cdot 3B_2O_3$ is thermodynamically less stable than the congruently melting borates $Na_2O \cdot 4B_2O_3$ and $Na_2O \cdot 2B_2O_3$ [19], and hence the presence of a large amount of the grouping $Na_2O \cdot 3B_2O_3$ in glasses is energetically unprofitable for the system. The above examples demonstrate the degree to which the structure of glasses, in terms of the mixture of superstructural units, is more complex than that of crystals.

According to Tables 4 and 5 in [17], the unit cells of various borate crystals include no more than two superstructural units, which co-exist, in a few cases, with single basic structural units, $B\emptyset_3$ or $B\emptyset_4^-$. From Fig. 5.5 it follows that the range of the superstructural units in glasses is noticeably larger. This is illustrated quantitatively in Table 5.2, for the diborate composition, using information on the types and quantities of the chemical groupings forming in this glass, together with information on the presence of superstructural units in the groupings (Table 5.1) and in crystalline sodium diborate, $Na_2O \cdot 2B_2O_3$ [17]. In terms of the energy, the difference between the vitreous and crystalline states is characterized by the enthalpy of crystallization (Fig. 5.4b).

Table 5.2: Comparison of the structure, at the superstructural units level, in crystalline and vitreous sodium diborate, $Na_2O \cdot 2B_2O_3$

Crystal	Glass, 800 K	
Superstructural units	Superstructural units	From what groupings
Di-pentaborate 50 %	Diborate* 70 %	$Na_2O \cdot 2B_2O_3$*
Di-triborate (NBO) 50 %	Triborate 20 %	$Na_2O \cdot 3B_2O_3$, $Na_2O \cdot 4B_2O_3$
	Cyclic metaborate 7 %	$Na_2O \cdot B_2O_3$
	Pentaborate 3 %	$Na_2O \cdot 4B_2O_3$

* The presence of the diborate superstructural unit introduced by the grouping $Na_2O \cdot 2B_2O_3$ into the glass is explained in section "Short-range order".

5.5 Short-range Order

5.5.1 Na$_2$O–B$_2$O$_3$ Glasses

Due to the knowledge of the equilibrium concentrations of the chemical groupings present and information on the types and numbers of the basic structural units introduced by these groupings into glasses, the short-range order in the structure can be calculated as a function of the glass composition and temperature. As an illustration, consider the structural changes in sodium borate glasses in terms of the fractions of the basic structural units, BØ$_4^-$, BØ$_3$, BØ$_2$O$^-$, and BØO$_2^{2-}$, as predicted by their chemical structure.

The model concentration dependencies of the fractions of the basic structural units at 800 K are shown in Fig. 5.6, together with the experimental data obtained by NMR [5, 24] and Raman spectroscopy [25]. For all of the units, the inaccuracy of calculations does not exceed ±0.05. It is important to note that, when the calculations were performed assuming that the Na$_2$O·2B$_2$O$_3$ chemical grouping introduces BØ$_2$O$^-$ asymmetric triangles into the glasses, as predicted by the crystal structure of α-Na$_2$O·2B$_2$O$_3$, the model fractions of BØ$_4^-$ units were noticeably lower than the experimental values [5, 24, 25]. The largest disagreement was observed at 33.3 mol% Na$_2$O, where the fraction of BØ$_4^-$ units was as low as ~ 0.34. The deviation from the highly precise experimental values reported in [5, 25] was equal to ~ 0.09, which was larger than the combined experimental and calculation errors. This suggests that the ratio of the basic structural units in the Na$_2$O·2B$_2$O$_3$ grouping differs from that in α-Na$_2$O·2B$_2$O$_3$,

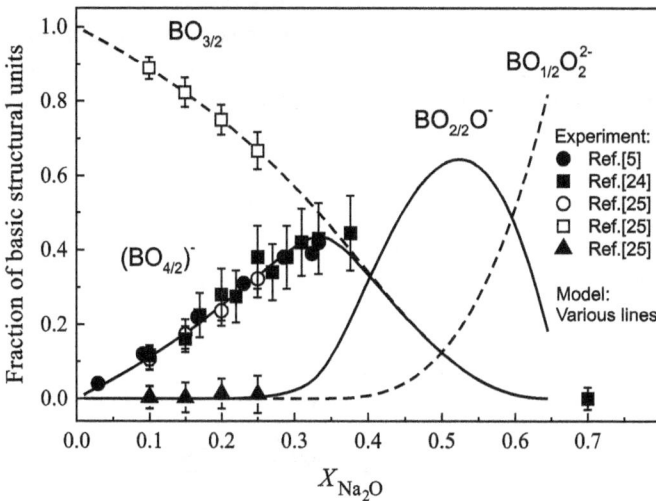

Fig. 5.6: Concentration dependencies of the basic structural units in sodium borate glasses at 800 K: model (lines) and experiment (symbols).

where it is $1.5B\emptyset_4 : 2B\emptyset_3 : 0.5B\emptyset_2O^-$ [17]. The calculations made on the assumption that the $Na_2O \cdot 2B_2O_3$ grouping contributes basic structural units in the ratio $2B\emptyset_4^- :$ $2B\emptyset_3$, as in the diborate superstructural unit, yields results that are in good agreement with the experimental data [5, 24, 25]. This indicates that the diborate grouping is unlikely to introduce asymmetric triangles into sodium borate glasses. The difference between the structures of vitreous and crystalline α-$Na_2O \cdot 2B_2O_3$ is also confirmed by NMR spectroscopy [26]. Note that, due to a negligibly small content of the groupings $Na_2O \cdot 9B_2O_3$, $3Na_2O \cdot 7B_2O_3$, $6Na_2O \cdot 13B_2O_3$ and $3Na_2O \cdot B_2O_3$ in the glasses at 800 K, their contribution to the structure tends to zero.

Fig. 5.5 shows that, over the region of 0–35 mol% Na_2O, the role of the groupings $Na_2O \cdot 5B_2O_3$ and $Na_2O \cdot 3B_2O_3$ in the chemical structure is insignificant as compared to that of the groupings $Na_2O \cdot 4B_2O_3$ and $Na_2O \cdot 2B_2O_3$. The content of two latter groupings in the glasses is very similar, whilst the fraction of 4-fold-coordinated boron atoms is higher in $Na_2O \cdot 2B_2O_3$. For this reason, a gradual replacement of tetraborate by diborate groupings, as the content of Na_2O in glasses increases up to 33 mol%, leads to an increase in the fraction of $B\emptyset_4^-$ units. This reaches the maximum value of ~ 0.43 in the composition with 33 mol% Na_2O and starts to decrease at a higher content of sodium oxide. This is explained by the formation of metaborate groupings, $Na_2O \cdot B_2O_3$, in the respective glasses. They do not include 4-fold-coordinated boron atoms but introduce non-bridging oxygen atoms. Metaborate groupings form concurrently with the rest of the species and, as soon as their presence in the system becomes noticeable (which is observed at a Na_2O content close to 30 mol%), the formation of the groupings containing 4-fold-coordinated boron atoms slows down, this process becoming more pronounced as the content of $Na_2O \cdot B_2O_3$ groupings increases. This slowing down explains why the maximum fraction of $B\emptyset_4^-$ in borate glasses is normally lower than that in corresponding crystals, where the maximum value of 0.5 is found in all of the known diborate crystals, except for the case of α-$Na_2O \cdot 2B_2O_3$ [17]. In this polymorph, the fraction of $B\emptyset_4^-$ is 0.375, whereas in $Na_2O \cdot B_2O_3$ glasses its maximum value is ~ 0.43.

The model dependencies, shown in Fig. 5.6, are calculated employing Eqs. (5.10)–(5.14). They present the ratios between the numbers of boron-oxygen polyhedra of each type (the numerators of Eqs. (5.10)–(5.14)) and the total number of all borate units, tetrahedra and triangles with and without non-bridging oxygen atoms present in a given glass (Eq. (5.14)).

$$[B\emptyset_4^-] = \frac{2(X_{Na_2O \cdot 5B_2O_3} + X_{Na_2O \cdot 4B_2O_3} + X_{Na_2O \cdot 3B_2O_3} + X_{Na_2O \cdot 2B_2O_3})}{\Sigma B_n}, \tag{5.10}$$

$$[B\emptyset_3] = \frac{2(X_{B_2O_3} + 4X_{Na_2O \cdot 5B_2O_3} + 3X_{Na_2O \cdot 4B_2O_3} + 2X_{Na_2O \cdot 3B_2O_3} + X_{Na_2O \cdot 2B_2O_3})}{\Sigma B_n}, \tag{5.11}$$

$$[B\emptyset_2O^-] = \frac{2X_{Na_2O \cdot B_2O_3}}{\Sigma B_n}, \tag{5.12}$$

$$[B\emptyset O_2^{2-}] = \frac{2X_{2Na_2O \cdot B_2O_3}}{\Sigma B_n}, \tag{5.13}$$

$$\Sigma B_n = 2X_{B_2O_3} + 10X_{Na_2O \cdot 5B_2O_3} + 8X_{Na_2O \cdot 4B_2O_3} + 6X_{Na_2O \cdot 3B_2O_3} \tag{5.14}$$
$$+ 4X_{Na_2O \cdot 2B_2O_3} + 2X_{Na_2O \cdot B_2O_3} + 2X_{2Na_2O \cdot B_2O_3}.$$

5.5.2 Li_2O–B_2O_3 Glasses and Melts

Using the system Li_2O–B_2O_3 as an example, it is shown that the concept of the chemical structure not only allows the distribution of the basic structural units in glasses and melts to be reliably calculated but also some specific features of the intermediate-range order in the glass/melt structure to be better understood.

Reference [25] yields the most extensive set of numerical experimental data on the short-range order in lithium borate glasses and melts as obtained by Raman spectroscopy over extended composition and temperature ranges. This work provides a unique chance to check quantitatively the reliability of the thermodynamic modeling at high temperatures against experimental results. The calculation of the chemical structure of lithium borate glasses and melts has been performed taking into account the presence of the following chemical groupings [19]: $Li_2O \cdot 3B_2O_3$, $3Li_2O \cdot 7B_2O_3$, $Li_2O \cdot 2B_2O_3$, $Li_2O \cdot B_2O_3$, $3Li_2O \cdot 2B_2O_3$, $2Li_2O \cdot B_2O_3$ and $3Li_2O \cdot B_2O_3$. The calculations have been carried out for six temperatures, 700, 800, 1000, 1200, 1300 and 1400 K, which cover the interval considered in [25]. However, for the sake of clarity, Fig. 5.7

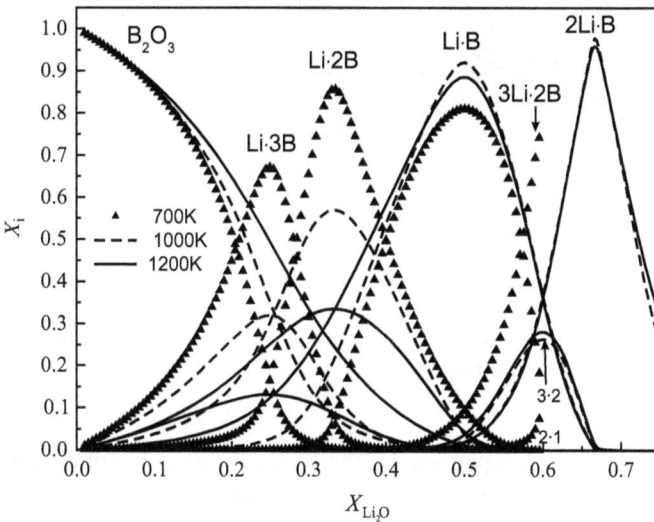

Fig. 5.7: Chemical structure of lithium borate glasses and melts, at 700, 1000 and 1200 K. The ratio of the oxides, Li_2O and B_2O_3, in the chemical groupings is denoted as mLi·nB.

only shows the results for 700, 1000 and 1200 K. Note that, at all of the temperatures considered, the content of the groupings $3Li_2O \cdot 7B_2O_3$ and $3Li_2O \cdot B_2O_3$ is so low (much less than 1 %) that their role in the chemical structure is negligible.

The knowledge of the chemical structure of glasses and melts, together with structural information from Table 5.1, enable the calculation of the content of the basic structural units present in them as a function of temperature and composition. The set of equations used for these calculations is similar to Eqs. (5.10)–(5.14). The inaccuracy of calculating the fractions of the basic structural units does not exceed ±0.05. Fig. 5.8 shows the results of calculations for $BØ_4^-$ units in glasses and melts containing 0–36 mol% Li_2O, together with the available experimental data obtained by NMR [5], Raman spectroscopy [25] and neutron diffraction [27]. It is seen that, at all of the temperatures considered, the calculated dependencies for glasses (700 and 800 K) and melts (1000–1400 K) are in good agreement with the experimental data. The largest deviation of the model value from the experimental result is observed only in the case of the melt containing 25 mol% Li_2O. However, even here the model dependence calculated at 1400 K agrees with the experimental point at 1373 K, within the limits of the calculation error.

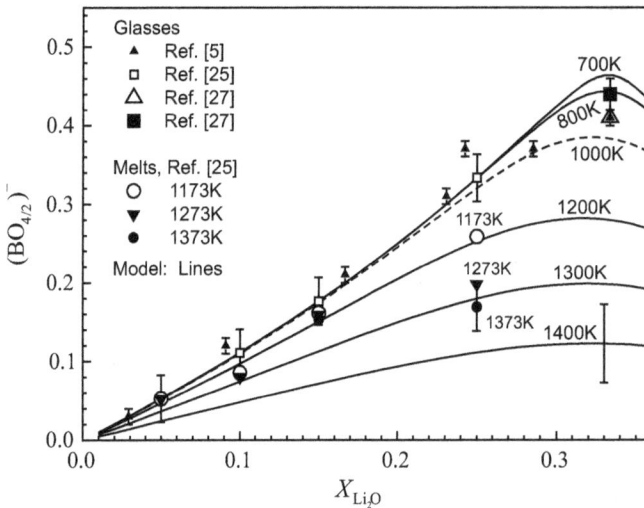

Fig. 5.8: Model and experimental fractions of 4-fold coordinated boron atoms in lithium borate glasses and melts: model (lines) and experiment (symbols).

As is seen from Fig. 5.8, the model dependencies reproduce the experimentally observed tendency of a decreasing fraction of 4-fold coordinated boron atoms with increasing temperature. This dependence can also be explained using the concept of the chemical structure. It should be remembered that, in any system, whose oxide components have different chemical nature, the formation of chemical groupings is an exothermic process, i.e. it proceeds with a release of heat (see Fig. 5.4a). From the experimental enthalpies of formation of lithium borates [16], it follows that, for example, lithium diborate forms from oxides, according to reaction

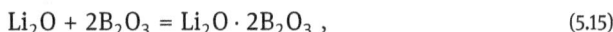

$$Li_2O + 2B_2O_3 = Li_2O \cdot 2B_2O_3 ,$$ (5.15)

the enthalpy of the process being equal to -272 kJ/mol. Note that the grouping can form in a different way, e.g. by interaction between other groupings, whose crystalline analogues are adjacent to $Li_2O \cdot 2B_2O_3$ in the phase diagram of the system:

$$Li_2O \cdot 3B_2O_3 + Li_2O \cdot B_2O_3 = 2 (Li_2O \cdot 2B_2O_3).$$ (5.16)

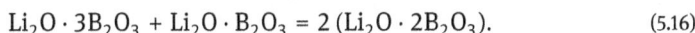

This process is noticeably less exothermic, its enthalpy being equal to approximately -42 kJ/mol [16]. It is known that, with increasing temperature, the equilibrium of exothermic processes shifts towards the initial components, either the oxides Eq. (5.15) or the groupings Eq. (5.16). Hence, in both cases, the content of the diborate grouping in the chemical structure decreases with increasing temperature. Fig. 5.7 shows that this is also true for the grouping $Li_2O \cdot 3B_2O_3$, which also brings $B\varnothing_4^-$ units into the glasses (melts). This explains the observed lower content of 4-fold-coordinated boron atoms at higher temperatures.

From the above, it follows that, as the temperature increases, the chemical groupings tend to dissociate either into the oxides or into some other groupings, the enthalpies of these processes being equal in value but opposite in sign to the enthalpies of the reactions Eqs. (5.15) and (5.16). The noticeable difference between these enthalpies points to the fact that one mechanism of dissociation is more probable than the other. As is indicated in [16], in the case of alkali borates, the difference between the standard Gibbs free energy and the standard enthalpy is insignificant. This enables the equilibrium constants of both processes to be estimated with a considerable degree of accuracy, using the equation:

$$K = \exp\left(-\frac{\Delta H^0}{RT}\right).$$ (5.17)

The estimation made at 800 K unambiguously indicates that it is the dissociation into the groupings that is more preferable for the system. The equilibrium constant for this process is about $1.8 \cdot 10^{-3}$, whilst that for dissociation into the oxides is much lower, being $1.6 \cdot 10^{-18}$. This fact strongly supports the suggestion made in [28] concerning the possibility of the interconversion of superstructural units as the temperature increases.

The knowledge of the chemical structure as a function of temperature provides an explanation for the different behavior of $B\varnothing_3$ units as compared to that of $B\varnothing_4^-$ and $B\varnothing_2O^-$. According to a high-temperature study of the sodium borate system by Raman spectroscopy [7], the only transformation of the basic structural units that takes place with increasing temperature is

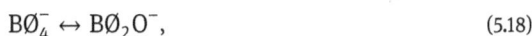

$$B\varnothing_4^- \leftrightarrow B\varnothing_2O^-,$$ (5.18)

whilst the content of $BØ_3$ units remains constant over an extended temperature interval. This fact is explained in [7] by the redistribution of $BØ_3$ triangles between boroxol rings, which are subject to the destruction as the temperature increases, and the random boron-oxygen network. This process is attributed to structural changes at the level of the intermediate-range order.

A typical picture of the temperature changes in the short-range order of borate glasses and melts is given in Fig. 5.9, by example of the lithium borate system. The model dependencies agree with the experimental observation that, as the temperature increases from 700 to 1200 K, the fraction of $BØ_2O^-$ units grows, at the cost of 4-fold coordinated boron atoms, while the content of the units $BØ_3$ remains constant. The reliability of the modeling is also confirmed by good quantitative agreement between the calculated dependencies at 1200 K and the experimental results at 1173 K, for the units $BØ_4^-$ and $BØ_2O^-$, as well as by the fact that the dependence for $BØ_3$ units is temperature insensitive and agrees with the available experimental data. The changes in the short-range order, given in Fig. 5.9, are determined by variations at the level of the intermediate-range order, as shown in Fig. 5.7, and, numerically, in Table 5.3. In the latter, the signs + and − denote an increase and decrease in the fractions of the chemical groupings with increasing temperature, respectively. It is seen that, for both compositions, the respective characteristic groupings, $Li_2O \cdot 3B_2O_3$ and $Li_2O \cdot 2B_2O_3$, dissociate

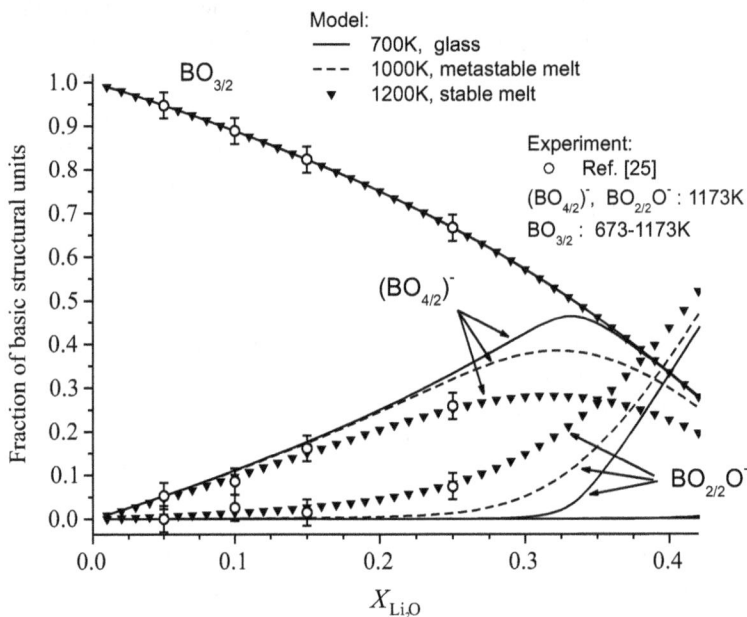

Fig. 5.9: Model and experimental fractions of various basic structural units in lithium borate glasses and melts.

Table 5.3: System $Li_2O-B_2O_3$. Influence of temperature on the structure at the superstructural unit level

Glass composition and temperature intervals	Changes in the fractions of the chemical groupings			
	$\Delta X_{B_2O_3}$	$\Delta X_{Li_2O\cdot3B_2O_3}$	$\Delta X_{Li_2O\cdot2B_2O_3}$	$\Delta X_{Li_2O\cdot B_2O_3}$
$0.25Li_2O\cdot0.75B_2O_3$ $700 \rightarrow 1000$ K	+ 0.189	−0.352	+ 0.138	+ 0.025
$0.33Li_2O\cdot0.67B_2O_3$ $1000 \rightarrow 1200$ K	+ 0.164	−0.045	−0.234	+ 0.115

giving rise to other groupings, i.e.,

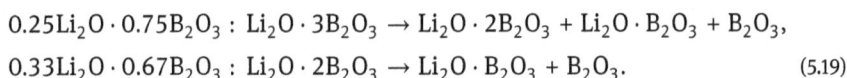

$$0.25Li_2O \cdot 0.75B_2O_3 : Li_2O \cdot 3B_2O_3 \rightarrow Li_2O \cdot 2B_2O_3 + Li_2O \cdot B_2O_3 + B_2O_3,$$
$$0.33Li_2O \cdot 0.67B_2O_3 : Li_2O \cdot 2B_2O_3 \rightarrow Li_2O \cdot B_2O_3 + B_2O_3. \tag{5.19}$$

Note that, at the second composition, the presence of the grouping $Li_2O\cdot3B_2O_3$ is neglected for simplicity, since the decrease in its content is of the order of the calculation error for the chemical structure, which is $\pm(0.03-0.05)$.

The data given in Table 5.3 numerically demonstrate to what an extent and into what species the groupings $Li_2O\cdot3B_2O_3$ and $Li_2O\cdot2B_2O_3$ dissociate with increasing temperature. Reactions Eqs. (5.19) show that, for both compositions, the metaborate grouping, $Li_2O\cdot B_2O_3$ (the source of asymmetric triangles, $B\emptyset_2O^-$) appears due to the dissociation of groupings that introduce $B\emptyset_4^-$ units. Hence, these reactions can be considered as a representation of the process, described by Eq. (5.18), in a molecular form. Special attention should be paid to the fact that, at both compositions, B_2O_3 is present among the products of dissociation. This points to the fact that the thermal behavior of $B\emptyset_3$ units is more complicated than it follows from experimental studies of the short-range order [7], i.e. the content of symmetric triangles in melts increases not only due to the thermal destruction of boroxol rings but also of other superstructural units. Since Raman spectroscopy and thermodynamic modeling (in the variant used in this case) do not differentiate $B\emptyset_3$ units included into boroxol or other superstructural units, both experimental and model results refer to the total number of symmetric triangles. Due to this, the concentration dependencies of $B\emptyset_3$ units become undistinguishable at different temperatures. This leads to the conclusion that the consideration at the superstructural unit level yields a more detailed and realistic picture of changes that occur in the glass (melt) structure as a function of temperature.

5.5.3 Na$_2$O–SiO$_2$ Glasses

The experimentally observed changes in the short-range order in the structure of sodium silicate glasses shown in Fig. 5.10 can also be explained using the concept of the chemical structure. According to the phase diagram of the system Na_2O-SiO_2 [30], the following polymorphs form: $3Na_2O \cdot 8SiO_2$, $Na_2O \cdot 2SiO_2$, $Na_2O \cdot SiO_2$, $3Na_2O \cdot 2SiO_2$ and $2Na_2O \cdot SiO_2$. Hence, the presence of the chemical groupings with similar stoichiometries can be expected in glasses. Fig. 5.11 shows that an increase in the Na_2O content in sodium silicate glasses leads to a decrease in the content of the unreacted silica, this process being accompanied by the formation of the groupings $3Na_2O \cdot 8SiO_2$ and $Na_2O \cdot 2SiO_2$ in the medium-alkali region (20−40 mol% Na_2O), which are replaced by $Na_2O \cdot SiO_2$, $3Na_2O \cdot 2SiO_2$ and $2Na_2O \cdot SiO_2$ groupings in the high-alkali region (over 40 mol% Na_2O). From Figs. 5.10 and 5.11, it is seen that the order in which the groupings replace each other in the chemical structure is in agreement with the order in which SiO_4 tetrahedra, Q^n, with decreasing numbers of bridging oxygen atoms form in sodium silicate glasses over the composition region of 0−60 mol% Na_2O. The fractions of various SiO_4 tetrahedra are calculated, with an error not exceeding ±0.05, using the following equations:

$$Q^4 = \frac{X_{SiO_2} + 2X_{3Na_2O \cdot 8SiO_2}}{\Sigma Q^n}, \tag{5.20}$$

$$Q^3 = \frac{2(3X_{3Na_2O \cdot 8SiO_2} + X_{Na_2O \cdot 2SiO_2})}{\Sigma Q^n}, \tag{5.21}$$

$$Q^2 = \frac{X_{Na_2O \cdot SiO_2}}{\Sigma Q^n}, \quad Q^1 = \frac{2X_{3Na_2O \cdot 2SiO_2}}{\Sigma Q^n}, \quad Q^0 = \frac{X_{2Na_2O \cdot SiO_2}}{\Sigma Q^n}, \tag{5.22}$$

$$\Sigma Q^n = X_{SiO_2} + 8X_{3Na_2O \cdot 8SiO_2} + 2X_{Na_2O \cdot 2SiO_2} + X_{Na_2O \cdot SiO_2}$$
$$+ 2X_{3Na_2O \cdot 2SiO_2} + X_{2Na_2O \cdot SiO_2}. \tag{5.23}$$

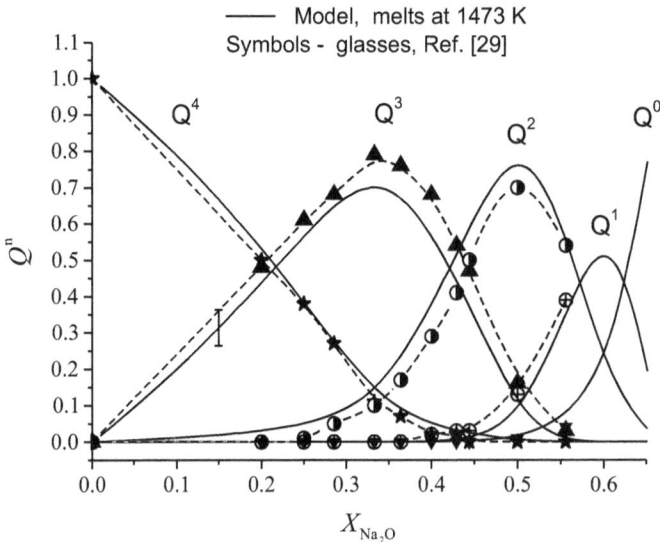

Fig. 5.10: Model and experimental fractions of Q^n species in sodium silicate glasses and melts.

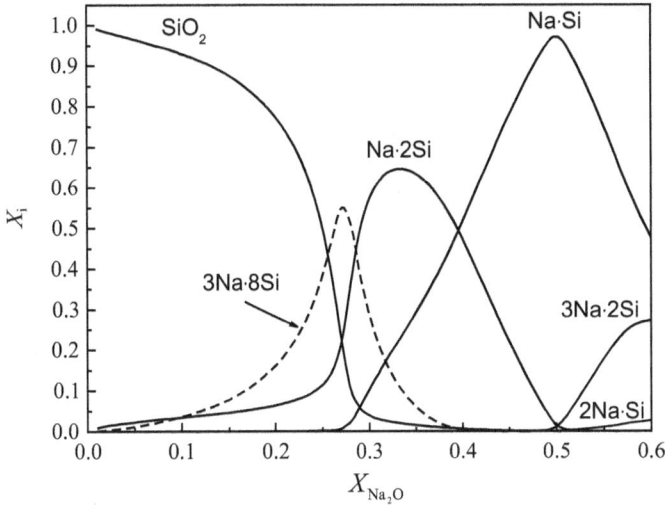

Fig. 5.11: Chemical structure of sodium silicate glasses. The ratio of the oxides, Na_2O and SiO_2, in the chemical groupings is denoted as mNa·nSi.

5.5.4 $Na_2O-B_2O_3-SiO_2$ Glasses

As it is shown in [14, 15], the approach considered also enables the structure of ternary glasses to be calculated. When borosilicate systems are concerned, the set of equations to be solved for determining the chemical structure has to include not only the equations for the formation of binary borate and silicate groupings but also those for the formation of ternary borosilicate groupings, e.g. in the case of sodium borosilicate glasses, these are the groupings $Na_2O·B_2O_3·2SiO_2$ and $Na_2O·B_2O_3·6SiO_2$. Fig. 5.12 shows the concentration dependence of 4-fold coordinated boron atoms in the cut with the constant content of Na_2O equal to 20 mol% [14], where the model dependence is in agreement with the available experimental data for the borate sub-network. In the same paper, Table 1 demonstrates that, in the cut with the equal content of the oxides Na_2O and B_2O_3, the calculated and experimental Q^n values agree within the limits of the total calculation and experimental error.

5.6 Intermediate-Range Order

It should be noted that the scale of the short-range order does not reflect all of the complexity of the structure of B_2O_3-containing glasses. A more detailed picture is given by consideration of the glass structure in terms of the superstructural units, i.e. beyond the first co-ordination sphere of the boron atoms. As is indicated in [17], "*superstructural units comprise well defined arrangements of the basic borate structural units, with*

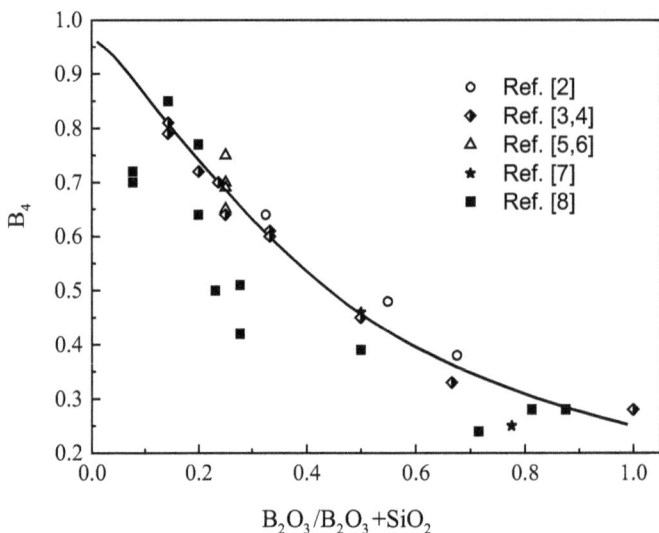

Fig. 5.12: Model and experimental fractions of 4-fold coordinated boron atoms in sodium borosilicate glasses, in the cut with the constant Na_2O content equal to 20 mol%: model (line) and experiment (symbols). After [14], where Refs. [2–8] are quoted.

no internal degrees of freedom in the form of variable bond or torsion angles". Various types of borate superstructural units, including those mentioned in Table 5.1, can be seen in Figs. 8–12 in [17], and two borosilicate rings, danburite and reedmergnerite ones, are shown in Fig. 1(e,f) in [23]. High-temperature Raman spectroscopy studies [7, 28] have revealed that, on quenching sodium borate melts, the re-arrangements at the level of the short-range order are "frozen-in" at temperatures noticeably higher than T_g, whilst structural changes at the level of the superstructural units become "frozen-in" at temperatures very close to T_g. This observation is very important, since it points to the fact that the structure of glasses, and hence their properties, are mainly determined by the superstructural units present. However, all of the experimental data, available in the literature, are qualitative and concern only the types of the units present and the way in which their content varies (an increase or decrease) with a variation in the glass composition or temperature. Hence, these data cannot be used for establishing the structure-property relationship.

To a noticeable extent, the limited value of the experimental results is due to the fact that in studies by Raman spectroscopy the breathing modes for the planar rings in the pentaborate and triborate groups, which comprise one $BØ_4$ tetrahedron plus two $BØ_3$ triangles, both occur at ~770 cm^{-1} [31–33], and so it is not possible to distinguish between the wavenumbers for the peaks due to these two superstructural units. The diborate group, on the other hand, does not incorporate any planar rings, and therefore does not give rise to an intense sharp line in the Raman spectrum but only a broad low-intensity peak at ~1100 cm^{-1} [31, 32], and so it is much more difficult to establish the presence of diborate groups using Raman spectroscopy. Similarly, NMR spectroscopy can distinguish between superstructural units that include zero, one or two BØ$_4^-$ tetrahedra but not separate pentaborate and triborate groups or di-pentaborate, diborate and di-triborate groups [34]. Although in favourable cases, it is still possible to gain some useful information concerning the presence of pentaborate, triborate and diborate groups, sometimes there is no unambiguous evidence as to what particular superstructural units are observed in the low-alkali region (pentaborate or triborate) or in glasses containing 30–35 mol% M$_2$O (di-pentaborate, diborate or di-triborate).

The concept of the chemical structure can be used for calculating the fractions of (super)structural units as a function of the glass composition, which describes the intermediate-range order in the glass structure. In Na$_2$O–B$_2$O$_3$-glasses at 800 K the set of the units includes not only various superstructural units but also triangles of two types, BØ$_3$ and BØO$_2^{2-}$, which are not included into superstructural rings. The calculation of the fractions of all (super)structural species are performed using Eqs. (5.24)–(5.30). They have been derived using the knowledge about the chemical structure of glasses (Fig. 5.5) together with information on the types and numbers of the superstructural units introduced into glasses by various chemical groupings (Table 5.1). These equations present the ratios between the numbers of (super)structural units of each type (the numerators of Eqs. (5.24)–(5.29)) and the total number of all superstructural units, denoted SSU, and independent triangles, denoted B$_n$, present in a given glass (Eq. (5.30)).

$$[\text{Boroxol}] = [\text{BØ}_3] = \frac{0.5X_{B_2O_3}}{\Sigma(SSU + B_n)}, \tag{5.24}$$

$$[\text{Pentaborate}] = \frac{2X_{Na_2O\cdot5B_2O_3} + X_{Na_2O\cdot4B_2O_3}}{\Sigma(SSU + B_n)}, \tag{5.25}$$

$$[\text{Triborate}] = \frac{2X_{Na_2O\cdot3B_2O_3} + X_{Na_2O\cdot4B_2O_3}}{\Sigma(SSU + B_n)}, \tag{5.26}$$

$$[\text{Diborate}] = \frac{X_{Na_2O\cdot2B_2O_3}}{\Sigma(SSU + B_n)}, \tag{5.27}$$

$$[\text{Cyclic metaborate}] = \frac{0.667X_{Na_2O\cdot B_2O_3}}{\Sigma(SSU + B_n)}, \tag{5.28}$$

$$[\text{BØO}_2^{2-}] = \frac{2X_{2Na_2O\cdot B_2O_3}}{\Sigma(SSU + B_n)}, \tag{5.29}$$

$$\Sigma(SSU + B_n) = X_{B_2O_3} + 2X_{Na_2O\cdot5B_2O_3} + 2X_{Na_2O\cdot4B_2O_3} + 2X_{Na_2O\cdot3B_2O_3}$$
$$+ X_{Na_2O\cdot2B_2O_3} + 0.667X_{Na_2O\cdot B_2O_3} + 2X_{2Na_2O\cdot B_2O_3}. \tag{5.30}$$

The results obtained are shown in Fig. 5.13. It is seen that in glasses containing 0–10 mol% Na$_2$O, boroxol rings and independent BØ$_3$ triangles are the dominating

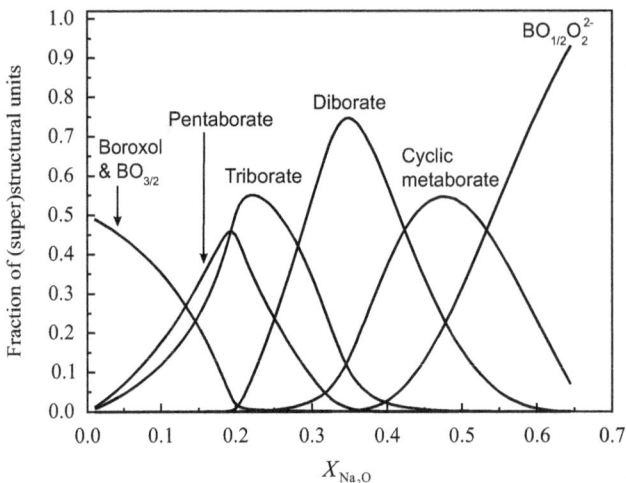

Fig. 5.13: Model fractions of superstructural units in sodium borate glasses.

structural units. Above the composition interval 0–20 mol% Na_2O both pentaborate and triborate units are present in glasses in practically equal amounts, which doubtlessly adds to the experimental difficulty of distinguishing between these units. In glasses containing from 20 to 30 mol% Na_2O triborate rings are the prevailing units, with a large contribution from pentaborate and diborate rings. At Na_2O content varying from 30 to 40 mol% Na_2O, the characteristic feature of the glass structure are diborate rings together with a considerable amount of triborate rings and cyclic metaborate anions. In the high-alkali glasses containing 40–55 mol% Na_2O the structure is dominated by cyclic metaborate anions with a decreasing amount of diborate superstructural units and a growing number of $BØO_2^{2-}$ pyroborate triangles. Above 55 mol% Na_2O, this triangle becomes practically the only structural unit in these glasses.

The above sequence of the superstructural units and anions as a function of the glass composition agrees with that reported in [32, 35], where the intermediate-range order in the structure of alkali borate glasses was studied by Raman spectroscopy. Fig. 5.13 also illustrates the experimental observation made in [32], according to which diborate superstructural units present can be divided into two categories: connected with the network (over the interval 15–45 mol% Na_2O) and "loose" (45–67 mol% Na_2O). As is seen from Fig. 5.13, in glasses containing 20–40 mol% Na_2O diborate rings find themselves in the well-connected network, where they are linked either to pentaborate and triborate units (20–30 mol% Na_2O), or to other diborate rings (30–40 mol% Na_2O). As the Na_2O content increases, the amount of cyclic metaborate anions exceeds that of diborate rings, which leads to the depolymerization of the network. Therefore, in the high-alkali region (> 40 mol% Na_2O) diborate rings are surrounded by anions: first, by cyclic metaborate groups, and then by $BØO_2^{2-}$ pyroborate triangles. These diborate units are called "loose" in [32].

Recall that the calculation of the chemical structure of sodium borate glasses has revealed a negligibly small amount of the grouping $3Na_2O\cdot7B_2O_3$, which is due to a low thermodynamic stability of the corresponding crystalline compound [19, 22]. This grouping comprises di-triborate superstructural unit [17], hence the presence of this unit in noticeable amounts in glasses can also hardly be expected. However, according to the high-temperature Raman studies [28, 36], di-triborate units are present instead of diborate rings in glasses containing 30–60 mol% Na_2O. This disagreement requires a further more detailed investigation.

5.7 Structure–Property Relationships

As it is indicated in [12, 13], the knowledge of the chemical structure of glasses, expressed in terms of the numbers of moles of the chemical groupings and of the unreacted constituent oxides, allows the calculation of a wide range of glass properties without use of adjustable parameters, as an additive function of the relevant property of the crystalline compounds that form in a given system. The molar volume of glass, V_{glass}, is one of such properties and is calculated as an additive function of the molar volume of the relevant crystalline compounds,

$$V_{glass} = \Sigma n_j \cdot V_j^0, \tag{5.31}$$

where V_j^0 represents the molar volume of the crystalline compounds and n_j represents the number of moles of the species j present in the glass. The index j refers both to the chemical groupings and to the unreacted oxides. On average, the inaccuracy of these calculations, estimated in terms of the density, does not exceed ± 0.01 g/cm^3, which is comparable to the uncertainty of reliable experimental measurements. This is confirmed by the good agreement between the calculated and experimental densities of glasses in such different systems as Na_2O–B_2O_3, Na_2O–SiO_2, Bi_2O_3–B_2O_3, Na_2O–B_2O_3–SiO_2 and Na_2O–CaO–SiO_2, shown in Fig. 5.14. Note that the reliability of calculating the density using Eq. (5.31) is entirely determined by the accuracy of the determination of the chemical structure and that of the densities of the compounds. The comparison of the calculation error for the chemical structure [13] with the scatter of the experimental data for the density of crystalline compounds made for a number of systems clearly shows that the latter is the major source of error.

Since the density is not an additive property, the partial contributions from the various structural units present have to be calculated in terms of the molar volume of glasses. The elimination of the contribution from the molar weight of glasses is particularly important when the molar weights of the constituent oxides are noticeably different (see Figs. 5.1 and 5.3). An example of the quantitative estimation of the structure-molar volume relationship is shown in Fig. 5.15 for sodium borate glasses. This figure presents the calculated molar volume of these glasses, V_{glass}, together with

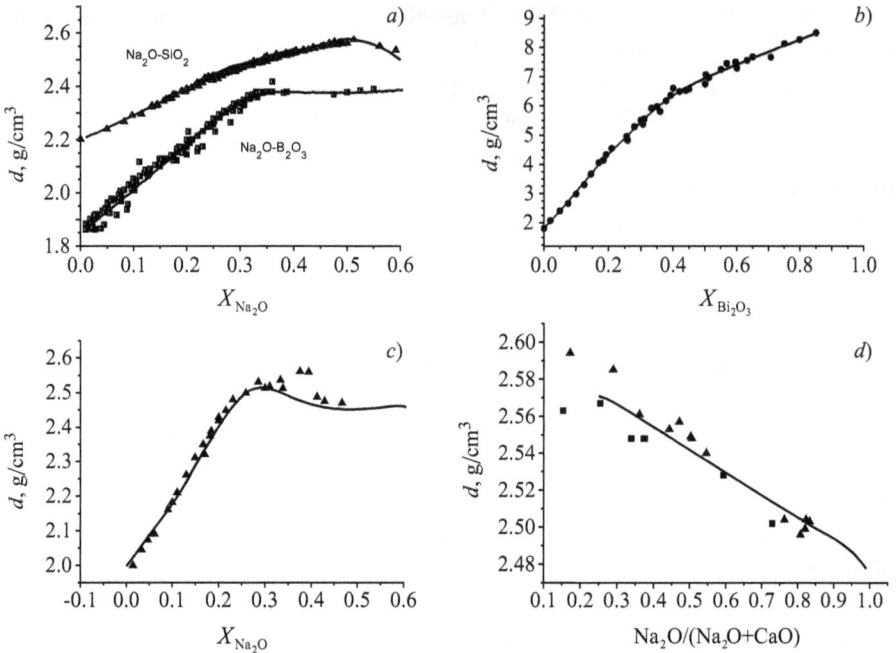

Fig. 5.14: Densities of glasses in the systems $Na_2O-B_2O_3$ and Na_2O-SiO_2 (a), $Bi_2O_3-B_2O_3$ (b), $Na_2O-B_2O_3-SiO_2$, the cut with K = 1 (c) and $Na_2O-CaO-SiO_2$, the cut with 70 mol% SiO_2 (d) : model (solid lines) and experiment (symbols [4]).

the contributions, $n_j \cdot V_j^0$, from the various chemical species present. These products are determined using Eq. (5.31).

Fig. 5.15 shows how the contributions from these species change as the content of Na_2O increases. The numbers, indicated by the vertical dashed lines, at the intervals of 10 mol% Na_2O represent the total contribution from all of the groupings present, i.e. the molar volume of the given glass. For example, at 10 mol% Na_2O, the value $\Sigma 34.1$ cm^3 is the sum of the contributions, $n_j \cdot V_j^0$, from the unreacted B_2O_3 and the groupings $Na_2O \cdot 5B_2O_3$ and $Na_2O \cdot 4B_2O_3$. This value is lower than $\Sigma 38.7$ cm^3 (the molar volume of pure vitreous B_2O_3) by 4.6 cm^3. As the Na_2O content grows, the differences between the compositions, marked by the adjacent dashed lines, become less and less pronounced (3.0 cm^3 between 10 and 20 mol% Na_2O, and 2.3 cm^3 between 20 and 30 mol% Na_2O). Over 30 mol% Na_2O, the decrease in these values slows down, and the difference between them becomes almost zero, varying from 0.2 to 0.4 cm^3 over the interval from 30 and 60 mol% Na_2O. Hence, the rate at which the molar volume of sodium borate glasses decreases changes at 30 mol% Na_2O.

On the other hand, the molar weight of the glasses decreases linearly over the entire composition interval. Since the density is equal to $M.Wt./V_{gl}$, it is the interplay between these two quantities that accounts for the shape of the density curve for sodium

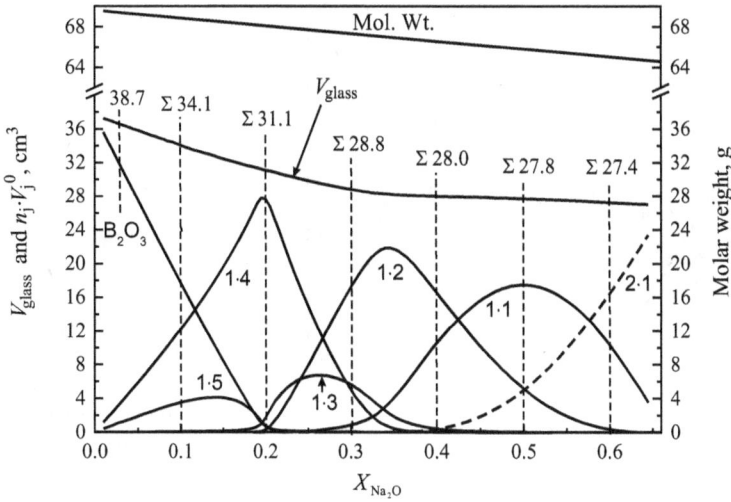

Fig. 5.15: Calculated molar volume of sodium borate glasses, V_{glass}, and the contributions from the various chemical groupings, $n_j \cdot V_j^0$.

borate glasses. As is seen from the concentration dependence of the density of sodium borate glasses (Fig. 5.14a), there is a pronounced change in gradient at 30 mol% Na$_2$O, due to which attempts to relate the shape of the density curve to the presence of 4-fold coordinated boron atoms have been made. However, Fig. 5.15 clearly shows that it is necessary to consider the glasses on the scale of the chemical groupings in order to explain the property (density) in terms of the structural changes. This conclusion is in agreement with experimental evidence [28] revealing that in sodium borate glasses the re-arrangements at the level of the superstructural units become "frozen-in" at temperatures very close to T_g. This observation points to the fact that the structure of borate glasses (and hence their properties) are mainly determined by the superstructural units present.

As follows from Fig. 5.14, this is also true for glasses in other binary and ternary systems, since the same approach is valid for calculating their densities. In particular, in [37], the structure-molar volume relationship is established for sodium silicate glasses (Fig. 5.16). This figure shows changes in the contributions, $n_j \cdot V_j^0$, from the various chemical groupings to the molar volume, V_{glass}, of these glasses as the content of Na$_2$O increases. From Fig. 5.16 it is seen that the minimum molar volume is observed for the composition with 50 mol% Na$_2$O. This is almost entirely due to the contribution from the grouping Na$_2$O·SiO$_2$, whose content in this glass is over 95 % (see Fig. 5.11), whilst the molar volume of the relevant crystalline compound is the lowest in the series of compounds involved. Finally, in [12] it is shown that other glass properties, such as the enthalpy of formation of lithium borate glasses and the heat capacity and molar

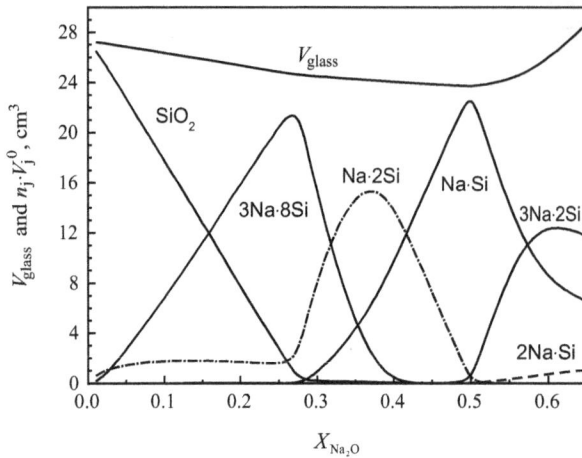

Fig. 5.16: Calculated molar volume of sodium silicate glasses, V_{glass}, and the contributions from the various chemical groupings, $n_j \cdot V_j^0$.

refraction of sodium silicate glasses, can be calculated as an additive function of the relevant property of the crystalline compounds.

5.8 Summary and Conclusions

The list of characteristic features of the concept of chemical structure, given in [37], can now be complemented with additional items resulting from the present more detailed comparison of glasses with crystals and melts. The concept of the chemical structure comprises a way of treating glasses and melts that is based on the understanding that (i) glasses (melts) are products of chemical interactions between their oxide components, (ii) the similarity between the energy and structure of the short-range order of glasses and crystals is an experimentally established fact, and (iii) only those structures that are thermodynamically stable yield the minimum Gibbs free energy for a given system. The advantages of viewing glasses (melts) in this way are as follows.

1. Calculations of the chemical structure of glasses (melts) are performed under the condition that all of the chemical reactions proceeding in a given system are taken into account, and that the requirement of the minimum of Gibbs free energy of the system is observed. Due to this, the concept of the chemical structure provides *a rigorous physical basis for understanding* the origin of various basic structural units or superstructural units in glasses (melts). In other words, this concept not only predicts *what* structural units are present in a given glass or melt but also explains *why* they form, in what *quantities*, and what their most probable *surroundings* might be.

2. The concept of the chemical structure enables the difference between the structure of glasses (melts) and crystals to be quantitatively determined at the superstructural units level.
3. The knowledge of the chemical structure as a function of temperature provides a more profound understanding of the temperature changes that occur at the levels of structural and superstructural units.
4. As is shown by numerous examples [12–15], this concept enables a wide variety of properties and the corresponding structural characteristics to be successfully calculated for glasses in binary and ternary systems. Since these calculations are performed on a unified basis, the approach allows the structure-property relationship to be determined *quantitatively*.
5. The approach is free of typical limitations of the models that calculate glass properties using information about their short-range order. For example, the questions of determining the properties of the basic structural units or of establishing the dependence of these properties on the content of modifying oxides, which inevitably requires the use of adjustable parameters, does not exist in the present approach. The knowledge of these vague details is unnecessary, as soon as the glass structure is considered at the scale of the superstructural units, and in terms of the chemical groupings. All specific features characterizing the short-range order in a given system are reflected in its energy. In this approach, these features are (implicitly) taken into account by the values of the standard Gibbs free energies of formation of the groupings. Due to this, a variety of glass properties can be calculated, without any adjustable parameters, as an additive function of the equivalent properties of the crystalline compounds that exist in a given system [12].

From the above, it follows that the concept of the chemical structure yields a comprehensive understanding of the structure of glasses (melts) at two levels (structural and superstructural units) and highlights the relationship between these levels. It also comprehensively explains structure-property relationships for glasses with any number of components. It may therefore be concluded that the concept of the chemical structure provides a sound basis for a better understanding of results of studies of both structure and properties. In this way, the concept of the chemical structure brings *a rigorous physical meaning* to the reasoning on the origin of the various basic structural units and superstructural units in glasses (melts), and establishes the relationship between these two levels in the glass (melt) structure. This concept also enables the difference between the structure of glasses (melts) and crystals to be quantitatively determined at the superstructural unit level. Since glass structures and properties are calculated on a unified basis, the concept allows the structure-property relationship to be determined *quantitatively*.

Acknowledgement: The authors are indebted to the late Boris Shakhmatkin for the computer programs employed in this work.

Bibliography

[1]   E.A. Porai-Koshits, *Proceedings XV International Congress on Glass*, Survey papers (Nauka, Leningrad, 1988, page 7).

[2]   M.M. Smedskjaer, J.C. Mauro, R.E. Youngman, C.L. Hogue, M. Potuzak, and Y. Yue, J. Phys. Chem. **B 115**, 12930 (2011).

[3]   H. Inoue, A. Masuno, Y. Watanabe, K. Suzuki, and T. Iseda, J. Amer. Ceram. Soc. **95**, 211 (2012).

[4]   SciGlass, Version 6.6, 1998–2006.

[5]   S. Kroeker, P.M. Aguiar, A. Cerquiera, J. Okoro, W. Clarida, J. Doerr, M. Olesiuk, G. Ongie, M. Affatigato, S.A. Feller, Eur. J. Glass Sci. Technology B: Phys. Chem. Glasses **47**, 393 (2006).

[6]   L. Cormier, G. Calas, and B. Beuneu, Eur. J. Glass Sci. Technology B: Phys. Chem. Glasses **50**, 195 (2009).

[7]   A.A. Osipov and L.M. Osipova, Russ. J. Glass Physics Chemistry **35**, 121 (2009).

[8]   M.M. Shultz, N.M. Vedishcheva, B.A. Shakhmatkin, and A.M. Starodubtsev, Sov. J. Glass Phys. Chem. **11**, 299 (1985).

[9]   B.A. Shakhmatkin and M.M. Shultz, Fiz. Khim. Stekla **4**, 271 (1978) (in Russ.).

[10]   B.A. Shakhmatkin and M.M. Shultz, Fiz. Khim. Stekla **8**, 270 (1982) (in Russ.).

[11]   B.A. Shakhmatkin and M.M. Shultz, Fiz. Khim. Stekla **6**, 129 (1980) (in Russ.).

[12]   B.A. Shakhmatkin, N.M. Vedishcheva, M.M. Shultz, and A.C. Wright, J. Non-Crystalline Solids **177**, 249 (1994).

[13]   B.A. Shakhmatkin, N.M. Vedishcheva, and A.C. Wright, *Proceedings XIX International Congress on Glass*, Invited papers (Soc. Glass Technology, Sheffield, 2001, p. 52).

[14]   N.M. Vedishcheva, B.A. Shakhmatkin, and A.C. Wright, J. Non-Crystalline Solids **345–346**, 39 (2004).

[15]   N.M. Vedishcheva, B.A. Shakhmatkin, and A.C. Wright, Phys. Chem. Glasses **46**, 99 (2005).

[16]   M.M. Shultz, N.M. Vedishcheva, and B.A. Shakhmatkin, in: *Fizika i Khimiya Silikatov (Physics and Chemistry of Silicates)*, Eds. M.M. Shultz and R.G. Grebenshchikov (Nauka, Leningrad, 1987, p. 5 (in Russ.)).

[17]   A.C. Wright, Eur. J. Glass Sci. Technol. B: Phys. Chem. Glasses **51**, 1 (2010).

[18]   A.C. Wright, G. Dalba, F. Rocca, and N.M. Vedishcheva, Eur. J. Glass Sci. Technol. B: Phys. Chem. Glasses **51**, 233 (2010).

[19]   R.S. Bubnova and S.K. Filatov, *Vysokotemperaturnaya Kristallokhimiya Boratov i Borosilikatov (High-temperature Crystal Chemistry of Borates and Borosilicates)* (Nauka, St. Petersburg, 2008, 760 p. (in Russ.)).

[20]   Ya. B. Zel'dovich, Zh. Fiz. Khim. **11**, 685 (1938) (in Russ.).

[21]   N.M. Vedishcheva, *Enthalpies of formation of alkali borate glasses and crystals*, Ph.D. Thesis, Leningrad (1989) (in Russ.).

[22]   A.C. Wright and N.M. Vedishcheva, European J. Glass Science Technology B: Phys. Chem. Glasses **54**, 147 (2013).

[23]   A.P. Howes, N.M. Vedishcheva, A. Samoson, J.V. Hanna, M.E. Smith, D. Holland, and R. Dupree, Phys. Chem. Chem. Phys. **13**, 11919 (2011).

[24]   P.J. Bray and J.G. O'Keefe, Phys. Chem. Glasses **4**, 37 (1963).

[25]   L.M. Osipova and A.A. Osipov, *Proc. XII Russ. Conf. Structure and Properties of Metal and Slag Melts*, **3** (Ekaterinburg, 2008, p. 41, (in Russ.)).

[26]   B. Chen, U. Werner-Zwanziger, M.L.F. Nascimento, L. Ghussn, E.D. Zanotto, and J.W. Zwanziger, J. Phys. Chem. **C 113**, 20725 (2009).

[27]   L. Cormier, G. Calas, and B. Beuneu, J. Non-Crystalline Solids **353**, 1779 (2007).

[28]   A.A. Osipov and L.M. Osipova, Russ. J. Glass Phys. Chem. **35**, 132 (2009).

[29]   H. Maekawa, T. Maekawa, K. Kawamura, and T. Yokokawa, J. Non-Crystalline Solids **127**, 53 (1991).

[30]   N.A. Toropov, V.P. Barzakovskii, V.V. Lapin, and N.N. Kurtseva, *Diagrammy Sostoyaniya Silikatnykh Sistem (Phase Diagrams of Silicate Systems)*, Ed. N.A. Toropov, Vol. 1 (Nauka, Leningrad, 1969 (in Russ.)).

[31]   B.N. Meera and J. Ramakrishna, J. Non-Crystalline Solids **159**, 1 (1993).

[32]   E.I. Kamitsos, M.A. Karakassides, and G.D. Chryssikos, Phys. Chem. Glasses **30**, 229 (1989).

[33]   E.I. Kamitsos and G.D. Chryssikos, J. Molec. Struct. **247**, 1 (1991).

[34]   D. Holland, private communication (2012).

[35]   W.L. Konijnendijk and J.M. Stevels, J. Non-Crystalline Sol. **18**, 307 (1975).

[36]   A.A. Osipov and L.M. Osipova, European J. Glass Science Technology B: Phys. Chem. Glasses **54**, 1 (2013).

[37]   N.M. Vedishcheva, B.A. Shakhmatkin, and A.C. Wright, In: *Glass – The Challenge for the 21st Century* (Advances Mater. Research **39–40**), Ed. M. Liška, D. Galusek, R. Klement and V. Petrušková (Trans Tech, Stafa-Zürich, 2008, p. 103).

Boris Z. Pevzner and Sergey V. Tarakanov

6 Bubbles in Silica Melts: Formation, Evolution, and Methods of Removal

Processes of formation and growth of bubbles in glass-forming melts and methods of their removal are analyzed. Part I is devoted to an overview on experimental data and basic mechanisms. Part II deals with general aspects of theoretical modeling and Part III is directed to the application of these methods to bubble evolution processes in silica melts.

Part I: Experimental Data and Basic Mechanisms

6.1 Introduction

The existence of bubbles in glasses is one of their main defects; it has an undesirable influence on the appearance of the manufactured glass articles as well as on their mechanical and optical properties. Bubbles under consideration result from the interaction between a glass-forming melt and gas products arising in different stages of glass manufacture. The formation of bubbles proceeds in various processes: the melt occludes some amount of the gases from the furnace atmosphere or from some of the batch components; some amount of the gases is physically and chemically dissolved in the melt and then isolated as bubbles. The opposite process, which consists in removing the bubbles from the melt, takes place due to buoyancy forces pushing the bubbles to the surface as well as by dissolving the gases physically and chemically in the melt.

The gases contained in bubbles have their origin in various sources: the batch components, the glass-furnace atmosphere, the glass-furnace lining, the physical-chemical processes in the melt which lead to the formation of the gaseous phase. Most manufactured glasses are many-component ones. For their manufacture a variety of minerals (quartz and quartz sands, feldspar, limestone, dolomite, nepheline, etc.) and a number of chemicals (for example, carbonates, nitrates, hydroxides, crystalline hydrates, sulphates, and peroxides) are used as the batch components. While heating the batch, the gases are formed in a wide temperature range, which is due to chemical reactions of dissociation and substitution. A comprehensive description of just mentioned reactions can be found in some manuals and monographs on glass melting, for example, in [1, 2]. The furnace atmosphere includes air components, products of combusting the fuel (coal, hydrocarbons, natural gas, or hydrogen) and the gases evolving from the batch. Synthesis of some glasses needs a special atmosphere (for example, vacuum, hydrogen, argon, nitrogen, or another). The fireproof furnace lining may contain some gaseous inclusions, which go over into the melt as a result of interaction with it. The chemical interaction of the melt with the melting-chamber walls may also result in formation of gaseous products. Some of the melt components may dissociate to form a gas phase or may give some volatile compounds. The main gases forming the bubbles are the following: N_2, O_2, CO_2, CO, SO_3, SO_2, H_2O, and H_2.

For each gas the mass exchange in the system "bubble – melt – gaseous furnace atmosphere" is determined both by the equilibrium constant in the solution (i.e. in the melt) and by that in the free state (i.e. in the gaseous phase). The kinetics of the process depends on the formation and diffusion rates of the gases in the melt. The process of removal of the bubbles due to their floating up is determined by the Archimedean force on the bubbles and by the melt viscosity. The final stage of removing the bubbles consists in rupturing the surface film by the bubble gas and its transfer into the furnace space. To summarize, the evolution of bubbles proceeds due to such physical-chemical properties of the melt as viscosity, density and surface tension, as well as to the solubility and diffusivity of gases in the melt. A number of technological approaches has been developed allowing one to decrease the bubble formation rate, to favor fast bubble removal and advancing the process closer to completion, for example, a special preparation of the batch, mixing of the melt, an advanced design of the melting chamber.

6.2 Sources of Bubbles in Silica Melt and Glass

6.2.1 Brief Account of the Technology of Silica Glass Production

Silica glass is the most remarkable one among all glasses. It has unique properties with respect to a variety of applications and exhibits very special features under manufacture. This is due especially to the high temperatures required for synthesizing the glass and processing the melt into wares as well as strict requirements to the homogeneity and chemical purity of the glass. An account of the properties and technology of silica glass production is given in the monograph by Leko and Mazurin [3] as well as in few courses for higher education [2, 4, 5]. No studies generalizing or advancing the matter to such an extent have been reported since then.

At the same time, data on various properties of silica glass are very numerous, significantly exceeding in quantity the data on other one-component glasses. Note that in January 2008 the electronic database Sci Glass [6] represented about 1000 papers, which describe various properties of glassy SiO_2 and its melts. Fig. 6.1 represents an integral and a differential curve that describe the time distribution of papers on silica glass and its melts over the period of 1900–2007. The maximum of publications appeared in the 1960s–1980s. The same period is also marked by the maximum of publications in the whole field of glasses. We briefly discuss here some aspects of manufac-

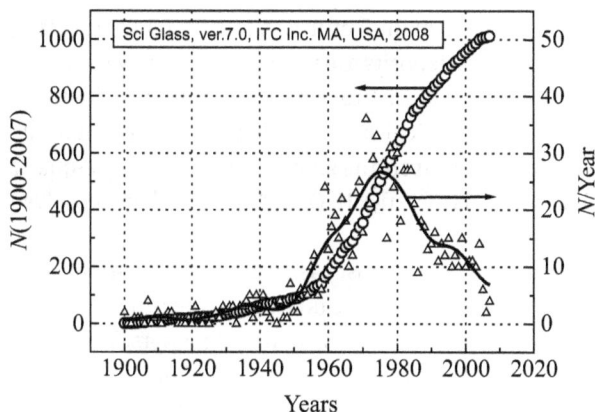

Fig. 6.1: Distribution of the number of papers on silica glasses and silica melts published over the period of 1900–2007.

turing transparent silica glasses, the ones connected with the formation, evolution, and removal of bubbles.

The distinctive features of manufacturing a silica glass are due to high refractoriness of crystal modifications of SiO_2 (quartz, tridymite, and cristobalite) and high viscosity of its melt. If we take the temperature ranges of melting compound glasses, the viscosities of their melts are found in the range from a few units to several dozens of Pa·s (10^0–10^2 Pa·s). The viscosity of silica melts in the conventional temperature range of melting has values within 10^3 – 10^7 Pa·s. Such a high viscosity excludes the application of conventional procedures employed in removing bubbles (mechanical and convective stirring). To produce a silica glass not containing any bubbles, one should optimize each stage of manufacture, which includes (i) selection and preliminary treatment of the raw materials; (ii) choosing the most appropriate heat source and heating technique; (iii) design of the furnace (selection of materials, design of heaters and shape-forming elements); (iv) optimization of the furnace atmosphere in melting; (v) optimization of the temperature-time mode of the fusion process. Depending on the preliminary treatment and other factors of the manufacturing process, one and the same raw material would lead to either a practically bubble-free glass or a glass containing a large number of bubbles, distributed either locally or over the whole sample.

6.2.2 Raw Materials as a Source of Bubbles

Natural or synthetic quartz, amorphous silica and some volatile compounds of silicon are commonly employed as the raw material for producing silica glass. Of importance in our case are the bubble-forming components of these materials and the bubble-forming impurities contained in them. When manufacturing a transparent silica glass, *rock crystals* and *vein quartz* from different deposits are used as *natural raw materials*. These materials contain two types of impurities: structural ones and those included in solid or gas-liquid inclusions. The most typical structural impurities are aluminum and alkaline metals, which isomorphically substitute silicon in the crystalline-quartz lattice, and also such gaseous impurities as CO_2, CO, H_2O, and H_2, which are dissolved in mineral quartz

Solid inclusions under consideration are, in most cases, particles of various minerals taken from hydrothermal solutions by growing quartz crystals; the just mentioned minerals are feldspars, mica, tiff, apatite, sphene and some others. These gas-liquid inclusions vary in composition within a very wide range – from almost pure water to very concentrated or pregnant solutions containing carbonates, chlorides, or sulphates of sodium, potassium, and calcium as well as, sometimes, suspended particles of these salts; as to the gas phase, it consists of gaseous water, carbonic-acid gas, ammonia, or hydrogen sulphide. The total content of solid and gas-liquid inclusions in vein quartz, granular and milk-white, lies within 1–10 mass percent. The concentration of these inclusions is much lower in rock crystals of high quality. Table 6.1 represents the content of gaseous CO_2 and H_2O in the main types of raw materials for silica glass manufacture, the measurements are made after enrichment [4, 5].

Table 6.1: Emission of gaseous CO_2 and H_2O in the 20–1200 °C range from different raw materials for silica glass manufacture, the data are obtained by gas chromatography analysis [4, 5]

| Raw material | Gas emission, mass% | |
	CO_2	H_2O
Rock crystal	$(0.08\text{–}0.6) \cdot 10^{-2}$	$(0.13\text{–}0.71) \cdot 10^{-2}$
Vein granular quartz	$(0.44\text{–}1.81) \cdot 10^{-2}$	$(1.3\text{–}4.5) \cdot 10^{-2}$
Vein white-milk quartz	$(2.6\text{–}4.7) \cdot 10^{-2}$	$(12\text{–}14) \cdot 10^{-2}$

When just mined, natural quartz is not suitable for obtaining a transparent silica glass and has to be preliminarily treated through many different processes. The preparation for melting includes breaking up and pounding in thermal and mechanical ways, magnetic separation, flotation enrichment, chemical enrichment (treatment in hot solutions of hydrochloric and hydrofluoric acids), washing, dehydration, and drying. The main aim of preliminary treatment is to remove the solid inclusions, to bring the gas-liquid inclusions to the outer ambience, and to get rid of the substances that belong to the latter ones and are capable of making gas. However, even such a substantial preliminary treatment cannot guarantee a thorough removal of all the possible sources of bubbles from silica raw materials, for there still remain some gases in the quartz lattice. The following sources can be distinguished: (i) *Man-made (synthetic) quartz* produced by the hydrothermal technique in the 350–400 °C range and the pressure range of 600–800 atm, includes some H_2O and CO_2, dissolved in the structure, and may contain gas-liquid inclusions, which are, however, much fewer than those in natural quartz. (ii) *Synthetic amorphous silicon* dioxide produced by hydrolysis and the subsequent polymerization of organic-silicon compounds, mostly tetraethyl-oxisilane-$Si(OC_2H_5)_4$, contains a lot of H_2O and CO_2 and are not suitable for direct manufacture of bubble-free silica glass. To remove the gas-forming components, one should treat this intermediate product in a thermal way for transforming it into cristobalite. (iii) *Volatile compounds of silicon*, for example $SiCl_4$, must be converted into silica by high-temperature hydrolysis or by high-temperature oxidation with oxygen. Gaseous silica formed in either of these ways condenses as amorphous particles of 0.1–1.0 µm, which may contain the physically and chemically dissolved initial gases and gaseous products formed during the treatment (H_2O, Cl_2, and HCl). If one conducts the fusing under certain conditions, these two groups may lead to bubbles in the melt. (iv) *Cristobalite* is the modification of crystalline silica whose lattice does not contain impurities and, therefore, sources of bubble formation in the melt. Though found in nature as an individual mineral, cristobalite is not used for melting of silica glass. Cristobalite as an intermediate product is obtained by a special-mode thermal treatment of synthetic amorphous dioxide or quartz grist; the material is thoroughly purified from gaseous impurities during the treatment, and the non-volatile impurities are extruded up to the grain surface. These nonvolatile impurities are either washed in a chemical way or sublimated and/or evaporated at high temperatures in vacuum. According to Leko [7], cristobalitization is the most optimal and economical method of producing a clear silica glass. Even if an initial bulk of quartz contains gas-liquid inclusions and dissolved gases in high concentrations, for example, as milk-white quartz does, we shall have the following. Rapid cristobalitization of this raw material, due to the alkaline admixtures contained in minute bubbles, especially in the form of chloride, under vacuum fusion yields a pure transparent silica glass without bubbles.

Rebrova and Shchekoldin [8] have analyzed the kinetics of cristobalitization of various vein quartz samples. These quartz samples were different in the amount of gas-liquid inclusions per cm³ (10^3-10^4 in vein granulated quartz, 10^6-10^7 in milk-white quartz), chemical composition of the solution ($Ca(HCO_3)_2$ and $CaSO_4$ in vein granulated quartz, $CaCl_2$ and NaCl in milk-white quartz), baking losses (0.005–0.01 % for granulated quartz, 0.1–0.2 % for milk-white quartz). Keeping the material to mature at 1500 °C for 10 or 20 h transformed 25–35 % of the granulated quartz and 90–100 % of the milk-white quartz into cristobalite.

6.2.3 Furnace Atmosphere as a Source of Bubbles

The gases of the furnace atmosphere can be physically captured by the powder layer of the batch being sintered as well as dissolve physically and chemically in the batch grains and melt layer. The composition of the furnace atmosphere depends on the source and manner of heating, the furnace design, and the fusion technique.

The gases of the furnace atmosphere arise from several predecessors: gases contained in the air or in the specially created gaseous medium, gases included in the gaseous fuel and combustion products, gases evolving from the batch, gaseous products generated by interaction of the batch and melt with the materials of heaters and shape-forming equipment. At present, the following techniques of heating are used: the thermal-electric method, the gas-plasma method, and the plasma manner.

The heat source for the *thermo-electric method in resistance furnaces* is a heater made of graphite, molybdenum, or tungsten. The induction furnace uses water-cooled tubular inductor as heater, made, as usual, of copper. It heats up the metallic (molybdenum or tungsten) or graphite crucible. The shape-forming equipment is made of the same materials. In some cases, the shape-forming equipment, represented by the tube and crucible, can function as a heater too. Development of the design involves keeping the melt from getting into a direct contact with the heater and the shape-forming equipment. One can achieve this by means of various combinations of the following design features: a tube or vessel of silica glass, a skull layer of quartz grist, and the gap between the surface under fusion and the heater (for example, we can use such a combination in a horizontal rotor furnace).

The thermal-electric method of heating is realized either in vacuum furnaces under continuous vacuum pumping, the pressure of the residual air gases not exceeding 10^{-2}–10^{-3} Pa at the beginning of the process, or in the medium of the gases He and/or H_2 quickly diffusing in the melt, the pressure being about 1 atm ($\sim 10^5$ Pa). Let us consider an appropriate vacuum furnace under heating. Gases moving into the vacuum system from the air, gases evolving in the batch and the gas products resulting from interaction between the batch (silica) and the materials of the heater and furnace equipment may enter the technological furnace zone. When the temperature of melting lies within 1800–2400 °C, the total gas pressure reaches 10^2–10^3 Pa. The conditions of melting are supposed to be reducing or neutral.

An electric furnace producing silica glass tubes contains an atmosphere of either hydrogen or its combination with nitrogen or argon, the combination composed of 20 % of the former and 80 % of the latter, and the total pressure is about 1 atm. It is very significant for the furnace atmosphere composition that the first stage under neutral or reducing conditions is the reaction of thermal dissociation of silica,

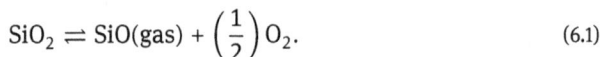

$$SiO_2 \rightleftharpoons SiO(gas) + \left(\frac{1}{2}\right)O_2. \tag{6.1}$$

Oxygen, while emitted, interacts with the materials of the heater and the equipment (C, Mo, and W). The following gaseous products are formed parallel to this: CO, MoO_3, and WO_3. The silicon monoxide partly dissolves in the melt, and partly reacts with graphite, molybdenum, and tungsten to give the corresponding gaseous products. The silica melt weakly wets graphite, thus, the reactions of oxidizing the graphite by the furnace atmosphere gases will keep going further. The silica melt readily wets molybdenum and tungsten, therefore, the reactions of oxidizing them by the furnace atmosphere gases will be over.

The melting space of a resistance or an induction furnace may include: gases of the initial atmosphere, namely, H_2, N_2, and He; gases evolving from gas-liquid inclusions of the charge, namely, O_2, N_2, H_2O and CO_2 vapors; gaseous products resulting from the interaction of the silica with the heaters and equipment, namely, SiO, CO, MoO_3, and WO_3. *The gas-flame method* involves an oxy-hydrogen flame as a heater. Sometimes natural gas is utilized instead of hydrogen. The pressure in the gas space of the furnace is about 1 atm. In most cases, the melting conditions are neutral or weakly reducing due to controlling the gas-oxygen ratio. The *plasma technique* uses a flame of high-frequency or arc plasma as heat source. Air, oxygen, nitrogen, argon, hydrogen, helium, or various combinations of them are applied as plasma-supporting gases. The redox conditions of melting are controlled by changing the concentrations of hydrogen and oxygen. The pressure in the gas space of the furnace can be varied within a wide range, namely, from a few percent to several atmospheres. While the arc plasma burns, the gaseous products of burning the graphite electrodes, namely, CO and CO_2, also come into the flame. With respect to the use of gas-flame or plasma heating methods in dependence of given raw materials, a few manufacture methods have been developed whose compositions of the furnace atmosphere are rather different.

While using quartz grist, this material is brought as a thin layer onto the melt surface being covered with the flame, the hot-patch temperature reaching 2100–2200 °C. The furnace chamber and the substrate, which goes down, as more and more silica glass is fused on it, are made of a refractory material not interacting with the gaseous furnace atmosphere – for example, zirconium dioxide. In case we have a different design, for example, a rotary furnace, centrifugal force spreads the grist as a layer of a few centimeters over the accessible part of the interior surface of a water-cooled metal tube. In other situations, a gas-flame or plasma flame may be initiated along the tube axis. The grist adjoining the tube surface forms a skull layer isolating the tube metal from the melt influence. Besides the gases forming the flame, the gaseous furnace atmosphere includes the gaseous products evolved by the batch.

If one synthesizes a silica glass in a vapor-phase way, the gas-flame or plasma flame is supplied with vapor of volatile silicon compounds, for example, the silicon tetrachloride $SiCl_4$. The gas-plasma way implies a high-temperature hydrolysis via the reaction

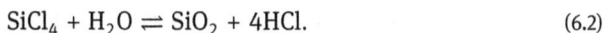

$$SiCl_4 + H_2O \rightleftharpoons SiO_2 + 4HCl. \tag{6.2}$$

The plasma method implies a high-temperature oxidation via the reaction

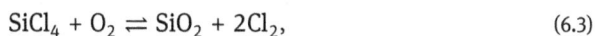

$$SiCl_4 + O_2 \rightleftharpoons SiO_2 + 2Cl_2, \tag{6.3}$$

the plasma including oxygen or nitrogen with oxygen. The gaseous silica, formed through either of the two reactions, condenses in the form of aerosol amorphous particles of 0.1–1 µm sizes, which settle on the surface of the melting silica glass block.

Besides the gases forming the flame, the furnace atmosphere contains HCl and Cl_2. The melt occludes the furnace atmosphere to a least possible extent under such a design which brings onto the melt surface a thin layer (ideally, a monolayer) of powder particles or melt drops.

6.2.4 Interaction of Heaters and Form-shaping Equipment with the Melt as Source of Bubbles

When silica melt gets in touch with graphite, there occurs a succession of reactions, to wit:

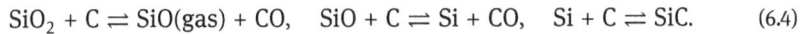

$$SiO_2 + C \rightleftharpoons SiO(gas) + CO, \quad SiO + C \rightleftharpoons Si + CO, \quad Si + C \rightleftharpoons SiC. \tag{6.4}$$

SiO dissolves in the melt, not yielding any bubbles. Under certain conditions the carbon monoxide forms numerous bubbles in the border zone. The determining factor for the rate of this formation is high quality of the graphite – a very moderate ash percentage, required to be as low as possible, and a high density of the material.

Molybdenum and tungsten are also capable of reducing the silicon in silica melt, through the reactions

$$3SiO_2 + Mo \rightleftharpoons 3SiO + MoO_3, \quad 3SiO_2 + W \rightleftharpoons 3SiO + WO_3, \tag{6.5}$$

to produce gaseous products (SiO, MoO_3, and WO_3). If the interaction goes on, it can lead to giving some molybdenum silicide and tungsten silicide. It will not provoke formation of bubbles in the melt.

The silica melt weakly wets graphite, thus the reactions of oxidizing the graphite by some gases of the furnace atmosphere will keep developing. The silica melt readily melts the molybdenum and tungsten, thus the reactions of oxidizing the metals by some gases of the furnace atmosphere will be over.

6.2.5 Concentrations of Impurities, Including Dissolved Gases, in Commercial Silica Glasses

In terms of technology and impurity concentrations, commercial silica glasses are classified into four groups, whose characteristics are represented in Table 6.2 [3]. Type I consists of non-hydroxyl silica glasses fused in vacuum. Type II consists of silica ones fused according to the gas-flame method. Type III consists of extra-pure hydroxyl-containing silica glasses obtained by high-temperature hydrolysis of $SiCl_4$. Type IV consists of extra-pure non-hydroxyl silica glasses. Table 6.3 represents data on the concentrations of a few gases in silica glasses made from different raw materials.

Table 6.2: Impurities in the main types of commercial silica glasses [3]

Glass type	Method of manufacture	Impurity concentrations, in mass%
I	Fusion of quartz grist in vacuum (thermal-electric method)	Summarized concentration of impurities is as much as $1 \cdot 10^{-2}$ for metals and no more than $5 \cdot 10^{-4}$ for OH-groups
II	Fusion of quartz grist in hydrogen-oxygen flame (gas-flame method)	Summarized concentration of impurities is $1 \cdot 10^{-2}$ for metals and no more than $(1.5 - 6) \cdot 10^{-2}$ for OH-groups
III	High-temperature hydrolysis of $SiCl_4$ in hydrogen-oxygen flame or in that of natural gas (vapor-phase method)	Summarized concentration of impurities is $1 \cdot 10^{-4}$ for metals, about 0.2 for OH-groups, and $(1 - 3) \cdot 10^{-2}$ for chlorine.
IV	(1) Oxidation of $SiCl_4$ with oxygen in high-frequency oxygen plasma (plasma-chemical method); (2) The two-stage technique consists in spraying the intermediate product by high-temperature hydrolysis of $SiCl_4$ and vitrifying the obtained material in dry atmosphere	Summarized concentration of impurities is $1 \cdot 10^{-4}$ for metals, $0.4 \cdot 10^{-4}$ for OH-groups, and as much as $6 \cdot 10^{-2}$ for chlorine

Table 6.3: Concentrations of a few gases in silica glasses made from different raw materials [4]

	Glass	Gas concentration, mol% $\times 10^3$					
		CO_2	CO	CH_4	H_2O	H_2	Cl_2
Thermal-electric method, rock crystal	KI	0.1	0.5	–	1.3	0.4	–
Thermal-electric method, vein granular quartz	KI	–	1.2	0.5	2.1	2.1	–
Gas-flame method, rock crystal	KU	0.1	0.2	–	127.6	10.6	–
Vapor-phase method, $SiCl_4$	KSG	–	–	–	460	15.5	5–50

6.2.6 Experimental Study of Formation and Evolution of Bubbles in Silica Melts

Table 6.4 demonstrates a number of data obtained on different raw materials and by different methods [9, 10]. During the repeated treatment above 1500 °C a number of bubbles can arise and will increase or decrease in size. This phenomenon can be caused by the following mechanisms: (i) emission of a gas physically or chemically dissolved in the melt if its concentration has exceeded the equilibrium one; (ii) dissolution and diffusion of some gases from the furnace space if the thermal treatment is carried out in a gaseous medium different from that of fusing.

Koryavin and Stepanchouk [11] have established the occurrence of a change in the chemical composition of gases in bubbles of a silica glass block if a corresponding material made by means of high-temperature hydrolysis of $SiCl_4$ (a glass of type III) is additionally kept within 1600–1780 °C in hydrogen (0.1 atm), argon (1.3 atm), and nitrogen (20 atm). The bubbles of the initial glass contained some of the gases (H_2 and

Table 6.4: Composition of gas in bubbles of different silica glasses, obtained by mass-spectrometry [9, 10]: (1–3) Synthesized quartz crystals; (4) Kyschtym quartz; (5) $SiCl_4$: Synthetic silicon dioxide; (6–8) SiO_2: Synthetic silicon tetrachloride

Sample number	Method of glass manufacture	Gas composition in bubbles, mass%
1	Gas-flame method	H_2 – 95; N_2 – 4; H_2O – 1
2	Method of oxygen-nitrogen plasma	O_2 – 39; N_2 – 60; NO – 1
3	Thermal-electric crucible method	CO – 70; N_2 – 30
4	Thermal-electric core method	CO – 88; CO_2 – 6; H_2O – 6
5	Method of oxygen-nitrogen plasma	O_2 – 61; N_2 – 26; CO_2 – 7; H_2O – 6
6	Method of nitrogen plasma	N_2 – 99; CO_2 – 1
7	Method of oxygen-nitrogen plasma	NO – 68; O_2 – 31; N_2 – 1
8	Ceramic method	N_2 – 97; CO – 2.5; $H_2O>0.5$; $H_2 \sim 0.5$.

H_2O) of the furnace atmosphere, in which the melt had been synthesized. After an additional thermal treatment of the glass for a long time there appeared N_2, CO, CO_2, and Ar. In addition, the chemical composition of the gas mixture in a given bubble depended on its size (from 0.5 to 5 mm) and disposition inside the block. The total gas pressure in the bubbles was as much as 7 atm.

Van der Steen and Papanikolau [12] carried out a lot of experimental work concerning the formation and evolution of bubbles in silica melt under controlled atmosphere. The initial material was Brazilian quartz containing Fe, K, Na, Ge, Li, and Al as components of the impurities, the granules from 30 to 200 μm in size. The granules were melted in a molybdenum crucible under heating by the high-frequency induction technique. The temperature of melting was 1950 °C, the procedure duration made up 15 min, the gaseous medium consisted of hydrogen under pressure of 1 atm. The obtained melt was cooled quickly. The samples did not include any detectable gas bubbles. The following heating up to 1700 °C and, keeping to mature for 15 min, led to the formation in the melt of an appreciable amount of bubbles from 960 to 187 μm in size, the calculated hydrogen pressure in them being found, respectively, from 2 to 9.5 atm at room temperature. The next series of experiments dealt with a gaseous medium including 99 % of He and 1 % of H_2, with the total pressure being 1 atm. The temperature of melting was 1950 °C, and the melting duration made up, correspondingly 5, 10 and 30 min. The obtained melt was cooled rapidly. When the melting duration was the shortest one, many bubbles, different in sizes, were observed in the glass. The longer was the melting, the smaller in number were the bubbles, and the larger was their average size.

Papanikolau and Wachters [13] have studied the growth of hydrogen bubbles in the course of hydrogen reboil in fused silica. Variation was made of the partial hydrogen pressure, the temperature from 1600 to 1900 °C, the mode of thermal treatment (an isothermal or non-isothermal one in dependence on the rate of cooling). The following properties have been analyzed: Size distribution of bubbles in dependence on dwelling time at different temperatures (the diameter ranging from 0.3 to 100 μm), partial hydrogen pressure in the bubbles, and growth and dissolution of hydrogen bubbles in dependence on bubble size, temperature, and hydrogen pressure in the gas phase. A theoretical model has been developed describing the formation of a bubble accounting for chemical reactions, diffusion, and viscous flow.

Boganov and others [10] systematically studied the mechanism of formation and growth of bubbles in silica glasses. Synthetic quartz (fused by a gas-flame method and in vacuum), vein quartz of the Kyschtym field (fused in a thermal-electric method in hydrogen and fused in oxygen-nitrogen plasma), and silicon tetrachloride (fused by the method of high-temperature hydrolysis) have been chosen as raw materials. Table 6.4 represents the composition of the gas in the bubbles. It is the fusing in vacuum that led to the least concentration of bubbles. The size distribution of bubbles is represented on Fig. 6.2, which demonstrates that most of the bubbles does not exceed 40–80 μm in size. The total concentration of gas in the bubbles is about $2.5 \cdot 10^{-8}$ weight percent. When the fusing was conducted in any of the mentioned gas media under atmospheric pressure, a considerable number of the bubbles were 1–2 mm in diameter, while the total content of gas in the bubbles was 2–3 orders of magnitude higher than at vacuum fusing.

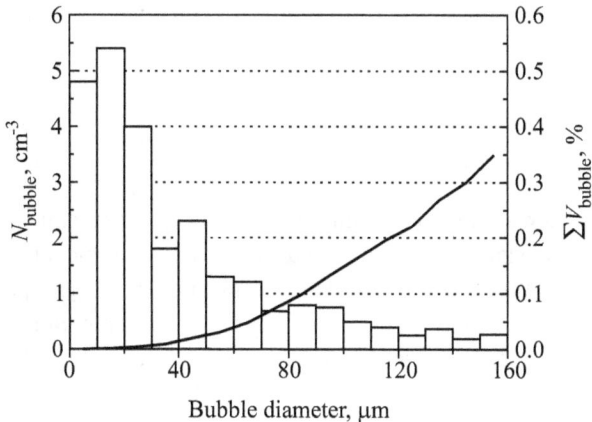

Fig. 6.2: Size distribution and the total volume of bubbles [10].

The authors of [14] have studied the removal of gases from the bubbles via physical dissolution and diffusion in the melt at 1800 °C and under external pressure of 5, 15, or 40 atm. The estimated time of dissolving the bubbles has been calculated in dependence on the diffusivity (D) and the solubility (S) of various gases in the melt. When bubbles under consideration contain gases of low D and S, for example, CO, N_2, or O_2, the time period of complete dissolution of bubbles of 2–200 μm in radius, is of the order of $10-10^5$ seconds, respectively. Bubbles including hydrogen and water exhibit higher values of D and S, whereas the period of dissolving bubbles of 20–2000 μm in radius makes up 1–10^4 seconds, correspondingly.

6.3 Physico-chemical Properties of Silica Melts Influencing the Formation and Evolution of Gas Bubbles

As has been noted above, the formation and evolution of gas bubbles are determined by the following physical-chemical properties of the melt: its surface tension, density, and viscosity as well as solubility and diffusion coefficients of the gases in the melt. Of particular importance are the values of each of just mentioned properties in the range of 1800–2400 °C, the typical temperatures of melting.

6.3.1 Surface Tension

There are only three papers devoted to surface tension in silica melts [15–17]. The glasses studied in [17] were found to have a surface tension coefficient of 0.305 ± 0.07 J/m^2 within 1950–2000 °C. There are not yet any reliable data concerning the dependence of the surface tension on temperature. The influence of ambience (vacuum, air, air with water vapor, nitrogen) has not been described either [15].

6.3.2 Density

The density of silica glass has been studied quite comprehensively. At room temperature the density does not depend significantly on the concentrations of typical impurities and is equal to 2.202 ± 0.005 g/cm^3. Only the concentration of OH-groups has some influence on the density: increasing this parameter at room temperature from a mass percentage of 0.0005 (glasses of type I) to that of 0.2 (glasses of type III) leads to decreasing the density by 0.006 g/cm^3. It is silica glass that has the lowest TEC (thermal expansion coefficient) among all the studied oxide glasses. The temperature dependence of the density – and, therefore, that is the case for TEC – exhibits an anomalous behavior and significantly depends on the type of a given glass, mainly on the content of OH-groups, and on the thermal history, but its absolute value varies within a very narrow range. Monograph [3] and the paper [18] have comprehensively discussed this property. Above the range 1600–1700 °C, the temperature dependence of density becomes of the usual type, i.e. the lower is the temperature the lower is the density. At 1700 °C the density is 2.190 g/cm^3. There is only one paper devoted to this dependence for the case when melting is conducted within 1900–2400 °C, i.e. in the equilibrium-melt region of the phase diagram of silica. This work is described in [19], the results are represented on Fig. 6.3.

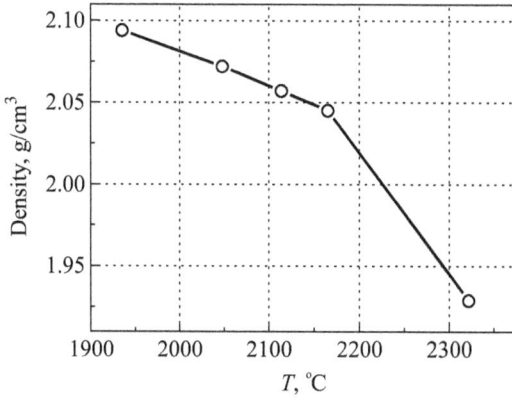

Fig. 6.3: Density of silica melt [19].

6.3.3 Viscosity

Researchers studied the viscosity of silica melts quite comprehensively. Paper [20] generalizes and interprets the scientific investigations on viscosity of silica melts. The paper includes a number of data concerning the influence of raw materials and manufacture techniques on the viscosity of the main types of commercial silica glasses [21].

As compared to other glass-forming melts, one should mention three features regarding the viscosity of silica melts. Firstly, it is the most high-melting one of all the studied glass-forming oxide melts. Secondly, within a rather wide temperature range (1200–2500 °C) a simple exponential function of Arrhenius's type fairly well describes the temperature dependence of viscosity. Thirdly, the viscosity of silica melt very strongly depends on the impurities and thermal history and can vary within a very wide range. In the 1000–1400 °C range, which covers the glass-transition temperatures of various silica glasses, the interval for viscosity makes up two or three orders of magnitude.

Fig. 6.4 demonstrates some data on the temperature dependence of viscosity exhibited by a number of silica melts of different types and from different manufactures. The dependence is considered from 1600 to 2500 °C, i.e. in an interval covering the region of fusing the batch and processing the melt. In this temperature range the viscosity values of silica melt do not go beyond an interval making up about one order of magnitude: from $10^7 - 10^8$ Pa·s at 1600 °C to $10^2 - 10^3$ Pa·s at 2500 °C. Most of the researchers have confined the temperature of measuring the viscosity by the upper limit within 2200–2300 °C. We have obtained the viscosity values up to 2500 °C with the help of extrapolation. There is only one work [22], which experimentally covers the temperature measurement interval of 1600–2480 °C. The data obtained in this range are located in the central part of the interval of experimental viscosity values corresponding to this temperature range.

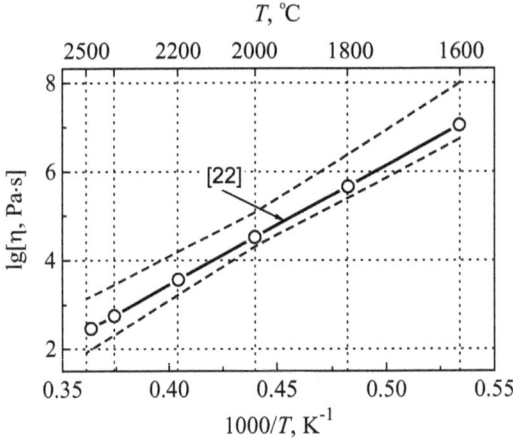

Fig. 6.4: Temperature dependence of viscosity of silica melts [20–22].

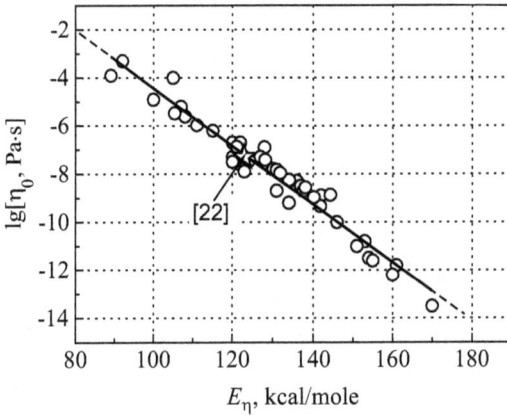

Fig. 6.5: Correlation between pre-exponential term $\lg \eta_0$ and viscous-flow activation energy E_η of silica melts.

As mentioned above, the temperature dependence of viscosity is described by the following equation:

$$\lg \eta = \lg \eta_0 - \frac{E_\eta}{2.303RT}.$$ (6.6)

Values of the pre-exponential term $\lg \eta_0$ and those of viscous-flow activation energy E_η exhibit a considerable discrepancy between one another when the results, obtained by different researchers, are compared. Fig. 6.5 represents the correlation between the two parameters, which corresponds to the so-called compensation rule: the higher is E_η, the lower is η_0. If one discusses the behavior in coordinates of $\lg \eta_0$ vs. E_η, all the points grouped rather tightly near the straight line of regression. The activation energy E_η varies from 80 to 180 kcal/mol, and the pre-exponential factor $\lg \eta_0$ [in Pa·s] from −3 to −14. The melts that correspond to glasses containing a minimum of impurities (type I) have a high value of viscous-flow activation energy and form a group in the right part of the graph. Glasses of types II and III are grouping in the left part. The point corresponding to the data of [22] is on the regression line, close to its middle.

6.3.4 Solubility and Diffusion of Gases

Concerning silica glasses and silica melts, one can divide the involved gases into two groups. The first group consists of those which do not chemically interact with silica – He, Ne, Ar, and Kr, i.e. noble gases, as well as N_2, a practically inert gas. The second one consist of the gases capable to interact chemically with silica at high values of temperature, to wit: oxygen, hydrogen, water vapor, chlorine and hydrogen chloride, carbonic gas and carbon monoxide as well as the gaseous products of chemical interaction between the melt and some of the materials of the electrodes and shape-forming equipment – graphite, molybdenum, tungsten. The first group of gases interacts with silica in the way of physical dissolution and diffusion. At certain conditions the second group can interact with silica apart from physical dissolution and diffusion, by acid-base and oxidation-reduction reactions. The present review will not consider one or another situation at different phase boundaries and inside the melt. The discussion will be restricted to the published data concerning the solubility and diffusion of some gases in silica melt.

Inert Gases

The physical solubility and diffusion of a gas in a silica melt is mainly determined by the ratios of the size of its molecule (atom) to a few characteristics of the glass (melt). The basic unit of silica glass and silica melt is the tetrahedron $SiO_{4/2}$. Tetrahedrons of $SiO_{4/2}$ are connected to one another by their vertices via the joint oxygen atoms, making rings. The atomic bonding of Si-O is definitely directed and to a more extent covalent than ionic. The ring under discussion can consist of four to eight tetrahedrons.

The capability of the atoms (molecules) of a given gas to dissolve physically and diffuse in glass is due to the following factors: the spatial distribution of the structural voids (solubility sites) in size, the lengths of jump paths and the dimensions of the doorways between adjacent solubility sites. In [23–30] the structural aspects of physical solubility and diffusion of the gases have been discussed, in particular, statistics of rings, topology of void subspace and migration ways. These data and information on such properties of the atoms (molecules) of a discussed gas as size, mass, and vibration frequency provide the basis for calculating its physical solubility and diffusion in silica glass. The calculation results are in satisfactory agreement with the experimental data. Table 6.5 represents the sizes of the atoms (molecules) of some gases for which enough data are available on their solubility and diffusion in silica glass.

Figs. 6.6 and 6.7 represent the temperature dependences of, respectively, the diffusion coefficient and of solubility of the noble gases in silica glass from various sources. The temperature dependencies corresponding to these properties are satisfactorily de-

Table 6.5: Sizes of the atoms (molecules) of some gases

| Gas | Sizes (diameter) of the molecules in nanometer | | | |
	[23, 24]	[26]	[27]	[31]
He	0.256	0.195	0.2	0.17, 0.196
Ne	0.275	0.24	0.24	0.32
Ar	0.341	0.32	0.32	0.384
Kr				0.396
H_2		0.25	0.25	
D_2		0.255	0.25	
N_2		0.34		
O_2	0.358	0.315	0.32	

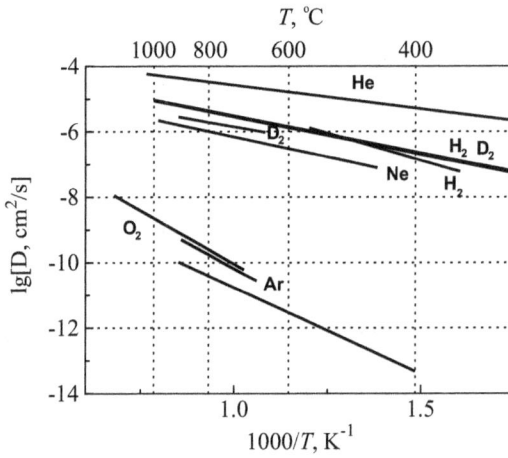

Fig. 6.6: Temperature dependence of physical (molecular) diffusivity of a gas in silica glass.

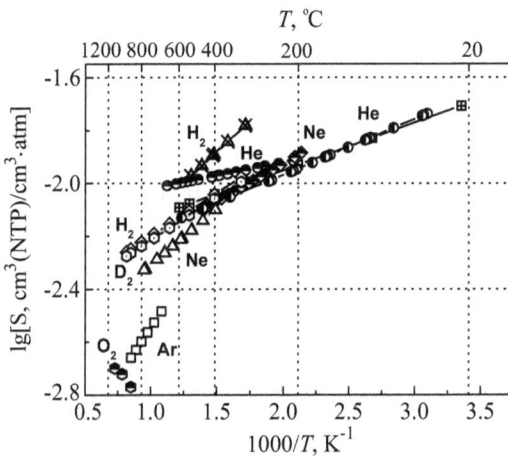

Fig. 6.7: Temperature dependence of physical (molecular) solubility of a gas in silica glass.

scribed by a simple exponential function of Arrhenius's type:

$$\lg D = \lg D_0 - \frac{E_D}{2.303RT},$$ (6.7)

$$\lg S = \lg S_0 - \frac{H_S}{2.303RT},$$ (6.8)

where D and S are the coefficients of, respectively, diffusion and solubility; E_D and H_S are, correspondingly, the diffusion activation energy and the dissolution heat of a gas; D_0 and S_0 are pre-exponential factors. Table 6.6 represents the corresponding values of D_0 and S_0, E_D and H_S. To evaluate the solubility coefficient S, the temperature dependence of which is represented in Fig. 6.7, we used either the data obtained from measuring the saturation and desorption of a given gas or the data calculated from the equation $K = DS$ where K is the permeability coefficient.

Table 6.6: Pre-exponential factors and activation energies for diffusivity (D), and pre-exponential factors and heat of solution (S) of gases in silica glass. Note (*), the concentration of dissolved chlorine is directly proportional to the fourth root of the chlorine pressure in the gaseous atmosphere (D_0 in cm^2/s], E_D in cal/mol; S_0 in cm^3(NTP)/s m^{3}atm; H_S, in cal/mol; temperature range in °C)

Gas	lg[D_0]	E_D	lg[S_0]	H_S	Temperature range	References
He	−3.131	6613	−2.13	−680	24–1034	[33]
	−3.26	6040	−2.309	−820	25–533	[34]
Ne	−3.657	11360	—	—	450–980	[35]
	−4.292	9550	—	—	630–984	[34]
	—	—	−2.725	−1913	50–443	[36]
	—	—	−2.503	−2000	208–616	[37]
Ar	−3.922	28670	−3.305	−3470	650–900	[34]
	−5.5	24100	−3.453	−2415	440–560	[38]
			—	—	400–900	[39]
H$_2$	−3.428	10370	−2.526	−1490	300–1000	[46, 48]
D$_2$	−3.300	10520	−2.534	−1470	300–1000	[46, 48]
H$_2$	−2.20	11400	—	—	200–800	[50]
D$_2$	−3,824	9260	—	—	25–550	[34]
H$_2$			−2.577	−2123	309–493	[36]
O$_2$	−3.553	23587			700–1100	[40]
−O−	−8.7	29060	—	—	830–1250	[41]
	0.415	108382	—	—	1200–1400	[44]
−OH	−6.0	18300			600–1200	[51]
	−5.818	19820			440–1150	[52]
CO, CO$_2$	−5.02	34000			1000–1400	[10]
Cl$_2$	−3.498	51000	1.836*	7390	1160–1700	[58]

We have to mention some important regularities: (i) Firstly, if taking any noble gas, the diffusion coefficient sharply gets larger with increasing temperature, whereas the solubility coefficient becomes lower, but to a less degree. (ii) Secondly, as the size of a diffusing particle becomes higher, the diffusion coefficient D abruptly decreases, and E_D significantly increases; at the same time, the solubility coefficient S gets lower.

Anderson and Stuart's [32] diffusion model stipulates the diffusion activation energy E_D to be directly proportional to r^2, where r is an atomic or molecular radius of a particle diffusing in a silica glass. The parameter E_D is estimated as the elastic energy required to widen the doorway of the radius r_d up to r. Fig. 6.8 represents the dependence of E_D on the radius of a diffusing particle employing data from different authors; the dependence demonstrates a fair agreement between the assumption and experimental data.

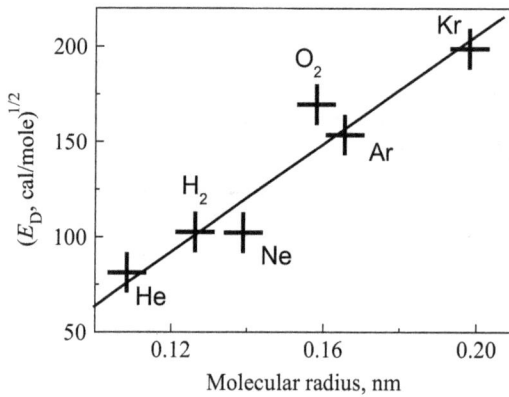

Fig. 6.8: Dependence of diffusion activation energy E_D on the radius of a diffusing particle in silica glass.

Oxygen, Hydrogen, and Water

The substances may not only dissolve in the glass and melt, they can also react with the silicon dioxide. This will have a significant influence on the formation and evolution of the bubbles containing these gases. One may a priori assume that in silica glass there are two diffusion mechanisms which involve *oxygen*. One of them is the diffusion of O_2 molecules in an interstitial way, the diffusion coefficient denoted by D_{O_2}; the other is diffusion of bridging oxygen atoms (ions) of the network, which can be defined as the self-diffusion of network oxygen. Besides, oxygen can diffuse in the form of non-bridging ions connected with the hydrogen ions, forming the groups of OH, and with the impurity cations. All the three mechanisms – the physical dissolution and the gas exchange by means of either the bridging (network) oxygen or non-bridging one – may act in the system "gas medium – glass".

Norton [40] seems to have been the first to measure the diffusion coefficient and solubility of molecular oxygen in silica glass. Since the molecules of O_2 and Ar do not differ much in diameter, the corresponding values being within 0.32–0.36 nanometers (see Tables 6.5 and 6.6 and Figs. 6.6, 6.7, and 6.8), the parameters of their physical solubility and diffusion will not show considerable divergences. Afterwards, the discussion was devoted to various mechanisms and some quantitative characteristics of oxygen in silica glass as well as in glassy films of silica. These topics were discussed in the papers by Williams [41], Revesz [42], Schaeffer [43], Mikkelsen [44]. It is appropriate to briefly set out the principal regularities. Different authors state that within 1000–1400 °C the value of D_{O_2}, the diffusion coefficient of molecular oxygen, is 5–7 orders of magnitude higher than that of D_O, the diffusion coefficient of network oxygen (see Figs. 6.6 and 6.9). At the same time the value of E_{DO_2}, the diffusion activation energy of molecular oxygen, the value being about 24 kcal/mol, is lower than that of E_{DO}, the diffusion activation energy of network oxygen. In such a way the paper of Mikkelsen [44] states that E_{DO} is about 108 kcal/mol, i.e. it is quite comparable with the energy of the bond Si–O. The value of S_{O_2}, the solubility coefficient of molecular oxygen in silica glass, is stated to be about $5 \cdot 10^{16}$ molecules/cm³·atm within 950–1000 °C [40], while the value of C_O, the concentration of network oxygen, is about $4 \cdot 10^{22}$ atoms/cm³, i.e. 6 orders of magnitude higher. The network oxygen mechanism and the molecular-oxygen mechanism provide diffusion fluxes comparable in value. According to [42], the fluxes are conjugate, which points to the possibility of the molecular oxygen migrating in the glass via its exchange with the network one.

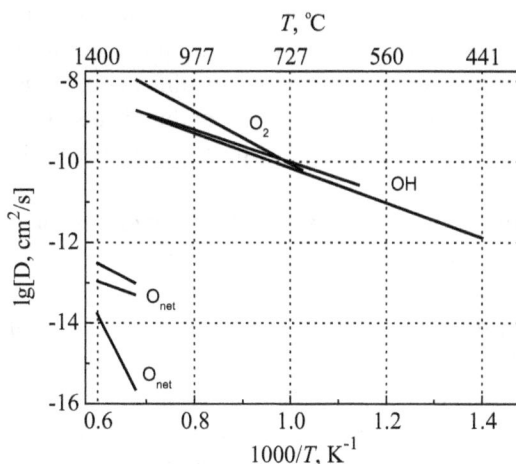

Fig. 6.9: Temperature dependence of the diffusivity of O_2, O_{net}, and –OH in silica glass.

The processes of transport of oxygen in silica melt are influenced by the reaction of thermal dissociation of SiO_2 and by the formation of silicon compounds not saturated with oxygen; the compounds can be represented as the structural groups $Si^{3+}O_{3/2}$, $S^{2+}O_{2/2}$ and so on. The initial stage of the reaction

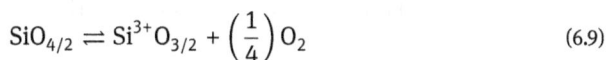

$$SiO_{4/2} \rightleftharpoons Si^{3+}O_{3/2} + \left(\frac{1}{4}\right)O_2 \qquad (6.9)$$

develops at the temperature as low as 1200 °C either in vacuum or under neutral conditions. If the temperature is above 1400 °C, the vacuum, the neutral or, especially, reducing gaseous atmosphere provides dissociation of the silicon dioxide to yield gaseous silicon monoxide in accordance with the reaction

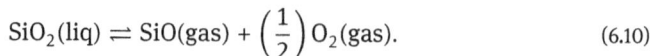

$$SiO_2(liq) \rightleftharpoons SiO(gas) + \left(\frac{1}{2}\right)O_2(gas). \tag{6.10}$$

Under the fusing temperature above 1800 °C and neutral conditions, namely, vacuum or an atmosphere of inert gases, the partial pressure of silicon monoxide gas is 1.5–2 orders of magnitude higher than that of silicon dioxide vapor. The solubility of SiO in the silicon dioxide melt is much higher than that of O_2. The silicon compounds not oxidized completely are stable only in the high-temperature region (above 1400 °C), while at lower values of temperature they tend to be fully oxidized. The oxidation takes place owing to either of the following process. The first is reducing some of the other oxides (for example, Na^+, Ca^{2+}, or some of the OH-groups) to the metals or hydrogen. The second is the disproportionation into SiO_2 and Si. It is clear that these chemical processes make their own contributions to the oxygen gas exchange between the silica glass or melt, the gaseous furnace atmosphere, and the bubbles.

Hydrogen and *water* can be present in different states either in the raw material or in the furnace atmosphere under fusing and treating thermally a silica glass (see above). The ratios between the various forms of hydrogen and water in a silica glass are determined quantitatively by the concentration and temperature conditions of some physical-chemical interactions in the system $SiO_2-H_2-H_2O$. Up to a temperature between 600–700 °C the molecular hydrogen physically dissolves and diffuses in a silica glass without appreciable chemical interaction. When the temperature is above the mentioned one, the substance begins to react, and the hydrogen dissolves and diffuses in a few chemically bounded forms.

The molecular water dissolves and diffuses in silica glass at a temperature of up to 600 °C and under high pressure, but some chemical interaction takes place concurrently. There is very hardly any dissolution and diffusion of the molecular water at higher temperature, the two processes develop in a few chemically bounded forms. In [45] the interaction between the water as overheated vapor and silica glasses of different types was comprehensively studied. The experimental conditions include the temperature from 350 °C to 600 °C, the pressure from 20 to 300 atm, and the duration from 10 to 300 hours. The molecular water can penetrate into the surface layer not deeper than 40–80 μm, with its concentration at the surface making up 1–3 %. The groups of -OH will advance under the same conditions as deep as 150 μm, whereas the concentration of them lies within 2–4 %. The higher is the temperature of molecular water, the deeper is the penetration while the lower is the concentration. With respect to the hydroxyl groups, the higher is the temperature, the more considerable are both the penetration depth and the concentration.

As reported by different authors [3, 33, 46–50] the molecular dissolution and diffusion of hydrogen or heavy-hydrogen are described within 25–1000 °C by the following quantities: the diffusion activation energy E_D is 9.3–10.4 kcal/mol, $\log[D_0, \text{cm}^2/\text{s}]$ lies within the range $-2.2 \div -3.8$, the solution heat H_S is $-1.4 \div -2.1$ kcal/mol, and $\log[S_0, \text{cm}^3(\text{NTP})/\text{cm}^3\cdot\text{atm}]$ lies within $-2.5 \div -2$ (see Table 6.6). With respect to molecular hydrogen, Figs. 6.6 and 6.7 represent the temperature dependencies of diffusivity and solubility, respectively. These temperature dependencies of hydrogen do not differ much from the corresponding dependencies of neon, which is due to only slight difference between the size of the hydrogen molecule and that of the neon molecule (see Table 6.5 and Fig. 6.8). The solubility of molecular hydrogen is directly proportional to its pressure in the gaseous phase. Some region above 700 °C exhibits a notable chemical interaction of water and hydrogen with silica, the interaction becoming the predominant factor in the dissolution and diffusion at the still higher values of temperature.

The following chemical reactions may take place between hydrogen or water vapor and silica:

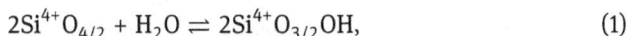

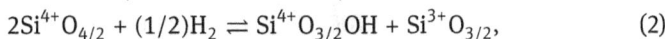

$$2Si^{4+}O_{4/2} + H_2O \rightleftharpoons 2Si^{4+}O_{3/2}OH, \tag{1}$$

$$2Si^{4+}O_{4/2} + (1/2)H_2 \rightleftharpoons Si^{4+}O_{3/2}OH + Si^{3+}O_{3/2}, \tag{2}$$

$$2Si^{4+}O_{4/2} + H_2 \rightleftharpoons Si^{4+}O_{3/2}OH + Si^{3+}O_{3/2}H, \tag{3}$$

$$3Si^{4+}O_{4/2} + H_2 \rightleftharpoons 2Si^{4+}O_{3/2}OH + Si^{2+}O, \tag{4}$$

$$Si^{4+}O_{4/2} + H_2 \rightleftharpoons Si^{2+}O + H_2O. \tag{5}$$

The silicon monoxide partly volatilizes, and dissolves to some extent in the melt, where it reacts with the silicon dioxide to form groups of $Si^{3+}O_{3/2}$. Thus, silica glass and silica melt can contain hydrogen and water in three different forms. The first form consists of molecules of hydrogen and water[1]. The second one includes the groups of Si–OH formed through reaction (1). The third form includes the groups of Si–H and those of Si–OH formed through reactions (2), (3), or (4), i.e. in the form of –OH and –H radicals connected with moderate part of the silicon atoms via valence bonds. Besides, some reducing groups including the trivalent silicon arise in silica glass. In [51–57] these problems are discussed, and Leko [3] has generalized various aspects of the dissolution and diffusion of hydrogen and OH-groups in silica glass and silica melt.

Our consideration is restricted to a concise account of the principal features of these reactions, since they are important for the matter under discussion: (i) Each of the reactions is endothermic, i.e. an increase in the temperature shifts the equilibrium to the right, and a decrease, to the left. (ii) The reactions take place on the silica surface, and the resulting hydrogen-containing groups diffuse deep into the glass or melt.

---

[1] To use the term "water" for –OH groups does not look quite appropriate, but it is often found in literature.

(iii) The reactions are reversible, i.e. changing the concentration of molecular hydrogen and/or water vapor in the gaseous atmosphere will accordingly vary the concentrations of hydrogen-containing groups in the glass or melt. The rate of the process is determined by the ratio of the chemical-reaction rate to those of diffusion processes as well as by the solubility of $-OH$, $-H$, H_2O and H_2 in the melt or glass. (iv) The groups of $Si^{3+}O_{3/2}$ and $Si^{+3}O_{3/2}H$ are reducing agents, and that of $Si^{4+}O_{3/2}OH$ is an oxidizer. As reported by Moulson [51] and Davis [52], the diffusion parameters of $\cdot OH$ groups in silica glass within 400–1200 °C are the following: D_0 is $1 \div 1.6 \cdot 10^{-6}$ cm^2/s, and E_D is 18.3–19.8 kcal/mole. If one takes an interval overlapping the mentioned one, the diffusion coefficient of $-OH$ groups is 3–4 orders of magnitude lower than that of molecular hydrogen (see Figs. 6.6 and 6.9).

The total hydrogen content in silica glass, including both the physically dissolved and chemically bounded hydrogen, is of importance for forming the bubbles. According to the data from Lee [46], Fig. 6.10 compares the solubility of molecular hydrogen and its total content in silica glass within 440–1000 °C. One can see that the higher is the temperature, the lower is the physical solubility of hydrogen while its total content gets higher at the expense of the chemically bounded hydrogen, the increase being most remarkable at values above 700 °C. In addition, the content of chemically bounded hydrogen is directly proportional to the square root of the pressure of hydrogen or water vapor in the gaseous phase. At 1000 °C this content exceeds 95 %.

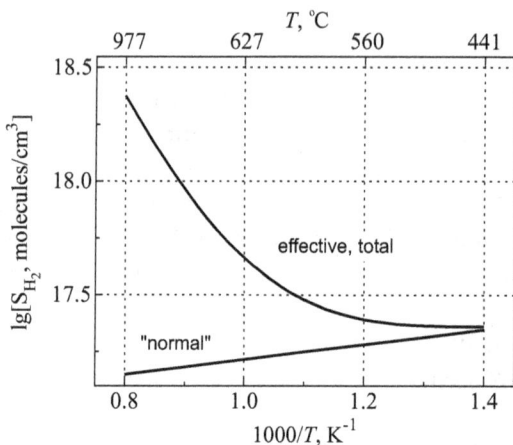

Fig. 6.10: Experimental (total) solubility for H_2 in fused silica as compared with the "normal" molecular solubility [46].

As described above, groups of $-OH$ in silica glass have different origins and form through different reactions. If taking a given silica glass, the total content of differently originated $-OH$ groups and the quantitative ratios between them are due to the manufacture way, the gaseous-atmosphere composition, and the water content in the raw material. Though the groups of $-OH$ are deprived of individuality when included in silica glass or melt, their further behavior is due to the ratios between the solubilities and the diffusivities of different hydrogen forms in the glass as well as the concentration of structural groups functioning as reducing agent.

To decrease the total concentration of (–OH)-groups in silica glass or melt, the reactions must be shifted to the left. One can realize it by decreasing temperature or concentration of hydrogen in the gaseous medium lower, or by conducting the two operations simultaneously. The rate of destructing the groups of –OH is limited by the rate of removing the molecular hydrogen from the melt in a diffusive way. Since the diffusion coefficient of hydrogen is 3–4 orders of magnitude higher than that of -OH groups, a thermal treatment in a gaseous medium without hydrogen (vacuum, dry air, or argon) will make reactions (2), (3), and (4) come to an end, and, therefore, the corresponding quantity of –OH groups will be removed from the glass. Some papers define these groups of –OH as "metastable" ones. In this connection the groups functioning as reducing agents will disappear in equivalent quantity. The excess amount of –OH groups, both formed through reaction (1) and inherited from the raw material, is much harder to remove from the glass. These groups of –OH are defined as "stable" ones.

Shifts in reactions (2), (3), and (4) towards the initial substances under decreasing temperature lead to supersaturating the melt with molecular hydrogen, which makes it possible for bubbles to arise and grow in the melt. It takes place if the furnace atmosphere contains hydrogen. In [12, 13] this phenomenon is studied experimentally. The nucleation of bubbles can be homogeneous or heterogeneous latter one proceeding either on the walls of the crucible or also in the bulk of the melt.

Chlorine

While silica glass is synthesized from volatile compounds of silicon, usually, $SiCl_4$, there is some molecular chlorine in the furnace atmosphere. Hermann [58] has studied the interaction between molecular chlorine and silica melt within 1160–1960 °C and under the pressure of chlorine from 0.5 to 13.5 atm. The dissolution and diffusion of chlorine in the melt must take place in the form of groups $Si^{4+}O_{3/2}Cl$. The process of dissolution and diffusion is described by the following values of the parameters:

$$D_0 = 3.18 \cdot 10^{-4} \text{ cm}^2/\text{s}, \qquad E_D = 51 \text{ kcal/mole}, \qquad (6.11)$$

$$S_0 = 1.836 \text{ cm}^3/\text{cm}^3 \cdot \text{atm}^{1/4}, \qquad H_S = 7.39 \text{ kcal/mole}. \qquad (6.12)$$

The concentration of the dissolved chlorine is directly proportional to the fourth root of the chlorine pressure in the gaseous atmosphere; the solution heat has a positive value, i.e. the higher is the temperature the higher is the solubility of chlorine. If the respective temperature values are about 1200 °C, the solubility of chlorine is two orders of magnitude higher than those of oxygen and argon, while the diffusion coefficient of chlorine is two orders of magnitude lower than those of the others.

6.4 Summary to Part I

There are three main technology sources of gas-containing bubbles in the melt: the starting materials, the furnace atmosphere, the materials of the heaters and shape-forming equipment. The properties having an influence on the formation and evolution of bubbles in the melt are density, surface tension, and viscosity of the melt as well as solubility and diffusion of the gases. In case the concentration of the gas dissolved in the melt has proved to be higher than the equilibrium one, the tendency to equilibrium can be realized in the two following ways: (i) The first way includes the diffusion of a given gas towards the free surface of the melt and, then, its passage through the surface. The two-stage process takes place when the concentration of the gas, i.e. partial pressure, in the furnace atmosphere is lower than the value corresponding to the equilibrium concentration of the gas in the melt. (ii) The second way includes formation of bubbles in the melt the gases diffusing into them.

The lower is the temperature, the higher are the physical solubility of gases in the glass or melt. Thus, while some values of temperature provide equilibrium, the decrease of temperature results in a concentration of a physically dissolved gas which is lower than the equilibrium one. In such cases, bubbles cannot be formed. If there is some chemical interaction, increasing the temperature, then, as a rule, the solubility of gases in the melt increases. In such cases, the melt becomes supersaturated with respect to this gas. Either decreasing the concentration (partial pressure) in the furnace space that was in equilibrium with the melt or glass or removing a given gas from this space also leads to such a supersaturation. In both cases the tendency to an equilibrium concentration is realized not only due to diffusion and evaporation of the gases through the free surface but also due to diffusion of them into bubbles. The formation of bubbles can proceed in a homogenous or heterogeneous way on some of the defects (for example, at the boundary with the solid wall of the fusion pot, on some of the inclusions, etc.). The equilibrium conditions in the system (the bubbles – the melt – the free surface) are different from those realized in the first case. For the second case, the capillary force determined by the curvature of the surface of the phase boundary "gas-melt" and the Archimedean force are effective making the bubbles floating up. Both factors are considered below.

Many-years' manufacture of the discussed production has led experts to various design and technology solutions providing yields of high-quality transparent silica glass. The solutions must be based on the following principles (regulations). The first rule is a preliminary treatment of raw materials that has to remove as much gas-forming components and gas-forming impurities as possible. The second implies the conduction of the process in vacuum or in a medium composed of gases having a rather high diffusion coefficient in silica melt. The third condition consists in a design of the fusion procedure and the temperature-time mode which removes gases from the batch as thoroughly as possible prior to the capsulation of the batch by the melt.

Part II: Theoretical Analysis and Computer Simulation of the Process

6.5 Introduction to Part II

The manufacturing of a silica glass includes a variety of stages, and experimental testing of them is very laborious and expensive. Some of the stages are impossible to reproduce under laboratory conditions. Theoretical analysis and computer simulation of some stages of the manufacturing process can help substantially in producing a bubble-free silica glass. The introduction to Part II is directed to the discussion of the following topics: (i) The main stages of the fusion of powdered silica under heating as well as the formation of a bubble structure. (ii) Selection of the parameters for the temperature dependence equations that describe the properties having an influence on the kinetics of the process.

6.5.1 Main Stages of Fusion of Powdered Silica under Heating and Evolution of Bubble Structure

When one produces a silica glass, the initial state of the starting material, i.e. batch, is a layer of powdered silica having some thickness (a powdered body). From one side, the layer is in contact with a solid substrate (the surface of the melting chamber), and with the gas of the furnace from the other side.

Phenomenologically, the fusion of the powder layer under heating is a physical process of transforming a material from a dispersed into a compact state. Each stage is characterized by its own degree of densification. While heating a layer of glass powder on a solid substrate, the formation of the melt layer goes through the following stages described in [59, 60]: (i) The first stage is rearrangement of the solid particles and formation of zones (non-isotropic zonal consolidation). (ii) The second one includes the onset of spheroidization of the particles, conglutination of them and formation of contact necks, adhesion of the particles to the substrate, formation of channels between the particles. (iii) The third one consists in the collapse of the channels. At this moment capsulation of the layer and formation of closed pores take place. The capsulation leads to the fact that the gas transfer via filtration through the layer of particles is substituted by that via diffusion through the melt layer containing closed pores. (iv) The fourth stage includes spheroidization of the pores or, in other words, formation of the bubbles. Evolution of the bubbles in vacuum leads to collapse of them while in gaseous atmosphere theirs size is stabilized, and floating up of the bubbles, and gas exchange in the system "furnace atmosphere ↔ melt ↔ bubble" takes place. (v) The fifth one consists in smoothing the surface of the melt.

As described above, manufacture of silica glass employs quartz and cristobalite crystal modifications of silica. At temperature values above 1400 °C, pure quartz starts melting slowly from its surface to give a highly viscous melt. Cristobalite melts at a temperature of about 1730 °C. Quartz and cristobalite are capable to exist when over-heated, i.e. at temperatures considerably exceeding that of melting (the difference lies within 200–300 K for quartz and makes up 40 K for cristobalite). Tridymite is not an intrinsic phase of crystalline silica, impurities of Na, K, and Ca (up to 1% in total) stabilize this phase. When heated above 1450 °C, tridymite begins to turn into cristo-balite and forms a high-viscosity silicate melt including the mentioned impurities; this phenomenon is defined as peritectic melting. Paper [61] has shed some light on the ki-netics of phase transformations of crystalline silica modifications.

Dispersed bodies may evolve via two main mechanisms: viscous flow and diffu-sion. The first one refers to the liquid state of a substance, a glass-forming melt in our case, while the latter concerns the solid state of a substance, i.e. a crystalline mate-rial. In the first case it is more correct to define the process as coalescence (merging) of drops, while in the second case as sintering of the powder. However, densification of powder under heating is often given the name of sintering regardless of the mech-anism involved.

When one manufactures silica glasses, the temperatures of fusion are commonly found within the range 1800–2400 °C, i.e. above the melting temperature of crys-talline silica modifications. Under actual conditions of manufacturing of a silica glass the particles of powdered crystalline silica begin to get sintered at a temperature about 1500 °C, thus the particles become non-moving with respect to one another. However, the packing density of the powder layer does not attain the stage of forming the occluded cavity until the temperature range 1700–1750 °C is reached. Thus, when one simulates fusion, it is quite permissible to take a silica glass as a powdered mate-rial and, therefore, to consider the viscous flow of the melt as the main mechanism of densification.

6.5.2 Selection of Parameters for the Temperature Dependence Equations that describe the Properties of the Silica Melt Affecting the Kinetics of the Process

Density, surface tension, and viscosity of silica melts, as discussed in Part I, have been measured in the region of fusion temperatures. The gas-diffusion properties have been measured experimentally only for the vitreous state of silica. In the glass transition re-gion most glasses change the kind of temperature dependencies of their properties. In addition, transport processes tend to increase the temperature coefficient in increas-ing the temperature. Thus, one has always to check the validity of the assumption that it is possible to use the temperature coefficient of a property, in particular, the diffusion activation energy and the dissolution heat of a given gas, to extrapolate the values of these properties to the high-temperature region.

It was experimentally demonstrated in [62] that it is permissible to extrapolate the temperature dependencies of solubility and diffusivity of helium in silica glass to the melt region, 1400–1960 °C. As mentioned above, the temperature dependence of the viscosity is described for each silica glass by an equation which includes E_η, the temperature-independent activation energy of viscous flow. Moreover, the density of silica melt does not differ much from that of silica glass. These facts lead to the conclusion that the concentration of structural voids or solubility sites does not change substantially in the glass transition region, more precisely from glass to melt. Thus, to estimate the gas-diffusion properties of silica melt with respect to noble gases, the diffusion activation energy and dissolution heat in silica melt can be taken to be close to those in silica glass. Carrying out a simulation, it is quite permissible to take the parameters of temperature dependence of diffusivity and solubility as constant. Table 6.7 demonstrates the results of our approximating the temperature dependencies of the silica melt properties that are important for simulating the process.

Table 6.7: Approximation of the temperature dependencies of a few properties of silica melts (* cm^3(gas, NTP)/[cm^3(glass)·atm]: gas volume at temperature 0 °C and pressure 1 atm; Dimension of E_η, E_D and H_S is kcal/mol)

Properties	Equations of temperature dependence	Dimension
Density, ρ	$-2.17851 + 4.39 \cdot 10^{-3}T\,(°C) -1.1286 \cdot 10^{-6}T^2\,(°C)^2$	g/cm^3
Surface tension, σ	0.305 ± 0.07	J/m^2
Viscosity, lg[η]	$-7.3 - 123/(2.303RT)$	Pa·s
Diffusivity and solubility of the gas		
Helium, lg[D]	$-3.2\ \ -6.32/(2.303\ RT)$	cm^2/s
Helium, lg[S]	$-2.21\ \ -(-0.75)/(2.303\ RT)$	$cm^3/(cm^3·atm)$ *
Argon, lg[D]	$-4.7\ \ -26.4/(2.303\ RT)$	cm^2/s
Argon, lg[S]	$-3.4\ \ -(-2.9)/(2.303\ RT)$	$cm^3/(cm^3·atm)$ *

6.6 Micro-rheological Model and Computer Simulation of the Process

The process of spontaneous densification of disperse particles is based on the following general principle. The driving force of reshaping each particle and forming an aggregate of them is the tendency of a discussed system to decrease its free surface energy. The resistance of a material to reshaping its particles is retarding the process. In [63], the main stages of developing the theory of sintering have been reviewed. According to Frenkel, whose paper [64] has proven to be the basic one in solving this problem, the work done by surface tension forces, while sintering the particles, is equal to the energy of dissipation under viscous flow. This statement can be expressed as the main sintering equation in the form

$$-\sigma_{MG}\Delta S_{MG} = W_\eta \Delta t. \tag{6.13}$$

One can supplement this equation with the main equation of spreading on a solid substrate, which is based on the same principles

$$(\sigma_{GS} - \sigma_{MS})\Delta S_{MS} - \sigma_{MG}\Delta S_{MG} = W_\eta \Delta t, \tag{6.14}$$

where σ and ΔS are, respectively, the surface tension and the change of the interface area of any one of the corresponding phase boundaries (M – melt, G – gas, and S – solid), W_η and Δt are, respectively, the power of viscous losses and the short time interval in which the interface area has changed by ΔS. Later a number of papers, reviews, and monographs were published devoted to this problem, for example, [65–72]. The foundation of the present work is a model developed by us earlier and published in [60].

6.6.1 The Micro-rheological Model of Powder Sintering and Structuring of a Porous Body

The model connects the following parameters. From one side, it reflects the initial characteristics of the system like shape and size of the particles, the manner of their packing, a few physical-chemical properties of the substance of the particles and properties at the phase boundaries. Form the other side, several characteristics of the arising melt layer including the gas phase are incorporated as well as the changes of these characteristics in the course of densification: the total volume of the layer under sintering, the size of the constituents of the layer (groups of particles, channels, cavity, the shape of the surface and so on). The following assumptions are made: (i) All the particles have the same initial shape (a ball) and size; the packing implies placing the particles in the nodes of a regular lattice; all the initial contacts between neighboring particles are points and equivalent; (ii) the substance of the particles is an incompressible Newtonian liquid (a glass-forming melt, in our case), the viscosity of which in the volume is equal to that at the phase boundaries; (iii) the mechanism of reshaping each particle, different groups of them, and all the ensemble is viscous flow; (iv) there is no chemical interaction between the substance of the particles and that of the substrate, the surface of the latter being taken solid, smooth, and isotropic; (iv) there is no gravitation; later we will substantiate this assumption, but still further the influence of gravitation will be considered; (v) the gas of the furnace atmosphere does not chemically interact with the substance of the particles, moreover the gas neither dissolves nor diffuses in it (later we will consider the case of physical dissolution and diffusion of the gas in the melt); (vi) all the constituents of the system have the same temperature at any instant of the simulation interval. Parallel to the terms "particle", "sintering", and "densification", we will use "drop", "merging", and "coalescence". The simulation of the process comprises three parts: a geometrical analysis, a consideration of the hydrodynamics (a calculation of the viscous losses), and of the process kinetics.

Geometrical Analysis

The geometrical analysis considers the following processes: the packing and shifting of particles (drops), reshaping of them during their merging, formation of the new structure blocks, changing of the interface areas, and the part of the melt that is under deformation.

Fig. 6.11 demonstrates the picture of the initial state of a powder multi-layer on a substrate (1). A few geometrical elements of the multi-layer are marked on the figure which describe the contributions of the various situations and stages to densification: (2) two contacting drops; (3) a group of drops inside the layer; (4) one drop on the substrate; (5) the monolayer of drops on the substrate, (6) the blanket of drops.

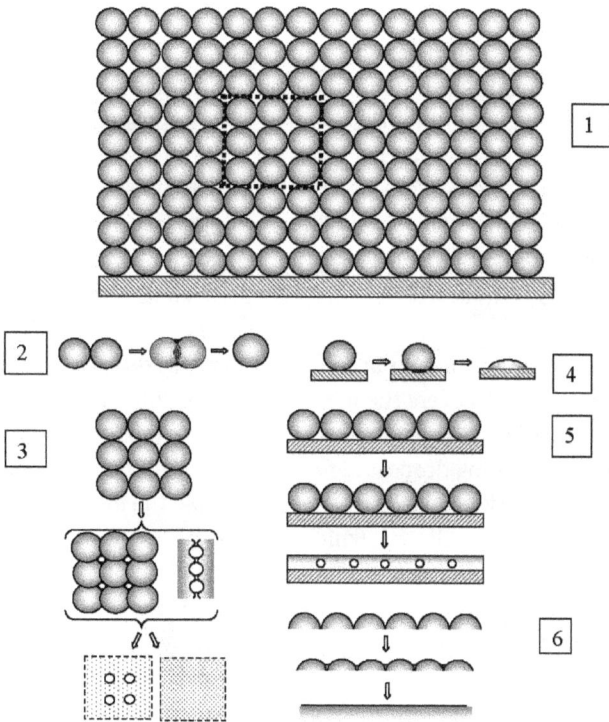

Fig. 6.11: Initial state of a powder multi-layer on a substrate and few geometrical elements of the multi-layer (see text).

Fig. 6.12 represents approximations of the shape and symbols for the characteristics of a single drop spreading on the substrate and of a pair of freely merging drops. Here are the symbols for the geometrical characteristics under discussion: (i) The radii of a drop under consideration are the initial radius R_0 and the current radius R. (ii) The contact

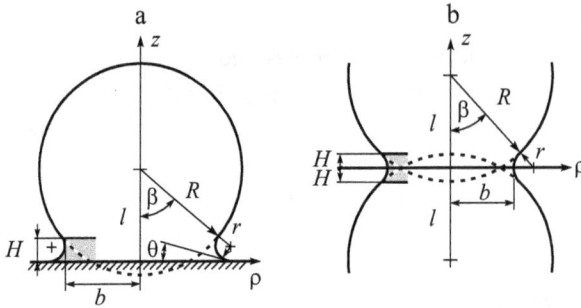

Fig. 6.12: Shape approximations of (a) single drop spreading on the substrate and (b) pair of freely merging drops.

neck, i.e. the place of merging the two drops, is described by the following characteristics: the height H and the two radii of curvature, b and r. The contact neck of the single drop spreading on the substrate is described by the height H, by b and r, the two radii of curvature, and by θ, the wetting angle of the substrate. (iii) The distances between the centers of the merging drops: the initial distance $2l_0$ and the current distance $2l$. (iv) The angle β formed by the line that connects the drop centers and the radius in the conjunction point of the contact neck surface and the drop surface. Elementary trigonometry makes it quite possible to associate these characteristics by a number of equations. Since the volume of the drop being deformed remains constant, one can calculate the volume of the contact neck and the change of the interface area per contact in dependence on β, r/R, and r/l. A comprehensive calculation is given in [60].

The process of merging of these regularly packed mono-disperse drops can be divided into three main stages: *Stage I* covers the interval from the point contacts between the drops to the critical value of the angle β, which is characterized by osculation of the contact necks of neighboring drops and formation of cylindrical channels having the radius r_{ch} in the narrowest place. *Stage II* proceeds from the instant of formation of the cylindrical channel to its collapse, which corresponds to the moment of capsulation of the multi-layer and formation of the closed cavity. *Stage III* starts with spheroidization of closed voids and formation of bubbles. The further development of stage III is due to the medium where the process proceeds, i.e., vacuum or gas atmosphere. In the former case the bubbles collapse and form a continuous bubble-free layer of melt. In the latter one the size of bubbles is stabilized with the help of equilibrating the gas pressure in a bubble by the external pressure of the gas, the capillary (Laplacian) pressure and the hydrostatic pressure of the liquid (the melt), the third factor being effective if considering gravitation. The number of the contacts per drop depends on the packing density of the particles. When the particles are arranged in the nodes of a hexagonal lattice, the characteristic K, i.e. the number of the contacts per particle, is equal to 12, while in case of a cubic lattice K is equal to 6.

Fig. 6.13 demonstrates the arrangement of the contacting drops at the instant of completing stage I and beginning stage II in different manners of packing. The critical

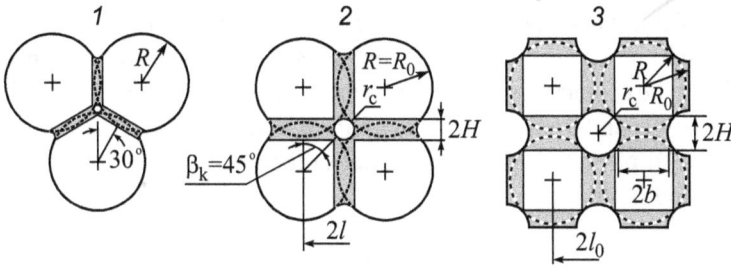

Fig. 6.13: Arrangement of the contacting drops at completing of stage I and beginning of stage II (formation of a channel) in different manners of packing: (1) hexagonal packing of particles, free merging; (2) cubic packing of particles, free merging; (3) cubic packing of particles, fixed centers of particles.

value of the angle β is equal to $30°$ at hexagonal packing, and to $45°$ at the cubic one. The space of the contact necks is shaded with dots. One can see that when the packing is hexagonal, the height of the contact neck and the radius of the channel are less than the corresponding values under cubic packing. While the drops are being freely merged at stage I, the densification in the volume of the multi-layer takes place at the expense of decreasing the center-to-center distance ($2l$) between these drops, and the drop radius far from the contact neck can be taken equal to the initial radius R_0. When the drops of the first monolayer are stuck to the substrate, the center-to-center distance is fixed at the level of $2l_0 = 2R_0$, whereas the densification takes place due to decreasing the drop radius R. Fig. 6.13 (2 and 3) visually distinguishes between the two situations for the cubic packing. One can see that when the center-to-center distance is fixed, the volume of the contact necks and the radius of the channel are higher than these characteristics under free merging of the drops. In addition, the drops of the first monolayer not only merge with one another and drops of the second layer, but also spread on the substrate surface.

The geometrical analysis of stages II and III in the volume of the multi-layer can be brought to calculating the decreases in areas of the interfaces while the channels and bubbles are being contracted, the calculation made from the initial values and those at the moment of collapse. A comprehensive consideration for the bottom and upper mono-layers of particles is presented in [60].

Hydrodynamics and Kinetics of Generation and Evolution of the Bubble Structure under Heating a Powdered Glass

The calculation of the viscous losses is based on a solution of the Navier–Stokes equation. The validity of the equation is restricted by $L \ll \eta^2/(\sigma\rho)$, where L is a scaling length ($\sim R_0$), η is the viscosity, σ is the surface tension, ρ is the density of the melt.

Let us consider merging of drops in the volume of the layer, i.e. under a mechanism of densification caused by approaching of the centers. A flow of the melt substance from both drops into the contact neck between them takes place at stage I. The neck can be represented as a cylinder having the radius r and the height $2H$ (see Fig. 6.12 b). In accordance with [74, p. 79, Eq. (16.3)], the power of the viscous losses in such a cylinder is given by the relation

$$W = 12\pi\eta H\left(\frac{db}{dt}\right)^2.$$

(6.15)

Some parts of the two drops, which are outside the neck, are also be involved in the deformation. In [75] it is shown that the power of viscous losses outside the contact neck does not differ much from that inside the neck. Thus, while merging of the two drops, the total power of viscous losses per contact neck can be is expressed as

$$W = 12\pi\eta H\left(\frac{db}{dt}\right)^2 \cdot 2.$$

(6.16)

Using Eq. (6.13) and the surface area equation for two merging drops, one can calculate the time pattern of merging the drops at stage I until the contact necks get in osculation. The computational algorithm looks as follows. We change the angle β by a small value, for example, $1°$, then calculate ΔS. In terms of Eq. (6.16) we can obtain W, the power of viscous losses, in a difference form; from Eq. (6.13) we can calculate then Δt, the time interval during which the mentioned changes occurred. Then, we repeat the procedure and perform a similar next step. The calculation is carried out up to $\beta = 30°$ for the hexagonal packing and up to $45°$ for the cubic one.

The further densification develops due to contracting the channels (stage II) and the voids (stage III), the contraction being accompanied by decreasing the distances between their centers. The equations describing the processes have been solved in an analytic way [65, 73]. The rate of contracting the infinite channel having the radius r_{chan} is given by

$$\frac{dr_{chan}}{dt} = \frac{-\sigma}{2\eta\,(1 - r^2_{chan}/l^2_{chan})},$$

(6.17)

where l_{chan} is half the distance between the axes of the channels. Two cases have to be distinguished with respect to the contraction of the cavities: the shaping in a vacuum and that in a gas, which neither dissolves nor diffuses in the matrix melt. The time required for the spheroidization is not taken into account. The rate of contracting a spherical pore (a bubble) having the radius R_b is calculated as follows:

$$\frac{dR_b}{R_b dt} = \frac{-1}{4\eta\,(1 - (R_b/l_b)^3)}\left(\frac{2\sigma}{R_b} - p + p^{(0)}\right),$$

(6.18)

where l_b is half the distance between the centers of adjacent bubbles, p and $p^{(0)}$ are, correspondingly, the gas pressure inside the bubble and the external pressure, these two values being equal to zero under the shaping in a vacuum. In terms of elementary geometrical relations for, respectively, the cubic and hexagonal packing, one can calculate the axis-to-axis distances and the densification due to contracting the channels as well as the center-to-center distances, the volume of the formed bubbles, and the densification due to contracting the pores. The respective equations are given in [60].

It is convenient to calculate the densification kinetics in normalized parameters. We normalize the time by $\tau = R_0 \eta / \sigma$, where $\tau = 1s$ at $R_0 = 1m$, $\eta = 1Pa \cdot s$, and $\sigma = 1J/m^2$; the linear dimensions should be normalized by R_0. Fig. 6.14 demonstrates the results of calculating the densification kinetics for the cubic packing of drops in an isothermal mode (η and σ are constant) in dimensionless coordinates. The time (in reduced units), $\alpha = t/\tau$, elapsed from the beginning of the process is given at the abscissa, parameters describing the densification are given on the ordinate, i.e., the relative densification $\Delta V / V_0$, the contact angle β, the relative radius of the channel r_{ch}/R_0, and the relative radius of the bubble R_b/R_0. A similar plot for hexagonal packing is given in [60, Fig. 4a]. The kink of the curve $\Delta V / V_0$ in the transition point between the two the stages is caused by the following reasons: Firstly, we do not take into account some overlapping of the stages in time, and, secondly, we approximate such shapes for the channels (a cylinder) and for the cavities (a sphere) that are more primitive than those of the real objects.

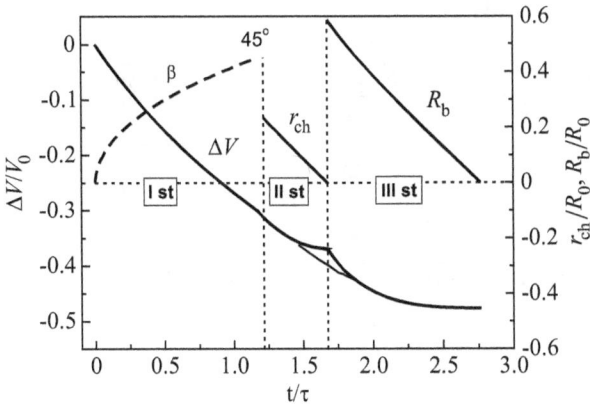

Fig. 6.14: Densification of drops multi-layer in dimensionless coordinates (cubic packing).

Table 6.8 includes design data on the structural parameters of the multi-layer at each stage of its formation for the hexagonal and cubic packing of spherical particles as well as the time of completion of each stage that we calculate from the beginning of the process. Since a hexagonal packing is denser than a cubic one, in the case of the hexagonal packing the time of having completed each stage and the whole process is significantly smaller than that for the cubic one. Table 6.8 also contains estimates of the structural parameters of forming the monolayer on the substrate. Since the centers of the particles are fixed, the densification proceeds much slower. We intend to refrain

Table 6.8: Structural parameters of the multi-layer and duration of its formation stages: m_{vol} is the volume concentration of the substance of particles; r_{ch} is the channel radius; V_b is the bubble volume; R_b is the bubble radius; l is the half of center-to-center distance; $\alpha = t/\tau$ is the time t, normalized by τ; the linear dimensions should be normalized by R_0

Number and essence of the stage	Parameter	Parameter value	
		Hexagonal packing, K = 12	Cubic packing, K = 6
Initial state	β, r, b, H, α	0	0
	R_0, l_0	1	1
	m_{vol}	0.740	0.524
1. In the volume the multilayer			
I stage Merging of particles and formation of channels	$\beta = \beta_c$	30°	45°
	r_{ch0}	0.090	0.254
	1	0.994	0.887
	b	0.455	0.663
	H	0.078	0.180
	$\Delta V/V_0$	0.158	0.302
	α	0.511	1.176
II stage Capsulation of the multilayer (channel collapse) and formation of vacuum pores	r_{ch}	0	0
	1	0.936	0.858
	V_{b0} (I)	0.039×2	0.864
	V_{b0} (II)	0.378	-
	R_{b0} (I)	0.210	0.591
	R_{b0} (II)	0.449	-
	$\Delta V/V_0$	0.179	0.369
	α	0.691	1.668
III stage Collapse of vacuum pores	$\Delta V/V_0$	0.260	0.476
	α (I)	1.111	2.768
	α (II)	1.591	-
2. The monolayer of particles on the substrate			
I stage Merging of particles and formation of channels	r_{ch0}	0.172	0.460
	R	0.983	0.955
	b	0.405	0.540
	H	0.149	0.325
	α	1.028	2.344
II stage The channel collapse	r_{ch}	0	0
	α	>1.372	>3.264

from considering the surface monolayer (see [60]). Coming to the factual values of the process parameters, we employ in the computations the real radius of the particle (in m), the real viscosity of the melt (in Pa·s), and the real surface tension (in J/m²).

Utilizing the value of the dimensionless parameter from Table 6.12 or from Fig. 6.14, it is easy to calculate its actual value.

To estimate the time corresponding to any of the stages of layer formation, in particular, the beginning and the end of any of the stages, under isothermal conditions it is convenient to use the following equation

$$\lg t_i = \lg \alpha_i + \lg R_0 + \lg \eta - \lg \sigma. \tag{6.19}$$

To estimate the real size of a given structural parameter, one should multiply the dimensionless value of this parameter by R_0. The model supplies us with the possibility to perform the computations also for non-isothermal conditions. When one knows or has approximated the temperature-time dependencies for, respectively, the viscosity $\eta(T, t)$ and the surface tension $\sigma(T, t)$, one comes to

$$\int_0^t \frac{dt}{\tau(t, T)} = \int_0^t \frac{\sigma(t, T)dt}{R\eta(t, T)} = \frac{t_i}{\tau} = \alpha_i. \tag{6.20}$$

Eq. (6.20) can be solved only numerically by dividing the process into small isothermal parts.

6.6.2 Influence of Some Technological Factors on Formation of Bubble Structure under Heating of Powdered Silica Glass: Computer Simulation of the Process

The proposed model and the calculation program developed in terms of this model allow us to perform the following steps. The first one is to vary in a wide range such process variables as temperature and rate of heating, pressure in the gas phase, type of the packing and initial radius of the drop, temperature dependence of viscosity and melt surface tension. The second step is to obtain the following parameters of the layer at any stage of its formation: the relative volume densification and the porosity; the size of such structure elements as contact necks, channels, bubbles, etc. as well as the pressure in the bubbles. We vary the process variables in the following range: (i) the furnace space temperature varies within 1800–2400 °C; (ii) the process is considered as an isothermal one or a constant heating rate is assumed within the range of 1–100 K/min; (iii) the gas pressure in the furnace space is varied within 0–10^6 Pa; (iv) the packing of particles is cubic; (v) the initial radius of the particles varies within 1–300 µm; (vi) the temperature dependence of viscosity and the surface tension are expressed via relations given in Table 6.7.

Isothermal mode
As an example, in Fig. 6.15 results of simulation of the kinetics of densification in vacuum of multi-layers with $R_0 = 100$ µm and cubic packing are given for a temperature

$R_0=100\mu m$, $T=1800°C$, $\lg[\eta,\ Pa\cdot s]=5.722$, $P_{ex}=0$

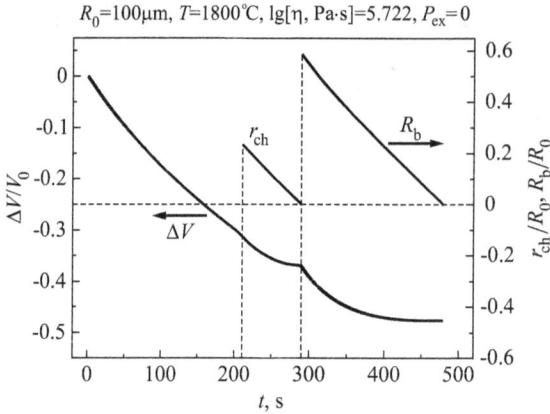

Fig. 6.15: Densification of multi-layer with $R_0 = 100$ μm and a cubic packing; at 1800 °C in vacuum.

1800 °C. The curves for the dependencies at vacuum conditions in real coordinates are identical in shape to those in dimensionless coordinates (see Fig. 6.14). Attention should be drawn to the fact that the channels and bubbles contract at a constant rate depending on temperature. The densification curves in vacuum coincide with those in the gaseous atmosphere until stage II is completed.

We discuss now the manner in which the pressure in the gaseous atmosphere influences the evolution of the forming bubbles. When considering the formation of the multi-layer at the pressure $p^{(0)}$, the main sintering equation, Eq. (6.13), is supplemented by the term corresponding to the conditions at the interface "melt – bubble"

$$-\sigma\Delta S = W\Delta t - (p - p^{(0)})\Delta V, \tag{6.21}$$

where p is the gas pressure in the bubble, ΔV is the change of the bubble volume during the time Δt. Fig. 6.16 shows how r_{ch}, the size of the channels, and R_b, the size of the

$R_0=100\mu m$, $T=1800°C$, $\lg[\eta,\ Pa\cdot s]=5.722$, $\sigma=0.3$ J/m^2

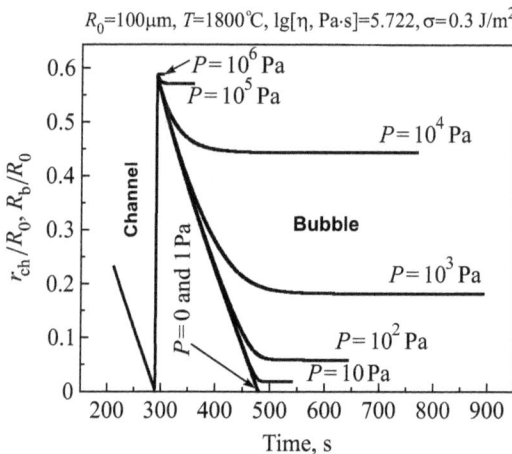

Fig. 6.16: Influence of pressure in the gas space, P_{ex}, on the kinetics of bubble formation.

$R_0 = 100 \mu m$, $T = 1800°C$, $\lg[\eta, \text{Pa·s}] = 5.722$, $\sigma = 0.3$ J/m^2

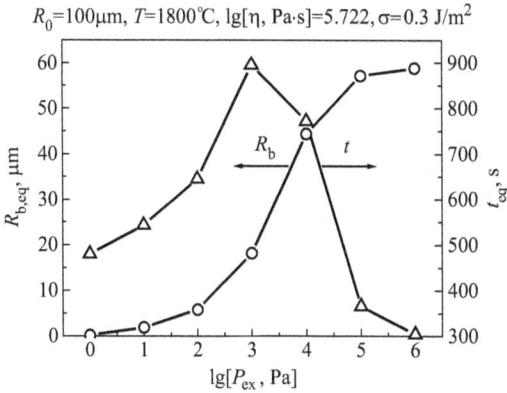

Fig. 6.17: Dependencies of the equilibrium bubble radius $R_{b,eq}$ and the stabilization time t_{eq} on the pressure in the gas space P_{ex} ($R_0 = 100$ μm, $T = 1800 °C$).

bubbles, depend on the time from the beginning of the process at different pressures in the gas space; the temperature is 1800 °C and R_0 is 100 μm. It is evident that the size of the channels does not depend on pressure. The size of a bubble decreases first with time but reaches then some constant value approaching $R_{b,eq}$. This state corresponds to equality between $P_{b,eq}$, the equilibrium pressure in the bubble, and the sum of P_{ex}, the external pressure, and P_{cap}, the capillary pressure, i.e., the relation

$$P_{b,eq} = P_{ex} + P_{cap} = P_{ex} + \frac{2\sigma}{R_{b,eq}} \tag{6.22}$$

holds here, where σ is the melt surface tension (see also Eq. (6.18)).

Strictly speaking, it will take an infinite time to establish equilibrium. In dependence on what aim is to be achieved, it is reasonable to specify a criterion for equilibrium when $R_{b,eq}$ is supposed to be attained and the rate of change of the radius is practically equal to zero. The time, t_{eq}, corresponding to establishment of this equilibrium state is denoted as time of stabilization of the bubble size. The higher is the pressure, the higher is $R_{b,eq}$. With an increase of pressure, the stabilization time first increases as well but then becomes lower. Fig. 6.17 represents the dependencies of the equilibrium bubble radius $R_{b,eq}$ and the stabilization time t_{eq} on the pressure in the gas space, P_{ex}.

Fig. 6.18 demonstrates the dependence of the equilibrium bubble radius $R_{b,eq}$, the equilibrium pressure in the bubble, $P_{b,eq}$, and its capillary contribution, P_{cap}, on pressure, P_{ex}. Finally, Fig. 6.19 presents the dependence of the pressure in the bubble $P_{b,eq}$ and its capillary contribution P_{cap} on the equilibrium bubble radius $R_{b,eq}$ at each pressure in the gas space P_{ex}. The analysis of these dependencies demonstrates that, when P_{ex} reaches values up to about 10^3 Pa, the equilibrium pressure in the bubble is largely determined by its capillary constituent. However, at higher values of P_{ex} the external-pressure contribution is predominant. It is the same external pressure that the maximum, t_{eq}, of the bubble size stabilization time corresponds to.

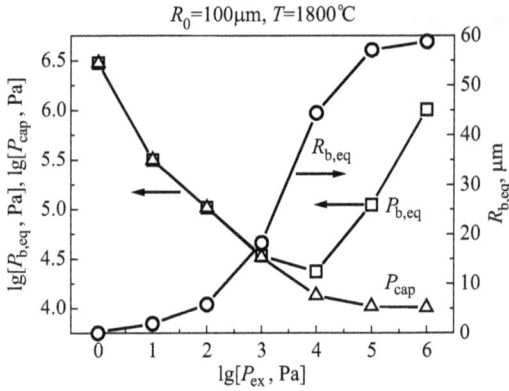

Fig. 6.18: Dependencies of the equilibrium bubble radius $R_{b,eq}$, the equilibrium pressure in the bubble $P_{b,eq}$, and its capillary constituent P_{cap} on the pressure P_{ex} (R_0 = 100 μm, T = 1800 °C).

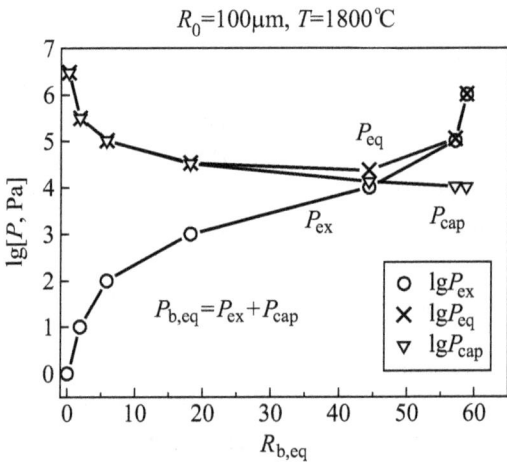

Fig. 6.19: Dependence of the pressure in the bubble $P_{b,eq}$ and its capillary constituent P_{cap} on the equilibrium bubble radius $R_{b,eq}$ (R_0 = 100 μm, T = 1800 °C).

Since the surface tension can be taken as independent of melt temperature, the equilibrium size of the bubble will not depend on temperature either. The temperature will determine the time of stabilization of its radius via its effect on the viscosity, only. Fig. 6.20 represents the dependence of the bubble radius stabilization time t_{eq} on the pressure in the gas space P_{ex} at temperatures 1800, 2000 and 2200 °C. One can see that all the curves are similar, and the stabilization time gets lower with decreasing melt viscosity.

Another question of interest is how the initial size of the particles, R_0, affects the bubble size stabilization time, t_{eq}. Fig. 6.21 demonstrates the dependence of the stabilization time on the pressure in the gas phase when the melt temperature is 2000 °C and the size of particles varies between 10 and 200 μm. The higher is the size of the particles, the higher is the stabilization time; the maximum of the stabilization time shifts from 10^4 Pa for the size within R_0 = 10–20 μm to 10^3 Pa for the larger particles.

Now we consider how R_0 affects the equilibrium bubble radius $R_{b,eq}$ and the residual porosity. The residual porosity can be represented as V_b/V_m or $V_b/(V_b + V_m)$,

$R_0 = 100\,\mu m$

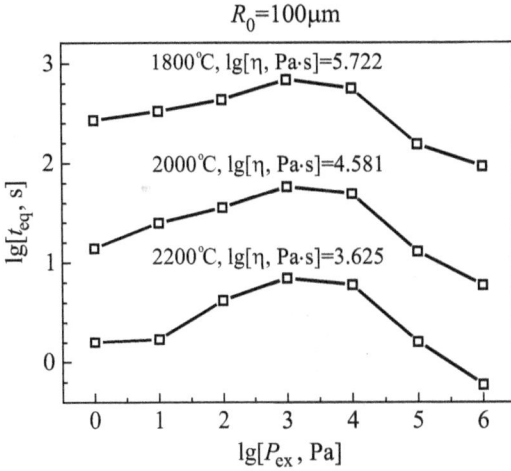

Fig. 6.20: Dependence of the bubble radius stabilization time t_{eq} on the pressure in the gas space P_{ex} at temperature values of 1800, 2000, and 2200 °C.

$2000°C$, $lg[\eta,\ Pa\cdot s] = 4.581$

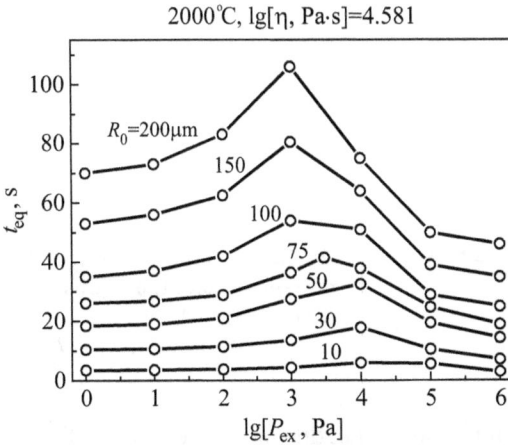

Fig. 6.21: Dependence of the stabilization time t_{eq} on the pressure in the gas phase and the different sizes of the particles (at $T = 2000\,°C$).

where V_b is the total volume of the bubbles and V_m is the melt volume. As an example, P_{ex} can be set equal to 1 atm. Fig. 6.22 presents the dependence of $R_{b,eq}$ on the particle radius R_0 and the dependencies of the corresponding values of $P_{b,eq}$ on the same parameter. $R_{b,eq}$ changes from 0.18 μm at $R_0 = 1$ μm to 195 μm at $R_0 = 333$ μm, and the value of $P_{b,eq}$ from 34 to 1.03 atm, respectively. Fig. 6.22 presents the dependence of $P_{b,eq}$ and the residual porosity on the initial particle radius R_0. Fig. 6.23 shows the porosity V_b/V_m, which increases from 0.6 % at $R_0 = 1$ μm to 19.9 % at $R_0 = 333$ μm.

If we consider the total gas volume in the bubbles at standard atmospheric pressure (1 atm), the total gas volume in the pores will amount to about 20.5 % of the melt volume regardless of the size of R_0 (the curve $Q_{gas,NP}/V_m$). This evident result is due to the fact that all the gas occluded at the moment of collapse of the channels, forms bubbles and, according to the assumptions adopted, neither dissolves in the melt nor is removed from it by diffusion.

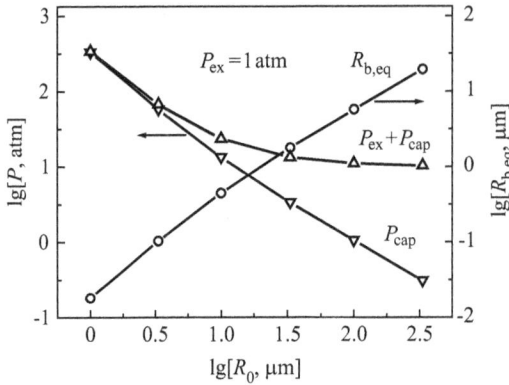

Fig. 6.22: Dependence of $R_{b,eq}$ and $P_{b,eq}$ on the particle radius R_0 (P_{ex} = 1 atm).

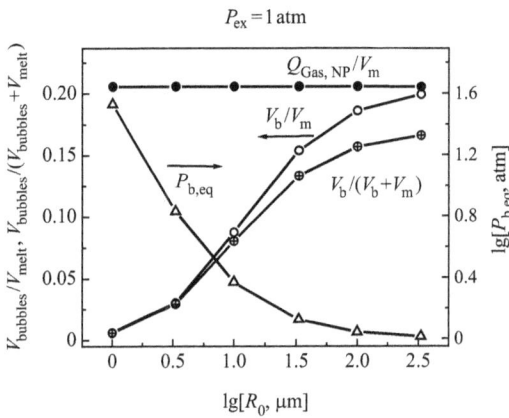

Fig. 6.23: Dependence of $P_{b,eq}$ and the residual porosity on the particle radius R_0 (P_{ex} = 1atm).

Non-isothermal Mode

Now we analyze the manner in which the rate of heating of the layer affects the formation of the bubble structure in it. As an example, Fig. 6.24 demonstrates results of simulating the kinetics of densification of the particle layer at R_0 = 100 μm, cubic packing, in vacuum, and a heating rate, q = 10 K/min. In contrast to the isothermal mode, we will have the temperature dependencies of the parameters of the forming layer and, respectively, the temperature dependencies of the different stages of the process. Note that the size of the channel and that of the bubble decrease practically linearly as the temperature becomes higher. Since the temperature gets higher at a constant rate, the rates of contracting, respectively, of the channels and bubbles remain actually constant, which is the case under the isothermal mode as well. Fig. 6.25 presents the curves of densifying the particle layer with R_0 = 100 at different heating rates in the range of 1–100 K/min. The same figure demonstrates the temperature dependence of the melt viscosity. As would be expected, an increase of the heating rate makes the curves shift towards the higher temperatures and, correspondingly, to

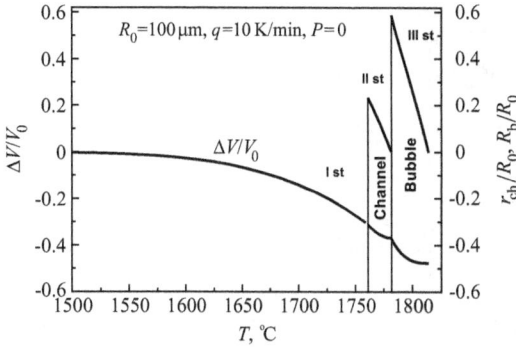

Fig. 6.24: Densification of the particle multi-layer at $R_0 = 100$ µm, cubic packing, vacuum, and heating rate $q = 10$ K/min.

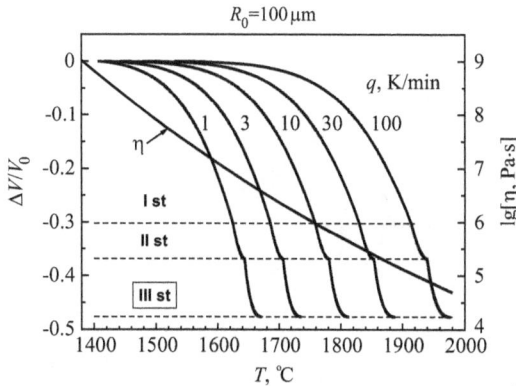

Fig. 6.25: Densification of the particles multi-layer at $R_0 = 100$ µm, cubic packing, vacuum, and different heating rates (q varies from 1 to 100 K/min).

lower viscosity values, which is valid for both the whole densification process and each stage.

Fig. 6.26 presents data on temperature and corresponding viscosity at each stage of densification of the layer. One can see that the dependencies $\lg \eta$ vs. $\lg q$ are described by straight lines, which directly results from the model under consideration. The dependence T vs. $\lg q$ exhibits modest deviations from straight lines. The magnitude of this deviation does not exceed 10 K in the discussed temperature range. Fig. 6.27 represents the curves of densification of the particle multi-layer at the constant heating rate $q = 10$ K/min and in varying the radius of particles R_0 from 1 to 300 µm. As expected, an increase of R_0 makes the curves shift towards higher temperatures and, correspondingly, the lower viscosity values, which is valid both for the whole densification process and for each stage.

Fig. 6.28 represents data on temperature and corresponding viscosity at each stage of densification of the multi-layer. One can see that the dependencies $\lg \eta$ vs. $\lg R_0$ are described by straight lines, which directly results from the model under consideration. The dependence T vs. $\lg R_0$ exhibit modest deviations from straight lines, the deviation magnitudes not exceeding 10 K in the discussed temperature range.

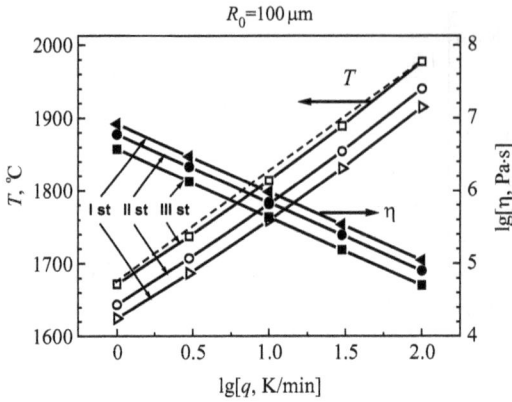

Fig. 6.26: Temperature and the corresponding viscosity at each stage of the densification of the multi-layer (R_0 = 100 μm, vacuum).

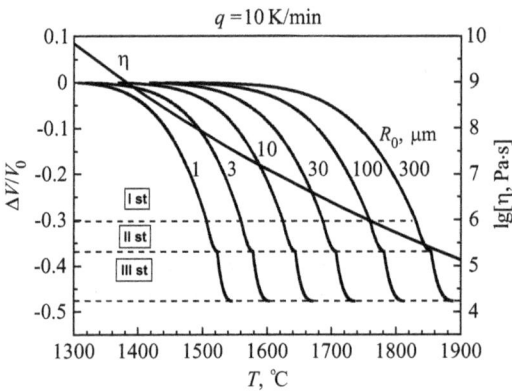

Fig. 6.27: Densification of the particles multi-layer at q = 10 K/min, cubic packing, vacuum, and different R_0 (R_0 ranging from 1 to 300 μm).

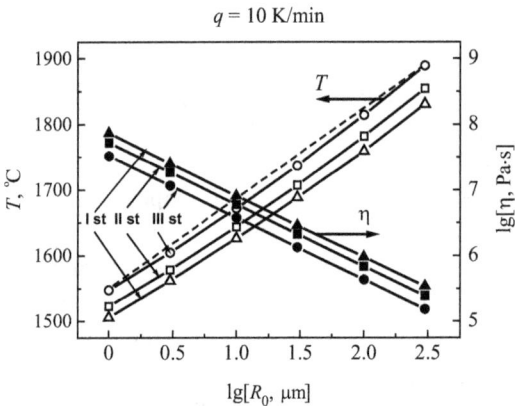

Fig. 6.28: Temperature and the corresponding viscosity at each stage of densification of the multi-layer (q = 10K/min, vacuum).

Further we analyze the way in which P_{ex}, the pressure in the gas phase of the furnace, affects the kinetics of densification of the powder layer. It is evident that the course of the process does not depend on the pressure until stage II is completed, i.e. the channels have been collapsed. As an example, we set here R_0 = 100 μm and

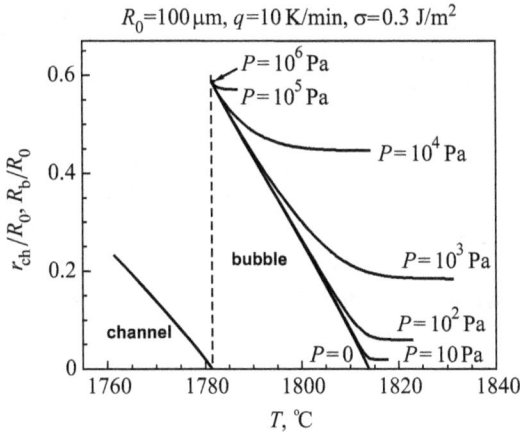

Fig. 6.29: Influence of pressure in the gas phase, P_{ex}, on bubble evolution ($R_0 = 100$ μm, $q = 10$ K/min).

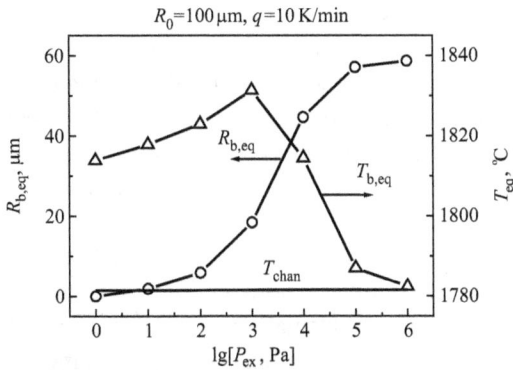

Fig. 6.30: Dependencies of the equilibrium bubble radius $R_{b,eq}$ and the stabilization temperature T_{eq} on the pressure in the gas phase P_{ex} ($R_0 = 100$ μm, $q = 10$ K/min).

$q = 10$ K/min. Fig. 6.29 demonstrates that, as the pressure increases, the bubble size stabilization temperature T_{eq} first increases but then decreases, again. The maximum of T_{eq} is found when P_{ex} is equal to 10^3 Pa·s (see Figs. 6.20 and 6.21). It should be recalled that the isothermal mode exhibits the maximum of the bubble size stabilization time t_{eq} at the same value of P_{ex}.

6.7 Summary to Part II

The proposed micro-rheological model and the computer program developed in terms of this model enables us to investigate the formation of bubbles under heating a multi-layer of silica glass particles and the further evolution of bubbles in the melt. The main assumptions of the model are the following: (i) The spherical mono-disperse particles of the glass are disposed in the nodes of a regular lattice under point contacts between the particles. (ii) The deformation of each particle and the whole layer develops in accordance with the mechanism of viscous flow of a Newtonian liquid (a glass-forming melt in our case), the process caused by the surface tension and the pressure of the furnace atmosphere gas. (iii) The furnace atmosphere gas neither dissolves nor diffuses in the melt. (iv) The gravitation is not taken into consideration.

The densification of the powder layer under heating includes the following three main stages: (i) the transformation of the multi-layer of contacting disperse particles (drops) of radius R_0 to the melt layer pierced by channels having the initial radius $r_{0\,channel}$; (ii) the contraction of the channels concluded by their collapse and the formation of a melt layer containing a lot of closed cavities having the initial radius $R_{0\,bubble}$; (iii) the contraction of the cavities concluded either by their collapse under heating in vacuum or by stabilization of the size of the bubbles under heating in a gaseous atmosphere.

The ranges for varying the process variables are the following: (i) R_0, the initial radius of the particle, varies within the range 1–300 μm. (ii) The mode of heating is an isothermal one or that at a constant heating rate within 1–100 K/min. (iii) The gas pressure in the furnace space, P_{ex}, is varied within the range 0–10^6 Pa. The estimated time is calculated for each stage as well as for the whole process. Such main parameters as relative densification, the size of the structure elements (the channels and bubbles), the pressure in the bubbles, and the porosity have been calculated for the powder layer and the bubble-containing melt layer formed from it, the parameters calculated for the various stages of the formation and evolution.

The following main regularities of the process have been established: (i) When the process takes place in vacuum, then it can be described in normalized (dimensionless) variables. The values of these parameters have been calculated and tabulated for each stage of the process. The initial radii of the channels and bubbles at cubic packing are, respectively: $r_{0\,channel} = 0.254R_0$, $R_{0\,bubble} = 0.59R_0$, where R_0 is the initial radius of the particle. The rates of contracting the channels and bubbles are constant. Under isothermal conditions, the times of bubble formation and collapse that are calculated from the beginning of the process and normalized in τ, make up 1.668 and 2.768, correspondingly $\tau = R_0\eta/\sigma = 1$s at $R_0 = 1$m, $\eta = 1$Pa·s, $\sigma = 1$J/m^2. The current time is obtained by multiplying the normalized time by τ for the current values of R_0, η and σ corresponding to the current temperature. (ii) In case heating is conducted in vacuum at a constant rate, calculation is made of the temperature intervals and viscosity ranges restricted by the moments of formation and collapse of the bubbles. These intervals are considered at different q and R_0. In terms of logarithmic coordinates the dependencies are described by the straight lines

$$\lg[\eta_q] = \lg[\eta_{1q}] - B_q \lg[q], \quad \lg[\eta_{Ro}] = \lg[\eta_{1Ro}] - B_{Ro} \lg[R_0]. \tag{6.23}$$

Here η_{1q} and η_{1Ro} are the viscosity values corresponding to the temperature at which the discussed structural change (the formation or collapse of the bubble) takes place under $q = 1$K/min and $R_0 = 1$ μm. B_q and B_{Ro} are the corresponding angular coefficients, whose values are about 0.94. (iii) For processes proceeding in a gas atmosphere at pressure P_{ex}, at the moment of channel collapse an isolated gas-filled cavity is formed, namely, a bubble having the effective radius $R_{0\,\text{bubble}} = 0.59R_0$. The pressure inside the bubble is equal to P_{ex} at this moment. Under the influence of the surface tension σ, the bubble contracts until its internal pressure becomes equal to the sum of the external pressure and the capillary pressure $P_{\text{cap,eq}} = 2\sigma/R_{\text{bubble,eq}}$, where $R_{\text{bubble,eq}}$ is the equilibrium radius of the bubble. Since the surface tension can be taken as practically independent of the melt temperature, the equilibrium size of the bubble will not depend on the temperature either. The stabilization time of the bubble radius depends on temperature via the temperature dependence of the viscosity. (iv) Quantitative estimates are given for the dependencies of the equilibrium bubble size $R_{\text{bubble,eq}}$ and the inside-bubble pressure $P_{\text{bubble,eq}}$ on the particle size R_0 and pressure in the gas atmosphere, P_{ex}. We also computed the bubble size stabilization time, t_{eq}. As R_0 increases at a constant value of P_{ex}, the equilibrium bubble radius $R_{\text{bubble,eq}}$ gets higher, while the capillary pressure $P_{\text{cap,eq}}$ and the inside-bubble pressure $P_{\text{bubble,eq}}$ becomes lower. When the external pressure P_{ex} increases at a constant value of R_0, the equilibrium bubble radius $R_{\text{bubble,eq}}$ gets higher, the capillary pressure $P_{\text{cap,eq}}$ becomes lower, whereas the total pressure $P_{\text{bubble,eq}}$ at first also decreases, then goes through the maximum at $P_{\text{ex}} = 10^3 - 10^4 \text{Pa}$, and further increases. In this connection, it is the capillary constituent that makes the main contribution to the total inside-bubble pressure $P_{\text{bubble,eq}}$ at $P_{\text{ex}} \le 10^3 \text{Pa}$, while, at higher pressure, the principal contribution is provided by the external pressure. At isothermal conditions and increasing external pressure, the values of $R_{\text{bubble,eq}}$ and t_{eq} at first get higher, then go through the maximum in the region of $P_{\text{ex}} = 10^3 - 10^4 \text{Pa}$, and further decrease. This shape of this dependence remains the same at different temperatures. (v) The layer porosity $V_{\text{bubble,eq}}/V_{\text{melt}}$, where V_b is the total volume of the bubbles after establishing the equilibrium and V_m is the melt volume, gets higher with increasing values of R_0 and P_{ex}. Since, in accordance with our assumption, the gas neither dissolves nor diffuses in the melt, its total amount in the bubbles remains constant under any possible following changes of their size and is defined by the amount of the gas captured at the moment of capsulation of the multi-layer.

Part III: Mathematical Modeling and Computer Simulation of the Behavior of Gas-filled Bubbles in Silica Melts

6.8 Introduction

In this third part we consider some problems of the mathematical modeling of the behavior of gas-filled bubbles in silica melts, not touching even the problem of bubble nucleation and assuming that all the bubbles are entrapped from surrounding space or formed already via nucleation processes in the system. The further behavior of the bubbles depends both on the state of the melt (temperature, pressure, gas solubility, diffusivity, etc.) and the parameters of the entrapped bubbles (number of bubbles, their sizes, amount of gas inside the bubbles, etc.).

Many well-written articles dealing with the mathematical modeling of the behavior of isolated or solitary bubbles can be found in the literature (see, for example [13, 76–87]). In [13, 75, 76, 78–80] the authors formulated mathematical models and studied the diffusion in the bubble neighborhood. They concentrated the attention to the derivation of more or less rigorous analytical solutions of the respective problem of bubble evolution. Moreover some questions of modeling of evolution of multi-component gas-filled bubbles were also considered in a number of publications (see [79, 81, 82]). Some problems of the floating of bubbles under the action of Archimedean force were discussed in [77, 87], and the transfer of the bubbles by glass melt flow was described in [84–86].

However, in reality one has to deal with populations of bubbles. As it is shown below, in this case the behavior of a bubble in the presence of neighboring bubbles may differ significantly (and, sometimes, principally) from that of a solitary bubble. Very interesting effects of such interdependence occur when the sizes of bubbles are different. For example, processes of coalescence or coarsening can take place then. Besides, with the growth of the volume of the bubbles in the melt, the effective density of the continuous glass melt changes and also an additional volume force appears as a result of floating of the bubbles. This latter factor evokes a flow of the melt. In addition, the presence of the bubbles changes the type of heat transfer in the melt, especially, its radiation contribution.

Existing publications on bubble evolution in silica melts are restricted to investigations of solitary bubbles, therefore, it is necessary to correct (at least partially) this omission. We have no pretension to arrive here at a complete description of all the complicated physical and chemical processes, which govern the behavior of gas-filled bubbles in the melt (see Part I), but we try to outline some systematic (as it seems to us) approach allowing one to develop mathematical models for calculations and computer simulations of the kinetics of behavior of gas-filled bubbles in the melt. In this analysis, we are interested in the time interval of the order of 1–10 hours representing the typical time scale of processes of fusion of silica glass. We limit the consideration here to bubbles of two inert gases, of helium and argon. Helium is an example of a well soluble and diffusing gas in silica glass. In contrast, argon is an example of a weakly soluble and diffusing gas. The values of parameters, typical for silica melts, are presented in Table 6.9. Having the goal to develop mathematical models for the description of the glass melt, containing a considerable quantity of gas-filled bubbles, we start with a study of the simplest case, the description of the evolution of a solitary isolated bubble of constant mass.

Table 6.9: Typical values of some parameters and properties of silica melts employed for the estimations of time scales of evolution of gas-filled bubbles. The solubility coefficient (Henry's constant) is dimensionless, the diffusion coefficient is given in m^2/s

Temperature of glass melt, °C		2400	2000
Dynamic viscosity coefficient, Pa·s		$6.5 \cdot 10^2$	$3.8 \cdot 10^4$
Diffusion coefficient	Helium	$1.9 \cdot 10^{-8}$	$1.5 \cdot 10^{-8}$
	Argon	$1.2 \cdot 10^{-11}$	$5.5 \cdot 10^{-12}$
Solubility coefficient	Helium	$7.1 \cdot 10^{-3}$	$7.3 \cdot 10^{-3}$
	Argon	$6.9 \cdot 10^{-4}$	$7.6 \cdot 10^{-4}$
Density of melt, kg/m^3		$2.2 \cdot 10^3$	
Radius of bubble, μm		100	
Pressure in continuum, Pa		10^5	
Surface tension coefficient, J/m^2		0.3	
@Gravity acceleration, m/s^2		9.8	

6.9 Behavior of Isolated Bubbles

We consider first a solitary gas-filled bubble of constant mass, i.e. the gas, contained in the bubble, cannot dissolve into the glass melt. We assume further that the melt is a viscous incompressible fluid. Then, the following equation can be employed for the description of the time dependence of the bubble radius (see [76])

$$\rho \left[r d_{tt}^2 r + \frac{3}{2} (d_t r)^2 \right] + 4\eta \frac{d_t r}{r} + 2\frac{\sigma}{r} + p = p_b, \qquad (6.24)$$

where t is time, r is the radius of the bubble, ρ is the density of the fluid (the glass melt), η is the dynamic viscosity, σ is the surface tension, p is the pressure inside the fluid (equal to the sum of hydrostatic pressure in the fluid and external pressure (pressure outside the fluid)), p_b is the pressure of the gas inside the bubble. Equation (6.24) expresses the equality of the sum of pressures of inertial fluid forces (due to the effect of virtual mass) $\rho \left[r d_{tt}^2 r + \frac{3}{2} (d_t r)^2 \right]$, viscosity stress $4\eta \frac{d_t r}{r}$, capillary pressure (surface tension) $(2\sigma/r)$, and pressure inside the fluid (outside of the bubble) p to the pressure of gas inside the bubble p_b.

At the typical temperature of glass fusion, the glass melt is a very viscous fluid and all inertial effects are not essential. Therefore, neglecting the terms inside the square brackets in Eq. (6.24) and additionally assuming the gas to be ideal we obtain:

$$4\eta \frac{d_t r}{r} + 2\frac{\sigma}{r} + p = \rho_b RT \equiv \frac{3 m_b}{4\pi r^3} RT, \qquad (6.25)$$

where ρ_b and m_b are the density and the mass of the gas inside the bubble $\left(m_b = (4\pi/3) r^3 \rho_b \right)$; T is the temperature; R is the universal gas constant. The stationary solution of Eq. (6.25) determines the value of the equilibrium radius of the bubble

$r = r(m_b, \sigma, T, p)$ as a function of the parameters enclosed in brackets. Note that there always exists a unique solution of this equation.

Let us study the problem of stability of equilibrium of an isolated bubble. Considering an infinitesimal deviation of the bubble radius from its equilibrium value $\tilde{r} = r - r_e$ and using Eq. (6.25), we get

$$d_t \tilde{r} = -\frac{3p + 4\frac{\sigma}{r_e}}{4\eta} \, \tilde{r}. \tag{6.26}$$

The solution of Eq. (6.26) is a decreasing exponential function $\tilde{r} = \tilde{r}_0 e^{-\frac{t}{\tau_i}}$ with a characteristic time

$$\tau_i \equiv \frac{4\eta}{3p + 4\frac{\sigma}{r_e}} = \frac{\frac{4\eta}{3p} \cdot \frac{\eta r}{\sigma}}{\frac{4\eta}{3p} + \frac{\eta r}{\sigma}} = \frac{\tau_p \tau_\sigma}{\tau_p + \tau_\sigma}. \tag{6.27}$$

Thus, the considered equilibrium state of the isolated bubble is stable.

There are two time scales in Eq. (6.27). The first one characterizes the time of collapse of a bubble governed by the capillary force: $\tau_\sigma = \frac{\eta r}{\sigma}$, and the second describes the relaxation time of a bubble governed by the pressure force: $\tau_p = \frac{4\eta}{3p}$. To estimate the numerical values of the time scales we assume (here and next) typical for silica melts values of the parameters as given in Table 6.9. Then we get

for $\quad T = 2400°C \quad \rightarrow \quad \tau_\sigma \approx 0.5$ s, $\quad \tau_p \approx 0.03$ s, $\quad \tau_i \approx \tau_p$,

for $\quad T = 2000°C \quad \rightarrow \quad \tau_\sigma \approx 25$ s, $\quad \tau_p \approx 1.5$ s, $\quad \tau_i \approx \tau_p$.

In comparison with the duration of silica glass fusion ($\sim$ 1–10 hours), the times τ_σ and τ_p are rather small. Evidently, the times τ_σ, τ_p and τ_i increase with decreasing temperature, because the viscosity increases as well.

The obtained ratio between both time scales τ_σ and τ_p, when τ_σ considerably exceeds τ_p, is typical for large bubbles and atmospheric pressure. For small bubbles, whose radii are much smaller than indicated in Table 6.9, or when the pressure is much smaller than the atmospheric one, the time scales τ_σ and τ_p become close to each other. Concerning the resulting relaxation time from Eq. (6.27) we can conclude that its value is never smaller as compared to the minimal value of both τ_σ and τ_p. The change of the size of an equilibrium bubble is due to the influence of ambient conditions such as pressure and temperature of the melt.

6.10 Behavior of Solitary Gas-filled Bubbles under Mass Exchange with the Melt

Mass exchange between bubbles and melt is very significant for the evolution of a gas-filled bubble. It occurs due to dissolution of gas at the surface of the bubble, and diffusion of gas molecules to the melt. The cause (driving force) of the mass exchange is the difference of density (concentration) of the gas dissolved in the melt at the bubble surface, ρ_s', and the density (concentration) of dissolved gas at some distance from the bubble surface, ρ_s. It was shown (see, for example, [88]) that the density of gas flow at the surface of a spherical bubble and the melt can be written as

$$j = \frac{D}{r}\left(\rho_s' - \rho_s\right),$$
(6.28)

where D is the diffusion coefficient. Strictly speaking this expression is valid only for a stationary distribution of concentration in the surrounding melt.

We assume here that the gas density inside the bubble, ρ_b, and the density of the dissolved gas at the surface of the bubble, ρ_s', obey Henry's law:

$$\rho_s' = h\rho_b,$$
(6.29)

where h is Henry's constant, i.e. there is a simple linear connection between ρ_b and ρ_s'. Then, the mass balance equation for the gas inside the bubble can be written as

$$d_t m_b = d_t\left(\rho_b V_b\right) = -\frac{3Dh}{r^2}V_b\left(\rho_b - \frac{\rho_s}{h}\right) = -\frac{V_b}{\tau_b^D}\left(\rho_b - \frac{\rho_s}{h}\right),$$
(6.30)

where V_b is the volume of the bubble $\left(V_b = \frac{4}{3}\pi r^3\right)$. In Eq. (6.30) we introduced the following notation

$$\tau_b^D \equiv \frac{r^2}{3Dh}.$$
(6.31)

The quantity τ_b^D has the dimension of time and it characterizes the time scale of mass exchange of the bubble with the melt. Along with the size of the bubble, the time τ_b^D is determined by such properties as solubility and diffusion of the gas in the melt. As it was mentioned above (see Part I), these properties are very different for various gases.

Estimating values of time τ_b^D for the bubble containing helium and using the data from Table 6.9, we get

$$\text{at} \quad T = 2400°C \quad \rightarrow \quad \tau_b^D \approx 25s,$$

$$\text{at} \quad T = 2000°C \quad \rightarrow \quad \tau_b^D \approx 30s.$$

Similar estimates for argon give values of mass exchange time, which are by four decimal orders larger then those for helium, i.e.,

$$\text{at} \quad T = 2400°C \quad \rightarrow \quad \tau_b^D \approx 5.0 \cdot 10^5 s \approx 108h,$$

$$\text{at} \quad T = 2000°C \quad \rightarrow \quad \tau_b^D \approx 8.0 \cdot 10^5 s \approx 220h.$$

The expression for mass flux in the right part of Eq. (6.30) is an essential simplification in comparison with those used in [76, 82].

As it was mentioned above, this expression is valid for stationary density profiles. The time of establishing of a steady-state in the melt near the bubble is approximately equal to $\tau_D = (r^2/3D)$. Comparing this time with the time scale of mass exchange of the bubbles τ_b^D, one concludes (see Eq. (6.31)): $(\tau_D/\tau_b^D) = h \ll 1$. Thus, the time of establishing of a steady-state in the melt near the bubble is even much less (for helium more than a hundred times) than the time of mass exchange. Therefore, the expression of mass flux in the right part of Eq. (6.30) is valid and it does not need any corrections.

Despite the fact that the time scales for the evolution of the bubbles of various gases are very different some common regularities can be observed. As it is seen from Eq. (6.30), at a rather large value of gas density inside the bubble, namely at $\rho_b > h^{-1}\rho_s$, the gas diffuses from the bubble into the melt, and the mass of gas inside the bubble decreases. On the contrary, if $\rho_b < h^{-1}\rho_s$, then the gas diffuses into the bubble and the mass of the gas in the bubble increases.

If the following condition is valid

$$\rho_b^e \equiv \rho_b = \frac{\rho_s}{h}, \tag{6.32}$$

then an equilibrium between the gas inside the bubble and the dissolved gas is established. At that value, the bubble radius is uniquely determined by Eq. (6.25) via

$$r^e = \frac{2\sigma}{\rho_b^e RT - p} = \frac{2\sigma}{\dfrac{(\rho_s RT)}{h} - p}. \tag{6.33}$$

An analysis shows that this steady state is not stable and if the radius of the bubble deviates from its equilibrium value by any random reason, then the bubble either collapses (if the radius becomes less than the equilibrium one) or, vice versa, unlimited growth of bubbles begins. The time constant of instability development is proportional to the time τ_b^D.

As illustrations, results of calculations (based on Eqs. (6.25) and (6.30)) of the evolution of solitary bubbles containing helium and argon at various values of the initial radius are shown on Figs. 6.31 and 6.32. For both gases the density of the dissolved gas is fixed so that for the bubble of radius $r^e = 100$ μm the equilibrium condition Eq. (6.33) holds. As it is seen from Figs. 6.31 and 6.32, if the initial radius of the bubble

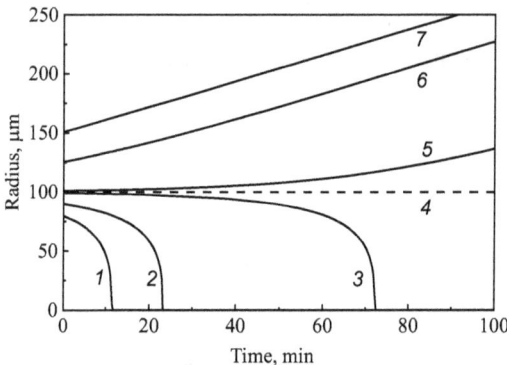

Fig. 6.31: Behavior of a solitary helium bubble for different values of the initial radius: $T = 2400\,°C$, $p = 10^5$ Pa, $\sigma = 0.3$ J/m^2, $\rho_s = 7.6 \cdot 10^{-5}$ kg/m^3, $r^e = 100$ μm. The numbers 1–7 correspond to the following values of the initial radius: $r_0 = 60, 80, 99, 100, 101, 125$ and 150 μm, respectively.

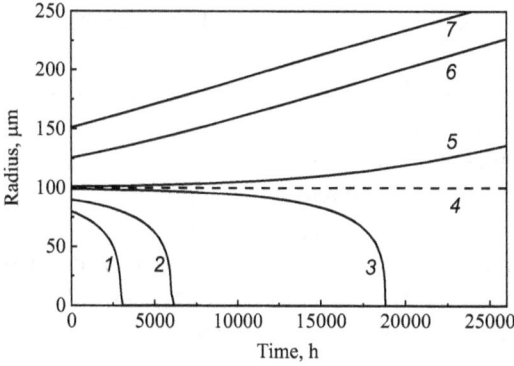

Fig. 6.32: Behavior of a solitary argon bubble for different values of the initial radius: $T = 2400\,°C$, $p = 10^5$ Pa, $\sigma = 0.3\,J/m^2$, $\rho_s = 7.4 \cdot 10^{-5}\,kg/m^3$, $r^e = 100\,\mu m$. Numbers 1–7 correspond to the following values of the initial radius: $r_0 = 60, 80, 99, 100, 101, 125$, and $150\,\mu m$, respectively.

is less than the equilibrium one, then the bubble collapses, and, vice versa, if the initial radius of the bubble is larger, then the bubble grows. From a qualitative point of view, the evolution kinetics of bubbles containing either helium or argon is absolutely similar.

A significant difference between the gases is the huge difference of time scales of the mass exchange processes. As a result, in practical technological processes of silica glass fusion, whose duration is a few hours, the bubbles, containing helium, evolve very actively: small bubbles, governed in its evolution by mass exchange, collapse, and the big ones grow. On the contrary, the bubbles with argon are of practically constant sizes because in realistic time scales mass exchange with the melt of the bubbles, containing argon, is negligible.

The field of possible application of the mathematical model based on Eqs. (6.25) and (6.30) is limited by the assumption that the density of the dissolved gas in the melt ρ_s stays invariable during the whole process. This requirement implies that the total amount of bubbles should be so small that the mass of the gas inside the bubbles could be remarkably less than the mass of the dissolved gas. In practice usually the reverse situation occurs when the largest amount of gas is contained exactly inside the bubbles. To describe a problem like this, the mathematical model of a solitary bubble, i.e. Eqs. (6.25) and (6.30), is not valid and it is necessary to further develop this model, taking into account the change of density of the dissolved gas in the melt ρ_s.

6.11 Two-phase Approach to the Description of Mono-disperse Ensembles of Bubbles

Let us consider the gas contained in the melt both in the dissolved form and in the form of gas-filled bubbles as a two-phase continuum. The first phase of it is the dispersed phase that consists of gas vesicles, i.e. bubbles. The second phase, which we denote further on as the dissolved phase, is the gas dissolved in the melt. Each phase is characterized by its own density and volume concentration.

At the beginning we consider a simplified situation when the dispersed phase is a population of bubbles of the same size. We denote the number density of bubbles in the melt by n_b, i.e. the number of bubbles per unit of the total volume of the melt. Then the volume concentration of the dispersed phase is equal to the product of the number density and the volume of a bubble i.e. $\alpha_b \equiv n_b V_b = n_b(4\pi/3)r^3$. Further we assume that bubbles neither appear nor disappear. In other words, we do not consider here nucleation processes. In addition, we suppose a fully dissolved (collapsed) bubble as existent with a radius equal to zero. Thus, if the bubbles do not move relatively to the melt, the number density of bubbles is constant,

$$n_b = n_{b0}. \tag{6.34}$$

Next we assume that the dissolved phase occupies the whole volume of the melt, which is free from bubbles. So, if we denote the volume concentration of the dissolved phase as α_s, then the product $\rho_s \alpha_s$ gives the density of the dissolved phase, i.e. the total mass of the dissolved gas per unit of the total volume of the melt. Then, as both phases occupy the whole volume of continuum, we have:

$$\alpha_b + \alpha_s = 1. \tag{6.35}$$

It is evident that the quantity $\rho_b \alpha_b$ is the density of the dispersed phase, i.e. the total mass of the gas inside a population of bubbles per unit of the total volume of the melt. An equation for the density of the dispersed phase can be found by multiplying the equation of mass balance for a gas-filled bubble Eq. (6.30) and the number density of bubbles, n_b. It yields

$$d_t\left(\rho_b \alpha_b\right) = -\frac{\alpha_b}{\tau_b^D}\left(\rho_b - \frac{\rho_s}{h}\right). \tag{6.36}$$

From this relation and accounting for mass balance of the gas in the melt we get the following balance equation for the density of the dispersed phase:

$$d_t\left(\rho_s \alpha_s\right) = \frac{\alpha_b}{\tau_b^D}\left(\rho_b - \frac{\rho_s}{h}\right). \tag{6.37}$$

For the case of constant number density of the dispersed phase, the system of Eqs. (6.35)–(6.37) jointly with Eq. (6.25) (latter one plays the role of the equation of state) describes the behavior of the gas-filled bubble population in the melt.

To determine the equilibrium parameters of the two-phase continuum from Eqs. (6.35)–(6.37), (6.25) we get:

$$n_b = \text{const.}, \qquad \rho_b = \frac{\rho_s}{h}, \qquad 2\frac{\sigma}{r} + p = p_b \equiv \rho_b RT, \tag{6.38}$$

$$\rho_b \alpha_b + \rho_s \alpha_s = \left(\rho_b \alpha_b + \rho_s \alpha_s\right)\big|_{t=0} = \rho_b^0 \alpha_b^0, \tag{6.39}$$

$$\alpha_b + \alpha_s = 1, \qquad \alpha_b = n_b \frac{4}{3}\pi r^3, \tag{6.40}$$

where the index '0' marks the parameters of the initial state of the continuum. Eqs. (6.39) can be reduced to the following equation, which determines the equilibrium radius of the bubbles:

$$\left(\frac{p_\sigma^0}{\xi} + p\right)\left[\xi^3 (1 - h) + \frac{h}{\alpha_b^0}\right] = p_b^0, \tag{6.41}$$

where $\xi = (r/r_0)$ is the dimensionless radius of the bubbles, p_b^0 is the initial gas pressure in the bubble, $p_\sigma^0 \equiv (2\sigma/r_0)$ is the additional pressure in the bubble at the initial moment of time resulting from capillary forces. The expression in the left part of Eq. (6.41) is a function of the dimensionless radius ξ. As for small and big values of ξ it becomes to be arbitrarily large, it implies, in particular, that the left part of Eq. (6.41) has a minimum. Evidently, a solution of Eq. (6.41) exists if this minimum is less than the value of the right part of Eq. (6.41). Otherwise, Eq. (6.41) has no solution. This latter case can be interpreted as an absence of bubbles in the melt. Let us find out when it occurs.

In order to do so, one needs to find out the conditions imposed on the parameters of the two-phase continuum, when the following inequality holds

$$\left(\frac{p_\sigma^0}{\xi} + p\right)\left[\xi^3 (1 - h) + \frac{h}{\alpha_b^0}\right] > p_b^0. \tag{6.42}$$

Differentiating Eq. (6.42) we obtain the equation to determine the value $\xi = \xi_{min}$, for which the left part of Eq. (6.42) has a minimum:

$$\left(2p_\sigma^0 + 3p\xi\right)\xi^3 = p_\sigma^0 \frac{h}{\alpha_b^0 (1 - h)}. \tag{6.43}$$

In particular, the inequality Eq. (6.42) shows that in order to get a melt without any bubbles the pressure of the gas in the bubbles, when they form and entrap a gas from surrounding space, should not exceed some critical value p_{b*}^0 that depends on surface tension, solubility, initial sizes, and volume concentration of the bubbles.

For example, in Figs. 6.33 and 6.34 results of the numerical solution of the problem based on Eqs. (6.42) and (6.43) are shown in terms of dependencies of critical value of helium pressure inside the bubbles on the initial bubble radius. The pressure of the continuum is a parameter for the graphics in Fig. 6.33. For some value of this pressure the area under the corresponding curve determines the pairs of initial bubble radii and initial pressure of helium in the bubble $\left(p_b^0, r_0\right)$ for which the bubbles in the melt disappear in the course of time. As it is seen from Fig. 6.33, the higher is the pressure of continuum the larger is the area. From a physical point of view, the disappearance of the bubble is caused by the action of the capillary pressure p_σ and the pressure of the continuum. They both set the pressure of the gas inside the bubbles and determine the gas flux from the bubbles to the melt. The higher are both pressures the better are the conditions for gas outlet. Note that the capillary pressure grows when the radius of the bubble decreases.

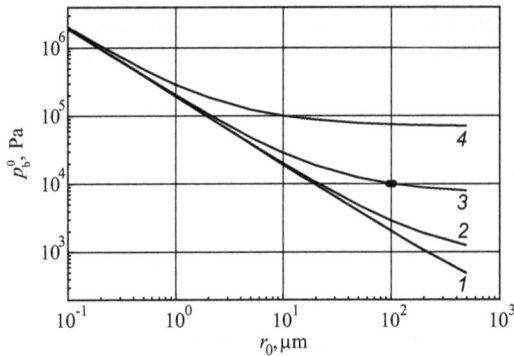

Fig. 6.33: Diagrams of critical initial values of helium pressure inside the bubbles and the radius of the bubbles in the melt. The parameters are: $T = 2400\,°C$, $\sigma = 0.3\,J/m^2$, $a_b^0 = 0.1$. The numbers 1–4 correspond to the following values of pressure in the continuum: $p = 10^3$, 10^4, 10^5, 10^6 Pa.

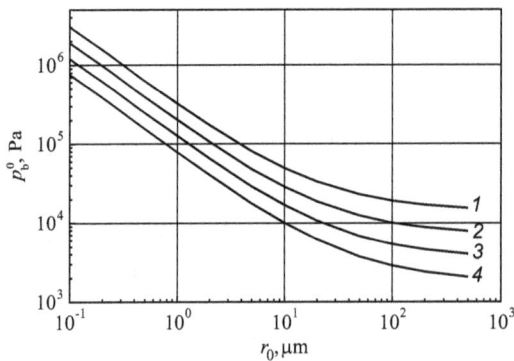

Fig. 6.34: Diagrams of critical initial values of helium pressure inside the bubbles and the radius of the bubbles in the melt. The parameter is the initial volume concentration of bubbles: $T = 2400\,°C$, $\sigma = 0.3\,J/m^2$, $p = 10^5$ Pa. Numbers 1–4 correspond to the following values of volume concentration of the bubbles: $a_b^0 = 0.05, 0.1, 0.2, 0.4$.

For Fig. 6.34 the initial volume concentration a_b^0 is chosen as the parameter. As it follows from the calculations, an increase of a_b^0 shifts the critical curves down, and the area of parameters $\left(p_b^0, r_0\right)$ for which there are no bubbles in the melt converges. For comparison the same critical diagrams for argon are shown in Fig. 6.35 (compare to Fig. 6.33). The situation for argon is qualitatively the same as for helium but the values of the critical initial pressure are approximately ten times less.

As a next step in the analysis, we model the non-stationary two-phase evolution based on Eqs. (6.25), (6.35)–(6.37). We study the kinetics of evolution of the bubbles starting from its initial state in establishing equilibrium in the melt. In Fig. 6.36 two time dependencies of the radii of the helium bubbles during the evolution for different initial conditions are shown. These initial conditions are marked as circles on the

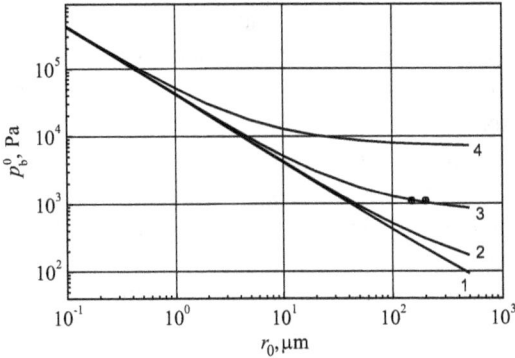

Fig. 6.35: Diagrams of critical initial values of argon pressure inside the bubbles and radius of bubbles in the melt. The parameter is the pressure of the continuum: $T = 2400\,°C$, $\sigma = 0.3\,J/m^2$, $\alpha_b^0 = 0.1$. The numbers 1–4 correspond to the following values of pressure: $p = 10^3, 10^4, 10^5$, 10^6~Pa.

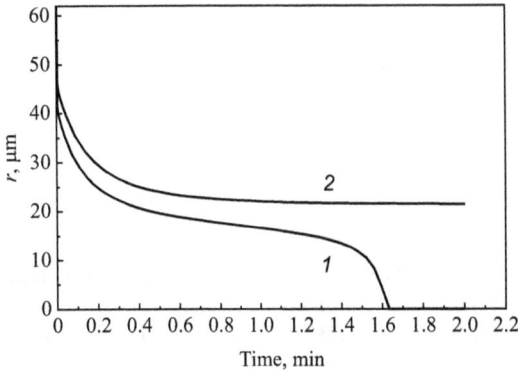

Fig. 6.36: Time dependencies of the radius of helium bubbles at subcritical (curve 2) and supercritical (curve 1) initial conditions: $T = 2400\,°C$, $\sigma = 0.3\,J/m^2$, $\alpha_b^0 = 0.1$, $p = 10^5$ Pa, $p_b^0 = 10^4$ Pa. The numbers 1, 2 correspond to the following values of the initial radius of the bubbles, $r_0 = 95$ and 105 μm.

diagram Fig. 6.33. At fixed initial parameters, shown on the caption to Fig. 6.36, the critical value of the initial radius of the bubbles is equal to $r_{0*} \cong 100$ μm. For the first case the bubble radius is $r_0 = 105$ μm and the point with the coordinates ($p_b^0 = 10^4$ Pa, $r_0 = 105$ μm) is located above the critical curve. Therefore, according to the performed analysis, the solution of the steady-state problem exists, i.e. the bubbles in their evolution do not vanish. For the second case $r_0 = 95$ μm and the point with the coordinates ($p_b^0 = 10^4$ Pa, $r_0 = 105$ μm) is located inside the supercritical area, then a steady-state of the bubbles in the melt does not exist. The latter statement means that all bubbles have to dissolve during its evolution. The graphics, shown in Fig. 6.36, confirms these conclusions completely. As it follows from the graphics, the bubbles, coming to be in continuum, where the pressure is ten times larger than the initial one inside the bubbles, abruptly shrink (for this case, approximately in half) and then the mechanism of mass exchange starts. As a result, the bubbles of subcritical sizes ($r_0 > r_{0*}$) become gradually stable. For bubbles of supercritical sizes ($r_0 < r_{0*}$) there could be only some partial stabilization of their sizes within some time interval.

For comparison on Fig. 6.37 similar time dependencies of radii of argon bubbles are shown. The initial conditions are marked as circles on the diagram in Fig. 6.35. At fixed initial data the critical initial radius of bubbles is equal to $r_{0*} \cong 175$ µm. As it can be seen from Fig. 6.37, the shape of evolution curves for argon and helium bubbles are similar and the whole difference is concluded in the time scales.

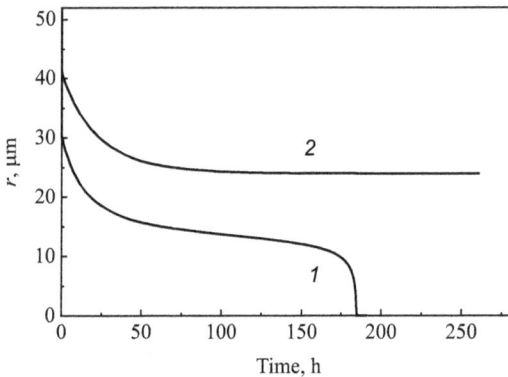

Fig. 6.37: Time dependencies of the radius of argon bubbles at subcritical (curve 2) and supercritical (curve 1) initial conditions: $T = 2400\,°C$, $\sigma = 0.3\,J/m^2$, $a_b^0 = 0.1$, $p = 10^5$ Pa, $p_b^0 = 1.1 \cdot 10^3$ Pa. Numbers 1, 2 correspond to the following values of initial radius of the bubbles: $r_0 = 150$ and 200 µm.

The mathematical model introduced above, describing a behavior of gas-filled bubbles in the melt, and the results, obtained on its basis, are valid for mono-disperse bubble populations. In practice populations of bubbles are usually poly-disperse, i.e. in each part of the melt there are bubbles of different sizes. For a mathematical description of the behavior of such poly-disperse ensembles a modification of the above-introduced mathematical model is required.

6.12 Two-phase Approach to the Description of Poly-disperse Ensembles of Bubbles

Let us characterize poly-disperse distributions of bubbles using a discrete distribution function. Namely, we assume that the bubble population – the disperse phase – contains a finite number of mono-disperse fractions of bubbles, N. Each fraction, for example, the i-th fraction, where $i \in [1, N]$, is specified by its own set of variables, which are the number density of bubbles (number of bubbles per unit volume), n_b^i, the radius of the bubbles, the density and the pressure of the gas inside the bubbles ρ_b^i, p_b^i, the volume concentration of bubbles $\alpha_b^i = n_b^i(4\pi/3)r_i^3$. The total volume concentration of the dispersed phase is equal to the sum of all volume concentrations of bubbles $\alpha_b = \sum_{i=1}^N \alpha_b^i$, and the total mass density of the disperse phase is equal to the sum of all mass densities, $\rho_b\alpha_b = \sum_{i=1}^N \rho_b^i\alpha_b^i$. Since the dispersed and dissolved phases occupy the whole volume of the continuum we have

$$\sum_{i=1}^N \alpha_b^i + \alpha_s = 1. \tag{6.44}$$

Eq. (6.36) is valid for the description of the mass exchange of bubbles of each fraction with the melt. For the i-th fraction this equation has the following form

$$d_t\left(\rho_b^i\alpha_b^i\right) = -\frac{\alpha_b^i}{\tau_{bi}^D}\left(\rho_b^i - \frac{\rho_s}{h}\right), \tag{6.45}$$

where $\tau_{bi}^D = (r_i^2/3Dh)$ is the typical time of the mass exchange of the i-th bubble fraction. The source of the right part of Eq. (6.45) is the mass of the gas (per unit volume) that passes from the bubbles of the i-th fraction to the melt in a unit of time or vice versa – from the melt into the bubbles of this fraction. Evidently, in the equation of the mass balance of the dissolved phase the sum of the same sources, taken with a minus sign, should appear. This requirement follows from the condition of conservation of the gas mass in the melt. Thus, we have:

$$d_t\left(\rho_s\alpha_s\right) = \sum_{i=1}^N \frac{\alpha_b^i}{\tau_{bi}^D}\left(\rho_b^i - \frac{\rho_s}{h}\right). \tag{6.46}$$

To close the system of Eqs. (6.44)–(6.46) one uses the equation of state Eq. (6.25) of each fraction of bubbles, i.e.:

$$4\eta\frac{d_t r_i}{r_i} + 2\frac{\sigma}{r_i} + p = \rho_b^i RT. \tag{6.47}$$

At equilibrium all derivatives with respect to time are equal to zero. Therefore, from Eq. (6.45) one gets that the gas density inside the bubbles is the same for every fraction, namely $\rho_b = \rho_b^i = \frac{\rho_s}{h}$, where $i \in [1, N]$. Besides, as it follows from Eq. (6.47), only one value of the radius of the bubbles corresponds to one value of the density ρ_b. In other words, the originally poly-disperse population of bubbles, having reached a steady-state, becomes mono-disperse.

To understand how this state is reached let us consider the solution of the non-stationary system of equations Eqs. (6.44)–(6.47). As an example the evolution of a poly-disperse population of helium bubbles of five fractions was calculated on the basis of Eqs. (6.44)–(6.47) for different values of the initial density of helium inside the bubbles and the pressure of continuum. In Figs. 6.38–6.40 the time dependencies of the parameters of state of the poly-disperse system during the evolution process are shown (at atmospheric pressure). The process of bubble evolution seems to proceed as follows. At the beginning the bubbles of all fractions shrink because of the influence of capillary forces. At the same time, the density of the gas inside the bubbles increases and becomes maximal for the bubbles of the smallest fraction (see Fig. 6.39). Therefore the diffusion of gas molecules from the bubbles to the melt starts. The intensity of mass flux is directly proportional (as it follows from Eq. (6.45)) to the difference of densities $\rho_b^i - \frac{\rho_s}{h}$ and is inversely proportional to the square of bubble radius. That is why the mass flux to the melt of the gas from the bubbles of the smallest fraction becomes dominating. As a result, the density of the dissolved gas in the melt increases (see Fig. 6.40) and the outflow of the gas from the melt into the bubbles of larger frac-

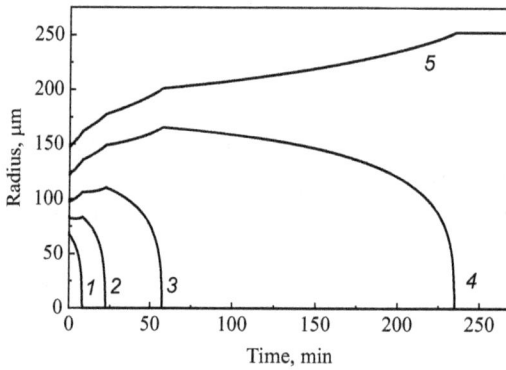

Fig. 6.38: Evolution of a poly-disperse bubble population: Radii vs. time: $r_{i0} = 70, 85, 100, 125, 150$; $\alpha_{b0}^i = 0.05$ at, $i = 1,\ldots,5$; $\rho_s^0 = 0$; $T = 2400\,°C$, $\sigma = 0.3\,J/m^2$, $p = p_b^0 = 10^5$ Pa. The numbers 1–5 correspond to the number of the fraction.

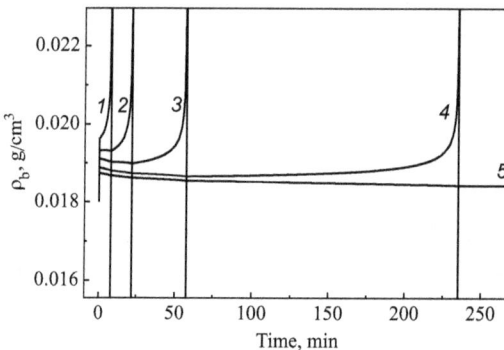

Fig. 6.39: Evolution of a poly-disperse bubble population: Helium densities inside the bubbles vs. time. The values of the parameters are the same as on Fig. 6.38, the numbers 1–5 correspond to the number of the fraction.

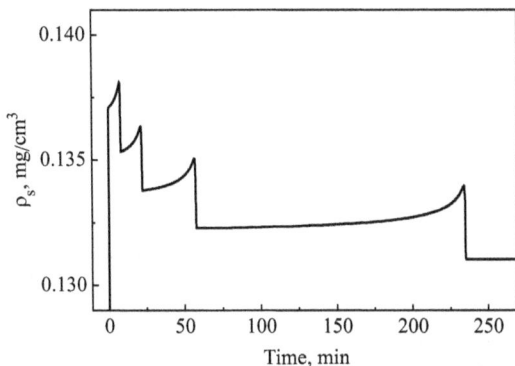

Fig. 6.40: Change of the density of dissolved helium during the evolution of a poly-disperse bubble population. Values of parameters are the same as in Fig. 6.38.

tions begins, i.e. these bubbles grow (see Fig. 6.38) and at the same time the sizes of the bubbles of the smallest fraction decrease. This decrease of the radius of the bubbles is accompanied by a growth of the capillary pressure and the density of the gas inside the bubbles. As a result, a transfer of gas molecules from the smallest bubbles to the melt and then into the bubbles of larger fractions becomes more intensive up to the moment when the smallest fraction of bubbles collapses (see Figs. 6.38–6.40). After that the growth of all remaining fractions of bubbles still continues during some time gap (see Fig. 6.40), but now it is accompanied by a strong decrease of the density ρ_s of the dissolved gas. It occurs while the difference $\rho_b^i - \frac{\rho_s}{h}$ for the smallest of the remaining fractions of bubbles stays negative. When this difference becomes positive, a new cycle of the evolution process starts, and it finishes again with a collapse of the smallest fraction of bubbles etc. Note that the time scale grows from cycle to cycle because for each new cycle the smallest fraction of bubbles is larger than for the previous one. Thus, the considered poly-disperse population at the mentioned conditions reaches equilibrium, and only the bubbles of the largest fraction survive. These bubbles accumulate the whole gas (with an exception of a small part that is dissolved in the melt) originally distributed inside the bubbles of all fractions.

The above described dynamics of evolution of an ensemble of gas-filled bubbles is typical for the stage of coalescence, the theory of which had been devised by I.M. Lifshitz and V.V. Slezov [89] in application to solid solutions. It was developed then by V.V. Slezov and others (see, for example [91]) to liquids supersaturated with gas. Nucleation and subsequent coarsening of the bubbles are caused by the initial supersaturation of the solution. In our case, the bubbles are entrapped from the surrounding space. For the mathematical description, Eqs. (6.44)–(6.47) are used. These relations have to be supplemented by the initial conditions, i.e., the sizes of bubbles, their volume concentration, initial pressure of gas inside the bubbles and density of dissolved gas. The concept of "supersaturation" of the solution or the melt is not required for the model given by Eqs. (6.44)–(6.47).

As an additional example, let us consider the same poly-disperse population formed at a reduced pressure $p_b^0 = 3.7 \cdot 10^3$ Pa. From Fig. 6.41 it follows that for this

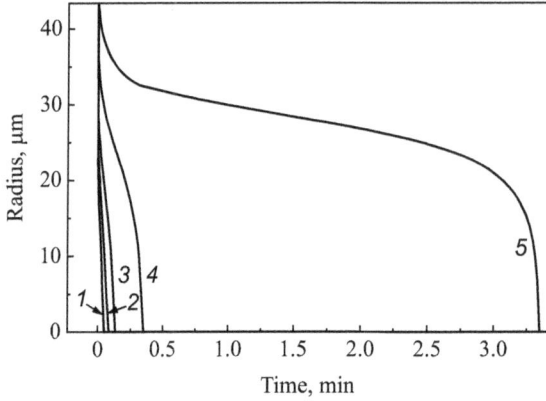

Fig. 6.41: Evolution of a poly-disperse bubble population at reduced initial pressure of helium inside the bubbles: Radii vs. time: $p = 10^5$ Pa, $p_b^0 = 3.7 \cdot 10^3$ Pa. The values of the parameters are the same as in Fig. 6.38. The numbers 1–5 correspond to the number of the fraction.

case even the largest fraction of bubbles collapses with time. This is similar to the situation from Section 6.11 for the mono-disperse population of bubbles, when the values of initial parameters are given on supercritical area. Note that the calculation of the critical values of the initial parameters for a mono-disperse population of bubbles, considered in Section 6.11, can be extended to the poly-disperse population of bubbles with a discrete size distribution function.

6.13 Diffusion of the Dissolved Gas in the Melt

In Sections 6.11 and 6.12, it was assumed in the mathematical models that the total amount of the gas in the two-phase continuum, that is equal to the sum of the dissolved gas and the gas inside the bubbles, is constant in time. At that we take into account a diffusive transfer of gas molecules between bubbles and glass melt neglecting a diffusion of gas to the surroundings. Such inconsequence can be excused by the essential difference between the time scales of these processes. For the first case a typical thickness of the diffusion layer is (in order of magnitude):

$$n_b^{-1/3} = \left(\frac{4\pi r^3}{3\alpha_b} \right)^{1/3} = r \left(\frac{4\pi}{3\alpha_b} \right)^{1/3}, \tag{6.48}$$

i.e. usually less than one millimeter. For the second case, a typical thickness is the thickness of the fused glass, i.e. a few centimeters and more. Nevertheless, for realistic systems some areas abutting on limits (where it is necessary to take into account diffusion processes in its full measure) may exist.

The development of a strict mathematical model for the description of diffusion in a melt with gas-filled bubbles is rather problematic. The authors of [90] succeeded to develop an approximate analytical solution for a limited number of the bubbles, which are placed in nodes of the cubic lattice. This analysis evidences a baffling complexity of the general problem one has to solve. A phenomenological approach, taking into account the difference of time scales of diffusion inside the neighborhood of the bubbles (i.e. in micro-volume and in the whole volume of the continuum, in the macro-volume) seems to be more constructive. This considerable difference of scales gives a possibility to suppose that the processes of diffusion in micro- and macro-volumes do not depend on each other. From the point of view of mathematical description it means that the equation of mass balance of the dispersed phase remains the same (see Eq. (6.45)) even under the condition of mass exchange with the surroundings. So that for a description of diffusion of the dissolved gas in the melt it is quite enough to add the usual diffusion term to the balance equation of the dissolved phase Eq. (6.46). As the result, instead of Eq. (6.46), we get:

$$\partial_t \left(\rho_s \alpha_s \right) = \sum_{i=1}^{N} \frac{\alpha_b^i}{\tau_{bi}^D} \left(\rho_b^i - \frac{\rho_s}{h} \right) + \nabla \cdot \left(\alpha_s D \nabla \rho_s \right), \tag{6.49}$$

where ∇ is the differential operator of gradient. The factor α_s at the diffusion coefficient in Eq. (6.49) takes into account (in a first approximation) that the dissolved phase occupies some part of the continuum volume only. Eq. (6.49) describes that a change of the density of dissolved gas is caused by mass exchange with bubbles of all fractions (diffusion in the micro-volume) and diffusion of dissolved gas in the macro-volume.

Let us discuss from a qualitative point of view the influence of mass exchange between the dispersed and the dissolved phases on the diffusion process of the gas dissolved in the melt. Summarizing all equations of mass balance of each fraction of the dispersed phase (Eq. (6.45) and Eq. (6.49)), we obtain the following equation for the total content of gas in the melt:

$$\partial_t \left(\sum_{i=1}^{N} \rho_b^i \alpha_b^i + \rho_s \alpha_s \right) = \nabla \left(\alpha_s D \nabla \rho_s \right). \tag{6.50}$$

It follows that the total amount of the gas (both dissolved and contained in the bubbles) is changed due to the diffusion of dissolved gas in the macro-volume, in scale of continuum only. Further let us assume, first, that the dispersed phase is in local equilibrium, i.e. the following relation $\rho_{bi} = \rho_b = \rho_s/h$ holds at every point of the continuum and, second, the volume fraction of bubbles is constant in the whole space. Then instead of Eq. (6.50) we have:

$$\partial_t \rho_s = \nabla \cdot \left(D_{eff} \nabla \rho_s \right), \tag{6.51}$$

where

$$D_{eff} = \frac{(1 - \alpha_b) h}{\alpha_b + (1 - \alpha_b) h} D \approx \frac{(1 - \alpha_b) h}{\alpha_b} D \tag{6.52}$$

is the effective (apparent) diffusion coefficient. In particular for helium and argon at conditions from Table 6.9 we get: $D_{eff}\big|_{He} \approx 0.06\, D\big|_{He}$, $D_{eff}\big|_{Ar} \approx 0.006\, D\big|_{Ar}$, i.e. the presence of the gas-filled bubbles reduces the diffusion of the gas in the melt by tens and even by hundreds of times. Note that the concept of an effective (apparent) diffusion coefficient for our two-phase system represents a rather rough idealization and it can be only used for estimations. For more accurate calculations, Eq. (6.49) should be used. The complete system of equations includes Eqs. (6.44), (6.45), (6.49) and the equation of state of each fraction Eq. (6.47).

In Figs. 6.42–6.45, some results of calculations of the evolution of a mono-disperse population of helium bubbles are shown for a rather thin layer of glass melt. It is assumed that above the upper surface the concentration of helium is equal to zero and dissolved helium can leave the glass melt freely. The lower surface of the layer is supposed to be impenetrable for helium, i.e. at the lower surface a diffusion flux is equal

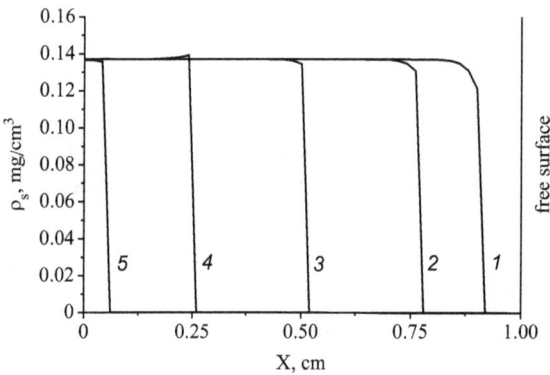

Fig. 6.42: Distribution of density of dissolved helium through the layer of the melt for appreciable volume content of bubbles (diffusion of dissolved gas): $p = p_b^0 = 10^5$ Pa, $\delta = 1$ cm, $\alpha_{b0} = 0.1$, $r_0 = 100$ μm. Moments of time (1–5): 0.14, 0.67, 2.8, 6.8, 10 h.

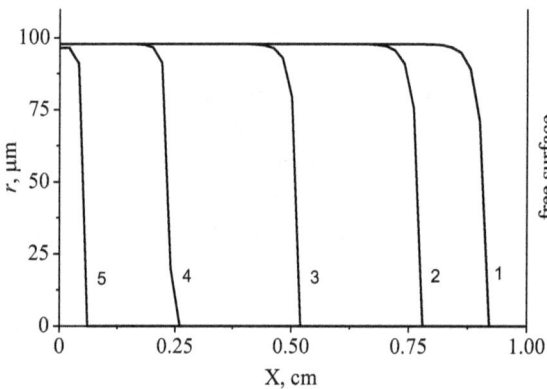

Fig. 6.43: Distribution of radii of helium bubbles through the layer of melt for appreciable volume content of bubbles (diffusion of dissolved gas): $p = p_b^0 = 10^5$ Pa, $\delta = 1$ cm, $\alpha_{b0} = 0.1$, $r_0 = 100$ μm. Moments of time (1–5): 0.14, 0.67, 2.8, 6.8, 10 h.

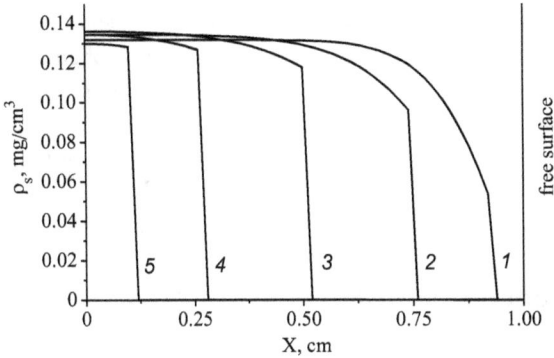

Fig. 6.44: Distribution of density of dissolved helium through the layer of the melt for an insignif-icant volume content of bubbles (diffusion of dissolved gas): $p = p_b^0 = 10^5$ Pa, $\delta = 1$ cm, $\alpha_{b0} = 0.001$, $r_0 = 100$ µm. Moments of time (1–5): 0.021, 0.055, 0.11, 0.18, 0.23 h.

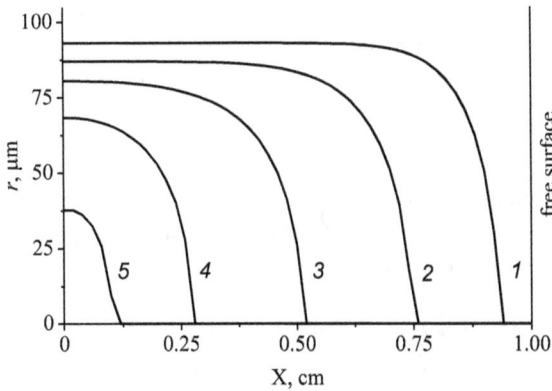

Fig. 6.45: Distribution of radii of helium bubbles through the layer of melt for an insignificant vol-ume content of bubbles (diffusion of dissolved gas): $p = p_b^0 = 10^5$ Pa, $\delta = 1$ cm, $\alpha_{b0} = 0.001$, $r_0 = 100$ µm. Moments of time (1–5): 0.021, 0.055, 0.11, 0.18, 0.23 h.

to zero. Further, at the initial moment of time the two-phase continuum is supposed to be in equilibrium. With time diffusion of dissolved helium to the upper surface oc-curs. A decrease of density of the dissolved gas disturbs the equilibrium and starts a mechanism of interphase mass exchange. As a result, the bubbles shrink, and some part of the gas flows from the bubbles to the melt. A flow of the gas from the bubbles to the melt reduces the rate of density decrease of the dissolved gas and the rate of interphase mass exchange. The more bubbles are in the melt, the more considerable is this reduction. This process may be considered as a motion of some wave (refining wave) from the upper surface (on graphics it is on the right) deep down the melt. For an appreciable volume content of bubbles the front of this wave is essentially steeper than for an insignificant volume content and the velocity of propagation is essentially less (compare the graphics in Figs. 6.42, 6.43 and 6.44, 6.45). This result agrees with the previous analysis of the influence of interphase mass exchange on the rate of gas diffusion in a melt, the effective diffusion coefficient. Particularly, in calculations, which results are shown on Fig. 6.42, 6.43 and 6.44, 6.45, at the initial volume concentration of bubbles equal to $\alpha_{b0} = 0.1$ the refining time is approximately 40 times larger than at $\alpha_{b0} = 0.001$.

6.14 Relative Motion of Bubbles in the Melt: Modification of the Mathematical Model

So far we assumed that the bubbles do not move in the melt. Strictly speaking this is not so because every bubble in the melt is subjected to the Archimedean force, which causes a motion of the bubble upwards, i.e., opposite to the direction of the vector of gravity acceleration. The velocity of bubble motion (velocity of floating-up) in the melt can be obtained setting equal Archimedean and Stokes forces. It yields

$$\vec{U} = -\frac{2}{9}\frac{\rho}{\eta}r^2\,\vec{g}, \tag{6.53}$$

where $\vec{U}$ is the velocity (vector) of floating-up of the bubble relative to the melt; $\vec{g}$ is the gravity acceleration. Estimating the value of the floating-up velocity (numerical values of parameters are taken from Table 6.9) we have

at $\quad T = 2400°C \quad \rightarrow \quad U \approx 7.4 \cdot 10^{-8}\text{m/s} \approx 0.3\text{mm/h}$,

at $\quad T = 2000°C \quad \rightarrow \quad U \approx 1.3 \cdot 10^{-9}\text{m/s} \approx 5\,\mu\text{m/h}$,

that indicates very slow floating of bubbles under the action of Earth's gravity field. Therefore, for these conditions, the rate of motion of the bubbles is negligibly small.

However, artificial gravity, caused by rotation, can be hundreds times larger than the usual gravity. Really, at the frequency of rotation of $v = 10$ Hz and a radius of $R = 0.5$ m the centrifugal acceleration is $a = (2\pi v)^2 R \approx 2 \cdot 10^3\text{m/s}^2$, that exceeds by two hundred times the usual gravitational acceleration. Accordingly, the velocity of floating-up increases by the same factors. The motion of the bubbles can essentially change the whole pattern of their evolution because, first, for a displacement of a bubble in the melt the conditions (for example, temperature, pressure etc.) near the bubble may vary, and, second, at a movement of bubbles relatively to the melt a new mass transfer of the gas from one point of the melt to another one appears. So, let us consider the behavior of the bubbles in the melt taking into account their motion due to Archimedean force. For simplicity we suppose here that the melt does not move. The flow of the melt is accounted for below in the next section.

The equations for the description of any fraction of the dispersed phase are modified in such case. They read

$$\partial_t n_b^i + \nabla \cdot \left(n_b^i \vec{U}_i \right) = 0,$$ (6.54)

$$\partial_t \left(\rho_b^i a_b^i \right) + \nabla \cdot \left(\rho_b^i a_b^i \vec{U}_i \right) = -\frac{a_b^i}{\tau_{bi}^D} \left(\rho_b^i - \frac{\rho_s}{h} \right),$$ (6.55)

$$\frac{4\eta}{r_i} \left(\partial_t r_i + \vec{U}_i \cdot \nabla r_i \right) + 2\frac{\sigma}{r_i} + p = \rho_b^i RT,$$ (6.56)

$$\vec{U}_i = -\frac{2}{9}\frac{\rho}{\eta} r_i^2 \vec{g}.$$ (6.57)

Eq. (6.54) describes the balance of the number of bubbles of the i-th fraction. If the bubble is at rest, i.e. $\vec{U}_i = 0$, Eq. (6.54) gives $\partial_t n_b^i = 0$, that means a constant number density of bubbles. It can change only due to the motion of the bubbles. The mass density of the i-th fraction of the dispersed phase is subjected to two factors: motion of bubbles and mass exchange between bubbles and melt (see Eq. (6.55)).

The following equation Eq. (6.56) defines the state of the bubbles of i-th fraction. The difference of Eq. (6.56) from Eq. (6.47) consists in the derivative with respect to time accounting for local and convective components. Note that the pressure from Eq. (6.56) is the sum of two pressures, the external (surrounding) pressure and the hydrostatic pressure. This latter condition of strong gravity can play an appreciable role. Finally, Eq. (6.57) defines the velocity of the floating-up bubbles and for the calculation of the density of the dissolved gas, Eq. (6.49) is valid.

Figs. 6.46–6.51 show some results of calculations (based on Eqs. (6.49), (6.54)–(6.57)) of the evolution of a poly-disperse helium bubble population in a melt layer of thickness $\delta = 10$ cm under a gravity field of $100g = 100 \cdot 9.8$ m/s^2. The diffusion of the dissolved gas to the upper (free) surface is also taken into account. It is supposed that at the initial moment of time three fractions of bubbles of different sizes are uniformly distributed inside the melt layer and the volume concentrations of these fractions are the same. The initial radii of the first, second, and third fractions amount to 150, 100, 50 µm, respectively. The initial pressure in the bubbles is equal to the atmospheric pressure. Besides, it is assumed that above the upper surface the concentration of helium is equal to zero and the lower surface of the layer is impenetrable for helium. For these initial and boundary conditions, the duration of the first stage is given by Eq. (6.27) and is less than a second. The bubbles shrink here to sizes determined according to the steady-state solution of Eq. (6.25). Maximum compression takes place near the lower surface of the layer, where hydrostatic pressure is the largest.

Simultaneously three slower processes start: interphase mass exchange, floating-up of bubbles, and diffusion of the dissolved gas to the upper (free) surface and then outside. Here the interphase mass exchange dominates (see Figs. 6.46 and 6.47). As a result, according to the mechanism described in Section 6.12, the dissolution of the lowest (third) fraction of bubbles and the growth of larger fractions occur. At that the

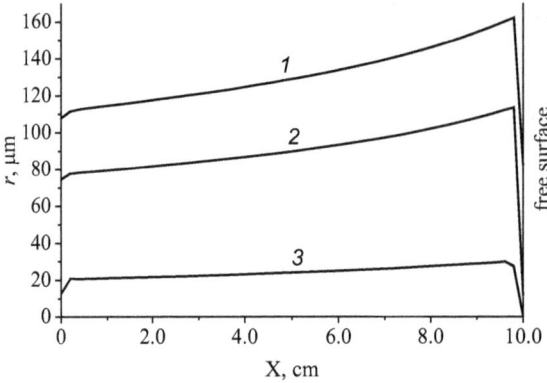

Fig. 6.46: Distributions of radii of helium bubbles of various fractions through the melt layer at $t = 3$ min (diffusion of dissolved gas and floating-up of bubbles): $p = p_b^0 = 10^5$ Pa, $\delta = 10$ cm, $\alpha_{bo}^I = \alpha_{bo} = 0.08$. The numbers 1–3 correspond to fractions with initial radii $r_{1-3}|_{t=0} = 150, \ 100, \ 50$ μm.

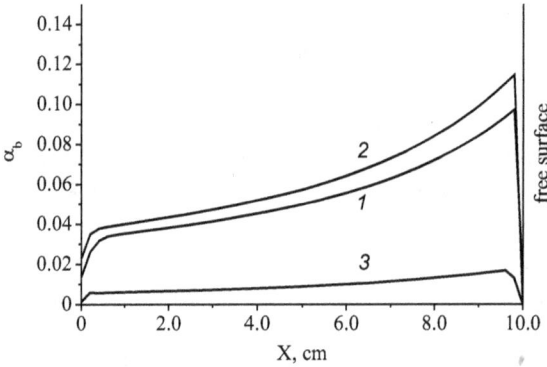

Fig. 6.47: Distributions of volume concentration of helium bubbles of various fractions through the melt layer at $t = 3$ min (diffusion of dissolved gas and floating-up of bubbles): $p = p_b^0 = 10^5$ Pa, $\delta = 10$ cm, $\alpha_{bo}^I = \alpha_{bo} = 0.08$. The numbers 1–3 correspond to fractions with initial radii $r_{1-3}|_{t=0} = 150, \ 100, \ 50$ μm.

bubbles of medium (second) fraction show a preferred growth. From Figs. 6.46 and 6.47 it is seen that after 3 minutes the size and the volume concentration of the smallest fraction is essentially decreased, whereas the size and the volume concentration of the largest fraction increases. Because of the diffusion of the dissolved gas to the upper surface some thin layer appears near this surface where the bubbles of all fractions quickly dissolve.

An influence of the floating-up of the bubbles becomes apparent somewhat later. After 45 minutes from the beginning of the process (see Fig. 6.48) some sinks of stratification of medium and largest fraction of bubbles are indicated (the lowest fraction has already had time to dissolve). The largest bubbles float to the upper part of the melt

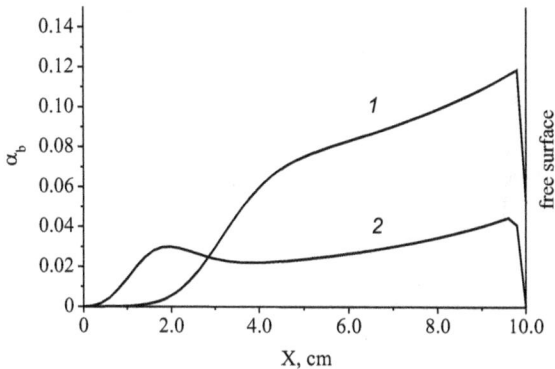

Fig. 6.48: Distributions of volume concentration of helium bubbles of various fractions through the melt layer at $t = 45$ min (diffusion of dissolved gas and floating-up of bubbles): $p = p_b^0 = 10^5$ Pa, $\delta = 10$ cm, $\alpha_{bo}^i = \alpha_{bo} = 0.08$. The numbers 1–3 correspond to fractions with initial radii $r_{1,2}|_{t=0} = 150, 100$ μm.

layer and the medium bubbles, whose velocity of floating-up is approximately half the velocity of the largest bubbles, remain behind. For all that the volume concentration of the bubbles of the largest fraction considerably increases because, staying here, the bubbles of the medium fraction shrink very intensively.

A complete stratification of bubbles is seen on Fig. 6.49. At the time $t = 1.5$ h the remaining bubbles of the largest fraction are concentrated near the upper surface (i.e. they practically leave the layer of the melt) and at the same time the bubbles of the medium fraction are located in the lower part of the layer. It is interesting to note, that the upper limit of the bubbles of medium fraction takes on the shape of a shock front, where the radii of the bubbles jump down to zero. An occurrence of such jump is caused by the fact that larger bubbles of the medium fraction catch up the lower ones.

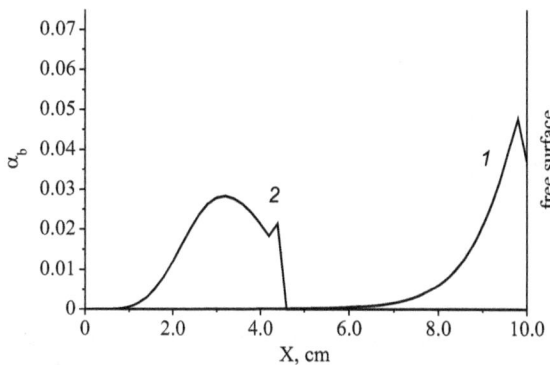

Fig. 6.49: Distributions of volume concentration of helium bubbles of various fractions through the melt layer at $t = 1.5$ h (diffusion of dissolved gas and floating-up of bubbles): $p = p_b^0 = 10^5$ Pa, $\delta = 10$ cm, $\alpha_{bo}^i = \alpha_{bo} = 0.08$. The numbers 1 and 2 correspond to fractions with initial radii $r_{1,2}|_{t=0} = 150, 100$ μm.

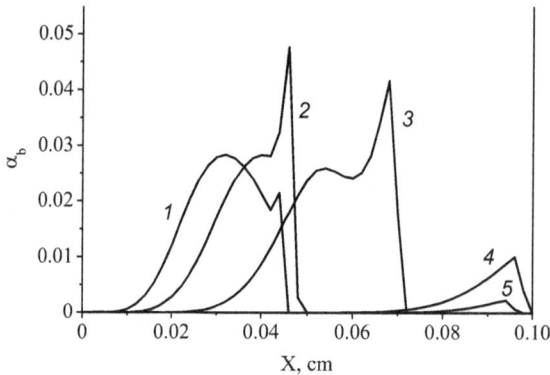

Fig. 6.50: Distributions of volume concentration of helium bubbles of medium fraction ($r_2|_{t=0}$ = 100 µm) through the melt layer at consecutive points of time (diffusion of dissolved gas and floating-up of bubbles): t_{1-5} =1.5, 2.0, 3.0, 5.9, 6.5 h. The remaining parameters are the same as in Fig. 6.47.

The further evolution of the bubbles is shown in Fig. 6.50. As it is seen, for a period of time from t = 1.5 h till t = 2 h, a closed area near to the front of the bubbles continues to develop, where the sizes of the bubbles are maximal. It occurs due to the presence of smaller bubbles (of the same fraction!), which collapse, raise the density of the dissolved gas in the melt (see Fig. 6.51) and therefore feed their own larger neighbors. The growth of the radius of the bubbles increases the velocity of floating-up and stretches the area of location of the bubbles.

The appearance of small bubbles is governed by diffusion of the dissolved gas in the melt: diffusion decreases the peak value of density of the dissolved gas on the

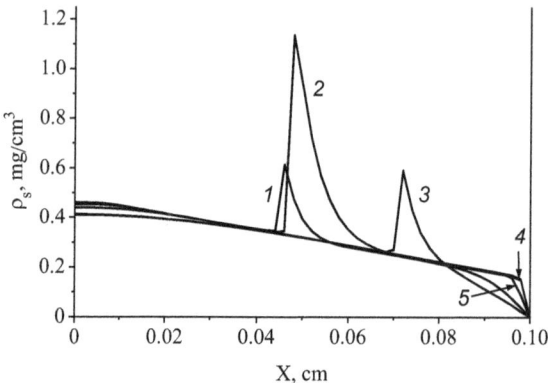

Fig. 6.51: Distributions of density of dissolved helium through the melt layer at consecutive points of time (diffusion of dissolved gas and floating-up of bubbles): t_{1-5} = 1.5, 2.0, 3.0, 5.9, 6.5 h. The remaining parameters are the same as on Fig. 6.47.

front. It causes the dissolution of the bubbles, a decrease of their sizes and a reduction of the velocity of motion. As a result, larger bubbles catch up smaller ones and absorb an abundance of dissolved gas. When the front of the bubbles approaches the upper surface, the role of diffusion increases, and the peak value of concentration of the dissolved gas decreases with time. Thus, simultaneous occurrence of interphase mass exchange, floating-up of bubbles, and diffusion of the dissolved gas give rise to the refining process which can obtain very intricate forms.

6.15 Flow of the Melt Governed by the Motion of the Bubbles: Complete System of Equations for Modeling of the Behavior of Gas-filled Bubble Ensembles in the Melt

The analysis of possible patterns of behavior of gas-filled bubble populations in a melt is now completed by the account of the flow of the melt. The flow of the melt could be initiated because of a floating-up of a bubble population, which might carry the melt.

Let us consider, from a hydrodynamic point of view, the population of bubbles in the melt and the melt as a multi-velocity continuum. Let $\vec{u}$ be the velocity of the melt, then the velocity of motion of the i-th fraction of bubbles of the dispersed phase is equal to $\vec{u}_b^i = \vec{u} + \vec{U}_i$, where $\vec{U}_i$ is the relative velocity of i-th fraction of bubbles (floating velocity). The mass averaged velocity of continuum is defined as

$$\vec{u}_m = \frac{(\rho + \rho_s)\,\alpha_s\,\vec{u} + \sum_i \rho_b^i \alpha_b^i\,\vec{u}_b^i}{(\rho + \rho_s)\,\alpha_s + \sum_i \rho_b^i \alpha_b^i}, \tag{6.58}$$

where ρ is the density of the melt. Because the gas density in the bubbles ρ_b^i and especially the density of the dissolved gas are considerably smaller than the density of the melt, it should be allowed that, first, the mass averaged velocity of continuum is equal to the velocity of the melt $\vec{u}_m \cong \vec{u}$ and, second, the density of continuum is expressed as $\rho_m \cong \alpha_s \rho$.

Summarizing, we can formulate a system of equations for the description of the behavior of the melt with the bubbles in the following way:

$$\partial_t n_b^i + \nabla \cdot \left[n_b^i \left(\vec{u} + \vec{U}_i \right) \right] = 0, \tag{6.59}$$

$$\partial_t \left(\rho_b^i \alpha_b^i \right) + \nabla \cdot \left[\rho_b^i \alpha_b^i \left(\vec{u} + \vec{U}_i \right) \right] = -\frac{\alpha_b^i}{\tau_{bi}^D} \left(\rho_b^i - \frac{\rho_s}{h} \right), \tag{6.60}$$

$$\frac{4\eta}{r_i} \left[\partial_t r_i + \left(\vec{u} + \vec{U}_i \right) \cdot \nabla r_i \right] + 2\frac{\sigma}{r_i} + p = \rho_b^i RT, \tag{6.61}$$

$$\vec{U}_i = -\frac{2}{9}\frac{\rho}{\eta} r_i^2 \vec{g}, \tag{6.62}$$

$$\partial_t \left(\rho_s \alpha_s \right) + \nabla \cdot \left(\rho_s \alpha_s \vec{u} \right) = \sum_{i=1}^{N} \frac{\alpha_b^i}{\tau_{bi}^D} \left(\rho_b^i - \frac{\rho_s}{h} \right) + \nabla \cdot \left(\alpha_s D \nabla \rho_s \right), \tag{6.63}$$

$$\alpha_s + \sum_i \alpha_b^i = 1, \quad \alpha_b^i = n_b^i \frac{4}{3} \pi r^3, \tag{6.64}$$

$$\partial_t \alpha_s + \nabla \cdot \left(\alpha_s \vec{u} \right) = 0, \tag{6.65}$$

$$-\nabla p + \nabla \cdot \overleftrightarrow{\sigma} + \alpha_s \rho \vec{F} = 0, \tag{6.66}$$

where $\overleftrightarrow{\sigma}$ is the tensor of viscous stresses with components $\sigma_{ij} = \eta_m \left(\partial_i u_j + \partial_j u_i \right)$; η_m is the effective viscosity coefficient. It can be expressed, in a first approximation, as $\eta_m = \alpha_s \eta$. $\vec{F}$ is the volume force affecting the melt and taking into account both natural and artificial gravity.

Eqs. (6.59)–(6.63) describe the behavior of every fraction of the dispersed phase, namely (i) Eq. (6.59), the balance of the number of the bubbles; (ii) Eq. (6.60), the balance of the mass of gas in the bubbles; (iii) Eq. (6.61), the dynamics of the change of the bubble size (equation of state); (iv) Eq. (6.62), the floating of bubbles in the melt; (v) Eq. (6.63), the mass balance of the dissolved gas, describes mass exchange within all bubble fractions, diffusion, and convective mass transfer. Eqs. (6.65) and (6.66) describe the flow of the melt: (i) Eq. (6.65) is the continuity equation of the melt; (ii) Eq. (6.66) is the equation of motion in Stokes form (smallness of Reynolds number is taken into account).

Because of the conditions of fusion of a glass, the process cannot be absolutely isothermal and the properties of the melt depend on temperature. For this reason, the system of equations Eqs. (6.59)–(6.66) has to be completed by the equation of energy balance,

$$\alpha_s \rho \, c \, (\partial_t T + \vec{u} \cdot \nabla T) = \nabla \cdot (\alpha_s \lambda \, \nabla T) - \nabla \cdot \vec{W}_r, \tag{6.67}$$

where T is the temperature, being the same for both phases; c is the heat capacity of the melt; λ is the conductivity of the melt; $\vec{W}_r$ is the density of radiation flux. Eq. (6.67) describes conductive, convective, and radiative transfer of energy. Note that for temperatures in the interval of silica glass fusion the radiative transfer usually dominates.

The radiation flux $\vec{W}_r$ is determined as an integral of the spectral intensity of radiation over the whole spectrum of radiation and the whole solid angle

$$\vec{W}_r = \int_0^\infty dv \int_{4\pi} d\Omega \, \vec{\Omega} I_v, \tag{6.68}$$

where v is the frequency of radiation; Ω is the solid angle; $\vec{\Omega}$ is the unit vector, which determines the direction of propagation of radiation; I_v is the spectral intensity of radiation. For the determination of the spectral intensity of radiation I_v the following equation of radiation transfer can be used:

$$\vec{\Omega} \cdot \nabla I_v = \kappa_v^{eff} \left(I_{ve} - I_v \right) = \left(\kappa_v^+ + \kappa_b^{eff} \right) \left(I_{ve} - I_v \right), \tag{6.69}$$

where the effective absorption coefficient of continuum is equal to the sum of the co-efficient of absorption of the melt and the dispersed phase of bubbles: $\kappa_v^{eff} = \kappa_v^+ + \kappa_b^{eff}$.

To illustrate the effect of carrying of the melt by floating-up bubbles the flow of silica melt inside a two-dimensional layer is considered. For such purpose, we employ Eqs. (6.59)–(6.69). It is assumed that the length of the layer is $L = 0.5$ m and the thickness is $\delta = 10$ cm. At the lower ($y = 0$) and the left ($x = 0$) boundaries of the layer the conditions of adhesion of the melt have to be accounted for

$$u|_{x=0} = v|_{x=0} = 0, \qquad u|_{y=0} = v|_{y=0} = 0. \tag{6.70}$$

At the upper ($y = \delta$) and the right ($x = L$) side, the boundary conditions of free glide have to be fulfilled,

$$u|_{x=L} = \partial_x v|_{x=L} = 0, \qquad \partial_y u|_{y=\delta} = v|_{y=\delta} = 0. \tag{6.71}$$

Besides, at the lower boundary of the layer the following distribution of volume concentration of bubbles is fixed:

$$\alpha_b|_{y=0} = \begin{cases} 0.09, & x < \dfrac{L}{2} \\ 0.11, & x > \dfrac{L}{2} \end{cases}. \tag{6.72}$$

Above studied processes of interphase mass exchange and diffusion of the gas in the melt are not considered here. Moreover, the size of the bubbles is supposed to be constant, $r = 100$ μm.

In Fig. 6.52 a stationary velocity field at a gravity of 160g due to rotation of the layer is shown and the appearing at the same time Coriolis acceleration is established. As

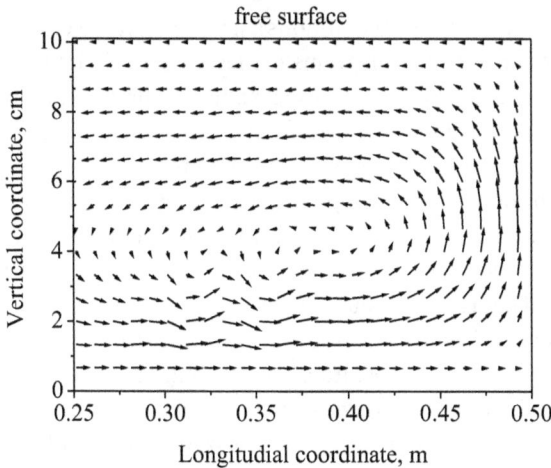

Fig. 6.52: Vector field of velocity of flow in a layer of glass melt, which is governed by a non-uniform distribution of bubbles (fragment): $T = 2400\,°C$, $r = 100$ μm, gravity $= 160$ g.

it follows from this figure, at accepted conditions a vortex-type flow in the layer of the melt develops. The center of the vortex is shifted to the right boundary, where the velocity of ascending flow is maximal.

6.16 Summary to Part III

The application of the two-phase approach for the description of gas-filled bubbles in a melt allows us to consider the population of the bubbles and the melt as one continuum and to formulate the system of equations Eqs. (6.59)–(6.66), which describes the behavior of this continuum. The system of equations Eqs. (6.59)–(6.66) gives the opportunity to calculate the characteristics of the following processes: flow of the continuum as a whole, relative motion of the different phases, their interaction, interphase mass exchange, and mass transfer (both convective and diffusive) in the continuum. Using Eqs. (6.59)–(6.66) (more precisely, its various forms) some special problems concerning the evolution of the bubbles in the melt were resolved. In particular, the following results are obtained: (i) An isolated bubble (which constant mass) always reaches equilibrium with the melt. At that the relaxation time (see Eq. (6.27)) is defined by the radius of the bubble, the coefficient of surface tension, pressure, and the viscosity of the melt, and does not depend on the kind of gas which fills the bubble. (ii) A solitary bubble under mass exchange with the melt never reaches equilibrium with the melt. It either grows with no limit or it collapses. For this process the time scale depends on the kind of the gas (see Eq. (6.31)). For example, for helium the time scale is small. However, for weakly soluble and diffusing gases (as argon) the time scale can be much larger than the time of glass fusion. Then, the mass of the gas inside the bubble may be assumed as constant. (iii) The behavior of a mono-disperse population of bubbles depends on its initial state (initial conditions). If initial parameters are larger than some critical values, defined by the diagrams in Figs. 6.33–6.35, then the population of bubbles during its evolution does not disappear and reaches an equilibrium. On the contrary, if the initial parameters are less than the critical ones, then the bubbles collapse. These diagrams may be useful in practice to find the conditions for refining the melt. (iv) The behavior of poly-disperse populations of bubbles also depends on their initial state. If the initial amount of gas inside the bubble population is small enough, then a stepwise collapse (from smaller to larger fractions) occurs (see Fig. 6.41). But if the initial amount of gas is large enough, then a process of coalescence or coarsening takes place. It is characterized by a consecutive consumption of smaller bubbles by larger ones. As a result, only the largest fraction of bubbles survives. The dynamics of this process is shown on the graphs (Figs. 6.38–6.40), where, along with the time dependencies of bubble sizes, the time dependencies of the density of the gas inside the bubbles and the density of the dissolved gas are presented. (v) The diffusion of the dissolved gas through the melt results in a change of the density of the dissolved gas and has, therefore, an influence on the dispersed phase. This influence appears on the graphs (Figs. 6.42–6.44) as a motion of some refining wave from the upper (free) surface deep down the melt. For an appreciable volume content of bubbles the front of this wave is essentially steeper than for insignificant volume content and the velocity of propagation is essentially lower. (vi) The relative motion of the bubbles in the melt is due to gravity influences on their evolution too. As it follows from the results outlined in Section 6.14, simultaneous occurrence of interphase mass exchange, floating-up of bubbles, and diffusion of the dissolved gas give rise to a refining process which can obtain very intricate forms. (vii) Ascending motion of bubbles with respect to the melt causes (at some well-defined conditions) a flow of the melt as a whole (see Fig. 6.52).

Evidently, all properties of the melt depend strongly on temperature. In addition, as a rule, glass fusion does not proceed isothermally. By this reason, it is quite necessary to have the possibility to determine the temperature fields. This task can be performed employing the energy balance equation Eq. (6.67). This equation, in combination with the equation of transfer of radiative energy Eq. (6.69), permits us to calculate temperature fields taking into account conductive, convective and radiative heat fluxes in continuum. However, the solution of this task is a problem, which goes beyond the limits of the present work.

Bibliography

[1]   Kitaigorodski, I.I., *Technologie des Glases* (Verlag Technik, Berlin, 1959).
[2]   Pavlushkin, N.M., *Chemical Technology of Glasses and Glass-ceramics* (Stroiizdat, Moscow, 1983).
[3]   Leko, V.K., Mazurin, O.V., *Properties of Quartz Glass* (Nauka, Leningrad, 1985).
[4]   Khalilev, V.D., Prokhorova, T.I., *Quartz Glass* (Leningrad, 1982).
[5]   Khalilev, V.D., Prokhorova, T.I., *Basic Principles of Silica Glass Technology* (Leningrad, 1983).
[6]   Sci. Glass, version 7.0, ITC.MA, USA, 2008.
[7]   Leko, V.K., private communication, March 2008.
[8]   Rebrova, K.P., Shchekoldin, A.A., *Peculiarities of Polymorphic Transformation of Vein Quartz into Cristobalite*. In: *Physical-Chemical Researches of Structure and Properties of Silica Glass*, 1974, Proceedings of the State Scientific Research Institute of Silica Glass, vol. 1, Moscow, pp. 119–125.
[9]   Leko, V.K., Komarova, L.A., *Crystallization of Silica Glasses in Various Gaseous Media*. In: Proceedings of the Academy of Sciences of the USSR, Series: Inorganic Materials **11**, 1115 (1975).
[10] Boganov, A.G., Rudenko, V.S., Cheremisin, I.I., The Soviet Journal of Glass Physics and Chemistry **10**, 208 (1984).
[11] Koryavin, A.A., Stepanchuk, V.I., Mechanical-Optical Industry **10**, 28 (1976).
[12] van der Steen, G.H.A.M., Papanikolau E., *Introduction and Removal of Hydroxyl Groups in Vitreous Silica. Part I. Influence of the Melting Conditions on the Hydroxyl Content in Vitreous Silica*, Phillips Res. Reports **30**, 103 (1975).
[13] Papanikolau, E., Wachters, A.J.H., *A Theory of Bubble Growth at Chemical Equilibrium with Application to the Hydrogen Reboil in Fused Silica*, Philips J. Res. **35**, 59 (1980).

[14]   Cheremisin, I.I., Rudenko, V.S., Boganov A.G., The Soviet Journal of Glass Physics and Chemistry **10**, 365 (1984).

[15]   Parikh, N.M., J. American Ceramic Society **41**, 18 (1958).

[16]   Kingery, W.D., J. American Ceramic Society **42**, 6 (1959).

[17]   Andreeva, N.A., Chernyshova, L.S., *Measurement of Surface Tension for Various Silica Glasses*. Theses 3rd All-Soviet Union Scientific and Technical Conference on Silica Glass. Moscow, 1973, pp. 80–81.

[18]   Shelby, J.E., J. Non-Crystalline Solids **349**, 331 (2004).

[19]   Bacon, J.F., Hasapis, A.A., Wholly, J.W., Phys. Chem. Glasses **1**, 90 (1960).

[20]   Leko, V.K., The Soviet Journal of Glass Physics and Chemistry **3**, 258 (1979).

[21]   Leko, V.K., Meshcheryakova, E.V., Gusakova, N.K., Lebedeva, R.B., Mechanical-Optical Industry **12**, 42 (1974).

[22]   Hoffmaier, G., Urbain, G., Sci. Ceram. **4**, 25 (1968).

[23]   Shackelford, J.F., J. Non-Crystalline Solids **30**, 127 (1978).

[24]   Shackelford, J.F., Brown, B.D., J. Non-Crystalline Solids **44**, 379 (1981).

[25]   Shackelford, J.F., J. Non-Crystalline Solids **49**, 299 (1982).

[26]   Norton, F.J., J. American Ceramic Society **36**, 90 (1953).

[27]   Mc Elfresh, D.K., Howitt, D.G., J. Non-Crystalline Solids **124**, 174 (1990).

[28]   Chan, S.L., Elliot, S.R., Physical Review B **43**, 4423 (1991).

[29]   Studt, P.L., Shackelford, J.F., Fulrath, R.M., J. Appl. Phys. **41**, 2777 (1970).

[30]   Boiko, G.G., Berezhnoi, G.V., Glass-Techn. Berichte **77 C**, 346 (2004). (Proc. N. Kreidl Memorial Conference – Slovakia).

[31]   Kratkaya Khimicheskaya Encyclopediya, Moscow, 1961.

[32]   Anderson, O.L., Stuart, D.A., J. Amer. Ceram. Soc. **37**, 573 (1954).

[33]   Swets, D.E., Lee, R.W., Frank, R.C., J. Chem. Phys. **34**, 17 (1961).

[34]   Perkins, W.G., Begeal, D.R., J. Chem. Phys. **54**, 1683 (1971).

[35]   Frank, R.C., Swets, D.E., Lee, R.W., J. Chem. Phys. **35**, 1451 (1961).

[36]   Shackelford, J.F., Studt, P.L., Fulrath, R.M., J. Appl. Phys. **43**, 1619 (1972).

[37]   Shelby, J.E., J. Appl. Phys. **47**, 136 (1976).

[38]   Flores, J.S., Shackelford, J.F., J. Non-Crystalline Solids **68**, 327 (1984).

[39]   Carroll, M.R., J. Non-Crystalline Solids **124**, 181 (1990).

[40]   Norton, F.T., *Gas Permeation Through the Vacuum Envelope*. In: *Transactions of VIII Vacuum Symposium*, vol. 1, 8 (Pergamon Press, New York, 1962).

[41]   Williams, E.L., J. Amer. Ceram. Soc. **48**, 190 (1965).

[42]   Revesz, A.G., Schaeffer H.A., J. Electrochem. Soc.: Solid-State Science and Technology **12**, 357 (1982).

[43]   Schaeffer, H.A., J. Non-Crystalline Solids **38&39**, 545 (1980).

[44]   Mikkelsen, J.C., Appl. Phys. Lett. **45**, 1187 (1984).

[45]   Bershtein, V.A., Emelyanov, V.A., Stepanov, Yu. A., The Soviet Journal of Glass Physics and Chemistry **4**, 475 (1978).

[46]   Lee, R.W., Frank, R.C., Swets, D.E., J. Chem. Phys. **36**, 1062 (1962).

[47]   Lee, R.W., J. Chem. Phys. **38**, 448 (1963).

[48]   Lee, R.W., Phys. and Chem. Glasses **5**, 35 (1964).

[49]   Lee, R.W., Fry, D.L., Physics and Chemistry of Glasses **7**, 19 (1966).

[50]   Saito, N., Motoyama, K., Vacuum Soc. Japan Journal **7**, 350 (1964).

[51]   Moulson, A.J., Roberts, J.P., Trans. Faraday Soc. **57**, 1208 (1961).

[52]   Davis, K.M., Tomozawa, M., J. Non-Crystalline Solids **185**, 203 (1995).

[53]   Drury, T., Roberts, J.P., Physics and Chemistry of Glasses **4**, 79 (1963).

[54]   Roberts, G.J., Roberts, J.P., Phys. Chem. Glasses **7**, 82 (1966).

[55]   van der Steen, G.H.A.M., Papanikolau, E., *Introduction and Removal of Hydroxyl Groups in Vitreous Silica. Part II. Chemical and Physical Solubility of Hydrogen in Vitreous Silica*. Philips Research Reports **30**, 192 (1975).
[56]   van der Steen, G.H.A.M., Papanikolau, E., *Introduction and Removal of Hydroxyl Groups in Vitreous Silica. Part III. Some Thermodynamic Data on the Reduction of Vitreous Silica*. Philips Research Reports **30**, 309 (1975).
[57]   Bell, R.I., Hetherington, G., Jack, K., Phys. Chem. Glasses **3**, 141 (1962).
[58]   Hermann, W., Rau, J., Ungelenk, H., Ber. Bunsenges. Phys. Chem. **89**, 423 (1985).
[59]   Pevzner, B.Z., Makhov, V.E., Borisenko, V.A., *Some Aspects of the Formation of Sintering Powder Coatings*. Proc. Conf. "Production and Application of Protective Coatings" (Nauka, Leningrad, 1987, pp. 27–34).
[60]   Pevzner, B.Z., Azbel, A. Yu., Glass Physics and Chemistry **19**, 82 (1993).
[61]   Polyakova, I.G., present volume.
[62]   Papanikolau, E., J. Non-Crystalline Solids **38–39**, 563 (1980).
[63]   Ristic, M.M., Science of Sintering **33**, 143 (2001).
[64]   Frenkel, Ya. I., J. Exp. Theor. Phys. **16**, 29 (1946).
[65]   Mackenzie, J.K., Shuttleworth, R., Proc. Phys. Soc. B **62**, 833 (1949).
[66]   Kingery, W.D., *Kinetics of High-temperature Processes* (MIT Press, New York, 1958).
[67]   Geguzin, Ya. E., *Physics of Sintering* (Nauka, Moscow, 1984).
[68]   Skorokhod, V.V., *Rheological Principles of Sintering Theory* (Naukova Dumka, Kiev, 1972).
[69]   Ivensen, V.A., *Phenomenology of Sintering* (Metallurgia, Moscow, 1985).
[70]   Scherer, G.V., J. Amer. Ceram. Soc. **60**, 236 (1977).
[71]   Scherer, G.V., Bachmann, D.L., J. Amer. Ceram. Soc. **60**, 239 (1977).
[72]   Scherer, G.V., J. Amer. Ceram. Soc. **60**, 243 (1977).
[73]   Azbel, A. Yu., Vasiliev, V.N., Khoruzhnikov, S.E., Glass Physics and Chemistry **14**, 749 (1988).
[74]   Landau, L.D., Lifshitz, E.M., *Hydrodynamics* (Nauka, Moscow, 1986).
[75]   Geguzin, Ya. E., Dokl. Akad. Nauk SSSR **20**, 1051 (1971).
[76]   Barlow, E.J., Langlois, W.E., *Diffusion of gas from a liquid into an expanding bubble*. IBM Journal, 1962, July, pp. 329–337.
[77]   Altmann, A., *Zum Verhalten von Blasen in Kieselglasschmelzen*, Silikattechnik **30**, 299–302 (1979).
[78]   Weinberg, M.C., Onorato, P.I.K., Uhlmann, D.R., *Behavior of bubbles in glass melts: Dissolution of a stationary bubble containing a single gas*, Journal of American Ceramic Society **63**, 175–180 (1980).
[79]   Nemec, L., *The behavior of bubbles in glass melts. Part 1. Bubble size controlled by diffusion*, Glass Technology **21**, 134–138 (1980).
[80]   Nemec, L., *The behavior of bubbles in glass melts. Part 2. Bubble size controlled by diffusion and chemical reaction*, Glass Technology **21**, 139–144 (1980).
[81]   Ramos, J.I., *Behavior of multi-component gas-filled bubbles in glass melt*, Journal of American Ceramic Society **69**, 149–154 (1986).
[82]   Cable, M., Frade, J.R., *Diffusion-controlled growth of multi-component gas-filled bubbles*. Journal of Materials Science **22**, 919–924 (1987).
[83]   Balkani, B., Ungan, A., *Numerical simulation of bubble behavior in glass melting tanks. Part 1. Under ideal conditions*. Glass technology **37**, 29–34 (1996).
[84]   Balkani, B., Ungan, A., *Numerical simulation of bubble behavior in glass melting tanks. Part 2. Dissolved gas concentration*, Glass technology **37**, 101–105 (1996).
[85]   Balkani, B., Ungan, A., *Numerical simulation of bubble behaviour in glass melting tanks. Part 3. Bubble trajectories*, Glass technology **37**, 137–142 (1996).

[86] Balkani, B., Ungan, A.: *Numerical simulation of bubble behaviour in glass melting tanks. Part 4. Bubble number density distribution*, Glass technology **37**, 164–168 (1996).

[87] Nemec, L., Tonarova, V., *The behavior of bubbles in glass melts under effect of the gravitational and centrifugal fields*, Ceramics-Silikaty **49**, 162–169 (2005).

[88] Frank-Kamenetzky, D.A., *Diffusia i teploperedacha v khimicheskoi kinetike* (Nauka, Moscow, 1987).

[89] Lifshitz, E.M., Pitaevsky, L.P., *Fizicheskaia Kinetika* (Nauka, Moscow, 1979).

[90] Zak, M., Weinberg, M.C., *Multi-bubble dissolution in glass melts*, Journal of Non-Crystalline Solids **38–39**, 533–538 (1980).

[91] Slezov, V.V., Abyzov, A.S., Slezova, Zh. V., *Nucleation and growth of gas-filled bubbles in low-viscosity liquids*. In: *Nucleation theory and applications*, Edited by J.W.P. Schmelzer, G. Röpke, V.B. Priezzhev. Joint Institute for Nuclear Research, Dubna, Russia 2005, pp. 222–266.

Victor K. Leko

7 Regularities and Peculiarities in the Crystallization Kinetics of Silica Glass

The present contribution gives a comprehensive overview on quantitative results of a systematic investigation of crystallization of different quartz glasses obtained at laboratory and industrial syntheses under conditions ruling out uncontrolled contamination of sample surfaces during heat treatment. It is shown that the process of crystallization proceeds through two stages: the stage of an induction period during which no crystallization traces are observed on the sample surface and the stage of crystalline layer growth. The value of the induction period can be taken to be a quantitative measure of crystallization stability of quartz glass. It was found that the crystallization kinetics of vitreous silica is strongly influenced by the content of structural water contained in vitreous silica, by the redox conditions on melting, and the degree of fusion penetration of batch material, quartz or cristobalite. It was also established that the crystallization kinetics of tubes made of quartz glass differs considerably from the crystallization kinetics of block-shaped glasses. Spontaneous crystallization on an ideally pure surface of quartz glass (on the surface of closed bubbles) is not observed. In order to explain these results the suggestion has been advanced that surface nucleation of vitreous silica is initiated by chemical reactions with the components of the gaseous medium (H_2O, O_2) or with components of contaminations attached to the sample surface. A study of the crystallization kinetics of quartz glasses in the atmosphere of different gases was carried out: in dry air, oxygen, argon, vacuum and water vapor. While studying samples of the same glass both crystallization rate and viscosity were measured. A comparison of the data on these two properties showed that the inversely proportional dependence between crystallization rate and viscosity (as discussed frequently in the literature) is not only not observed in actual practice, but there is an opposite tendency: the factors increasing the crystallization rate simultaneously increase glass viscosity. A possible mechanism allowing one to explain this phenomenon is proposed.

7.1 Introduction

Quartz glass as a high-temperature material is used in a variety of applications in modern engineering, mainly in semiconductor industry and lighting engineering where quartz glass products are in operation at high temperatures for an appreciable length of time. Under such conditions products start crystallizing. Crystallization begins from the surface and a resulting transparent poly-crystalline layer of cristobalite gradually increases its thickness. A crystallizing product is a laminated composite material whose outer layers are formed by cristobalite and the inner layer is formed by the highly viscous SiO_2-melt. At high temperatures partially crystallized products are visually indistinguishable from not crystallized ones. However, the mechanical properties of a partially crystallized product differ significantly from those of a not crystallized product.

At temperatures above 1000 °C the density and thermal expansion coefficient of the SiO_2-melt are very close to those of cristobalite. Therefore crystallization of the respective products is not accompanied by their fracture or changes of spatial dimensions. Drastic temperature changes in this region result in the appearance of weak fast-relaxing stresses on the melt-cristobalite interface. Unlike the melt, cristobalite does not possess fluidity; hence, viscous deformation of the products slows down with an increasing degree of crystallization and is eventually fully terminated. Such products can be kept at high temperatures without further significant deformation practically for infinite times.

When the temperature is decreased below the range of about 1000 °C, the coefficient of linear thermal expansion of cristobalite starts growing rapidly ($\alpha_{(250-800)°C}$ = 80 · 10^{-7} K^{-1} [1]), while the SiO_2-melt transforms into brittle glass with a very low coefficient of linear thermal expansion ($\sim 5 \cdot 10^{-7}$ K^{-1}). Hence cooling of partially crystallized products is accompanied by the development of tensile strength in a cristobalite layer, which results in its cracking. Cooled to ~250 °C the product looks transparent but it is covered by a net of coarse cracks. At temperatures less than 250 °C, high-temperature α-cristobalite is transformed gradually into the low-temperature β-cristobalite. Each step is accompanied by a 4 % change of the volume and by the formation of a thick net of micro-cracks in the cristobalite layer. The crystallized layer becomes white, while the products loose their strength, transparency and become practically not suitable for further use.

Depending on the product operating conditions crystallization can play both positive and negative roles. At a long-term continuous operation at temperatures above 1000 °C, a rigid cristobalite layer appears on the products; this layer drastically reduces irreversible deformation, which increases the life-time for service of the product. For products operating under periodic mode, crystallization is a very undesirable phenomenon resulting in a quick failure of a product. Crystallization is inadmissible in conducting bending and sintering processes, as well as in studying some properties of quartz glass (viscosity, electrical conduction, gas permeability, etc.) over the range of high temperatures. Extremely undesirable is crystallization when annealing is carried out.

To control the process of crystallization, so that the life-time of service of a product is kept as long as possible and to properly manage the technological processes it is necessary to have full information on the specific features of crystallization of quartz glasses. It is essential to understand what the quantitative measure of crystallization stability is, what factors affect it and what should be done to prevent premature crystallization of products. Apart from these practical aspects, a detailed investigation of the crystallization process of vitreous silica is also of obvious general scientific interest. This is due to a number of reasons: (i) Quartz glass is a single-component glass and simple in its composition. (ii) The glass and the product of its crystallization – cristobalite – are identical in composition. (iii) Quartz glass is crystallized from the surface (there is no spontaneous bulk crystallization), the thickness of the resulting layer is easily measured. (iv) Some properties of single-component vitreous silica are very sensitive to impurities. A study of this problem with respect to crystallization can supply us with a deeper insight into the specific features of phase transitions in the silica system.

The commercial production of quartz glass goes back to the early 20th century. The first investigations of quartz glass crystallization also appeared at that time. Quartz glass was a rare and hard-to-produce material till the middle of the last century. Of general interest at that time were investigations of the dependence of the properties on compositions of multi-component glasses, as well as the possibility to change their chemical compositions within wide ranges; against this background a systematic study of the properties of the simplest in chemical composition quartz glass was not thought to be a currently central problem. For a long time, studies in this direction were either totally absent or of random character. This statement is particularly true for the studies of crystallization of quartz glass. Among the results of the first few studies, special attention should be devoted to those directed at the investigation of the sensitivity of the quartz glass crystallization process to the composition of gas medium [2, 3] and to surface contamination by alkali traces [4] (a review of the first papers on crystallization of quartz glass is given in [5]).

The second half of the last century was marked by a rapid growth of new fields of science and technology: semiconductor industry, lighting engineering, atomic and laser technologies, rocket engineering, precision and optical instrument engineering, metrology. In connection with this there arose a demand for quartz glass as a material with a variety of unique properties (high heat resistance, thermal stability, transparency, low dielectric and mechanical losses, high resistance to electromagnetic irradiation and particle bombardment, as well as to deleterious chemical agent actions). Rapid improvement of the old and development of new technologies of quartz glass production was evident.

The first experience of a widespread commercial use of quartz glass showed that many of its important properties (viscosity, crystallization, spectral properties, radiation and optical stability) can change over unexpectedly wide ranges. The information on quartz glass properties available at that time was absolutely inadequate for an efficient solving of the appearing practical problems and a proper understanding the phenomena observed. This is the reason why intensive studies of various properties of vitreous silica were carried out at research centers of the leading industrial countries (USA, Germany, Great Britain, France) during the second half of the last century. For such purposes, in the USSR in the 1960s a specialized research and development center – the State Scientific-Research Institute of Quartz Glass – was set up. The institute both developed technological processes of production of quartz glasses and quartz glass articles and investigated various properties including crystallization of all types of quartz glasses on record.

At present quartz glasses are obtained in different ways from various raw materials. To be better guided in the variety of transparent commercial glasses it was proposed in [6] to classify them by four types varying in the synthesis technique and content of impurities of metals and OH-groups. Table 7.1 gives a list of initial raw materials, conditions of production, impurity content and, as examples, commercial brands for each of the four types of glasses.

Table 7.1: Types of silica glass. Examples of trademarks for different types of glasses are: (I) GE-105, GE-125, GE-204 (USA); IR-Vitreosil (Great Britain); Infrasil (Germany); Pursil-453 (France); 2020, 2030 (Japan); KI (Russia); (II) GE-102, GE-104 (USA); OG-Vitreosil (Great Britain); Herasil (Germany); Pursil optique (France); 1030, 1320 (Japan); KV (Russia); (III) Corning 7940 (USA); Spectrosil (Great Britain); Suprasil (Germany); Tetrasil (France); 4040 (Japan); KU (Russia); (IV) Spectrosil WF (Great Britain); Suprasil W (Germany); KUVI (Russia)

Type	Raw material	Melting condition	Impurity content
I	Quartz, cristobalite	Vacuum, dry gases	Σmetals $< 1 \cdot 10^{-2}$; OH$< 1 \cdot 10^{-3}$
II	Quartz, cristobalite	In H_2+O_2 flame	Σmetals $< 1 \cdot 10^{-2}$; OH$\sim (2 \div 6) \cdot 10^{-2}$
III	$SiCl_4$	In H_2+O_2 flame	Σmetals $< 1 \cdot 10^{-4}$; OH$\sim (0.1 \div 0.2)$
IV	$SiCl_4$	In O_2 plasma	Σmetals $< 1 \cdot 10^{-4}$; OH$< 1 \cdot 10^{-4}$

With respect to crystallization of quartz glasses, particularly intensive studies were conducted in the (1960–1970)s mainly in the USA and USSR. The results of these studies were published in English and Russian in the respective scientific journals. In the following years the studies in this direction and the corresponding publications on this subject practically stopped. At the same time the analysis of the papers devoted to the properties of vitreous silica published in recent years shows that in the papers dealing with crystallization problems references are provided actually only to papers written in English. Thus, the results of the studies of various aspects of crystallization processes for all types of quartz glasses obtained at the specialized research center of the USSR – the Institute of Quartz Glass – and published in Russian language are practically unknown to today's English-language readers. The aim of this review is to introduce the reader (as far as possible) to the considerable amount of information on the problem of crystallization of quartz glass including the papers originally published in the Russian language.

7.2 Literature Review

When discussing the problems connected with surface crystallization, it is necessary to know the general phenomenological equations that describe quantitatively the dependence of crystalline layer thickness on time and temperature. The first attempt in this direction was made by Brown and Kistler [7]. The authors investigated crystallization kinetics of five different glasses: Corning company commercial glass obtained by high-temperature hydrolysis of $SiCl_4$ and four glasses of laboratory vacuum melting from synthetic silicon dioxide Cab–O–Sil–0 (containing about 0.03 wt% of impurities about 0.02 wt% of which are alkali impurities). Three of these glasses contained different additions of Al_2O_3.

Rods of about 2 mm in diameter and 5 cm in length, drawn in oxyhydrogen flame, were used as samples. The samples were placed into a vertical tubular furnace heated up to a specified temperature. It was noted that the exposure of such samples at high temperature resulted in nonuniform crystallization of the surface. To obtain uniform crystallization the samples were pre-soaked in water for no less than 48 hours. During heat-treatment of these samples a linear growth of a crystalline layer with time was observed (see Fig. 7.1). It is seen from this figure that the nucleation process proceeds very fast and crystallization starts practically immediately after the beginning of heat treatment. The rate of crystallization (V) increases exponentially with temperature. The values of the activation energy of crystalline layer growth were calculated from the slopes of the (log V vs. 1/T)-curves.

Paper [8] describes a study of crystallization kinetics of GE 204A glass of type I. Sections of tubes were used as samples. Crystallization was conducted in a tubular electric furnace with an interior platinum-rhodium heater allowing one heating up to 1900 °C. The authors confirmed a high sensitivity of quartz glass crystallization with respect to minor amounts of surface impurities. For example, a drop of distilled water

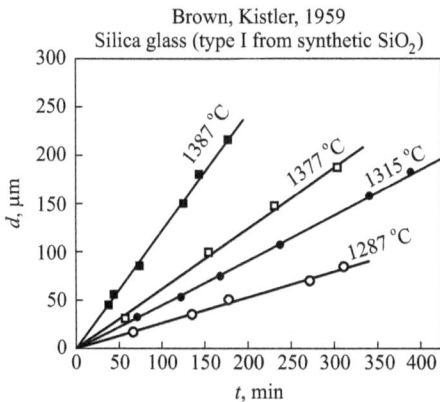

Fig. 7.1: Dependence of crystalline layer thickness on temperature (from data of [7]).

Ainslie et al., 1962
Silica glass I type GE204 (tube)

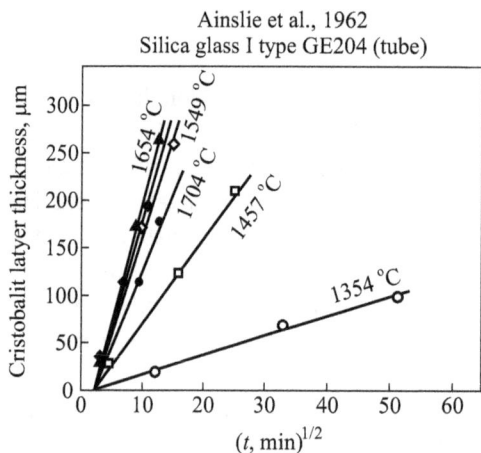

Fig. 7.2: Growth parabolas for GE 204A fused silica corresponding to several temperatures [8].

applied to flame-polished surface of a sample after its heat-treatment at 1400 °C during 30 minutes initiated traces of crystallization, whereas a drop of triply distilled water at the same heat treatment did not initiate crystallization traces. The authors found that, when quartz glass tubes were crystallized in air, the thickness of the crystalline layer increased linearly as a function of the square root of the heat-treatment time, i.e. a parabolic dependence was observed (Fig. 7.2). The fact that the parabolas do not pass through the origin of the coordinate system is explained by them by the necessity of a 15–20 minute warm-up of the sample in order to reach thermal equilibrium. Since at parabolic growth of a crystalline layer the growth rate decreases with time, the authors judge a crystalline layer growth rate from the slope of the curve representing layer thickness vs. the square root of time (Fig. 7.3). The "crystallization rate" adopted by the authors passes through a maximum at ~1655 °C, while at a temperature of ~1710 °C the rate is zero (melting point of cristobalite).

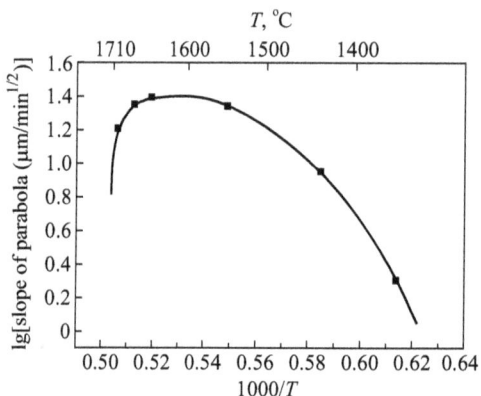

Fig. 7.3: Slopes of the parabolas plotted in Fig. 7.2 as a function of reciprocal temperature [8].

To the opinion of the authors, the parabolic dependence of crystalline layer thickness on time indicates that some materials attached to the glass surface catalyze the process of crystallization, while the rate of motion of a glass-cristobalite interface depends on the rate at which these materials diffuse through the cristobalite layer. Since the process of crystallization is carried out in the ordinary atmosphere, the material catalyzing crystallization could be one of the air components: nitrogen, oxygen or water vapor. To find out which of the air component catalyzes the crystallization process, experiments on heat-treatment of samples sealed in quartz glass ampoules were carried out containing a certain gas: superpure nitrogen, spectrally pure oxygen, water vapor, chemically pure argon. The result was assessed by the thickness of the crystalline layer which was formed in the course of heat-treatment. It was established that both oxygen and water have the same effect on crystallization in ordinary atmosphere, whereas crystallization in argon proceeded considerably more slowly. Thus, it was found that water vapor and oxygen are catalysts of quartz glass crystallization. It was also noticed that the presence of reducing agents (hydrogen, graphite, chromium, and germanium) in sealed ampoules resulted in a rapid decrease of the crystallization rate. Films of silicon or pyrolytic graphite on the surface of quartz glass also decreased the rate of its crystallization. The authors of [8] observed glass crystallization not only starting from the surface, but also the so-called internal or bulk crystallization occurring on solid impurities present in quartz glass. The authors noted that the observed crystal growth in the interior of glass proceeded approximately three times faster than on the surface.

In [9], crystallization of four different glasses was investigated both in water vapor and in oxygen. Three of the four glasses were glasses of a reducing synthesis (GE-201, GE-204 type I and Corning 7943 obtained by the two-stage method: formation of a porous blank glass by high-temperature hydrolysis of $SiCl_4$ and vitrification of this blank in the atmosphere of dry hydrogen). The fourth glass was a glass of a laboratory synthesis. It was obtained by dehydration of Corning glass 7940 of type III by powdering it and exposing to vacuum at 1110 °C during 247 hours. The dehydrated powder was fused then at 1750 °C during 40 minutes. The resulting glass practically did not contain OH-groups and it was fused in the absence of any reducing agent. Cubes of about 3 mm edge length were cut from all glasses except GE-204. The surface of edges was being polished by diamond dust. Glass GE-204 was used in the form of tubes. Samples of the glasses GE-201 and GE-204 had a high enough nucleation rate; hence, their surfaces were not pre-treated. Nucleation of samples of Corning 7943 and glass of a laboratory synthesis was sensitive to impurities during the process of sample polishing, which resulted in nonuniform nucleation. To achieve a more uniform nucleation the samples were held in a weak solution of sodium silicate or in tap water. Crystallization was conducted in a vertical tubular furnace with heaters of molybdenum disilicide. The samples were placed on a platinum substrate that was suspended by a platinum wire inside the furnace.

The authors came to the conclusion that crystallization of glasses of a reducing synthesis in the atmosphere of water vapor or oxygen proceeds under the parabolic law, in the atmosphere of water vapor crystallization proceeding at a higher rate than in the atmosphere of oxygen (Fig. 7.4). In the authors' opinion, the parabolic form of the time-dependence of crystalline layer thickness implies a diffusion-controlled mechanism of crystallization. Glasses of a reducing synthesis have a deficit of structural oxygen: the more reducing the synthesis mode is, the higher is the deviation of the glass composition from a stoichiometric one. Cristobalite has a stoichiometric composition, so the authors believe that in order to be transformed into cristobalite the glass should be oxidized. The authors suppose that such oxidation occurs because of the diffusion of oxygen atoms or water vapor from the ambient gas medium through the growing layer of cristobalite. From Fig. 7.4 it follows that water vapor accelerates crystallization to a greater extent than oxygen would do. The authors explain this by the fact that water is not only a supplier of oxygen, but also a material that breaks the three-dimensional network of glass and decreases its viscosity.

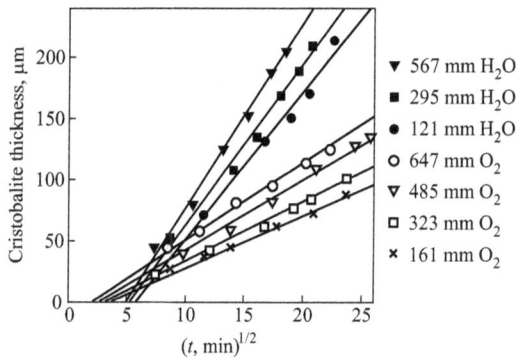

Fig. 7.4: Parabolic growth rates for type 204A glass at 1508 °C in H_2O and O_2-atmosphere [9].

At the same time a particular laboratory glass (the authors call it stoichiometric) demonstrates a linear time-dependence of the growth of crystalline layer thickness (Fig. 7.5), while the rate of its crystallization is the same both in the atmospheres of oxygen and of nitrogen (Fig. 7.6). This result, in their opinion, confirms the validity of the hypothesis of the diffusion-controlled mechanism of crystallization.

In the paper by Wagstaff and Richards [10], an attempt was made to compare the crystallization kinetics of a stoichiometric glass and that of a glass with a certain deficit of oxygen. The synthesis of glasses was performed in the following way. A powder from a very pure glass Corning 7940 was prepared and heated in vacuum at 1080 °C during 300 hours to remove hydroxyl groups. The dehydrated powder was divided into two parts and 70 ppm of metal silicon powder were added to one of them. Melting was performed in vacuum in a tube of quartz glass at 1850 °C. When heating a mixture of SiO_2 and Si-powders gaseous SiO evolved; to prevent bubbling in the glass an expo-

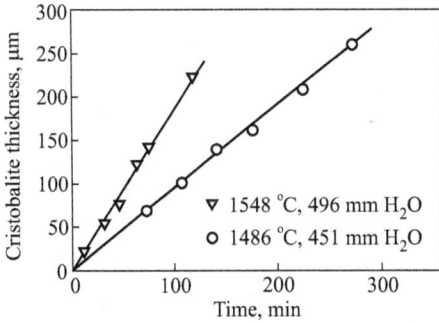

Fig. 7.5: Growth rates for de-watered fused SiO_2 showing a non-parabolic time-dependence in H_2O-atmosphere [9].

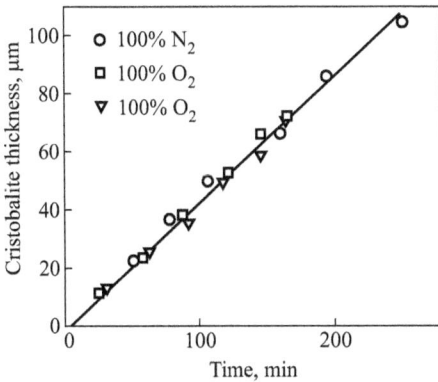

Fig. 7.6: Comparison of growth rates in N_2 and O_2 for de-watered fused SiO_2 at 1486 °C [9].

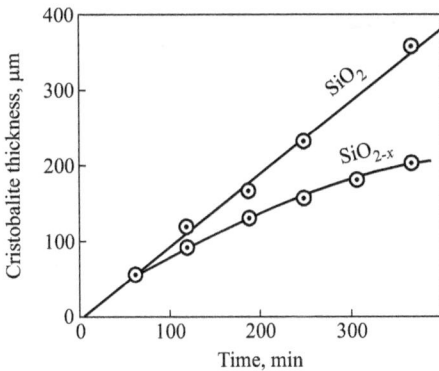

Fig. 7.7: Growth rates at 1460 °C (478 Torr H_2O-atmosphere) for stoichiometric and oxygen-deficient vitreous silica [10].

sure at 1300 °C was performed during 2 hours. The gaseous SiO condensed on the cold parts of the vacuum system in the form of a brown deposit. The authors believed that in this way they succeeded in obtaining two glasses, which differ from each other: one of them was stoichiometric and of formula SiO_2, while the other has a deficit of oxygen and was of formula SiO_{2-x}. No additional check of this assumption was given in the paper. The results of the measurement of the crystallization kinetics for both glasses are shown in Fig. 7.7. As it is seen from the figure, the stoichiometric glass is crystallized according to a linear in time law, the glass with the deficit of oxygen is crystallized via a parabolic law, and its crystallization proceeds more slowly than that of stoichiometric glasses. Based on these results, Wagstaff and Frank received a patent in 1968 for the method of production of crystallization-stable quartz glass [11].

In their next paper, Wagstaff and Richards [12] published the results of their investigation of stoichiometric glass crystallization in different gaseous media (water vapor, nitrogen, argon) and vacuum (see Figs. 7.8 and 7.9). The greatest influence on the crystallization rate is exerted by water vapor. The crystallization rate becomes higher with an increase of the partial pressure of water vapor. The influence of argon on the crystallization rate is practically indistinguishable from that of nitrogen. The lowest crystallization rate was observed in vacuum. Basing on the temperature dependencies of crystallization rates the authors calculated the activation energies for the process of

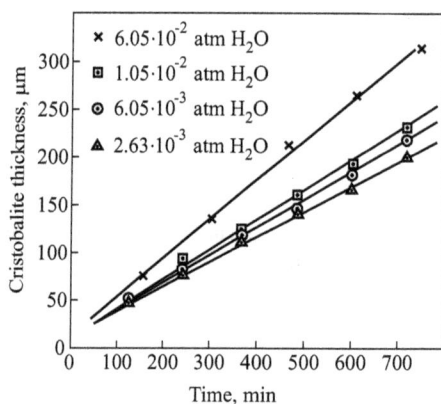

Fig. 7.8: Linear growth rates of stoichiometric SiO_2-glass at 1426 °C in H_2O-vapor atmosphere [12].

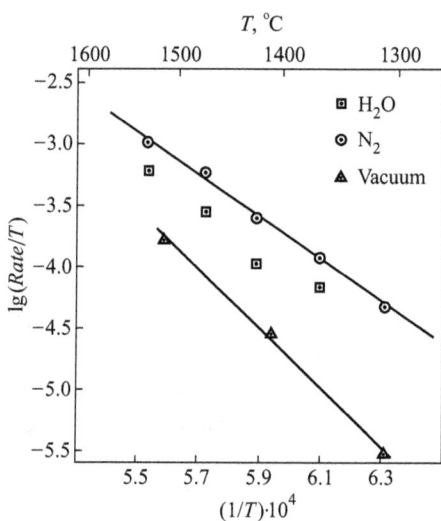

Fig. 7.9: Comparison of crystallization rates of stoichiometric SiO_2-glass in H_2O, N_2, and vacuum [12].

stoichiometric glass crystallization in the atmosphere of water vapor and in vacuum; these proved to be equal to 77 kcal/mol and 134 kcal/mol, respectively.

In [13], Wagstaff studied internal crystallization of type I quartz glass, while in [14] he analyzed the melting kinetics of cristobalite appearing via internal crystallization. Internal crystallization usually originates on solid microscopic inclusions of mineral origin that were inherited from natural quartz. These cristobalite crystals grow inside the silica melt not contacting the atmosphere surrounding the sample.

The specific feature of internal crystallization is the absence of $(\alpha \leftrightarrow \beta)$-transformations in cristobalite when cooling a sample to room temperature. This makes it possible to conduct repeated experiments using the same sample. The samples for the investigation were cut from a glass plate that was first heat-treated at a temperature of 1483 °C during 70 hours in order to form interior crystals of cristobalite. After that cubes of about one centimeter edge length were cut from this plate. The kinetics of crystallization and melting was studied under the microscope by measuring the changes in crystal sizes after a specified heat treatment. At the discussed here internal crystallization, a linear time-dependence of the growing crystal sizes is observed (Fig. 7.10). Fig. 7.11 shows the temperature dependence of the radial growth rate of

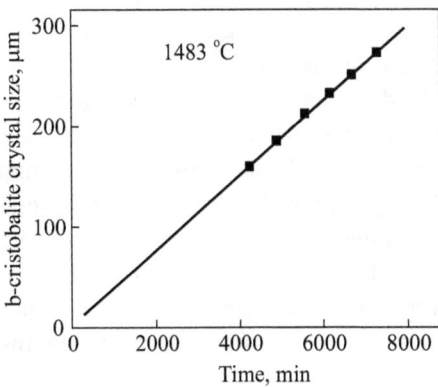

Fig. 7.10: Growth rates at 1483 °C for β-cristobalite in type I fused quartz [13].

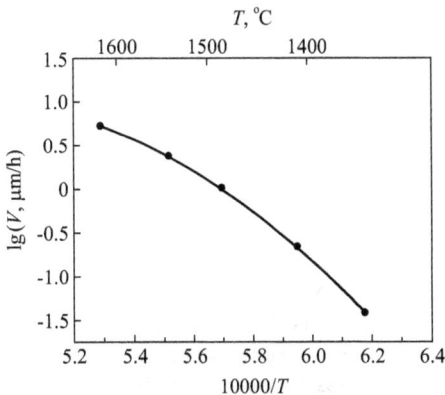

Fig. 7.11: The internal crystallization of type I glass.

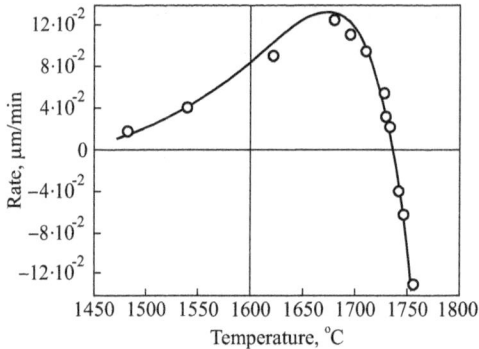

Fig. 7.12: Growth rates as a function of temperature [14].

cristobalite crystals. The dependence was plotted from the data given in the table [13]. In [14] Wagstaff investigated both the kinetics of internal crystallization and the kinetics of crystal melting in a silica melt. The results are demonstrated in Fig. 7.12. It is seen that the maximum rate of crystallization was observed at 1670 °C, while the melting temperature of cristobalite was 1734 °C.

In the papers by Hlavach and Vashkova [15, 16], surface crystallization of tube samples of type I glass produced in Czechoslovakia and France was investigated. The samples under study were treated with HF-solution to remove a surface layer 6–8 microns thick. After that procedure, they were soaked in distilled water during 48 hours. The authors also drew the conclusion that the dependence of crystal layer thickness on time is described by a curve resembling a parabola. On closer examination they found out that the obtained experimental curve can be divided into 3 parts. The first short part is linear and it starts from the origin of the coordinate system. The second part is parabolic and it is attended by a gradual decrease of crystallization rate with time smoothly transforming then to a linear part (Fig. 7.13). The activation energy values calculated from the temperature dependencies of the rates in the final (linear) part were found to be equal to 57 kcal/mol for the French glass and 53 kcal/mol for the Czechoslovakian glass.

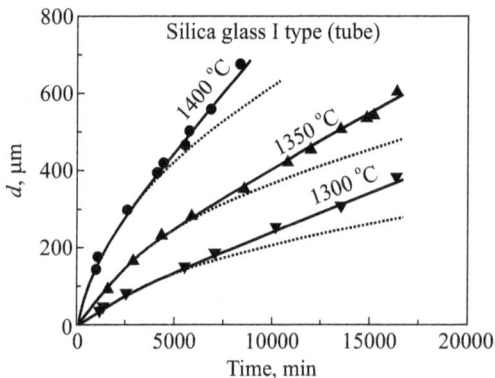

Fig. 7.13: Time-dependence of the thickness of the crystalline layer for a type I glass (French production) [15].

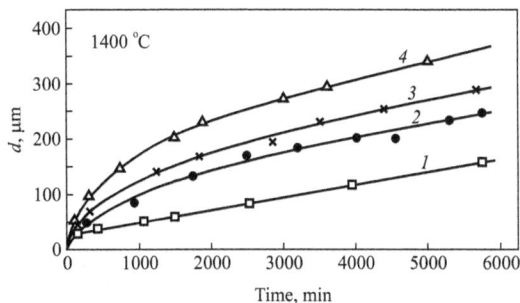

Fig. 7.14: Time-dependence of the thickness of the crystalline layer when measurements are per-formed under different conditions at 1400 °C [16]: (1) argon medium, (2) air medium (samples in a platinum crucible put into a closed tube of quartz glass and placed into a furnace), (3) air medium, crystallization in a corundum tube placed into a furnace, (4) air medium, samples placed immedi-ately into the core of the internal winding of a furnace.

The authors investigated crystallization of identical samples in air and in argon, as well as under different conditions: in a platinum crucible, in a crucible of quartz glass, in a tube of corundum and in an open furnace. They showed that the conditions of crystallization significantly affect the kinetics of crystallization (Fig. 7.14). It is seen from the figure that the dependence of crystalline layer thickness on time for sam-ples of the same glass is determined to a large extent by particular conditions of heat treatment of each sample. At crystallization in air these dependencies are curvilinear (resembling a parabola), they differ from one another by their initial crystallization rate that gradually decreases with time, reaching a constant value. For samples crys-tallizing under different conditions approximately the same rate of crystalline layer growth was finally established. From the temperature dependence of the growth rate constant the authors calculated the values of activation energy in air (as 55.5 kcal/mol) and in argon (as 51.2 kcal/mol).

Boganov et al. [17] described an investigation of crystallization of type I quartz glass in different gaseous media: in dry oxygen, in commercial nitrogen, in nitrogen freed from oxygen and water vapor, in dry hydrogen and in air. The investigation was conducted in a special apparatus where two furnaces were located one over another in vacuum. In the upper furnace glass melting was conducted in vacuum at 1780 °C during 2 hours. The lower furnace was heated to 1300 °C. Upon completion of melting, the upper furnace was cooled down to a temperature of 1300 °C, the sample was re-leased into the lower furnace, and a particular gas was supplied into the system. Heat treatment of samples was conducted at a temperature of 1300 °C during 72 hours. The influence of gaseous medium on crystallization was judged from the thickness of the resulting crystalline layer. It was found that heat treatment in the atmosphere of dry oxygen, commercial nitrogen or air resulted in the development of a crystalline layer 100–250 microns thick, while heat treatment in purified nitrogen or hydrogen did not

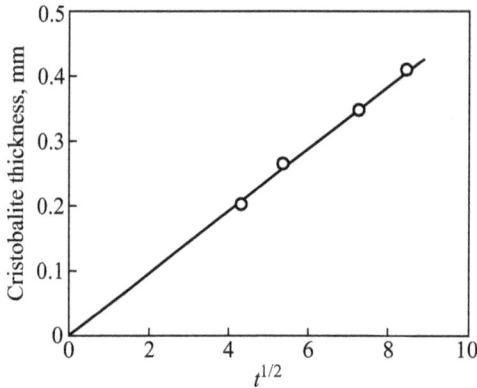

Fig. 7.15: Time-dependence (in hours) of crystalline layer thickness for a sample of dry commercial glass crystallized in the air at 1400 °C [17].

result in crystallization of the sample. No traces of crystallization were observed when the glass was heat-treated in vacuum at the same temperature during 50 hours. In the course of crystallization of a commercial glass of type I in air the authors obtained a parabolic dependence of crystalline layer thickness on time (Fig. 7.15), which in the authors' opinion resulted from a diffusion-controlled process. The information obtained gave the authors grounds to formulate their point of view on the nature of crystallization of quartz glass: according to their point of view the process of crystallization of quartz glass is of a chemical nature. The authors state that "crystallization occurs when glass, being a stoichiometric product, receives the necessary chemical agent, i.e. oxygen, so that a more sophisticated (with respect to its stoichiometry) cristobalite can be formed. Crystallization does not occur when the stoichiometric formula SiO_2 cannot be reached by one means or another ... Hence vitreous silica is not a metastable form of stoichiometric silicon dioxide, as is generally believed".

Above discussed well-known and frequently cited papers are practically the only references to the problems under consideration referred to in the English-language literature. The interest to the problem of quartz glass crystallization was satisfied to a considerable degree by these papers. After this analysis, till the present time further detailed investigations in this direction have not been carried out with one exception consisting in the comprehensive research conducted in the USSR in the 1970s at the Institute of Quartz Glass in St. Petersburg. However, these papers were published in Russian and, judging from the references in the literature, they till now remain widely unknown to English-language readers.

Mentioned research on quartz glass crystallization was initiated at the Institute of Quartz Glass in the late 1960s. The background for this decision was the following: Despite the existence of a number of published papers, an acute lack of information was observed on various aspects of practical importance concerning the process of quartz glass crystallization. In particular, almost all of the published papers were devoted to the investigation of glasses of type I. In addition, here only the second stage of the process – the growth rate of the already developed layer of cristobalite – was studied.

The nucleation stage was not analyzed at all. The investigators even tried to accelerate it by intentionally contaminating the glass surface, keeping glass in a weak solution of alkalis or in water over prolonged periods. In practice, in most cases specialists try to avoid such type of contaminations. It was also impossible to find out from the published information what was meant by the term crystallization stability and whether it could be described quantitatively. For example, in the above mentioned patent by Wagstaff and Frank [11], by crystallization stability the rate of crystalline layer growth was understood, i.e. a parameter was employed which is related to a state of the sample (or a product) when crystallization has already occurred. At the beginning of the present review, we mentioned cases when the appearance of even the first evidence of crystallization on the sample surface could not be tolerated. Therefore such use of the term "crystallization stability" is inapplicable for practical purposes. In addition, crystallization of glasses of types II and IV was not analyzed at all.

Having in mind such state of affairs, it is evident that in the published literature one could not find answers to many questions of practical importance: which of the four main types of glass is the most crystallization-stable, what factors affect crystallization stability, what should be done in order to increase the service life-time of quartz glass products at high temperatures.

7.3 Development of Experimental Techniques

Preliminary experiments on crystallization of vitreous silica samples with a thoroughly cleaned surface, performed at the Institute of Quartz Glass under various conditions (in crucibles of different materials, in an open furnace on supports of different materials, in different furnaces, etc.), showed that a particular type of dependence of crystalline layer thickness was determined not only (and not so much) by the type of quartz glass, but also by particular test conditions. The first experiments confirmed the observations of different authors, for example, those of [12, 16] that the furnace atmosphere may have a profound effect on crystallization of quartz glass. Such effect was never observed in crystallization of multi-component glasses. This feature is determined by the simultaneous action of two factors: (i) by the extreme sensitivity of quartz glass crystallization to infinitesimal contamination of the sample surface. Such contamination may occur (ii) mediated by the gas phase by extraneous agents (especially by alkali metals and platinum impurities) at the high temperatures of crystallization. At temperatures above ~1000 °C furnace materials (refractory materials, heaters, supports, screens) are no longer chemically inert: alkali impurities start to volatilize from ceramic materials, various transport reactions can proceed between materials of different parts of a furnace [18]. These reactions are accompanied by transport of matter through the gaseous phase from hotter parts of the system to colder ones.

After a series of preliminary experiments it became obvious that reproducible results on quartz glass crystallization in the absence of impurities penetrating through a gaseous phase can be obtained only when the samples under investigation are isolated from these material flows in the gaseous phase. In this case one can hope to observe the intrinsic crystallization process depending on the conditions of glass production and the composition of the gaseous medium. The simplest way to achieve this is heat treatment of samples that are inside a tube of quartz glass (as it is achieved upon conducting oxidation and diffusion reactions when producing semiconductor devices). However, high temperatures of crystallization (>1400 °C) and long term exposures (dozens of hours) cause fast deformation of such tubes. That is why other means had to be looked for.

To begin with, it was decided to investigate surface crystallization of a sample surrounded on all sides by the surface of the same glass. This was achieved by using samples with an artificial cavity: cubes with ~20 mm edge length were sawed out of a block of a glass under investigation. A plate ~2 mm thick was cut from this cube, and in the center of one of the cube faces a hole ~8 mm in diameter at the depth of ~ 12 mm was drilled. The flat surfaces of the cubes and the plates were polished, then the cubes and the plates were treated with HF-solution to remove the contaminated surface layer ~ 30 microns thick, whereupon they were triply thoroughly washed with distilled water. A flat plate closed the blind hole of the sample from above with resulting development of a cavity with a pure surface, the cavity being not tight, but sufficiently well isolated from contamination of volatile impurities from the furnace. Alternatively to this procedure another way of protection of the sample surface from impurities, penetrating from the furnace, was tested: a pure sample was placed into a small crucible of quartz glass, the small crucible being covered from above. Prior to testing, the crucible with the cover as well as the samples under investigation, were treated with HF-solution to remove a surface layer ~30 microns thick. The photograph in Fig. 7.16 demonstrates both ways of protection of sample surfaces from contamination during heat treatment.

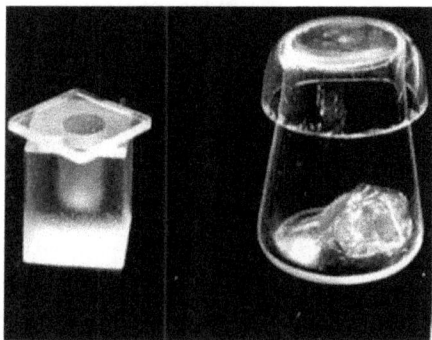

Fig. 7.16: Samples of protected surfaces of quartz glass to be crystallized [20].

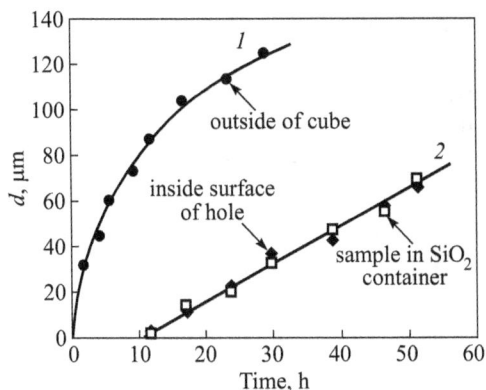

Fig. 7.17: Crystallization of samples of a type I glass with unprotected (outer surface of a tube) and protected (hole surface of a cube) surface of a sample in SiO_2-container surfaces at $T = 1430\,°C$ [20].

Fig. 7.17 shows the results of crystallization of samples of the same glass (type I) with surfaces unprotected and protected by two ways. It is seen that irrespective of the way of protection, both types of samples (surface of a hole in a tube and pieces of glass in a crucible) with a protected surface are equally crystallized. The attempts to improve protection by placing a cube with a hole into a crucible did not involve a change of crystallization results and henceforth all experiments were conducted with samples placed into uptight closed containers (crucibles) of quartz glass. This procedure was used in investigations of the intrinsic crystallization of various types of quartz glass, i.e. of crystallization not affected by random impurities penetrating the surface. By special experiments it was demonstrated that not all types of quartz glass are suitable for preparation of protective crucibles: type I glasses, obtained under reducing conditions, are not applicable for this purpose since crystallization of the interior walls of the crucible causes fast crystallization of a sample placed there [19]. In other words, crystallization of any samples started practically immediately after crystallization of the interior walls of a crucible of type I glass. In the subsequent production of protective containers a type II glass was used, for which a similar phenomenon was not observed [19].

Different alternative versions of sample manufacturing were tested: in the form of cubes or plates with polished or ground surfaces, as well as in the form of glass fragments of arbitrary shape. The experiments showed that the surface of samples after mechanical treatment were found to be severely contaminated and started to crystallize rapidly. No usually employed means of washing provided the removal of the impurities that were concentrated in surface micro-cracks developed during the process of mechanical treatment. Only a removal of the contaminated surface layer by treating it with HF-solution produces the required result. In this case a great variety of samples (samples with polished and ground surfaces or mere glass fragments) crystallized identically. For secure removal of the contaminated surface layer the samples were treated with 20 % HF during 2 hours and then washed in distilled and triply distilled water and dried. All operations with the samples were performed using a forceps with tips of quartz glass [20].

The data, shown in Fig. 7.17, indicate that intrinsic crystallization (curve 2) differs fundamentally from crystallization of samples in an open furnace (curve 1): crystallization of a sample in an open furnace starts so rapidly that one fails to observe any induction period and the time-dependence of crystalline layer thickness is described by a curve resembling a parabola. In contrast, the intrinsic crystallization shows up after a considerable period of time since the beginning of heat treatment, and the dependence of crystalline layer thickness on time is described by a linear law. It will be shown below that in many cases a crystallization rate acceleration is observed after the linear part, which was never described earlier in any scientific paper devoted to crystallization of vitreous silica. All this cast doubt on the hypothesis of the diffusion-controlled mechanism of crystallization and determined a considerable scope of studies on the regularities of quartz glass crystallization under conditions preventing random contaminations of sample surfaces.

7.4 Basic Phenomenological Features of the Crystallization Processes

We will first consider the simplest case of crystallization of a very-high-purity type III Corning 7940 glass [21]. The time-dependencies of crystalline layer thickness measured at different temperatures are demonstrated in Fig. 7.18. It is seen that for each temperature the beginning of the growth of the crystalline layer is preceded by an induction period. The value of the induction period was determined by an extrapolation of the linear part of the time-dependence of crystalline layer thickness, i.e. by its intersection with the abscissa axis. Prior to the completion of the induction period, it was not possible to reveal a crystalline layer on the sample surface under the microscope (the minimum thickness of a crystalline layer that could be determined by this

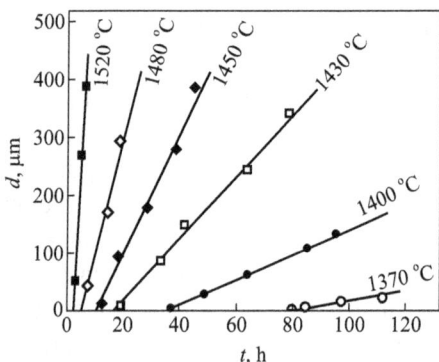

Fig. 7.18: Crystallization of a type III Corning 7940 glass [21].

method was ~3 microns). The length of the induction period decreased rapidly with an increase of temperature.

The value of the induction period is a natural and very convenient measure of crystallization stability of glasses; it is during an induction period that a glass surface remains free of crystallization traces. During the induction period processes of nucleation take place, the nuclei growth proceeds along the surface and into the depth of the sample, separate microscopic nuclei coalesce and a continuous crystalline layer is formed, its thickness growing to ~ 3 microns. The mathematical formulation of such process is given in [22, 23].

Upon termination of the induction period, the growth of the resultant crystalline layer starts. It is surprising that crystallization proceeds via a linear law until large enough cristobalite thickness is reached (up to ~400 microns). This result is surprising because the investigated very pure glass, carrying as few as possible metallic impurities, contains a large amount (up to 0.2 wt%) of structural water. In the crystallization process, structural water is not incorporated into the cristobalite and is accumulating in the vicinity of the cristobalite-glass interface [24]. Strangely enough, this fact does not practically affect the rate of crystallization, the rate of crystallization does not change with time. The crystallization rate grows with an increase of temperature. Figs. 7.19 and 7.20 show the temperature dependencies of the induction period and crystallization rates of Corning 7940 glasses [21]. It is seen that the temperature dependencies of induction period and crystallization rates can be well described by simple exponential equations.

Fig. 7.21 shows the dependencies of crystalline layer thickness on time when crystallizing type II glasses are hold at different temperatures [25]. It follows from the figure that a linear rate of crystallization is observed until the crystalline layer thickness reaches the value of ~200 microns, thereupon the crystallization rate starts increasing with time. Fig. 7.22 shows the same dependencies for a type I glass obtained under reducing conditions [26]. It is seen, firstly, that this glass is much less stable to crys-

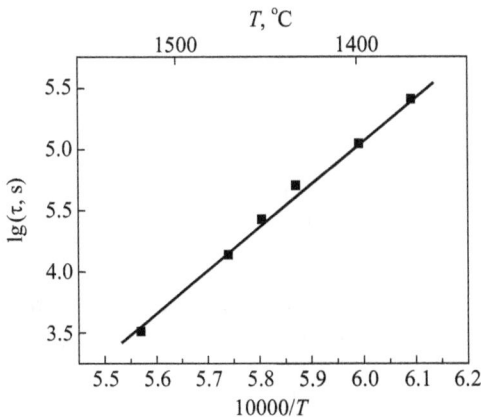

Fig. 7.19: Temperature dependence of the induction period of crystallization of Corning 7940 glass [21].

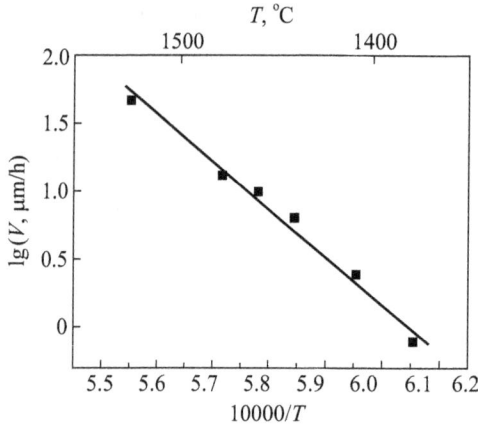

Fig. 7.20: Temperature dependence of the rate of crystalline layer growth of Corning 7940 glass [21].

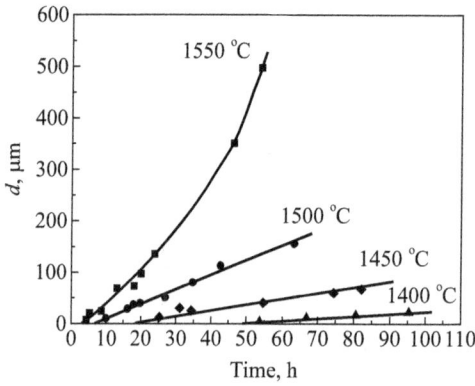

Fig. 7.21: Crystallization of type II glass (KV) [25].

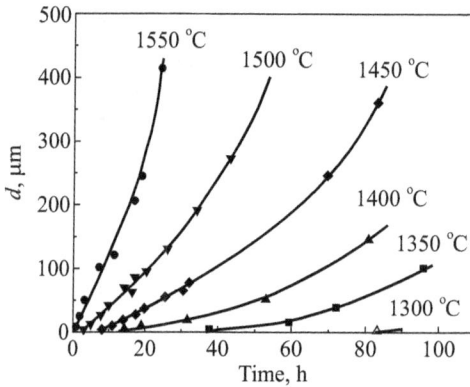

Fig. 7.22: Crystallization of type I glass (KI) [26].

tallization as compared to type II and III glasses and, secondly, its linear portion of crystallization is relatively short, after which the crystallization rate starts increasing rapidly.

The cited examples show that the intrinsic crystallization is characterized by two parameters, (i) the value of the induction period (τ) and (ii) the initial rate (V, the rate of growth in the linear in time range) of crystalline layer growth. As an example, Figs. 7.23 and 7.24 show the temperature dependencies of the induction period and rates of crystalline layer growth (over the range of temperatures lower than 1500 °C) for four types of quartz glasses of Soviet production [27]. Both dependencies are described by simple exponential equations

$$\tau = \tau_0 \exp\left(\frac{E_\tau}{RT}\right), \tag{7.1}$$

$$V = V_0 \exp\left(-\frac{E_V}{RT}\right). \tag{7.2}$$

Table 7.2 presents the values of the "activation energies" E_τ and E_V, as well as the values of τ and V for a temperature of 1400 °C computed employing the data shown in Figs. 7.19, 7.20, 7.23 and 7.24. From the data given in the table it follows that (i) the presented glasses differ from one another by their crystallization behavior; (ii) $E_\tau \approx E_V$ holds within the accuracy of the measurements; (iii) out of the four types of commer-

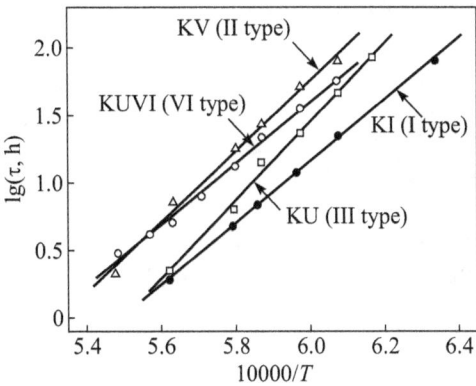

Fig. 7.23: Temperature dependence of the induction period of crystallization for commercial glasses of Soviet production [27].

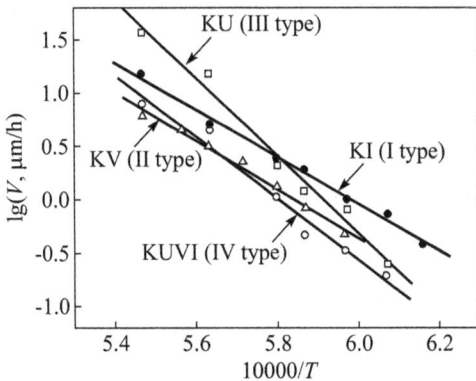

Fig. 7.24: Temperature dependence of the rate of cristobalite growth for commercial glasses of Soviet production [27].

Table 7.2: Quantitative characteristics of crystallization of commercial silica glasses

Type of silica glass	Trade mark	E_τ kcal/mol	E_V kcal/mol	τ_{1400} hours	V_{1400} μm/h
I	KI	103 ± 4	101 ± 7	11	1.3
II	KV	123 ± 7	120 ± 8	54	0.3
III	KU	150 ± 7	158 ± 10	20	0.8
III	Corning 7940	161 ± 9	154 ± 9	30	2.3
IV	KUVI	106 ± 5	109 ± 9	35	1.5

cial quartz glasses, the type II glass is the most stable one with respect to crystallization (at $T = 1400\,°C$).

Crystallization of vitreous silica can be influenced by different factors, such as reducing conditions of melting, fusion penetration degree of quartz grains, concentration of structural impurities, composition of ambient gaseous medium, process of tube drawing. Let us consider, now, the influence of each of the listed factors separately.

7.5 Influence of the Degree of Silica Reduction

At heating silica to high temperatures ($>1500\,°C$) a reaction of silica dissociation into SiO and O_2 starts to proceed at an appreciable rate [28]

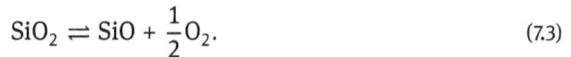

$$SiO_2 \rightleftharpoons SiO + \frac{1}{2}O_2. \tag{7.3}$$

Silicon monoxide (SiO) at high temperatures is found in a gaseous state. On the one hand, gaseous silicon (SiO) reacts with melting silica powder, dissolving in it to yield a partially reduced form of SiO_{2-x}, i.e.,

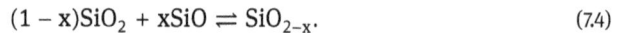

$$(1 - x)SiO_2 + xSiO \rightleftharpoons SiO_{2-x}. \tag{7.4}$$

On the other hand, while condensing on cold parts of the vacuum system it separates into silicon and silica yielding a characteristic yellow-brown condensate. This results from the fact that SiO in the form of an individual compound and being in a gaseous state is stable only over the range of high temperatures. When cooled over the temperature range $< 1500\,°C$ it becomes thermodynamically unstable and decomposes according to the reaction

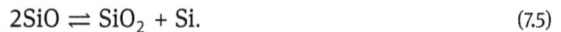

$$2SiO \rightleftharpoons SiO_2 + Si. \tag{7.5}$$

The resulting minute particles of silicon tint the condensate into yellow-brown colour.

A partially reduced form of silica SiO_{2-x} behaves in a similar way: it is stable over the temperature range above $1500\,°C$ and becomes unstable over the range of lower temperatures. Its transition to a stable state over the range of low temperatures can be performed by two ways: either by reduction of impurity ions according to the reaction of the type:

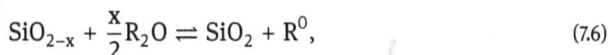

$$SiO_{2-x} + \frac{x}{2}R_2O \rightleftharpoons SiO_2 + R^0, \tag{7.6}$$

or by the reaction

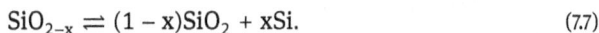

$$SiO_{2-x} \rightleftharpoons (1-x)SiO_2 + xSi. \tag{7.7}$$

If during the process of melting a reducing agent (C, Si, Mo, H_2, etc.) is present in the system, the resulting by reaction Eq. (7.3) oxygen binds with a reducing agent and reaction Eq. (7.3) shifts rapidly to the right towards intensive formation of SiO. As this takes place, the reaction of interaction of silica with a reducing agent (for example, graphite) proceeds through a gaseous phase without direct contact of reacting materials. The resulting by reaction Eq. (7.4) nonstoichiometric vitreous silica with oxygen deficit (SiO_{2-x}) contains oxygen vacancies $\equiv$Si-Si$\equiv$ that determine an absorption band at 248 nm and a luminescence band over the region of 280 nm [29].

Silica containing oxygen vacancies is stable over the region of high temperatures ($> 1500\,°C$) and becomes unstable with cooling. Its transition to a stable state can be affected at the expense of elimination of oxygen vacancies by reduction of impurities (reaction Eq. (7.6)) while with a high degree of silica reduction (with high values of x in SiO_{2-x}) it can be affected at the expense of silica release by the reaction Eq. (7.7). In the first case, a reduction of impurities to atoms takes place, which results in a change of properties of vitreous silica that are very sensitive to impurities, for example, it leads to a change of absorption spectra when there are traces of copper impurity or to an increase of viscosity when there are alkali impurities [30]. In the second case the process results in the so-called thermal darkening of glass: glass becomes yellow-brown because of light scattering by colloid particles of silicon [31].

The degree of silica reduction is characterized by the value x that is determined by melting conditions. Thus, on melting of a type II glass in a hydrogen-oxygen flame the value of x ranges between $3 \cdot 10^{-5}$ and $1 \cdot 10^{-4}$ (or $(1.5 - 5) \cdot 10^{-3}$ mol% O) [32], for glasses, melted in vacuum in graphite crucibles (type I glasses), oxygen deficit amounts to 0.01–0.02 mol% O [33]. Type II glasses contain 0.02–0.06 wt% (or 0.06–0.18 mol%) of OH-groups, that is approximately by a factor of 10^2 more than the content of oxygen vacancies or alkali impurities, therefore a reducing reaction (type Eq. (7.6)) of partially reduced silicon will proceed almost completely at the expense of structural water with the release of gaseous hydrogen whose rate of diffusion removal from a sample will determine the rate of the reducing reaction [34]. Due to the high rate of hydrogen diffusion, in vitreous silica the reaction of "self-oxidation" of a type II glass is completed fast enough even at the initial stage of heat treatment and the glass becomes stoichiometric.

A different situation arises with type I glasses obtained under reducing conditions. Such glasses practically do not contain OH-groups and the concentration of oxygen vacancies in them is commensurable or greater than the concentration of metal oxide impurities, therefore such glasses even after completion of type Eq. (7.6) reactions will be non-stoichiometric. That is the reason why the problem of the influence of non-stoichiometry degree on crystallization of quartz glasses refers almost exclusively to type I glasses.

Elucidation of the influence of reducing conditions of melting was carried out with glasses of a laboratory synthesis and a commercial glass, melted in vacuum in a graphite crucible [35, 39]. Laboratory melting of "anhydrous" glasses under neutral and reducing conditions was performed in the following way. A cylindrical ampoule of quartz glass 50–60 mm in diameter and 100–200 mm in height, its wall thickness being ≤ 1 mm, with a narrow hole for filling a batch was filled with quartz grit and placed into a molybdenum crucible. Melting was performed in vacuum electric furnaces with tungsten and molybdenum fixtures at a residual gas pressure in the furnace not less than 10^{-4} Torr. During the process of melting, "structural" gases H_2O, CO_2, CO, H_2, N_2 [36, 37] are released from quartz, the gases permeating the melting silica powder. The gases include both oxidizers (H_2O, CO_2) and reducers (CO, H_2); therefore such melting conditions can be considered as neutral. To create reducing conditions, 0.1 wt% of chemically pure silicon powder was added to quartz powder. During melting silicon reacts with silica yielding gaseous SiO which, solving in melting silica, determines its non-stoichiometry. Similar conditions exist when melting quartz glass in industrial crucible furnaces where silica powder is permeated by gaseous silicon monoxide that is formed on interaction of melting silica grit with graphite of the crucible. Absorption spectra of glasses obtained by the described methods are shown in Fig. 7.25. It is seen that commercial and laboratory glasses synthesized under reducing conditions have a well-defined absorption band at 248 nm determined by oxygen vacancies [38]. For a glass synthesized under neutral conditions such band was not observed.

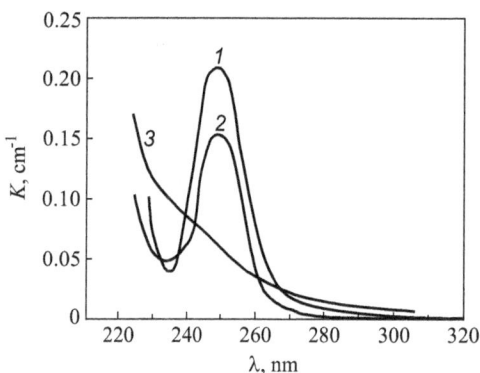

Fig. 7.25: Absorption spectra of type I glasses melted under different conditions: (1) melting synthetic quartz in a graphite crucible; (2) melting synthetic quartz + 0.1 % Si in a SiO_2 bulb; (3) melting synthetic quartz in a SiO_2 bulb [20].

The crystallization kinetics of these glasses is shown in Fig. 7.26. It is seen that (i) contrary to existing opinions [10, 11, 17], crystallization stability of glasses of a reducing synthesis is considerably lower than that of glasses of a neutral synthesis; (ii) in the case of glasses of a reducing synthesis, for the time-dependence of crystalline layer

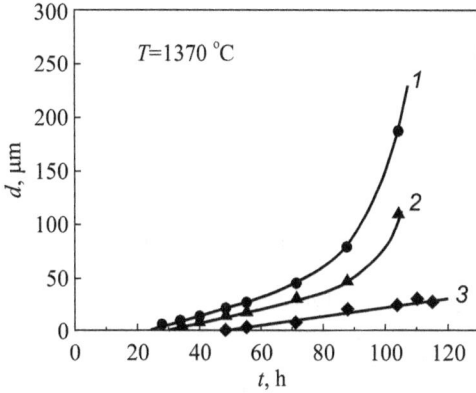

Fig. 7.26: Crystallization kinetics of glasses of neutral and reducing synthesis at a temperature of 1370 °C [35, 39]: (1) melting synthetic quartz in a graphite crucible; (2) melting synthetic quartz + 0.1 % Si in a SiO₂ bulb; (3) melting synthetic quartz in a SiO₂ bulb.

thickness a short linear part is detected, followed by acceleration of crystallization, while for glasses of a neutral synthesis a lengthy linear part is observed; (iii) it is seen by the slope of the linear parts that the crystallization rate of reduced glasses is higher than that of glasses of a neutral synthesis. Figs. 7.27 and 7.28 show the temperature dependencies of the induction period and the rate of crystalline layer growth for the same glasses. Consideration should be given to the fact that transition from neutral conditions of synthesis to reducing ones does not practically affect the slopes of the respective curves. In other words, the decrease of the induction period of crystallization and the increase of the rate of growth by Eqs. (7.1) and (7.2) occurs not because of the change of activation energies but due to the change of the pre-exponential factor.

Considerable changes in redox melting conditions, and hence in stoichiometry of melted glass, are possible only with synthesis of type I glasses. The redox conditions on melting type II glasses can be changed only to a small extent by changing the

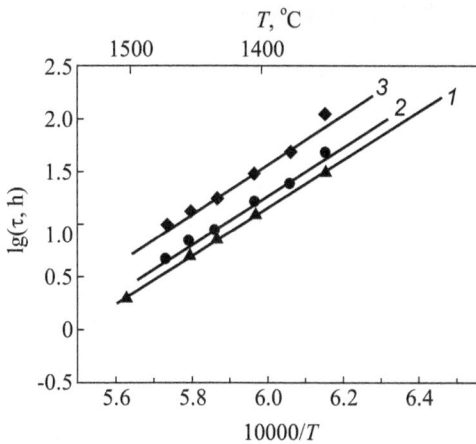

Fig. 7.27: Temperature dependence of induction periods for type I glasses of different syntheses [39]: (1) melting synthetic quartz in a graphite crucible; (2) melting synthetic quartz + 0.1 % Si in a SiO₂ bulb; (3) melting synthetic quartz in a SiO₂ bulb.

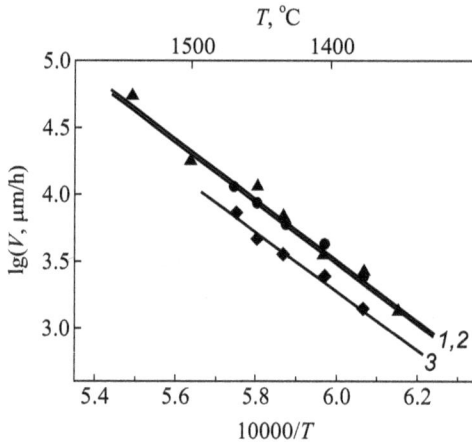

Fig. 7.28: Temperature dependence of the rate of crystalline layer growth for type I glasses of different syntheses [20]: (1) melting synthetic quartz in a graphite crucible; (2) melting synthetic quartz + 0.1 % Si in a SiO_2 bulb; (3) melting synthetic quartz in a SiO_2 bulb.

H_2/O_2-ratio in a hydrogen-oxygen flame. Oxygen vacancies developing by the reaction

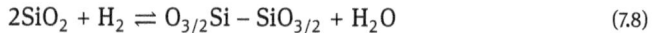

$$2SiO_2 + H_2 \rightleftharpoons O_{3/2}Si - SiO_{3/2} + H_2O \tag{7.8}$$

can exist only in a chilled glass, since sample exposure during the process of crystallization results in a shift of reaction Eq. (7.8) to the left increasing the degree of stoichiometry of the glass [34]. The attempts to correlate the value of hydrogen in relation to oxygen in the furnace flame and the kinetics of crystallization were not successful, although a pronounced correlation between H_2/O_2-ratio and the intensity of the band of 248 nm was observed. As the melting mode changed, the concentration of structural water changed as well. All variations in the crystallization kinetics at changes of the melting mode were correlated only with changes in OH-group concentration [20].

7.6 Influence of Concentration of "Structural Water"

Experimental data on the kinetics of crystallization of type I and II glasses obtained from various initial raw materials [20, 40] show that among glasses of both type I (obtained in graphite crucibles) and type II there is a certain (about ±20 %) scattering of data in the values of the induction period and the rate of crystalline layer growth. This result may indicate that impurities introduced into a glass from initial raw materials affect the kinetics of crystallization. However, this influence is considerably less than the influence of structural water. The content of structural water decides to which type of glasses, type I or type II glasses, the particular system has to be assigned to.

The study of the influence of the hydroxyl-group concentration on the kinetics of crystallization was conducted with two series of glasses: a series of laboratory glasses with different concentrations of structural water and a series of specially selected commercial glasses (melted in open gas furnaces and plasma furnaces from synthesized quartz crystals and type III glasses differing in concentration of OH-groups).

Laboratory melting was carried out in vacuum furnaces with metal accessory, in quartz glass ampoules placed in a molybdenum crucible. A powder of type III glass with a maximum content of structural water was used as an initial batch. During melting over the temperature range 1100–1200 °C different times of exposure were used for partial removal of structural water, after which the temperature was raised to 1800 °C and was held so during 30 minutes. Depending on temperature and duration of an intermediate exposure, it was possible to obtain a series of glasses practically free from metal impurities and with various contents of OH-groups. The concentration of OH-groups in glasses was determined by the value of absorption at 2.73 μm [41] as

$$C_{OH}[wt\%] = 0.1 \left(\frac{D_{2.73}}{d} \right), \qquad (7.9)$$

where $D_{2.73}$ is the optical density at 2.73 microns, expressed in natural logarithm, and d is the sample thickness in millimeters.

The dependence of the induction period and crystallization rate at a temperature of 1370 °C on the content of OH-groups is demonstrated in Figs. 7.29 and 7.30 [39]. It is seen that at a temperature of 1370 °C the maximum of crystallization stability and, respectively, the minimum of crystallization rates are observed in glasses containing 0.02–0.04 wt% of structural water; such amount of structural water is usually contained in type II glasses. The temperature dependencies of the induction period and the rate of crystalline layer growth of a given series of glasses are well described by Eqs. (7.1) and (7.2). It is interesting to note that the values of activation energies E_τ and E_V grow smoothly with an increase in concentration of hydroxyl groups (see Fig. 7.31).

Apart from the series of laboratory glasses containing a variable content of hydroxyl groups, the influence of structural water on crystallization was investigated on a series of commercial glasses of types I, II and III with various content of OH-groups [39]. Fig. 7.32 shows the dependence of the induction period on concentration of structural water for a series of such glasses at a temperature of 1430 °C. The concentration dependence of the induction period shown in this figure corresponds qualitatively to

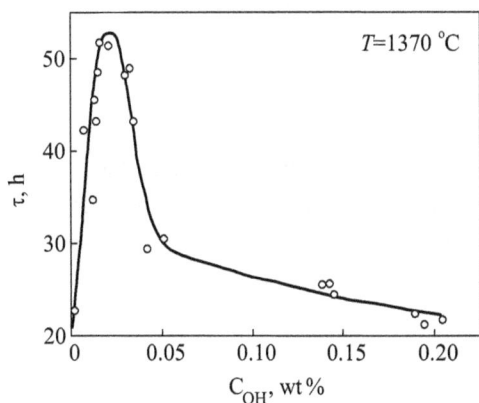

Fig. 7.29: Dependence of the induction period on concentration of OH-groups [39].

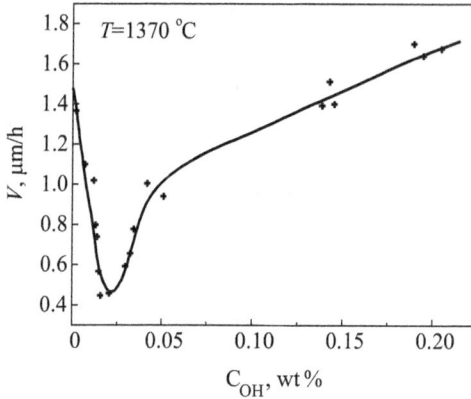

Fig. 7.30: Dependence of the rate of crystalline layer growth on concentration of OH-groups [39].

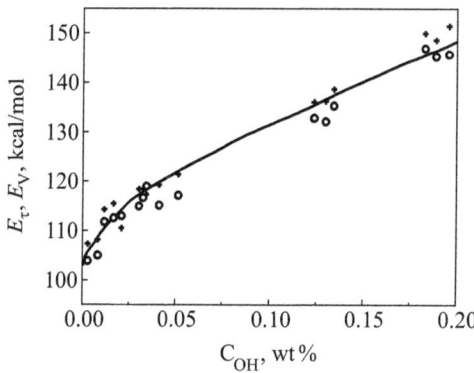

Fig. 7.31: Dependence of the activation energies E_τ and E_V on the concentration of hydroxyl groups in vitreous silica [39].

a similar dependence obtained for a series of laboratory glasses (Fig. 7.29), which indicates a profound effect of structural water on crystallization of vitreous silica. The large scattering in the data for glasses with a very small content of hydroxyl groups (Fig. 7.32) seems to be determined by the fact that another factor – the extent of silica reduction – affects crystallization in such glasses. Owing to the high stability of hydroxyl-containing glasses to crystallization it is possible to study the concentration dependencies of the induction period and crystallization rate experimentally only at comparatively high temperatures (~1400 °C). However, the respective such information for lower temperatures can be obtained through calculations by Eqs. (7.1) and (7.2) basing on experimental data obtained for the high-temperature range [40] (Figs. 7.33 and 7.34). It is seen that, at temperatures lower than 1300 °C, crystallization stability grows smoothly, while the rate of crystalline layer growth decreases with an increase in concentration of OH-groups. At temperatures higher than 1300 °C both dependencies pass through an extremum at a concentration of OH-groups of about ~0.03 wt%.

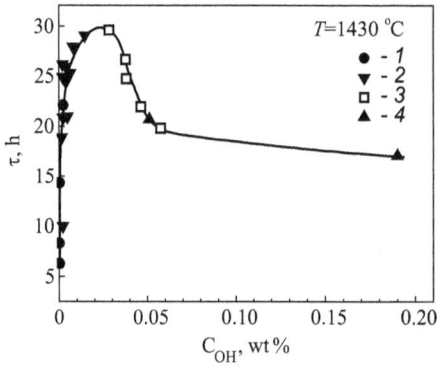

Fig. 7.32: Dependence of induction periods of crystallization at 1430 °C on the content of hydroxyl groups in commercial quartz glasses melted [39]: (1) in vacuum from quartz powder; (2) in high-frequency plasma from quartz powder; (3) in oxygen-hydrogen flame from quartz powder; (4) in oxygen-hydrogen flame from SiCl$_4$.

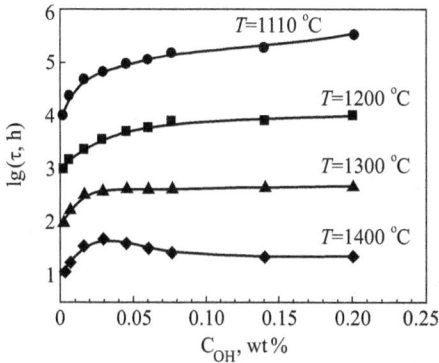

Fig. 7.33: Dependence of the induction period on concentration of OH-groups at different temperatures [40].

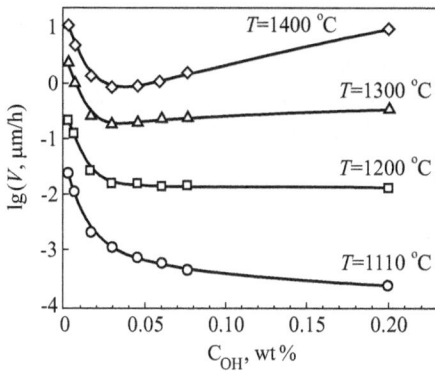

Fig. 7.34: Dependence of the rate of crystalline layer growth period on concentration of OH-groups at different temperatures [40].

7.7 Influence of the Degree of Fusion Penetration of Quartz or Cristobalite Particles on Crystallization of Quartz Glasses

Type I and II glasses are obtained by melting of particles of crystalline modifications of silica, quartz or cristobalite. A special feature of melting of these crystalline materials is the low rate of motion of the melt-crystal interface at temperatures several hundreds degrees (for quartz) and several dozen degrees (for cristobalite) higher than their respective thermodynamic melting temperatures [14, 42–44]. Slow melting of silica grit results in the fact that the melting time at a specified temperature may be not large enough for complete melting of the crystalline particles. In this case different techniques are used to reveal glass micro-inhomogeneities in the form of not completely melted micro-crystals of quartz and cristobalite [44, 45], the structure and some properties of such glasses depending on the initial raw material, quartz or cristobalite [44, 46, 47].

For example, on the curve describing the time-dependence of micro-hardness for a glass melted from quartz at a temperature of 1860 °C a maximum at 570 °C (the temperature of the $\alpha \leftrightarrow \beta$-transformation in quartz) was observed, and only on melting quartz grit at 1900 °C during 1 hour this maximum did not show up any more in the molten glass. Neither showed it up in glasses melted from cristobalite under any operating conditions. In addition, it was found that quartz does not immediately transform into the usual melt: at first, over the range of temperatures lower than 1670 °C a dense highly viscous phase is formed, its refractive index being ~1.476; the phase then gradually transforms into usual vitreous silica with the refractive index equal to 1.458 [44]. The presented data suggest that incomplete melting of the initial crystalline grit may also affect the kinetics of crystallization of molten glass.

Glasses employed for the investigation of the influence of the degree of melting on the crystallization kinetics were prepared in laboratory electric furnaces under neutral conditions from powders of different granulometric compositions at different temperatures and different periods of time [48]. Fig. 7.35 shows the results of investigation of the influence of grain fineness of synthesized quartz crystals on crystallization of glasses melted at 1750 °C during 40 minutes. It is seen from the plot that glass crystallization kinetics at the same temperature is strongly dependent on the powder fraction. The investigators succeeded in obtaining a glass which is stable enough to crystallization. Such glass was melted under the conditions mentioned above from a fine fraction of batch (0.09–0.14 mm). It crystallizes in the form of a homogeneous crystalline layer. Most likely such glass was sufficiently penetrated.

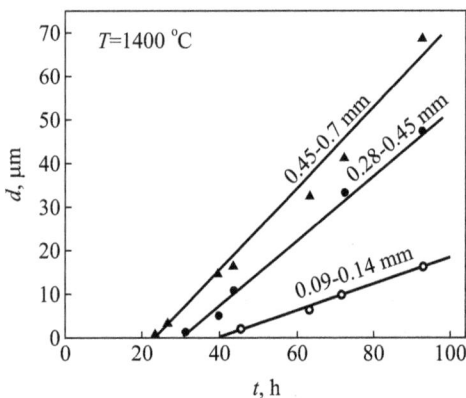

Fig. 7.35: Crystallization kinetics of glasses melted in a laboratory vacuum electric furnace from different fractions of synthesized quartz powder under exposure at 1750 °C during 40 minutes [48].

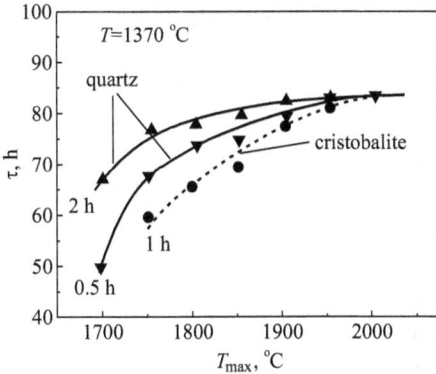

Fig. 7.36: Temperature dependence of induction periods on the maximum temperature and time of melting for glasses obtained from quartz or cristobalite (fraction 0.28–0.45) [48].

Completely different is the behavior of glasses obtained from coarse-grained fractions. Here crystallization starts not in the form of a uniform crystalline layer, but in the form of separate local zones of indefinite shapes, that are randomly distributed over the sample surface. With time these zones grow over the surface and penetrate deeply into the sample. Later a uniform crystalline layer is formed on the sample surface free from local crystalline zones; this layer starts growing penetrating deeply into the glass. Fig. 7.35 demonstrates the induction period and crystallization rate precisely for an uniform crystalline layer. The induction period for crystallization in the form of local zones is shorter and the rate of their penetration deep into the sample is higher. Fig. 7.36 shows the influence of the maximum temperature of melting on the value of the induction period of glass crystallization at 1370 °C. The experiments are performed with three different exposure times (2 hours, 1 hour and 30 minutes), the glasses being melted from quartz or cristobalite.

Fig. 7.37 shows the dependence of the rates of crystalline layer growth on the maximum temperature and time of melting of glasses obtained from synthesized quartz (fraction 0.28–0.45). It is seen that on melting at a temperature lower than 1900 °C glass crystallization depends both on exposure time at the maximum temperature and

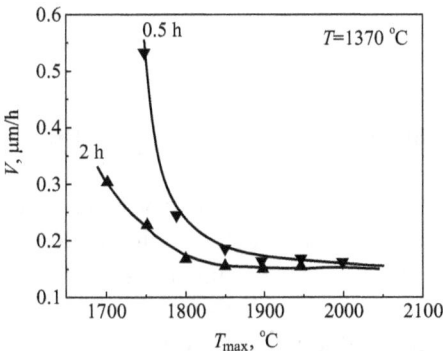

Fig. 7.37: Dependence of the rate of crystalline layer growth on the maximum temperature and time of melting for quartz glasses obtained from synthesized quartz (fraction 0.28–0.45) [48].

on batch material. At the same time, crystallization of glasses melted at 2000 °C with melting times of one hour and more is no longer dependent both on the type of the initial material of the batch (quartz or cristobalite) and the increase in the duration of melting. It is believed that such glasses are completely melted.

The problem of the influence of fusion penetration degree of quartz particles on crystallization of melted glass seems to be related mainly to type I glasses. Type II glasses are melted in the hot point of a hydrogen-oxygen flame at a temperature of 2100–2200 °C. In a hydrogen-oxygen flame, the melting rate of quartz is much higher than in air. Thus, for example, a quartz particle 0.4 mm in size is completely melted during the period of ~1 second at a flame temperature of 2100 °C, while in air at the same temperature such particle does not melt even in several minutes [49]. In the course of melting of type II glasses, quartz particles are hold out at melting conditions for a sufficiently large enough period of time, so that they have time to melt, merge and spread over the surface of a melted block. As a rule, type II glasses are crystallized in the form of a continuous crystalline layer.

Incomplete penetrating of quartz particles is a particular characteristic of type I glasses melted under reducing conditions, for example, in graphite crucibles in vacuum. At the temperature of melt formation from the crystalline grit an intensive interaction of silica with graphite takes place, yielding silicon monoxide accompanied by volatilization of silica and blockage of the vacuum system with condensed silicon monoxide. The process of interaction of silica with graphite accelerates drastically with increasing temperature. That is why, in order to avoid excess losses and premature blockage of a vacuum system, technologists try to conduct the process of melting at lower temperatures (~1800 °C) and during as short times as possible. Such conditions are not favorable for complete melting of crystalline silica powder.

Moreover, for melting of type I glass, a powder of a wide-range granulometric composition is employed where along with finer particles coarse grains are present, which is time consuming for their complete melting. Therefore, such glasses will be certainly heterogeneous with respect to their degree of melting. Based on the results, shown in Fig. 7.35, it should be expected that local regions of the sample surface, containing insufficiently melted particles, will start crystallizing earlier and at a higher rate, i.e. crystallization will be spatially inhomogeneous.

7.8 Influence of Surface Contamination on Crystallization Kinetics

The techniques for the investigation of quartz glass crystallization, developed at the Institute of Quartz Glass in St. Petersburg, made it possible to obtain well-reproducible results and to establish important regularities concerning the influence of different factors on crystallization kinetics. The main requirements of this technique are high purity of the surface of initial samples and prevention of any contamination of this surface during heat treatment. In the papers published earlier [7, 8–12, 15–17] either both these conditions or one of them were not met, that is why the data of these papers differed considerably from the data on the intrinsic crystallization of quartz glasses obtained under pure conditions.

During the process of development of this technique it was found that reproducible results could be obtained only after removing a thin surface layer of the sample by treating it with HF-solutions followed by thorough washing in triply distilled water [20]. Cleaning of polished or ground surfaces of samples by other methods (washing with alcohol, treating with chrome mixture, etc.) did not yield reproducible results. However, the treatment of samples with HF-solutions is not always desirable, as it entails deterioration of polishing and breaking of bubbles and capillaries located near a surface. To understand the mechanism of crystallization of samples with a contaminated surface and to establish the link with the results of the previous studies it is important to compare crystallization of samples of the same glass with pure and controlled contaminated surfaces.

With these considerations in mind two versions of experiments were carried out [50]: (i) Crystallization of samples with pre-contaminated surfaces under conditions preventing further contamination during the process of heat-treatment (samples were placed into small crucibles of type II glass). (ii) Crystallization of samples with pure surfaces under conditions of controlled contamination of sample surfaces during the process of heat treatment. The samples employed for the first version of experiments were soaked in 10 % solution of NaOH during 3 days. This procedure was followed by thorough washing in distilled and triply distilled water. After such washing, sodium ions, chemically bound with the SiO_2 surface, remained on the glass surface. The experiments of the second version were carried out using samples with pure surfaces, the samples being placed into platinum boxes to crystallize. To obtain reproducible results the boxes were fabricated out of the same platinum plate. Prior to the experiments, the boxes were first boiled in hydrochloric acid over a long period of time and then calcinated in vacuum during several hours. The investigation was carried out with type I, II and III glasses. Results of qualitatively the same kind were obtained for all glasses. As an example, Figs. 7.38 and 7.39 show the data on crystallization of a type II glass.

From the data of Fig. 7.38 it is seen that the induction period of crystallization for glasses with a contaminated surface is considerably lower than the one for samples with a pure surface, and the dependence of crystalline layer thickness on time has a form of a curve resembling a parabola. It is evident that, in the course of crystallization, the impurities from the surface are partially captured by cristobalite and partially displaced into the silica melt near the cristobalite-melt interface. As the crystalline layer thickness increases, the concentration of impurities decreases, gradually approaching zero. Correspondingly, the rate of crystalline layer growth has its maximum at the initial stage of crystallization and, gradually decreasing, it reaches a constant value that is equal to the crystallization rate of a sample with a pure surface.

Silica glass (II type)
Samples with dirty surface:

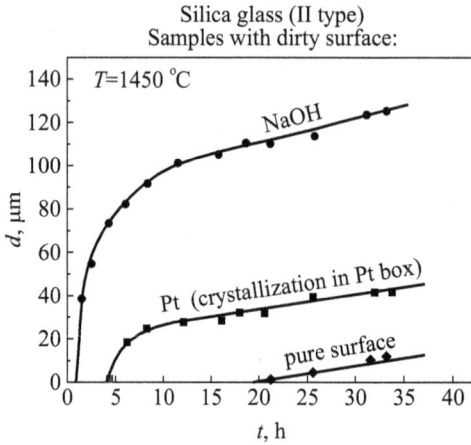

Fig. 7.38: Influence of sample surface contamination on the time-dependence of crystalline layer thickness for type II quartz glass [50]: (1) samples with a pure surface; (2) the surface is contaminated with traces of platinum during the process of heat treatment; (3) the surface is contaminated with traces of alkali.

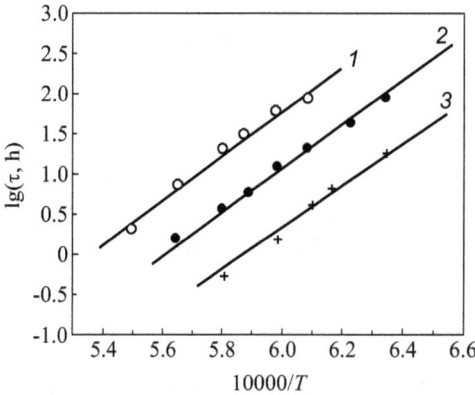

Fig. 7.39: Influence of sample surface contamination on the value of the induction period of type II quartz glass [50]: (1) samples with a pure surface; (2) the surface is contaminated with traces of platinum during the process of heat treatment; (3) the surface is contaminated with traces of alkali.

Fig. 7.39 shows the temperature dependencies of induction periods of crystallization of samples with different degrees of their surface contamination. It is seen that the presence of contaminations on sample surfaces does not practically change the slope of the temperature dependencies, the value of the activation energy E_τ in Eq. (7.1), but affects only the value of the pre-exponential factor. Similar results were also obtained for other types of glasses, and it was found that with a similar degree of sample surface contamination of different glasses their induction period decreases approximately by the same factors, not changing the relative stability of the different glasses [50].

7.9 Influence of the Composition of the Gas Medium on Crystallization of Quartz Glass

7.9.1 Introductory Comments

Quartz glass is almost the only glass whose crystallization is sensitive to the composition of the ambient medium. This feature led researchers to develop the hypothesis of the diffusion-controlled mechanism of crystallization of silica glass. It was put forward in the 1960s and was adopted by many investigators.

As it was already noted earlier in the present brief review, the influence of the composition of the gas medium on crystallization of quartz glasses was investigated in a number of studies [8, 12, 16, 17]. However crystallization of samples in these studies was carried out under such complicated conditions (inadequate surface purity of initial samples, surface contamination during heat treatment) that it is very difficult to draw correct conclusions basing on these studies. Hence, experiments were carried out aimed on the elucidation of the mechanism of influence of different gases on crystallization of quartz glass samples with a pure surface in the absence of uncontrollable impurities on their surfaces during the process of heat treatment [51, 52].

Two series of experiments were conducted: (i) Crystallization of sample surfaces in contact with the ambient gas and (ii) crystallization of sample surfaces in contact with gases which were evaporated from a melt into bubbles. For the experiments of the first series the following gases were chosen: dry argon, dry oxygen, dry air, vacuum ($\sim 10^{-4}$ Torr), and air with different partial pressures of water vapor. The experimental procedure is described in [20]. Type I, II and III glasses were investigated. These experiments were characterized by the fact that the sample surface was in contact with an "external" gas whose composition was specified by the investigators. In the experiments of the second series, the gas contained in closed glass bubbles was in equilibrium with the melt. In the experiments of this series, crystallization of bubble surfaces containing a specified gas mixture was studied. The results of the experiments on crystallization in an "external" gas showed that crystallization in dry gases and vacuum differs significantly from this process as proceeding in moist media.

7.9.2 On Crystallization in Dry Gas Media

In dry gas media, glasses containing structural water (type II and III glasses) crystallize differently as compared with hydroxyl-free glasses of type I; crystallization of type II and III glasses does not depend on the composition of the gaseous medium and proceeds in the form of a continuous layer (Fig. 7.40). Crystallization of type I glasses starts in the form of separated local zones of an indefinite shape. They are randomly distributed over the glass surface and are characteristic for insufficiently melted glasses (see Fig. 7.35). However, at crystallization in dry argon or vacuum no

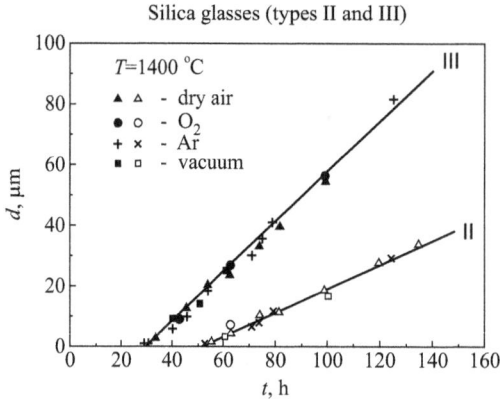

Fig. 7.40: Crystallization kinetics of type II and III glasses in dry gaseous media at 1400 °C [51].

continuous uniform layer between separate crystallization zones is formed. It is possible to observe inter-growing crystallization zones and patches of pure glass surfaces between them over a long period of time. Due to the indefinite shapes of the crystallization zones and the difference in their sizes it is not only difficult to measure their growth rates, but this procedure also involves averaging over the sizes of numerous separate zones. In this case the maximum time of heat treatment during which local crystallization zones are not formed on the sample surface is taken as the duration of the induction period. Despite the mentioned difficulties of measurement it can still be stated that on heat-treatment of samples in the atmosphere of dry oxygen or in dry air the induction period is lower and the crystallization rate is higher than at heat-treatment in the atmosphere of argon or in vacuum (Fig. 7.41). However, no significant differences of crystallization in vacuum or argon are observed. It is also impossible to detect differences in crystallization kinetics in dry oxygen or in dry air, though the partial pressure of oxygen in dry air is five times lower than in the atmosphere of dry oxygen.

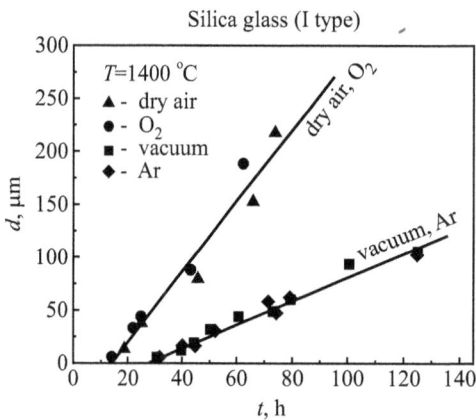

Fig. 7.41: Crystallization kinetics of type I glasses in dry gaseous media at 1400 °C [51].

7.9.3 Experiments on Crystallization in an Atmosphere Containing Water Vapor

The respective experiments showed that in all cases and for all the investigated glasses the formation of a continuous crystalline layer is observed. For type I glasses melted under reducing conditions on crystallization in the air with a low content of water vapor (~8 Torr) both types of crystallization (uniform layer and local zones) can be detected, whereas with increasing partial pressure of water vapor (for example, at pressures ≥190 Torr) the crystallization only in the form of a uniform layer can be observed. Fig. 7.42 shows the time-dependencies of crystalline layer thickness when type I glasses crystallize in air at different partial pressures of water vapor. It is seen that with increased partial pressure of water vapor the induction period decreases and the crystallization rate grows. Consideration should be given to the fact that at a low partial pressure of water vapor the linear part of crystallization of type I glass is comparatively short, thereafter the crystallization rate increases with time. When crystallizing samples of the same glass are kept in an atmosphere with a high content of water vapor only linear crystallization takes place.

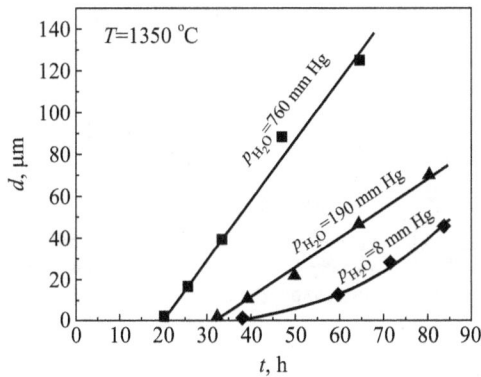

Fig. 7.42: Crystallization kinetics of type I glasses at 1350 °C in the atmosphere with different partial pressures of water vapor [52].

Figs. 7.43 and 7.44 demonstrate the temperature dependencies of the induction period and crystallization rate of type II glass at three different partial pressures of water vapor. It is seen from the plots that the slope of the lines describing the temperature dependencies does not practically change when the concentration of water vapor in the ambient gaseous medium changes. The changes of both the induction period and the growth rate occur due to the change of the pre-exponential factor. Similar dependencies are also observed for type III glasses.

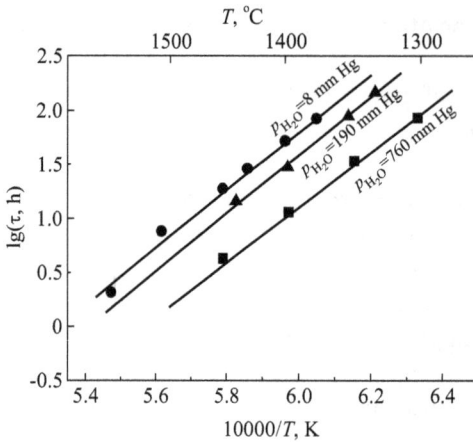

Fig. 7.43: Temperature dependencies of the induction period of crystallization of type II glass at different partial pressures of water vapor [52].

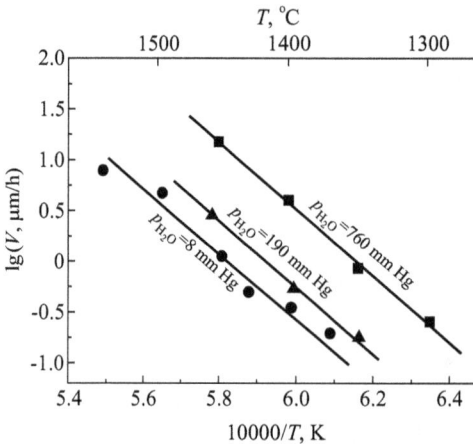

Fig. 7.44: Temperature dependencies of crystalline layer growth rates for type II glasses at different partial pressures of water vapor [52].

7.9.4 Crystallization of Quartz Glass in the Atmosphere of Gases in Equilibrium with the Melt

Experiments in this direction were conducted with glass samples containing closed bubbles. Glasses obtained under various conditions and thus containing bubbles with various gas compositions were selected for the experiments. The compositions of the gases in the bubbles, determined by mass-spectrometric methods, are presented in Table 7.3 [51]. Glasses have been selected whose bubbles contain various gases, including H_2O and O_2, actively affecting crystallization of silica glass. Samples were cut from these glasses taking care that each of the samples contains at least one closed bubble. When conducting all the experiments described above the heat-treatment was performed for extended periods of time (several hundred hours) at temperatures 1500–1600 °C (some samples remained in the furnace at all experiments). In none of the cases under investigation, any evidence of crystallization of the surface of the

Table 7.3: Results of the mass-spectrometric analysis of gas composition in bubbles [51]

Glass type	Initial material	Method of synthesis	Composition of gas bubbles, wt%
II	Synthetic quartz crystals	Gas-flame	$H_2 - 95; N_2 - 4; H_2O - 1$
IV	Synthetic quartz crystals	Oxygen-nitrogen plasma	$O_2 - 39; N_2 - 60; NO - 1$
I	Synthetic quartz crystals	Electrothermal crucible	$CO - 70; N_2 - 30$
I	Kyshtym quartz	Electrothermal (rod furnace)	$CO - 88; CO_2 - 6; H_2O - 6$
IV	$SiCl_4$	Oxygen-nitrogen plasma	$O_2 - 61; N_2 - 26; CO_2 - 7; H_2O - 7$
I	Synthetic SiO_2 (cristobalite)	Nitrogen plasma	$N_2 - 99; CO_2 - 1$
I	Synthetic SiO_2 (cristobalite)	Oxygen-nitrogen plasma	$NO - 68; O_2 - 31; N_2 - 1$
I	Synthetic SiO_2 (cristobalite)	Ceramic method	$N_2 - 97; CO - 2.5; H_2O > H_2 \sim 0.5$

closed bubble was revealed. At the same time, a thick layer of cristobalite grew on the outer surface of the samples. These experiments indicate that on the surface of closed bubbles spontaneous crystallization does not occur (within reasonable observation times).

An interesting regularity is observed when crystallization of bubbles (8–10 mm in diameter), opened immediately prior to an experiment, is carried out in air: for all glasses, except the glass melted under severe reducing conditions (the third from the top in Table 7.3), surface crystallization of an opened bubble does not practically differ from crystallization of an outer surface of the sample. Whereas the opened bubbles of glasses melted under severe reducing conditions were crystallized considerably earlier, the rate of crystalline layer growth on bubble surfaces being considerably higher (Figs. 7.45 and 7.46). The fact that the surface of closed bubbles does not crystallize and starts crystallizing after their opening can indicate that the process of crystallization is initiated by surface chemical reactions with the components of the gaseous medium with respect to which the surface of an opened bubble is in non-equilibrium. Out of the possible reactions, the reactions of hydration, dehydration and oxidation seem to be the most probable of them. Very fast crystallization of an opened bubble in a glass melted under reducing conditions is probably determined by the reaction of oxidation of a heavily reduced surface of an opened bubble.

As it is shown by the data of the mass-spectrometric analysis (see Table 7.3), the gas phase in the bubbles in this glass consists of CO (70 %) and N_2 (30 %). During the process of melting, CO is formed by the reaction

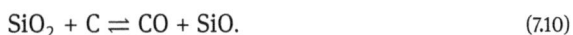

$$SiO_2 + C \rightleftharpoons CO + SiO. \tag{7.10}$$

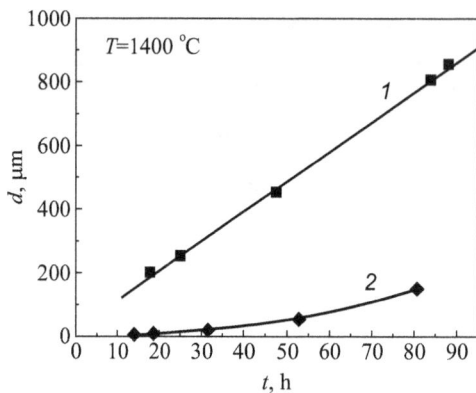

Fig. 7.45: Dependence of crystalline layer thickness on time for type I glass melted under reducing conditions: (1) crystallization of an opened bubble surface; (2) crystallization of the sample surface.

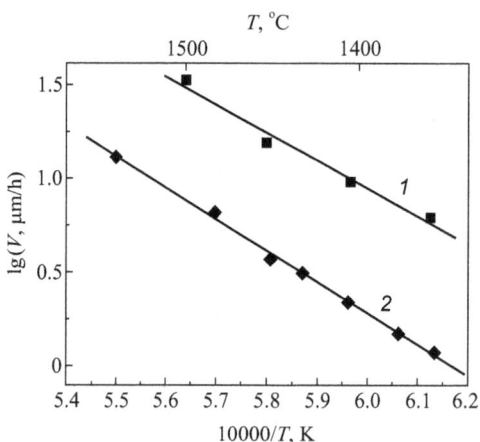

Fig. 7.46: Temperature dependence of the growth rate of the crystalline layer for a type I glass melted under reducing conditions: (1) crystallization of an opened bubble surface; (2) crystallization of the sample surface.

SiO is also formed in the same quantity. SiO is in a gaseous state at the temperatures of glass melting. When going over into the closed cavity, SiO starts reacting with the cavity surface layer yielding non-stoichiometric silica SiO_{2-x} by the reaction Eq. (7.4). On cooling of the melt, a thin layer of reduced (non-stoichiometric) silica containing (Si–Si)-bonds is formed on the surface of the bubble. In the presence of oxygen, non-stoichiometric silica oxidizes. The change of (Si–Si)-bonds for Si–O–Si stimulates the change of the surface layer structure. Structural changes of the surface layer initiate the processes of nucleation and crystal growth. The example of crystallization of opened bubbles in the glass obtained under reducing conditions can be a further indication of the fact that the reduced (non-stoichiometric) form of silica is considerably less stable to crystallization in air than a more stoichiometric form with a smaller value of x in the formula SiO_{2-x}.

7.10 Influence of the Drawing Process on the Crystallization Kinetics of Tubes of Quartz Glasses

Up to now, we were dealing with crystallization of quartz glass samples cut from a block. Yet, in actual practice quartz glass, as a high-temperature material, is used in the majority of cases in the form of tubes. During the process of drawing two additional factors affect the properties of quartz glass: (i) The drawing process is carried out at high temperatures when transport reactions accompanied by mass transfer through a gas phase proceed quite intensively, so that it is very difficult to avoid contamination during the process of tube surface molding. (ii) In the process of drawing a glass, beginning with the melt and ending with the final product (tube, rod), the system is affected by uniaxial tensile stress which may have some effect on glass structure as well.

Tubes are often used as samples for investigation of physico-chemical properties of quartz glasses. For example, a large amount of published data on crystallization of quartz glasses was obtained on samples having the shape of tubes [8, 9, 15, 16]. Therefore, a comparison of the crystallization features of glass blocks and tubes, respectively, is important both from practical and scientific points of view. There are three main techniques of drawing tubes of quartz glass: (i) Drawing a melt through an annular hole in the bottom of a graphite crucible. (ii) Drawing a melt through an annular hole in the bottom of a molybdenum crucible (in the atmosphere of a shielding gas containing hydrogen). (iii) Drawing from a face of a tubular bar (usually denoted as similitude method). The most general of the described methods is the first one of the listed above. Using this method it is possible to draw tubes of any profile (oval, triangular, rectangular, etc.) of vitreous silica of any type. Hence crystallization of tubes drawn from a graphite crucible was investigated in most detail.

In the course of the process of drawing, a quartz glass is subjected to a number of factors that considerably change its crystallization behavior. Among these the following factors have to be mentioned, in particular: (i) Contamination of tube surface layers by impurities from molding elements; (ii) partial reduction of surface layers resulting from the interaction with reducers (graphite, molybdenum); (iii) possible orientation of structural elements of the glassy network. These factors can result in considerable differences of tube crystallization from glass block crystallization.

The experiments on comparative studies of glass block samples and samples in the form of tubes were conducted in the following way. A plate (12–15) mm thick, out of which the samples were prepared, was cut from a block of quartz glass. The remaining part of the block was drawn into a tube. Glass block samples and tube samples were placed each in its container of type II silica glass and were crystallized simultaneously. The surface of glass block samples was cleaned by the standard technique: samples were treated with 20 % HF during 2 hours and then washed in distilled and triply distilled water. Surface treatment of tubes was performed by two ways: (i) The surface was cleaned by carbon tetrachloride and then thoroughly washed by alcohol

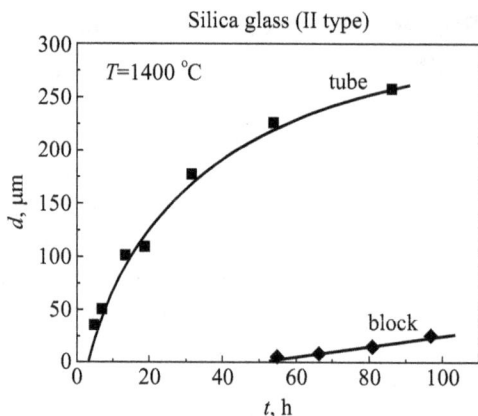

Fig. 7.47: Time-dependence of crystalline layer thickness for samples of type II glass for a tube and a block, respectively, at 1400 °C [26].

and triply distilled water; (ii) samples were treated with HF-solutions to remove a surface layer of a specified thickness.

Fig. 7.47 shows the time-dependencies of crystalline layer thickness for samples of a tube and type II glass block when crystallized in the air at 1400 °C [26]. It is seen that the initial surface of the tube is crystallized much faster than the surface of the block. The small value of the induction period and curvilinear dependence (resembling a parabola) points to the fact that during the molding process of the tube its surface was contaminated. In order to check this assumption, an analysis of the number of impurities in the layers of one micron thickness was carried out in dependence on the distance of these layers from the surface of the tube [53]. The results obtained are demonstrated in Fig. 7.48. It is seen from the figure that the content of impurities near to the surface by far exceeds the one in the interior layers.

A removal of the contaminated and partially reduced surface layer by treatment with HF-solution results in a drastic change of crystallization kinetics of the tube [20, 54] (Fig. 7.49): the induction period increases and the dependence of crystalline

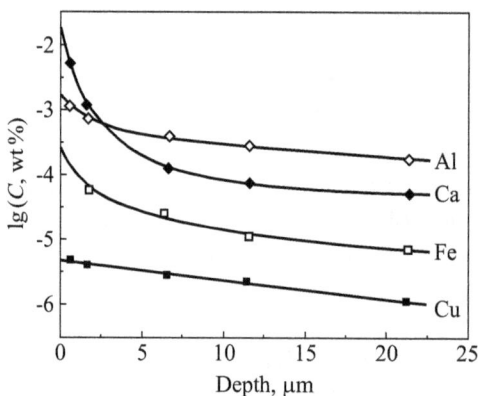

Fig. 7.48: Distribution of impurities in dependence on the distance to the surface of a tube of type II glass [53].

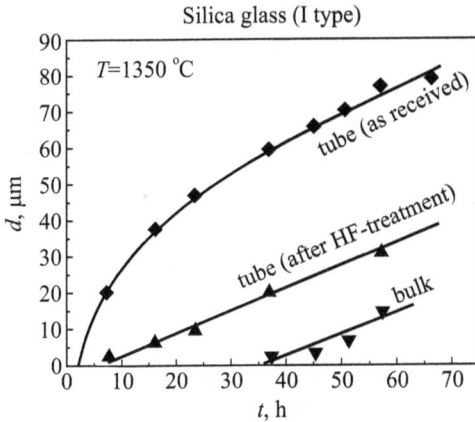

Fig. 7.49: Effect of treatment with HF-solution (to remove the contaminated surface layer ~ 30 μm thick) on crystallization of a tube of type I glass. Comparison with crystallization of block-shaped glass [20].

layer thickness on time becomes linear. It is seen that the crystallization rate of the initial (not treated with HF-solution) tube after formation of a crystalline layer is very high; it decreases with time and eventually becomes linear and equal to the rate of tube crystallization with a removed surface layer. A removal of the contaminated and partially reduced layer of the tube ~30 microns thick results in a drastic increase in the crystallization stability of the tubes. However, tubes treated with HF-solutions are considerably less stable with respect to crystallization as compared to the corresponding block-shaped glass.

The thickness of the contaminated layer of a tube depends both on the way of drawing the tube and the quality of graphite used as the molding element and the quality of the heater. Most contaminated are the internal and external surfaces of tubes drawn through an annular hole of a graphite crucible. The degree of surface contamination in this case depends both on the ash content of graphite and the time of its contact with the surface of the drawn tube, i.e. on the rate of drawing: the lower is the ash content of graphite and the higher is the rate of drawing, the more stable to crystallization is the initial tube [20, 55]. Due to these reasons, the crystallization behavior of initial tubes is poorly reproducible. Complete removal of the contaminated surface layer in HF-solution makes it possible to obtain reproducible results that are practically independent of the listed above factors. Generally, contaminations in tubes, obtained by the use of graphite accessories, are localized in a layer of finite thickness. Therefore, a further increase in the thickness of the removed layer beyond mentioned amount of 20–30 micron does not affect the kinetics of quartz tube crystallization. To attain the same effect on tubes fabricated by molybdenum molding path or on tubes drawn by the similitude method it is sufficient to remove a surface layer of about 5–7 microns thickness [54].

Fig. 7.50 demonstrates the temperature dependencies of induction periods of crystallization of glass tubes (with initial and clean surfaces) and type II glass block samples. It is seen that over the whole range of temperatures the crystallization stability

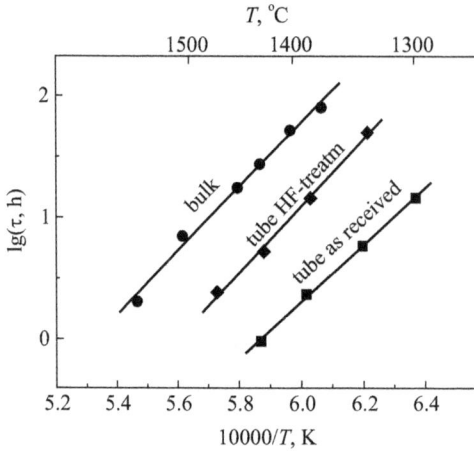

Fig. 7.50: Temperature dependence of the induction period of crystallization of block-shaped glass and tubes drawn from it [20].

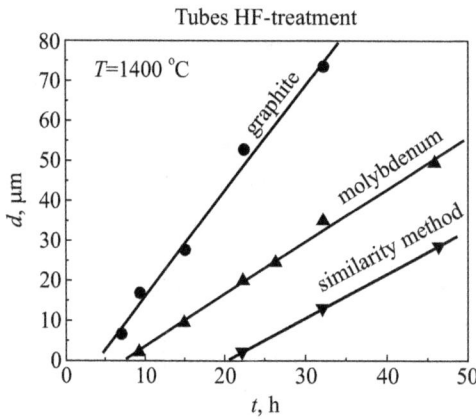

Fig. 7.51: Time-dependence of crystalline layer thickness for tubes drawn by different methods and treated with HF-solutions to remove a surface layer 30 microns thick [20].

of tubes with removed contaminated surface layer is significantly lower than the crystallization ability of block-shaped glass samples, out of which the tube under review was drawn. Fig. 7.51 illustrates the dependencies of crystalline layer thickness on time of heat treatment for tubes of type I glass drawn by different techniques and treated with HF-solution [20]. It is seen that the tubes drawn by the similitude method were the most stable ones with respect to crystallization and the tubes drawn through a graphite nozzle were the least stable ones. Figs. 7.52 and 7.53 demonstrate the temperature dependencies of induction periods and crystallization rates for the same tubes [54].

Crystallization of tubes has one significant distinction from crystallization of glass blocks: the activation energy of the induction period, E_τ, for glass block samples is practically equal to the activation energy of the rate of growth E_V (see Table 7.3), whereas for tubes the value E_τ is considerably higher than the value of E_V. Owing to these circumstances the activation energy for crystallization rates of block samples of

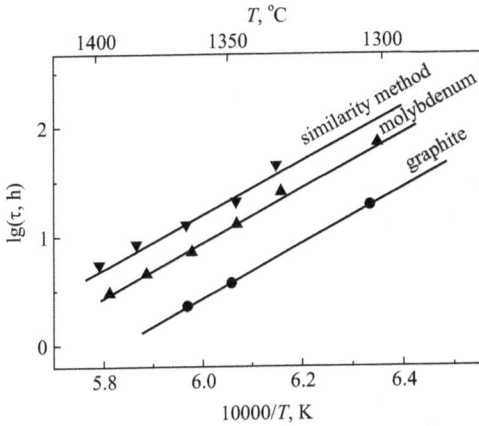

Fig. 7.52: Temperature dependence of induction periods for tubes drawn by different methods from glasses of type I and treated with HF-solutions to remove a surface layer 30 microns thick [54].

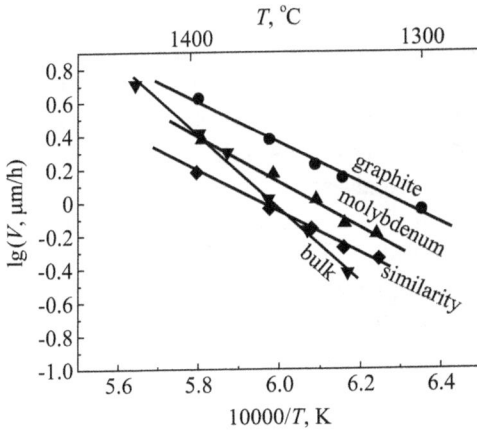

Fig. 7.53: Temperature dependence of the crystalline layer growth rate for tubes drawn by different methods from glasses of type I and treated with HF-solutions to remove a surface layer 30 microns thick [54].

quartz glass is almost twice as high as the activation energy for crystallization rates of tubes drawn from the same glass. The quantitative characteristics of the process of crystallization of tubes and block glass are demonstrated in Table 7.4. So significant

Table 7.4: Values of the activation energy of crystallization of glass blocks and tubes

Type of glass	Raw material	Method of tube production	E_τ, kcal/mol		E_V, kcal/mol	
			Tubes	Blocks	Tubes	Blocks
I	Granular quartz	From a graphite crucible	97	105	56	94
I	Granular quartz	From a molybdenum crucible	114	—	61	—
I	Granular quartz	Similarity method	110	113	55	98
II	Synthetic quartz	From a graphite crucible	112	121	57	123
III	SiCl₄	From a graphite crucible	121	151	86	154

differences in the crystallization behavior of glass block-shaped samples and tubes are indicative of the subtle differences between the structure of block-shaped glass samples and glass drawn from the melt and cooled down to room temperature under the action of uniaxial tensile stress.

7.11 Summary of Results and Discussion

7.11.1 Introductory Remarks

The above presented experimental data, obtained at the Institute of Quartz Glass in St. Petersburg, illustrate how sensitive the process of crystallization of quartz glass is with respect to technological features of their production and processing. Such studies became possible owing to the fact that the investigators at this institute had the opportunity to monitor the overall production process of quartz glasses of different types at different stages: from preparation of the batch to the production of the completed items they had the chance to specify the required modes of melting, drawing, treating with HF-solutions and obtaining data on the results of chemical analysis. This unique situation enabled the investigators to reveal a variety of physico-chemical factors affecting any property of interest (in this case, crystallization) and to direct the attention to the necessity of a more detailed analysis of the influence of these factors. In addition, a wide spectrum of various quartz glasses of all types was available to the investigators, which made it possible to study crystallization of a great variety of glasses simultaneously. This spectrum of glasses included glasses being obtained under both industrial and laboratory conditions. The fact that the investigators had the opportunity to receive the required quantity of crucibles with covers of type II quartz glass to perform crystallization of samples in them contributed considerably to the successful realization of such studies.

Colleagues working at other research centers throughout the world did not have such comprehensive package of opportunities. Quartz glass is usually produced by specialized companies and goes on sale in the form of different products: tubes, chemical utensils, optical crude products, etc. The detailed information on the production process is a trade secret; hence it is commonly not published. Without knowledge of some important details of quartz glass production it is impossible to make an unambiguous conclusion about the nature of quartz glass crystallization. Hence it is not surprising, for example, that the interesting hypothesis of an diffusion-controlled mechanism of quartz glass crystallization [8, 17] did not find an experimental confirmation in a more detailed investigation.

7.11.2 Influence of Surface Reactions on Crystallization

From a theoretical point of view, most important with respect to the understanding of crystallization is the answer to the following question: Does the process of silica melt crystallization occur spontaneously or only when it is stimulated by some external factors? Unsuccessful persistent attempts to crystallize the surface of closed bubbles indicate that spontaneous surface crystallization of quartz glass over a reasonable range of heat-treatment times (up to several thousand hours) is not observed. After bubbles (or capillaries) are opened and reheated, their surface crystallizes quite fast. As was noted above, these facts may indicate that the surface nucleation of quartz glass is stimulated by various chemical reactions proceeding at high temperatures. These are mainly the reactions of establishing chemical equilibrium between components of the gaseous medium and those dissolved in the silica melt. The typical reactions are the reactions of hydration (interaction of silica surface with water vapor), dehydration (reaction of removal of structural water from a glass surface) and reactions of oxidation of partially reduced silica by oxygen (in the case of crystallization of non-stoichiometric glass). A profound effect on the process of nucleation is exerted by different metal (Pt, Mo, W, etc.) particles and by alkali impurities transferred through the gaseous phase.

Many distinguishing features of quartz glass crystallization (compared to crystallization of other glasses) are determined by the fact that, simultaneously with this process and over the same temperature range, the processes of interaction of silica melt with an ambient gaseous medium proceed along with intensive transport reactions that determine the transfer of various materials through the gaseous phase from one part of the system into another. Crystallization of multi-component glasses occurs, as a rule, at significantly lower temperatures at which chemical reactions of the melts with an ambient gaseous medium as well as transport processes are frozen-in. The investigators, who studied crystallization of quartz glasses (mainly type I glasses), dealt with the simultaneous influence of several factors (the use of samples with inadequately cleaned surfaces, contamination of sample surfaces during treatments, chemical reactions of surfaces with water vapor and oxygen) and thus obtained kinetic dependencies that were not observed on other glasses. The attempts to interpret the obtained results came up with the appearance of the hypothesis of diffusion-controlled crystallization of quartz glass.

The following two factors give the basis for the development of the hypothesis of a diffusion-controlled mechanism of crystallization: (i) Crystallization of quartz glass is accelerated in the atmosphere of oxygen or water vapor; (ii) the dependence of crystalline layer thickness on time has the shape of a curve resembling a parabola. However, both these facts can be debated with respect to their evidence in support of the validity of this hypothesis by the following considerations. First, the authors of [8, 17] dealt only with samples of drawn items (tubes and rods) of reduced type I glass from whose surface a technologically contaminated layer was not removed. Second, no attempts were made to protect sample surfaces from contamination during heat treatment. Hence, the "parabolic" dependencies obtained by the authors are determined by uncontrollable surface contamination during the process of drawing tubes (Fig. 7.49) or during crystallization (Fig. 7.17), and not by the distinguishing features of the intrinsic crystallization of quartz glass. It should be also noted that a parabolic form of the kinetic dependence is not necessarily the evidence of a diffusion-controlled mechanism of crystallization.

The following general rule is valid: to any simple mechanism of a process there must be a correspondence to a certain type of kinetic dependence describing its evolution in terms of this mechanism. However, this rule fails if reversed: one cannot derive an unambiguous conclusion about a given mechanism of some process basing on the shape of an experimentally determined kinetic curve [56]. The evidence of the absence of the dependence of the crystallization rate on oxygen diffusion is the disappearance of the parabolic dependence of crystallization kinetics after a contaminated surface layer is removed and contamination of a sample surface is prevented during crystallization (Figs. 7.17 and 7.49). Acceleration of the crystallization process in the atmosphere of oxygen relates only to a type I reduced glass (Fig. 7.41) and is not observed with glasses of other types (Fig. 7.40).

Surface chemical reactions cannot only contribute to an acceleration of nucleation, but, in contrast, they may also suppress this process. A drastic decrease in the rate of quartz glass crystallization or even its complete suppression in the presence of reducers (H_2, C, Cr, Ge) was noted in a number of papers [8, 9, 17]. The phase transition quartz-cristobalite also drastically slows down in the presence of hydrogen [44]. The observed phenomena of this kind are evidently connected with chemical reactions, for example, the reactions Eqs. (7.10) or (7.11) yielding volatile silicon monoxide:

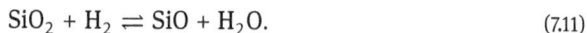

$$SiO_2 + H_2 \rightleftharpoons SiO + H_2O. \tag{7.11}$$

Such interaction occurs mainly in the places of new phase nucleation, resulting in a destruction of nuclei (the so-called Hedvall effect [57]).

The rate of quartz glass crystallization in a reducing medium may depend on the interaction kinetics of silica with a reducer. Two versions of interaction can be observed here: (i) The rate of interaction with a reducer is lower than the rate of nucleation, in this case deceleration of melt crystallization will be observed; (ii) the rate of interaction with a reducer is equal or higher than the rate of nucleation, in this case no crystallization will be observed at all.

The issue of the influence of water on crystallization of quartz glasses is also of interest. Two opposing tendencies are observed here: an increase of water concentration in the atmosphere results in a decrease of the induction period and an increase of the rate of quartz glass crystallization with constant values of activation energies (Figs. 7.42–7.44), whereas an increase of structural water concentration in glass results in an increase of the induction period and a decrease of the rate of crystallization over the range of temperatures lower than 1300 °C, as well as in the appearance of a corresponding maximum and minimum on the respective dependencies on hydroxyl group concentration over the range of higher temperatures (Figs. 7.32–7.34). Let us consider these issues in greater detail.

As was mentioned above, the presence of water vapor in gas bubbles (see Table 7.3) does not result in their crystallization, since the composition of the gaseous medium is in equilibrium with respect to the melt, and there is no chemical interaction between water vapor and bubble surfaces. The external surface of a sample is not in equilibrium with the atmosphere. Chemical reactions on a glass surface result in the change of the surface layer structure thus initiating the processes of nucleation. An increase of water vapor concentration in the atmosphere surrounding the sample increases the rate of surface reactions at the expense of the increase in the number of sites where a reconfiguration takes place thus increasing the rate of nucleation, which results in a notable decrease of the induction period (Figs. 7.42 and 7.43).

Along with the decrease of the induction period the rate of crystalline layer growth increases. In the literature [8, 9, 12], this increase of the rate of crystallization is related to the breakage of bridge bonds Si-O-Si and to a diffusion-controlled occupation in the glassy network of oxygen vacancies by oxygen atoms, their source being, as the authors believe, water vapor. It is hard to accept this explanation as satisfactory for two reasons: first, it is known that the diffusion of oxygen atoms through a cristobalite layer proceeds by several orders of magnitude slower than through the glass [58, 59]; second, the dependence of crystalline layer thickness on time is not parabolic (Fig. 7.42).

A series of experiments with a change of the gas medium during heat treatment was carried out to verify how effective a diffusion-controlled mechanism can act as a factor determining the rate of crystallization [52]. The experiments were conducted in two versions: (i) Samples of type I glasses obtained from synthesized quartz crystals were exposed in dry air at a temperature 1450 °C for a time in which a crystalline layer ~50 microns thick was formed on their surface (the layer thickness was determined from a control sample removed from the furnace before changing the atmosphere). Thereafter, the air was replaced by water vapor. (ii) Samples of the same glass before forming a crystalline layer ~ 50 microns thick were crystallized at the same temperature in water vapor. At the completion of this process, the water vapor atmosphere was replaced by dry air. The obtained results were compared with the results of control experiments on crystallization of samples of the same glass at the same temperature contained either completely in air or completely in an atmosphere of water vapor. In the case of a diffusion-controlled mechanism of crystallization, in both versions a significant deviation of the obtained dependencies from the control curves should have been observed soon after the change of gas medium: in the first version, an acceleration of cristobalite growth should take place, in the second version, a decrease in the rate of crystalline layer growth should be observed. However the experimental data, shown on Fig. 7.54, indicate that the change of the gas medium after formation of a continuous crystalline layer no longer affects the further progress of crystallization. A decrease of the induction period of crystallization with increasing partial pressure

Change of gaseous medium

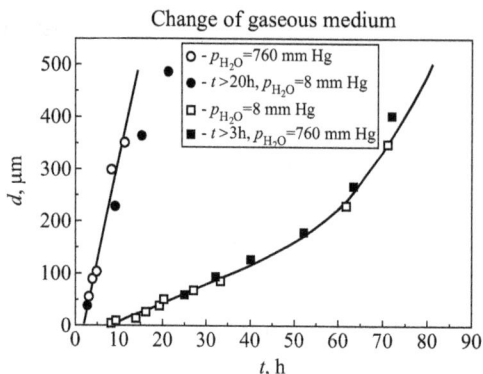

Fig. 7.54: Crystallization of type I silica glass obtained from a synthetic crystal at $T = 1450\,^{\circ}C$ [52].

of water vapor in the gas medium is not in contradiction with the assumption that the nucleation process in this case is initiated by the reaction of silica hydrolysis on the sample surface. An increase of partial pressure of water vapor results in the acceleration of the reaction of hydrolysis and in the increase of the nucleation intensity expressed via the decrease of the induction period (Figs. 7.42–7.43).

The fact that the activation energy values of the processes, proceeding during the induction period and crystalline layer growth, do not change with changing water vapor concentration in the gas phase indicates that the increase of nucleation intensity with growing water vapor content is determined by an increase of the number of sites on the surface where crystals nucleate. It should be noted that the factors, affecting the induction period, decrease simultaneously with an increase of the crystallization rate. This fact is illustrated, for example, both by the influence of water vapor partial pressure on crystallization (Figs. 7.42–7.44) and by the influence of oxygen (being present in the atmosphere of the furnace) on crystallization (Fig. 7.41), where no indications of a parabolic growth of the crystalline layer are found. The nature of the close relation between the value of the induction period and the rate of crystalline layer growth is not yet quite clear.

An increase of structural water concentration in glass results in an increase of the activation energy of both the induction period and the rate of crystalline layer growth (Fig. 7.31). At a first glance, this would seem strange. Indeed, an increase in structural water content results in an increase of the number of broken bridge bonds Si-O-Si and in a weakening of the silicon-oxygen glass network thus decreasing the activation energy of the viscous flow process. The discussion of this problem is closely related to the more general problem, the problem of the interrelation between viscosity and crystallization.

7.11.3 Relation Between Crystallization Rate and Viscosity

Viscosity and crystallization refer to the processes of transfer, the transfer of momentum and, respectively, mass transfer. In glass-forming melts both processes are affected by switching covalent chemical bonds [60], that is why for already several decades investigators have been trying to establish a quantitative interrelation between crystallization processes and viscous flow of glass-forming melts.

As far back as 1932, Frenkel, by theoretical considerations, arrived at the conclusion that the linear rate of crystallization of liquids in a supercooled region must be inversely proportional to viscosity [61]. A quantitative interrelation between crystallization rate and viscosity for a number of multi-component silicate glasses was established for the first time in the 1940s by Leontyeva. She suggested the following empirical equation for the rate of crystal growth, V [62, 63, 64]

$$V = \frac{k_1}{\eta} + k_2 \log \eta.$$ (7.12)

In 1947, Swift [65] established for sodium-calcium-silicate glasses the following relation between viscosity and crystallization rate

$$V = C\frac{\Delta T}{\eta}.$$ (7.13)

In Eqs. (7.12) and (7.13), k_1, k_2 and C are empirical constants, $\Delta T = (T_m - T)$ is the value of the supercooling, T_m is the melting temperature.

A number of theoretical papers, for example [66, 67], appeared later in which the authors advance equations describing the rate of motion of the crystal-liquid interface and establish its relation with viscosity. When discussing the relation between crystallization rate and viscosity, as applied to quartz glass [8, 13, 14], consideration is given to the equation of the rate of motion of the crystal-liquid interface derived in [66]. In this approach, out of the broad spectrum of factors influencing the rate of motion of this interface (process of mass transfer, temperature of the interface, presence of impurities, etc.), only one factor, the mass transfer, was taken into consideration in [66] when deriving the equation of interrelation between crystallization rate and viscosity. Heat release during the process of melt crystallization can considerably increase the temperature of the interface, which occurs on crystallization of a number of liquids with a great thermal effect and high rate of motion of the interface [8].

Since crystallization of quartz glasses proceeds slowly enough and is accompanied by a moderate heat release only (~2500 cal/mol) accompanied by a high radiative thermal conductivity, the released heat of crystallization is quickly dissipated not causing a considerable rise of the interface temperature. In the estimation of the authors of [68] the temperature of the interface differs from the sample temperature by a negligibly small amount of the order ~0.1 °C. Thus, the factor of interface temperature rise can be neglected for quartz glasses. But, on the other hand, the rate of quartz glass crystallization is strongly influenced by impurities. High sensitivity of both crystallization rate and viscosity of quartz glasses to impurities is one of the main peculiarities that distinguishes this glass from multi-component glasses for which the same properties are determined by their main composition and depend only weakly on impurities.

The equation describing the rate of motion of the interface, according to [66], has the following form

$$V = f \frac{D}{\delta} \left[1 - \exp\left(-\frac{\Delta g}{k_B T} \right) \right].$$ (7.14)

Here Δg is the crystallization free energy change per molecule, δ is the length of a molecule's jump when crossing the interface, f is the fraction of sites on the interface available for molecules ($f \cong 1$), D is a kinetic constant with the dimension $cm^2 s^{-1}$, which the authors of [66] identify with the self-diffusion coefficient. For the case $\Delta g \ll k_B T$, Eq. (7.14) reads

$$V = f \frac{D}{\delta k_B T} \frac{\Delta h}{T_m} \Delta T,$$ (7.15)

where Δh is the enthalpy of melting.

For silica melts, in this equation all quantities except D are known. Its value cannot be calculated from theoretical considerations, hence, it is postulated that D can be expressed in terms of viscosity by the Stokes-Einstein equation,

$$D = \frac{k_B T}{3 \pi \delta \eta}.$$ (7.16)

Inserting the expression Eq. (7.16) for D into Eq. (7.15) we obtain

$$V = \frac{f \Delta h}{3 \pi \delta^2 T_m} \frac{\Delta T}{\eta}.$$ (7.17)

Taking into consideration that the temperature dependence of viscosity of quartz glasses over a wide range of temperatures is described by a simple exponential dependence [70]

$$\eta = \eta_0 \exp\left(\frac{E_\eta}{k_B T} \right),$$ (7.18)

the temperature dependence of interface rate of motion can be written (using Eqs. (7.17) and (7.18)) as

$$V = f \frac{\Delta h}{3 \pi \delta^2} \frac{\Delta T}{T_m} \frac{1}{\eta_0} \exp\left(-\frac{E_\eta}{k_B T} \right).$$ (7.19)

The temperature dependence of the diffusion coefficients of the atoms, for example, of silicon or oxygen are also described by an exponential equation [69, 70]. If we substitute such relations into Eq. (7.15), we obtain another relation for the temperature dependence of the rate of interface motion

$$V = f \frac{\Delta h}{\delta k_B T} \frac{\Delta T}{T_m} D_0 \exp\left(-\frac{E_D}{k_B T}\right),$$ (7.20)

where E_D is the activation energy of the diffusion process.

For silica melts for sufficiently high supercooling ($> 150\,°C$), the temperature dependence of the crystallization rate is practically completely determined by E_D. In this limiting case, Eq. (7.20) becomes a special case of Eq. (7.2). Eq. (7.2) describes experimental data on the crystallization rate in dependence on temperature, with $E_D = E_V$. Eq. (7.19) describes the crystallization rate in dependence on temperature, the calculations made from viscosity. If the postulate on the relation between diffusion coefficient D and viscosity η is true, the values of crystallization rates, calculated by Eq. (7.19), and the experimental values should agree. In this case, firstly, the crystallization rate should be inversely proportional to viscosity and, secondly, the values of activation energies E_V and E_η should be equal. Thus, one of the important criteria of applicability of Eqs. (7.17) and (7.19), interrelating crystallization rate with viscosity, is the fulfillment of the condition

$$E_V \approx E_\eta \quad \text{or} \quad \frac{E_V}{E_\eta} \approx 1.$$ (7.21)

Due to the strong dependence of crystallization rate and viscosity of quartz glass on impurities and melting conditions a reliable check of the applicability of Eqs. (7.17) and (7.19) can be performed only if measurements are performed on a series of glasses with a regular change of impurity composition, the temperature dependencies of both crystallization rate and viscosity being measured on the samples of the same glass.

The first paper in which measurements were performed of the crystallization rate and viscosity on samples of the same glasses with changing impurity composition was the already cited paper by Brown and Kistler [7]. This paper meets all above listed requirements which should be fulfilled for the answer to the question if there is realized in experiment on quartz glasses the theoretically assumed relation between crystallization rate and viscosity, i.e. whether condition Eq. (7.21) is fulfilled for them or not. Let us recall that in this paper an investigation was carried out on crystallization and viscosity of five different glasses: a type III (Corning) glass obtained when melting synthetic silicon dioxide in vacuum and three glasses melted from synthetic SiO_2 with different additions of Al_2O_3. The data on crystallization of these glasses are presented in Fig. 7.1. The data concerning the temperature dependencies of crystallization rate and viscosity, obtained and presented in the form of a separate table in [7], are given here in Table 7.5.

From the data of Table 7.5 it is seen that Eq. (7.21) is not fulfilled for any of the investigated glasses. For example, for very-high-purity glasses of type III the activation energy of crystalline layer growth was found to be by almost 30 % higher than the activation energy of viscous flow, while for glasses with additions of aluminum oxide the

Table 7.5: Interrelation between crystallization rate and viscosity from the data by Brown and Kistler [7]

Glass	$(Al/Si)\cdot10^2$	E_V, kcal/mol	E_η, kcal/mol	E_V/E_η
Corning fused silica	$1\cdot10^{-7}$	118.6	92.9	1.277
Cab-O-Sol-0	0.0118	109.5	119.6	0.916
Al-3	0.225	74.3	108.7	0.684
Al-1	0.634	66.1	94.6	0.698
Al-2	1.097	83.4	119.0	0.701

value was, conversely, by about 30 % lower. From this result it follows that Eq. (7.17), describing the relation between viscosity and crystallization rate, is not applicable to quartz glass. The analysis of the experimental data on crystallization rate and viscosity, published in this paper, shows that the factors leading to viscosity growth (decrease of structural water concentration, increase of aluminum admixture concentration) also simultaneously contribute to an increase of the crystallization rate, which is contradictory to the theoretical expectations. To our regret, the authors of [7] refrained from discussing the data obtained. A paradoxical situation resulted here: while the paper [7] was one of the most quoted papers devoted to separate analysis of the problems of quartz glass crystallization and the problems of viscosity of this glass, the results of a combined investigation of viscosity and crystallization obtained in [7] were not discussed in the literature at all. Such omission can be traced also in the papers where the problem of the relation between crystallization rate of quartz glass and its viscosity received primary consideration [8, 13, 14, 68, 69].

The authors of [8, 13, 14] compared their data on glass crystallization with the published data on viscosity and obtained a reasonable agreement between the experiments and calculations, which gave them the basis to state that equations of the type of Eq. (7.17) are applicable for the description of the process of crystallization. After more than forty years after the appearance of above mentioned papers, the authors of [68] returned to a discussion of this problem and arrived at the same conclusions. They correlated the data on internal crystallization of quartz glasses from [13, 14] with the data of [70] on the investigation of silicon diffusion coefficients and viscosity in quartz glass thus confirming the conclusions made in [8, 13, 14]; in addition, they concluded that the value of D in Eq. (7.14) is the silicon diffusion coefficient.

The main drawback of the mentioned papers is the fact that their conclusions were drawn based on the comparison of the data obtained on measuring different glasses by different authors. From the analysis as presented here it is evident how sensitive the crystallization kinetics of quartz glasses is both with respect to impurities and to the conditions of their production. Similarly sensitive to impurities is viscosity. The scattering of viscosity values over the range of crystallization temperatures can exceed two orders of magnitude [73]. Thus, the conclusions about the interrelation between crystallization rate and viscosity, which have been drawn based on viscosity measurements of glasses different from the glasses for which crystallization measurements were made, significantly depend on the choice of glasses employed in the analysis and cannot be considered, consequently, as sufficiently correct and convincing.

The most preconceived approach to the problem of interrelation between viscosity and crystallization can be found in the recent paper devoted to this problem [69]. Using the database on glass properties SciGlass 5, the authors constructed plots of the crystallization rate logarithm against temperature and plots of viscosity logarithm against inverse temperature for each type of quartz glass, covering practically all the published data. The authors took notice of the wide scatter of data on both viscosity and crystallization rate that reached two orders of magnitude. To analyze such an extensive amount of information the authors had at least two possibilities: (i) To study the sources closely and select the data on viscosity and crystallization rate measured on the same glasses (glasses of the same melt), (ii) to try to generalize the available data without going into detail. The authors did chose the second possibility. They suggested that there must be a correspondence between the data on high crystallization rates and the data on low values of viscosity, and, vice versa, glasses with high viscosity must have a low crystallization rate. Such assumption is equivalent to an *a priori* acknowledgement of the correctness of Eq. (7.19). It is, therefore, not surprising that the authors confirmed the applicability of Eqs. (7.17) and (7.19) to quartz glass. At the same time they neglected the data analysis of a number of papers [7, 21, 71], in which the data on viscosity and crystallization, obtained for glasses of the same melt, are presented. In particular, in the middle of the 1970s an investigation on the interrelation between crystallization and viscous flow of vitreous silica was carried out at the Institute of Quartz Glass in St. Petersburg. These investigations have been performed on glasses of the same melts obtained by different methods and containing different amounts of impurities [21, 35, 71, 72].

One of the factors affecting crystallization rate and viscosity of vitreous silica is the content of structural water in glass. The crystallization rate changes in a complicated way (Fig. 7.34). Over the temperature range < 1300 °C it grows with increasing

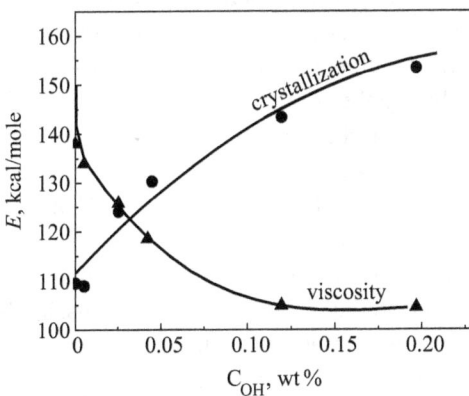

Fig. 7.55: Dependence of the activation energies of viscous flow and rate of crystalline layer growth on the content of structural water in glasses of laboratory synthesis [71].

concentration of OH-groups, while viscosity decreases [74]. Fig. 7.55 shows the concentration dependence of the activation energy of viscosity and crystallization rate on structural water content for a series of glasses of a laboratory synthesis, obtained by vacuum remelting of type III glass powder [74]. It is seen that with increasing concentration of structural water the activation energy values of viscosity and crystallization rate change in the opposite directions. At the same time, the values of viscosity and crystallization change in the same direction, but not in inverse proportion, as it is predicted by Eq. (7.17). A similar tendency is observed not only for glasses of the laboratory synthesis, but also for commercial glasses (Fig. 7.56).

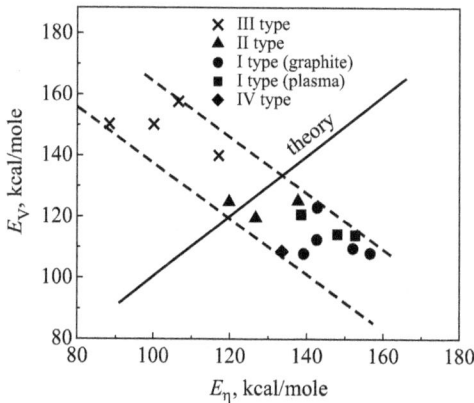

Fig. 7.56: Interrelation between the activation energies of viscous flow processes and crystallization rate for different types of commercial glasses.

From the plots, presented in Figs. 7.55 and 7.56, it is apparent that Eq. (7.21) does not hold for vitreous silica. It is seen that for glasses with a high content of structural water (type III glasses) the ratio $(E_V/E_\eta) > 1$ (for different glasses it varies between 1.2 and 1.6). This result correlates well with the results of Brown and Kistler [7] (see Table 7.5), where the ratio $(E_V/E_\eta) = 1.28$ was obtained for high-water Corning glass. It should be noted that when investigating similar Corning glass 7940 at the Institute of Quartz Glass in St. Petersburg, the ratio $(E_V/E_\eta) = 1.5$ [21] was obtained. With decreasing content of structural water the ratio (E_V/E_η) decreases. For water-less glasses, it becomes less than one. Thus, the ratio (E_V/E_η) for vitreous silica can vary from ~ 1.6 to ~ 0.7 and only in a particular case it can be equal to 1. With $E_V \neq E_\eta$, the calculated temperature dependencies of crystallization rate will differ significantly from the experimental ones as it is shown in Fig. 7.57 by the example of type III glass.

Both the crystallization rate (see Figs. 7.26 and 7.28) and viscosity [73, 75] of type I glasses are sensitive to redox conditions of melting: glasses melted under reducing conditions have a higher viscosity and they are crystallized at a higher rate compared to those glasses melted from the same batch material under neutral conditions. It follows that, also in this case, under changing redox melting conditions viscosity and crystallization rate change in the same direction.

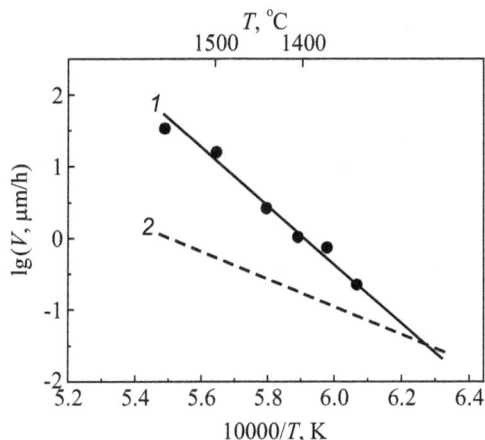

Fig. 7.57: Temperature dependence of the crystallization rate of a type III glass: (1) experiment, (2) calculation by Eq. (7.19).

Here is one more example demonstrating the absence of the postulated by Eq. (7.17) interrelation between crystallization rate of quartz glass and its viscosity. The glass viscosity of a tube does not differ, or differs only moderately, from the glass viscosity of a block from which this tube was drawn. For type I and II glasses this difference is determined by additional volatilization of alkali impurities in the course of the process of drawing. An increase of viscosity as a result of drawing depends on the content of alkali impurities in the initial glass: if it is $< 10 \cdot 10^{-4}$ mol%, then after reheating and drawing the viscosity (at $1200\,^{\circ}$C) grows by 0.1–0.3 orders of magnitude without any noticeable change in the activation energy [73]. At the same time the temperature dependence of crystallization rate reduces rapidly for tubes compared to that for the initial glass (see Table 7.4). A comparison of the corresponding values of activation energy for three commercial glasses is given in Table 7.6. It is evident from the table that the activation energy values of crystalline layer growth for tubes are much smaller (by 1.5–2 times) than the values of activation energy of viscous flow. If it is considered that crystallization of quartz glasses was investigated not only on block glass samples, but also in many cases on tube samples [8, 9, 15, 16], then the data of Table 7.6 present additional evidence that Eq. (7.21) is not suited for the description of quartz glass.

Table 7.6: Comparison of the values of the activation energy of viscous flow E_η (in kcal/mol) and the activation energy of crystallization rate E_V (in kcal/mol) for samples obtained from blocks and tubes [71]

Glass	Type	E_η	E_V (bulk)	E_V (tubes)
KI	I	132	102	48
KV	II	137	123	57
KU	III	110	154	86

The above presented facts indicate that the theoretically postulated interrelation between viscosity and crystallization rate (Eq. (7.17)) is not confirmed for vitreous silica. In contrast, the opposite tendency is observed: with increasing viscosity the crystallization rate increases as well. Such conclusion should not seem unexpected, since the considered processes are related to different types of transfer phenomena. However, the obtained regularities demand a thorough and in-depth theoretical study. At present, we can merely try to qualitatively explain the tendency of the changes of the discussed properties basing on the assumption that the processes of crystallization and viscous flow are affected by switching covalent chemical bonds [60] and the kinetics of the considered processes depends on the properties of these bonds.

A wide distribution of the angles of (Si–O–Si)-bonds over the range of values from 120 °C to 180 °C is found in the network of vitreous silica [76]. The change of the angle of a Si–O–Si bond results in the change of the electronic structure of the Si–O bond. The occurring redistribution of electronic density of the Si–O bond with an increasing angle of Si–O–Si is accompanied by a change in all characteristics of a bond, including its length, strength, force coefficient and effective charges on atoms of silicon and oxygen (these issues are considered in greater detail in the review [77]). However, the most profound change of the parameters of Si–O bonds occurs on breaking the network, which is determined by the entry of impurities into the network. The electron distribution in a SiO_4 tetrahedron strongly depends on to which second neighbor an oxygen atom is bound. If, instead of a silicon atom, it is bound to a positively charged ion, for example, H^+ or Na^+, a redistribution of electron densities in a SiO_4-tetrahedron will develop with the result that an effective negative charge on a non-bridging oxygen atom will decrease, the length of the Si-O-bond will reduce, while the strength and force coefficient will increase. Simultaneously the lengths of Si-O(Si)-bonds increase and their strength and force coefficients decrease [78, 79, 80]. The change of electron density in one SiO_4-tetrahedron causes a certain change of electron density in the neighboring tetrahedra as well. Thus, introduction of impurity cations into vitreous SiO_2 results in a differentiation of the Si–O-bonds causing the appearance of stronger and weaker bonds in the glassy network.

On crystallization, a transfer of particles through an interface by switching Si–O chemical bonds is performed. To transform melt into cristobalite all available bonds must be involved in the process of switching. The kinetics of the crystallization process must be limited by switching the strongest bonds: the greater is the portion of the strong bonds, the slower is the growth of the crystalline layer. A transfer of an amount of motion is sensitive to the presence of the weakest Si–O bonds. The weaker are the chemical bonds, the more intensive is this process and the lower is glass viscosity. With viscous flow it is not necessary to switch all, even the strongest, bonds. Small fragments of the network, containing strong bonds, can be retained in a network on viscous flow. Thus, the process of bond differentiation must result in a decrease of the crystallization rate and the decrease in viscosity and, vice versa, a reduction in the differentiation degree must increase viscosity and crystallization rate.

7.12 Conclusions

The process of surface crystallization of quartz glasses is not a spontaneous process; it is stimulated by surface chemical reactions between silica and impurities located on the surface or components of the gaseous medium. At the temperatures of quartz glass crystallization, numerous chemical processes can occur in the furnace space where the sample is located, the processes being accompanied by a transfer of impurities through the gas phase. Contacting the surface of the sample, which is under investigation, and contaminating it, such impurities have a profound effect on the process of crystallization of vitreous silica, distorting the process considerably. Hence, correct results on the intrinsic crystallization of quartz glass can be obtained only then if appropriate measures are undertaken to protect the surfaces of the samples to be investigated from contamination through the gas medium during heat treatment.

Heat treatment of quartz glasses under pure conditions showed that the process of their crystallization proceeds in two stages: at the first stage (during an induction period) there are no traces of crystallization on the surface of the sample for an appreciable length of time, thereafter the stage of crystalline layer growth starts. The value of the induction period decreases exponentially with increasing temperature and the growth rate exponentially increases. For the samples of block-shaped glass the activation energy values of the processes proceeding during the induction period and the process determining the growth rate are equal. The value of the induction period is a quantitative measure of crystallization stability of quartz glass.

Crystallization stability of type I glasses is strongly dependent on redox melting conditions that determine the degree of their non-stoichiometry. The more reduced is the glass, the lower is its crystallization stability. The degree of melting of quartz crystals has a profound impact on the crystallization stability of glasses of this type. Glasses with completely melted crystals have a maximum crystallization stability. The fusion penetration degree of crystalline particles depends on the dimension of crystalline particles, and on the temperature and time of melting: all factors being equal, the degree of melting increases with decreased particle sizes and increased temperature and/or time of melting. Crystallization stability of type II and III glasses depends on the content of structural water. Crystallization stability of glasses at temperatures below 1300 °C increases with an increased concentration of hydroxyl groups; over the range of higher temperatures type II glasses containing 0.02–0.04 wt% of structural water have the maximum crystallization stability. Surface contaminations decrease the induction period and increase the crystallization rate. These changes occur at the expense of the changes in the values of the pre-exponential factors (see Eqs. (7.1) and (7.2)), the activation energy values therewith not changing.

Dry gases – oxygen, air, argon, as well as vacuum – have no impact on crystallization of type II and III glasses. Type I glasses, melted under reducing conditions in the atmosphere of oxygen or air, are crystallized faster than in argon or vacuum. Water vapor accelerates the process of crystallization of glasses of all types. With an increase of the partial pressure of water vapor the value of the induction period decreases, while the growth rate increases. As with surface contamination, all changes proceed at the expense of the change in the pre-exponential factors.

The process of drawing tubes strongly affects crystallization. Surface layers of a tube are found to be contaminated with impurities volatilizing from molding elements and the heater. The degree of contamination depends on the purity of the materials (graphite, molybdenum, tungsten) used for production of molding elements and the heater. The presence of a contaminated surface layer drastically decreases crystallization stability of tubes. Removal of this layer by treating it with a solution of hydrofluoric acid considerably increases the crystallization stability of the tube, however it still remains significantly lower than the crystallization stability of the initial block glass. Crystallization of tubes after removal of the contaminated layer is described by the same exponential laws Eqs. (7.1) and (7.2) as the ones for crystallization of block-shaped glass, but, unlike block-shaped glasses, the value of E_V for tubes is by (1.5–2) times lower than the one of E_r.

In a number of theoretical papers, devoted to the investigation of quartz glass, it is stated that the crystallization rate of vitreous silica over the region of considerable supercooling is inversely proportional to viscosity. In order to confirm this statement, experimental data on the crystallization rate were compared with experimental data on viscosity, the data being obtained by different authors and on different glasses. Considering the strong dependence of both viscosity and crystallization rate of quartz glasses on a variety of factors, such comparison cannot be considered as correct. The measurements of the crystallization rate and viscosity of samples of the same melt showed that the mentioned dependence is experimentally not observed, but even an opposite tendency is found, the factors influencing the increase of the crystallization rate increase viscosity as well and vice versa.

Bibliography

[1]   W. Johnson, K.W. Andrews, *An X-ray study of the inversion and thermal expansion of cristobalite*, Transact. Brit. Ceram. Soc. **55**, 227 (1956).

[2]   F. Thomas, *Siloxid, ein höherwertiger Ersatz des Quarzglases*, Chem. Zeitung **36**, 25 (1912).

[3]   R. Rieke, K. Endell, *Über die Entglasung von Quarzgläsern*, Silikat Zeitschr. **1**, 6 (1913).

[4]   F.A. Kurlyankin, *Thermal properties of articles from a transparent quartz glass*, Ceramics and Glass **2**, 27 (1936).

[5]   R.B. Sosman, *The Properties of Silica* (Chemical Catalog Company, New York, 1927).

[6]   G. Hetherington, K.H. Jack, M.W. Ramsay, *The high-temperature electrolysis of vitreous silica*, Phys. Chem. Glasses **6**, 6 (1965).

[7]   S.D. Brown, S.S. Kistler, *Devitrification of high-SiO₂ glasses of the system $Al_2O_3 - SiO_2$*, J. Amer. Ceram. Soc. **42**, 263 (1959).

[8]   N.G. Ainslie, C.R. Morelock, D. Turnbull, *Devitrification Kinetics of Fused Silica*. In: *Symposium on Nucleation and Crystallization in Glasses and Melts* (Columbus, 1962, pgs. 97–107).

[9]   F.E. Wagstaff, S.D. Brown, I.B. Culter, *The influence of H_2O and O_2 atmospheres on the crystallization of vitreous silica*, Phys. Chem. Glasses **5**, 76 (1964).

[10] F.E. Wagstaff, K.J. Richards, *Preparation and crystallization behavior of oxygen-deficient vitreous silica*, J. Amer. Ceram. Soc. **48**, 382 (1965).

[11] F.E. Wagstaff, E. Frank, *Crystallization Resistant Vitreous Silica formed by the Addition of Silicon to Silica* (Pat. USA N 3370921, 1968).

[12] F.E. Wagstaff, K.J. Richards, *Kinetics of crystallization of stoichiometric SiO_2 glass in H_2O atmospheres*, J. Amer. Ceram. Soc. **49**, 118 (1966).

[13] F.E. Wagstaff, *Crystallization kinetics of internally nucleated vitreous silica*, J. Amer. Ceram. Soc. **51**, 449 (1968).

[14] F.E. Wagstaff, *Crystallization and melting kinetics of cristobalite*, J. Amer. Ceram. Soc. **52**, 650 (1969).

[15] J. Hlavach, L. Vashkova, *Crystallization of reduced silica glass in an oxidizing atmosphere*, Silikaty **9**, 213 (1965).

[16] L. Vashkova, J. Hlavach, *Crystallization of quartz glass*, Silikaty **13**, 211 (1969).

[17] A.G. Boganov, V.S. Rudenko, G.L. Bashnina, *Patterns of crystallization and the nature of quartz glass*, Neorg. Mater. **2**, 363 (1966) (in Russian).

[18] H. Schaefer, *Chemische Transportreaktionen* (Verlag Chemie, Weinheim, 1961).

[19] V.K. Leko, L.A. Komarova, *The effect of crystallization induction of silica glasses*, Fizika i Khimiya Stekla **1**, 335 (1975) (in Russian).

[20] L.A. Komarova, *Issledovanie Zakonomernostei Kristallizatsii Kvartsevykh Stekol* (Thesis, Leningrad, 1973 (in Russian)).

[21] O.V. Mazurin, V.K. Leko, L.A. Komarova, *Crystallization of silica and titanium oxide-silica corning glasses (codes 7940 and 7971)*, J. Non-Crystalline Solids **18**, 1 (1975).

[22] O.M. Todes, *Kinetics of topochemical reactions*, Zh. Fiz. Khim. **14**, 1224 (1940) (in Russian).

[23] K. Mampel, *Zeitumsatzformeln für heterogene Reaktionen an Phasengrenzen fester Körper*, Z. Phys. Chem. **A 187**, 43, 235 (1940).

[24] Kim Dong-Lae, M. Tomozawa, *Water concentration profile in silica glasses during surface crystallization*, J. Non-Crystalline Solids **279**, 170 (2001).

[25] V.K. Leko, L.A. Komarova, O.V. Mazurin, *An investigation of the crystallization rate of silica glasses*, Neorg. Mater. **8**, 1125 (1972) (in Russian).

[26] V.K. Leko, L.A. Komarova, *Investigation of the induction period of crystallization of silica glasses*, Neorg. Mater. **7**, 2240 (1971) (in Russian).

[27] V.K. Leko, L.A. Komarova, O.V. Mazurin, *Untersuchung der Kristallisationskinetik von Kieselgläsern in Luftatmosphäre*, Silikattechnik **25**, 81 (1974).

[28] H.L. Schick, *A thermodynamic analysis of the high-temperature vaporization properties of silica*, Chem. Reviews **60**, 331 (1960).

[29] A.V. Amosov, *Oxygen Vacancies in a Network of Quartz Glasses and Their Spectroscopical Displays*. In: *The Glassy State* (Leningrad, 1983, p. 155–159 (in Russian)).

[30] V.K. Leko, *Effect on the properties of vitreous silica of the physicochemical processes occurring during synthesis and heat treatment*, Fizika i Khimiya Stekla **8**, 129 (1982) (in Russian).

[31] V.K. Leko, I.A. Stepanova, *Investigation of a thermal darkening of electromelted quartz glasses*, Opt. Mekh. Prom. **6**, 44 (1973) (in Russian).

[32] T. Bell, G. Hetherington, K.H. Jack, *Water in vitreous silica*, Phys. Chem. Glasses **3**, 141 (1962).

[33] I.M. Vasserman, M.P. Nikitina, A.V. Amosov, *About non-stoichiometry of vitreous silica*, Doklady Akademii Nauk SSSR **203**, 99 (1972) (in Russian).

[34] G. Hetherington, K.H. Jack, *The oxidation of vitreous silica*, Phys. Chem. Glass **5**, 147 (1964).

[35] V.K. Leko, *Investigation of Viscosity and Crystallization of Silica Glasses*. In: *Novye neorganicheskie materialy i pokrytiya na osnove stekla i tugoplavkikh soedinenii* (Moscow, 1973, pgs. 9–11 (in Russian)).

[36] N. Saito, K. Motoyama, *Evolution of gas from fused silica*, J. Vacuum Soc. Japan **7**, 350 (1964).

[37] I.I. Cheremisin, A.G. Boganov, *Investigation of the gaseous impurities evaluated at transformation of quartz in glass*, Fizika i Khimiya Stekla **3**, 87 (1977) (in Russian).

[38] A.V. Amosov, *The new concept of the mechanism of formation of the radiating paramagnetic color centeres in quartz glasses*, Fizika i Khimiya Stekla **9**, 569 (1983) (in Russian).

[39] V.K. Leko, L.A. Komarova, *Effect of concentration of oxygen vacancies and hydroxyl groups on crystallization of silica glasses*, Neorg. Mater. **11**, 1864 (1975) (in Russian).

[40] V.K. Leko, O.V. Mazurin, *Svoistva Kvartsevogo Stekla* (Nauka, Leningrad, 1985 (in Russian)).

[41] G. Stephenson, K. Jack, A.J. Moulson, J.P. Roberts, *Water in silica glass*, Transact. Brit. Ceram. Society **59**, 397 (1960).

[42] N.G. Ainslie, J.D. Mackenzie, D. Turnbull, *Melting kinetics of quartz and cristobalite*, J. Phys. Chem. **65**, 1718 (1961).

[43] G. Scherer, P.J. Vergano, D.R. Uhlmann, *A study of quartz melting*, Phys. Chem. Glasses **11**, 53 (1970).

[44] J.D. Mackenzie, *Fusion of quartz and cristobalite*, J. Am. Ceram. Soc. **43**, 615 (1960).

[45] F. Oberlies, *Inhomogeneity in Quartz Glass*, Naturwissenschaften **44**, 488 (1957).

[46] J.H. Westbrook, *Hardness-temperature characteristics of some simple glasses*, Phys. Chem. Glasses **1**, 32 (1960).

[47] V.A. Florinskaya, R.S. Pechenkina, *Spectra of simplified glasses in infrared range and their connection with the structure of a glass*. In: *Stroenie Stekla* (Nauka, Moscow-Leningrad, 1955, pgs. 70–95 (in Russian)).

[48] V.K. Leko, L.A. Komarova, *Effect of temperature and melting period on crystallization of silica glass*, Neorg. Mater. **11**, 2106 (1975) (in Russian).

[49] E.V. Gurkovskii, I.M. Levin, *Gas-flame melting a quartz glass*. In: *Fiziko-Khimicheskie Issledovaniya Struktury i Svoistv Kvartsevogo Stekla* (Moscow, 1974, pgs. 147–150 (in Russian)).

[50] V.K. Leko, L.A. Komarova, *Study of the influence of surface additions on crystallization kinetics of silica glasses*, Neorg. Mater. **10**, 1872 (1974) (in Russian).

[51] V.K. Leko, L.A. Komarova, *Study of crystallization of silica glasses in various gaseous media*, Neorg. Mater. **11**, 1115 (1975) (in Russian).

[52] V.K. Leko, L.A. Komarova, *Effect of aqueous vapors on crystallization of silica glasses*, Neorg. Mater. **11**, 2046 (1975) (in Russian).

[53] V.K. Leko, L.A. Komarova, *Investigation of distribution of impurities in surface layers of tubes from a quartz glass*, Opt. Mekh. Prom. **6**, 33 (1974) (in Russian).

[54] V.K. Leko, L.A. Komarova, *Features of crystallization of quartz glasses*. In: *Fiziko-Khimicheskie Issledovaniya Struktury i Svoistv Kvartsevogo Stekla* (Moscow, 1974, No. 1, pgs. 97–103 (in Russian)).

[55] V.K. Leko, L.A. Komarova, *Crystallisation of tubes from silica glass*, Steklo i Keramika **8**, 10 (1974) (in Russian).

[56] A. Ya. Rozovskii; *Kinetics of Topochemical Reactions* (Moscow, Khimiya, 1974 (in Russian)).

[57] J.A. Hedvall, *Einführung in die Festkörperchemie* (Vieweg, Braunschweig, 1952).

[58] R.M. Barrer, *Mechanism of activated diffusion through silica glass*, J. Chem. Soc. **136**, 378 (1934).

[59] R.M. Barrer, *Diffusion in and through Solids* (Univ. Press, Cambridge, 1941).

[60] R.L. Myuller, *Nature of activation energy and experimental data of fluidity of refractory glass-forming substances*, Zh. Prikl. Khimii **28**, 363 (1955) (in Russian).

[61] Ya. I. Frenkel, Sov. Phys. **1**, 498 (1932) (in Russian).

[62]   A. Leontyeva, *The rate of crystallization of some silicate glasses as a function of viscosity*, Acta Physicochim. URSS **13**, 423 (1940) (in Russian).

[63]   A. Leontyeva, *Relation between the linear rate of crystallization and viscosity for* Na_2O-SiO_2 *glasses*, Acta Physicochim. URSS **14**, 245 (1941) (in Russian).

[64]   A. Leontyeva, *Linear rate of crystallization of potassium, sodium and lithium disilicates*, Acta Physicochim. URSS **16**, 97 (1942) (in Russian).

[65]   H.J. Swift, *Effect of magnesia and alumina on rate of crystal growth in some soda-lime-silica glasses*, Amer. Ceram. Soc. **30**, 170 (1947).

[66]   W.B. Hillig, D. Turnbull, *Theory of crystal growth in undercooled pure liquids*, J. Chem. Phys. **24**, 914 (1956).

[67]   J.W. Cahn, W.B. Hillig, G. Sears, *Molecular mechanism of solidification*, Acta metallurgica **12**, 1421 (1964).

[68]   M.L.F. Nascimento, E.D. Zanotto, *Mechanisms and dynamics of crystal growth, viscous flow, and self-diffusion in silica glass*, Phys. Rev. **B 73**, 024209 (2006).

[69]   M.L.F. Nascimento, E.D. Zanotto, *Diffusion processes in vitreous silica revisited*, Phys. Chem. Glass: Eur. J. Glass Sci. Technol. **B48**, 201 (2007).

[70]   G. Brebec, R. Seguin, C. Sella et al., *Diffusion du silicium dans la a silice amorphe*, Acta metall. **28**, 327 (1980).

[71]   V.K. Leko, *Interrelationship between viscosity and crystallization of silica glasses*, Neorg. Mater. **12**, 99 (1976) (in Russian).

[72]   V.K. Leko, *Simultaneous investigation of viscosity and crystallization of quartz glasses*. In: *Fiziko-Khimicheskie Issledovaniya Struktury i Svoistv Kvartsevogo Stekla* (Moscow, 1974, No. 1, pp. 90–97 (in Russian)).

[73]   V.K. Leko, *Viscosity of quartz glasses*, Fizika i Khimiya Stekla **5**, 258 (1979) (in Russian).

[74]   V.K. Leko, N.K. Gusakova, *The characteristics of viscous flow of vapor-synthetic silica glasses*, Fizika i Khimiya Stekla **3**, 226 (1977) (in Russian).

[75]   V.K. Leko, E.V. Meshcheryakova, *Investigation of specific features of the influence of reduction-oxidation conditions of melting on viscosity of electromelted silica glasses*, Fizika i Khimiya Stekla **1**, 264 (1975) (in Russian).

[76]   R.L. Mozzi, B.E. Warren, *The structure of vitreous silica*, J. Appl. Cryst. **2**, 164 (1969).

[77]   V.K. Leko, *The structure of silica glass*, Fizika i Khimiya Stekla **19**, 673 (1993) (in Russian).

[78]   M.G. Voronkov, *About interatomic distances and the nature of Si-O bonds in silicates*, Doklady Akademii Nauk SSSR **138**, 106 (1961) (in Russian).

[79]   D.W. Cruickshank, *The role of 3d-orbitals in p-bonds between (a) silicon, phosphorous, sulphur, or clorine and (b) oxygen and nitrogen*, J. Chem. Soc. **12**, 5486 (1961).

[80]   A.N. Lazarev, *Nature of Si-O-Si bonds and values of valence angles of oxygen*, Izv. Akademii Nauk SSSR (Ser. Chem.) **2**, 235 (1964) (in Russian).

Vladimir M. Fokin, Alexander Karamanov, Alexander S. Abyzov,
Jürn W.P. Schmelzer, and Edgar D. Zanotto

8 Stress-induced Pore Formation and Phase Selection in a Crystallizing Stretched Glass

In the present chapter, we describe results of experimental investigations and theoretical analysis of phase selection and nucleation of pores in small samples of undercooled diopside liquid when it is enclosed by a solid crystalline surface layer. The formation of the surface crystalline layer starts with nucleation and growth of highly dense diopside crystals. At the moment of impingement of these crystals on the sample surface, the crystallization pathway switches from diopside to a wollastonite-like (WL) phase. The origin of such switch can be explained by the fact that the formation of the WL-crystal produces less elastic stress energy than the same amount of diopside. This difference is due to the lower (as compared to diopside) density of the WL-crystal phase, which is closer to the liquid density. The relative content of the two crystalline phases can be changed by varying the sample size. Due to the density misfit the growth of the WL-crystalline layer leads to uniform stretching of the encapsulated liquid. This negative pressure leads finally to the formation of one small pore, which rapidly grows up to a size that almost eliminates the elastic stress and, therefore, dramatically reduces the driving force for further pore nucleation. The nucleation process of the pore is experimentally found to occur in a very narrow range of the relative widths of the surface layer (compared to the sample size) and, consequently, of negative pressures. We consider this fact as an indication that pore nucleation proceeds via homogeneous nucleation. The above-given qualitative explanation of the observed phenomena is corroborated by detailed theoretical calculations of elastic stress fields and their impact on phase selection and pore nucleation. Good qualitative and partly even quantitative agreement between experiment and theory is found. An overview on other systems with similar or related properties is included as well. The findings of this research are quite general because the densities of most glasses significantly differ from those of their iso-chemical crystals. By this reason, the studied phenomena are of high technological significance for the development of different types of glass–ceramic materials and the understanding and control of sinter-crystallization processes. The latter problem is also considered in the present chapter.

8.1 Introduction

In crystallization processes of glass-forming liquids elastic stresses may arise due to the difference between the densities of the original glass and newly formed crystals. These stresses may strongly affect the kinetics of phase transitions in condensed systems [1]. In more detail, the influence of internal elastic stresses on crystal nucle-ation and growth rates in glasses is considered from both theoretical and experimental points of view, for example, in [2–7]. A detailed overview on different investigations on the effect of elastic stresses on segregation and crystallization processes in glass-forming melts is given in [8]. As shown in the cited references, elastic stresses evolving in crystallization result in the reduction of the effective thermodynamic driving force for crystallization. In case of the same composition of melt and crystal (iso-chemical crystallization), elastic stresses can be of significance only at temperatures that are lower than the so-called decoupling temperature, i.e. when the viscosity (determining stress relaxation) increases with decreasing temperature faster as compared to the rate of decrease of the effective diffusion coefficient determining crystallization and hence stress formation.

In the present chapter, by considering a model system first we focus upon two aspects of elastic stress effects. We demonstrate that elastic stresses may trigger the formation of new phases that do not commonly develop at the given temperature and normal pressure (cf. also [2]). In addition, it is shown as well that stresses can induce vacuum pore (void) nucleation, i.e., they may result, similarly to cavitation processes in liquids, in the formation of *pores* in crystallizing glass samples. These pores are formed in crystallization to compensate, at least partly, the elastic stress caused by the density difference between glass and crystal phase. This process of pore formation is typical for glass crystallization if isolated samples of a residual glass crystallize, e.g., in the case of surface crystallization during sintering of glass powders. Isolated parts of the residual melt can form also at the advanced stages of phase transformation due to percolation of crystals distributed in the glass volume. In glass production it is also known that vacuum bubbles are formed at stretching of glass-forming liquids due to the fast cooling of glass surface layers resulting in tensile stresses in the interior of the vitreous melt (see, e.g. [9]).

Crystallization of isolated glass samples may result in pore formation also via an-other mechanism similar to segregation processes in solids. The solubility of gases in the crystalline phase is commonly lower as compared to the respective liquid. For this reason, crystallization may lead to an increase of gas concentration in the resid-ual melt and, as a result, to the formation of gas-filled bubbles. However, opposite to vacuum pores, this kind of pores commonly evolves at the advanced stages of crys-tallization when the supersaturation of the dissolved gas achieves the critical value for bubble nucleation (see e.g. Fig. 8.1). However, this topic is not discussed in de-tail in the present chapter. It should be also mentioned that internal residual stresses may arise in glass-ceramics upon cooling down from crystallization temperature due to the thermal expansion and the elastic mismatch between the crystalline and glassy phases (see e.g. the review [10]).

All the mentioned aspects make the effects of elastic stresses of great importance for the development and manufacture of glass-ceramics (cf. also [12]), especially, for sinter-crystallization processes. It should be noted that due to sintering a part of the

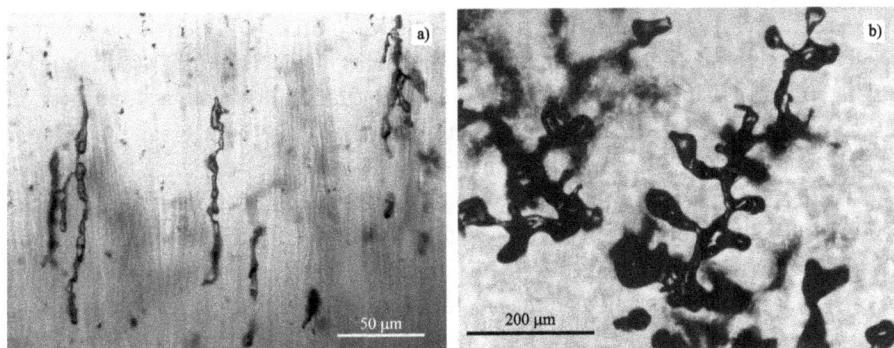

Fig. 8.1: Gas-filled pores in crystallized lithium calcium silicate (a) and sodium calcium silicate (b) [11] glasses.

glass sample surface could be removed from the crystal nucleation process. It results in an increase of average distance between growing crystals and hence in a delay of formation of closed glass structures or crystalline skeletons. Thus, the condition for formation of elastic stresses and accumulation of stress energy, finally resulting in pore nucleation, will arise at higher volume fraction of the crystalline phase than in the case of crystallization of single glass particles. However the latter case allows one to analyze in detail the effect of crystallization on pore formation. The relevance of the present study in glass technology problems will be considered in application to different systems in Section 8.3 of the present chapter for glass powder sintering-crystallization processes.

8.2 Stress Induced Pore Formation and Phase Selection in a Crystallizing Stretched Glass of Regular Shape

8.2.1 The Model

In the subsequent analysis we consider the situation when a crystalline layer is formed on the surface of a glass particle, i.e., we consider particles of glass powder or other particles with isometric form, such as spheres or cubes. The formation of such crystalline surface layer can fix, like a nutshell, the total volume of the system that then leads (in further crystallization) to uniform stretching of both crystal and residual glass that is enclosed by the crystalline layer. The crystalline layer does not allow the stresses to relax and thus elastic stress energy accumulates in such a stretched system. The elastic energy reduces the thermodynamic driving force for crystallization. In this way, the evolution of elastic stresses inhibits and even may fully terminate crystal growth.

There exists, however, another possible way of evolution by which the system may react to elastic stress fields: by the formation of pores inside the liquid regions encapsulated by the crystalline layer. In the course of development of a crystalline layer on the sample surface, the remnant liquid is uniformly stretched. As a result, similar to cavitation processes in liquids, pores may spontaneously evolve. Hence, in this case, the elastic stress energy in the glass is the origin of pore nucleation. Thus we will be dealing with nucleation of pores in a stretched liquid. According to the principle of le Chatelier-Brown this process reduces the intensity of elastic stresses (which arise due to the density difference between the crystalline and liquid parts of the system).

In the described crystallization pathway, further growth of the crystalline layer is accompanied by an increase of the pore volume. As the result, elastic stresses will then decay. Thus a distinctive feature of this nucleation process is that, in the general case, a second pore does not form since the first one almost fully eliminates the stretching of the residual liquid and the thermodynamic driving force for pore formation. Therefore, a theoretical treatment of pore nucleation can be performed in terms of determining the waiting time for the appearance of the first pore as, e.g., in the case of crystallization of metal droplets [13] or boiling of liquids [14] following, with some modifications, the basic ideas of the classical nucleation theory and/or its extensions [13–16]. A second pore can appear only if the crystalline layer forms on the surface of the first pore and thereby arrests its growth resulting in a renewed stretching of residual liquid.

It should be noted that classical nucleation theory has been successfully applied to the description of pore formation in elastic solids under load, which leads to cracks and destruction of the material [17, 18], i.e. to effects which by their physical origin are similar to the processes discussed here. However, to the best of our knowledge, the nucleation approach has not been applied so far to pore formation caused by crystallization of glass-forming melts. The aim of the present first part of the chapter is to fill this gap and to apply nucleation theory to the analysis of experimental data on pore nucleation.

The subsequent analysis presents the results obtained for glass samples with regular shape crystallizing from the surface. The assumption of a regular shape allows us to obtain reproducible data and strongly simplifies its quantitative treatment. The main qualitative conclusions of the description of phase transformation in isolated glass samples with fixed volume are expected to be valid also for glass particles of any arbitrary shape including isolated glass particles formed in sinter-crystallization processes. This direction of the analysis will be advanced in connection with different technological applications in Section 8.3 of the present chapter.

8.2.2 Experiments

8.2.2.1 Material and Methods

Diopside (CaO·MgO·2SiO$_2$) glass was chosen as a model for the analysis due to the very high difference between the densities of diopside crystal, ρ_{cr}, and diopside glass, ρ_{gl}. This density difference corresponds to the value of the density misfit parameter, δ, equal to

$$\delta = \frac{\rho_{cr} - \rho_{gl}}{\rho_{gl}} \cong 0.16. \tag{8.1}$$

This parameter determines the magnitude of the elastic stresses in the crystallization process under investigation.

The glass was melted in a platinum crucible at about 1500 °C in an electric furnace in air. Analytic grade carbonates of calcium and magnesium and anhydrous amorphous silicon dioxide were used for the synthesis. The melt was cast into a massive steel plate. To eliminate a few bubbles detected by optical microscopy in the polished glass plates, the melting procedure was repeated at 1550 °C for 5 hours. After re-melting and annealing at temperatures T close to the glass transition temperature T_g, no bubbles were observed within the resolution limit of an optical microscope (~ 1 μm). This glass was used to study pore formation.

Opposite to the glass obtained via single melting, after proper heat-treatment above T_g, the re-melted glass revealed a few spherulites in its interior, with a number density not exceeding 0.2 mm^{-3}. However, it is well-known that diopside glass crystallizes only from the surface as indeed observed for the single melted glass. This means that during re-melting some impurities entered the melt acting as active centers for crystal nucleation. Nevertheless, neither the growth rate of the crystalline surface layer nor the glass transition temperature differs from that of single-melted glass. Thus, we neglected the above mentioned impurities.

However, luckily these (unknown) impurities gave us the unique possibility to measure the growth rate of the crystalline phase in the glass interior. Since we employed glass samples with a volume not smaller than 8 mm^3, the total volume of the crystals inside the sample can be neglected compared to that of the crystalline surface layer, at least, during mean expectation time of pore formation. One of the possible reasons of the absence of detectable volume nucleation in pure diopside glass may be the strong elastic stresses which evolve in crystallization. These stresses decrease the thermodynamic driving force for crystallization and suppress homogeneous bulk nucleation in the glass (cf. e.g. [3, 4]).

The compositions of the studied glass, before and after re-melting, are shown in Table 8.1. Within the accuracy of the analysis (about 0.3–0.4 mol%) they are close to each other and to that of stoichiometric diopside. The glass samples were cut by a diamond saw as cubes with sides a equal to about 2, 3, and 4 mm, respectively. Such cubes were then dropped into a previously stabilized vertical box furnace. Opposite to the crystallization behavior of any polished surface, crystalline surface layers on the

Table 8.1: Glass compositions (in mol%) by analysis and nominal one

	SiO_2	CaO	MgO
Origin glass	50.9	24.4	24.7
Glass after remelting	50.5	25.8	23.7
$CaO \cdot MgO \cdot 2SiO_2$	50	25	25

cube surfaces are quickly formed due to the presence of many active centers for nucleation. After a given period of time the samples were quenched to room temperature. Then the top and bottom surfaces of the cubes were eliminated by grinding and polishing to study their interior and to measure the crystalline layer thickness using an optical microscope (Leica DMRX coupled with a Leica DFC490 CCD camera). Optical microscopy and X-ray analysis were employed to identify the crystalline phases. X-ray diffraction measurements were carried out on the powdered samples using a Siemens D5005 X-ray diffractometer operating at 40 mA and 40 kV. CuK_α (1.5406 Å) was employed as incident radiation. A Netzsch 404 Differential Scanning Calorimeter (DSC) was used to detect glass-crystal and crystal-crystal transitions.

8.2.2.2 Results
Crystalline Phases
As already mentioned, the dominating type of crystallization of the studied glass is surface crystallization. Two crystal morphologies were observed by optical microscopy on the cross sections of the large bulk (regular shaped) and the small (powdered) samples. The first type of crystals are square faceted (Fig. 8.2a and b, (1))

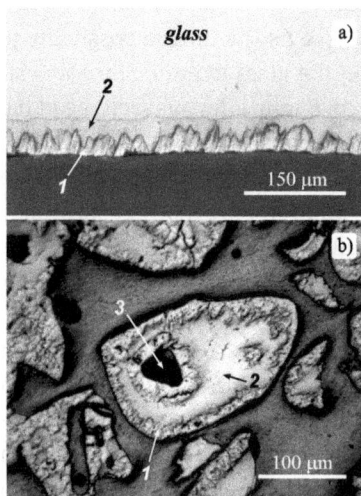

Fig. 8.2: Reflected light optical micrographs of crystallizing bulk (a) and powder (b) diopside glass samples heat-treated at 870 °C for 90 min and at 850 °C for 50 min, respectively. Arrows (1) and (2) refer to diopside and wollastonite-like crystals, arrow (3) shows the location of a pore.

whereas the second formed a relatively dense layer (Fig. 8.2a and b, (2)) with the crystal/glass interface parallel to the external boundaries of the sample. The latter feature is clearly seen in the samples with planar surfaces (Fig. 8.2a).

The growth of the first type of crystals was terminated by the formation of the second type ones. The sequence of occurrence of the different crystalline phases is illustrated in Fig. 8.3. This fact allowed us to vary the ratio between these kinds of crystals by changing the shape and size of the glass samples prior to crystallization. It is reasonable to expect that crystallization of a bulk glass sample produces more of the second type of crystals than crystallization of a glass powder.

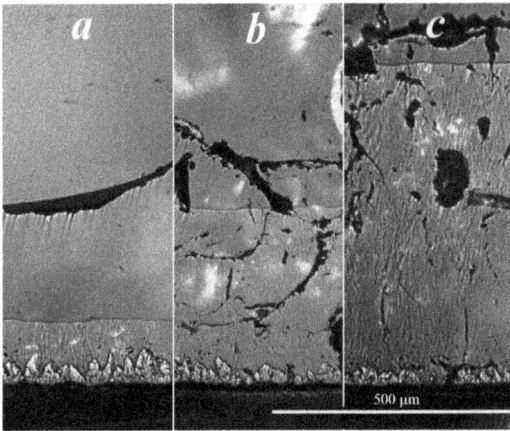

Fig. 8.3: Reflected light optical micrographs of cross sections of bulk diopside glass samples heat-treated at 870 °C for 7 (a), 14 (b), and 24 (c) hours.

Fig. 8.4 shows X-ray diffraction spectra of a glass powder with an average size of about 60 µm and (3 · 1.5 · 7mm) bulk glass crystallized at 850 °C for 24 and 120 hours, respectively. This bulk sample revealed only surface crystallization. It should be noted that 120 hours of heat treatment at 850 °C was not sufficient for full crystallization of the monolithic sample (please, see the weak halo of the amorphous phase in Fig. 8.4b). The spectrum of the crystallized glass powder matches that of diopside [75-1072], while the spectrum of the crystallized bulk glass is similar to that of wollastonite [72-2284] with the distinction that the peaks shift to higher angles. Thus, we can suppose that the second phase is a solid solution with wollastonite structure, where half of Ca is substituted by Mg. This phase can be treated as a phase having a structure similar to that of the low or high temperature forms of wollastonite with the following cell parameters: low temperature triclinic

$a/b/c$ 7.605/7.049/6.822 Å, $\alpha/\beta/\gamma$ 90.39/95.08/103.16(°), $Z = 3$,

and high temperature monoclinic

$a/b/c$ 14.80/7.048/6.8221 Å, $\alpha/\beta/\gamma$ 90/95.08/90(°), $Z = 6$.

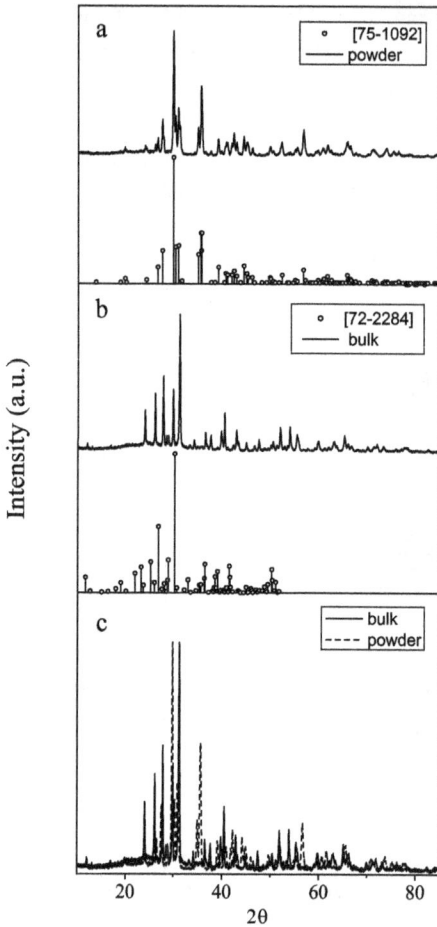

Fig. 8.4: *X*-ray diffraction spectra of diopside powder (a) and bulk (b) glass crystallized at 850 °C for 24 h and 120 h, respectively. Both spectra are presented in Fig. 8.3c.

The *X*-ray densities of these supposed structures were calculated as

$$\rho = 1.6602 \ ZM/V, \tag{8.2}$$

where *M* is the molar weight of diopside and *V* is the volume of a cell given by

$$V = abc \left[\sqrt{1 - \cos(\alpha)^2 - \cos(\beta)^2 - \cos(\gamma)^2 + 2\cos(\alpha)\cos(\beta)\cos(\gamma)} \right]. \tag{8.3}$$

The respective values of ρ are:

$$\rho_{tric} = 3.042 \ \text{g/cm}^3 \ , \qquad \rho_{mono} = 3.058 \text{g/cm}^3.$$

As we mentioned before, some samples of the re-melted glass revealed a few spherulites in the interior after proper heat treatment. According to an *X*-ray analysis, these spherulites are also a wollastonite-like phase as the second phase in the

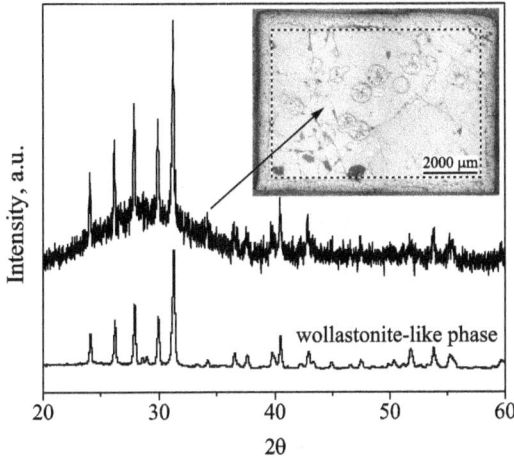

Fig. 8.5: *X*-ray diffraction spectra of the internal part (see inset of sample cross section) of monolithic samples (4 · 7 · 9 mm) of remelted diopside glass crystallized at 870 °C for 8 h (upper spectrum) and wollastonite-like phase (lower spectrum).

crystalline surface layer. The *X*-ray diffraction spectrum of the internal part of the diopside glass sample (see inset of Fig. 8.5a), crystallized at 870 °C for 8h and including a few spherulites, is shown in Fig. 8.5 together with the diffraction spectrum of the wollastonite-like phase taken from Fig. 8.4b.

The sequence of crystal phase formation and the effect of sample shape on phase composition were corroborated by DSC analysis. Fig. 8.6 shows the heating and cooling DSC runs of bulk and powder glasses (Fig. 8.6a) and samples preliminary crystallized using bulk and powder glass (Fig. 8.6b). The DSC-curve of the powdered glass has only one exothermic peak, which, according to *X*-ray data, refers to diopside crystallization. We recall that diopside crystals are the first phase forming on the glass surface, and in the case of small powder particles it is the main or unique crystalline phase. Opposite to powdered glass, the DSC curve of the bulk glass re-

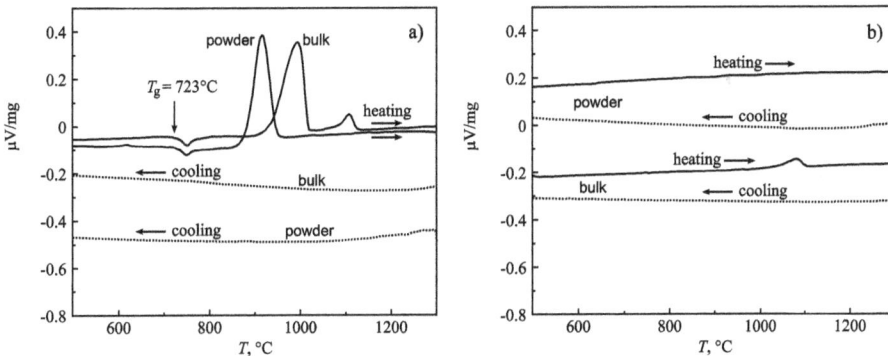

Fig. 8.6: DSC heating and cooling curves: The experiments were performed at $|C| = 10 \text{K/min}$ for (a) bulk ($\cong$ 4 mm) and powder ($\cong$ 200 µm) glass and (b) crystallized powder (< 60 µm) and bulk (7 · 5 · 2 mm) glass at 850 °C for 24 h and 219 h, respectively.

veals two exothermic peaks. As was shown in Figs. 8.3 and 8.4, the main crystalline phase in the crystallized bulk sample is represented by the second type of crystals (wollastonite-like phase). Hence we could interpret the first peak as crystallization of the wollastonite-like phase, and the second peak, close to 1100 °C, as its transformation into diopside. Since diopside is the stable phase in a non-stretched system, the cooling curves of both powder and bulk glasses do not reveal any thermal effects.

The above discussed impact of sample shape and size on the type of the crystal phases which evolve is also confirmed by DSC-measurements of preliminary crystallized powder glass and bulk glass (Fig. 8.6b). The curve of the crystallized powder glass does not reveal any thermal effects since the latter consists of diopside crystals, while the curve of a monolithic piece, previously crystallized at a relatively low temperature, shows an exothermic peak at about 1100 °C that is caused by the transition of a wollastonite-like phase into diopside. But a DSC-scan of a sample, preliminary crystallized at 1200 °C, does not reveal any peaks independent of its shape.

Fig. 8.7 shows the thickness of the crystalline surface layer, h, versus time at $T = 870\,°C$; vacuum pore (void) formation was studied at this temperature. As we already mentioned, with a proper heat-treatment, the re-melted diopside glass reveals sometimes a few internal wollastonite-like crystals with spherulitic form, which have no effect on the results of the present analysis and, therefore, were not studied in detail. However, we used the sample of re-melted glass which revealed some spherulites in the interior to measure the size evolution of a given spherulite with heat-treatment time. Fig. 8.8 presents the radii of nine spherulites together with h as a function of time of heat-treatment at 870 °C. One can see that the rate of crystal growth of the crystalline layer is practically equal to that of the spherulites. This result is explained by X-ray analysis data; both are wollastonite-like crystals. It should be also noted that spherulite growth generally starts somewhat later than that of the crystalline layer.

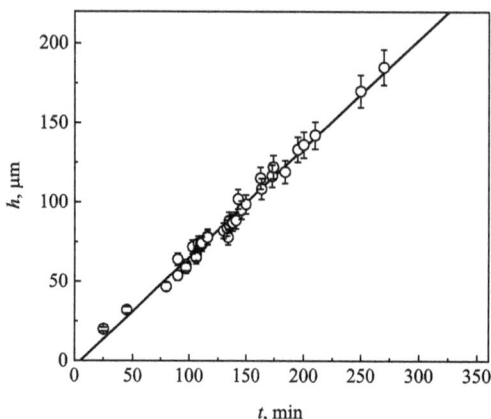

Fig. 8.7: Thickness of the crystalline layer as a function of heat-treatment time at 870 °C.

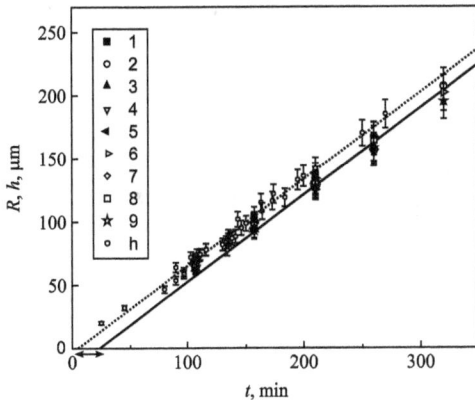

Fig. 8.8: The radii of spherulites (*1-9*) versus time of heat-treatment at 870 °C. The solid line is a linear fit of the respective data, while the dotted line is a linear fit of the $h(t)$-data, shown also in Fig. 8.7.

Pores

Cubic samples with sides of size a of about 2, 3, and 4 mm were cut from the re-melted diopside glass and heat-treated at 870 °C for different times. Then the crystalline layers were removed from two parallel sides by grinding and polishing to measure the layer thickness, h, and to check whether pores had been formed in their volume. The following dimensionless parameter X was used to characterize the condition of pore nucleation. This parameter is defined as

$$X = \frac{a - 2h}{a} = 1 - \frac{2h}{a}. \tag{8.4}$$

The values of X vary from 1 (absence of a crystalline layer) to zero (fully crystallized sample). The latter case was never realized in our experiments due to the formation of pores. As was shown in [19, 20], the value of the parameter X determines the degree of elastic stresses in a finite system of the considered geometry.

The growth of the crystalline layer results in stretching the residual liquid inside the cube and finally in the formation of pores via nucleation and fast growth up to a volume that compensates the density difference between the amorphous and crystalline phases. Thus it is reasonable to suppose that pore nucleation is the rate-limiting process. The peculiarity of this nucleation process is that, as a rule, a second pore does not appear since fast growth of the first one eliminates the elastic stresses. Thus, the nucleation experiment is reduced to the detection of the first pore.

To show the statistical results of pore detection we plotted the reduced thickness of crystalline layer ($2h/a$) versus heat-treatment time at $T = 870$ °C (cf. Fig. 8.9). Each point corresponds to a single heat-treatment for the given time. Filled and empty triangles refer to samples without and with a pore, respectively. These data refer to cubes of different sizes denoted close to proper lines. According to Fig. 8.9, nucleation of a pore occurs in a very narrow interval of ($2h/a$)-values. The average value of ($2h/a$), at which a pore nucleates, is marked by ($2h_*/a$). We interpret the existence of such

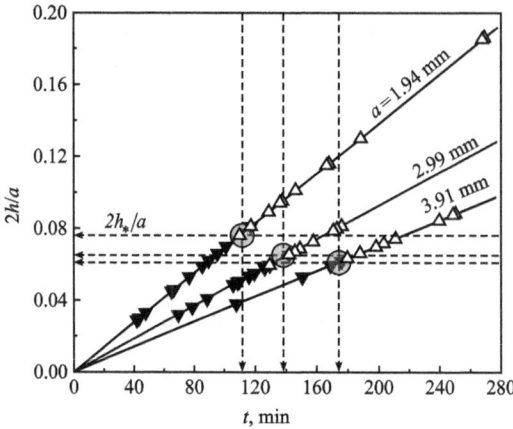

Fig. 8.9: Reduced thickness of crystalline layer versus heat-treatment time at $T = 870\,°C$ for cubic samples of different sizes denoted close to respective lines. Each point corresponds to single heat-treatment for given time. Fill and empty triangles refer to samples without and with a pore, respectively.

a narrow interval of $(2h_*/a)$-values for the occurrence of pores as a strong indication that pores nucleate homogeneously [22].

In such an interpretation we connect the possibility of pore observation with its nucleation and growth. Alternatively, one could suppose that at low values of h, i.e. at $0 < (2h/a) < (2h_*/a)$, or $(X_* < X < 1)$, the pore has a size that is too small for detection by optical microscopy. To eliminate this doubt we calculated the diameter of a pore, D, which has to compensate the density mismatch for different X. The respective value can be easily obtained taking into account that the volume of the pore is equal to the volume of the crystalline phase multiplied be the misfit parameter, δ. We then get the following dependence for the pore diameter D as a function of X:

$$D = a \left\{ \frac{6}{\pi} \delta \left[1 - X^3 \right] \right\}^{1/3}. \tag{8.5}$$

According to Fig. 8.10, the expected diameter of a pore at values of X of the order of 0.93–0.97 should be larger than the resolution limit of optical microscopy. This result indeed implies that pores do not form until X ($2h/a$) achieves some critical value, $X_*(2h_*/a)$. Fig. 8.11 shows that the value of X_* grows with an increase of the cube volume. Some (hypothetically diopside) crystals are occasionally formed on the pore surface (Fig. 8.12).

8.2.3 Theoretical Interpretation: Classical Nucleation Theory

The following main experimental results will be discussed now theoretically: (i) the switch from one to another crystalline phase during phase transition and the possi-

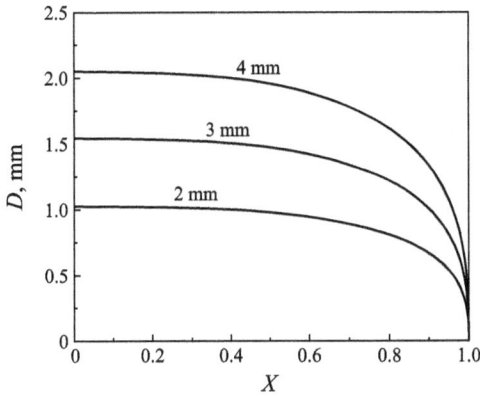

Fig. 8.10: Diameter of a pore calculated by Eq. (8.3) and using the density of the wollastonite-like phase (see Table 8.2) in dependence on the parameter X for cubes with sides 2 mm, 3 mm, and 4 mm.

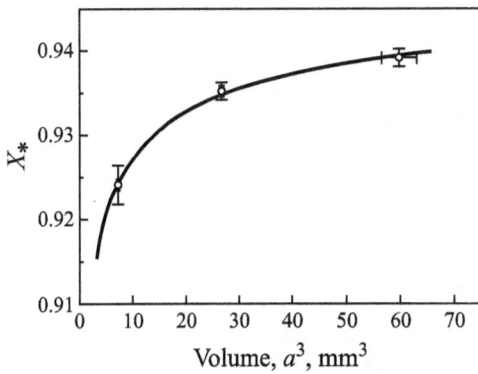

Fig. 8.11: Critical value of X for $T = 870\,°C$ versus volume of the sample. Circles show experimental data. Curve is calculated for sphere with $R_3 = (a/2)$.

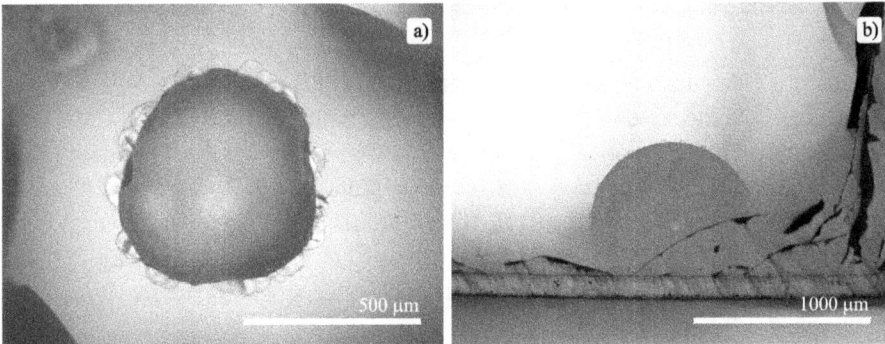

Fig. 8.12: Transmitted light optical micrographs of diopside glass, heat-treated at 870 °C for 200 min (a) and at 890 °C for 64 min (b).

bility to affect the type of phase developing by changing the size of the glass samples, and (ii) the formation of pores at a relatively early stage of crystallization within a narrow interval of sizes of the crystal mantle, X $(2h/a)$. Both groups of results will be interpreted in terms of the influence of elastic stresses.

8.2.3.1 Sequence of Appearance of Crystalline Phases

As discussed in the outline of the experimental results, diopside first forms and is then replaced by a wollastonite-like phase. In order to give an interpretation of the origin of this sequence of formation of different crystalline phases, the following fact is very important: the density of the wollastonite-like phase is considerably lower than that of diopside crystals. Thus, the density misfit and the resulting elastic stress effect in the crystallization process for the wollastonite-like phase are lower than that for diopside.

Table 8.2 shows the respective values of ρ, δ, and δ^2. Since the total energy of elastic deformation is proportional to δ^2, the effects of elastic stresses on nucleation and growth of the wollastonite-like crystals must be less than those for diopside by a factor of about five. The elastic stress energy Δf increases with the growth of the crystalline layer and hence with a decrease of the parameter X in the case of diopside faster than in the case of the wollastonite-like phase. Thus the effective thermodynamic driving force for crystallization of diopside has to decrease with decreasing X faster than that for the wollastonite-like phase. This difference determines the switching at $X = X_{d/w}$ of the type of crystalline phases growing in the system.

Table 8.2: Densities (in g/cm^3) and density misfits for different crystalline phases

Phase	Density	δ	δ^2
Diopside crystal	3.29	0.1584	0.025
Wollastonite-like crystal, triclinic	3.04	0.0704	0.005
Wollastonite-like crystal, monoclinic	3.06	0.0775	0.006
Diopside glass	2.84		

To illustrate this effect, we employ the analytical results obtained in [19, 20] for the evolution of elastic stresses, Δf, of the residual amorphous phase in surface crystallization of a finite spherical domain with constant values of the external radius. Fig. 8.13 shows the dependence of Δf on the parameter $r = R_1/R_3$, where R_3 is the radius of the spherical sample and R_1 is the radius of its internal amorphous core. The parameter r is similar to the parameter X for a cubic shape given by Eq. (8.4). At $r > X_{d/w}$ and at $r < X_{d/w}$ the curve $\Delta f(r)$ refers to diopside and wollastonite-like phases, respectively.

The model employed in [19, 20] treats the layer of diopside crystals as a smooth one. This approach is reasonable for the second layer consisting of the wollastonite-like phase or more exactly for its internal surface. But the interface between the diopside crystal and the melt or the wollastonite-like phase is rough (see e.g. Fig. 8.2a). Such property of the interface is caused by the relatively large distance between crystals nucleated on surface defects. At a time corresponding to the coalescence of diopside crystals (formation of the crystalline layer) a fast increase of the density f_{cr} of elastic deformation energy can be expected to occur due to restrictions on elastic

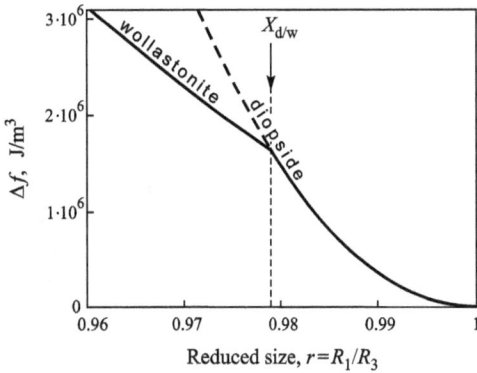

Fig. 8.13: Elastic stress energy of residual amorphous phase as a function of the parameter r, calculated for the case of a sample of spherical shape. R_1 is the radius of the sphere enclosing the melt and $R_3 = 1$mm is the external radius of the sample. $X_{d/w}$ corresponds to switching of phases.

stress relaxation. The change of f_{cr} can be sufficiently large to trigger the growth of the wollastonite-like phase, producing smaller values of f_{cr}. Thus we can suppose that the moment of coalescence corresponds to $X \cong X_{d/w}$. It could be also expected that, since the growth of single crystals nucleated on glass surface is three-dimensional, the average thickness of the diopside layer at the moment of its formation depends mainly on the average crystal size determined by the typical distance between the diopside crystals. By varying the size of the glass sample (from a fine powder to monolithic pieces) one can change the ratio between the amount of diopside crystals and the wollastonite-like phase in the crystallized sample, as we showed here earlier. The growth of the wollastonite-like phase also results in melt stretching, but to a less extent as compared to diopside. Nevertheless, such combined crystal growth finally triggers the formation of a pore, which is followed by fast growth up to a volume that, at each given moment of time, partially compensates the elastic stresses induced by the density misfit between the crystalline and glassy phases.

It should be emphasized that sometimes new crystals (hypothetically diopside) form on the pore surfaces (see the crystals located around pore (3) on Fig. 8.2b and Fig. 8.12). Since the formation of a pore eliminates elastic stresses in the melt, we can suppose that these crystals represent diopside. Moreover, their shape is similar to that of diopside crystals on the polished surface of diopside glass giving additional support to this suggestion.

8.2.3.2 Nucleation of Pores

To theoretically treat the results presented in Fig. 8.9, first we employ here classical nucleation theory (CNT). The following equation can be used for the determination of the nucleation rate, I, of pores:

$$I = I_0 \exp\left(-\frac{W_* + \Delta G_D}{kT}\right),$$ (8.6)

$$I_0 \approx N_1 \frac{kT}{h}.$$ (8.7)

Here N_1 is the number density of "structural units" in the ambient phase, h is Planck's constant, and T is the absolute temperature. ΔG_D is the activation free energy of attachment of "structural units" to the new phase, and W_* is the thermodynamic barrier for nucleation or the work of critical cluster formation.

Employing CNT, the work of critical cluster formation (in application to pore nucleation) can be written as [14, 15, 16]

$$W_* = \frac{16\pi}{3} \frac{\sigma^3}{p^2},$$ (8.8)

where p is the negative pressure in the stretched melt, and σ is the specific surface free energy of pore surface, i.e. of the melt-vacuum interface. Here it is additionally assumed that the vapor pressure in the cavitation bubble p' is small as compared to the negative external pressure p in the liquid ($|p'| \ll |p|$), and thus we neglect it. If one replaces, as usually done, ΔG_D by the activation energy of viscous flow, Eq. (8.6) can be rewritten as [26]

$$I = I_0 \frac{h}{4l^3} \frac{1}{\eta} \exp\left(-\frac{W_*}{kT}\right),$$ (8.9)

where η is the viscosity of the melt and l is a size parameter of the order of $2 \cdot 10^{-10}$m.

To apply the above equations to nucleation of pores one must specify the structural units, i.e. we have to specify at the expense of what units in the ambient liquid phase pore formation and growth occurs. In the case of crystalline materials, pores are considered as "negative" crystal units that grow due to the attachment of vacancies [27]. These vacancies serve as "void atoms" and their number density increases under load. As opposite to a crystalline phase, the definition of vacancies in glasses, which have a structure similar to that of the liquid, is different. In this case, the holes in lattice-hole models of simple liquids are treated as vacancies. Hereby the molar fraction of holes is taken to be equal to the relative free volume of the liquid [8]. We will use Eq. (8.9) assuming, in a first approximation, that the number of holes does not strongly differ from that of the structural units in the liquid, and its diffusivity can be estimated via viscosity. Some arbitrariness in the definition of the pre-exponential term, I_0, connected with these assumptions, will not strongly affect the nucleation rate.

The following equation can be written then for the number of pores nucleated in the stretched melt in a period of time t

$$N(t) = \int_0^t I(t')V(t')dt',$$ (8.10)

where I and V are the pore nucleation rate and the volume of the stretched melt, respectively. Since the negative pressure p determines, to a large extent, the thermodynamic barrier for nucleation (see Eq. (8.8)), its increase with increasing thickness h of the crystalline layer leads to a strong increase of the nucleation rate with heat-treatment time, t, while V weakly decreases.

Generally, only *one* pore appears in a stretched melt since its fast growth eliminates the negative pressure and terminates further nucleation. This nucleation event occurs at some moment of time t_1, determined via

$$N(t_1) = 1. \tag{8.11}$$

Here t_1 corresponds to the critical value of X, detected in experiment,

$$X_* = \frac{a - 2Ut_1}{a}, \tag{8.12}$$

where U is the growth rate of the crystalline layer.

To estimate I and then perform calculations by Eqs. (8.10)–(8.11) one has to know the value of the negative pressure, p. In [19, 20] this problem was solved for a model that treats a spherical layer of diopside as a smooth one that forms early in the crystallization process. However, as was noted in Section 8.2.3.1, in fact, the diopside crystals practically do not participate in melt stretching. At the moment of formation of the diopside layer, when it could stretch the melt, a switch to the wollastonite-like phase occurs. Therefore, here we present results of slightly modified calculations of negative pressure for the case that the melt is stretched only by the layer of the wollastonite-like phase. Hereby the thickness of the diopside crystal layer was considered as independent of the size of the sample. Fig. 8.14 shows the results of these calculations for different radii of the sphere estimated as $R_3 = a/2$ versus $r = R_1/R_3$. After approaching some critical value of pressure, p_*, that corresponds to X_* i.e. formation of the pore, negative pressure drops rapidly.

In addition to the computed values of $p(r)$, experimental data on the specific surface energy, σ, and viscosity, η, were taken from [25, 28], respectively. These data al-

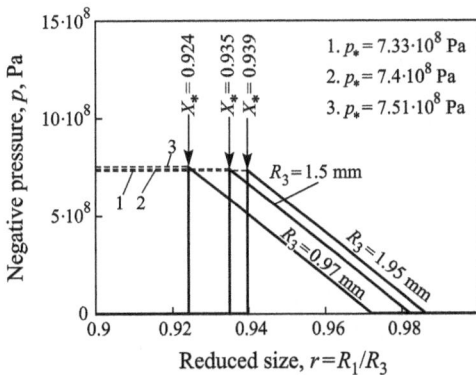

Fig. 8.14: Negative pressure versus reduced size for different sizes of spherical samples, R_3.

lowed us to estimate the thermodynamic barrier for pore nucleation, W_*, the nucleation rate I and then the function $N(r)$ by employing CNT in the described way. The condition of pore formation is given then by the relation $N(X_*) = 1$.

8.2.3.3 Discussion

The comparison between theory and experiment shows that pore nucleation experimentally occurs at lower values of pressure or a smaller width of the crystalline layer than estimated theoretically. Consequently, CNT overestimates the work of critical bubble formation. In order to arrive at a satisfactory agreement of experimental values of X_* with theoretical predictions, we have to reduce the work of critical bubble formation W_* by a factor 0.544. The results obtained in such a way are shown in Figs. 8.11 and 8.15. So, the remaining question to be answered is how such reduction of the theoretically estimated thermodynamic barrier for nucleation of the pore can be explained.

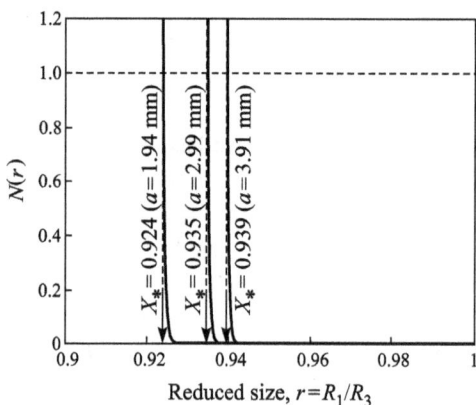

Fig. 8.15: Number of pores calculated by Eq. (8.10) for spherical samples with size $R_3 = (a/2)$ heat treated at $T = 870\,°C$ versus parameter r.

Employing classical nucleation theory, such deviations may be explained assuming that the specific surface energy is size-dependent, i.e., a dependence $\sigma = \sigma(R)$ is assumed. Recall that in the computations we employ the value of σ measured for a planar interface, i.e., $\sigma = \sigma_\infty$ was utilized. Since the specific surface energy in Eq. (8.8) refers to pores of critical radius

$$R_* = \frac{2\sigma}{p} \, , \tag{8.13}$$

the size dependence of surface energy must be taken into account. Proceeding in such a way, the work of critical cluster formation is changed according to

$$W_*(\sigma(R)) = W_*(\sigma_\infty) \left[\frac{\sigma(R)}{\sigma_\infty} \right]^3 . \tag{8.14}$$

It follows that in order to reconcile theory and experiment

$$\sigma(R) = \sigma_\infty (0.544)^{1/3} = 0.816\sigma_\infty \qquad (8.15)$$

must be assumed. The reduction of the thermodynamic barrier by a factor 0.544 can be realized via the reduction of the specific surface energy from $\sigma_\infty = 0.377 J/m^2$ [28] to $\sigma(R_*) = 0.308 J/m^2$.

To a first approximation, let us employ – with the necessary precaution (cf. e.g. [29] and the discussion below) – Tolman's equation

$$\sigma(R) = \sigma_\infty \left(1 + \frac{2\delta}{R}\right)^{-1} \qquad (8.16)$$

for the description of the size dependence of the surface tension. Here the Tolman parameter, δ, characterizes the width of the interfacial region between the coexisting phases (it has to be of the order of atomic dimensions). The required reduction of the thermodynamic barrier corresponds at a critical radius $R_* \approx 10^{-9}$m to the reasonable value of Tolman's parameter, equal to $\delta = 0.112 R_* \approx 10^{-10}$m.

However, as was shown in [30–33], in contrast to Tolman's equation, the curvature dependence of surface tension in processes of condensation and boiling or segregation in solutions is determined by term of the order $(1/R)^2$ in the expansion of σ with respect to cluster size, i.e.

$$\sigma(R) = \sigma_\infty \left(1 + \frac{B}{R^2}\right). \qquad (8.17)$$

Assuming $B < 0$ leads to a decrease of the work of critical cluster formation by a constant value as compared with the results obtained via CNT and assuming the capillarity approximation, i.e., size-independence of the surface tension. A widely similar result has been recently obtained in the analysis of crystal nucleation of certain classes of glass-forming melts [34]. Thus, in the framework of CNT, we can connect the reduction of the thermodynamic barrier for pore nucleation with the reduction of the specific surface energy due to its size dependence. However, the question about the foundation of the underlying assumptions like $\delta > 0$ or $B < 0$ remains to be answered.

It should be also recalled that our experiments were performed on cubes and the theoretical analysis was performed here for spheres. The difference in the shapes of the samples, studied theoretically and experimentally, may be another origin for the quantitative deviation of theoretical and experimental results.

An alternative general approach to reconcile experiment and theory can be given by employing the generalized Gibbs' approach for the determination of the work of critical cluster formation [7]. In this approach it is shown that the bulk properties of the critical clusters considerably deviate from the properties of the newly evolving macroscopic phases. As a consequence, the specific surface energy is also necessarily size-dependent. Moreover, within this approach it can be demonstrated that – under very weak assumptions concerning the systems under consideration – CNT (employing the capillarity approximation) overestimates the work of critical cluster formation.

Consequently, since we employed CNT here, the theoretical values of the work of critical cluster formation are (by necessity) too high. A detailed analysis of the interpretation of pore formation in terms of the generalized Gibbs' approach will be presented in the next section.

8.2.4 Theoretical Interpretation: Generalized Gibbs Approach

8.2.4.1 Thermodynamic Aspects: Equation of State of the Stretched Diopside Fluid
In order to apply the generalized Gibbs approach to the description of nucleation [15, 16, 36], the thermal equation of state of the system where the process of nucleation takes place has to be known. Lacking suitable data for the particular systems we are analyzing here, we employ (at least, as a qualitatively appropriate expression) the reduced form of the van der Waals equation of state [37–39] for the specification of the thermal equation of state of the diopside melt. In such an approach, we get

$$\Pi(\omega) = \frac{8\theta}{3\left(\omega - \frac{1}{3}\right)} - \frac{3}{\omega^2}, \tag{8.18}$$

$$\Pi = \frac{p}{p_c}, \qquad \omega = \frac{v}{v_c}, \qquad \theta = \frac{T}{T_c}. \tag{8.19}$$

Here p is the pressure, v is the specific volume, T is the temperature, while p_c, v_c, and T_c refer to the critical point. These critical parameters can be determined via density, ρ, Young's modulus, E, and bulk thermal expansion coefficient, β,

$$\rho(\theta) = \frac{\rho_c}{\omega(\theta)}, \qquad E(\theta) = -\omega p_c \left(\frac{d\omega}{d\Pi}\right)^{-1}, \qquad \beta(\theta) = \frac{1}{\omega T_c}\frac{d\omega}{d\theta}. \tag{8.20}$$

For diopside glass [28]

$$\rho_0 = 2.84 \text{ kg/m}^3, \qquad E_0 = 10^{11}\text{J/m}^3 \text{ (at } T = 20°C),$$

$$\beta = 11.73 \cdot 10^{-5}\text{K}^{-1} \text{ (at } T = 870°C),$$

the solution of the system of equations yields

$$p_c = 102 \text{ MPa}, \quad \rho_c = 971 \text{ kg/m}^3, \quad v_c = 1.03 \cdot 10^{-3} \text{ m}^3, \quad T_c = 3590 \text{ K}. \tag{8.21}$$

The chemical potential of the molecules in a van der Waals fluid can be written generally as [35]

$$\frac{\mu}{p_c v_c} = -\frac{8\theta}{3}\ln(3\omega - 1) + \frac{8\theta\omega}{3\omega - 1} - \frac{6}{\omega} + \chi(\theta). \tag{8.22}$$

Here $\chi(\theta)$ is some well-defined function of temperature.

Here we consider pore formation processes proceeding via nucleation and growth. Such processes occur for homogeneous initial states of the system (diopside melt) located in the region between binodal (the boundary between stable and metastable regions) and spinodal (the boundary between metastable and unstable regions) curves.

All further calculations will be performed here for a heat-treatment temperature, $T = 870°C$. It corresponds to a reduced temperature, $\theta = 0.318$. In this particular case, the position of the binodal curves is given by $w_b^{(left)} = 0.373$ and $w_b^{(right)} = 1663$, the respective parts of the spinodal curves are located at $w_{sp}^{(left)} = 0.445$ and $w_{sp}^{(right)} = 6.343$, correspondingly. Thus, we consider initial states located between the left hand side branches of the spinodal and binodal curves, respectively, i.e., initial states in the range of reduced volumes $w_b^{(left)} < w < w_{sp}^{(left)}$. It corresponds to the interval of tensile stresses $p_{sp} < p < p_b$ ($p_{sp} \approx -769$ MPa, $p_b \approx -0.06$ MPa). Isotherms for the diopside melt, according to Eq.(8.22), for different values of the reduced temperature $\theta = 0.318, 0.6, 0.84, 1$ are shown in Fig. 8.16, dashed and dashed-dotted curves refer to binodal and spinodal curves, correspondingly.

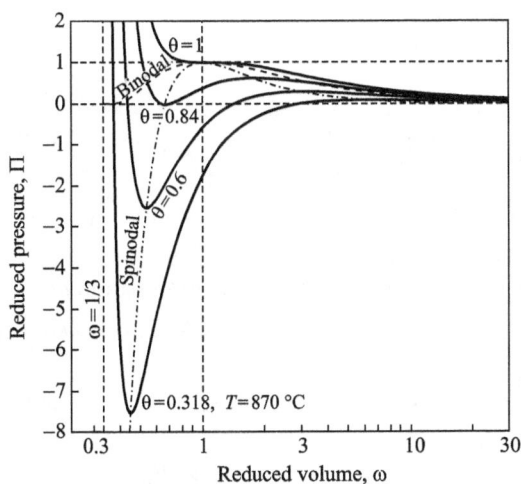

Fig. 8.16: van der Waals's isotherms adopted for the description of the diopside melt for different values of the reduced temperature, $\theta = 0.318, 0.6, 0.84$, and 1. The first value corresponds to the temperature of the experiment [40], the latter three curves are given for illustration of the general behavior.

One can see that there are two classes of isotherms: for the first group of isotherms ($\theta \geq \theta_s$), the inequality $p \geq 0$ holds, and for the second class, ($\theta < \theta_s$), the pressure may be both positive and negative. Only in this temperature range the melt can exist in a stretched state. This parameter θ_s, separating the behavior, is determined via the equation

$$\Pi_l(w_{sp}) = 0, \tag{8.23}$$

which yields $\theta_s \approx 0.844$ and $T_s \equiv T_c\theta_s \approx 3029$ K. Let us note that the parameters T_c and T_s are rather formal quantities here. The diopside melt can decompose partially at high temperatures. For this reason, the temperature values may not be reached in reality. They are employed only for the specification of the equation of state, which is used then exclusively in the physically realistic range of temperatures and pressures.

8.2.4.2 Work of Critical Pore Formation and Nucleation Rate

In order to determine the work of critical pore formation, governing the nucleation process, both the thermodynamic driving force for nucleation and the value of the surface tension have to be known. For the surface tension we choose here an equation of the form [16, 36]

$$\sigma(w_g, w_m, \theta) = \Theta(\theta) \left(\frac{1}{w_m} - \frac{1}{w_g} \right)^{\delta}, \qquad \delta = 2. \tag{8.24}$$

Here w_g is the specific volume of the gas in the pore, w_m is the specific volume of the melt, the function $\Theta(\theta)$ is determined via

$$\Theta(\theta) = A \left(\frac{1}{w_b^{(\text{left})}} - \frac{1}{w_b^{(\text{right})}} \right)^{4-\delta}. \tag{8.25}$$

The parameter A is defined via $A = 0.0333 \text{J/m}^2$, which corresponds to the experimental value of the specific surface energy, $\sigma = 0.377 \text{J/m}^2$ [28].

Similarly to [16, 36], the basic equations employed for the determination of the work, W_*, of formation of a pore of radius R in the generalized Gibbs approach read

$$\frac{\Delta g(r, w_g, w_m, \theta)}{k_B T} = 3 \left(\frac{1}{w_m} - \frac{1}{w_g} \right)^{\delta} r^2 + 2f(w_g, w_m, \theta) r^3, \tag{8.26}$$

$$f(w_g, w_m, \theta) = \Pi(w_g, \theta) - \Pi(w_m, \theta) + \frac{1}{w_g} \frac{\mu(w_m, \theta) - \mu(w_g, \theta)}{p_c v_c}, \tag{8.27}$$

where the driving force of pore formation, Δg, and the critical pore size, r, are given in dimensionless form. In detail, the following notations have been used

$$\Delta g \equiv \frac{\Delta W_*}{\Omega_1}, \qquad \Omega_1 = \frac{16\pi}{3} \frac{1}{p_c^2 k_B \theta T_c} \Theta(\theta)^3, \tag{8.28}$$

$$r \equiv \frac{R}{R_\sigma}, \qquad R_\sigma = \frac{2}{p_c} \Theta(\theta). \tag{8.29}$$

The dependencies of the scaling factors Ω_1 and R_σ on temperature are shown in Fig. 8.17. For the chosen temperature, $T = 870 \,^{\circ}C$, their values are equal to $\Omega_1 = 14.665$ and $R_\sigma = 1.027$ nm. These values are specified in the figure by dashed curves. These parameters tend to zero at $T = T_c$, but this limiting case is far beyond the physically interesting temperature interval between the glass transition, T_g, and melting temperature, T_m. So, the analysis can be performed similarly with analogous results for other temperatures within the range $T_g < T < T_m$.

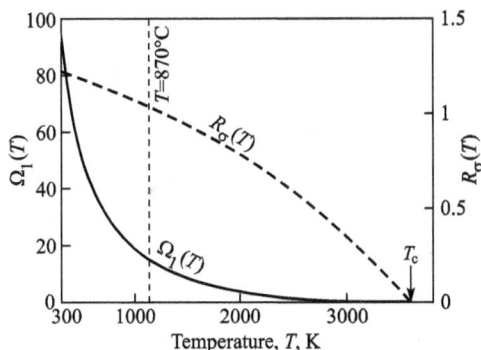

Fig. 8.17: Dependencies of the scaling factors, Ω_1 (left) and R_σ (right), on temperature (see text).

The Gibbs free energy surface for the metastable initial state has a typical saddle shape (see Fig. 8.18). This saddle point position – supplying us with the work of formation and the size of the critical pore – is determined by the set of equations

$$\frac{\partial \Delta g(r, \omega_g, \omega_m, \theta)}{\partial r} = 0, \qquad \frac{\partial \Delta g(r, \omega_g, \omega_m, \theta)}{\partial \omega_g} = 0. \qquad (8.30)$$

The dependencies of work of formation, size and gas density in the critical pore on negative pressure are shown in Figs. 8.19a–c. The full curves correspond to the generalized Gibbs' approach, while the dashed curves refer to computations performed in the framework of CNT [14, 16, 35].

Once the work of critical pore formation is known, we can employ Eq. (8.9) for the determination of the rate of nucleation, J, of pores. As in Section 8.2.3.2, we used $l \approx 2 \cdot 10^{-10}$ m (the size parameter of the diffusing building molecules, which is equivalent to the jump distance or the lattice parameter – parameters commonly employed in

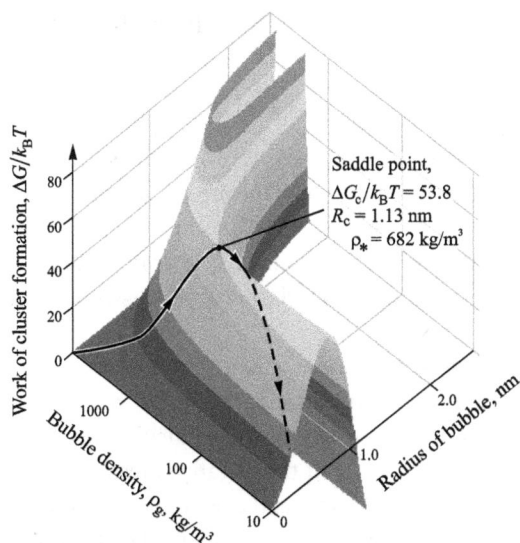

Fig. 8.18: Dependence of work of pore formation, $\Delta G/k_B T$, on the radius of the pore and the gas density in the pore.

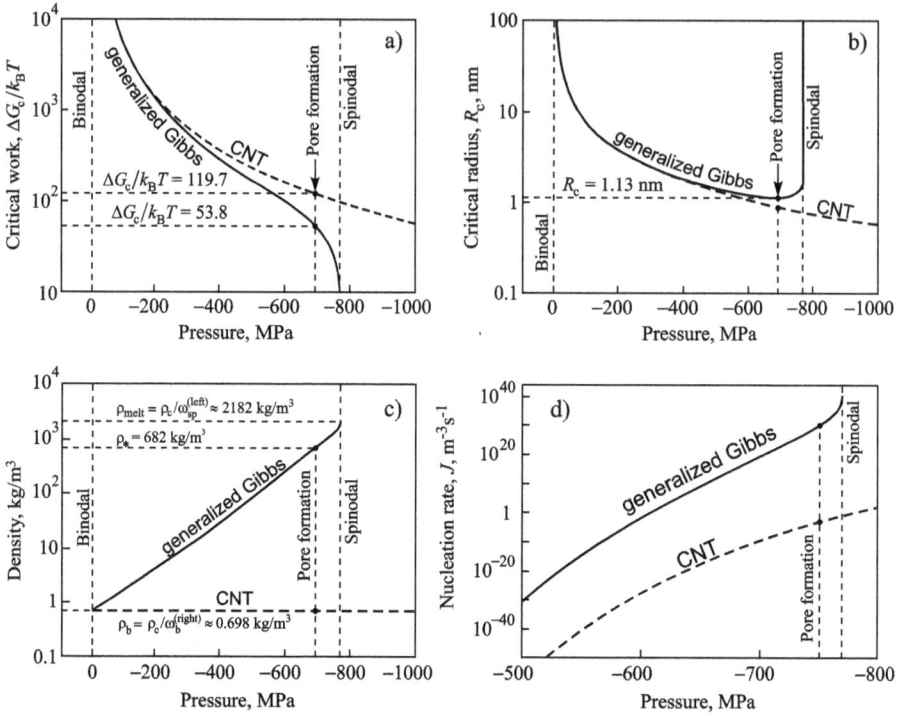

Fig. 8.19: (a) Work of the critical pore formation, $\Delta G_c/k_B T$, as a function of negative pressure; (b) radius of the critical pore, R_c, as a function of negative pressure; (c) gas density in the critical pore as a function of negative pressure; (d) nucleation rate, J, as a function of negative pressure (full curves – generalized Gibbs' approach, dashed curves – CNT).

such kinetic analysis (see, e.g. [41, 42])) and the viscosity of the diopside melt [25] well described by the following VFT equation

$$\eta(T) = 10^{[-4.27+3961.2(T-750.9)^{-1}]}, \ [\text{in Pa} \cdot \text{s}]. \tag{8.31}$$

According to Eq.(8.31), at the heat-treatment temperature $T = 1143$ K the viscosity is equal to $6.74 \cdot 10^5$ Pa s. Some uncertainty in the definition of the pre-exponential term, J_0, will not strongly affect the nucleation rate, the value $J_0 = 10^{41} \text{s}^{-1} \text{m}^{-3}$ has been used for the calculation [14]. In Fig. 8.19d, dependencies of the nucleation rates, obtained via the generalized Gibbs' approach (full curve) and CNT (dashed), on pressure for the same values of temperature are shown.

As we already noted, generally, only one pore appears in a stretched melt since its fast growth eliminates the negative pressure and terminates further nucleation. In order to estimate the nucleation rate and then to perform the calculations using Eqs. (8.9)–(8.17), one needs to know the dependence of the negative pressure, p, on the position of the crystal-melt interface, X. As was noted, at the beginning of the crystallization process the diopside crystals practically do not participate in melt stretch-

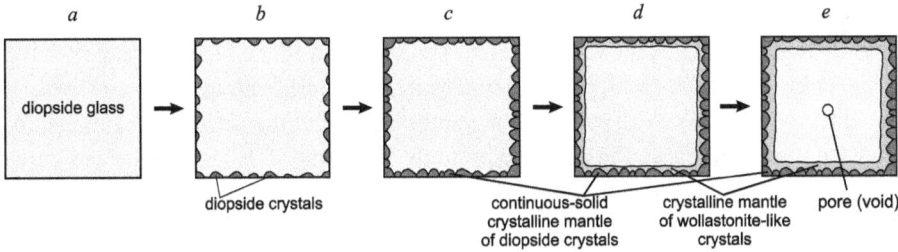

Fig. 8.20: Sketch of experimental results showing the switch of the crystallizing phase in dependence on the width of the crystalline layer and stress induced pore formation: (a) cubic sample of diopside glass; (b) formation of diopside crystals at the surface of the sample; (c) formation of a continuous solid crystalline layer; (d) formation of a wollastonite-like crystalline layer; (e) growth of the wollastonite-like phase and pore formation.

ing, and only at the moment of formation of a continuous diopside layer, latter one can stretch the melt, and a switch to formation of the wollastonite-like phase occurs (see Fig. 8.20c). Therefore the thickness of the diopside crystal layer was considered as independent of the size of the sample.

For the computations of the evolving elastic fields, similar to Section 8.2.3.2 we considered a sample of spherical size. Fig. 8.21 (dashed curve) shows the results of these calculations versus $r_a = R_a/R_s$ for a radius of the sphere estimated as $R_s = a/2$ (here R_s is the radius of the spherical sample and R_a is radius of the amorphous core). After approaching some critical value of pressure, p_*, that corresponds to X_*, i.e. to the size of the layer at the moment of formation of the pore, negative pressure drops rapidly. Taking into account Eq. (8.18), one can rewrite the condition of pore formation Eq. (8.17) as $N(X_*) = 1$. A comparison between the predictions of the value of X_* via CNT and experiment (in agreement with earlier mentioned general considerations [43]) shows that CNT overestimates the work of critical pore formation. In contrast,

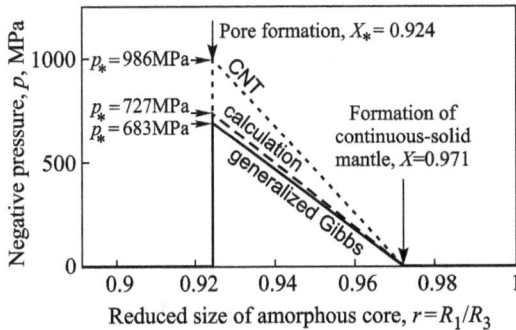

Fig. 8.21: Dependence of the negative pressure on reduced size of the amorphous core. Full curve – dependence as it is predicted for nucleation of a pore according to the generalized Gibbs' approach, short-dashed curve – according to CNT, dashed line – calculation of [20].

the generalized Gibbs approach gives results which are much nearer to experiment, it slightly underestimates the work of critical pore formation, i.e., a pore is created at a lower value of negative pressure than that found in experiment. In order to arrive at a satisfactory agreement of experimental values of X_* with theoretical predictions, for CNT we have to increase the negative pressure by a factor 1.356 (short-dashed curve in Fig. 8.21), and for the generalized Gibbs' approach we have to reduce the negative pressure by a factor 0.939 (full curve in Fig. 8.21). Such reduction of the theoretically estimated pressure can be easily explained by the difference in the shapes of the samples studied theoretically [20] and experimentally [40].

Fig. 8.11 shows the dependence of X_* on the volume of the sample, circles show experimental data, the curve is calculated by Eqs. (8.14)–(8.18). The thickness of the diopside crystal layer, $X_{d/w}$, was used here as a fit parameter, the best result is obtained for a value $X_{d/w}$ = 27.1 µm, which is in good agreement with the experimental data.

8.2.4.3 Conclusions

The generalized Gibbs approach leads to much smaller values of the work of critical pore formation (ΔG_c = 56.8$k_B T$), as compared to CNT, being nearly identical to the value required for a quantitatively exact prediction of the experimental results. The result obtained via CNT (ΔG_c = 121.9$k_B T$; see Fig. 8.19a) is much higher, leading to a huge difference of the values of the steady-state nucleation rates obtained via the two different methods (Fig. 8.19d). So, the generalized Gibbs approach provides us with a much more adequate description of the process of pore nucleation as compared to classical nucleation theory and allows one to interpret pore formation in the considered elastically stretched liquids as cavitation-like processes caused by elastic stresses also in a quantitatively correct way.

The switch of surface crystallization of diopside melts from diopside crystals to a wollastonite-like crystalline phase and pore formation in the glass-forming diopside melt can be described both qualitatively and quantitatively as the result of elastic stresses caused by crystallization. In contrast to CNT, the generalized Gibbs approach has shown to be capable of giving not only a qualitative but even a quantitatively correct interpretation of this process. In this way, phase switch and pore formation due to crystallization is a general phenomenon, which has to be properly taken into account in any processes like sintering, fabrication of glass-ceramic materials involving partial crystallization of glass powders.

8.3 Sintered Diopside-albite Glass-ceramics Forming Crystallization-induced Porosity

8.3.1 Introduction

Usually glass-ceramics, produced by bulk nucleation and crystal growth, are described as non-porous materials, and micro-voids in their structure are explained as defects from the parent glass or cracks formed due to significant difference between the thermal expansions of crystal and amorphous phases present in the considered materials [44, 45]. At the same time, the observed voids in sintered glass-ceramics were explained only as residual inter-granular porosity; similarly to ceramic materials, this porosity is considered as a negative factor, reducing the quality of the mechanical properties. However, in some studies it was noted that another type of porosity might be formed in the glass-ceramics due to the crystallization process. This phenomenon is not well studied and explained from both experimental and theoretical points of view yet. First attempts for a systematic experimental work, related to the formation of this *crystallization induced porosity*, P_{CR}, were carried out in the last decade [46–52]. In these studies sintered glass-ceramics belonging to the pseudo-binary system albite-diopside, which form different percentages of diopside, were used. Later P_{CR} formation was confirmed in other sintered glass-ceramic systems [53–62], as well as in some new ceramics with high crystallinity [63–65].

During sinter-crystallization, processes of densification and crystal phase formation take place in the same temperature interval. When the sintering completes before the actual beginning of phase formation its kinetics may be explained by models for viscous flow sintering. However, in the major part of sintered glass-ceramics densification and crystallization take place simultaneously. When phase formation is characterized by a relatively high velocity, a rapid increasing of the apparent viscosity, η_{app}, is observed; this phenomenon reduces the sintering rate and may even terminate further densification. A somewhat similar phenomenon is well-known from the production of bulk glass-ceramics, where the increase of apparent viscosity avoids the deformation during heating between the nucleation and the crystallization steps [44, 45]. It is recommended to carry out the crystallization treatment always at η_{app} higher than 10^{10-11} dPa s (i.e. at viscosity values similar to the dilatometric softening point). It is interesting to note that at similar viscosities the densification rate in the sintered glasses or glass-ceramics is negligible.

Various models might be used to estimate the relationship between apparent viscosity and effective volume of a rigid phase in a slurry. Notwithstanding of the different theoretical approaches, all relations predict a relatively small η_{app} variation of up to one "critical" percentage of rigid phase, followed by a rapid increasing of apparent viscosity by 10^3–10^8 dPa s. This means that a completed sintering could be obtained only if it is finished prior to the formation of this "critical" percentage of crystalline phase; otherwise, the densification will be inhibited by the high apparent viscosity.

The increase of η_{app} not only diminishes sintering but can also influence the crystallization shrinkage, which is a consequence of the volume variations during crystallization, ΔV_{CR}. It can be assumed that only that kind of phase formation, taking place at low apparent viscosity, leads to a crystallization shrinkage; the subsequent crystallization, going at high η_{app}, firstly can create tensile stresses in the material and after that can provoke the formation and growth of *crystallization induced pores*.

The amount of P_{CR} mainly depends on the percentage and kind of the formed crystal phase. Since the melt density is an additive function of its chemical composition, while the density of crystal phases depends on their different structures and packing, ΔV_{CR} varies significantly. Its value is lower at loose crystal structures and higher at denser crystal arrangement. Very high ΔV_{CR} (at about 16 vol%) is observed during the crystallization of diopside, which is one of the most "popular" crystal phases in the glass-ceramics. This phase is typical for many glass-ceramics from industrial wastes [44, 45] as well as for several sintered glass-ceramics [66]. In the present analysis a part of the above-mentioned results from diopside ($CaO \cdot MgO \cdot 2SiO_2$)-albite ($Na_2O \cdot Al_2O_3 \cdot 6SiO_2$) system are summarized together with some data unpublished up to now. The phase diagram of this pseudo-binary system [46] shows a very large crystallization field of diopside and a very narrow crystallization field for albite. This indicates that practically only diopside is formed, while the final residual glasses retain a composition similar to albite. In addition, the viscosity-temperature curves of the investigated diopside-albite compositions are very similar [46]. This justifies the evaluation of the experimental results mainly as a function of the different crystallization trends.

8.3.2 Experimental

The theoretical compositions of the used three model glasses (labeled **G1, G2,** and **G3**) are reported in Table 8.3. The melting procedure, the experimental results for the glass compositions, obtained by XFR-analysis, and the experimental conditions, realized to prepare the different glass fractions, are reported in previous papers.

Table 8.3: Chemical compositions of parent glasses

	G1	G2	G3
SiO_2	54	58	62
Al_2O_3	2	4	6
CaO	21	17	13
MgO	21	17	13
Na_2O	2	4	6

"Green" samples with initial size 10/10/(8–10) mm^3 were prepared by mixing the parent glass powders with 7.5 % PVA solution and by un-axial pressing at 100–150 MPa. After drying and a 30 min holding at 270 °C (to eliminate the PVA) the samples were heated to different temperatures in the range 650–1000 °C at a heating rate of 5 °C/min and then sinter-crystallized for 1–10 hours. The degree of sintering was evaluated by the variation of total porosity, P_T, and closed porosity, P_C, by measuring apparent density, ρ_a, skeleton density, ρ_s, and absolute density of the glass-ceramics samples, ρ_{gc} through the following relationships:

$$P_T = 100\frac{\rho_{gc} - \rho_a}{\rho_{gc}}, \tag{8.32}$$

$$P_C = 100\frac{\rho_{gc} - \rho_s}{\rho_{gc}}, \tag{8.33}$$

where ρ_a was measured by a dry flow pycnometer (GeoPyc 1360), while ρ_s and ρ_{gc} – by He displacement Pycnometer (AccuPyc 1330). Firstly, skeleton density was measured and then the absolute density after crashing and milling the samples below 26 µm. The experimental associated errors to the evaluation of ρ_a, ρ_{gc} and ρ_s were estimated as ±0.013 g/cm^3, ±0.003 g/cm^3 and ±0.005 g/cm^3, respectively. The experimental error associated to P_T and P_C was evaluated as ±0.6 % and ±0.3 %, respectively. The open porosity P_T was estimated as the difference between P_T and P_C.

The crystalline fraction, x (wt%), was also evaluated by density measurements through the expression:

$$x = 100\left(\frac{\frac{1}{\rho_g} - \frac{1}{\rho_{gc}}}{\frac{1}{\rho_{g(cr)}} - \frac{1}{\rho_{cr}}}\right), \tag{8.34}$$

where ρ_g is the absolute density of parent glass and ρ_{gc} is the absolute density of glass-ceramic, $\rho_{g(cr)}$ is the density of a hypothetical glass with the composition of the formed crystal phase and ρ_{cr} is the density of the crystal phase. In the case of diopside, it was assumed that ρ_{cr} and $\rho_{g(cr)}$ have values of 3.27 and 2.75 g/cm^3, respectively. The experimental associated error of the ρ_g and ρ_{gc} values was evaluated as ± 0.003 g/cm^3, which corresponds to an experimental error of ±1 % on the evaluation of the amount of the formed diopside. A detailed description of the methodological approach is given in [67]. The structures of samples were studied by Scanning Electron Microscopy (Philips XL30CP). Both polished fractured samples and their surfaces were observed. The samples were treated with 2% HF solution for 5 seconds.

8.3.3 Results and Discussion

8.3.3.1 Density and Porosity Evaluations

Figs. 8.22–8.24 show the variations of open and closed porosities as well as the percentages of formed diopside after 1 hour of heat-treatment at different temperatures for compositions **G1**, **G2** and **G3** (fraction 75–125 µm). At low temperatures, up to 750 °C, the porosity is only open, at 800–850 °C the transformation from open into closed porosity takes place, while in the range of 850–1000 °C the porosity becomes closed. Diopside formation starts at 750 °C for **G1** and **G2** and at 800 °C for **G3**; above 900 °C

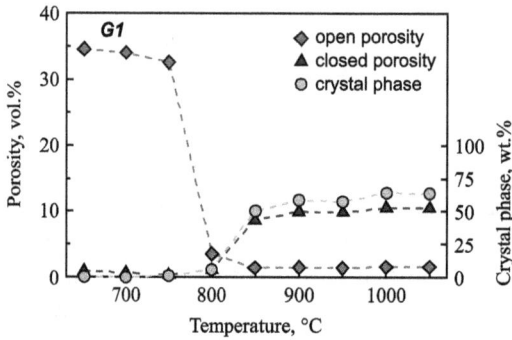

Fig. 8.22: Open and closed porosities and percentage of crystal phase in sintered **G1** glass after 1 hour at different temperatures (fraction 75–125 µm).

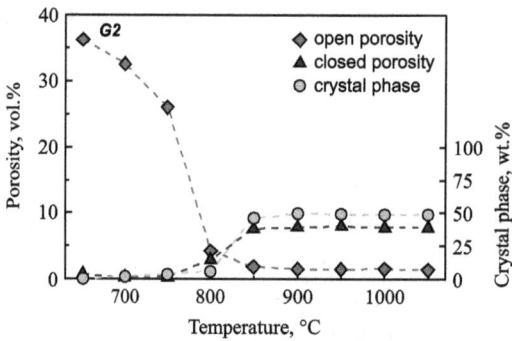

Fig. 8.23: Open and closed porosities and percentage of crystal phase in sintered **G2** glass after 1 hour at different temperatures (fraction 75–125 µm).

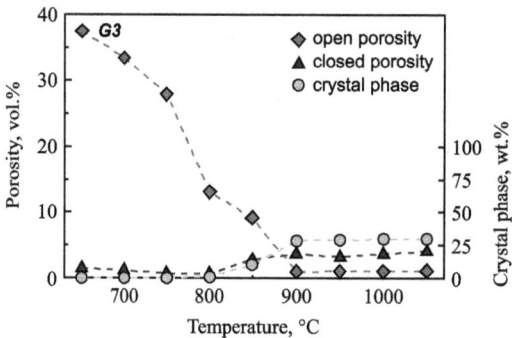

Fig. 8.24: Open and closed porosities and percentage of crystal phase in sintered **G3** glass after 1 hour at different temperatures (fraction 75–125 µm).

the amounts of formed crystal phase practically remain constant. It is evident that the percentages of formed crystal phase are well related to the amounts of closed porosity.

However, it is incorrect to assume that the entire closed porosity is crystallization-induced because some residual closed porosity, P_R, always remains in sintered glass powders. In order to estimate the value of P_R identical experiments with a non-crystallizing SiO_2–CaO–Na_2O container glass were carried out. The results are presented in Fig. 8.25. They highlight that after sintering at $700\,°C$ the porosity becomes only closed and P_R value is $3.6 \pm 0.6\,\%$. This result is comparable with the values obtained by other authors [68, 69]. The lower sintering temperatures of samples from container glass are in a good agreement with its inferior glass transition temperature, T_g: in the studied diopside-albite glasses T_g values are found within the range of 660–680 °C [48], while the container glass has glass transition temperature of ~560 °C.

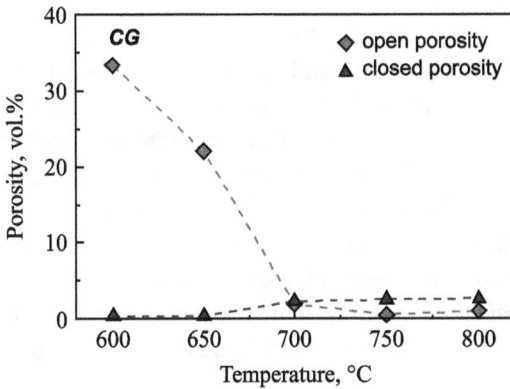

Fig. 8.25: Open and closed porosities in sintered container glass after 1 hour at different temperatures (fraction 75–125 μm).

Table 8.4 summarizes the density and porosity results obtained after 1 hour at 900 °C. It it obvious from these data that the closed porosity significantly increases with the amount of the formed crystal phase. Assuming that the residual porosities in **G1, G2,** and **G3** are similar to their values in the sintered container glass the following results for P_{CR} are obtained: ~6.5 % in **G1**, ~4.1 % in **G2** and ~0.1 % in **G3**. Using thus obtained P_{CR} values and combining Eqs. (8.33) and (8.34) the percentages of crystal

Table 8.4: Porosity and crystal phase in **G1, G2** and **G3** (1 hour at 900 °C and fraction 75–125 μm)

	glass density (g/cm^3)	apparent density (g/cm^3)	skeleton density (g/cm^3)	absolute density (g/cm^3)	crystal phase (wt%)	total porosity (wt%)	closed porosity (vol%)
G1	2.74	2.69	2.73	3.04	58.8	11.5	10.1
G2	2.69	2.65	2.69	2.92	49.4	9.2	7.9
G3	2.56	2.55	2.58	2.68	28.6	4.8	3.7

phase, leading to formation of crystallization induced porosity (labeled as **Cr-P**), can be estimated. The other part of crystal phase (labeled as **Cr-S**) can be related to the crystallization induced shrinkage. The results of these approximations, together with the ratio **Cr-P/Cr-S**, are summarized in Table 8.5.

Table 8.5: Amounts of crystal phase, leading to crystallization induced porosity, **Cr-P**, and crystallization induced shrinkage, **Cr-S**, for **G1**, **G2** and **G3** (1 hour at 900 °C and fraction 75–125 μm)

	Cr-P	Cr-S	Cr-P/ Cr-S
G1	40 ± 3	19 ± 3	2.1
G2	27 ± 3	22 ± 3	1.2
G3	4 ± 3	25 ± 3	0.2

Cr-P notably increases with the crystallization trend, while the percentage of **Cr-S**, on the contrary, decreases. This phenomenon leads to significant variations of the **Cr-P/Cr-S** ratio and highlights that the relative amount of crystal phase, leading to huge increasing of apparent viscosity and inhibiting of the crystallization shrinkage, notably decreases with the rise of crystallization trend. It might be assumed that for the fractions of 75–125 μm the crystallization shrinkage in **G1** stops after formation of 5–10 μm surface crystal layers (which corresponds to ~ (1/3) of the total amount of the crystal phase), whereas in **G3** the shrinkage continues even after formation of a crystal front of 50 μm (i.e. **Cr-P** is negligible). These differences elucidate that a dense crystal core can be formed in **G1**, while in **G3** the crystal front can be more lax and can allow some flow between the crystals.

The ratio **Cr-P/Cr-S** also depends on the size of the used glass particles. If the crystallization shrinkage at different fractions stops at similar thickness of the crystal front (for each of the compositions), it follows that the finer are the powders the higher is **Cr-S** and the lower is **Cr-P**. This hypothesis was confirmed using fractions 26–32 μm, 40–53 μm and 75–125 μm of the **G1** glass. In order to guarantee a completed sinter-crystallization the samples were heat-treated for 2 hours at 900 °C. The obtained density and porosity results are summarized in Table 8.6, whereas the corresponding values for **Cr-P, Cr-S,** and **Cr-P/Cr-S** are shown in Table 8.7. It is assumed that the percentage of 3.6 ± 0.6 % for residual porosity might be also used for the fractions of 26–32 μm and of 40–53 μm.

The results for these three fractions show similar crystallinity, while **Cr-P** and **Cr-S** percentages vary significantly. In the fraction of 26–32 μm two thirds of the formed crystal phase lead to shrinkage and only one third – to crystallization induced porosity (i.e. contrary to the results for the fraction of 75–125 μm). In the fraction of 40–53 μm the amounts of diopside, leading to shrinkage and porosity, are similar. If it is assumed that the crystallization shrinkage in the fractions of 26–32 μm and 40–53 μm also stops

Table 8.6: Porosity and crystal phase in **G1** after 2 hours at 900 °C

Fraction size (μm)	apparent density (g/cm^3)	skeleton density (g/cm^3)	absolute density (g/cm^3)	crystal phase (wt%)	total porosity (vol%)	closed porosity (vol%)
26–32	2.81	2.83	3.06	62.9	8.2	7.5
40–53	2.75	2.78	3.06	62.9	10.0	9.0
75–125	2.70	2.73	3.05	61.1	11.3	10.3

Table 8.7: Amounts of crystal phase, leading to crystallization induced porosity, **Cr-P**, and crystallization induced shrinkage, **Cr-S**, for **G1** after 2 hours at 900 °C

Fraction size, (μm)	Cr-P	Cr-S	Cr-P/ Cr-S
26–32	22 ± 3	41 ± 3	0.5
40–53	32 ± 3	31 ± 3	1.0
75–125	41 ± 3	20 ± 3	2.1

after formation of about 5–10 μm surface crystal layers the estimations for **Cr-S** parts in fact are ~(2/3) and ~(1/2), respectively.

In order to evaluate the simultaneous formation of crystal phase and crystallization induced porosity various treatments at 800 °C for different times (between 1 min to 10 h) were also carried out. The results, obtained for the composition **G1** and the fraction of 40–53 μm, are plotted in Fig. 8.26. This figure demonstrates that after 1 min the porosity is only open and the amount of crystal phase (formed during the heating) is negligible. After 1 hour holding the crystallinity increases to ~ 10 % and open porosity still remains. After that, up to 4 h heat-treatment, the open porosity is transformed into closed residual porosity and diopside percentage reaches ~25 %. Then, the increasing of the amount of the crystal phase is accompanied by the formation of additional crystallization induced porosity. Finally, after 10 hour holding the sinter-crystallization is completed by the formation of ~60 % diopside and ~9 % closed porosity.

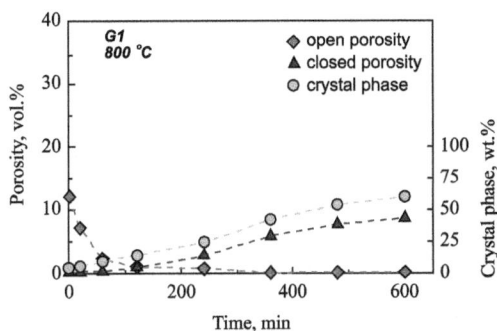

Fig. 8.26: Open and closed porosities and percentage of crystal phase in sintered **G1** glass after different times (fraction 40–53 μm).

8.3.3.2 SEM-observations of G1-glass-ceramics

The structures of **G1** glass-ceramics (both surfaces and polished fractures), obtained after 1, 4 and 7 hours holding at 800 °C (for fraction 40–53 μm), were studied with Scanning Electron Microscopy. A part of the results are summarized in Figs. 8.27–8.31. The fracture of the sample after 1 hour holding (Fig. 8.27) elucidates a well sintered body, where the surface crystallization is in its initial stage and as a result the separated grains are well distinguished. The residual intergranular pores are characterized by a smooth surface, while no additional intragranular pores are observed in the volume of the grains.

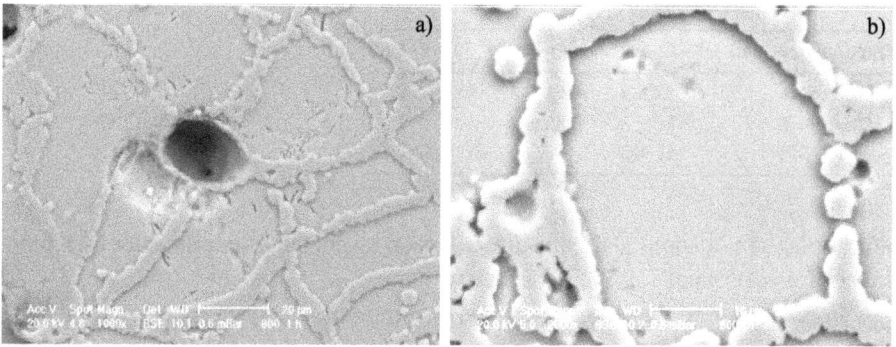

Fig. 8.27: BSE-SEM images of polished fracture after 1 hour at 800 °C.

Fig. 8.28 shows some details of the sample, heat-treated for 1 hour, obtained at higher magnification with SE-SEM technique and after Au-metallization. Fig. 8.28a highlights the contact zone between two sintered grains and shows the beginning of crystal growth in both particles. Fig. 8.28b presents the surface of a closed residual pore (i.e. non-sintered part of a grain) with crystals growing inside of the particle.

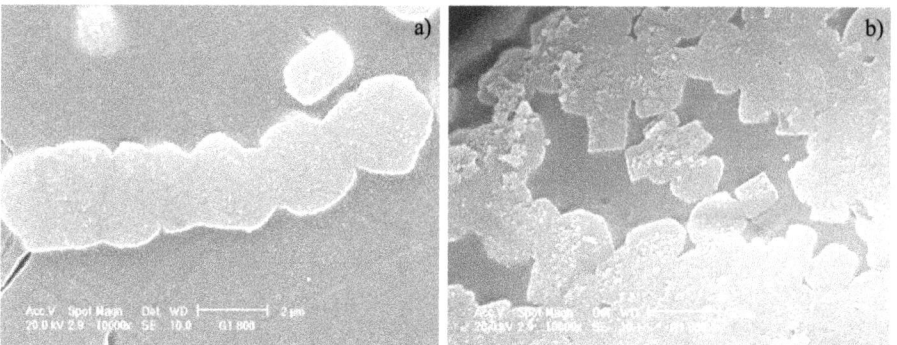

Fig. 8.28: SE-SEM images of fracture after 1 hour at 800 °C (details).

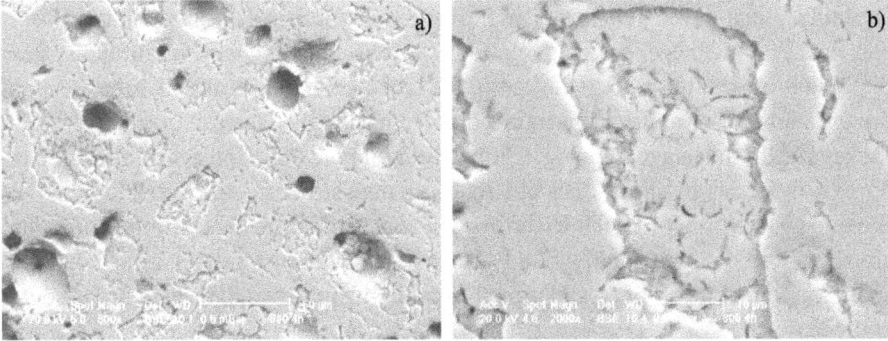

Fig. 8.29: BSE-SEM images of polished fracture after 4 hours at 800 °C.

Both images elucidate that the width of crystals in the beginning of phase formation is 1–3 μm.

Fig. 8.29 presents the polished fracture of the sample, obtained after 4 hours hold-ing. The image at lower magnification (Fig. 8.29a) demonstrates an increased closed porosity and the co-existence of two different kinds of pores. The bigger part of poros-ity is presented by intergranular residual pores with smooth surfaces. However, some intragranular pores with poly-crystalline surface can also be identified. The photos from Fig. 8.29b show an individual grain without crystallization-induced pore and the increase of crystal front up to 6–8 μm. The comparison with the image from Fig. 8.27b clearly demonstrates that the central amorphous part is more affected by the chemical attack (i.e. the chemical durability of glassy phase decreases during the crystalliza-tion). This interesting feature might be explained by formation of tensile stresses due to the additional crystallization. In fact, the un-annealed glasses (i.e. glasses under tensile stress) demonstrate significantly lower chemical resistance.

Fig. 8.30 highlights the structure after 7 hours of sinter-crystallization. The crystal-lization process is completed in the major part of the particles and the pores are mainly

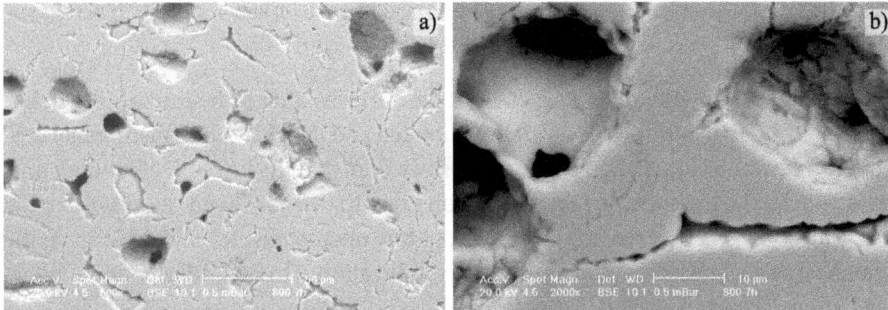

Fig. 8.30: BSE-SEM images of polished fracture after 7 hours at 800 °C.

crystallization induced and located in the centers of grains. The residual glassy phase, observed in some bigger grains, is characterized by a better chemical durability than that one formed after 4 hours holding. The latter difference probably indicates that the formation of crystallization-induced pores leads to a decrease of the stresses in the sample. In the image from Fig. 8.30b a residual pore (left) and a crystallization induced pore (right) are shown. The differences in their structures are well illustrated: P_R shows smooth surface (similar to the one in Fig. 8.27b) where the beginning of crystal growth is well distinguished, while P_{CR} is characterized by a rough surface. It corresponds to the final part of crystal growth completing in the void, formed in the center of the particle.

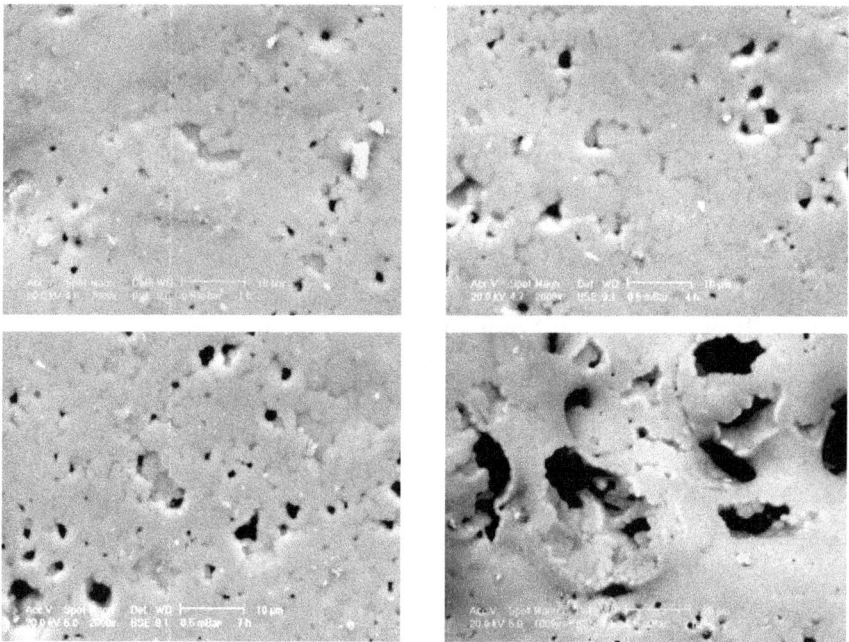

Fig. 8.31: BSE-SEM images of surface after 1 (a), 4 (b), and 7 (c) and (d) hours at 800 °C (from left top to right bottom).

Finally, Fig. 8.31 summarizes images from the surfaces after different holding times. The sintering in the samples is entirely completed and it is difficult to distinguish separated grains or residual open pores. It is also evident that the increase of holding time does not change the crystals morphology on grain's surface (i.e. no re-crystallization is found). However, the rise of the amount of tiny open pores, formed directly on the grains facade, is observed during sinter-crystallization. This peculiarity might be related to continued crystallization shrinkage or/and increasing gas pressure in the crystallization-induced pores. Very interesting images (Fig. 8.31d), showing rupture of the crystal core, were made on the edge of a sample heat-treated 7 hours. This behavior will be investigated and explained in future reports.

Bibliography

[1] J.W. Christian, *The Theory of Transformations in Metals and Alloys*, Part 1 (Pergamon, Oxford, 1981).

[2] J. Möller, J. Schmelzer, and I. Gutzow, Z. Phys. Chemie **204**, 171 (1998).

[3] J.W.P. Schmelzer, R. Pacova, J. Möller, and I. Gutzow, J. Non-Cystalline Solids **162**, 26 (1993).

[4] J.W.P. Schmelzer, J. Möller, I. Gutzow, R. Pascova, R. Müller, and W. Pannhorst, J. Non-Crystalline Solids **183**, 215 (1995).

[5] J.W.P. Schmelzer, O.V. Potapov, V.M. Fokin, R. Müller, and S. Reinsch, J. Non-Crystalline Solids **333**, 150 (2004).

[6] J.W.P. Schmelzer, E.D. Zanotto, I. Avramov, and V.M. Fokin, J. Non-Crystalline Solids **352**, 434 (2006).

[7] V.M. Fokin, E.D. Zanotto, J.W.P. Schmelzer, and O.V. Potapov, J. Non-Crystalline Solids **351**, 1491 (2005).

[8] I.S. Gutzow and J. W. P. Schmelzer, *The Vitreous State: Thermodynamics, Structure, Rheology, and Crystallization* (Springer, Berlin, 1995; Springer, Berlin-Weinheim, 2013).

[9] V.T. Slaviansky, *Gases in Glasses* (Oboronprom Publishers, Moscow, 1957).

[10] F.C. Serbena and E.D. Zanotto, J. Non-Crystalline Solids **358**, 975 (2012).

[11] Oscar Peitl Filho, *Vidro-Ceramica Bioativa de Alto Desempenho Mecanico* (PhD Thesis, UFS-car, Sao Carlos, Brazil, 1995).

[12] I. Gutzow, R. Pascova, A. Karamanov, and J. Schmelzer, J. Materials Science **33**, 5265 (1998).

[13] V.P. Skripov and V.P. Koverda, *Spontaneous Crystallization of Super-cooled Liquids* (Nauka, Moscow, 1984 (in Russian)).

[14] V.P. Skripov, *Metastable Liquids* (Wiley, New York, 1974).

[15] J.W.P. Schmelzer and J. Schmelzer Jr., *Kinetics of Bubble Formation and the Tensile Strength of Liquids*. In: J.W.P. Schmelzer, G. Röpke, and V.B. Priezzhev, *Nucleation Theory and Applications* (Joint Institute for Nuclear Research Publishing Department, Dubna, Russia, 2002, 88ff).

[16] J.W.P. Schmelzer and J. Schmelzer Jr., J. Atmospheric Research **65**, 303 (2003).

[17] S.A. Kukushkin, J. Applied Physics **98**, 033503 (2005).

[18] T. Yokobori, J. Chem. Phys. **22**, 951 (1954).

[19] A.S. Abyzov, J.W.P. Schmelzer, and V.M. Fokin, *On the Effect of Elastic Stresses on Crystallization Processes in Finite Domains*. In *Nucleation Theory and Applications*, Eds. J.W.P. Schmelzer, G. Röpke, and V.B. Priezzhev (Joint Institute for Nuclear Research Publishing Department, Dubna, Russia, 2009, pages 379–398).

[20] A.S. Abyzov, J.W.P. Schmelzer, and V.M. Fokin, J. Non-Crystalline Solids **356**, 1670 (2010).

[21] R.S. Roth, M.A. Clevinger, and D. McKenna (Eds.), *Phase diagrams for ceramists* (United States National Bureau of Standards, American Ceramic Society, Columbus, 1984).

[22] V.G. Baidakov, *Explosive Boiling of Superheated Cryogenic Liquids* (WILEY-VCH, Berlin-Weinheim, 2007).

[23] W. Hinz, *Silikate* (Verlag für Bauwesen, Berlin, 1970).

[24] S. Reinsch, PhD thesis, Technische Universität Berlin, 2001.

[25] M.L.F. Nascimento, E.B. Ferreira, and E.D. Zanotto, J. Chem. Phys. **121**, 8924 (2004).

[26] E.D. Zanotto and V.M. Fokin, Phil. Trans. Royal Society London **A 361**, 591 (2003).

[27] P.G. Cheremskoi, V.V. Slyozov, and V.I. Betekhtin, *Pores in the Solid State* (Energoatomizdat, Moscow, 1990 (in Russian)).

[28] Sciglass database: http://www.esm-software.com/sciglass.

[29] A.I. Rusanov, *Thermodynamic foundation of mechano-chemistry* (Nauka, St. Petersburg, 2006 (in Russian)).

[30] V.G. Baidakov and G.Sh. Boltachev, Phys. Rev. E **59**, 469 (1999).

[31] V.G. Baidakov, G.Sh. Boltachev, and J.W.P. Schmelzer, J. Colloid Interface Science **231**, 312 (2000).

[32] M.P.A. Fisher and M. Wortis, Phys. Rev. **29**, 6252 (1984).

[33] R. McGraw and A. Laaksonen, Phys. Rev. Lett. **76**, 2754 (1996).

[34] V.M. Fokin, E.D. Zanotto, and J.W.P. Schmelzer, J. Non-Crystalline Solids **356**, 2185 (2010).

[35] J.W.P. Schmelzer and J. Schmelzer Jr., J. Chem. Phys. **114**, 5180 (2001).

[36] J.W.P. Schmelzer and J. Schmelzer Jr., J. Atmospheric Research **65**, 303 (2003).

[37] J.D. van der Waals, Thesis. Leiden (1873).

[38] J.D. van der Waals, *Die Kontinuität des gasförmigen und flüssigen Zustandes* (2nd edition, Johann-Ambrosius-Barth Verlag, Leipzig, 1899–1900).

[39] J.D. van der Waals and Ph. Kohnstamm, *Lehrbuch der Thermodynamik* (Johann-Ambrosius-Barth Verlag, Leipzig und Amsterdam, 1908).

[40] V.M. Fokin, A.S. Abyzov, J.W.P. Schmelzer, and E.D. Zanotto, J. Non-Crystalline Solids **356**, 1679 (2010).

[41] M.L.F. Nascimento, E.B. Ferreira, and E.D. Zanotto, J. Chem. Phys. **121**, 8924 (2004).

[42] J.W.P. Schmelzer, J. Non-Crystalline Solids **356**, 2901 (2010).

[43] J.W.P. Schmelzer, G. Sh. Boltachev, and V.G. Baidakov, J. Chemical Physics **124**, 194503 (2006).

[44] Z. Strnad, *Glass-Ceramic Materials* (Elsevier, Amsterdam, 1986).

[45] W. Höland and G. Beall, *Glass-Ceramics Technology* (The American Ceramics Society, Westerville, 2002).

[46] A. Karamanov, L. Arrizza, I. Matecovetc, and M. Pelino, *Properties of sintered glass-ceramics in the diopside-albite system*, Ceramics International **30**, 2129 (2004).

[47] A. Karamanov and M. Pelino, *Sinter-Crystallization in the System Diopside-Albite, Part I. Formation of Induced Crystallisation Porosity*, J. European Ceramic Society **26**, 2511 (2006).

[48] A. Karamanov and M. Pelino, *Sinter-Crystallization in the System Diopside-Albite, Part II. Kinetics of Crystallization and Sintering*, J. European Ceramic Society **26**, 2519 (2006).

[49] A. Karamanov and M. Pelino, *Induced Crystallization Porosity and Properties of Sintered Diopside and Wollastonite Glass-Ceramics*, J. European Ceramic Society **28**, 555 (2008).

[50] A. Karamanov, I. Georgieva, R. Pascova, and I. Avramov, *Pore Formation in Glass Ceramics: Influence of the Stress Energy Distribution*, J. Non-Crystalline Solids **356**, 117 (2010).

[51] A. Karamanov, I. Avramov, L. Arrizza, R. Pascova, and I. Gutzow, *Variation of Avrami parameter during non-isothermal surface crystallization of glass powders with different sizes*, J. Non-Crystalline Solids **358**, 1486 (2012).

[52] A. Karamanov, *Influence of micro- and nano-induced crystallization porosity in sintered glass-ceramics on their structure and properties*, Nanoscale Phenomena and Stucture, 2008, Edited by D. Kashchiev, Prof. M. Drinov, Academic Publishing House, p. 101–104.

[53] E. Bernardo, J. Doyle, and S. Hampshire, *Sintered feldspar glass–ceramics and glass–ceramic matrix composites*, Ceramics International **34**, 2037 (2008).

[54] Wei Yi Zhang, Hong Gao, and Yu Xu, *Sintering and reactive crystal growth of diopside–albite glass–ceramics from waste glass*, Journal of the European Ceramic Society **31**, 1669 (2011).

[55]   D.U. Tulyaganov, S. Agathopoulos, J.M. Ventura, M.A. Karakassides, O. Fabrichnaya, and J.M.F. Ferreira, *Synthesis of glass–ceramics in the* $CaO-MgO-SiO_2$ *system with* B_2O_3, P_2O_5, Na_2O *and* CaF_2 *additives*, Journal of the European Ceramic Society **26**, 1463 (2006).

[56]   Huizhi Yang, Changping Chen, Hongwei Sun, Hongxia Lu, and Xing Hu, *Influence of heat-treatment schedule on crystallization and microstructure of bauxite tailing glass–ceramics coated on tiles*, Journal of materials processing technology **197**, 206 (2008).

[57]   E. Bernardo and R. Dal Maschio, *Glass–ceramics from vitrified sewage sludge pyrolysis residues and recycled glasses*, Waste Management **31**, 2245 (2011).

[58]   I.K. Mihailova, P.R. Djambazki, and D. Mehandjiev, *The effect of the composition on the crystallization behavior of sintered glass-ceramics from blast furnace slag*, Bulgarian Chemical Communications **43**, 293 (2011).

[59]   J.K.M.F. Daguano, K. Strecker, E.C. Ziemath, S.O. Rogero, M.H.V. Fernandes, and C. Santos, *Effect of partial crystallization on the mechanical properties and cytotoxicity of bioactive glass from the* $3CaO \cdot P_2O_5 - SiO_2 - MgO$ *system*, Journal of the Mechanical Behavior of Biomedical Materials **14**, 78 (2012).

[60]   E. Bernardo, L. Esposito, E. Rambaldi, A. Tucci, Y. Pontikes, and G.N. Angelopoulos, *Sintered esseneite–wollastonite–plagioclase glass–ceramics from vitrified waste*, Journal of the European Ceramic Society **29**, 2921 (2009).

[61]   Weiyi Zhang and He Liu, *A low cost route for fabrication of wollastonite glass–ceramics directly using soda-lime waste glass by reactive crystallization–sintering*, Ceramics International **39**, 1943 (2013).

[62]   L. Schabbach, F. Andreola, E. Karamanova, I. Lancellotti, A. Karamanov, and L. Barbieri, *Integrated approach to establish the sinter-crystallisation ability of glasses from secondary raw material*, J. Non-Crystalline Solids **357**, 10 (2011).

[63]   A. Karamanov, L. Arrizza, and S. Ergul, *Sintered Material From Alkaline Basaltic Tuffs*, J. European Ceramic Society **29**, 595 (2009).

[64]   E. Karamanova, G. Avdeev, and A. Karamanov, *New Building Ceramics based on Blast Furnace Slag*, J. European Ceramic Society **31**, 989 (2011).

[65]   L.M. Schabbach, F. Andreola, L. Barbieri, I. Lancellotti, E. Karamanova, B. Ranguelov, and A. Karamanov, *Post-treated incinerator bottom ash as alternative raw material for ceramic manufacturing*, J. European Ceramic Society **32**, 2843 (2012).

[66]   R.D. Rawlings, J.P. Wu, and A.R. Boccaccini, *Glass-ceramics: Their production from wastes-a review*, J. Materials Science **41**, 733 (2006).

[67]   A. Karamanov and M. Pelino, *Evaluation of the Degree of Crystallization in Glass-Ceramics by Density Measurements*, J. European Ceramic Society **19**, 649 (1999).

[68]   W.D. Kingery, H.K. Bowen, and D.R. Uhlmann, *Introduction to Ceramics* (John Wiley & Sons, New York, 1975).

[69]   E.D. Zanotto and M. Prado, *Isothermal sintering with concurrent crystallization of monodispersed and polydispersed glass particles. Part 1*, Phys. Chem. Glasses **42**, 191 (2001).

Vladimir G. Baidakov

9 Crystallization of Undercooled Liquids: Results of Molecular Dynamics Simulations

The present contribution is devoted to molecular dynamics (MD) simulations modeling of the kinetics of spontaneous crystal nucleation in under-cooled one-component Lennard-Jones liquids and detailed comparison with the basic assumptions and results of classical nucleation theory (CNT). In the MD-computations the following spectrum of properties of the respective nucleating systems under consideration is determined: nucleation rate, J, diffusion coefficient of the crystal clusters in cluster size space, $\mathcal{D}$, non-equilibrium Zeldovich factor, Z, size of the critical crystal nucleus, n_*, pressure inside the critical crystal nucleus, p_*. Based on these data, the interfacial energy density of the critical crystal nucleus is determined. Simultaneously, the interfacial energy density is computed by molecular dynamics methods for the planar interface liquid-crystal. It is found that for typical sizes of the critical nuclei in the range of 0.7–1.0 nm the value of the effective specific interfacial energy differs from that of the planar interface by less than 15 %. A comparison of the molecular dynamics results with the classical nucleation theory shows that for the considered case of crystallization of one-component liquids MD simulation results are in good agreement with the classical nucleation theory not only with respect to the final result, the nucleation frequency, but also with respect to the parameters $\mathcal{D}$, Z, n_*. Consequently, the results of molecular dynamics simulations of crystallization in one-component liquids demonstrate the validity of the basic assumptions and the final results of CNT for this particular case of phase formation.

9.1 Introduction

The process of crystallization of liquids starts with formation and subsequent growth of critical clusters of the crystalline phase. If the homogeneous liquids do not contain heterogeneous nucleation cores, and the system is not exposed to external disturbances initiating nucleation, the clusters of the newly evolving crystalline phase are developing spontaneously due to thermal fluctuations. This process is called homogeneous nucleation. However, as a rule, the formation of nuclei of the new phase starts at the walls of the container the liquid is embedded in or at solid particles, dissolved in the liquid. In such cases, crystallization may start at considerably lower under-cooling.

In order to develop a theoretical description of nucleation-growth processes, first one has to develop a theory of homogeneous nucleation. Specific features of heterogeneous nucleation can be described then in the next step, appropriately modifying the basic relations of homogeneous nucleation theory. In this way, the theoretical description of homogeneous nucleation is a complex and fundamental problem itself, and also as giving the basis for the treatment of heterogeneous nucleation. For this reason, it has been attracting the attention of scientists for a long time. Historically, it was developed taking the basic studies of Gibbs as the starting point [1]. In its thermodynamic part (going back to the work of Gibbs), homogeneous nucleation theory is of universal nature, specific features of nucleation for different systems have to be incorporated into the theory via differences in the kinetics, as shown first by Volmer and Weber [2], Farkas [3], Kaischew and Stranski [4], Zeldovich [5], Frenkel [6] and others, leading to formulation of the classical nucleation theory (CNT).

For an experimenter, who wants to use homogeneous nucleation theory in application to experiment, the task consists in creating pure conditions in the system under investigation. Numerous experiments with the use of different methods point to the feasibility of fluctuation formation of a crystal phase in actual experimental conditions [7–10]. Experimental data, at least in final results, agree with CNT. Nevertheless, in experimental investigations of the kinetics of spontaneous crystallization of a supercooled liquid there are some fundamental points one has to take into account: (i) Practically all the investigations have been conducted at pressures close to atmospheric. (ii) The appearance of a fluctuation nucleus in a meta-stable phase is a random event. In many papers (including the frequently cited reference [11]), however, a statistical analysis is not made. (iii) At present, there do not exist reliable methods of measuring the surface free energy, γ, at the boundary between liquid and crystal phases. Therefore, when studying crystallization, one is forced to regard γ as a "free" parameter. This complicates the comparison with experiment of the theory of homogeneous formation of crystal nuclei. (iv) With the use of different investigation techniques [7] one can trace with increasing supercooling variations in the nucleation rate, J, from 10^2 to 10^{25} s^{-1}m^{-3}. However, it is problematic to advance into regions with nucleation rates, $J > 10^{25}$ s^{-1}m^{-3} in experiments. (v) With the nucleation rate, defined as a temperature function, experiments do not allow one measuring directly such nucleation characteristics as the work of formation of a critical nucleus, its size and the rate of the crystal transition through the critical size.

Methods of computer simulation (molecular dynamics (MD) and Monte-Carlo methods) open up new opportunities in the investigation of homogeneous nucleation. The use of computer models makes it possible to extend the temperature and nucleation rate ranges considerably in elucidating the applicability of CNT. The small dimensions of computer models allow one advancing into the region of very high nucleation rates $J = (10^{30} - 10^{32})$ s^{-1}m^{-3} [11–16], and the use of special computational methods, such as the umbrella sampling method [17], the transition path sampling [18], meta-dynamics [19], makes it possible to study nucleation even at low supersaturation, when the energy barriers are high. Besides, computer simulation provides detailed microscopic information about the initial stage of phase transitions. This microscopic information enables one determining (along with the nucleation rate) the size of the critical nucleus, the work required for its formation, the equilibrium and steady-state size distributions of nuclei, and the rate of particle attachment to the critical nucleus.

The radius of the critical nucleus and the work of its formation largely depend on the value of the interfacial free energy. When comparing the results of experiments on the observation of spontaneous liquid crystallization with CNT, one can evaluate the solid–liquid interfacial free energy. This approach was first realized in [20] for organic liquids and in [21] for metals. The subsequent more detailed experimental investigations of spontaneous crystallization kinetics [7] made it possible to specify and extend data on the interfacial free energy of metals, water and organic substances. All these investigations, however, refer to atmospheric pressure or pressures close to it, and therefore give the value of γ only at one point of the melting line.

The solid–liquid interfacial free energy may also be determined in computer experiments. Three methods of calculating the interfacial free energy at a planar solid–liquid interface via molecular simulation have been suggested in recent years: the cleaving potentials method [22–24], the capillary fluctuation method [25–28], and the Gibbs-Cahn integration technique [29, 30]. The first calculations of the solid–liquid interfacial free energy were performed by Broughton and Gilmer [22] for the systems of particles whose interaction was described by a modified Lennard-Jones potential. These investigations showed a weak anisotropy in γ.

A meta-stable system losing stability against heterophase changes of state (nucleation process) retains its reducing reaction to infinitesimal (homophase) perturbations. It is usually assumed that with a deeper penetration into the meta-stable region approaching the spinodal, the phase loses stability against small perturbations of state as well. However, for a supercooled liquid the spinodal is evidently absent [31]. The absence of the spinodal of a supercooled liquid may be connected with the same fundamental reason that causes the absence on the melting curve of a singular point similar to the liquid–gas critical point. Nevertheless, as shown by computer experiments [32, 33], in a simple one-component system there is a point of termination of crystal-liquid phase equilibrium, which is located in the region of negative pressures and is the point of contact of the meta-stable extension of the melting curve with the spinodal of a tensile-stressed (superheated) liquid.

In the present contribution, our own results of molecular dynamics modeling of homogeneous crystal nucleation in one-component meta-stable (under-cooled) Lennard-Jones liquids are reviewed and compared with CNT. Such computations have been performed by us for both positive and negative pressures, For this purpose, in Section 9.2, basic assumptions and results of classical theory of crystal nucleation in under-cooled liquids are briefly summarized. In Sections 9.3 and 9.4, the basic models and methods of molecular dynamics simulations of nucleation employed are discussed. In Section 9.5, the methods of determination of the specific interfacial free energy crystal-liquid for planar interfaces are described and the results of computations are outlined. The particular attention devoted to this quantity is caused by the fact that it represents one of the basic material parameters determining the work of critical crystal cluster formation and the spontaneous nucleation rate. In addition, the attention to this parameter for crystal formation is further motivated by the fact that – in contrast e.g. to condensation and evaporation or segregation in solutions – direct experimental methods of determination of this quantity for the interface liquid-crystal are lacking. Sections 9.6 and 9.7 are devoted to the outline of the results of molecular dynamics simulations of crystallization. In particular, the temperature and pressure dependencies of the nucleation frequency and of the specific interfacial free energy of critical crystal nuclei are established and analyzed. In addition, the results of molecular dynamics simulations are compared with classical nucleation theory. A summary and the discussion of the results, given in Section 9.8, complete the chapter.

9.2 Thermodynamics and Kinetics of Crystal Formation

In CNT it is assumed that new-phase nuclei originate as a result of heterophase fluctuations [34]. At given temperature, T, and pressure, p, in a supercooled liquid the work of formation of a crystal nucleus is determined by the variation of the Gibbs thermodynamic potential, $\Delta\Phi$. In the variables determining the volume, the shape of a crystal nucleus and the pressure in it, the function $\Delta\Phi$ is described by a hyper-surface, whose saddle point corresponds to the unstable equilibrium of a crystal nucleus in the supercooled liquid. In the vicinity of the saddle point we can write

$$\Delta\Phi(V,A) = \rho_{s_*} \left[\mu(p,T) - \mu_{s_*}(p_{s_*},T) \right] \int_V dV + \int_A \gamma dA, \tag{9.1}$$

where μ and μ_s are the chemical potentials of the liquid and crystal phases, ρ_{s_*} and p_{s_*} are the density and pressure inside the critical crystal nucleus, V and A are the volume and surface area of the nucleus.

The work of formation of a critical nucleus W_* corresponds to the value of the thermodynamic potential difference at the saddle point of the surface, $\Phi(V,A)$, at a given chemical potential difference, $\Delta\mu = \mu - \mu_s$. The relation between the volume of a crystal nucleus and the surface area of a nucleus with an equilibrium shape may be obtained from an extended formulation of the Gibbs-Wulff theorem [35]

$$3V_* = \frac{2}{\rho_{s_*}(\mu - \mu_{s_*})} \int_{A_*} \gamma dA, \tag{9.2}$$

where A_* is the area of the surface of a critical-sized crystal having an equilibrium shape. From Eqs. (9.1) and (9.2) we have

$$W_* = \frac{1}{3} \int_{A_*} \gamma dA, \tag{9.3}$$

i.e. the work of formation of a critical nucleus is equal to one third of its surface energy. With Eqs. (9.2) and (9.3) the work of formation of a crystal nucleus can be presented as a function of its volume V_* in just the same form as for a nucleus of a spherical shape, i.e.,

$$W_* = \frac{1}{2} V_* \rho_{s*} (\mu - \mu_{s*}) = \frac{1}{2} V_* (p_{s*} - p). \tag{9.4}$$

The calculation of the work of formation of a crystal-phase critical nucleus is considerably simplified with introduction of the effective specific surface energy γ_e [7]

$$\gamma_e = \frac{1}{(4\pi R_0^2)} \int\limits_{A_0} \gamma dA. \tag{9.5}$$

The integral in Eq. (9.5) has been taken over the surface A_0 of the crystal in its equilibrium shape, R_0 is the radius of the sphere with the same volume as the crystal in its equilibrium shape with the surface area A_0.

The introduction of the effective specific surface energy, γ_e, allows one to treat a crystal nucleus in terms of the spherical approximation. In this case, by virtue of Eqs. (9.3) and (9.4), the spherical approximation does not lead to any changes in the difference of chemical potentials, $\mu - \mu_s$, the nucleus volume V and the total surface energy. However, the effective specific surface energy γ_e, determined by Eq. (9.5), will differ from the averaged over the different crystal faces crystal surface energy $\bar{\gamma}$, which is usually calculated as $\bar{\gamma} = \sum \gamma_i A_i / \sum A_i$. Since the surface of the sphere has the smallest area as compared with all other shapes of bodies of the same volume, it is evident that $\gamma_e \geq \bar{\gamma}$. It should also be noted that the tips and edges of an unstable equilibrium crystal are smeared out by thermal fluctuations [36], therefore the form of a critical nucleus may differ from the macroscopic equilibrium shape of a crystal.

For the work of formation of a crystal-phase critical nucleus in a supercooled liquid according to Eqs. (9.2), (9.4), and (9.5) we have

$$W_* = \frac{16\pi}{3} \frac{\gamma_e^3}{\rho_{s*}^2 (\mu - \mu_{s*})^2} = \frac{16\pi}{3} \frac{\gamma_e^3}{(p_{s*} - p)^2}. \tag{9.6}$$

At small supercooling $\Delta\mu$ may be presented as

$$\Delta\mu = (s_l - s_s)\Delta T - \left(\frac{1}{\rho_l} - \frac{1}{\rho_s}\right)\Delta p, \tag{9.7}$$

where s is the specific entropy, $\Delta T = T_m - T$ and $\Delta p = p_m - p$. T_m and p_m are the equilibrium temperature and pressure of melting, respectively. In the process of isobaric cooling, we have

$$\Delta\mu = (s_l - s_s)\Delta T = \Delta h\left(\frac{\Delta T}{T_m}\right), \tag{9.8}$$

where Δh is the melting enthalpy along the crystal–liquid equilibrium coexistence curve. For the alternative case of an isothermal penetration into the region of supercooled states, we get instead the relation

$$\Delta\mu = \left(\frac{1}{\rho_l} - \frac{1}{\rho_s}\right)\Delta p. \tag{9.9}$$

The work of formation of a critical nucleus largely determines the nucleation rate, J. At stationary conditions the value of J is equal to the average number of viable nuclei forming in a unit volume of the liquid in a unit of time. According to CNT, the expression for J may be presented as [5]

$$J = \rho \mathcal{D} Z \exp\left(-\frac{W_*}{k_B T}\right),$$ (9.10)

where $\mathcal{D}$ is the diffusion coefficient in the space of the sizes of nuclei, k_B is the Boltzmann constant, Z is the non-equilibrium Zeldovich factor

$$Z = \left(\frac{W_*/k_B T}{3\pi n_*^2}\right)^{1/2} = \left(\frac{4R_*^2 \gamma_e}{9 k_B T n_*^2}\right)^{1/2}.$$ (9.11)

Here n_* is the number of molecules in a critical nucleus of radius R_*.

Turnbull and Fisher [37] have obtained the following expression for the diffusion coefficient of nuclei in cluster size space in the vicinity of the saddle point of the potential barrier

$$\mathcal{D} = i_* \left(\frac{k_B T}{h}\right) \exp\left(-\frac{E}{k_B T}\right),$$ (9.12)

where i_* is the number of molecules at the surface of a critical nucleus, $k_B T/h$ is the molecular vibration frequency at the crystal surface, E is the activation energy of the process of molecule transition from the liquid into the crystal phase, h is the Planck constant. In the calculations E is usually taken equal to the activation energy of viscous flow [38]. With such assumption, we have

$$\mathcal{D} = \frac{i_* k_B T}{3\pi d_0^3 \eta}.$$ (9.13)

Alternatively, E can be identified with the activation energy of self-diffusion in a melt [39], in such case one gets

$$\mathcal{D} = \frac{24 D n_*^{2/3}}{l^2}.$$ (9.14)

In Eqs. (9.13) and (9.14), η is the viscosity of the liquid, d_0 is the effective diameter of a molecule, D is the self-diffusion coefficient, l is the characteristic diffusion length.

Another approach to determining the value of $\mathcal{D}$, different from that of Turnbull and Fisher [37], was suggested by Zeldovich [4]. Since the growth of supercritical nuclei proceeds by macroscopic laws, then, using hydrodynamic equations, one can determine the growth rate of such nuclei. The growth rate of nuclei is related to $\mathcal{D}$ by a relation which follows from the condition of vanishing of the flow of nuclei at stable equilibrium. Examining the isothermal growth of an isotropic crystal in an unlimited volume of a viscous liquid, we arrive in such approach at [40]

$$\mathcal{D} = \frac{3}{4}\left(\frac{k_B T \rho n_*}{\eta}\right).$$ (9.15)

A rigorous treatment of the kinetics of evolution of crystallization centers requires taking into account the anisotropy of the interfacial free energy of a crystal nucleus even at the stage of obtaining the appropriate kinetic equations describing their evolution [41]. However, the use of the Gibbs-Wulff principle in the thermodynamic analysis of the conditions of equilibrium of a small crystal within a supercooled melt and the use of the spherical approximation for the crystal nucleus with an effective surface energy in calculating the minimum work of nucleation make it possible to employ a one-parameter description of the nucleation kinetics. The system of kinetic equations for the determination of the size and shape distribution function of nuclei given in [41], when going over to the spherical approximation of the nuclei, coincides with the Zeldovich equation [5].

9.3 Description of the Systems under Investigation in the Present Study

9.3.1 Models

To study nucleation we employ the standard method of molecular dynamics (MD) simulations in *NVE*, *NVT*, and *NPT* ensembles [42, 43]. The systems under investigation contain $N = 2048$, 8788, 32000, 108000, 256000, and 1000188 interacting particles, respectively. The particles were located in a cubic cell with periodic boundary conditions. The interaction between particles was described:

– by the Lennard–Jones cutoff pair potential (cLJ model)

$$u(r) = \begin{cases} 4\varepsilon \left[\left(\frac{\sigma}{r}\right)^{12} - \left(\frac{\sigma}{r}\right)^{6} \right], & r < r_c \\ 0, & r > r_c \end{cases} \tag{9.16}$$

where ε is the unit of energy, σ is the unit of length. The potential cutoff radius r_c was taken to be equal to half the length of the cell edge for the system with $N = 2048$ particles at the largest value of the liquid density, and was kept constant with changes in the number of particles. In the investigated temperature range, we have $r_c = (6.27 - 6.58)\,\sigma$.

– Lennard–ones potential in the form proposed by Broughton and Gilmer [22] (mLJ-model):

$$\phi(r) = \begin{cases} 4\varepsilon \left[\left(\frac{\sigma}{r}\right)^{12} - \left(\frac{\sigma}{r}\right)^{6} \right] + c_1, & r \leq 2.3\sigma \\ c_2 \left(\frac{\sigma}{r}\right)^{12} + c_3 \left(\frac{\sigma}{r}\right)^{6} + c_4 \left(\frac{\sigma}{r}\right)^{2} + c_5, & 2.3\sigma < r < 2.5\sigma \\ 0, & 2.5\sigma \leq r \end{cases} \tag{9.17}$$

where

$$c_1 = 0.016132\varepsilon, \qquad c_2 = 3136.6\varepsilon, \qquad c_3 = -68.069\varepsilon,$$

$$c_4 = -0.083312\varepsilon, \qquad c_5 = 0.74689\varepsilon.$$

This potential is constructed in such a way that both the potential and the force become equal to zero at $r = 2.5\sigma$.

From here on all the calculated quantities are given in dimensionless form and marked by the superscript (*). The reduction units are the particle mass m and the parameters σ and ε of the Lennard-Jones potential. We define the reduced distance as $r^* = r/\sigma$, and the reduced potential energy as $u^* = u/\varepsilon$. For the reduced pressure we have $p^* = p\sigma^3/\varepsilon$, density $\rho^* = \rho\sigma^3$, surface free energy $\gamma^* = \gamma\sigma^2/\varepsilon$, and temperature $T^* = k_B T/\varepsilon$. The unit of time, τ^*, is defined as $\tau^* = \sqrt{m\sigma^2/\varepsilon}$. For argon $m = 6.63\cdot10^{-26}$ kg, $\sigma = 0.3405$ nm, $\varepsilon/k_B = 119.8$ K and, accordingly, $\tau^* = \sqrt{m\sigma^2/\varepsilon} = 2.15$ ps. In integrating the equations of particle motion the time step was chosen as $\Delta\tau^* = 0.0023$, which corresponds to $5\cdot10^{-3}$ ps. In modeling the planar crystal-liquid interface, the particles were assumed to be located in parallelepiped shaped cells with edge lengths L_x, L_y, and L_z. Periodic boundary conditions were superimposed on the cell boundaries. The number of particles, N, in it varied from 80 000 to 250 000 depending on the method of calculating the interfacial free energy and the orientation of the crystal phase. In the framework of the described model the thermodynamic and kinetic properties of a Lennard-Jones fluid in stable and meta-stable states were calculated as discussed first in [44–46].

9.3.2 Phase Diagram

The properties of the Lennard-Jones fluid and, consequently, the location of the phase-equilibrium curves in the phase diagram essentially depend on the cutoff radius of the intermolecular potential. For the mLJ model the phase diagram in p, T and T, ρ coordinates in the region of states adjacent to the triple point is shown in Fig. 9.1. The lines of phase equilibria, i. e. the melting line DT_t and the saturation line AT_t have been found from the condition of equality of temperature, pressure and chemical potentials of coexistent phases from the data given in [32, 33]. The parameters of the triple point are

$$T_t^* = 0.692, \quad p_t^* = 0.0012, \quad \rho_{t,l}^* = 0.847, \quad \rho_{t,cr}^* = 0.962.$$

Each of the lines of phase equilibrium has also been determined at temperatures lower than the temperature of the triple point, where they correspond to the equilibrium of meta-stable phases (dotted lines). As distinct from the saturation line, whose meta-stable extension (owing to the absence of the spinodal in a supercooled liquid) has been determined down to $T = 0$, the equilibrium coexistence curve of a crystal and liquid ends at the point of contact K_m of the melting curve with the spinodal of a tensile-stressed liquid. The point K_m is located in the region of negative pressures. At this point

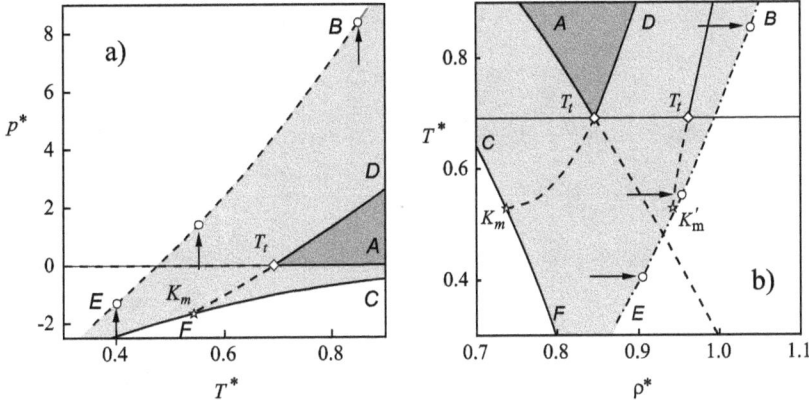

Fig. 9.1: Phase diagram of the Lennard-Jones system close to the triple point and at negative pressures in p^*, T^* (a) and T^*, ρ^* (b) coordinates. T_t is the triple point; DT_t, AT_t are melting and saturation lines; CF is the spinodal of a tensile-stressed (superheated) liquid; K_m is the end point of crystal–liquid phase equilibrium; dotted lines are meta-stable extensions of melting and saturation lines; arrows show the temperatures at which nucleation in a super-cooled liquid was investigated; dashed-dotted line is the line of attainable super-cooling ($J^* = 5 \cdot 10^{-7}$).

the critical (spinodal) state is achieved only for one (liquid) phase. The crystal phase in this case remains stable against infinitesimal perturbations of the state parameters. The parameters of the end point of the crystal–liquid phase equilibrium are

$$T_K^* = 0.529, \quad p_K^* = -1.713, \quad \rho_{K,l}^* = 0.737, \quad \rho_{K,cr}^* = 0.942.$$

Being meta-stable (supercooled), the liquid can also exist at temperatures below T_K^*.

9.4 Methods of Modeling of Spontaneous Crystallization

9.4.1 Mean Life-time Method

The formation of a new-phase nucleus in a meta-stable system is a stochastic event. Both the place and the times of appearance of phase-transition centers must satisfy certain distributions.

Under time-independent boundary conditions the probability of formation of a supercritical nucleus in a short time interval τ, $\tau + \Delta\tau$ is directly proportional to the length of this interval $\lambda\Delta\tau$, where λ is a certain constant parameter. For fixed values of p and T the quantity λ is constant. This parameter has the meaning of the density of probability for formation of a supercritical nucleus, i.e. $\lambda = JV$. The probability for the absence of formation of phase-transition centers in the interval, τ, $\tau + \Delta\tau$, will then be equal to $1 - \lambda\Delta\tau$. The probability $P(m,\tau)$ of appearance of m supercritical nuclei in the time τ in the sample is determined by the Poisson distribution [47]. We have

$$P(m, \tau) = (m!)^{-1}(\lambda\tau)^m \exp(-\lambda\tau).$$ (9.18)

Hence, the probability of absence of supercritical nuclei in a sample in the time τ is

$$P(0, \tau) = \exp(-\lambda\tau),$$ (9.19)

and the probability of appearance of one supercritical nucleus in the time τ is

$$P(1, \tau) = \lambda\tau \exp(-\lambda\tau).$$ (9.20)

At high supersaturations, the growth rate of supercritical nuclei is usually very high. In this case, for a phase transition to proceed in the whole sample, the appearance of only one supercritical nucleus is sufficient. The distribution of nucleation events in repeated experiments with one sample may be found as the ratio of the number of nucleation events k in the time interval, $\tau, \tau + \Delta\tau$, to the total number of experiments $\mathcal{N}$. In this case from Eqs. (9.19) and (9.20) we have [48]

$$\frac{k}{\mathcal{N}} = P(1, \Delta\tau) = \lambda\Delta\tau \exp(-\lambda\tau).$$ (9.21)

The mean expectation time of occurrence of the first supercritical nucleus is

$$\bar{\tau} = \sum_j \frac{\tau_j}{\mathcal{N}},$$ (9.22)

where τ_j is the life-time of the meta-stable sample in the j-th experiment. According to Eq. (9.21), we get then

$$\bar{\tau} = \int_0^\infty \tau\lambda \exp(-\lambda\tau)d\tau = \lambda^{-1} = \frac{1}{(JV)}.$$ (9.23)

Another approach to the determination of the distribution function of nucleation events is connected with the registration of the number of nucleation events in the time τ. The ratio of the number of nucleation events i in the time τ to the total number of experiments $\mathcal{N}$ determines the nucleation event distribution function

$$\frac{i}{\mathcal{N}} = P(\geq 1, \Delta\tau) = 1 - \exp(-\lambda\tau).$$ (9.24)

Eqs. (9.21) and (9.24) may be immediately employed for finding the nucleation rate without involving Eq. (9.22).

In a real experiment the moment of nucleation is generally registered by some secondary indication (phase transition heat release, jump of volume, pressure, etc., caused by the formation of a sufficiently large amount of the new phase) rather than by the appearance of a supercritical nucleus. Thus, the time τ, registered in experiment, includes (besides the time of expectation of a supercritical nucleus, τ_n) the time of its growth τ_g to macroscopic dimensions. The value of $\tau - \tau_g$ may also contain the delay connected with the establishment of both a stationary flow of nuclei [5] and the thermodynamic parameters p and T after the conversion of a meta-stable system to them. It is quite easy to take into account the delays, indicated in the distributions Eqs. (9.21) and (9.24), as the process described by the Poisson law possesses the property of independence of the probability of the onset of an individual event from the beginning of the time registration.

In real experiments the method of life-time measurements was first used in investigating the kinetics of spontaneous boiling-up of superheated liquids [48]. In computer simulations this method was used in studying the melting of a superheated crystal [49] and cavitation in tensile-stressed liquids [50, 51]. A typical histogram of experiments, i.e., distribution of the number of events of crystallization, k, within the interval $\tau, \tau + \Delta\tau$, is shown in Fig. 9.2. Crystallization was registered by the pressure drop in the fluid in the process of crystal growth. The smooth curve has been built based on Eq. (9.21) with the use of the value $\lambda = 1/\bar{\tau}$, obtained in the course of the computer simulation. In determining $\bar{\tau}$ via Eq. (9.22) from the time measured in experiment we subtracted the value of the delay time $\tau_0^* = 50$, which exceeded knowingly the duration of the process of transition of the system under consideration to a prescribed state.

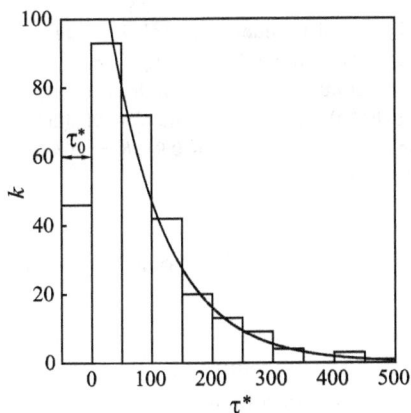

Fig. 9.2: Histogram of experiments on crystallization of the super-cooled Lennard-Jones liquid ($T^* = 0.55, \rho^* = 0.96, \mathcal{N} = 303, \bar{t}^* = 121$, $N = 2048$). The smooth curve is drawn according to Eq. (9.21) with $\bar{t}^* = 121$ and $\tau_0^* = 50$.

The expectation time for crystallization in every individual experiment is presented in Fig. 9.3 in the form of a histogram. At $\tau^* = 200$ the formation of a crystal phase has been registered in 270 experiments out of 303. Thus, we have $P(\tau) = 0.89$. The probability of distribution of expectation times for crystallization is illustrated in Fig. 9.4. The solid line in the figure shows the result of calculating $P(\tau)$ by Eq. (9.24) with $1/\lambda = \bar{\tau} = 121$. The delay time τ_0^* is here equal to 60, which is in good agreement

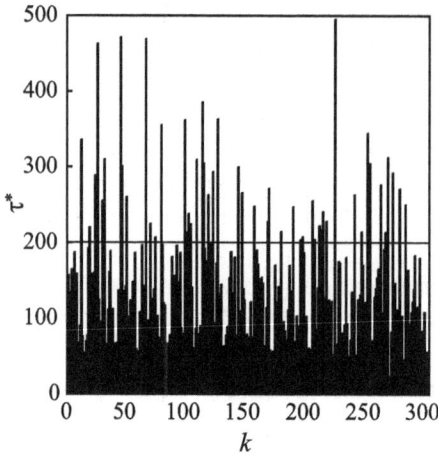

Fig. 9.3: Expectation time for crystallization in a set of $\mathcal{N} = 303$ experiments ($T^* = 0.55$, $\rho^* = 0.96$, $N = 2048$).

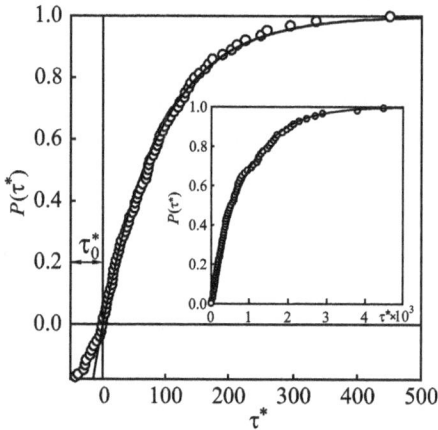

Fig. 9.4: Distribution of expectation times for crystallization ($T^* = 0.55$, $\rho^* = 0.96$, $N = 2048$, $\mathcal{N} = 303$). Smooth curve – by Eq. (9.24) with $\bar{\tau}^* = 121$ and $\tau_0^* = 50$. (In the inset – $T^* = 0.55$, $\rho^* = 0.95$, $N = 2048$, $\mathcal{N} = 218$).

with the processing of data of computer simulations with the Poisson distribution (see Fig. 9.2).

A decrease of super-cooling (super-compression) results in a decreasing slope of the dependence $P(\tau)$ at short times. With a decrease in the reduced liquid density by a value of 0.01 the mean expectation time for crystallization increases by an order of magnitude. In this case, $\tau_0^* \ll \bar{\tau}$ holds. The form of distribution of expectation times for this case is shown in the inset to Fig. 9.4. The results of calculation of the nucleation rate by Eqs. (9.21) and (9.24) are in agreement with each other within an error of 5 %.

9.4.2 Mean First-passage Time Method

A more comprehensive information about the kinetics of initiation of new phase formation, as compared with the mean life-time method, can be obtained by utilizing another well-known notion from the theory of random processes, "the time of the first passage beyond some region boundary" [52]. In application to nucleation processes, this characteristic determines the time during which a new-phase nucleus of size x_0 in the process of its growth reaches the prescribed size, x_1. There is a certain distribution of such times, and yet often it will be sufficient to limit ourselves to the determination of the average value of this time, $\tau(x_0, x_1)$.

Locating the region boundary x_1 at the point corresponding to the vertex of the activation barrier, i.e. assuming $x_1 = x_*$, where x_* is the size of the critical nucleus, and supposing that a new-phase nucleus on reaching the vertex with probability α will transfer the whole system into a two-phase state, for the nucleation rate we may write the following simple formula

$$J = \frac{\alpha}{\bar{\tau} V}. \tag{9.25}$$

Since a critical-sized nucleus has equal probabilities of growth and dissolution, we have to set $\alpha = 1/2$. For a supercritical nucleus $(x_1 > x_*)$, $\alpha \approx 1$ holds.

In computer simulations, for the characteristic size of a new-phase nucleus at every moment of time it is convenient to choose the size of the maximum cluster out of those existing in the system. The size will be determined here by the number of particles in the cluster, i.e., $x = n_{max}$. Knowing the τ-dependence of n_{max}, with a large number of observations one can calculate the mean expectation time for appearance of a cluster of any prescribed size.

The function $\tau(x_0, x_1)$ may be determined theoretically. This was first done in [53], resulting in the expression

$$\tau(x, x_1) = \int_x^{x_1} dy \mathcal{D}(y) \exp\left[\frac{\Delta\Phi(y)}{k_B T}\right] \int_{-\infty}^{y} dz \exp\left[-\frac{\Delta\Phi(y)}{k_B T}\right], \tag{9.26}$$

where $\mathcal{D}(x)$ is the generalized diffusion coefficient, which characterizes the probability of transition of a new-phase nucleus into the neighboring state on the x-axis, $\Delta\Phi(x)$ is the activation barrier (the work of formation of a new-phase nucleus of size, x), which separates the region of heterogeneous fluctuations from the two-phase region.

For a high activation barrier $(\Delta\Phi(x_*)/k_B T \gg 1)$, when the initial distribution in the meta-stable ambient phase is close to the equilibrium one, and the deviation from equilibrium is observed only in a narrow $(|\Delta\Phi(x_*) - \Delta\Phi(x)| \approx k_B T)$ vicinity of the barrier vertex, the integrals in Eq. (9.26) are calculated by the steepest descent method [54] and one obtains

$$\tau(x) = \frac{\bar{\tau}}{2}\left\{1 - \mathrm{erf}\left[Z\sqrt{\pi}(x - x_*)\right]\right\}, \tag{9.27}$$

where erf(x) is the error function, $\bar{\tau}$ is the mean expectation time of appearance of a supercritical nucleus, and Z is the non-equilibrium Zeldovich factor. Thus, in approximating data of computer simulations on the dependence $n_{max}(\tau)$ by Eq. (9.27) one can calculate (along with the nucleation rate and the size of a critical nucleus) the Zeldovich factor, which determines the relative excess of the number of passages $n_{max,*} \rightarrow n_{max,*} + 1$ over the passages, $n_{max,*} \rightarrow n_{max,*} - 1$.

The mean life-time (MLT) and mean first-passage time (MFPT) methods are efficient when the nucleation barrier is small enough and nucleation rates are of the order of $J^* = 10^{-5}-10^{-9}$. In computer experiments, the MFPT method was previously used in studying condensation of supersaturated vapors [54, 55]. The number of particles in a cluster of maximum size was chosen as the characteristic variable (reaction coordinate).

To isolate crystalline clusters in a supercooled liquid, we used a method developed by Frenkel et al. [56], based on the analysis of the ordered bonds between the Steinhardt particles [57]. In this method, a complex vector $q_{6m} = \sum_j Y_{6m}(\hat{r}_{ij})$ is calculated for every particle, i, where Y_{6m} is the spherical harmonic function of sixth order, $\hat{r}_{ij}$ is the unit normal vector that determines the direction of the bond between particle i and its neighbor. The summation is performed with respect to all nearest particles, i.e. those within the distance 1.4σ from particle, i. We introduce the unit complex vector

$$\hat{q}_{6m}(i) = \frac{q_{6m}(i)}{\left[\sum_{m-6}^{6} |q_{6m}(i)|^2\right]^{1/2}} \tag{9.28}$$

and the value of $\hat{q}^*$ complex conjugate to $\hat{q}$. If the scalar product $\sum_{m=-6}^{6} \hat{q}_{6m}(i)\hat{q}_{6m}^*(j)$ between two nearest particles exceeds the value 0.5σ, particle j is a bound neighbor of particle i. When particle i has at least 11 bound neighbors, it is considered to be crystal-like. Two crystal-like particles are a part of the same crystalline cluster if they are bound (and not just neighbors).

In the MFPT-method in every experiment, after the liquid is transferred to the prescribed value of pressure p, after every 100 steps of integration of the equations of particle motion the configuration of particles in the cell was retained, and in every retained configuration one could find the largest crystalline cluster. The time dependence of the number of particles in the maximum-sized cluster for one of the experiments is shown in Fig. 9.5. The time of the first appearance of a crystalline cluster of a definite size in a supercooled liquid is determined in every experiment. Then it is averaged over the number of the experiments conducted.

Fig. 9.6 presents the mean time of the first appearance of a crystalline cluster with a definite number of particles in the liquid. At every given value of n_{max} the distribution of expectation times for the appearance of a crystalline cluster has the Poisson form (see Fig. 9.2). The appearance of a supercritical (viable) nucleus in the system leads to an irreversible growth of the crystal phase (see Fig. 9.5). The growth rate of supercritical nuclei is, in a first approximation, constant. This distinguishes the molecular

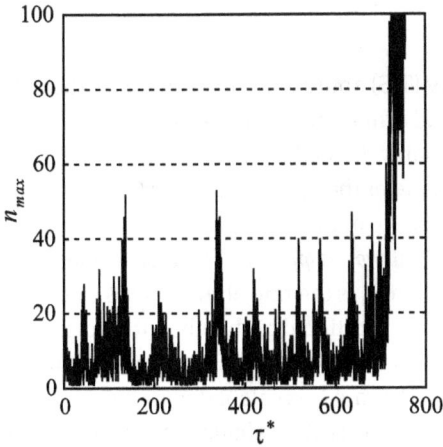

Fig. 9.5: Number of particles in a maximum-sized cluster as a function of time.

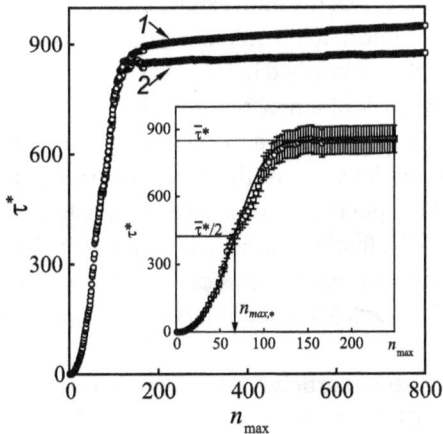

Fig. 9.6: Mean expectation time for appearance in the system of a crystalline maximum-sized cluster as a function of this size: (1) without correction for the supercritical nuclei growth rate, (2) with an introduced correction ($T^* = 0.55$, $\rho^* = 0.95$, $N = 2048$, $\mathcal{N} = 218$). For approximation of the results of computer simulation by Eq. (9.27) see the inset to the figure (solid line).

dynamics calculation of the dependence $n_{max}(\tau)$ from Eq. (9.27), in which it is assumed that supercritical nuclei are removed from the system, i.e. they have an infinitely high growth rate.

The Poisson law of distribution of expectation times for a crystalline cluster makes it possible to allow one to account for the final growth rate of a crystalline cluster rather easily by introducing an appropriate correction, whereupon the dependence $n_{max}(\tau)$ takes the form shown in the inset of Fig. 9.6. Approximating the dependence $n_{max}(\tau)$ by Eq. (9.27) along with the nucleation rate $J = 1/\bar{\tau} V$, where $\bar{\tau}^* = 850 \pm 60$, we obtain the value of the non-equilibrium Zeldovich factor as $Z = 0.015 \pm 0.001$ and the number of particles in a critical nucleus as $n_{max,*} = 67 \pm 2$.

9.4.3 Transition Interface Sampling

The idea of the transition interface sampling (TIS) approach is to decompose the path connecting an initial meta-stable state (A) and a final (two-phase) state (B) into a series of successive intermediate paths along a suitably defined order parameter φ [58]. The order parameter must be a function of the state of the system that evolves monotonically from the initial to the final state. In the space of the order parameter a number of interfaces for the values of φ_i may be defined with $\varphi_A < \varphi_i < \varphi_B$. The nucleation rate is written then in the form of $J = J_0 P(\varphi_n|\varphi_0)$, where J_0 is the rate of transitions through a certain base (first) surface, and $P(\varphi_B|\varphi_0)$ is the conditional probability of reaching the final state, i.e. the crossing probability for the last of the surfaces under consideration after the transition through the base surface. The rate J_0 is calculated in the same way as in the MLT-method. The crossing probability $P(\varphi_B|\varphi_0)$ is calculated by factorizing it into a product of crossing probabilities $P(\varphi_{i+1}|\varphi_i)$ with subsequent computing of each of these probabilities from trajectories that come from the initial state, cross the interface i, and then either cross the interface $i+1$ or return to the initial region [58, 59]. For improved sampling, the parallel paths swapping, described in [60], was used.

When a critical nucleus is formed, the growth of a new phase is thermodynamically irreversible, and the choice of φ_B has no effect on the nucleation rate as long as the inequality $\varphi_B > \varphi_*$ holds, where φ_* is the value of the order parameter for the critical nucleus. The maximum value of the parameter φ_A is given by the level of homophase fluctuations, and the position of the first interface φ_0 by the level of heterophase fluctuations. The distance between the interfaces $\delta\varphi = \varphi_i - \varphi_{i-1}$ was chosen from the condition $P(\varphi_{i+1}|\varphi_i) \approx 0.1 - 0.2$. This procedure requires the generation of $10^3 - 10^4$ paths for every interface [58, 59].

When investigating nucleation, the number of particles in the largest nucleus is usually chosen as the order parameter [61, 62]. We have somewhat modified such a choice of the order parameter by introducing the smoothing function

$$f(r) = \begin{cases} \exp\left(1 - \dfrac{r_0^2}{r_0^2 - r^2}\right), & r < r_0 \\ 0, & r \geq r_0 \end{cases} \tag{9.29}$$

where r_0 is the smoothing parameter. The function $f(r)$ determines the magnitude of the contribution of some particle i, located at the distance $r = |\vec{R} - \vec{r}_i|$ from the marker point $\vec{R}$.

In the case of cavitation a virtual cubic lattice with a period a was assigned in the volume of a MD-cell. The period was taken approximately equal to the average distance between the particles of the liquid. The lattice sites were determined by the vector $\vec{R}/a = l_1\vec{i} + l_2\vec{j} + l_3\vec{k}$, where l_1, l_2, l_3 are integer numbers. For every site we calculated the sum $S = \sum_i f\left(|\vec{R} - \vec{r}_i|\right)$, where the summation was performed with respect to all particles falling within the region of validity of Eq. (9.29). The smoothing parameter r_0 was chosen approximately equal to the radius of the critical nucleus. The order parameter of cavitation was determined as the minimum value of S of those calculated for every lattice site, i.e. $\varphi = \min S$.

In crystallization the order parameter must distinguish the crystal-phase nuclei as compared to the liquid, and also describe "the degree" of their crystallinity. As in the MFPT-method, we employed the method developed by Frenkel and co-workers [56] and the Steinhardt bond-ordering analysis [57] after modifying it. For every particle i, with the help of function $f_1(r)$ we determined a smoothed complex vector $q_{6m}(i) = \sum_j Y_{6m}(\hat{r}_{ij})f_1(|\vec{r}_{ij}|)$, where Y_{6m} is the spherical harmonic function of order six, $\hat{r}_{ij}$ is the unit normal giving the direction of the bond between particle i and its neighbor j, and the summation was carried out over all neighboring particles i. The quantitative measure of correlation between the structures of environment of two particles is the scalar product of normalized vector $\vec{q}_6$ of the nearest particles i and j, i.e.

$$S_{ij} = \sum_{m=-6}^{6} \hat{q}_{6m}(i)\hat{q}_{6m}^*(j)f_2(|\vec{r}_{ij}|),$$

where $\hat{q}_{6m}(i)$ is given by Eq. (9.28), again, and $\hat{q}^*$ is the complex conjugate of $\hat{q}$. The quantity S_{ij} was smoothed by the function $f_2(|\vec{r}_{ij}|)$ within its nearest environment $S_i = \sum_j S_{ij}f_2(|\vec{r}_{ij}|)$, and "collective coherence" for every particle i was determined as $SS_i = \sum_j S_i f_3(|\vec{r}_{ij}|)$. The maximum value of the smoothed function SS_i was taken as the order parameter of crystallization, i.e. $\varphi = \max SS_i$. Smoothing functions f_1, f_2, f_3 are determined by Eq. (9.29), where for f_1, f_2 we have $r_0 \approx 2.5 - 3.0$, and for f_3 the value of r_0 is approximately equal to the radius of a critical nucleus ($r_0 \approx 5.0$).

The order parameter φ has also to distinguish stable and meta-stable states. But the efficiency of the methodology does not dramatically depend on the "quality" of a chosen reaction coordinate [63]. The probability of crossing, $P(\varphi|\varphi_0)$, i.e. the conditional probability of crossing the interface to which the value of the order parameter φ corresponds, is shown in Fig. 9.7. The figure refers to the case of liquid crystallization in an NpT-ensemble at a temperature $T^* = 0.4$ and three initial values of pressure

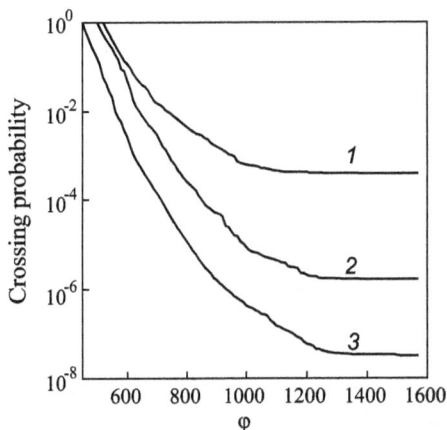

Fig. 9.7: Conditional probability of intersection, $P(\varphi|\varphi_0)$, as a function of the order parameter φ at $T^* = 0.4$ and pressures: $p^* = -1.666$ (1), $p^* = -1.781$ (2), and $p^* = -1.931$ (3).

$p^* = -1.666$, $p^* = -1.781$, and $p^* = -1.931$, respectively. In all three cases the probability of crossing becomes constant at the achievement of a certain value of the order parameter. All crystal nuclei with larger values of the order parameter will eventually result in complete liquid crystallization.

9.5 Temperature Dependence of the Interfacial Free Energy Density Crystal-liquid for Planar Interfaces

9.5.1 Triple Point

The crystal–liquid interfacial free energy γ, which is determined as the reversible work required for the formation of a unit area of the interface between liquid and crystal, plays a key role in the kinetics of crystal phase nucleation from the melt [64]. To calculate the interfacial free energy at the triple-point of a cLJ system we used the method of cleaving potential [22, 24]. The triple point temperature in reduced units is $T_t^* = (k_B T_t/\varepsilon) = 0.692$. The densities of the crystal and liquid phases are equal to

$$\rho_{s,t}^* = \rho_{s,t}\sigma^3 = 0.9619(6), \qquad \rho_{l,t}^* = 0.8469(7),$$

respectively [33, 34]. Simulation was carried out in an NVT – ensemble. In the direction of the z–axis the length of the cell edges L_z was approximately three times larger than in the directions of x and y. The length of the cell edges L_x, L_y was about five cutoff radii of the potential, r_c, and the sectional area $L_x L_y$ exceeded the sectional area of the cell used in [24] by 8 to 9 times. Calculations started with the formation of homogeneous crystal and liquid phases. The crystal phase was given in the form of an ideal FCC crystal. With respect to the cell plane (x, y) the crystal was oriented by crystallographic planes (100), (110), (111).

    The procedure of calculation of the crystal–liquid interfacial free energy included several stages: cutting of the homogeneous crystal and liquid phases by introduction of cleaving walls, superposition of the surfaces of different phases, removal of the cleaving walls. For cleaving walls we used monolayers of an FCC crystal with an appropriate orientation for each of the three formed crystals. The potential of interaction of the wall with the media was given in the form as described in [24]. The repulsive potential of the walls separated the homogeneous system into two subsystems. In the process of separation of homogeneous phases the thermodynamic parameters did not change. The knowledge of the potential of interaction of the wall particles with the particles of the system in the cleaving process allows us to calculate the work $w_{1,2}$ performed for it. Here the indices 1 and 2 refer to the liquid and crystal phases, respectively.

    At the beginning of the second stage the interaction between the particles through the cleaving plane was smoothly "switched off". For this purpose, during integration of the equations of motion of the particles the depth of the potential well was changed with a step 0.001 from 1 to 0 (in terms of ε). Next, the halves of the crystal phase were joined together, so that the surfaces formed as a result of the cell cutting were on the outer sides. The liquid phase cut was joined to these sides in a similar way. As a result of such a procedure, in the cell $2L_z$ long there are formed two crystal–liquid interfaces. It is essential that in the two-phase system, obtained owing to periodic boundary conditions, the interaction between the particles at the place of contact of the homogeneous phases was not disturbed.

After the formation of the compound cell the interaction between the particles of the different phases was smoothly "switched on" and the work w_3 done at the second stage of formation of a two-phase system was calculated by the method of thermodynamic integration. At the final stage, the cleaving walls were removed from the system. The work performed in doing so is w_4. The interfacial free energy was calculated as

$$\gamma = \frac{1}{(2A)} (w_1 + w_2 + w_3 + w_4), \tag{9.30}$$

where $A = L_x L_y$ is the area of one of the two crystal–liquid interfaces in the cell. In the system with the truncated LJ-potential, the values of the interfacial free energy for three crystal orientations are

$$\gamma^*(100) = 0.430(4), \qquad \gamma^*(110) = 0.422(4), \qquad \gamma^*(111) = 0.408(5).$$

The orientation-average value of the interfacial free energy is $\gamma^*_{0,t} = 0.420(5)$.

At the triple point of the system with the LJ-potential cut-off at $r_c^* = 6.78$ the crystal–liquid interfacial free energy exceeds that at the triple point of a system with a mLJ potential [22, 24, 27]. According to the data of Broughton and Gilmer [22], one gets $\gamma^*_{0,t} = 0.35(2)$, of Davidchack and Laird [24], $\gamma^*_{0,t} = 0.359(3)$, of Morris and Song [27], $\gamma^*_{0,t} = 0.362(8)$. With such a comparison one should bear in mind that the temperature of the triple point in a system with a mLJ potential is equal to $T^* = 0.617$, whereas with the cLJ potential it is $T^* = 0.692$. Despite the difference in the values of γ, as given in [27, 24], compared to those presented here, there is good agreement in the anisotropy of γ. As in the papers by Davidchack and Laird [24], Morris and Song [27], we find $\gamma_t^*(100) > \gamma_t^*(110) > \gamma_t^*(111)$. The value of $\gamma_t^*(100) - \gamma_t^*(111) = 0.022(5)$ agrees (within the calculation error) with the data of [24], where this difference is equal to 0.024(4).

9.5.2 Melting Line

The Gibbs–Cahn thermodynamic integration method [29, 30] has been used to determine the temperature dependence of the interfacial free energy along the melting line. The basic equation of this method has the form [30]

$$\frac{d(\gamma_g/T)}{dT} = -\rho_s^{-2/3} \left[\frac{e}{T^2} + \frac{2\tau}{3\rho_s T} \left(\frac{d\rho_s}{dT} \right) \right], \tag{9.31}$$

where ρ_s is the crystal number density, $\gamma_g = \rho_s^{-2/3}\gamma$ is the interfacial free energy per interface particle, e is the excess interfacial energy, and τ is the excess interfacial stress. The derivatives are taken along the line of phase coexistence. The excess interfacial quantities in Eq. (9.31) are calculated as related to the respective Gibbs dividing surface. Knowing γ at one point of the line of phase coexistence, and integrating Eq. (9.31), one can determine the interfacial free energy at any other point of the melting line where ρ_s, e, and τ are known.

At the formation of a two-phase system the crystal phase with a spatial extension $L_z/2$ was located at the center of the cell. Sections of the liquid phase with a length $L_z/4$ were located on two sides of the crystal. The interfaces are perpendicular to the z-axis. Three initial configurations of particles were formed, in which the crystal phase was oriented to the crystal–liquid interface by crystallographic planes (100), (110) and (111). The parameters of the crystal–liquid phase equilibrium (ρ_s, ρ_l, p) and the values of e, τ were calculated for $T^* = T_t^* = 0.692$ and temperatures above $(T^* = 0.85, 1.0, 1.2)$ and below $(T^* = 0.65, 0.625, 0.6)$ the respective value at the triple point. Using the value of the interfacial free energy at the triple point calculated by the method of cleaving potential, and integrating Eq. (9.31), we found the value of γ at all the mentioned temperatures on the phase equilibrium curve.

The profile of the density of the number of particles in a two-phase LJ crystal–liquid system at $T^* = 1.0$ for the crystal phase orientation (111) is presented in Fig. 9.8. In the crystal phase and the transition layer the dependence $\rho(z)$ shows large oscillations. To obtain a monotonic function $\rho(z)$ and to determine the properties of the bulk crystal phase, we used the filter suggested by Davidchack and Laird [65]. The result of its application to the profile of the density of the number of particles is shown in Fig. 9.8 as a smooth line.

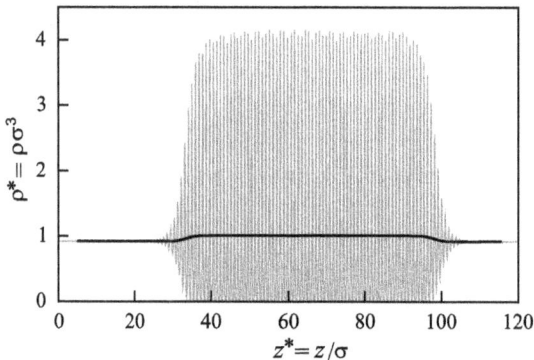

Fig. 9.8: Distribution of the number density in an equilibrium crystal–liquid system at $T^* = 1$ and a crystal-phase orientation to the interface (111). Lines with large oscillations show directly the results of calculations with partition of the cell into layers with the thickness $\Delta z^* = 0.05$ and subsequent averaging over $5 \cdot 10^5$ steps. The smoothed line shows the result of using the profile filtration procedure.

If $\rho_{U,s}$ and $\rho_{U,l}$ are, respectively, the densities of potential energy of the crystal and the liquid phases, the excess interfacial energy with respect to the Gibbs separating surface is

$$e = \frac{1}{A} \left[U_{int} - \rho_{U,s}^A L_e - \rho_{U,l}^A (L_{int} - L_e) \right], \tag{9.32}$$

where U_{int} is the potential energy of the interfacial layer, L_{int} is its width, L_e is the location of Gibbs' dividing surface. The excess interfacial energy as a function of temperature for three crystal orientations with respect to the interface is shown in Fig. 9.9. With decreasing temperature, the magnitude of $e(T)$ increases. If for the orientations (110) and (111) at any temperature the values of the excess interfacial energy are closely similar, at $T^* = 1.0$ the quantity $e(100)$ is approximately two times smaller than $e(110)$ and $e(111)$.

The excess interfacial stress, τ, was determined as the integral of the difference in the normal, P_N, and tangential, P_T, components of the pressure tensor

$$\tau = \int (P_N - P_T) dz, \tag{9.33}$$

where $P_N = P_{zz}$ and $P_T = (P_{xx} + P_{yy})/2$. The results of calculating the interfacial stresses are shown in Fig. 9.10. As distinct from the excess interfacial energy, the interfacial stress is negative. With decreasing temperature, the value of $\tau(100)$ decreases, whereas the values of $\tau(110)$ and $\tau(111)$ increase, demonstrating the approach of the maximum.

Fig. 9.11 presents the results of calculating the interfacial free energy for three orientations of the crystal phase. An increase in the temperature leads to a monotonic increase of γ. The anisotropy of γ in this case varies only slightly, and the relations between the values of $\gamma(100)$, $\gamma(110)$ and $\gamma(111)$ are retained. The derivative $d\gamma/dT$ decreases at the approach to the triple point and with penetration into the region of

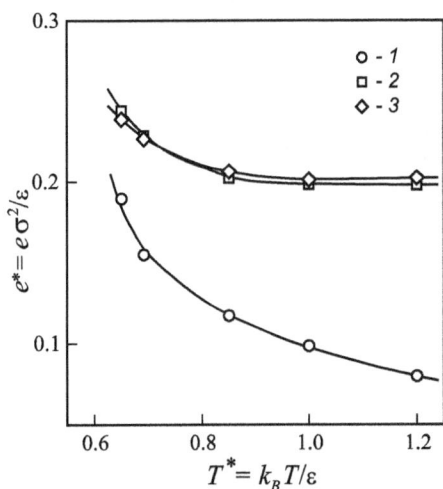

Fig. 9.9: Excess interfacial energy at three crystal orientations to the interface: 1 – (100), 2 – (110), 3 – (111).

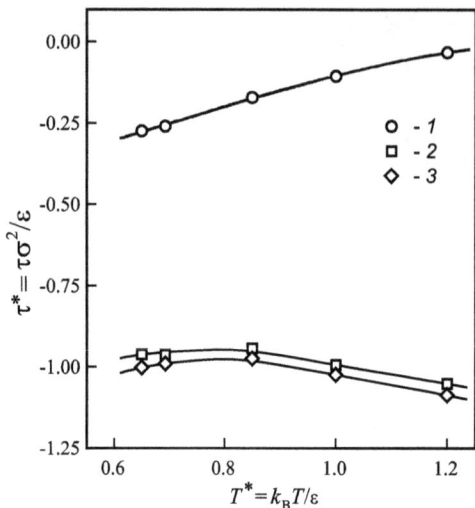

Fig. 9.10: Excess surface tension as a function of temperature for three crystal orientations with respect to the interface: 1 – (100), 2 – (110), 3 – (111).

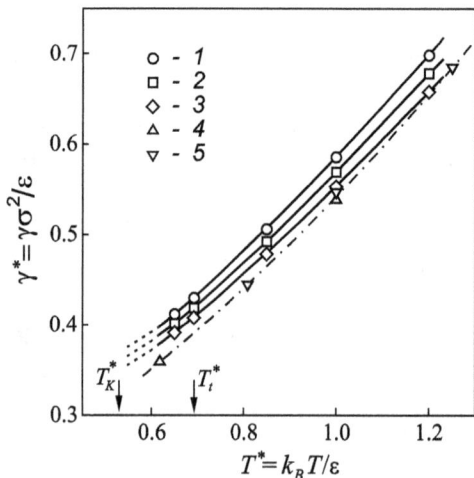

Fig. 9.11: Temperature dependence of the interfacial free energy. Our data for three crystal orientations with respect to the interface are: 1 – (100), 2 – (110), 3 – (111). Orientation-averaged values of γ_0 of Davidchack, Laird et al.: 4 – [24], 5 – [30]. T_t^* and T_K^* are temperatures at the triple point and the critical endpoint. The data of [24, 30] have been obtained for a mLJ-potential [22] with a smaller value of the cutoff radius of the potential than in the present paper.

meta-stable states. The values of γ of Davidchack and Laird [24], Laird, Davidchack et al. [30], obtained for a system with the Broughton and Gilmer [22] mLJ potential, are systematically lower than ours (see Fig. 9.11).

9.6 Kinetics of Crystallization in a cLJ-system

9.6.1 Crystallization Parameters

Fig. 9.12 shows the time dependencies of pressure for several realizations of the molecular dynamics experiments pertaining to temperatures $T^* = 0.865$ and 0.4. Each value of pressure has been obtained by averaging over intervals, $\Delta\tau^* = 0.232$. The dependencies $p^*(\tau^*)$ have three clearly distinguished sections. The initial section corresponds to a homogeneous meta-stable state of a liquid. Throughout the lifetime of a meta-stable phase, crystals of subcritical size appear and disappear in it. The formation of a supercritical nucleus in a liquid leads to a pressure decrease. The pressure and internal energy established after crystallization of the liquid, as a rule, exceed those in the state of complete crystalline ordering (Fig. 9.12a).

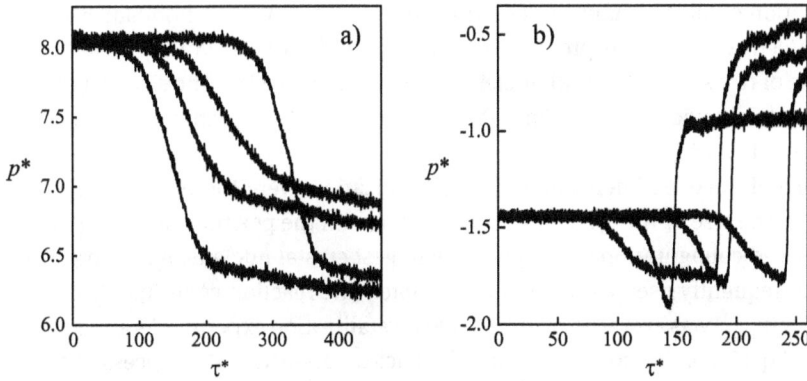

Fig. 9.12: Time dependence of pressure for several molecular dynamics experiments on crystallization of a super-cooled liquid at temperatures $T^* = 0.865$ ($a: \rho^* = 1.029, N = 32000$) and $T^* = 0.4$ ($b: \rho^* = 0.895, N = 32000$).

The character of liquid crystallization is qualitatively different at temperatures $T^* = 0.865$ and $T^* = 0.4$. At $T^* = 0.4$ the initial pressure drop in the process of crystallization is then followed by its subsequent rise (Fig. 9.12b). The latter is connected with the formation of cavitation pockets in the crystallizing liquid [66].

It has been established that at given temperature and pressure in systems with different numbers of particles the condition of inverse proportionality of the mean lifetime $\bar{\tau}$ to the volume (number of particles) of the system ($\bar{\tau} \sim 1/V$) is met. This property indicates that the process proceeds with the same probability in any of the volume elements and that the assumed periodic boundary conditions do not affect the process (Fig. 9.13). The possible effect of periodic boundary conditions on the crystallization of a super-cooled liquid with a number of particles in the system less than 1500 was previously established in [67].

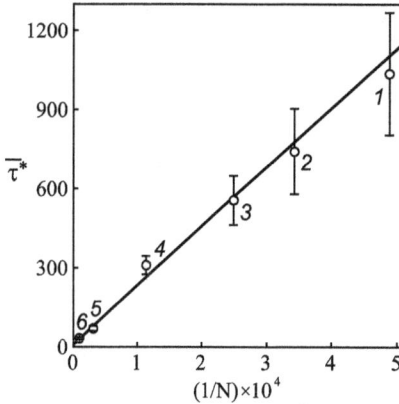

Fig. 9.13: Dependence of the mean lifetime of a super-cooled liquid on the inverse value of the number of particles in the cell at $T^* = 0.865$, $\rho^* = 1.029$: (1) $N = 2048$, (2) 2916, (3) 4000, (4) 8788, (5) 32000, (6) 108000.

A detailed comparison of CNT and computer simulation presupposes an independent determination of all the quantities entering Eq. (9.10). The MD-method makes it possible to directly calculate the pre-exponential factor in Eq. (9.10), which is determined by the rate of transition of crystal nuclei through the critical size. The size of a critical nucleus and Zeldovich factor in an MD-simulation may be determined by using the MFPT-method [54, 55].

In each of the experiments on liquid crystallization every 100 or 1000 steps, depending on the size of the system, we saved a file with the positions of particles and in every saved configuration determined the largest crystal nucleus, n_{max}. This value n_{max} is subsequently used as the order parameter (the reaction coordinate). The dependence $\tau(n_{max})$ was averaged over 100–300 crystallization experiments and approximated by Eq. (9.27). The configurations of particles, retained in the process of calculation of $\bar{\tau}(n_{max})$ and containing near-critical nuclei, were subsequently used to determine the coefficient of diffusion of nuclei in cluster size space. Following Auer and Frenkel [68], we determine the coefficient of nuclei diffusion near the critical size as

$$\mathcal{D} = \frac{1}{2}\frac{\langle \Delta n^2_{max}(t)\rangle}{t}, \qquad \Delta n^2_{max}(t) = [n_{max}(t) - n_{max}(0)]^2, \qquad (9.34)$$

where $\Delta n^2_{max}(t)$ is the mean-square change of the number of particles in a crystal nucleus. In calculating $\mathcal{D}$, 100 configurations with n_{max} close to n_* ($n_* \simeq n_{max} \pm 5$) were selected, which we subsequently used as seeds for new MD-configurations, with the changed initial particle velocities corresponding to the prescribed temperature. We re-determined the zero time whenever n_{max} was close to n_*. The simulation was terminated if the crystalline cluster contained less than 20 or more than $2n_*$ particles. The slope of the curve $\langle [n_{max}(t) - n_{max}(0)]^2 \rangle$ (Fig. 9.14) determines the value of $\mathcal{D}$.

The effective surface free energy γ_e at the crystal nucleus–liquid interface was calculated from the data on p_{cr_*} and R_*^* from the pressure balance relation $p_{cr_*} - p = 2\gamma_e/R_*^*$. The data obtained indicate a rather weak pressure (size) dependence of γ_e and in the subsequent discussion this dependence is ignored. The values of surface

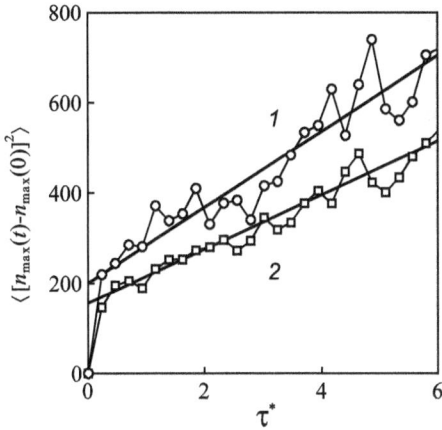

Fig. 9.14: Mean-square change in the number of particles in the critical nucleus at $T^* = 0.6185$ and densities $\rho^* = 0.960$ (1) and $\rho^* = 0.975$ (2). The solid line corresponds to the best approximation of calculation data (circles and squares).

free energy at the critical crystal–liquid interface, obtained in such an approximation, are shown in Fig. 9.15. The value of γ_e is a monotonically increasing function of temperature. As distinct from the surface free energy at a crystal nucleus–liquid interface, the surface free energy γ_∞ of a planar interface is determined only down to the temperature ($T_K^* = 0.529$) of the terminal critical point of the melting curve, where the crystal–liquid phase equilibrium ceases to exist.

Another quantity in Eq. (9.14) that requires determination is the parameter l. The value of l can be found by calculating the rate of addition of particles to a critically sized nucleus [68]. We determine l by referring to values of J obtained in a computer experiment. At a fixed temperature and γ_e which does not depend on the nucleus size the value of l may be assumed as constant (with an accuracy of $\pm 7\,\%$). A decrease in the temperature results in decreasing values of l. In the whole investigated range of temperatures and pressures the values of J, calculated with an error close to the error

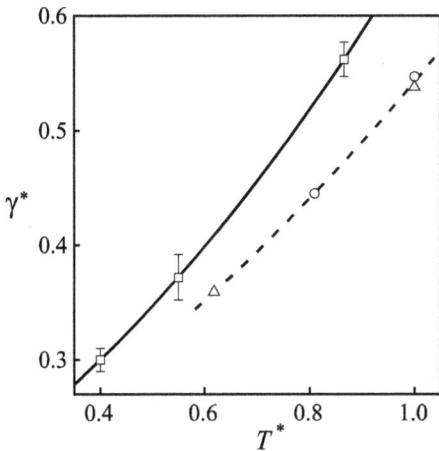

Fig. 9.15: Temperature dependence of the surface free energy at the crystal nucleus–liquid interface ($\square$) and the flat crystal–liquid interface ($\triangle$, $\bigcirc$) [24, 30].

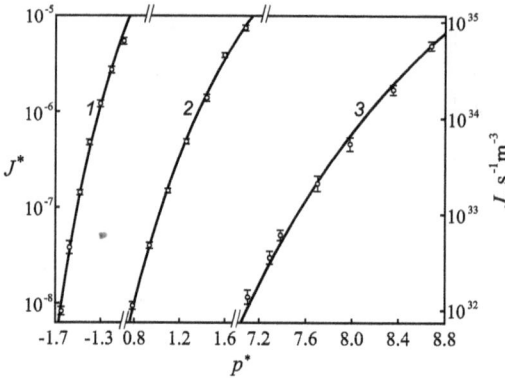

Fig. 9.16: Pressure dependence of the nucleation rate in the super-cooled Lennard-Jones liquid at temperatures: $1 - T^* = 0.4$, $2 - 0.55$, $3 - 0.865$. Solid lines present the calculation by homogeneous nucleation theory, Eqs. (9.6), (9.10)–(9.14).

of their determination, may be described in the framework of homogeneous nucleation theory in the approximation of $\gamma_e = \text{const}$, $l = \text{const}$ at $T = \text{const}$. The results of such an approximation are shown in Fig. 9.16 by solid lines. The nucleation rate has been calculated from the data on the mean expectation time of liquid crystallization [66]. The pressure dependence of the logarithm of the nucleation rate is shown in Fig. 9.16. The minimum and maximum values of the nucleation rate were 10^{-8} and 10^{-5} in reduced units and 10^{32} and 10^{35} $\text{s}^{-1}\text{m}^{-3}$ in dimensional ones, respectively. The increase in supercooling and temperature is accompanied by the decrease in the slope of the pressure dependence of J. The error bars in Fig. 9.16 represent the mean-square deviation of the nucleation rate.

The crystallization of a super-cooled Lennard-Jones liquid is studied in [12, 72, 73]. Wang et al. [72] investigated crystal nucleation by Monte Carlo computer simulations. The height of the activation barrier was evaluated as the difference of the excess free energies of a critical n_* and an equilibrium n_{min} cluster in a meta-stable liquid, i.e. $W_* = \Delta F(n_*) - \Delta F(n_{\text{min}})$. At $\rho^* = 0.95$ and $T^* = 0.55$ from data of [72] one obtains $n_{\text{min}} = 7$, $n_* \simeq 22$, $W_*/k_B T \simeq 0.8$. To study the crystallization of Lennard-Jones liquids, Lundrigan and Saika-Voivod [73] used the MFPT-method [54]. By the data of [73] at $\rho^* = 0.95$ and $T^* = 0.55$ the number of particles in a crystal nucleus is $n_* = 65$, the reduced height of the nucleation barrier equals $W_*/k_B T = 15.74$, and the nucleation rate is $J^* = 9.4 \cdot 10^{-8}$. In our calculations at $\rho^* = 0.95$ and $T^* = 0.55$ we obtained

$$n_* = 48, \qquad \frac{W_*}{k_B T} = 14.0, \qquad J^* = 4.9 \cdot 10^{-7}.$$

We did not observe any distinct differences in the kinetics of crystallization of supercooled Lennard-Jones liquids at temperatures above and below the temperature of the critical end point of the melting curve. As distinct from the surface free energy at a planar crystal–liquid interface, which is not determined at $T < T_K$, the surface free energy at a curved crystal nucleus–liquid interface has quite a definite value here. No peculiarities of the crystallization process which might show that a super-cooled liquid has a spinodal and which were previously discussed in [72, 75] have been revealed up to a nucleation rate $J^* = 10^{-5}$ as well. At a fixed nucleation rate a decrease in the temperature of the liquid is accompanied by a decrease in the height of the nucleation barrier W_* and an increase in the value of $W_*/k_B T$, which is a measure of stability of a meta-stable system against the nucleation process.

9.6.2 Nucleation Rate

As evident from Fig. 9.12, the pressure clearly traces the beginning of the process of crystallization in a super-cooled liquid. If a phase transition initiates only the first nucleus, a certain arbitrariness in the choice of this beginning, by virtue of the Poisson law Eq. (9.21), may not lead to considerable differences in the values of the mean expectation time for crystallization. An increase in super-cooling is accompanied by a decrease in the particle mobility owing to the increase of the liquid viscosity and the decreasing height of the nucleation barrier. This results in a decreasing growth rate of supercritical nuclei. Fig. 9.17 shows the time dependence of the number of particles in the largest crystalline cluster along with the dependence $p^* (\tau^*)$. As can be seen from Fig. 9.17, the moments of the appearance of a supercritical nucleus in the system and the pressure decrease correlate well. The constancy in the first approximation of the values of $dp/d\tau$ and $dn_{max}/d\tau$ after the beginning of crystallization ($T^*, p^* = $ const) and the Poisson character of nucleation make it possible to vary the moments of fixation of the liquid crystallization over sufficiently wide limits. The values of $\bar{\tau}$ calculated with 3, 5 and 7 percent pressure decreases lie within the error of determination of $\bar{\tau}$.

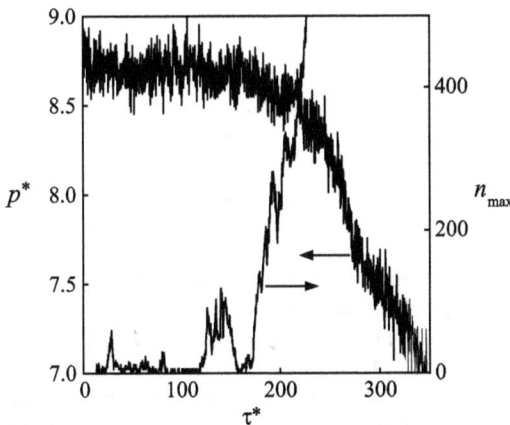

Fig. 9.17: Time dependence of the pressure and the number of particles in the largest crystalline cluster by the data of one experiment at $T^* = 0.865, p^* = 1.040, N = 2048$.

9.6.3 Comparison of Homogeneous Nucleation Theory with Computer Simulation

The results of the computer experiment were compared with homogeneous nucleation theory with the use of Eqs. (9.6), (9.10)–(9.14). Such quantities as density ρ, pressure p, which appear in them, were calculated in the process of simulating crystallization of a super-cooled liquid. The pressure p_{cr_*} in the critical crystal nucleus was determined from the condition of equality of the chemical potentials of the liquid and crystal phases by the data from [32, 33]. The self-diffusion coefficient D in the super-cooled Lennard-Jones liquid was calculated by molecular dynamics methods [45]. The volume, V_*, of the critical nucleus was determined by the nucleation theorem with allowance for the dependence of the self-diffusion coefficient and the pressure in the critical nucleus on pressure in the liquid [69–71] via

$$V_* = \frac{k_B T \left(\dfrac{\partial \ln J}{\partial p} - \dfrac{\partial \ln D}{\partial p} \right)}{\dfrac{\partial p_{cr_*}}{\partial p} - 1}. \tag{9.35}$$

The number of particles in the crystal nucleus was found to be equal to $n_* = \rho_{cr_*} V_*$, where ρ_{cr_*} is the density in the critical crystal nucleus.

An increase of pressure in the liquid leads to a decrease in the derivative $\partial \ln J / \partial p|_T$, and, consequently, in the size of a critical crystal. In the investigated range of nucleation rates and temperatures the number of particles in the critical crystal nucleus varied from 30 to 80, and their effective radius was $R^* = 1.9 - 2.6$.

If in an experiment the pressure or the temperature is kept constant, the dependence of the stationary rate of formation of crystal nuclei in a super-cooled liquid on temperature and pressure, respectively, will be dome-shaped (Fig. 9.18). The decrease in the rate of formation of crystallization centers at high pressures is connected with the decrease of the self-diffusion coefficient (increasing viscosity). The very fact of

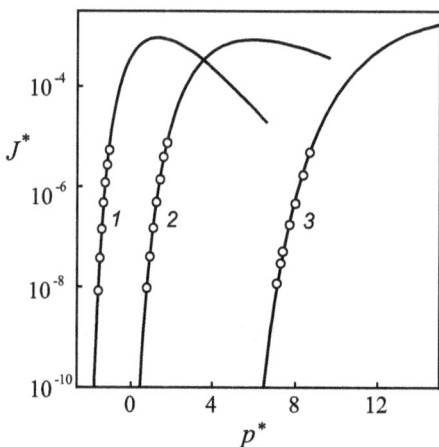

Fig. 9.18: Pressure dependence of the nucleation rate in the super-cooled Lennard-Jones liquid: (1) $T^* = 0.4$, (2) $T^* = 0.55$, (3) $T^* = 0.865$. Solid lines present the calculation by homogeneous nucleation theory, Eqs. (9.6), (9.10)–(9.14). Circles show the results of molecular-dynamics simulations.

the presence of such a maximum on the dependencies $J(p)$ or $J(T)$ points to the fundamental possibility of realization of high super-cooling in liquids. A passage over the maximum of the dependence $J(T)$ was observed in studying the crystallization of droplets of liquid germanium 0.01–5 μm in diameter [76]. Droplets with diameters smaller than 0.3 μm did not crystallize and were transformed into a glassy state at temperatures lower than that corresponding to the maximum of the dependence $J(T)$.

9.6.4 Nucleation in the Region Below the Endpoint of the Melting Line

In normally melting substances a liquid phase below the temperature T_t is always meta-stable. In the region between meta-stable extensions of the melting and saturation lines, which mainly corresponds to negative pressures, nucleation may proceed both with the formation of a crystal phase and by means of liquid cavitation. Negative pressures may be realized in molecular systems, where along with repulsive forces there act forces of intermolecular attraction, i.e. in liquids, solids, but not in gases. The equilibrium coexistence of a meta-stable liquid and a meta-stable crystal at negative pressures ceases at the critical endpoint T_K, i.e. the point of contact of the melting line and the spinodal of a tensile-stressed liquid [33]. Below T_K in the absence of actions initiating phase transitions the processes of crystal-phase nucleation and cavitation compete at nucleation rates that may be realized in computer experiments.

We have used MD-methods to investigate the kinetics of spontaneous cavitation and crystallization in the cLJ-model at temperatures below the temperature of the critical endpoint of the melting line [77]. Two methods of determining the nucleation rate have been employed, the MLT-method [66, 74] and the TIS-method [58]. The models under investigation contained from 2048 to 500 000 particles. By the MLT-method nucleation rates were calculated in an NVE–ensemble at temperatures $T^* = 0.35$ and 0.4. The TIS-method was realized in NVE and NpT–ensembles at $T^* = 0.4$.

Fig. 9.19 presents density dependencies of the crystallization and cavitation rates of the liquid. As the liquid is stretched, the crystallization rate decreases, and the cavitation rate increases. After the formation of a viable cavity the continuity of the liquid phase is disrupted. The crystallization of the liquid also is terminated with the appearance of cavities in it. The formation of crystal nuclei leads to tensile stresses in the parent phase, and it cavitates. In the MLT-method at $T^* = 0.35$ the signal of appearance of the first crystal nucleus at the approach to the point where $J_{cav} = J_{cr}$ holds becomes very weak, whereas cavitation is registered often enough even after the passage through this point. At $T^* = 0.4$ in the TIS-approach one can observe the approach of a state where $J_{cav} = J_{cr}$ both in the case of cavitation and in the case of crystallization of a liquid. The data of the TIS- and MLT-methods are in good agreement in the overlapped density range. A good agreement is also observed for the values of J obtained in NVE- and NpT-ensembles, which indicates a weak effect of initiation conditions on nucleation.

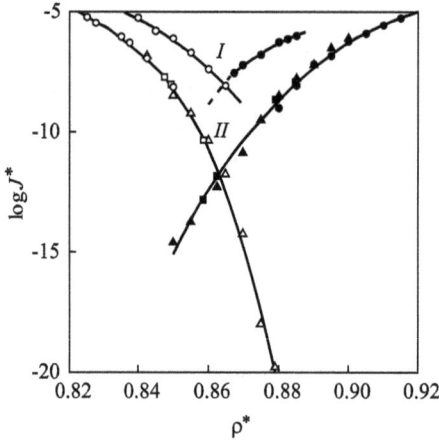

Fig. 9.19: Rate of cavitation (white symbols) and crystallization (black symbols) of a liquid at temperatures $T^* = 0.35$ (I) and $T^* = 0.4$ (II). Calculations have been made by the MLT method in NVE-ensemble ($\circ$), the TIS method in an NVE-ensemble ($\square$) and the TIS method in an NpT-ensemble ($\triangle$). At the point where $J_{cav} = J_{cr}$ at $T^* = 0.35$ we have: $\rho^* = 0.864$, $p^* = -2.32$, $J^* \approx 1.0 \cdot 10^{-8}$; at $T^* = 0.4$: $\rho^* = 0.863$, $p^* = -1.92$, $J^* \approx 1.3 \cdot 10^{-12}$.

Let us discuss the obtained data in the context of CNT. Despite the seemingly qualitative difference in the processes of crystal nucleation and cavitation, the formulas describing these phenomena are identical. The stationary rate of both crystal nucleation and cavitation is given by Eq. (9.10).

Critical cavities at $T < T_t$ contain no gas-phase particles. Therefore, the growth of near-critical cavities in the framework of the CNT traditional scheme may be presented as the transfer of elementary holes from the liquid into a cavity. The main idea is that the growth of a near-critical bubble at cavitation in the framework of the traditional scheme of homogeneous nucleation theory corresponds to the transfer of holes from the liquid rather than to the transfer of molecules into the vapor phase. The coefficient $\mathcal{D}$ has in this case the meaning of the frequency of the appearance of holes at the cavity surface. Every such transition is accompanied by overcoming the activation barrier E', for which we may assume $E' \approx E$. Usually E is considered to be equal to the activation energy of viscous flow or of the self-diffusion coefficient in the liquid [7].

Estimates show that the product $i_* Z$ during crystallization and cavitation is close to unity. Thus, at $T^* = 0.4$ in the case of crystallization $i_* Z \approx 1.3$, and in the case of cavitation $i_* Z \approx 1.1$. Therefore, from Eqs. (9.10) and (9.12) we have

$$J \approx \rho \left(\frac{k_B T}{h} \right) \exp \left(-\frac{E}{k_B T} \right) \exp \left(-\frac{W_*}{k_B T} \right). \tag{9.36}$$

According to Eq. (9.36), at the point where the rate of crystallization is equal to that of cavitation $W_{S*} \approx W_{*V}$, the work of formation of a critical nucleus is given by Eq. (9.6), where we have

$$\frac{\gamma_{LV}}{\gamma_{LS}} \simeq \left(1 - \frac{p_{S*}}{p}\right)^{-2/3}. \tag{9.37}$$

Here n_* is the number of particles in a critical crystal (crystallization) or the number of elementary holes (cavitation), γ is the liquid–solid (γ_{LS}, crystallization) or liquid–void (γ_{LV}, cavitation) interfacial free energy, p_{S*} is the pressure in a critically sized crystal. The pressure in a critical cavity is given by $p_{V*} = 0$. Since $p_{S*} > p$ holds, from Eq. (9.37) we have $\gamma_{LV} > \gamma_{LS}$.

The pressure in a critical-sized crystal nucleus p_{S*} is determined by the condition of equality of the chemical potentials of the particles in the nucleus and the ambient phase, i.e. $\mu_L(p, T) = \mu_S(p_{S*}, T)$. In [44] the MD-method is used to calculate the thermodynamic properties of the Lennard-Jones liquid and crystal in meta-stable states. With these data we have determined the chemical potentials and found the value of p_{S*}. For $T^* = 0.35$, at the point, where $J_{cav} = J_{cr}$, $\gamma_{LV} \simeq 5.4\gamma_{LS}$. For $T^* = 0.4$ we have $\gamma_{LV} \simeq 4.5\gamma_{LS}$. According to MD calculations the surface tension of the Lennard-Jones fluid, along the meta-stable extension of the saturation line for a planar interface, is equal to 1.96 at $T^* = 0.35$ and 1.83 at $T^* = 0.4$. Thus, the solid–liquid interfacial free energy at $T^* = 0.4$ is evaluated as 0.41. By the data of [66] the surface free energy of a crystal nucleus in the Lennard-Jones system at $T^* = 0.4$ is equal to 0.30 and a higher nucleation rate $J^* = 5 \cdot 10^{-7}$ is found.

From the nucleation theorem [70, 78] it follows that

$$\frac{dW_*}{d(p_* - p)} = -V_*$$

holds, where V_* is the volume of the critical nucleus defined by choosing the equimolecular dividing surface. At the point of intersection of the isotherms of the rates of crystal nucleation and cavitation

$$\left|\frac{d\ln J}{d(p_{S*} - p)}\right|_S > \left|\frac{d\ln J}{d(-p)}\right|_V.$$

Hence, $V_{S*} > V_{V*}$ holds. At $T^* = 0.4$, we have $V_{V*} \simeq 30.5$, $R_{V*} \simeq 1.9$. For a critically sized crystal nucleus at this temperature

$$V_{S*}^* \simeq 110, \qquad R_{S*}^* \simeq 3.0, \qquad n_{S*} \simeq 107$$

holds. In [66] at the same temperature but with a higher nucleation rate ($J^* = 5 \cdot 10^{-7}$) the results

$$V_{S*}^* \simeq 45.5, \qquad R_{S*}^* \simeq 2.2, \qquad n_{S*} \simeq 46$$

are reported.

So, liquid states below the triple point located in between the meta-stable extensions of the saturation and melting lines allow the simultaneous formation of cavities and crystals. What will form directly from a tensile-stressed liquid (disperse phase-crystal or cavity (void)) depends on the mechanism (homogeneous or heterogeneous) of nucleation. In the case of homogeneous nucleation the knowledge of the nucleation rate makes it possible to draw certain conclusions about the predominance of transitions in one or another direction, to distinguish the regions of temperatures and pressures with an increase of nucleation of a crystal phase or voids. Since there exists a simple relation between the rate and the work of nucleation (see Eq. (9.36)), a comparison may be made with respect to the work of critical cluster formation, W_*. At the point where the rates of cavitation and crystal nucleation are equal one can observe simple relations between the volumes of nuclei and their surface free energies.

9.7 Kinetics of Crystallization in the mLJ-system and Free Energy of the Clusters of the Crystalline State

9.7.1 Pressure Dependence of the Nucleation Rate

The equilibrium melting temperature T_m is the starting point in studying the phenomena of liquid super-cooling. We have determined this temperature in mLJ-models with a planar liquid–crystal interface containing 25 966 ($T_m^* = 0.6175$ at $p^* = -0.011$) and 122 778 ($T_m^* = 0.6180$ at $p^* = -0.007$) particles in much the same way as it was done in [79]. The obtained value of the equilibrium melting temperature is in good agreement with the results of Broughton and Gilmer ($T_m^* = 0.617$ at $p^* = 0$) [22], Morris and Song ($T_m^* = 0.620$ at $p^* = -0.01$) [27] and Davidchack and Laird ($T_m^* = 0.617$ at $p^* = -0.02$) [24].

We have also calculated the p, ρ, T–properties and the internal energy in stable and meta-stable states of an mLJ-liquid and an mLJ-crystal. Calculations were made for states along isotherms in the range of reduced temperatures from 0.4 to 0.7 with a temperature step 0.1. The data obtained are described by the thermal and caloric equations of state, from which the entropy, the chemical potential and the enthalpy of the liquid and the crystal phase have been calculated. Parameters of the melting line of an mLJ-system have been determined from conditions of diffusion ($\mu_l = \mu_S$) and mechanical ($p_l = p_S$) equilibria [80]. At $p^* = 0$ we obtained

$$T_m^* = 0.6180 \pm 0.0005, \qquad \rho_l^* = 0.8288, \qquad \rho_s^* = 0.9456,$$

$$\Delta h^* = 1.0339, \qquad \Delta s^* = 1.6730.$$

The dependence of the crystal nucleation rate on the value of $(\Delta p^*)^{-2} = (p_{s_*}^* - p^*)^{-2}$ at the temperature of 0.6185±0.0005 is presented in Fig. 9.20. The dependence of $\log J^*$ on $(\Delta p^*)^{-2}$ is close to a linear one. The rigorous linearity of this dependence indicates the fulfillment of the relations $\rho^* \mathcal{D}^* Z = \text{const}$ and $\gamma_e^* = \text{const}$. The best agreement with the results of MD-simulations is observed with

$$\gamma_e^* = 0.404 \pm 0.002, \qquad \rho^* \mathcal{D}^* Z = 1.17 \pm 0.06.$$

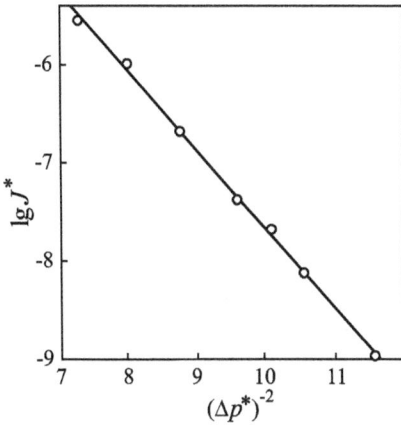

Fig. 9.20: Nucleation rate as a function of $(\Delta p^*)^{-2} = (p_{s_*}^* - p^*)^{-2}$ at $T^* = 0.6185$.

The number of particles in a crystal, calculated from the condition of mechanical equilibrium of a critical nucleus and a super-cooled liquid at $\rho^* = 0.960$ and $\rho^* = 0.975$, is approximately equal to the average value of n_* obtained by the MFPT-method.

The values of the self-diffusion coefficients D and the shear viscosity η at $T^* = 0.6185$ were calculated in [45]. The values of d_0^* and l^* were determined by the data on D, η and $\mathcal{D}$ from Eqs. (9.13) and (9.14). Both d_0^* and l^* are less than unity. For the value of l^* this fact was previously noted in [73]. Eq. (9.15) gives for $\mathcal{D}$ a value which is 20–30 times smaller than the one obtained by a direct MD calculation with the use of Eq. (9.34). It should be mentioned that Eqs. (9.13) and (9.14) lead to the same values of $\mathcal{D}$ employing Eq. (9.34) with $l^* \simeq 1.2$, $d_0^* \simeq 1$.

Each of the parameters in the complex $\rho \mathcal{D} Z$ is a function of the density. The values 0.625 ($\rho^* = 0.960$) and 0.600 ($\rho^* = 0.975$) have been obtained from the results of MD calculations of $\mathcal{D}$ and Z for the complex $\rho^* \mathcal{D}^* Z$. Thus, the value of $\rho \mathcal{D} Z$ rather weakly depends on the density (pressure), but differs considerably (by approximately 35 %) from the result of its determination by the data on the pressure dependence of the nucleation rate ($\rho^* \mathcal{D}^* Z = 1.17$). If we take into account the density dependence of the complex $\rho \mathcal{D} Z$ in Eqs. (9.6) and (9.10), then γ_e will depend on the curvature of the separating surface of a spherical crystal nucleus. At $\rho^* = 0.960$ we obtain $\gamma_e^* = 0.399$, $R_*^* = 2.47$, and at $\rho^* = 0.975$ we have $\gamma_e^* = 0.396$, $R_*^* = 2.14$. Thus, in the pressure range under investigation the dependence $\gamma_e(R)$ is rather weak, the effective surface energy of crystal nuclei at $T = $ const being higher than at a planar interface (Fig. 9.21) [81].

9.7.2 Temperature Dependence of the Nucleation Rate

Limiting super-cooling of liquids is usually investigated at isobaric conditions at pressures close to atmospheric one. In experiments on spontaneous crystallization of liq-

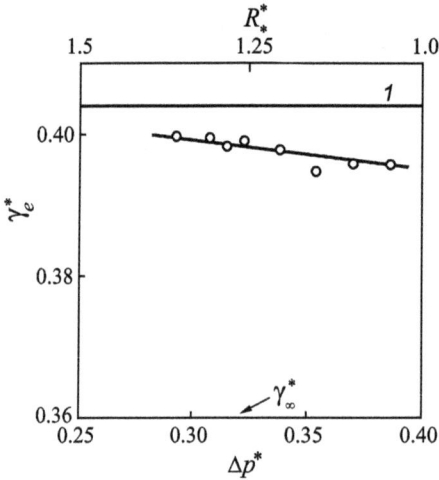

Fig. 9.21: Effective surface energy of crystal-phase critical nuclei as a function of pressure and radius at temperature $T^* = 0.6185$ (circles). (1) values of γ_e^* in approximation of linear dependence of $\ln J^*$ on $(\Delta p^*)^{-2}$. The surface free energy at the planar liquid–crystal interface is equal to $\gamma_\infty^* = 0.360$ [24]. The dependence of γ_e^* on R_*^* is shown qualitatively with the size uncertainty of ± 0.03.

uid metals in particles of micron and nano-sizes it has been noted [7] that the logarithm of the nucleation rate $\ln J$ in a first approximation is a linear function of the complex $T^{-1}(T_m - T)^{-2}$. For tin this is fulfilled in the range of $J = 10^5 - 10^{19}\ \mathrm{s}^{-1}\mathrm{cm}^{-3}$, which in dimensionless units, used in the present work, corresponds to the range of $J^* = 10^{-29} - 10^{-15}$.

In MD-simulations we have calculated the temperature dependence of the crystal nucleation rate at $p^* = 0.00 \pm 0.01$ in the range of $J^* = 10^{-10} - 10^{-5}$. The dependence of $\ln J^*$ on $T^{*-1}(T_m^* - T^*)^{-2}$ is presented in Fig. 9.22. It has a weakly expressed curvature.

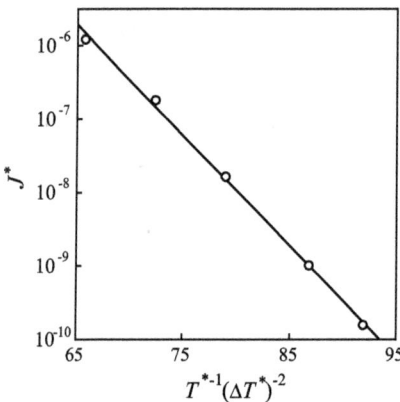

Fig. 9.22: Nucleation rate as a function of $T^{*-1}(T_m^* - T^*)^{-2}$ at $p^* = 0$.

Approximation of the obtained data by a linear function in accordance with Eqs. (9.6), (9.8) and (9.10) gives $\rho^* \mathcal{D}^* Z \approx 1.1 \cdot 10^4$, $\gamma_e^* \approx 0.373$. The value of γ_e^* differs from the data of [24, 27] on γ_∞^* by approximately 4 %.

Just as it is done in the previous section, we have calculated the temperature dependencies of $\mathcal{D}$ and Z, determining in such a way the pre-exponential factor in Eq. (9.10). By such a calculation, we get $\rho^* \mathcal{D}^* Z \approx 0.15$–$0.55$, which is by five orders of magnitude less than the value obtained in approximating the dependence of $\ln J^*$ on $T^{*-1}(T_m^* - T^*)^{-2}$ by a linear function (Fig. 9.21). Note that the value of $\rho^* \mathcal{D}^* Z \approx 0.15$–$0.55$ is close to that obtained in experiments on the crystallization of metal melts ($\rho \mathcal{D} Z = 10^{32}$–$10^{33} \mathrm{s}^{-1}\mathrm{cm}^{-3}$, which corresponds to $\rho^* \mathcal{D}^* Z = 10^{-2}$–$10^{-1}$) and calculated from CNT [7]. Eq. (9.15) gives a value of $\mathcal{D}$ which is 10–20 times smaller than that of Eq. (9.34). The values of d_0^* and l^*, calculated from Eqs. (9.13) and (9.14) based on the results of MD-simulation of the coefficient $\mathcal{D}$ (Eq. (9.34)), are less than unity.

As indicated earlier, Eq. (9.8), which was used traditionally in estimating the values of the pre-exponential factor and the effective surface energy from Eq. (9.10) in experimental research of the spontaneous crystallization of liquids, is valid only at low super-cooling. Fig. 9.23 presents the values of the pressure in a critical crystal (at zero ambient pressure), calculated from the condition of equality of the chemical potentials of a liquid and a crystal nucleus and equation $p_{s_*} = \rho_s \Delta h (\Delta T / T_m)$, which follows from Eq. (9.8) at $p = \mathrm{const}$. For $T^* = 0.43$ the discrepancy between the data of these approaches is 15 %.

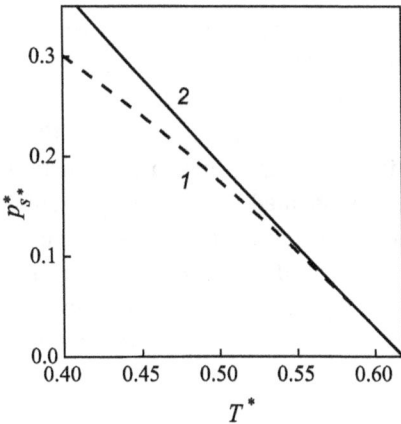

Fig. 9.23: Pressure in a crystal–phase critical nucleus at $p^* = 0$: (1) from the condition of equality of chemical potentials of liquid and critical nucleus $\mu_s(p_{s_*}, T) = \mu(p, T)$; (2) according to Eq. (9.8).

On the isobar $p^* = 0$ the dependence of $\ln J^*$ on $(\Delta p^*)^{-2} = p_{s_*}^{*-2}$ is close to rectilinear if $p_{s_*}^*$ has been determined from the condition of equality of the chemical potentials of a liquid and a crystal nucleus. In approximating this dependence by a linear function in accordance with Eq. (9.10) we obtain $\rho^* \mathcal{D}^* Z \approx 1.01 \cdot 10^3$, $(16\pi \gamma_e^{*3})/(3T^*) \approx 1.448$. With such an approximation the effective surface energy of a crystal nucleus is an

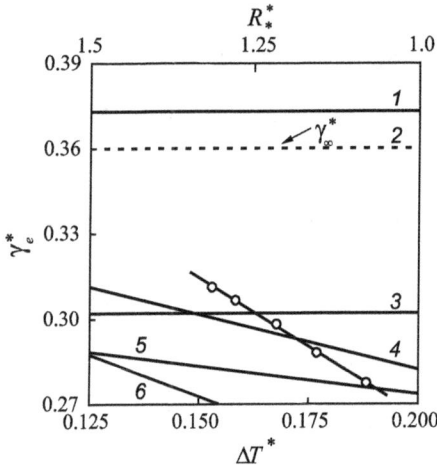

Fig. 9.24: Effective surface energy of crystal–phase nuclei as a function of temperature and radius at $p^* = 0$ (circles). (1) the value of γ_e^* in approximation of linear dependence of $\ln J^*$ on $T^{*-1}(\Delta T^*)^{-2}$; (2) surface free energy at a planar crystal-liquid interface γ_∞^* at $T_m^* = 0.618$ [24]; (3) results of the work of Bai and Li [82] when the pressure in a critical crystal is determined via the approximation Eq. (9.8); (4) calculation of γ_e^* by the relation $\gamma_e(T) = C_T \rho_s^{2/3} \Delta h(T)$, where $C_T = 0.3687$; (5) results of the work of Bai and Li [82] when the pressure in the critical nucleus is determined from the condition of equality of the chemical potentials of the liquid and the crystal nucleus; (6) calculation of γ_e^* by the relation $\gamma_e(T) = \gamma_\infty(T/T_m)$. The dependence of γ_e^* on R_*^* is shown qualitatively with the size uncertainty of ± 0.03.

increasing function of the temperature (the nucleus size), and the pre-exponential factor exceeds the result of the direct MD-simulation by four orders of magnitude.

The effective surface energy of crystal nuclei, calculated from the data of molecular dynamics determination of all the parameters involved in Eqs. (9.6), (9.10), viz. J, $\mathcal{D}$, Z, is presented in Fig. 9.24. Along the isobars, γ_e is an increasing function of temperature. In the range of state parameters investigated the value of γ_e is smaller than that of γ_∞ at T_m^* [81]. Fig. 9.24 also presents the values of the surface energy of crystal nuclei obtained by Bai and Li [82] by the method of implantation of a crystal into a super-cooled liquid in the approximation Eq. (9.10) and with allowance for a temperature dependence of this parameter. In [83] the following expression for the temperature dependencies of the interfacial free energy is proposed: $\gamma_e(T) = \gamma_\infty T/T_m$ and $\gamma_e(T) = C_T \rho_s^{2/3} \Delta h(T)$, which is a modification of Turnbull's formula [11], taking into account the temperature dependence of the heat of phase transition. The plots of such temperature dependencies $\gamma_e(T)$ are also presented in Fig. 9.24.

9.8 Discussion and Conclusions

Crystallization is a well-known phenomenon, which has been investigated for a long time. At small super-cooling, it commonly proceeds by the mechanism of heterogeneous nucleation. With a further increase of the degree of under-cooling, the size of the critical nucleus and the work of formation of the critical cluster decrease. As a consequence, the probability of formation of viable nuclei by thermal fluctuations in the system increases also in the absence of heterogeneous nucleation sites. A variety of experimental data, dealing with crystallization in one-component liquids, demonstrate the possibility of homogeneous formation of viable crystallites in real systems.

Employing thermodynamic approximations for the description of the clusters of the newly evolving phase and elementary kinetic considerations one can arrive at an, at least, qualitatively satisfactory description of homogeneous nucleation, as realized in Classical Nucleation Theory (CNT). The degree of quantitative agreement between experiment and theory is a reflection (in the final result) of the level of correspondence of the real process with its model description in terms of CNT.

Computer modeling of the nucleation process, as realized e.g. in molecular dynamics and Monte Carlo simulations, supplies us with a tool for testing to which degree the basic assumptions of CNT (both with respect to thermodynamic and kinetic aspects) are realized in the real nucleation-growth process, at least, for the model systems under consideration. These methods allow one to determine the whole set of quantities, entering the final expression derived in the framework of CNT, Eq. (9.10), in an independent way.

Each single process of stochastic generation of a supercritical crystal is connected with an overcoming of the thermodynamic activation barrier. As shown for the first time by Gibbs, the existence of such barrier to nucleation is caused by the appearance of surface or interfacial contributions to the thermodynamic potentials of the system crystallite–meta-stable initial phase. The interfacial energy is hereby a function of the crystallographic orientation of the crystal faces.

In classical homogeneous nucleation theory, the task of determination of the height of the activation barrier is simplified by the introduction of an effective interfacial free energy density, γ_e. Since direct experimental methods of determination of this quantity, γ_e, for the crystal-liquid interface do not exist, this quantity is used in the theory as a free or fit parameter. Molecular dynamics simulations allow one to calculate the crystal-liquid free energy density for a planar interface crystal-liquid, to estimate its anisotropy, to directly determine the height of the thermodynamic potential barrier to nucleation, the value of the effective interfacial free energy density, γ_e, and the bulk excess free energy (Δh) of the crystal phase. The molecular dynamics computations, as performed in the present contribution, have demonstrated that (i) the highest under-cooling which can be reached corresponds to the critical cluster size of 0.7–1 nm, (ii) the value of the effective interfacial free energy density, γ_e, at an isothermal penetration into the meta-stable state differs from its value at the planar interface by not more than 15 % (at $T^* < 0.9$) and not more than 22 % at isobaric cooling ($p^* = 0$).

The results of molecular dynamics computations give a verification of the correctness of the assumption to treat the formation of clusters as a sequence of attachment and separation of single units (atoms, molecules) and its adequate description via the diffusion model of nucleation. The diffusion coefficient of the clusters in the critical region of the activation barrier is computed via molecular dynamics methods, and it is shown to be in agreement with theoretical estimates within 1–2 orders of magnitude. Similarly, the molecular dynamics estimates of the non-equilibrium Zeldovich factor turn out to be in good agreement with analytical expressions.

The development of multi-scale methods of modeling of molecular systems, passing high activation barriers in their evolution [17–19], allows one to investigate the spontaneous crystallization of under-cooled liquids at low super-cooling, i.e. also for the cases where nucleation experiments can be and are performed in real systems. In future this would imply the possibility of a direct comparison of experimental data with the results of molecular dynamics computer model investigations, and not merely computer simulations and predictions of classical nucleation theory.

Acknowledgement: The author would like to express his gratitude to the Deutsche Forschungsgemeinschaft (DFG SCHM 937/16-1) and Dr. Jürn W.P. Schmelzer for the possibility to design this chapter in the course of a stay at the Institute of Physics of the University of Rostock, Germany, and the Russian Foundation for Basic Research (Project number 12-08-00467) as well as the Program of the Presidium of the Russian Academy of Sciences No. 2 (Project 12-П-2-1008) for funding the research, the results of which are outlined in the present chapter.

Bibliography

[1]   J.W. Gibbs, *On the equilibrium of heterogeneous substances*, Trans. Connecticut Academy of Sciences 3, 108, 343 (1875–1879); J.W. Gibbs, The Collected Works, vol. 1, *Thermodynamics* (Longmans, New York – London – Toronto, 1928).

[2]   M. Volmer, A. Weber, Z. Phys. Chem. **119**, 277 (1926).

[3]   L. Farkas, Z. Phys. Chem. **125**, 236 (1927).

[4]   R. Kaischew, I.N. Stranski, Z. Phys. Chem. **B 26**, 317 (1934).

[5]   Ya. B. Zeldovich, Zh. Eksp. Teor. Fiz. **12**, 525 (1942).

[6]   Ya. I. Frenkel, *Kinetic Theory of Liquids* (Oxford University Press, Oxford, 1946).

[7]   V.P. Skripov, V.P. Koverda, *Spontaneous Crystallization of Supercooled Liquids* (Nauka, Moscow, 1984).

[8]   L. Bosio, Met. Corros.-Ind. **40**, 421 (1965).

[9]   G. Wood, A.G. Walton, J. Appl. Phys. **41**, 3027 (1970).

[10] J.H. Perepezko, D.H. Rasmussen, AIAA Pap. **30**, 1 (1979).

[11] D. Turnbull, R.E. Cech, J. Appl. Phys. **21**, 804 (1950).

[12] W.C. Swope, H.C. Andersen, Phys. Rev. **B 41**, 7042 (1990).

[13] P.R. ten Wolde, M.J. Ruiz-Montero, D. Frenkel, J. Chem. Phys. **104**, 9932 (1996).

[14] P.R. ten Wolde, D. Frenkel, J. Chem. Phys. **109**, 9901 (1998).

[15] C. Valeriani, E. Sanz, D. Frenkel, J. Chem. Phys. **122**, 194501 (2005).

[16] E. Mendez-Villuendas, I. Saika-Voivud, R.K. Bowles, J. Chem. Phys. **127**, 154703 (2007).

[17] G.M. Torrie, J.P. Valleau, Chem. Phys. Lett. **28**, 578 (1974).

[18] P.G. Bolhuis, D. Chandler, C. Dellago, P.L. Geissler, Annual Rev. Phys. Chem. **53**, 291 (2002).

[19] A. Laio, M. Parrinello, Proc. Natl. Acad. Sciences USA **99**, 12562 (2002).

[20] D.G. Thomas, L.A.K. Staveley, J. Chem. Soc. **12**, 4569 (1952).

[21] D. Turnbull, J. Appl. Phys. 21, 1022 (1950).

[22] J.Q. Broughton, G.H. Gilmer, J. Chem. Phys. **84**, 5759 (1986).

[23] R.L. Davidchack, B.B. Laird, Phys. Rev. Lett. **85**, 4751 (2000).

[24] R.L. Davidchack, B.B. Laird, J. Chem. Phys. **118**, 7651 (2003).

[25] J.J. Hoyt, M. Asta, A. Karma, Phys. Rev. Lett. **86**, 5530 (2001).

[26] J.R. Morris, Phys. Rev. B **66**, 144104 (2002).

[27] J.R. Morris, X. Song, J. Chem. Phys. **119**, 3920 (2003).

[28] Y. Mu, X. Song, J. Phys. Chem. B **109**, 6500 (2005).

[29] T. Frolov, Y. Mishin, J. Chem. Phys. **131**, 054702 (2009).

[30] B.B. Laird, R.L. Davidchack, Y. Yang, M. Asta, J. Chem. Phys. **131**, 114110 (2009).

[31] V.P. Skripov, V.G. Baidakov, High Temp. **10**, 1102 (1972).

[32] V.G. Baidakov, S.P. Protsenko, Dokl. Akad. Nauk **402**, 754 (2005).

[33] V.G. Baidakov, S.P. Protsenko, Phys. Rev. Lett. **95**, 015701 (2005).

[34] V.G. Baidakov, S.P. Protsenko, J. Exp. Theor. Phys. **103**, 876 (2006).

[35] L.D. Landau, E.M. Lifshitz, *Statistical Physics*, Vol. 1 (3rd edition, Pergamon Press, Oxford, 1980).

[36] A.A. Chernov, In: *Sovremennaya kristallizatsiya*, Volume 3, *Obrazovanie Kristallov* (Nauka, Moscow, 1980).

[37] D. Turnbull, J.C. Fisher, J. Chem. Phys. **17**, 71 (1949).

[38] D. Turnbull, R.L. Cormia, J. Chem. Phys. **34**, 820 (1961).

[39] K.F. Kelton, Solid State Physics **45**, 75 (1991).

[40] V.G. Baidakov (unpublished).

[41] A. Ziabicki, J. Chem. Phys. **48**, 4368 (1968).

[42] M.P. Allen, D.J. Tildesley, *Computer Simulation of Liquids* (Oxford University Press, Oxford, 1989).

[43] D. Frenkel, B. Smit, *Understanding Molecular Simulation: From Algorithms to Applications* (Academic Press, San Diego, 2002).

[44] V.G. Baidakov, S.P. Protsenko, Z.R. Kozlova, Fluid Phase Equil. **263**, 55 (2008).

[45] V.G. Baidakov, Z.R. Kozlova, Chem. Phys. Lett. **500**, 23 (2010).

[46] V.G. Baidakov, S.P. Protsenko, Z.R. Kozlova, J. Chem. Phys. **137**, 164507 (2012).

[47] A.T. Bharucha-Reid, *Elements of the Theory of Markov Processes and Their Application* (McGraw-Hill Book Company, New York, 1960).

[48] V.P. Skripov, *Metastable Liquids* (Wiley, New York, 1974).

[49] G.E. Norman, V.V. Stegailov, Dokl. Akad. Nauk **386**, 328 (2002).

[50] T. Kinjo, M. Matsumoto, Fluid Phase Equlibria **144**, 343 (1998).

[51] V.G. Baidakov, S.P. Protsenko, Dokl. Akad. Nauk **394**, 752 (2004).

[52] C.W. Gardiner, *Handbook of stochastic methods for physics, chemistry and the natural sciences* (Springer, Berlin, 1997).

[53] L.A. Pontryagin, A.A. Andronov, A.A. Vitt, Zh. Eksp. Teor. Fiz. **3**, 165 (1933) [translated by J.B. Barbour and reproduced in *Noise in Nonlinear Dynamics*, edited by F. Moss and P.V.E. McClintock (Cambridge University Press, Cambridge, 1989), vol. 1, p. 329].

[54]   J. Wedekind, R. Strey, D. Reguera, J. Chem. Phys. **126**, 134103 (2007).

[55]   J. Wedekind, G. Chkonia, J. Woelk, R. Strey, D. Reguera, J. Phys. Chem. **131**, 114506 (2009).

[56]   D. Frenkel, B. Smit, *Understanding Molecular Simulation: From Algorithms to Applications* (Academic Press, San Diego, 2002).

[57]   P.J. Steinhardt, D.R. Nelson, M. Ronchetti, Phys. Rev. B **28**, 784 (1983).

[58]   T.S. van Erp, D. Moroni, P.G. Bolhuis, J. Chem. Phys. **118**, 7762 (2003).

[59]   T.S. van Erp, P.G. Bolhuis, J. Comp. Phys. **205**, 157 (2005).

[60]   T.S. van Erp, Phys. Rev. Lett. **98**, 268301 (2007).

[61]   D. Moroni, P.R. ten Wolde, P.G. Bolhuis, Phys. Rev. Lett. **94**, 235703 (2005).

[62]   S. Jungblut, C. Dellago, J. Chem. Phys. **134**, 104501 (2011).

[63]   T.S. van Erp, J. Chem. Phys. **125**, 174106 (2006).

[64]   D.R. Woodruff, *The Solid-Liquid Interface* (Cambridge University Press, London, 1973).

[65]   R.L. Davidchack, B.B. Laird, J. Chem. Phys. **108**, 9452 (1998).

[66]   V.G. Baidakov, A.O. Tipeev, K.S. Bobrov, G.V. Ionov, J. Chem. Phys. **132**, 234505 (2010).

[67]   J.D. Honeycutt, H.C. Andersen, Chem. Phys. Lett. **108**, 535 (1984).

[68]   S. Auer, D. Frenkel, J. Chem. Phys. **120**, 3015 (2004).

[69]   V.G. Baidakov, A.M. Kaverin, I.I. Sulla, Teplofiz. Vys. Temp. **27**, 410 (1989).

[70]   D. Kashchiev, J. Chem. Phys. **125**, 014502 (2006).

[71]   G. Wilemski, J. Chem. Phys. **125**, 114507 (2006).

[72]   H. Wang, H. Gould, W. Klein, Phys. Rev E **75**, 031604 (2007).

[73]   S.E.M. Lundrigan, I. Saika-Voivod, J. Chem. Phys. **131**, 104503 (2009).

[74]   V.G. Baidakov, A.O. Tipeev, Thermochim. Acta **522**, 14 (2011).

[75]   F. Trudu, D. Donadio, N. Parrinello, Phys. Rev. Lett. **97**, 105701 (2006).

[76]   Z. Vucic, D. Subasic, Z. Ogorelic, Phys. Stat. Sol. a **47**, 703 (1978).

[77]   V.G. Baidakov, K.S. Bobrov, A.S. Teterin, J. Chem. Phys. **135**, 054512 (2011).

[78]   J.W.P. Schmelzer, J. Colloid. Interface Sci. **242**, 354 (2001).

[79]   V.G. Baidakov, S.P. Protsenko, A.O. Tipeev, J. Non.-Crystalline Solids **356**, 2923 (2010).

[80]   S.P. Protsenko, V.G. Baidakov, A.O. Tipeev, Thermophysics and Aeromechanics **20**, 95 (2013).

[81]   V.G. Baidakov, A.O. Tipeev, J. Chem. Phys. **136**, 074510 (2012).

[82]   X.M. Bai, M. Li, J. Chem. Phys. **124**, 124707 (2006).

[83]   L.J. Peng, J.R. Morris, R.S. Aga, J. Chem. Phys. **133**, 084505 (2010).

Gyan P. Johari and Jürn W.P. Schmelzer

10 Crystal Nucleation and Growth in Glass-forming Systems: Some New Results and Open Problems

In this chapter, we describe four aspects of nucleation and crystal growth processes in highly viscous glass-forming melts. In the first aspect, we critically analyze the relation between the typical sizes of supercritical nuclei *vis a vis* the sizes of co-operatively rearranging regions (CRR) of the configurational entropy theory and of the domains of heterogenous dynamics (DHD) envisaged in the structure of ultraviscous melts. We argue that stochastic structural fluctuations in a melt that produce supercritical nuclei are irreconcilable with those that produce the co-operatively rearranging regions and the dynamic heterogeneity domains – the nuclei have structural order and the other mentioned above regions do not. Since the nanometer size of such nuclei is the same as that of the disordered regions and domains, it seems improbable that nuclei would form inside these regions and domains. In the second aspect, we describe the crystallization kinetics of a metal-alloy glass studied by calorimetry under, (a) isothermal conditions and (b) heating at a fixed rate. Only one prominent exotherm appears in procedure (a), and two exotherms in procedure (b). The Kolmogorov–Johnson–Mehl–Avrami relation for the fraction crystallized [$x(t) = 1 - \exp(-kt^m)$] and the corresponding relation for rate-heating fit the results from both studies until x begins to approach one and the experimentally measured crystallization rate becomes lower than the calculated rate. This deviation is attributed to the slower crystallization of inter-granular melt. The rate coefficient value yields an activation energy of 169 kJ/mol. On annealing the ultraviscous melt, the first peak vanishes but the second persists. On reheating it also vanishes, indicating certain effects of prior thermal history on crystallization. The slower kinetics of the second exothermic peak indicates that the melt crystallized partially to a metastable phase, which transforms on heating. In the third aspect, we consider how buoyancy-induced mass transfer of the melt due to Rayleigh–Bénard convection and due to Marangoni-flow are expected to affect the crystallization process and thereby the diffusion-controlled crystallization kinetics and its activation energy. Both types of convection mechanisms are affected by the difference between the densities of the melt and the crystal phase, by the heat evolved in crystallization, by the composition difference in the various regions of the crystallizing melt and by the surface tension gradient within the melt and the surface tension of the ambient gas-melt interface. These effects can be resolved by studies performed in microgravity conditions or by subjecting a crystallizing melt to a centrifugal force and by studying the effects of pressure on crystallization kinetics. In the fourth aspect, which is entirely based on the studies of Schmelzer and coworkers, we consider thermodynamic aspects in the derivation of the correct expression for the work of critical cluster formation. In order to describe theoretically the kinetics of nucleation-growth processes, the so-called work of cluster formation – the change of the thermodynamic potential due to the formation of a cluster of a given size and composition – has to be known. This quantity is conventionally determined in the framework of Gibbs' thermodynamic theory of heterogeneous systems employing certain additional assumptions. However, Gibbs restricted his analysis exclusively to "equilibrium states of heterogeneous substances" and, already by this limitation, does not supply us with a fully satisfactory solution of mentioned problem. Generalizing Gibbs' method, recently a thermodynamic description of non-equilibrium states consisting of clusters of arbitrary sizes and composition in the otherwise homogeneous ambient phase was developed by one of us. This approach leads not only to a sound foundation of the thermodynamic aspects of the theoretical description of cluster growth processes but also to a variety of principally new insights into the course of nucleation-growth or spinodal decomposition processes, in general. In particular, it leads to a different set of thermodynamic equilibrium conditions for the determination of the properties of the critical clusters as compared with Gibbs' classical treatment. It supplies us further with a tool allowing one to determine the bulk and interfacial properties of the evolving clusters in dependence on their sizes. In addition, a generalization of the Skapski–Turnbull relation for the determination of the specific interfacial energy is developed for aggregates of the newly evolving phase not being in thermodynamic equilibrium with the ambient phase. The approach further allows one to give a novel answer to the question at what conditions the evolution of the cluster ensemble proceeds via a saddle point trajectory passing the critical cluster and under which conditions ridge crossing is preferred. These and some of the further consequences of the new approach in application both to nucleation and growth-dissolution processes are briefly reviewed. Finally, we complete the analysis with a discussion of prehistory effects in phase formation and suggest possible ways to incorporate them into the theory of crystal phase formation.

10.1 Introduction

Monographs on the subject of crystallization of melts (cf. [1, 2] for an overview) remind us of two major aspects of chemical physics of solidification: the roles of (i) the free energy of a bulk material relative to its surface energy, as given by Gibbs [3] for size-dependent net free energy of a material, and (ii) the sequence of events in the process of crystallization, i.e. the evolution of the crystal phase proceeding first by crystal nuclei formation and then by crystal growth, as discussed in detail already by Tammann [4, 5]. Gibbs' concept has led us to expressing the conditions under which clusters of atoms or molecules can produce nuclei with atomic arrangements similar to but not identical to those of a crystal phase. These conditions constitute the basis for the different approaches to the theoretical modeling of crystal nucleation and growth. It is implemented into the Kossel [6], Kaischew–Stranski–Krastanov [7], and Turnbull's approaches [8] to critical nuclei formation and the subsequent crystal growth in (pure) molten metals as a function of their supercooling below the freezing point – the view employed for all mono-component systems and also for most multi-component systems. This approach has led us to the construction of the time-temperature-transformation plots of a host of materials which is presently an integral part of the undergraduate courses in materials science [9]. The subject is also referred to as a particular mechanism of first-order phase transformations.

It is recognized that, at a molecular level, stochastic fluctuations, meaning that a system's subsequent state is determined both by the process's predictable actions and by random elements, cause clustering of atoms and molecules. Nuclei formation is one aspect of that clustering process. Briefly, stochastic fluctuations produce a variation of inter-molecular distances from site to site in the bulk of a melt, and these fluctuations occasionally lead to clustering (self-association, or aggregation) not only at random sites in the structure but also randomly in time. It is also understood that the self-diffusion coefficient of individual molecules within the bulk of a melt varies with time, as does the position of molecules relative to others, i.e. some molecules diffuse much faster than the others, amounting to a distribution of kinetic energies. The subject has been discussed by several workers and an up to date list of references on these topics may be found both in the earlier and the recent versions of the monograph by Gutzow and Schmelzer [1]. Its developments have also appeared in several recent reviews on the thermodynamic aspects of nuclei formation and the growth of critical size nuclei to an identifiable crystal phase [10–13].

According to our current understanding of solidification of melts, when a melt is cooled towards its freezing point, molecular (or atomic) clusters form stochastically in the bulk and they vary in size, apparently with a distribution of sizes about some mean value. The clusters decay (uncluster, dissolve or disaggregate) in size ultimately to their original molecular or atomic state in the bulk of a melt. At temperatures above the freezing point, none of the sizes in this distribution is large enough to be regarded as having the size of a critical nucleus. Moreover, above the freezing point, the Gibbs potential is a monotonically increasing function of cluster size and a critical cluster size does not exist at all. In contrast, below the freezing temperature, critical nuclei have a finite size and nuclei can be formed being large enough that their free energy ($G = H - TS$) reaches a maximum value, G^*, in terms of the sum of the thermodynamic surface and bulk free energy contributions [1, 2]. When a melt is supercooled, the probability of the clusters to form larger size nuclei increases with increasing degree of supercooling, the average of their size distribution shifts toward larger sizes and in part of this distribution some of the nuclei are equal to or larger than the critical nuclei. The size of the critical nuclei normally also decreases with increasing supercooling as well.

The number of the critical nuclei produced in a unit volume per unit time is called the nucleation rate J. It increases first as the temperature T is decreased during supercooling. In this view there is no indication as to what conditions of temperature are needed for formation of molecular or atomic clusters that can grow to form nuclei in a given melt, and one requires prior knowledge of the equilibrium freezing (or crystal melting) point, T_m, to obtain the temperature range in which clusters may not grow to the critical size nuclei and the temperature range in which they may grow to critical size nuclei. At T infinitesimally below T_m, the probability of formation of critical size nuclei increases from zero to a finite value, but it cannot be explained straightforwardly (by theoretical argumentation) how many such nuclei should form before crystal growth can be sustained and complete crystallization occurs. One of the reasons is that neither the specific surface energy of the nuclei-bulk interface nor their shape is known. As a rule, the crystal nuclei, taken to be of the order of nanometers in dimensions, are not expected to be spherical.

Theories of nucleation and crystallization have been part of the effort for understanding why melts of low-viscosity, typically below 100 Pa·s, occasionally supercool or do not crystallize at temperatures below their equilibrium melting or freezing points. Since the degree of supercooling is small, i.e. the magnitude of $T_m - T$ is small, in the development of the respective theories it is supposed as a rule that the surface tension and enthalpy of freezing do not depend upon temperature T in this narrow range. One expects that such theories would not apply when an ultraviscous melt crystallizes at T far below T_m. The theories provide a rationale by suggesting that for melts such as glycerol, that do not crystallize even when the melt is kept 10 to 50 K below its freezing point, supersaturation does not reach high enough values to initiate homogeneous nucleation and for melts such as water and pure metals that do not readily supercool such a supersaturation is reached quickly.

There seems to be no *a priori* manner of knowing whether a melt would crystallize on cooling or not. It is also known that nucleation and crystallization can in some cases be promoted by increasing the pressure on a melt held at a fixed temperature, and that some melts that do not crystallize on supercooling can be crystallized by application of pressure, i.e. they crystallize when "superpressed" (pressure higher than the equilibrium value) at a fixed temperature. In such cases, the change in free energy involves contributions from a decrease in volume expressed via compressibility and not by change in the temperature and the associated decrease in volume expressed by thermal expansion coefficient. There is also the possibility that the change of pressure may result in the transfer of the system to states where crystalline phases of different space groups can be formed ([14, 15], cf. also Chapters 2, 3, 4, and 8 of the present book).

There are several other factors that also affect the kinetics of crystal growth in a melt. The magnitude of these effects is determined by the rate of heat transfer in the melt away from a growing crystal front. Briefly, crystallization is an exothermic process and as a result the melt in the vicinity of the crystal front becomes warmer than the melt away from the front. Therefore, a crystal front would advance and the crystal would grow only when this local rise in temperature vanishes by removal of thermal energy by heat conduction or convection mechanisms in the melt kept at a fixed macroscopic temperature. While the heat transfer by conduction involves phonons, the convection involves molecular diffusion.

There are also two macroscopic mechanisms both of which involve heat transfer by convection. In the first the convection is driven by the higher buoyancy (lower density) of the melt in the vicinity of the crystal growth front, an effect that is influenced by the gravitational field – crystal growth rate in microgravity conditions differs from that in the Earth's gravity. This effect is often described in terms of Rayleigh-Bénard convection. In the second the convection is driven by the difference between the surface tension of the warmer melt in the vicinity of the crystal's growth front and the somewhat cooler melt elsewhere, an effect that is influenced by adding surfactant molecules into the melt. This effect is often described in terms of Marangoni flow. In addition, the question has been analyzed as to whether the viscosity or effective diffusion coefficients determine the kinetics of crystal formation and growth (possible decoupling of diffusion and viscosity) or whether this process is realized by alternative mechanisms not studied in detail so far (e.g. [16], cf. also Chapters 7 and 9).

Although the time evolution of the size distribution of the nuclei can be determined by SAXS and SANS experiments and may be described as discussed e.g. in [17–20], there is a difficulty in taking into account the change in the dynamics with temperature caused by the above-mentioned clustering of different kinds. On the other hand, a variety of experimental methods (like calorimetry) allows one to determine exclusively the total amount of the newly evolving crystal phase. In order to describe the evolution of the crystal phase in time, thermodynamics and kinetics of nucleation process is usually dealt with separately from the kinetics of crystal growth. The separate knowledge of these factors allows one then to combine them and to advance theoretically the formalism for overall crystallization kinetics, known as the Kolmogorov-Johnson-Mehl-Avrami (KJMA) equation [1, 2, 21–23].

In the present chapter, we consider the consequences of some of the above sketched, so far less recognized, effects generally, and also particularly for crystallization of glass-forming metal alloy melts in ultraviscous states. We also discuss the limitations on the models of crystallization when phase separation occurs, when the occurrence of crystallization continuously changes the melt composition, and we use metallic alloys with varying average atomic size, as examples. Since atoms do not rotate, diffusive motions in metal-alloy glasses and melts are only translational and there are no internal degrees of freedom or steric hindrance for orientation of molecular groups favorable for crystal formation.

Numerous studies have appeared on crystallization of metal-alloy glasses (cf. also Chapter 2), and calorimetry has been commonly used for this purpose (cf. also Chapter 1). Most such studies have been performed by rate-heating or cooling during which it is usually the ultraviscous melt that crystallizes. Analysis of the data obtained from such studies often does not distinguish the various kinetic processes. Such studies also do not in all cases use different heating rates to resolve the various thermal effects that appear in such results. Moreover, assumptions made in analyzing the crystallization kinetics overlook features related to the material's structure, and often the parameters obtained by fitting a formalism to the data for crystallization have little resemblance with the quantities that determine nucleation and growth.

As an example of these features observed on crystallization, we describe here a study of crystallization of ultraviscous melt of $Mg_{65}Cu_{25}Tb_{10}$-glass performed by using two procedures, (i) by keeping the samples at fixed temperatures, and (ii) by heating the samples at different fixed rates. Studies of such metal-alloy glass-forming melts are particularly advantageous in the analysis of glass-formation, nucleation and crystal growth. There are no hydrogen-bonds or covalent-bonds interactions in such melts, but they do have an entropy of mixing, which is retained when crystalline alloys form. Some of the results given here have been already published [24]. A parallel study of the quaternary composition $Pd_{40}Ni_{10}Cu_{30}P_{20}$-ultraviscous melt has shown remarkably similar crystallization features. These results will be published later.

Finally, we describe some of the thermodynamic aspects of nucleation and crystallization and analyze a set of problems encountered in the application of the basic ideas of classical nucleation theory to crystallization processes in glass-forming melts (cf. also Chapter 9). The analysis covers the properties of critical clusters (whether they are identical to the properties of the evolving macroscopic phases or not), and, resulting from the respective assumptions, we describe the consequences with respect to critical cluster size and the work of critical cluster formation. In this context we also describe the limitations of Gibbs' classical approach to the thermodynamic description of heterogeneous systems. In particular, this discussion will be directed to its consequences with respect to the predictions of the properties of the critical clusters (cf. also [1, 11]). Also covered are the following problems: (i) Determination of the number of nucleation sites extending previous analyses of the size of the structural units in the expressions both for crystal nucleation and growth in the framework of the classical theory of nucleation and growth [16]. (ii) The problem is analyzed whether nucleation always proceeds (as in most applications assumed) via the saddle-point of the thermodynamic potential surface. (iii) Appropriate expressions for the specific interfacial energy (generalizing the Skapski-Turnbull relation to systems in thermodynamic non-equilibrium states) are derived. (iv) The problem is analyzed as to what extent the thermodynamic and kinetic properties of glass-forming melts can be considered as a function of the actual thermodynamic state parameters of the systems under consideration. Thus, we describe approaches to the implementation of the effects of prehistory on crystallization provided recently [25] and discussed also in Chapter 1 of this book.

10.2 Consequences of Stochastic Structural Fluctuations in Ultraviscous Melts

10.2.1 Structure Fluctuations, Nucleation and Distribution of Relaxation Times

There are a number of features of molecular diffusion that are common to melts and some of these distinguish a liquid-crystal transformation from liquid-liquid transformations (including both nucleation-growth processes and spinodal decomposition) and from crystal-crystal transformations. While the free energy plays an important role in all transformations, specific features may be of more or less importance in particular occurrences of phase transformation processes. For example, compositional change plays an exceptional role in spinodal decomposition (frequently also in nucleation-growth processes in multi-component systems but may be also fully absent in latter cases) and the mechanical strain energy and lattice diffusion of atoms may play a vital role in crystal-crystal transformation. Theoretical details of spinodal decomposition have been widely discussed starting with the work by Hillert [26], Cahn and Hilliard [27, 28]. The broad subject of crystal-crystal transformations has been reviewed in detail, for example, in the monographs on crystallization and solid-solid transformations [29, 30] and with particular emphasis on metals by Christian in [31].

In regard to the liquid-crystal transformation, the postulate is that local fluctuations of thermal energy lead to clustering and unclustering of molecules in the melt's structure. In this stochastic process, some clusters grow to a size larger than the critical size in the supercooled state of a melt, and their number formed per unit time per unit volume is called the homogeneous nucleation rate. This rate increases when temperature is decreased until the decrease of the work of critical cluster formation due to the decrease of the critical cluster size is not overcompensated by the decrease of mobility or increase of viscosity with decreasing temperature. Supercritical size nuclei grow to form crystals at a rate that depends upon the self-diffusion coefficient of the entities.

There are theories for nucleation and crystallization going back to Turnbull and Fisher [8], and these have been reviewed in detail with reference to vitreous solids by Gutzow and Schmelzer [1] and Kelton and Greer [2]. There have been conferences on the subject of nucleation and crystallization of ultraviscous melts and a paper has reviewed this subject and proposed a scenario similar to spinodal decomposition [27, 28] for the description of nucleation and growth of metastable phases in multi-component systems [32–34]. Application of the basic ideas of this theory to ultraviscous multi-component melt crystallization, accounting quantitatively even for pre-history effects, was recently described by Schmelzer and Schick [25]. The latter topics will be briefly described below.

There are two types of nucleation, homogeneous and heterogeneous [1, 2, 35, 36]. Homogenous nucleation occurs predominantly in the bulk of a melt and the free energy increase in this type of nucleation is high and the probability of its occurrence is relatively low at low supercooling. Heterogeneous nucleation occurs on the walls of a container and on solid particles in the melt; the free energy increase for nuclei formation is relatively low and the probability of its formation is relatively high. Although both homogeneous and heterogeneous nucleation occur in normal laboratory conditions, in container-less processing in the near zero-gravity of space only homogeneous nucleation can occur in a melt devoid of any impurity solid particles. Cooling favors formation of lower energy states, e.g., H-bond formation in hydrogen bonded melts, and promotes both homogeneous and heterogeneous nucleation. The conditions for nucleation in melts mathematically described by Gibbs are referred to as conditions for formation of a *heterogeneous system*. Therefore, it seems important to avoid confusion between the terms Gibbs' heterogeneous system, heterogeneous nucleation and nucleation in multi-component melts and therefore we do not use here the term heterogeneous system for a homogeneously nucleated state.

There is another long known aspect of molecular diffusion, namely the existence of a distribution of viscoelastic and other relaxation times (cf. e.g. [37, 38]). One expects that this well-known, characteristic feature would also have an effect on nucleation and crystal growth in ultraviscous melts. As this distribution becomes broader on cooling or pressurizing, its effect on nucleation and crystallization would become significant on cooling or pressurizing a melt.

10.2.2 Structure Fluctuations and the Notion of Disordered Cluster Formation

In the theoretical description of glass formation, ultraviscous melts are considered to possess another kind of clustering (spatial regions in the melt) which are disordered, and a non-Arrhenius, Vogel-Fulcher-Tammann type, temperature dependence of viscosity, η, and of self-diffusion coefficient, D, is explained in terms of formation of such clusters that are known as co-operatively rearranging regions (CRR) [39]. Briefly, for small undercooling or at high temperatures, D varies according to the Arrhenius relation,

$$D \propto \exp\left(-\frac{E_A}{RT}\right),$$

were E_A is the Arrhenius activation energy and R is the universal gas constant. Its temperature derivative, as it is the case for self-diffusion in ordinary crystals, is seen to be independent of the changing structure of the melt with varying T. But at low temperatures in the ultraviscous state, the clusters grow in size and their growth causes the viscosity, η, and the self-diffusion coefficient, D, to become structure-dependent. In such cases, η and D vary with T not according to the Arrhenius equation any more but according to dependencies with a temperature dependent activation energy, frequently appropriately described by the Vogel-Fulcher-Tammann equation. Such clustering or formation of CRR is the basis of the configurational entropy theory for glass formation [39].

In relaxation measurements, the long-range motions of atoms or molecules appear as the non-Arrhenius variation of the characteristic time of the α-relaxation process, which thus depends upon the structure of the melt. In contrast to critical size nuclei that also form as a result of structural fluctuations, CRR's have a disordered arrangement. However, the formation of CRR in this theory does not explain why there is a distribution of relaxation times, nor does it explain the occurrence of faster modes of motion or local mobility, known as the β- or the JG-relaxation process, in ultraviscous melts [37, 40, 41] and of the persistence of this mobility in the glassy state as well as that of the Arrhenius variation of this mobility. These localized modes of motions are also recognized as the source of the unexpectedly rapid nucleation and growth in the glassy and ultraviscous states of small molecule organic substances [37, 42–50]. There have been indications that the overall crystallization rate is most rapid in a viscous melt far below its freezing point, in the temperature range in which cooling causes the α-relaxation to evolve from the JG-relaxation process and then to grow in strength on further cooling, as the number of molecules involved in the α-relaxation increases and the number of those involved in the JG-relaxation decreases [37, 38, 51].

Since the entropy theory and the formation of CRR do not explain the distribution of relaxation times and the occurrence of the JG-relaxation, in recent years computer simulations have been performed to explain it. These studies have been focused mainly on the origin of the distribution of relaxation times in ultraviscous melts. Computer simulations using spheres for atoms and molecules have indicated that in the structure of an ultraviscous melt there are dynamically heterogeneous domains (clusters, self-organized or associated structures) that form as a result of thermal fluctuations in local mass density [52, 53], composition, and energy in the structure of melts [54–63]. These (disordered) domains form and decay stochastically both in (positional) space and macroscopic time. Each distinct region has a single (exponentially-decaying) relaxation process, with time constants that differ significantly from one domain to the other. The heterogeneity is referred to as spatial heterogeneity. When there are alterations in the dynamics of the system or a given spatial region within the system over time, the heterogeneity is referred to as temporal heterogeneity.

Computer simulations of simple models of melts demonstrated that some particles jump frequently and move over large distances while the others remain relatively immobile, and particles with similar relaxation times cluster together [64, 65]. Multiplicity of such time scales led to the inference that dynamic heterogeneity is the source of the distribution of relaxation times (for limitations of this view, see [66]). The non-exponential response is supposed to originate from a sum of parallel exponential processes and these processes may be spatially-separated or temporally-separated. Thus stochastic diffusion leads (i) to density and free volume fluctuations, (ii) to the formation and disappearance of co-operatively rearranging regions, (iii) to the formation and vanishing of dynamically heterogeneous domains, and, in the context of this chapter, iv) to the formation of crystal nuclei in the melt. Therefore one is faced with three types of cluster formation in ultraviscous melts: (a) nucleation of ordered structures, (b) formation of structurally disordered domains of CRR, and (c) formation of dynamically and/or structurally heterogeneous domains, i.e., formation of spatially and temporally different domains (domains of heterogeneous dynamics (DHD) without walls, or spatial fluctuations in the local dynamic behavior). The length scales of structurally disordered domains of CRR and of the domains of heterogeneous dynamics (DHD) have been estimated to be of the order of 1–5 nanometers [37, 56, 57, 59–63] in organic melts.

In the context of structural inhomogeneity at a given moment of time in an ultraviscous melt, we should note also that Bokov [67], who originally, based on his light scattering experiments, had independently concluded that there are domains of structural heterogeneity has now found that the earlier observations were adversely affected by thermal gradients in the sample [68], thus putting into question the earliest experimental evidence of structural heterogeneity in melts. A thorough discussion of the subject is given in [66]. Interestingly, the size of the critical cluster appearing in *low-viscosity* melts at temperatures below the freezing point of a liquid is also estimated to be a few nanometers.

Moreover, there exist numerous studies of apparent nanocrystals, ranging in size from 2 to 80 nm, either in isolated states or in a state formed by freezing of a liquid in a nanoporous matrix. As these nanoscale dimensions of the regions, domains and nuclei are comparable albeit different in structures, they become relevant to nucleation and crystallization in the ultraviscous melt. As mentioned earlier, they are not equally relevant to nucleation and crystal growth in low-viscosity melts for which the concepts developed by Gibbs and Tammann were used originally to formulate the theory of nucleation and crystal growth (for a detailed analysis of the pre-conditions of applicability of the classical nucleation-growth concepts see also the detailed analysis given in [30, 35] and here in Chapter 9).

The overall structure of an ultraviscous melt is expected to contain temporally and spatially different regions of different energies, densities, relaxation times, etc. They may be held together by the van der Waals forces or by hydrogen bonds, giving no well-defined shape to the regions, and they may have an ordered or disordered structure. Their size grows on cooling and this growth is seen to decrease the configurational entropy whether they are disordered [39] or ordered. Such clusters have not been isolated, although their chemical composition in a multicomponent melt has been studied by SANS technique (as reported in [69] and described theoretically in [70, 71] (see the discussion of Figs. 2.4 and 4.1 there)). It is also not known whether homogeneous nucleation would occur as likely in CRR's of large size and in slowly relaxing dynamically heterogeneous domains as in the regions between the CRR's. Yet their formation and growth is regarded as the basis for the commonly accepted theory of melt crystallization. While these occurrences of self-association or aggregation of nanoscale dimensions remain stochastic both spatially and temporally, crystal growth from critical size nuclei after both homogeneous and heterogeneous nucleation is deterministic.

It is also known that the size of the supercritical nuclei decreases as a melt is cooled, or increases as the melt is heated from the glassy state. On cooling the size of CRR and/or of DHD increases, or this size decreases on heating a melt from the glassy state. Thus, one expects (at least within these notions) that a temperature can be reached on cooling at which the size of the supercritical nuclei would be the same as the size of the CRR and/or DHD, and on further cooling, the size of the CRR and/or DHD would be greater than the size of the supercritical nuclei (cf. also the discussion of similar aspects by Kauzmann [72] considering such effect as one possible solution of certain problems denoted later as Kauzmann paradox (see e.g. [1])). This would imply a crossing over of the CRR or DHD and of the supercritical nucleus at some temperature, and likely at some pressure of a melt. Stevenson and Wolynes have modeled this view [73] by arguing that in low viscosity melts at high temperatures the mechanism suggested by Turnbull may be tenable, but in ultraviscous melt (at a temperature approaching T_g), i.e. below the presumed cross-over temperature, a new crystallization mechanism may emerge, enabling rapid development of a large scale web of sparsely connected crystallinity. In Section 10.4 here we describe other effects that play an equally important role in crystallization kinetics and which do not involve such model considerations.

The incorporation of the above-mentioned aspects of the structure of a liquid has also consequences for the description of melts in terms of the potential energy landscape model. This energy landscape consists of a surface in a $3N$-dimensional space containing minima of different depths [74–77]. The state point of a liquid that represents its structure explores these minima. The number of such minima determines the configurational entropy of the liquid while the curvature of the minima determines its vibrational entropy. The entropy is related to viscous flow through the configurational entropy theory [39] for glass formation. There is only one deep minimum in the potential energy surface for the state point of a crystal, and several such deep minima for the state point of several different crystal states. But when a mono-component liquid begins to crystallize, interaction of atoms and molecules with the crystal interface moves the state point of the mixture to another part of the potential energy landscape, which contains the energy minima associated with the arrangement of the molecules at the crystal-melt interface. When crystallization of a multi-component melt occurs, the composition of the melt changes. This further causes the state point of the melt to move to a different part of the energy landscape.

It is well-known that when a melt is cooled towards its T_g, its transport coefficients decrease by many orders of magnitude. At high temperatures and low viscosity, there is neither a distribution of relaxation times (the response is exponential) nor a non-Arrhenius temperature dependence of diffusion coefficient or of viscosity, which is consistent with the Fick's diffusion law employed in the nucleation and crystal growth theories. But in the ultra-viscous state at low temperatures and/or at high pressures there is a distribution of relaxation times (the response is non-exponential) and the temperature dependence of diffusion as seen in the relaxation (relaxation times, τ) and flow (viscosity, η) experiments is non-Arrhenius.

The classical (including the formulation by Turnbull) theory for nucleation and growth implies that the time-scale for formation of a supercritical nucleus as a result of structural fluctuations is separated from the time scale of disordered clustering envisaged to explain the non-Arrhenius temperature dependence of η. Here the disordered clusters are CRR when only the non-Arrhenius dependence of the viscosity η is considered, and the disordered clusters are DHD when in addition to a non-Arrhenius dependence there is a distribution of viscoelastic and of other α-relaxation times. This separation of the time scale for formation of supercritical nuclei from the time scale of disordered clustering may make it possible to distinguish that part of the potential energy landscape that contains the free energy maximum (saddle point) for formation of a supercritical nuclei, from that part which contains the free energy minima that represent the state point of an ultraviscous melt.

Localized diffusion of the JG-relaxation in the structure of a glass can also lead to formation of nuclei, but their growth is expected to be slower as the long range diffusion of atoms and molecules required for crystal growth is slow and therefore crystallization starting from the bulk of glass substantially below its glass transition temperature, T_g, is extremely slow. However, depending upon the rate of nucleation and crystal growth, a glass may contain an insignificant population of nuclei if the nucleation rate near the vitrification temperature is low, or the glass may contain a large population of such nuclei if the crystallization rate is low. These so-called "athermal nuclei" grow progressively more rapidly as the temperature is increased, thus causing the glass to partially crystallize and thereafter the melt to fully crystallize. In addition, new nuclei form and grow slowly in the same ultraviscous melt.

It is now known that nuclei may also form in regions of localized fast motions, i.e. in the regions where the JG-relaxation occurs. In these regions they grow to form crystallites at random sites in the structure of a glass and of an ultraviscous melt. It has been suggested that nucleation and growth in these regions occur by self-diffusion over distances which are shorter than the usual inter-atomic distances. Studies of some ultraviscous melts have exhibited the so-called induction period for crystallization of some inorganic glasses (cf. [1, 2, 78]; for one of the first, the Zeldovich description of the induction time [79] in the process of crystallization of melts and glasses, the readers may consult [1] in the reference list to this chapter and also the discussion by I. Gutzow, D. Kashchiev, and I. Avramov [80]. An overview on further developments is given also in [1] and, in particular, in [78]). This time can be identified in one of its possible interpretations with the time required for establishing a steady-state distribution of nuclei up to critical cluster sizes, the unusually slow growth of the atomic or molecular clusters to a critical size and in some cases to indicate early appearance of metastable phases of the material and its components.

When the one-component melt has only one crystal form, the crystal size in the microstructure may vary depending upon the ratio of the nucleation rate to crystallization rate. But when different crystal types of the same composition are produced as in a mono-component melt, or of different compositions are produced as in a multi-component melt, micro-structural details of the ultimately formed solid vary because the crystal type, size and composition vary during the course of nucleation and growth. This variation leads also to unexpected thermal effects.

It is obvious that there is a basic difference between homogeneous nucleation and self-structuring that yields CRR or else DHD. The process of nucleation occurs according to a rate equation, but no such rates of self-structuring to CRR or DHD are known. In the current theories of nucleation and growth, the steady-state homogeneous nucleation rate is given by

$$J = J_0 \exp\left(-\frac{W_c}{k_B T}\right), \tag{10.1}$$

where J_0 is the – according to the classical theory – widely temperature-independent upper limiting value of the nucleation rate, J, k_B is the Boltzmann constant and W_c is the work of critical cluster formation, which is equal to the minimum change of the appropriate thermodynamic potential of the system (glass-forming melt here) required to form a cluster of critical size capable of further deterministic growth [1]. For a fixed temperature, there is an (at least partial) equilibrium between clustered and unclustered states, otherwise, the structure of a melt would contain only nuclei if a long time is allowed, and this equilibrium would be affected by the change in T for a mono-component system and also by any change in the composition of a multi-component system. As the term W_c is affected also by the pressure, P, one expects that this equilibrium would be additionally affected by the change in P (cf. also [1, 14]).

It has been found that the conventional equations of hydrodynamics are not valid for ultraviscous melts (cf. [1, 81]) and that, in addition, on cooling a melt towards T_g the various transport properties "decouple" from one another, and on heating a melt from the glassy state, the same properties are initially decoupled and then become coupled. To describe this effect briefly, the viscosity, η, the translational diffusion coefficient, D_{trans}, and temperature, T, are linked by the hydrodynamic Stokes-Einstein relation,

$$D_{trans}\eta \propto T.$$

However, this relation is not valid for ultraviscous melts, for which the product $D_{trans}\eta$ is found to be by orders of magnitude larger than the result expected from the Stokes-Einstein relation. In this respect we note that NMR-experiments on a molecular liquid, o-terphenyl, have shown that translational diffusion in its ultraviscous state has a weaker temperature dependence than the viscosity or its rotational correlation time [82]. In such measurements, the translational diffusion in units of $m^2 s^{-1}$ is by orders of magnitude less than the rotational diffusion in units of s^{-1}. The two have the same temperature dependence at $T > 290$ K, some 40 K above the glass transition temperature, T_g, of o-terphenyl [82]. This finding has been variously seen as consistent with the presence of dynamically heterogeneous regions in the structure of an ultraviscous melt.

Assuming the reliability of these data, the simple argument is that, if the structural fluctuations were homogeneous (absence of CRR and/or DHD), the apparent decoupling of the translational diffusion from rotational diffusion in a temperature or in a viscosity plane would mean that the relaxation of the structure on application of a stress, or equivalently the viscosity, is apparently not determined by the diffusion of molecules. Thus, it seems that the notions of CRR and DHD and of the decoupling of the temperature dependence of the translational and rotational diffusion coefficient on cooling towards T_g should be interrelated, so that the discussion related to the consequence of one would be equivalent to the discussion related to the consequence of the others. This interdependence suggests that an experimental test for either one of the notions would be an experimental test for all the three of them.

It is not known whether or not dynamic heterogeneity and crystallization kinetics are linked. One can only deduce that domains of correlated mobility are most likely to crystallize and hence attempt to predict crystallization sites and times from the spatiotemporal structure of the melt. This means that, as crystallization progresses, the population of such domains would decrease, causing the crystallization to be retarded by the continuing loss of such domains. It is conceivable that part of the reason for which the rate coefficient of isothermal crystal growth may not be proportional to the viscosity of the melt may be due to the fact that the heat generated in the vicinity of a growing crystal front decreases the viscosity of the melt itself, and, therefore, lack of this proportionality may not be caused by the presumed decoupling. It may not indicate that the viscosity does not control the growth rate, because the heat generated at the interface of the crystal growing front may lower the viscosity of the melt significantly below the macroscopic value. This aspect of the crystal growth and its feedback effects are described in terms of the Rayleigh-Bénard convection and Marangoni effect in Section 10.4 here.

However, there is another effect which would be caused by contraction in local volume around the growing crystal in most melts. If the volume does not relax rapidly enough, this occurrence would amount to a negative pressure on the melt or to an effective increase in its volume, thus reducing the viscosity. One test of such a view is that for materials in which the crystals are less dense than the melt (such as ice, Ga, etc.) an opposite effect to that found for common crystals would be observed. If so, the view of decoupling of the viscosity for common materials would be inappropriate for materials whose crystals are less dense than the melt. If it is found to be the case then there would be no need to assume that there is a decoupling of the (macroscopic) viscosity-determined relaxation processes and the nucleation and crystal growth-determining self-diffusion coefficient. The view of the effect of elastic stresses on crystallization kinetics can still be reconciled with the possibility of a negative pressure developing on a short time scale in the melt in the vicinity of a growing crystal front.

It is well known that (i) lattice strain caused by external deformation, lattice mismatching and internal stresses in a crystal raise its free energy and makes it more reactive, (ii) nano-size particles are more chemically reactive than larger particles, and (iii) surfaces of all solids are more reactive than the bulk. The same applies to their melting process, (i) a poly-crystalline mass of small size crystals coexists with the melt at a lower temperature than single crystals with the melt, (ii) nano-size particles have a lower melting point than micron-size particles, and (iii) nano-size particles melt from the surface inwards. Moreover, if microscopic crystals, containing different amounts of defects, are formed, the macroscopic free energy may vary with the amount of such defects. It is also known that an external stress causing strain in a glassy polymer [83] increases the diffusion coefficient. Some lithium silicate glasses [84] show the same effect, namely that on heating they lose the stored energy at temperatures below their T_g and some glassy polymers and silicate glasses crystallize at $T < T_g$ [85]. It is not finally clear how these effects would add to the observed nucleation and crystallization rates of ultraviscous melts.

10.3 A Case Study: Crystallization Kinetics of a Typical Metal Alloy Melt

10.3.1 General Considerations

Most studies of crystallization are performed by measuring the heat released as crystallization progresses at fixed temperature and pressure, P, or when the material is heated with time at a fixed rate at a fixed pressure. The technique employed is usually differential scanning calorimetry (DSC) which allows one to measure the rate of heat flow as a result of the change in the specific heat and occurrence of exothermic or endothermic processes in terms of Watts ($1\,W = 1\,J\,s^{-1}$ (cf. e.g. Chapter 1)). The measured value in Watt per gram of the sample is converted then to $JK^{-1}mol^{-1}$ by dividing the value of the thermal energy in Watt by the heating rate $q_h = (dT/dt)$ in Ks^{-1} and multiplying it by the molecular mass.

Crystallization is usually studied in temperature scanning mode by heating the glassy sample at a fixed rate to $T > T_g$, and measuring the heat released with increasing T. In the terminology used for the scanning mode, the temperature for onset of crystallization is usually denoted as T_x and the temperature of the exothermic minimum in the heat flow rate or exothermic peak is T_p. In such studies, x, the volume fraction of the sample crystallized at different temperatures is obtained from the relation

$$x(T) = \frac{1}{\Delta H_{cryst}} \int_{T'}^{T} \frac{1}{q_h}\left(\frac{dH}{dt}\right) dT, \qquad (10.2)$$

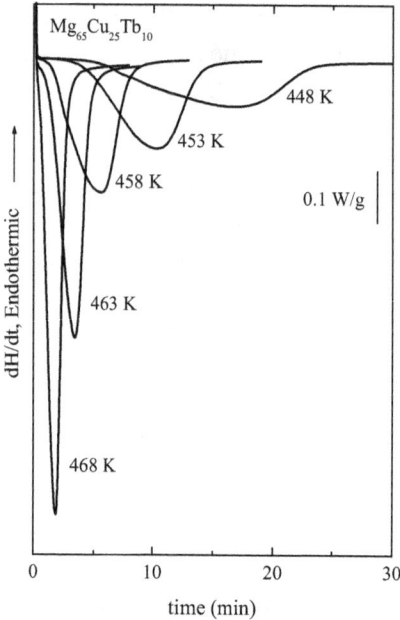

Fig. 10.1: Plots of (dH/dt) against time of isothermal crystallization of the ultraviscous melt at 448, 453, 458, 463, and 468 K, respectively. For each isothermal study, a new sample of the as-cast glass was heated from 363 K at 100 K/min rate to the above-mentioned selected temperatures, T_{cryst}, of 448, 453, 458, 463, and 468 K and kept at that temperature for a predetermined time period. After crystallization appeared to have been completed, the sample was heated at 20 K/min to a temperature where crystallization could occur. None of the samples thus treated showed an exothermic minimum. Their featureless DSC scans are not shown here. The plots are taken from [24].

where ΔH_{cryst} here is equal to the total peak area in the plot of $(dH/dt)/q_h$ in the temperature plane, and T' is a reference temperature. The rate of crystallization is obtained from

$$\frac{dx}{dt} = \frac{dx}{dT}\frac{dT}{dt}. \tag{10.3}$$

In rare cases, crystallization is also studied in the isothermal mode, i.e. by keeping the sample at a fixed temperature T_{cryst} usually in the melt state at $T > T_g$ and measuring the heat released with time. For the isothermal mode, T_{cryst} is kept sufficiently below T_p so that thermal effects of heat flow can be measured with reasonable accuracy. This method also shows a peak at a certain temperature and the partial area under the peak is proportional to x, at time, t, at the relevant temperature,

$$x(t) = \frac{1}{\Delta H_{cryst}} \int_{t=0}^{t} \left(\frac{dH}{dt}\right) dt, \tag{10.4}$$

where ΔH_{cryst} is equal to the total peak area. At the initial time, which is taken as zero, $x(t = 0) = 0$ holds and at a time near to the end of the experiment, $x = 1$.

To investigate whether crystallization on rate-heating produces the same calorimetric state as does crystallization isothermally, DSC-scans of the crystallized samples from the two studies are obtained and compared. If the two DSC-scans are found to be identical within experimental errors, and if these scans do not show the characteristic glass transition endotherm and/or crystallization minimum, it is assumed that the specific heats of the ultimate state formed on isothermal crystallization and on crystallization by rate heating are the same. However, it does not indicate whether the ultimately formed states in the two procedures are the same or whether the ultimate state has been reached by the same process of transformation.

10.3.2 One Experimental Example

As an example of the crystallization kinetics of metal-alloy glasses, we include here data on crystallization of a system with ternary composition, $Mg_{65}Cu_{25}Tb_{10}$-ultraviscous melt. The description is taken almost entirely from our previous study [24]. The usual DSC heating scan for a heating rate of 20 K min^{-1} yielded its glass softening temperature T_g of 414 K and an increase in the specific heat of 13.3 J/ mol^{-1}K^{-1}. Its crystallization was studied in isothermal mode by keeping the sample at a fixed temperature $T_{cryst} > 414$ K and also analyzed in the usual scanning modes by heating the glassy sample at a fixed rate from $T < 414$ K. The plots of (dH/dt) against t are shown in Fig. 10.1, where T_{cryst} is specified. The plots show only one exothermic peak, which is due to crystallization at high temperatures. As expected, increasing the sample's T_{cryst} shifts the peak to a higher T, increases its height, the peak becomes narrower and sharper, and the time taken for full crystallization is reduced. But at the lowest temperature of 448 K, the beginning of the broad peak appears to have become distorted. We will discuss this change here after the description of the data analysis.

As an example, Fig. 10.2 shows the $((dH/dt)$ vs. $T)$-plots obtained during rate heating at q_h of 5, 10, 20, 40, and 80 K/min. It is observed that there are two exothermic peaks due to crystallization, $T_{p,1}$ and $T_{p,2}$ in the plots, with their onset temperatures, $T_{x,1}$ and $T_{x,2}$. The low temperature peak is much more prominent than that at the high temperature. This feature indicates that there are two crystallization processes occurring at different rates. As is expected, when q_h is increased, $T_{p,1}$, $T_{p,2}$, $T_{x,1}$, $T_{x,2}$ as well as the height and the half-width of the peaks increase.

The quantity x, calculated from the plots in Fig. 10.1, is plotted against t in Fig. 10.3a. The corresponding dependencies of the rate of crystallization are plotted in Fig. 10.3b. Since this rate, (dx/dt), is proportional to $-(dH/dt)$, the (dx/dt)-plots are equivalent to the inverted plots of Fig. 10.1. The quantities x and (dx/dt), determined for peak 1 in the plots in Fig. 10.2, are shown against T in Fig. 10.4a and b, and the corresponding quantities for peak 2 are given against T in Fig. 10.5a and b.

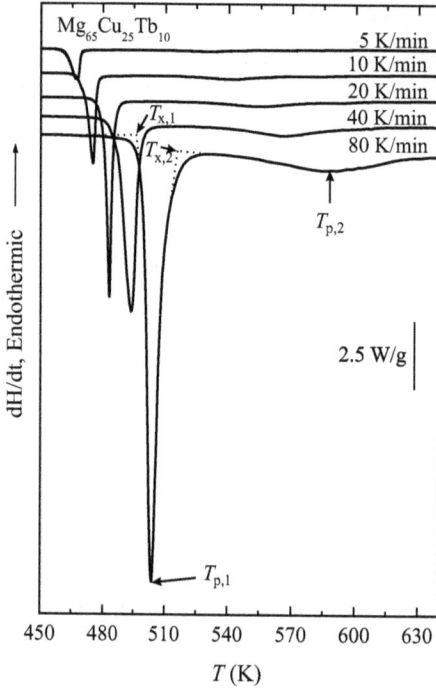

Fig. 10.2: Plots of (dH/dt) against T of the glass during heating at rates of 5, 10, 20, 40, and 80 K/min. $T_{x,1}$ and $T_{x,2}$ are the onset crystallization temperatures for peak 1 and peak 2, respectively. $T_{p,1}$ and $T_{p,2}$ are the peak temperatures for peak 1 and peak 2, respectively. For each study, the as-cast sample was heated at a fixed rate from 363 K to a certain temperature, T_{max}, and its DSC-scan was obtained. The heating rates for this study were 5, 10, 20, 40 and 80 K/min. During the first heating, the characteristic crystallization feature appeared as the exothermic minimum in the (dH/dt) against T plot. The sample was then cooled to 363 K at the same rate as in heating, and then rescanned to the same T_{max} of 693 K in order to determine whether or not the sample is fully crystallized. The plots are taken from [24].

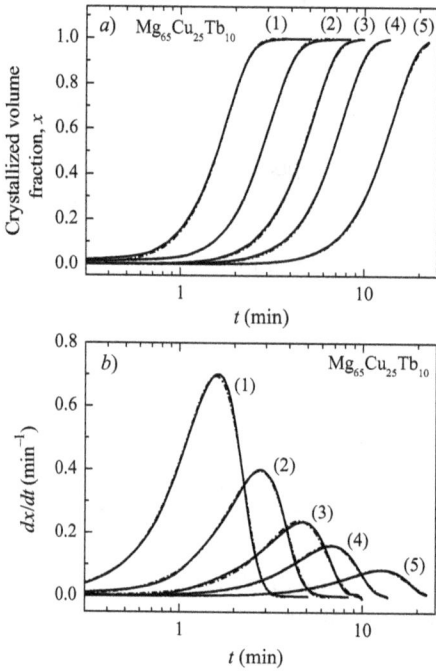

Fig. 10.3: (a) Crystallized volume fraction plotted against time and the fitting of the KJMA equation to data at (1) 468 K, (2) 463 K, (3) 458 K, (4) 453 K, and (5) 448 K. (b) The corresponding rate of crystallization. Solid lines are the experimental data and the dashed lines are the fitting curves. The plots are taken from [24].

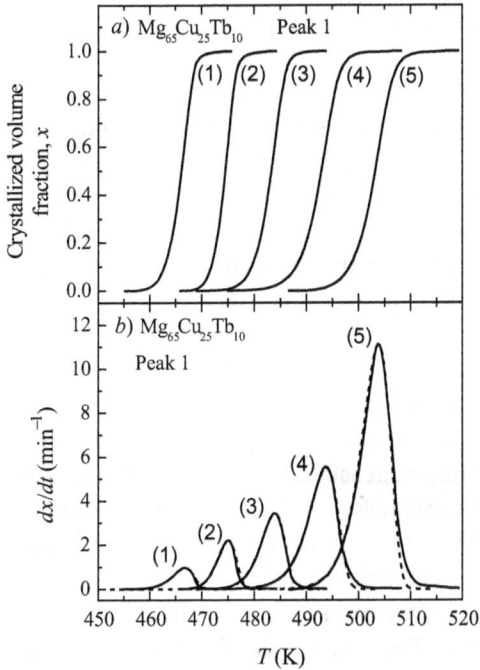

Fig. 10.4: (a) Crystallized volume fraction from peak 1 is plotted against temperature. The data were obtained for heating at (1) 5 K/min, (2) 10 K/min, (3) 20 K/min, (4) 40 K/min, and (5) 80 K/min. (b) The corresponding plots of the crystallization rate against temperature. Solid lines are the experimental data and the dashed lines are the fitting curves. The plots are taken from [24].

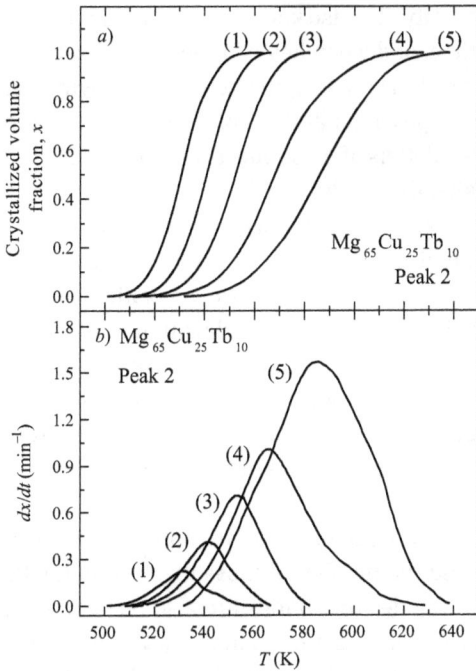

Fig. 10.5: (a) Crystallized volume fraction from peak 2 plotted against temperature. The data were obtained for heating at (1) 5 K/min, (2) 10 K/min, (3) 20 K/min, (4) 40 K/min, and (5) 80 K/min. (b) The corresponding plots of the crystallization rate against temperature. The plots are taken from [24].

10.3.3 Theoretical Interpretation in Terms of the KJMA-approach

According to the thermodynamic theory of crystal nucleation and growth in melts supercooled only by few degrees below their freezing/melting point, T_m, the rate of homogeneous nucleation is written in terms of two quantities, (i) the degree of supercooling (denoted as ΔT) below the equilibrium freezing/melting point T_m (i.e., $\Delta T = T_m - T$) and (ii) the excess entropy of the supercooled melt over the crystal phase. The kinetic equation for the nucleation rate requires also the knowledge of the solid-melt interfacial energy and the free energy barrier for molecular diffusion across the interface. As the melt is cooled, the term $(T_m - T)^2$ that appears in the denominator of the negative exponential term for the nucleation rate (see [1, 2] and here Chapter 2) increases rapidly, and hence the nucleation rate increases from a very slow to a very large value over a small temperature range. For deep supercooling to T far below T_m, the excess entropy becomes significantly less and this decrease has a significant effect on the nucleation rate. According to a simplified terminology of thermodynamic and kinetic driving forces for crystallization, the effect of increase in $(T_m - T)^2$ is relatively small at T just below T_m and large in the ultraviscous melt, particularly as T approaches the vitrification range. The effect of lessening of the excess entropy is large at T just below T_m and small in the ultraviscous state as T approaches the vitrification range. The nucleation rate is zero at T_m and crystallization rate is vanishingly small in the fully vitrified state due to the high values of viscosity. It is also known that some amorphous solids in storage crystallize slowly with time, and when heated their crystallization is rapid even before their T_g is reached. Plots of the nucleation and crystal growth rates against T show peaks. Their widths and shapes differ and the plots often partly overlap. Reviews on this subject have provided details of these effects [1, 2, 11, 12, 86].

It is not generally recognized (as already briefly mentioned here earlier) that crystallization of an ultraviscous melt and glass is distinguished from that of a low-viscosity melt. In the ultraviscous melt, molecular dynamics is of two types, (i) co-operative dynamics of slow diffusion, that shows up as the α-relaxation process and viscous flow exhibiting a non-Arrhenius variation of the relaxation time, τ, and the viscosity, η, with temperature, T, and whose activation energy decreases as T is increased, and (ii) faster dynamics of localized motions that show up as the JG-relaxation, and whose characteristic time varies with temperature, T, according to the Arrhenius equation. In the glassy state, the co-operative dynamics is too slow and only the JG-relaxation dynamics is observed.

It has been generally found that, on cooling, the number of molecules involved in the α-relaxation dynamics increases and the number of those involved in the JG-relaxation dynamics decreases. Localized modes of motions are also recognized as the source of the unexpectedly rapid nucleation and growth in the glassy and ultraviscous states of small molecule organic substances. There have been indications that the overall crystallization rate is most rapid in a viscous melt far below its freezing point, at the Donth temperature [87], where on cooling, the α-relaxation process of a melt evolves from the JG-relaxation process [88, 89]. In contrast, self-diffusion coefficient, D, and its temperature derivative are independent of the changing structure of the melt with changing temperature, i.e. the self-diffusion coefficient D, as for an ordinary crystal, varies according to the Arrhenius relation, $D \propto \exp(E_A/RT)$, with E_A being the Arrhenius activation energy, R is the universal gas constant. Because we are unable to take these aspects into account, we use, as others did, the Kolmogorov-Johnson-Mehl-Avrami (KJMA) equation [1, 2] for the overall crystallization kinetics, and henceforth discuss our results in terms of this formalism.

As mentioned earlier here, Tammann [4] proposed that crystallization occurs in two consecutive stages. In the first stage crystallization centers or nuclei form and in the second one the nuclei grow. Therefore, if a procedure could be found by which the time scale or temperature range of the two processes could be made substantially different, one can expect to observe the two processes and then investigate them separately by methods other than calorimetry (cf. e.g. [10]). The plots of the nucleation and crystal growth rates against T do show their different shape peaks at different temperatures and the plots partly overlap. For example, the relative position of these plots in the temperature plane are used for producing glass ceramics of different microstructures and properties. In this process, both homogeneous and heterogeneous nucleation are utilized.

When an ultraviscous melt obtained by heating a glassy state crystallizes, the crystal nuclei, if already present, are expected to grow before new nuclei may form in the bulk of an ultraviscous melt and in the localized regions of the JG-relaxation in the glassy state. Recent studies of crystallization of metallic glasses by Ichitsubo et al. [48–50], who used radio-frequency ultrasonic energy absorption, have shown that mechanical stability caused by the resulting shear leads to crystallization by motions in local regions in the structure of a glass where JG-relaxation occurs. In the melt, crystallization kinetics would be determined only by the molecular diffusion rate, the rate of the α-relaxation process or the viscosity. During this occurrence new nuclei may form concurrently with crystal growth, and these may also grow. The crystallization kinetics in such a case would be determined by both the nucleation and crystal growth rates, but the slower of the two processes would dominate the observed kinetics and thermal effects.

It is obvious that nucleation and growth occur in the same volume of a melt. Therefore, as crystallization progresses, the volume of the melt, available for nucleation, decreases and this decrease plays a major role in the crystallization kinetics. Kolmogorov calculated the probability of nucleation in a certain volume that remains available at a certain time after crystallization has begun [21]. Gutzow and Schmelzer [1] provided an interesting discussion of the subject, in which they compared Kolmogorov's probability calculation with Poisson's probability treatment of a mathematically similar problem. They also discussed the formal analysis of the volume available for nucleation in a partially crystallized melt by Johnson and Mehl [22], and by Avrami [23]. Their description may be consulted for further details. Burbelko et al. [90] have also reviewed the progress of the Kolmogorov–Johnson–Mehl–Avrami (KJMA)-approach since the time when Kolmogorov's original paper was published. The subsequently developed outline here follows the way as given in the monograph by Gutzow and Schmelzer [1].

Accordingly, the ratio of crystallized volume to the total volume of the system, x_n, can be written (for constant values of the nucleation, J, and linear growth rates, v, and this restriction – as will be discussed later in more detail – leads to serious restrictions in the applicability of this relation to rate-heating crystallization) as

$$x_n(t) = 1 - \exp\left(-Y_n(t)\right), \qquad Y_n(t) = \omega_n J v^n \int_0^t (t - t')^n dt', \tag{10.5}$$

where Y_n is the (hypothetical) volume of the new phase formed until time t provided the different crystalline aggregates would not affect each other. J is the nucleation rate, v is the linear growth velocity and ω_n is a geometrical shape factor equal to $4\pi/3$ for spherical clusters. The parameter n has different values for different nucleation and growth mechanisms and dimensions of space in which the transformation occurs. These values have been listed in Table 10.1 in [1]. After integrating and substitution of the value of the integral $Y_n(t)$, the final equation is obtained now known as Kolmogorov-Johnson-Mehl-Avrami (KJMA) equation, i.e.

$$x_n(t) = 1 - \exp\left(-\frac{\omega_n}{(n + 1)} J v^n t^{n+1}\right) \tag{10.6}$$

or

$$x_n(t) = 1 - \exp\left(-k_n t^{n+1}\right), \qquad k_n = \frac{\omega_n}{(n + 1)} J v^n. \tag{10.7}$$

The sum $(n + 1)$ is usually written as $m = n + 1$, and m is denoted as the KJMA-coefficient for the phase transformation. The quantity k_n is known as KJMA-kinetic coefficient. Macroscopically, it is the temperature-dependent rate constant for crystallization in such a study. The value of m is usually an integer, which depends upon the dimensionality and morphology of crystal formation and growth.

With mentioned notations, in practice the form of the KJMA-equation used for fitting the overall crystallization kinetics data is given by

$$x(t) = 1 - \exp\left(-kt^m\right). \tag{10.8}$$

By differentiating Eq. (10.8) with respect to time, t, one obtains the condition $(dx/dt) = 0$ at which the rate of crystallization reaches a maximum value, i.e.

$$\frac{dx}{dt} = mkt^{m-1} \exp\left(-kt^m\right) = 0. \tag{10.9}$$

The rate of enthalpy decrease or heat release in the crystallization process is, therefore, given by

$$\Delta H\left(\frac{dx}{dt}\right) = \Delta H\left[mkt^{m-1} \exp(-kt^m)\right], \tag{10.10}$$

or, equivalently (employing Eq. (10.8)), by

$$\Delta H \left(\frac{dx}{dt} \right) = \Delta H m k^{1/m} \left[-\ln(1-x)^{(m-1)/m} (1-x) \right].$$

(10.11)

By fitting Eqs. (10.8) and (10.9) to the plots of x against t in Fig. 10.3a and b one obtains the values of m and k at different temperatures as listed in Table 10.1. The fitted and experimental curves of x and (dx/dt) are shown in Fig. 10.3a and b.

Table 10.1: Enthalpy of crystallization and KJMA-parameters of the crystallization kinetics of an $Mg_{65}Cu_{25}Tb_{10}$-ultraviscous melt at isothermal crystallization. The data are taken from [24]

T_{cryst} (K)	ΔH_{cryst} (kJ/mol)	k (min^{-m})	m	E (kJ/mol)
448	2.81	$2.6 \cdot 10^{-4}$	3.11	
453	2.92	$1.7 \cdot 10^{-3}$	3.15	
458	2.96	$5.5 \cdot 10^{-3}$	3.17	169
463	3.23	0.026	3.21	
468	3.61	0.15	3.25	

For further tests, these k-values in turn may be used to determine the rate constant k_0 and the activation energy E for the relation,

$$k = k_0 \exp \left(-\frac{E}{RT} \right),$$

(10.12)

treating the process similar to chemical reaction kinetics (cf. e.g. [91, 92]). The data for $\ln(k)$ obtained from isothermal crystallization studies (circles) are plotted against $(1/T)$ in Fig. 10.6, where the values obtained from the results of fixed heating rate experiments are shown simultaneously by a smooth line.

In the formalism for nucleation and crystal growth in molten metals, interactions between atoms have a spherical symmetry, i.e. there is no directional bias, and in molecular melts interactions are also taken to be spherical but in terms of the van der Waals forces. Any change in these interactions may be seen as a process occurring concurrently to crystal growth and this change is expected to show up as a decrease in the enthalpy. A discussion of this aspect is given in [24].

10.3.4 Crystallization on Rate Heating

The parameters for the crystallization kinetics, observed on heating, are also obtained frequently by fitting the KJMA-equation modified for this purpose. In such cases, the nucleation rate is expected to decrease as the melt is heated. There are several methods for such an analysis, including analytical ones. Meisel and Cote [93] reviewed

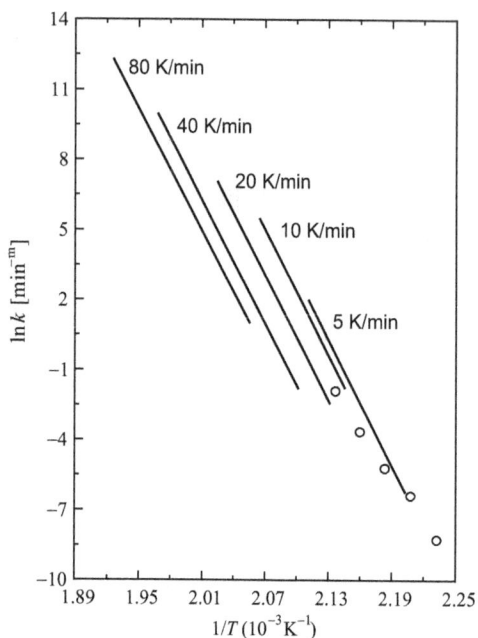

Fig. 10.6: Plots of $\ln(k)$ against the reciprocal temperature. Circles are data from isothermal kinetics, and lines from the rate-heating kinetics. The plots are taken from [24].

some of the difficulties in the analysis and provided a different approach to the use of the formalisms for the non-isothermal rate kinetics involving metastable phases, particularly those that form on heating metal-alloy glasses. A critical discussion of such analytical approaches was provided in detail by Yinnon and Uhlmann [94]. These authors came, however, to the conclusion that analytical methods lead to serious uncertainties. According to them, the reason for the uncertainties is that even though the approximations made in the methods used for such analyses are different, the KJMA-equation is always used as a basis in all methods. However, as evident from the derivation of this relation as sketched by us here above, this relation in the form given by Eq. (10.8) holds strongly only for constant nucleation and linear growth rates. Yinnon and Uhlmann [94] concluded further that, "All these methods are based on the Avrami treatment of transformation kinetics and define an effective crystallization rate coefficient having an Arrhenian temperature dependence." They continued then [94] "Most are shown to be based on an incorrect neglect of the temperature dependence of the rate coefficient", and "Thus, in general, non-isothermal transformation cannot be treated analytically. A detailed description of non-isothermal transformation can, however, be obtained by numerical methods."

In view of the above-mentioned reasons, we will employ here numerical methods in the analysis of rate heating crystallization. From the spectrum of approaches employed in this direction, we employ the one suggested by Kelton and Greer [2] and Greer [95] as it seems to us to be the most appropriate for use here. Greer assumed that crystallization at different T will vary only in the time scale, i.e., that the transformation is iso-kinetic and its rate at any instant depends only on T and the value of x at that instant. It is independent of prior thermal history of the sample, as would be the case for a melt in which only one type of crystals of only one composition are formed. In such an analysis, direct use is made of the isothermal KJMA-equation by approximating the linear heating profile as a series of short isothermal anneals.

To describe the method briefly in more detail, consider a material transformed at T_1 for a time of t_1. The transformed fraction at T_1 is $x = f_1(t')$, where t' is the time from the start of the transformation at T_1. At the end of the transformation at T_1, the transformed fraction is $x = f_1(t_1)$. The transformation continues at T_2 for a time of t_2 at which $x = f_2(t'')$, where t'' is the time from the start of transformation at T_2. The course of the transformation at T_2 is precisely the same as if the initial transformed fraction at $T_2, f_1(t_1)$, had been formed at T_2. If $t_{1'}$ is the time it would have taken at T_2 to produce $x = f_1(t_1)$, i.e., $f_1(t_1) = f_2(t_{1'})$, the transformation at T_2 is given by $x = f_2(t + t_{1'} - t_1)$, where t is the time from the start of the transformation at T_1 (or T_2). At the end of the transformation at T_2, $x = f_2(t_2 + t_{1'} - t_1)$. The calculations were performed by using a computer program for anneals at uniformly incremented temperature. At the end of the rate-heating, values of x are obtained and (dx/dt) is evaluated at the midpoints of temperature step. For the ith increment, (dx/dt) at $T = (T_i + T_{i+1})/2$ is taken to be $(x_{i+1} - x_i)/(t_{i+1} - t_i)$.

Fig. 10.4b shows the result of fitting of the (dx/dt)-plots for crystallization peak 1 at different heating rates. The fitted parameters for the crystallization kinetics are listed in Table 10.2. The activation energies obtained by the two methods differ, but both values are considerably high. Still these values are lower than those observed for molecular ultraviscous melts, e.g., the activation energy for isothermal crystallization for syndiotactic poly(styrene) is 792 kJ/mol [96].

In earlier studies, the activation energies of several relatively stable metal-alloy glasses, $Pd_{77}Cu_6Si_{17}$ and $Pd_{48}Ni_{32}P_{20}$, were determined [97] by using crystallization kinetics data obtained by rate-heating. These values were found to be equal to the

Table 10.2: Heating rate, onset and peak temperatures, enthalpy of crystallization, and KJMA-parameters of crystallization kinetics on rate heating the $Mg_{65}Cu_{25}Tb_{10}$-glass for the first exothermic feature. Also listed are the onset and peak temperatures, and the enthalpy of crystallization for the second exothermic feature. The fitting method used is similar to the one used by Kelton and Greer [2] and Greer [95]. The data are taken from [24]

q_h (K/min)	$T_{x,1}$ (K)	$T_{p,1}$ (K)	$\Delta H_{cryst,1}$ (kJ/mol)	$\ln k_0$	m	E (kJ/mol)	$T_{x,2}$ (K)	$T_{p,2}$ (K)	$\Delta H_{cryst,2}$ (kJ/mol)
5	461.9	466.9	3.47	191.90	3.30		469.5	530.7	0.76
10	469.8	475.2	3.51	190.10	3.28		477.5	541.6	0.85
20	478.2	483.2	3.58	187.01	3.26	246	487.1	553.5	0.89
40	484.3	493.9	3.74	183.25	3.23		498.8	565.9	0.97
80	496.1	503.8	4.05	180.30	3.20		515.7	585.9	0.99

activation energy for viscous flow which decreased when the crystallization temperature was increased. This equality led to the conclusion that the crystallization rate is controlled by the viscosity. It was also suggested that crystallization in the amorphous state may occur by a diffusion-less mechanism ([98], cf. also Chapter 7 and [30]) and in such a case the activation energy is low, and hence the crystallization rate is small. This presents an interesting situation when one realizes that the activation energy for localized motions of the JG-relaxation is low and not that of the motions that involve viscous flow, indicating thereby that crystallization may involve localized motions which occur over a range which is long enough to produce nuclei and crystal growth. However, it is now established that the processes of crystallization, as well as of structural relaxation, in metal-alloy glasses are diffusion-controlled.

10.3.5 Differences Between Isothermal and Rate-heating Crystallization

One finding for the $Mg_{65}Cu_{25}Tb_{10}$-melt here is that its crystallization occurs only in one step isothermally at high temperatures and in two steps on rate-heating (only one peak is observed in Fig. 10.1 and two are observed in Fig. 10.2). But also a small broadening to the left of the already broad peak appears at isothermal crystallization at 448 K in Fig. 10.1. Crystallization is almost complete in less than 10 min isothermally at 468 K in Fig. 10.1, and this temperature is considerably lower than the 510–580 K range at which $T_{p,2}$ appears in Fig. 10.2. Therefore, it seems necessary to determine (by extrapolation) the temperature at which the $T_{p,2}$-peak might appear at a formally zero heating rate. This temperature is 525 K. The exponent, m, is interpreted in terms of the manner in which nuclei grow to form crystals, and its magnitude is seen to indicate the mechanism of crystal growth [31]. The value of m for isothermal crystallization obtained here varies from 3.11 to 3.25 (Table 10.1) and for rate-heating crystallization from 3.2 to 3.3. This would suggest that crystallization occurs by interface-controlled growth with decreasing nucleation rate. The validity of these interpretations can be ascertained by *in situ* X-ray diffraction and electron microscopy studies.

The plots in Figs. 10.3 and 10.4 show that both the measured values of x and the rate of crystallization do not agree entirely with the values calculated from the KJMA-equation. The deviation seems to be the highest at long times and high temperatures. To investigate its source, we determined the ratio of the measured rate of crystallization to that calculated for both isothermal and rate-heating crystallization. This ratio is plotted in Figs. 10.7a and b, respectively. It shows significant deviations at short times when crystallization occurs at 468 K. For other temperatures, the deviation increases with t and with T. It is at most 86 %, indicating that crystallization of the melt becomes slower than expected as $x \rightarrow 1$.

There are several occurrences that may produce this feature, but the main one seems to be that the crystals formed are small with a large fraction of inter-granular melt (cf. also [99] and here Chapter 8). This melt crystallizes more slowly than the bulk,

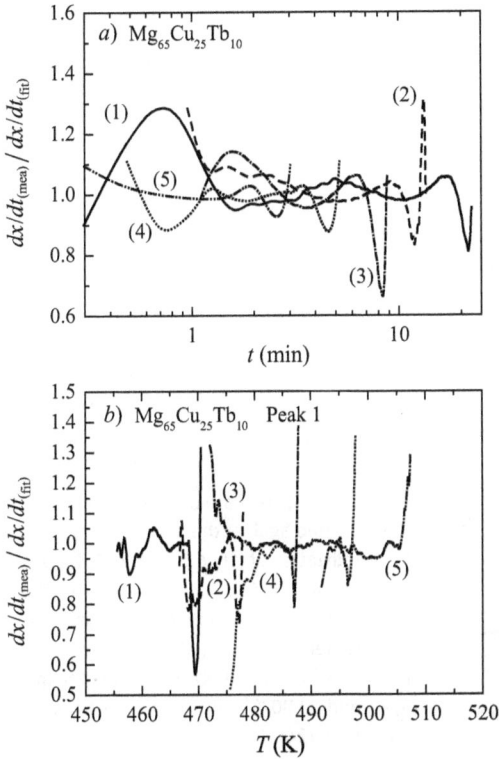

Fig. 10.7: (a) The ratio of the measured rate of crystallization to that calculated from fitting the KJMA equation is plotted against time. The data are taken from Fig. 10.3b. (b) The corresponding plots against temperature from data in Fig. 10.4b. Curves 1, 2, 3, 4, and 5 refer to the data at temperatures, (1) 468 K, (2) 463 K, (3) 458 K, (4) 453 K, and (5) 448 K, as in Fig. 10.3. The plots are taken from [24].

despite its viscosity being lower. Alternatively, it may be one of the melt composition that is different from the original composition and whose crystallization to another phase occurs more slowly than that of the parent phase. Finally, some uncertainty may be introduced as well by the method of computation employed. When the temperature is increased, the critical cluster size is increased. As a consequence, not all clusters can grow at these temperatures but only those having sizes above the new value of the critical cluster size. Clusters with sizes in the range between the new and the original critical cluster size will, as a rule, decay. Consequently, not the whole amount of the newly evolving phase, formed at a temperature T, is available for further growth at a temperature $T + \Delta T$ [1, 100].

It is worth investigating how the crystallization rate, $(dx/dt)_T$, changes with x on crystallization at a fixed temperature T and on crystallization during heating at a fixed rate. Fig. 10.8a shows plots of this rate against x isothermally and Fig. 10.8b shows some of them on rate-heating. According to Eq. (10.8), a plot of $(dx/dt)_T$ against t would show a peak only when $m > 1$, which, of course, is evident from the sigmoid-shaped plots of x against t in Fig. 10.3a. However, the plots of $(dx/dt)_T$ against x in Fig. 10.8a show a hump-like feature whose peak shifts with x from $x = 0.521$ at 448 K to $x = 0.528$ at 468 K, as indicated by the arrows. The shift in the peak is attributable mainly

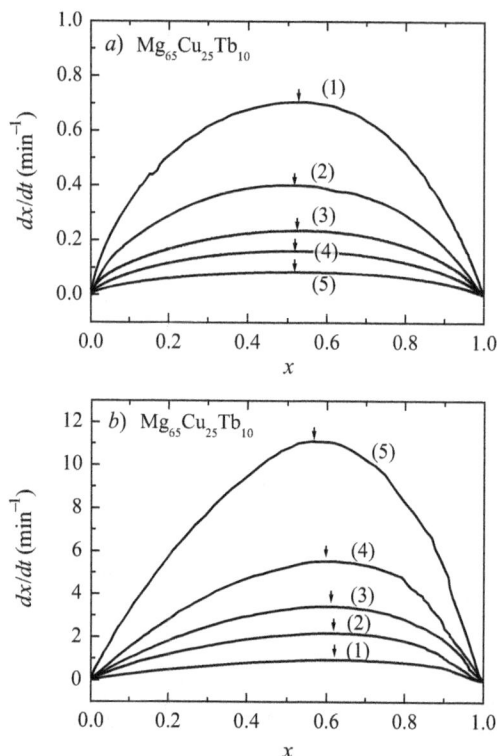

Fig. 10.8: (a) Rate of crystallization is plotted against the degree of crystallization at (1) 468 K, (2) 463 K, (3) 458 K, (4) 453 K, and (5) 448 K. (b) The corresponding plots from the data obtained for heating at (1) 5 K/min, (2) 10 K/min, (3) 20 K/min, (4) 40 K/min, and (5) 80 K/min. The plots are taken from [24].

to change in m with T_{cryst}. In contrast, a broad peak is observed in Fig. 10.8b, and this peak shifts from $x = 0.62$ at 5 K/min to $x = 0.56$ at 80 K/min, as is indicated by the arrows.

10.3.6 Origin of the Second Peak for Crystallization on Rate-heating

A broad and low-intensity exothermic peak at high temperatures in Fig. 10.2 appears for each of the five heating rates, which indicates an additional irreversible process in the metastable melt. When q_h is increased, it shifts to higher temperatures, as expected for a thermally-activated process. For a one-component system, the second exotherm may indicate that a small fraction of the melt persists after the first crystallization and this fraction crystallizes at a lower rate. It may alternatively indicate that the crystals formed on first crystallization are of metastable structure that transforms to stable crystal modifications on heating i.e. solid-solid transformation, as expected on the basis of the Ostwald rule of stages (cf. e.g. [12, 14, 101, 102]). But this would not be the case for a ternary melt of $Mg_{65}Cu_{25}Tb_{10}$. Alternatively, the melt initially crystallizes to form an inter-metallic compound, leaving behind a melt rich in one component. Nucleation and growth in this melt is slow and its crystallization appears as a broad peak at higher temperatures. It is also conceivable that the second peak is a reflection of the grain growth of some of the fine structure or nanocrystalline phases formed as a result of the first crystallization.

The enthalpy decrease, associated with the second exotherm, that appears as peak 2 is 0.89 kJ/mol for a heating rate of 20 K/min. This is about 25 % of the enthalpy release of the first crystallization peak. This seems to be a substantial effect, although it does not refer to the relative amounts of the states formed on the second and first crystallization. The second peak observed on rate heating also appears to be too broad to be described by the KJMA-formalism with a reasonable value for m. Therefore, we consider that the second peak indicates occurrence of one or several of the following processes: (i) Crystals produced on fast crystallization are metastable and these transform to another structure. (ii) The glass does not completely crystallize and the crystals which persist within the melt are of a different composition, which crystallize with a much slower kinetics to another phase at high temperatures. (iii) Some of the crystals formed on first crystallization are small and they grow in the solid state to form larger grains. The heat evolved in cases (i) and (ii) is due to the usual liquid-solid and solid-solid phase transformations and in case (iii) it is due to grain growth. Calorimetric data of these glasses cannot unambiguously determine which of these occurrences is more plausible. The structure of the phases, formed during this crystallization process, and their chemical compositions can be determined by using X-ray diffraction and electron microscopy at high temperatures. Such studies may also reveal new crystalline alloys in the binary and ternary compositions of the Mg-Cu-Tb system.

To resolve the origin of the second exothermic peak, seen in Fig. 10.2, we performed a further experiment. A new as-cast sample was heated at 20 K/min from 300 K to 458 K ($\sim$ 44 K above the glass transition temperature, T_g, being of the order of 414 K) and then kept isothermally at 458 K for 10 min. Thereafter, the sample was heated at 20 K/min to 613 K. The curve obtained is shown as curve 1 in Fig. 10.9. It shows no specific features until a temperature T of about 480 K is reached, and thereafter an exotherm appears. This result indicates that annealing at 458 K for 10 min does not remove the second exothermic peak. Since the annealing was performed at $\sim$ 44 K above its T_g, it is unlikely that a significant amount of the melt would have persisted (in comparison against the plot at 458 K in Fig. 10.1, this amounts to 10 more minutes at 458 K). The sample was then cooled back to 495 K at 20 K/min and reheated at the same rate, and curves 2 and 3 in Fig. 10.9 were obtained. These curves show no indication of a further exotherm. The vanishing of the exotherm may indicate that crystallites growing at different sites produced a solid, containing extremely small grains, and hence a large surface area, and that the annealing time was insufficient for considerable grain growth by Ostwald ripening [13, 103]. The grain growth occurred only on heating, and reduction in surface energy produced an exotherm. It is also possible that a metastable phase had formed on annealing that transformed irreversibly to a stable phase on heating. The total heat, released on heating, is 1.40 kJ/mol, which seems to be too large to be caused by surface energy reduction. We conclude that the

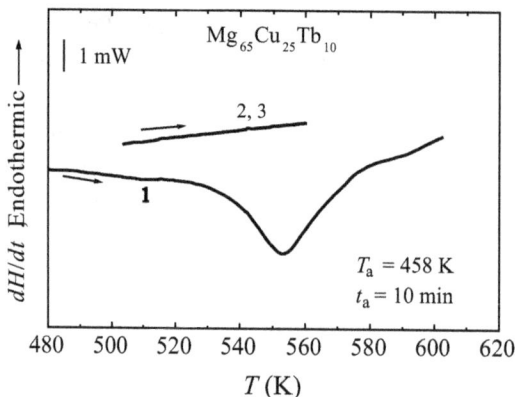

Fig. 10.9: Plots of (dH/dt) against T of the glass that was annealed at 458 K for 10 min and then heated at 20 K/min, as shown by curve 1. The scan did not show the first peak but the second peak persisted. The curves 2 and 3 were obtained by reheating the sample to 613 K. These curves showed no such features. The plots are taken from [24].

melt contained a metastable phase that grew on annealing at 458 K and transformed to a stable phase on subsequent heating.

10.4 Thermal Effects of Crystallization on Its Kinetics

10.4.1 General Remarks

When a melt begins to crystallize, a number of physical processes occur, and these processes, which are often overlooked, contribute to the rate of crystallization, the types of crystals formed, their density, their composition in a multi-component melt and the morphology of the crystals. We consider that there are at least four effects resulting from these processes: (i) the density of the crystal is generally higher than that of the melt and the fractional density change (or the crystal-melt volume misfit parameter), $(\rho_{crystal} - \rho_{melt})/\rho_{crystal}$, is positive. As a consequence, the melt flows towards the crystallizing region and fills the regions of volume shrinkage. Note that in the case of ice, Ga, and several other metals, for which $\rho_{crystal} < \rho_{melt}$, this fractional change and the volume misfit parameter are negative, and in those cases the melt flows away from the crystallizing region as the crystal growth tends to push the melt away from its crystal-melt interface. (ii) The latent heat of crystallization that is liberated at the growing front of a crystal produces hot spots at the crystal-melt interface. This heat is transported away from the interface by thermal conduction due to phonons and due to thermal convection. The latter arises as a result of the density gradient due to the temperature change. (iii) In a multi-component melt the various components with their own diffusion coefficients become redistributed as crystallization occurs incongruently. The melt is thus either richer or deficient in, at least, one of the components. This effect results into a continuously changing composition of both the crystal phase and the melt, which in turn results into a continuous change in the fractional density or the volume misfit parameter, the enthalpy of melting and the thermal effect on the fractional density. In this case, the product of crystallization is inhomogeneous in composition. (iv) The spatial temperature distributions, resulting from the above-given processes, cause different values of surface tension, which then causes the melt to flow from one crystal interface to another, an effect known as Marangoni flow. Thus the ratio of the buoyancy driven effect to viscosity driven effect may be caused by exothermal crystallization alone or by solute difference alone or by a combination of the two. If the surface area of the melt is much larger than its depth, the effects related to the melt-gas interface may become more important than the effects in the bulk. In addition, such large surface effects may cause a metastable crystal state to crystallize. An example of such crystallization is formation of cubic ice on freezing of thin films [104]. Moreover, the diffusion coefficient of an atom or molecule varies as the inverse of its radius according to the Stokes-Einstein equation, $D_{AB} = k_B T/(b\pi R_0 \eta_B)$, where A refers to the molecule diffusing in the continuum B of viscosity η_B, k_B is the Boltzmann constant, R_0 is the radius of the diffusing molecules, and the constant b depends on the size of the diffusing molecules. The relation $b = 6$ appears to hold for molecules larger than those of the base substance; $b = 4$ for identical molecules; and b can be less than 4 for smaller molecules.

To quantitatively analyze the above-mentioned effects of heat and mass transport, one needs to solve the appropriate kinetic equations accounting for the boundary values for conservation of mass, momentum and energy in the geometry of crystal-melt regions. In the usual approach one formulates these conservation laws in terms of a dimensionless variable or a number, which may have values between zero and infinity. Two such effects that have been formulated in such terms are described in the sections below.

10.4.2 Rayleigh–Bénard Convection Effects

Although phonon conduction plays a role in transporting heat from the crystal-melt interface to both the crystal and the melt, and the phonon conduction is affected by both the temperature and the continuously changing constituents of the crystal and the melt, the convection plays the dominant role in the heat transfer during the crystallization process. A review of the subject of the buoyancy-driven or Rayleigh–Bénard convection was published by Bérge and Dubois [105]. Here we use the information in that review as a basis for relating the buoyancy effect to crystallization.

Briefly, the ascending motion of a hot and less dense element and descending motion of a cold and denser element of a melt in the Earth's gravity produces a bifurcation between the steady state and the convective states, thus producing cellular patterns in a liquid [105]. Such pattern formation in plasma melting of glasses was also discussed by Tseskis et al. [106]. Although the buoyancy driven convection is mainly due to difference in the temperature of the melt, there is an additional effect on heat transport as the generally heavier crystals sediment in the melt or lighter crystals, ice, Ga, etc., rise to the surface of a melt and (except in the case of mono-component or congruently melting systems) there is usually segregation of one or more components at the crystallizing interface, resulting in a boundary layer of different composition at the crystal-melt interface. If the metal density depends upon the composition, this would also contribute to the thermal convection. This raises an interesting situation when the crystallizing interface rejects components higher in density than the bulk melt, as may occur on crystallization of ultraviscous metal alloys in which one component is much heavier than the remaining ones. The fluid next to the interface is made heavier by the rejected component but lighter by the higher temperature due to heat released on crystallization. In summary, there is a host of complicated factors that make the crystal growth process to be limited by the molecular transport mechanism (kinetics of attachment of entities to the growing crystal surface), and, when the melt is ultraviscous, molecular transport appears to be determined by the co-operatively rearranging regions or dynamically heterogeneous domains of nanometer size dimensions. This presents not only a conflict between the theories of molecular transport for ultraviscous melts and the mechanism by which supercritical nuclei may form at a fixed rate during the crystal growth process, but also a time dependent microscopic size crystal growth as a result of time-dependent buoyancy convection effects.

Considering the buoyancy convection effects in the presence of the gravitational force, when the gravitational vector and temperature gradient are aligned, the dimensionless Rayleigh number [105, 107] is

$$R_a = g\frac{\alpha \Delta T d^3}{\nu \kappa}, \tag{10.13}$$

where g is the acceleration due to gravity, α is the thermal expansion coefficient of the melt, ΔT the vertical temperature difference imposed or caused by the occurrence of some process, d is the melt depth corresponding to ΔT, ν is the kinematic viscosity ($\nu = \eta/\rho$), and κ is thermal diffusivity. It is found that only in the case when R_a exceeds a certain value characteristic of a melt, the convective motion begins. Thus changes in any of the variables in Eq. (10.13) can cause conditions in which convection may or may not occur.

As an example, the critical temperature difference, ΔT_c, corresponding to R_{ac} for the threshold motion for $d = 1$ cm is 0.1 K for water and 2.2 K for silicone oil at room temperature [105]. When the quantity g is zero, R_a is zero, there are no convective currents because there is no buoyancy and the growth of crystals occurs slowly, because there is no transfer of the heat evolved away from the crystal-melt interface. This is true for both mono-component and multi-component systems, and the crystal growth can be spheroidal. However, if crystallization of a melt is forced to occur in a container spinning at a high rate, the buoyancy effect is high and the crystals grow highly dendritically on cooling even when the viscosity of the melt is only a few dPa·s, as found in laboratory studies of crystallization of $Sn_{99.7}Cu_{0.3}$-melts on cooling [108] and in the crystallization studies of AgI-AgPO$_3$-ultraviscous melts in microgravity conditions, created onboard NASA's KC135 flights, performed in 1987–1988 by G.P. Johari.

So, although it is acknowledged that buoyancy-induced convection influences the microstructure of a solidified material from its melt, most models assume a steady-state, purely diffusive rate of heat transfer during solidification, and it is not certain how inclusion of this source of heat transport by mass transfer due to the buoyancy convection can affect the observed crystallization rate. Nevertheless, there is little doubt that when this convection effect begins to dominate, the generally used potential energy barrier for viscous flow would become less relevant for crystallization kinetics. It is also understood that the *apparent* energy barrier to crystal growth is decreased by the buoyancy convection. It is conceivable that the low activation energy, listed in Table 10.1, is also partly due to this effect. Crystallization experiments performed on mono-component systems under different gravitational conditions, where the buoyancy convection may be largely reduced, or different centrifugal forces, where it can be considerably increased, may be able to resolve these effects on the activation energy.

10.4.3 Marangoni or Thermo-capillarity Convection Effect

The Marangoni effect refers to the mass transfer of a melt along an interface caused by a difference in the surface tension. This process may result into homogenization of the surface tension, σ_t, as a melt flows from regions of low surface tension to the region of higher surface tension. The topic was treated already by Gibbs [3]. The difference in surface tension may be caused by the temperature difference resulting from the exothermic effect of crystallization and/or by compositional differences in the two regions of the melt when incongruent crystallization occurs at different rates in different regions. It is understood that the Marangoni flow dominates over the buoyancy-driven flow in the Earth's gravity. Therefore, the Marangoni number, a dimensionless parameter, can be used to characterize the relative effects of surface tension and viscous forces.

A detailed mathematical treatment of this effect based on the Navier-Stokes equations and the equations of thermodynamics is given by Chandrasekhar [109]. Briefly, the Marangoni number is given by [107]

$$M_a = \left(\frac{\partial \sigma_t}{\partial T} \right) \frac{\Delta T d}{\rho \nu \kappa}, \tag{10.14}$$

where the terms have the same meaning as in Eq. (10.13). Originally σ_t was referred to the liquid-gas interface, but here it refers to solid-liquid interface. Thus, the ratio of

M_a to R_a is given by

$$\frac{M_a}{R_a} = \left(\frac{\partial \sigma_t}{\partial T}\right)\frac{1}{\alpha \rho\, d^2 g},$$

(10.15)

so that for high values of g, R_a dominates the convection flow, and for zero or low values of g, Marangoni flow dominates. Also for small systems as for crystallization of a melt, d^2 is small and the Marangoni convection of heat dominates.

As mentioned earlier here, if the surface area of the melt is much larger than its depth, the effects related to the melt-gas interface may become more important than the effects in the bulk. In such a case the effects of the Marangoni flow can be examined by changing the gas in contact with the melt and thereby alter the surface tension. Since DSC-studies are performed on a sample that has a large surface area exposed to the purge gas and to the metal of the DSC-pan in relation to the depth of the liquid, it is likely that these effects may be observed by varying the purge gas, the rate of heating or by crystallizing at different low temperatures over a long period of time.

10.5 Classical and Generalized Gibbs' Approaches to Cluster Formation and Growth

10.5.1 Basic Ideas

In addition to the analysis of the peculiarities of the motion of the particles in the liquid (cf. also [37]) and its effects on crystal formation and growth, we would like to briefly review here also some new developments connected with the thermodynamics of cluster formation and growth processes.

In his fundamental analysis [3], first published in the period of 1875–1878, J.W. Gibbs extended classical thermodynamics to the description of heterogeneous systems consisting of several macroscopic phases in thermodynamic equilibrium. As one application, he analyzed thermodynamic aspects of nucleation phenomena and the dependence of the properties of critical clusters – aggregates being in unstable thermodynamic equilibrium with the ambient phase – on supersaturation. Regardless of the existing impressive advances of computer simulation techniques and density functional computations, the method developed by Gibbs is predominantly employed till now in the theoretical interpretation of experimental data on nucleation-growth phenomena (cf. e.g. [1, 2, 30, 35, 36, 86, 110]). However, as evident from the title of his work [3] (“*On the equilibrium of heterogeneous substances*”), Gibbs directed his analysis exclusively to equilibrium states of thermodynamically heterogeneous systems. It follows immediately as a consequence that Gibbs' thermodynamics in its original form is not applicable – without developing more or less founded additional assumptions – to the description of growth and dissolution processes of clusters of super- and subcritical sizes, i.e. of small aggregates not being in thermodynamic equilibrium with the ambient phase. The description of the properties of clusters of such arbitrary sizes in the ambient phase is not covered by Gibbs' classical method, restricting its strict applicability to nucleation-growth processes to the specification of the properties of critical clusters (or clusters in stable equilibrium with the ambient phase) and, of course, to macroscopic equilibrium states. However, even in the application to the description of the properties of critical clusters problems occur as will be evident from the following considerations.

In application to the analysis of phase equilibria of macroscopic systems, Gibbs' theory served so well that it is considered frequently as being equivalent to the basic laws of thermodynamics or even as being a consequence of them. Such a point of view is not correct as can be traced easily following Gibbs' derivations. In addition, such interpretation contradicts Gibbs' own point of view considering his theory merely as one of the possible methods of description of thermodynamically heterogeneous systems but, of course, a good one. He wrote (cf. [111]): "Although my results were in a large measure such as had been previously obtained by other methods, yet, as I readily obtained those which were to me before unknown or vaguely known, I was confirmed in the suitableness of the method adopted". Mentioned point of view about the equivalence of Gibbs' approach to the basic laws of thermodynamics is also in contrast to different attempts to modify or replace Gibbs' treatment. However, these alternative approaches have their own limitations as mentioned partly by the authors themselves or as it became evident in their further discussion in the scientific community. In most cases, these and further alternative approaches (if correct) turned out to be widely equivalent in their consequences to the results of Gibbs' theory.

Restricting the analysis to equilibrium states, Gibbs considers exclusively variations of the state of heterogeneous systems proceeding via sequences of equilibrium states. For such quasi-stationary reversible changes of the states of a heterogeneous system, Gibbs' theory leads to the consequence that the surface tension depends on the state parameters of one of the coexisting phases merely. This limitation is not restrictive with respect to the analysis of macroscopic equilibrium states and quasi-stationary processes proceeding in between them. For such cases, the properties of one of the phases are uniquely determined via the equilibrium conditions by the properties of the alternative coexisting phase. However, the situation becomes very different if Gibbs' theory is applied to the description of cluster nucleation and growth.

Critical clusters, determining the rate of nucleation processes, correspond to a saddle point of the appropriate thermodynamic potential. In order to search for saddle or other singular points of any thermodynamic potential surface, we have to know the values of the potential function first for any possible states of the system. In application to cluster formation and growth, we have to know also the thermodynamic functions of a cluster or an ensemble of clusters not being, in general, in equilibrium with the otherwise homogeneous ambient phase. Only having this information, we can search for singular points by well-established rules. Since Gibbs restricts his analysis from the very beginning to equilibrium states, his theory does not allow us – strictly speaking – to apply the common methods of search for saddle points. And here we come to the basic limitation of Gibbs' theory in application to cluster formation and growth processes: In the search for the critical cluster we have to compare not different equilibrium states but all possible states including non-equilibrium states of the heterogeneous system under consideration. For the different non-equilibrium states considered, the surface tension has to depend, in general, on the state parameters of both coexisting phases. Gibbs' classical approach does not allow us, in principle, to account for such dependence and has to be generalized to open the possibility to incorporate this essential new ingredient into the thermodynamic description.

An extension of Gibbs' original thermodynamic treatment of heterogeneous systems to include non-equilibrium states along the lines as discussed above was initiated by one of the present authors in cooperation with Ivan Gutzow [112] (by formulating the (as we denoted it) generalized Ostwald's rule of stages) and then further advanced into a comprehensive thermodynamic theory in cooperation with Vladimir G. Baidakov and Grey Sh. Boltachev (both Yekaterinburg, Russia [113]), Jörg Möller (Dresden, Germany) and Alexander S. Abyzov (Kharkov, Ukraine) [28, 33, 34, 71] and applied to the interpretation of nucleation-growth processes in glass-forming melts in cooperation with Vladimir M. Fokin and Edgar D. Zanotto (St. Petersburg, Russia & São Carlos, Brazil) [10, 12, 32]. This – as we denote it – generalized Gibbs' approach employs Gibbs' model as well. However, Gibbs' fundamental equation for the superficial or surface quantities is generalized (extending previous approaches of one of the authors [114], assuming certain well-defined constraints to prevent irreversible flow processes) allowing one to introduce into the description the essential dependence of the surface state parameters (including the surface tension) on the bulk state parameters of both coexisting phases. Then the thermodynamic potentials for the respective non-equilibrium states are formulated. After this task is performed, the equilibrium conditions are derived. Similarly to Gibbs' classical theory, the critical cluster corresponds to a saddle point of the characteristic thermodynamic potential. It is a maximum with respect to variations of the cluster size at fixed intensive state parameters of both cluster and ambient phase and a minimum with respect to variations of the intensive bulk state parameters of the cluster. However, the state parameters of the critical clusters are different as compared with the predictions of the classical Gibbs' approach. These and further consequences are briefly sketched below.

10.5.2 Application to Nucleation

The thermodynamic state parameters – size, composition, and structure – of a critical cluster are determined in Gibbs' classical approach via a subset of the well-known thermodynamic equilibrium conditions (equality of temperature and chemical potentials of the different components) identical in Gibbs' classical treatment to those obtained for the description of phase equilibrium of macroscopic systems. Employing these dependencies it turns out that – following Gibbs' classical approach – the bulk properties of the critical clusters are widely the same as the respective properties of the newly evolving macroscopic phase (at least, as far as the formation of condensed phases is considered). Once this is the case, one can then assume that the properties of sub- and supercritical clusters deviate also only slightly from the respective parameters of the newly evolving macroscopic phases. This assumption is commonly employed in the description of crystal growth and dissolution processes [2, 115]. However, the above mentioned result of Gibbs' theory concerning the properties of critical clusters is in contradiction to results of molecular dynamics and density functional computations of the respective parameters as demonstrated first in detail by Hillert [26], Cahn and Hilliard [27]. In such approaches it can be shown that the properties of critical clusters, in general, significantly deviate from the properties of the newly evolving macroscopic phases, in particular, for large values of the supersaturation. In this way, the question arises what the origin of such discrepancies is and how they can be removed eventually.

As already mentioned, restricting the analysis to equilibrium states, Gibbs considers exclusively variations of the state of thermodynamically heterogeneous systems (in Gibbs' notations) proceeding via sequences of equilibrium states. For such quasi-stationary reversible changes of the states of a thermodynamically inhomogeneous system, Gibbs' adsorption equation is valid (which is reduced for isothermal conditions to Gibbs' adsorption isotherm frequently employed in applications). This equation describes (in the framework of Gibbs' theory) the dependence of the surface tension on the state parameters of the system under consideration. It leads to the consequence that the surface tension has to depend (in the simplest case) on $(k + 1)$ independent thermodynamic parameters, where k is the number of components in the system. This result implies that – according to Gibbs' original approach – the surface tension depends on the state parameters of one of the coexisting phases merely.

This limitation is not restrictive for equilibrium states and quasi-stationary processes proceeding in between them. For such cases, the properties of one of the phases are uniquely determined via the equilibrium conditions by the properties of the alternative coexisting phase. However, in the search for the critical clusters (or for the saddle point of the appropriate thermodynamic potential) we have to compare not different equilibrium states but different non-equilibrium states of the heterogeneous system under consideration. For the considered different non-equilibrium states, the surface tension has to depend, in general, on the state parameters of both coexisting phases, i.e. on $2(k + 1)$ independent parameters. Gibbs' adsorption equation does not allow one, in principle, to account for such dependence. Since Gibbs' adsorption equation is a consequence of Gibbs' fundamental equation for the thermodynamic parameters, describing the contributions of the interface to the thermodynamic functions, the latter relation has to be changed in order to allow us to develop a thermodynamic description of thermodynamic non-equilibrium states of the considered type and to allow the surface tension to depend on the sets of state parameters of both ambient and cluster phases. This is the essence of the generalization of Gibbs' description to non-equilibrium states as it was performed recently [113].

The generalized Gibbs' approach employs Gibbs' method of dividing surfaces as well. However, Gibbs' fundamental equation for the superficial or surface quantities is generalized allowing one to introduce into the description the essential dependence of the surface tension on the state parameters of both coexisting phases. In this theory, first the thermodynamic potentials for the respective non-equilibrium states are formulated. After this task is performed, the general equilibrium conditions are derived. They have the following form (in application to the surface of tension)

$$(T_\alpha - T_\beta)s_\alpha + (p_\beta - p_\alpha) + \sigma \frac{dA}{dV_\alpha} + \sum_{j=1}^{k} \rho_{j\alpha}(\mu_{j\alpha} - \mu_{j\beta}) = 0, \tag{10.16}$$

$$(\mu_{j\beta} - \mu_{j\alpha}) = \frac{3}{R}\left(\frac{\partial \sigma}{\partial \rho_{j\alpha}}\right)_{\{\rho_{j\beta}\},T_\beta}, \qquad (T_\beta - T_\alpha) = \frac{3}{R}\left(\frac{\partial \sigma}{\partial s_\alpha}\right)_{\{\rho_{j\beta}\},T_\beta}.$$

Here T is the temperature, p is the pressure, σ is the surface tension, A is the surface area, V is the volume, s is the entropy density, ρ_i are the particle densities, and μ_i are the chemical potentials of the different components, R is the radius of the critical cluster referred to the surface of tension, the index α specifies the parameters of the cluster while β refers to the ambient phase. The equilibrium conditions coincide with Gibbs' expressions for phase coexistence at planar interfaces ($R \to \infty$) or when (as required in Gibbs' classical approach) the surface tension is considered as a function of only one of the sets of intensive state variables of the coexisting phases, either of those of the ambient or of those of the cluster phase. In such limiting cases, Gibbs' equilibrium conditions

$$(p_\beta - p_\alpha) + \sigma \frac{dA}{dV_\alpha} = 0, \qquad \mu_{j\alpha} = \mu_{j\beta}, \qquad T_\alpha = T_\beta \tag{10.17}$$

are obtained as special cases from Eqs. (10.16). In the generalized Gibbs' approach, not only the pressure in the cluster is different as compared to the value in the ambient phase but also chemical potentials [28] and temperature [116, 117]. The generalized Gibbs' approach predicts thus as one consequence the existence of additional thermal effects on nucleation beyond those discussed here earlier.

The expression for the work of critical cluster formation W_c remains the same in the generalized Gibbs' approach as in the classical Gibbs' treatment

$$W_c = \frac{1}{3}\sigma A, \tag{10.18}$$

provided in both approaches the surface of tension is chosen as the dividing surface. However, since the parameters of the critical clusters are determined in a different way in both the classical and generalized Gibbs' approaches, Eq. (10.18) leads, consequently, also to different results for the work of formation of clusters of critical sizes and other characteristics of the nucleation process. A detailed analysis shows [113] further that the classical Gibbs approach employing the capillarity approximation i.e. assuming that the surface tension is equal to the respective value for a planar coexistence of both phases at planar interfaces overestimates the work of critical cluster formation as compared with the generalized Gibbs approach. Independent of the number of components in the system and the application discussed, the work of critical cluster formation – computed via the classical Gibbs approach – is larger as compared with the results of the generalized Gibbs method. This results can possibly give also the key to the understanding of one of the problems in the theoretical interpretation of nucleation in metallic glass-forming melts discussed in Chapter 2, the unexpected in classical terms high crystal nucleation rates.

Assuming that the kinetic pre-factor, J_0, in the expression for the steady-state nucleation rate, J

$$J = J_0 \exp\left(-\frac{W_c}{k_B T}\right), \tag{10.19}$$

does not depend significantly on the chosen path, it follows as a direct consequence that the classical Gibbs approach to the determination of the steady-state nucleation rate leads to too low values of the nucleation rate and related quantities like the limit of accessible supersaturations (i.e., the value of the supersaturation at which nucleation proceeds at a measurable at normal experimental time scales rate). Consequently, experimentalists interpreting their data on nucleation in terms of the classical theory should be aware that the process proceeds in reality as a rule with higher nucleation rates and/or earlier as predicted by classical theory.

These ideas have been recently extended to the analysis of heterogeneous nucleation [118]. In particular, it is shown that, in the generalized Gibbs' approach, contact angle and catalytic factor for heterogeneous droplet nucleation become dependent on the degree of metastability (undercooling or superheating) of the fluid. For the case of formation of a droplet in supersaturated vapor on a hydrophobic surface and bubble formation in a liquid on a hydrophilic surface the solid surface has only a minor influence on nucleation. In the alternative cases of condensation of a droplet on a hydrophilic surface and of bubble formation in a liquid on a hydrophobic surface, nucleation is significantly enhanced by the solid. Effectively, the existence of the solid surface results in a significant shift of the spinodal to lower supersaturations as compared with homogeneous nucleation. Qualitatively, the same behavior is observed now near the new (solid surface induced) limits of instability of the fluid as compared with the behavior near to the spinodal curve in the case of homogeneous nucleation. An extension of this approach to segregation and crystal formation is in progress.

Completing the analysis of the consequences of the generalization of Gibbs' method with respect to nucleation, we would like here to underline the following major consequences one obtains from the generalized Gibbs' approach as compared to the classical Gibbs' treatment of thermodynamically heterogeneous systems (cf. also [120]): (i) The critical clusters will have different (as the rule, significantly different) properties (density, composition, structure etc.) as compared to the properties of the newly evolving macroscopic phases. It follows that the commonly employed relations for the determination of the driving force of crystallization like [1, 2]

$$\Delta\mu = \Delta s_m(T - T_m) \qquad (10.20)$$

may not be appropriate for the description. This and also more general relations of similar type or even expressions based on the measurements of the respective properties of the two macroscopic phases whose chemical potentials are compared presuppose that the bulk properties of the cluster phase are identical to the properties of the newly evolving macroscopic phase, an assumption, which is in general incorrect. (ii) The classical method of adjustment of experiment and theory – the introduction of a curvature dependence of the surface tension in Gibbs' classical approach – is as a rule not the appropriate method of correcting the theory. A curvature dependence of the surface tension remains to be accounted for but it is caused basically by the size dependence of the bulk properties of the clusters. Consequently, the basic size dependence entering the description is the size-dependence of the bulk properties of the clusters of the newly evolving phase generating, of course, a size dependence of the surface properties (Comment: The surface tension is treated in classical nucleation theory as size dependent, and, in addition, as crystal face-dependent for an anisotropic crystal. So if the nuclei have shapes similar to an anisotropic crystal, the surface tension would not only vary with the size but also according the facet of the nuclei. But also in such more general cases, the primary size dependence is the dependence of the bulk properties on the size of the crystallites reflected then in a size dependence of the surface parameters). (iii) The general scenario of cluster formation and growth is not correctly reflected – at least, for segregation in solutions [11, 32, 33] – by the classical picture. Cluster formation and growth do not occur – as assumed in classical theory of nucleation and growth processes – by growth of clusters with nearly constant bulk properties but by a process of, in general, simultaneous change of cluster size and properties (cf. Fig. 10.10). Hereby in certain stages of the process the change of the bulk properties at nearly constant cluster size may dominate (cf. [12]). (iv) As already mentioned, under relatively weak additional assumptions, the general result can be derived [113] that the work of critical cluster formation obtained via the classical Gibbs' approach involving the capillarity approximation is – as a rule – larger than the respective value obtained via the generalized Gibbs' approach. The possibility opened in the latter method to account for and predict the possible changes of the parameters of the critical clusters and the choice of the most suitable ones in order to minimize the thermodynamic barrier for the transition is the origin of such result. It follows that, employing the classical Gibbs' approach and the capillarity approximation, one gets as a rule an upper limiting value for the work of critical cluster formation. As a consequence, the steady-state nucleation rates determined theoretically in such a way will be, as a rule, lower than observed in experiment. (v) Similarly to homogeneous nucleation, in heterogeneous or activated nucleation possible changes of state of the cluster phase as compared to the classical picture have to be accounted

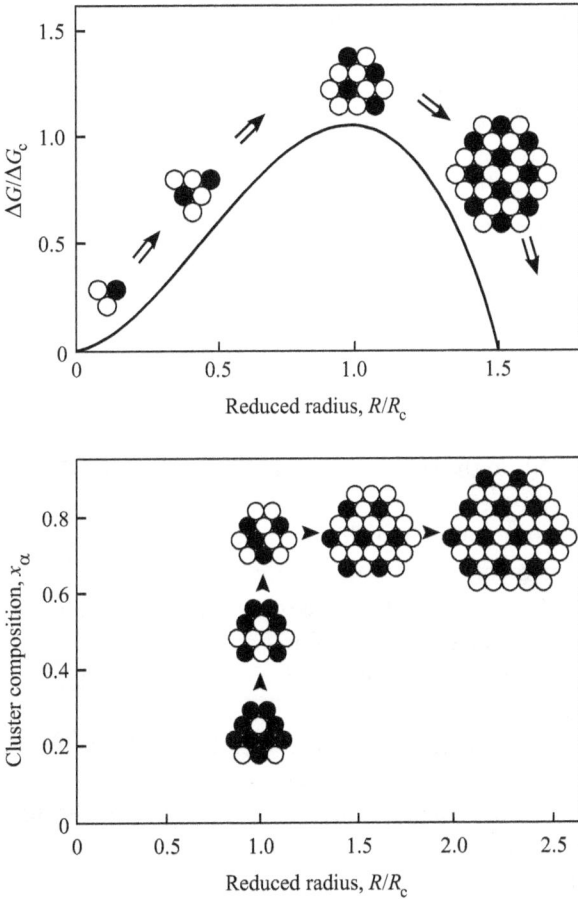

Fig. 10.10: Comparison of the classical model of phase separation in multi-component solutions (top) with the scenario as developed based on the generalized Gibbs' approach (bottom). A widely similar illustration has been presented for the first time in [121].

for, in general, as well in addition to the introduction of an activity parameter, Φ, for heterogeneous nucleation (cf. e.g. [1, 118]). The form of the expression for the work of critical crystalline cluster formation,

$$W_c \cong \frac{16\pi}{3} \frac{\sigma^3}{(\rho_\alpha \Delta\mu)^2} \Phi, \tag{10.21}$$

entering the expression for the steady-state nucleation rate for heterogeneous nucleation, remains formally the same as in classical nucleation theory but has to be supplemented by an account of the dependence of the driving force of phase formation on supersaturation and on the properties of the critical cluster. (vi.) One further typical feature of the dependence of critical cluster properties on supersaturation in application to segregation in solutions consists in the following: the work of critical cluster formation decreases with increasing supersaturation. Similarly, the critical cluster size (referred to Gibbs' surface of tension) computed via the classical Gibbs' approach decreases with increasing supersaturation. In contrast, the critical cluster size, computed via the generalized Gibbs' approach, after an initial decrease in response to increasing molar fraction of the solute in the ambient phase, increases, again, in approaching the spinodal curve. This difference is also the basic origin of differences in the results of the investigation of finite-size effects in phase transition performed via the classical and generalized Gibbs' approaches, respectively [119, 120].

10.5.3 Application to Cluster Growth Processes

The above performed analysis leads to the consequence that clusters of critical sizes have properties which are widely different, in general, from the properties of the newly evolving macroscopic phases. By this reason, the properties of sub- and supercritical clusters have also to depend, in general, both on supersaturation and cluster size. In order to develop an appropriate description of the course of the phase transitions, as a next step one has to develop consequently a method to establish the dependence of the bulk state parameters (density, composition, structure) of clusters of arbitrary sizes on the mentioned set of parameters.

For the determination of the most probable trajectory of evolution of the clusters in the space of thermodynamic state variables, recently we proposed that the preferred most probable path of evolution of the clusters is determined by the deterministic equations of cluster growth and dissolution starting with initial states slightly above and below the critical cluster size [33, 34]. In its simplest tentative version [32], the behavior of the system resembles then the motion of a mass in a viscous fluid (i.e. with a velocity proportional to the force acting on the mass) in some force field determined by the shape of the thermodynamic potential surface. These methods allow one to straightforwardly determine the most probable path of evolution and the bulk cluster properties as a function of the size of the aggregates independent of the particular kind of phase transformation considered.

The change of the composition of the clusters in dependence on their sizes leads to a size-dependence of almost all thermodynamic (in particular, the driving force of cluster growth and surface tension) and kinetic parameters (diffusion coefficients and growth rates), determining the course of the phase transition (for the details, see again [32–34]). Some first experimental analyses confirming these theoretical predictions are summarized in [1, 12]. Taking into account such size dependence, it can be easily explained why thermodynamic and kinetic parameters, obtained from nucleation experiments, may not be appropriate for the description of growth or dissolution and vice versa. Following the thermodynamic analysis in the framework of the generalized Gibbs' approach, as analyzed here, and the method of determination of the most probable trajectory of evolution employed we come to the conclusion that the kinetics of nucleation and growth in solutions or melts does not proceed according to the classical picture but exhibits features typical for spinodal decomposition. Moreover, essential features of the process of spinodal decomposition and the phase transformation kinetics in the vicinity of the classical spinodal curve can be interpreted in terms of the generalized Gibbs' approach as well [28].

10.5.4 Thermodynamics versus Kinetics: Ridge Crossing

In the analysis of nucleation-growth processes in glass-forming melts, it is commonly assumed that nucleation processes proceed along a trajectory passing the maximum or, more generally, the saddle point of the thermodynamic potential surface, the critical cluster. Hereby the properties of the critical clusters are identified as a rule with the properties of the newly evolving macroscopic phases in line with Gibbs' classical theory of heterogeneous systems. Extending Gibbs' classical theory to the description of heterogeneous systems in non-equilibrium states, we have re-analyzed in recent publications the process of segregation in solutions from thermodynamic [28] and kinetic [19, 20] points of view by analytical methods and by solving numerically the set of kinetic equations describing nucleation and growth processes. The starting point of the analysis and the problem to be analyzed can be described as follows:

The critical cluster size, R_c, and the work of critical cluster formation, ΔG_c, in dependence on the initial solute concentration, x, in the ambient phase can be represented in a form as shown in Fig. 10.11. While in the respective dependencies obtained via the classical Gibbs' approach, $R_{CNT(\sigma_\infty)}$ and $\Delta G_{CNT(\sigma_\infty)}$ (employing the capillarity approximation) no peculiarities occur in the vicinity of the spinodal curve, the critical cluster – computed via the generalized Gibbs' approach – diverges here and the work of critical cluster formation tends to zero remaining equal to zero also in the region of unstable initial states. Provided we introduce in the classical Gibbs' approach a curvature dependence of the interfacial tension in such a way that the work of critical cluster formation tends – as it should be the case – to zero at the spinodal, then the critical cluster size tends to zero at the spinodal in such approach as well. In the subsequent derivation, we will not consider the latter case but employ the capillarity approximation as usually done in the classical theory of nucleation and growth. Note as well that in the framework of the generalized Gibbs' approach – in contrast to the classical Gibbs' method of description – a critical cluster size can be determined also for unstable initial states, it corresponds to the lower limit of the size of the region where spontaneous density or composition amplification may be realized according to the Cahn-Hilliard theory of spinodal decomposition. So, the question we would like to address here is: Will the evolution of the system to the new phase proceed via some of the specified saddle points – determined either by the classical or generalized Gibbs'

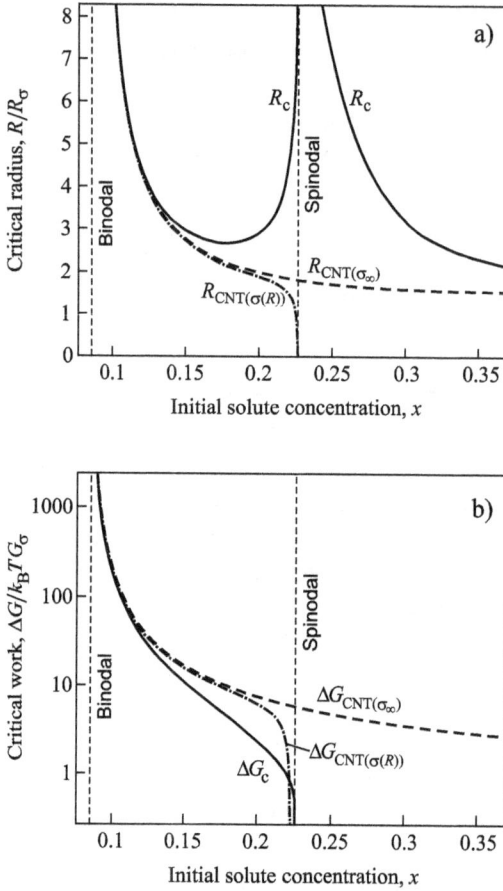

Fig. 10.11: Critical cluster size and work of critical cluster formation according to the classical (employing the capillarity approximations, dashed curve, $R_{\text{CNT}(\sigma_\infty)}$, $\Delta G_{\text{CNT}(\sigma_\infty)}$) and the generalized (full curve, R_c, ΔG_c) Gibbs' approaches (scaling parameters, R_σ and G_σ, for the cluster radius and the work of critical cluster formation are introduced (cf. [19, 20])). If in the classical Gibbs' approach a curvature dependence of the interfacial tension is introduced in such a way that the work of critical cluster formation tends to zero at the spinodal curve, then the critical cluster size approaches zero as well (dashed-dotted curve, $R_{\text{CNT}(\sigma(R))}$, $\Delta G_{\text{CNT}(\sigma(R))}$). Figures are taken from [20].

approaches – or will there occur deviations from the saddle-point trajectory of cluster evolution and, if this is the case, why.

Following earlier suggestions by other authors (cf. [36]) and based on the solution of the set of kinetic equations, describing nucleation and growth in solutions, the most probable path of evolution of the cluster ensemble in nucleation and growth processes has been specified in dependence on the initial supersaturation (cf. Fig. 10.12 for an illustration). Hereby, on one side, the classical Gibbs' approach is employed for the description of the thermodynamic properties of the system (utilizing the capillarity

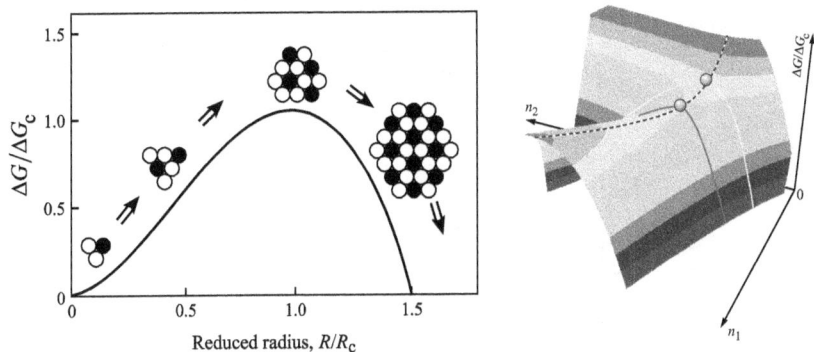

Fig. 10.12: Change of the Gibbs free energy, ΔG, with radius, R (in reduced units), or the number of particles, n, in the crystal cluster formed in the ambient melt. On the left hand side, the common classical picture is shown when one parameter is sufficient to describe the state of the cluster. On the right hand side, it is assumed that the cluster consists of two different components, n_1 and n_2. The dark curve shows the path of evolution via the saddle. This is the path which is usually supposed to be the main path of evolution to the new phase. The light line indicates evolution via a ridge of the thermodynamic potential. $W_c = \Delta G_c$ is the value of ΔG at the saddle point, while W_j refers to the respective value at some given ridge point with a size R_j. Figures are taken from [25].

approximation). As an alternative method, the classical Gibbs' method of description is replaced by the generalized Gibbs' approach. It is shown that significant deviations from the saddle point trajectory of evolution are found only if in the thermodynamics of cluster evolution the generalized Gibbs' approach is employed allowing one to account for and to determine changes of the state of the clusters in dependence on supersaturation and cluster size. The results are illustrated in Fig. 10.13. They can be summarized as follows.

Employing the generalized Gibbs' approach, the work of critical nucleus formation in solutions is found generally to have lower values as compared with the result obtained by the classical Gibbs' approach when the capillarity approximation is employed. Therefore the nucleation rate computed via the generalized Gibbs' approach is, as a rule, considerably larger. These results are a consequence of the possible variations of bulk properties of the critical clusters accounted for and determined in the generalized Gibbs' approach. For small supersaturation, the results of the classical and generalized Gibbs' approaches lead to widely equivalent results. This is the range, where the classical Gibbs' method underlying classical nucleation theory is directly applicable. For moderately metastable states, the thermodynamic analysis in the framework of the generalized Gibbs approach can be applied in order to determine the flux via the saddle point dominating here the nucleation process. However, for both metastable and unstable initial states near to the spinodal curve the evolution to the new phase can and will, as a rule, proceed not via the saddle point, but via the ridge of the appropriate thermodynamic potential relief. The saddle point tra-

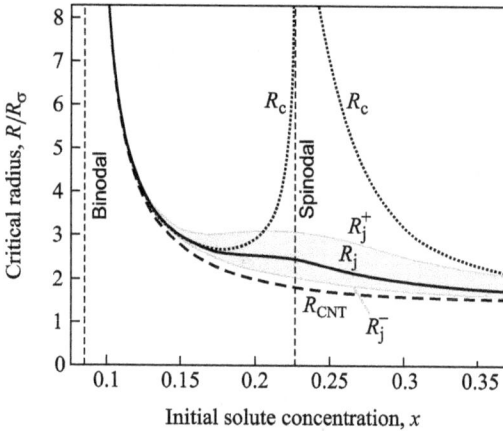

Fig. 10.13: Dependence of the ridge flux parameters R_j (location of the maximum of the flux, full curve), R_j^- and R_j^+ (lower and upper values of the size along the ridge where intensive flow processes of clusters to the new phase are observed) on the initial solute concentration, x. Critical cluster sizes, R_c, for nucleation via the saddle point obtained via the generalized Gibbs' approach (dotted curve), and R_{CNT} for CNT (dashed curve) are shown for comparison. Figure is taken from [20].

jectory would result here in values of the nucleation rate near to zero. In this range of supersaturation, only the analysis of the kinetics of nucleation and growth based on the solution of the set of kinetic equations employing for the thermodynamic description the generalized Gibbs approach is valid. For low interfacial tension values, the process of nucleation proceeds visually similar to CNT – critical size and work of critical cluster formation are near to the values predicted by CNT. Nevertheless, the physical nature of the process is very different: in the framework of CNT, nucleation proceeds via the saddle point, but in the generalized Gibbs' approaches (i) the saddle point is as a rule located at another place, and (ii) nucleation proceeds via the ridge of the thermodynamic potential relief. It follows that nucleation does not proceed necessarily via the saddle-point trajectory and, in particular, in crystallization in multi-component systems such deviation from the saddle-point trajectory of evolution may have to be accounted for (for some recent investigations of particular features in crystal formation in multi-component systems see also [122]).

Finally, we would like to analyze why the path of evolution via the saddle is switched to evolution proceeding via the ridge of the thermodynamic potential surface when supersaturation increases. The origin of such kind of behavior is the following: as discussed here already in connection with Fig. 10.10, the mechanism of nucleation in solutions does not consist (as assumed in the classical picture) in the growth of the cluster in size with more or less given composition. In contrast, nucleation is characterized by an initial amplification of density fluctuations in a region of the ambient phase with a radius of the critical cluster size. The nucleation rate of clusters evolving via the saddle can therefore be represented as being proportional to

$$J(R_c) \propto \phi_c(R_c) = N(R_c) \exp\left(-\frac{W_c}{k_B T}\right) = \left(\frac{R_\sigma}{R_c}\right) \exp\left(-\frac{W_c}{k_B T}\right). \tag{10.22}$$

Here R_σ is some well-defined length parameter [20], $N(R_c)$ is proportional to the number of nucleation sites in the system when nucleation proceeds via the saddle of the thermodynamic potential surface. With an increase of the supersaturation, the work of critical cluster formation via the saddle, W_c, decreases and the exponential term increases. However, the pre-factor being proportional to the number of nucleation sites, $(R_\sigma/R_c)^3$, in the expression of the steady-state nucleation rate overcompensates this effect as soon as R_c as determined via the generalized Gibbs' approach starts to increase with increasing supersaturation (cf. Fig. 10.11). By this reason, interpreted in terms of the generalized Gibbs' approach, the steady-state nucleation rate at initial states near to the spinodal curve tends to zero provided the evolution to the new phase proceeds via the saddle.

In the alternative case, assuming that the process proceeds similarly via a ridge, for the ridge nucleation rate, we get the following analogous expression

$$J(R_j) \propto \phi_j(R_j) = N(R_j) \exp\left(-\frac{W_j}{k_B T}\right) = \left(\frac{R_\sigma}{R_j}\right) \exp\left(-\frac{W_j}{k_B T}\right). \tag{10.23}$$

For such path, the nucleation rates may be larger as compared with the results given by Eq. (10.22) as soon as R_c starts to increase with increasing supersaturation, again. The potential barrier which has to be overcome is larger but this effect is overcompensated by the much larger number of nucleation sites (for the details cf. [20]). Consequently, employing the generalized Gibbs' approach it turns out that the pre-factor, J_0, in the expression for the steady-state nucleation rate is as a rule not – as in the classical treatment – more or less a constant but may vary significantly as well determining the switch from saddle point evolution to evolution via the ridge of the thermodynamic potential hyper-surface.

The switch from the saddle point evolution path to ridge crossing allows the system to realize higher nucleation rates, and by this reason, higher rates of change of the characteristic thermodynamic potential. This switch in the preferred path of evolution can be considered in this way as a special realization of the principle of maximum entropy production (or here the Gibbs' free energy decrease) as formulated in [36] as a criterion of selection of the most probable among several possible reaction paths.

As it is evident from this brief overview, the generalized Gibbs approach is able to lead to a variety of further new insights into the course of first-order phase transformations. In addition to the mentioned results, it allows one, for example, to reconcile Gibbs' and van der Waals' approaches to the interpretation of thermodynamically inhomogeneous systems [112], to give a new interpretation of the problem of existence or non-existence of metastable phases in crystallization of different glass-forming melts and the evolution of bimodal cluster size distributions for intermediate stages of segregation processes (cf. [32] and Chapter 2), a new approach to the description of spinodal decomposition and a variety of other phenomena [10, 12, 71, 112, 118]. This way, we believe that the further development of the generalized Gibbs' approach in application to the description of phase formation processes in future may serve – combining the simplicity of the classical Gibbs' approach with the accuracy of density functional approaches and computer simulation methods – as a quite powerful new and generally applicable tool in order to resolve problems in the comparison of experimental results and theoretical predictions which have not found a satisfactory solution so far.

10.6 Specific Interfacial Energy and the Skapski–Turnbull Relation

10.6.1 General Approach to the Determination of the Specific Interfacial Energy: Taylor Expansion

In order to apply thermodynamic concepts to cluster formation and growth processes, the specific expression for the surface tension or specific interfacial energy and the bulk properties of both the ambient and newly evolving phases have to be known. In particular, these expressions have to be at one's disposal not only for the case that both phases are in equilibrium but also for cases when they are not. In the present section, we consider some new developments concerning possible ways of determination of the surface tension of clusters of the newly evolving phase not being in equilibrium with the ambient phase going over then in the next section to the discussion of bulk properties of the ambient phase and their possible dependence on prehistory.

According to the general results of the theoretical description of thermodynamically heterogeneous systems, both in the classical and generalized Gibbs approaches the interfacial tension depends primarily on the intensive bulk state parameters of the coexisting phases as far as the surface of tension is chosen as the dividing surface [1, 3, 113]. The difference in both approaches consists in the fact that in the classical Gibbs' approach (due to the assumption of full thermodynamic equilibrium) the surface tension depends on the bulk state parameters of only one of the coexisting phases while in the generalized Gibbs' approach (treating initially systems in non-equilibrium states) both the intensive state parameters of the ambient and the newly evolving phase affect the value of the surface tension. Taking into account these results of the generalized Gibbs' approach, the dependence of the surface tension on the intensive state parameters, $\{\phi_i\}$, of the coexisting phases in non-equilibrium states can be expressed in the simplest form as (cf. [113])

$$\sigma = \sum_{i,j=1}^{k+1} \Theta_{ij} \left(\phi_{i\alpha} - \phi_{i\beta} \right) \left(\phi_{j\alpha} - \phi_{j\beta} \right) \qquad \text{with} \qquad \Theta_{ij} = \Theta_{ji}. \tag{10.24}$$

In the derivation of Eq. (10.24) it is merely assumed that the surface tension has to depend basically on the differences of the intensive state parameters of both coexisting phases, i.e., it is assumed that the dependence of the surface tension on the state parameters can be expressed as

$$\sigma\left(\{\phi_\beta\}, \{\phi_\alpha\}\right) = \sigma\left(\{\phi_\beta^{(\infty)}\}, \{\phi_{i\alpha} - \phi_{i\beta}\}\right).$$

The first non-vanishing term in the Taylor expansion with respect to the differences $(\phi_{i\alpha} - \phi_{i\beta})$ results then in Eq. (10.24). Particular versions of this expression we employed e.g. in [71, 112, 113, 117, 118, 123]. Note that due to symmetry considerations only even terms have to be retained in the expansion. Moreover, the surface tension becomes equal to zero if the parameters of both the ambient and the newly evolving phases become identical (here we get the limit of a one-phase state and interfacial contributions have to vanish).

The coefficients of the expansion, Θ_{ij}, have to obey the conditions that the quadratic form as given by Eq. (10.24) is positive semi-definite. In application to the interpretation of phase formation in real or model systems, the values of the coefficients, Θ_{ij}, in the expansion Eq. (10.24) can be determined either from measurements or from statistical-mechanical model computations of the values of the surface tension for equilibrium phase coexistence at planar interfaces (specified by the superscript ∞) knowing the values of the state parameters of the coexisting phases and the value of the surface tension for these particular states. By this reason, the coefficients, Θ_{ij}, may depend on the state parameters of the ambient phase coexisting in equilibrium with the newly evolving phase at a planar interface, i.e., $\Theta_{ij} = \Theta_{ij}\left(\{\rho_\beta^{(\infty)}, T_\beta\}\right)$.

Expressions for the surface tension of the type as given by Eq. (10.24), we denote as self-consistent capillarity approximation. It is assumed in agreement with the general theoretical result that the bulk properties of the ambient and cluster phases determine the value of the interfacial tension. It follows that Eq. (10.24) has to be considered as an approximation only as far as higher-order terms in the expansion with respect to the differences of the intensive state parameters of both coexisting phases are omitted. In contrast, employing the classical capillarity approximation it is supposed that the value of the interfacial tension is determined by the bulk properties of the coexisting phases they have for the particular state of an equilibrium coexistence of the respective phases at planar interfaces. Since the properties of the critical clusters and the ambient phase change with supersaturation, latter approximation has to be considered as theoretically not consistent with the basic equation both in the classical and generalized Gibbs' approaches. Such account of changes of the state parameters, in particular, of the critical clusters is especially important having in mind the possibility of significant variations of the state parameters of the critical clusters in dependence on the supersaturation.

10.6.2 Stefan's Rule and Skapski–Turnbull Relation: Some Interpretation and Extension to Thermodynamic Non-equilibrium States

Eq. (10.24) supplemented by described above method of determination of the expansion coefficients supplies us with a generally applicable tool of specification of the dependence of the surface tension on the state parameters of both the ambient and newly evolving phases. Going beyond this method, one can employ also similar but alternative dependencies derived directly from experiment. Some examples in this respect can be found in mentioned previous analysis.

With respect to crystallization, a widely employed method of specification of the specific interfacial energy consists in its determination via the latent heat (or enthalpy) of the phase formation, a proposal which goes back to a suggestion by Stefan [124] (Stefan's rule). In application to melt crystallization, this method is widely employed and denoted commonly as Skapski-Turnbull relation [1, 2]. In the present section, we will discuss this relation in order to develop an alternative general method of specification of the dependence of the specific surface energy on the state parameters required in the application of the generalized Gibbs' approach. First, we give a brief derivation of the Skapski-Turnbull expression supplementing it by an analysis of the possible values of some essential fit parameter entering the respective expression. In a next step, the Skapski-Turnbull relation is generalized to non-equilibrium states. Finally, some inter-relations of the Skapski-Turnbull relation and another similar dependence developed by Faizullin [125], Skripov and Faizullin [30] for liquid-vapor interfaces are briefly discussed.

Going over to the first part of the analysis, let us consider one-component systems with a number of particles, n_1, on the surface of a cluster consisting of $n = N_A$ (N_A is Avogadro's number) particles. The volume, V, of the cluster and its surface area, A, are equal to

$$V = \frac{4\pi}{3}R^3, \qquad A = 4\pi R^2, \tag{10.25}$$

respectively, where R is the radius of the cluster. If we denote with v_1 and r_1 the volume and the radius of one particle, then we can write similarly for the volume, v_1, and the surface area, a_1, occupied by one particle the relations

$$v_1 = \frac{4\pi}{3}r_1^3, \qquad a_1 = 4\pi r_1^2. \tag{10.26}$$

The number of particles in the cluster is given then via

$$n = \frac{V}{v_1} = \frac{4\pi}{3}\frac{R^3}{v_1}, \tag{10.27}$$

while the number of particles on the surface of the cluster, n_1, can be determined via

$$n_1 = k\frac{4\pi R^2}{4\pi r_1^2} = k\left(\frac{R}{r_1}\right)^2 = k\left(\frac{\frac{3}{4\pi}(nv_1)}{\frac{3}{4\pi}v_1}\right)^{2/3} = n^{2/3}k, \tag{10.28}$$

where k is the thickness of the surface layer. In other words, k is a measure of the number of mono-layers in the surface required to generate the surface layer.

Now, we make – following Stefan [124] – the basic assumption that the specific surface energy, σ, is proportional to the enthalpy difference, ΔH, or, in application to crystallization or melting, to the heat of melting required to remove the surface layer divided by the area of the interface, i.e.,

$$\sigma \propto \frac{n_1 q_1}{A} = \frac{n_1 q_1 N_A}{A N_A} = \frac{n^{2/3} k q_m}{N_A 4\pi \left(\dfrac{3n}{4\pi} \dfrac{v_m}{N_A}\right)^{2/3}} = \frac{k}{(36\pi)^{1/3}} \frac{q_m}{N_A^{1/3} v_m^{2/3}}, \tag{10.29}$$

where q_m is the molar heat of melting and v_m is the molar volume, i.e.,

$$q_m = q_1 N_A, \qquad v_m = v_1 N_A, \tag{10.30}$$

and q_1 being the latent heat of melting per particle. As the result we get the Skapski-Turnbull relation in the form (cf. [1])

$$\sigma = \chi \frac{q_m}{N_A^{1/3} v_m^{2/3}}, \qquad \chi = \frac{k}{(36\pi)^{1/3}} = 0.4 - 0.6 \quad \text{for} \quad k = 2 - 3. \tag{10.31}$$

Typical values of the parameter χ are found in experiment to be in the range $\chi \cong 0.4 - 0.6$. Based on above sketched considerations, this fact can be interpreted in such a way that the thickness of the surface layer varies from two to three mono-layers.

Employing this result, Eq. (10.31), as the starting point, we can now formulate an expression for the specific interfacial energy of a cluster in the ambient phase not being necessarily in thermodynamic equilibrium. For that purpose, we have to express the change in the bulk contributions to the enthalpy of phase formation in the general form as developed in the framework of the generalized Gibbs' approach, when a certain amount of matter in the ambient phase (mole number n) is transferred into the appropriate alternative phase. With the general relations

$$H_\alpha = T_\alpha S_\alpha - \sum_i \mu_{i\alpha} n_{i\alpha}, \qquad H_\beta = T_\beta S_\beta - \sum_i \mu_{i\beta} n_{i\beta}, \tag{10.32}$$

we get

$$\Delta H = H_\alpha - H_\beta = T_\alpha S_\alpha - T_\beta S_\beta + \sum_i \left(\mu_{i\beta} n_{i\beta} - \mu_{i\alpha} n_{i\alpha}\right), \tag{10.33}$$

$$\Delta H_m = T_\alpha s_\alpha - T_\beta s_\beta + \sum_i \left(\mu_{i\beta} x_{i\beta} - \mu_{i\alpha} x_{i\alpha}\right), \tag{10.34}$$

$$s_\alpha = \frac{S_\alpha}{n}, \qquad s_\beta = \frac{S_\beta}{n}.$$

Here $x_{i\alpha}$ and $x_{i\beta}$ are the molar fractions of the given amount of the sample in the different states, i.e., in the generalization performed the Skapski–Turnbull relation is formulated for the case of multi-component systems. With Eqs. (10.31) and (10.34), the specific interfacial energy is given then by

$$\sigma = \chi \frac{\Delta H_m}{N_A^{1/3} V_m^{2/3}}. \qquad (10.35)$$

Assuming that the composition of the both phases coincides (i.e. $x_{i\alpha} = x_{i\beta}$), we get as a special case

$$\Delta H_m = T_\alpha s_\alpha - T_\beta s_\beta + \sum_i \left(\mu_{i\beta} - \mu_{i\alpha} \right) x_{i\alpha}. \qquad (10.36)$$

Employing further the classical Gibbs' (or the macroscopic) equilibrium conditions $(T_\alpha = T_\beta, \mu_{i\alpha} = \mu_{i\beta})$, we arrive at $\Delta H_m = T(s_\alpha - s_\beta) = q_m$ as a special case, again, i.e. arrive at the classical expression for the Skapski–Turnbull relation.

Eqs. (10.34) and (10.35) can be used as an alternative to Eq. (10.24) starting point for the determination of the specific interfacial energy in dependence on the state parameters of both bulk phases. However, we have to draw here the attention also to the fact that the supposed in Stefan's law (and in the Skapski–Turnbull relation) linear relationship between enthalpy of the transformation and the specific surface energy is not as obvious as commonly believed. Indeed, Faizullin [125], Skripov and Faizullin [30] showed by a direct approximation of experimental results on liquid-vapor phase coexistence that the relation between surface tension and enthalpy of phase transformation is not necessarily expressed by a linear dependence. Instead they derived a relation of the form

$$\sigma_{LV} \cong A \left(\frac{\Delta H_{LV}}{V_L} \right)^m, \qquad m = 2.15. \qquad (10.37)$$

Here ΔH_{LV} is the change of the enthalpy, when the volume V_L of the liquid is transferred into the gas phase. The parameter A can be expressed hereby via the respective parameters for an appropriately chosen reference state,

$$A = \sigma_{LV} \left(t_{0.6} \right) \left(\frac{V_{LV} \left(t_{0.6} \right)}{\Delta H_{LV} \left(t_{0.6} \right)} \right)^m, \qquad (10.38)$$

taken by the authors to refer to the reduced temperature $t = (T/T_c)$, $t_{0.6} = 0.6$. This relation can be described quite satisfactorily also as

$$\sigma_{LV} \cong A_1 \left(\frac{\Delta H_{LV}}{V_L} \right)^2 + A_2 \left(\frac{\Delta H_{LV}}{V_L} \right)^4 + \cdots . \qquad (10.39)$$

This relation shows that the specific surface energy (surface tension) for liquid-vapor coexistence can be represented as a truncated Taylor expansion of σ with respect to $(\Delta H_{LV}/V_L)$ containing only even terms. Note as well that deviations from the linearity occur mainly in the vicinity of the spinodal curve when the entropy of the liquid-gas transformation tends to zero. Taking into account the statement of the absence of a spinodal curve in congruent melt crystallization as formulated first by Skripov and Baidakov in [126] and reconfirmed in [30], the linearity in the dependence of specific surface energy and enthalpy of crystal-melt transformation could be considered then as a consequence of this experimental fact. However, very recent molecular-dynamics computations of the crystal-liquid interfacial free energy in a Lennard-Jones system show [127] that that the parameter χ in Eq. (10.31) depends on the state parameters; it is found to vary in the range between $\chi \cong 0.35-0.45$ for the range of state parameters where the computations have been performed. So, even for this simple system, the linearity of crystal-liquid interfacial energy in dependence on melting enthalpy is not confirmed. As it turns out the respective problem of the relation between specific surface energy and enthalpy of the respective phase transformation deserves further detailed consideration.

10.7 Dependence of Crystal Nucleation and Growth Processes on Pre-history

10.7.1 Introductory Comments

In the previous sections, we have shown that clusters of the newly evolving phases have as a rule properties different from the properties of the final macroscopic phases they evolve to. However, there exists also another circle of problems, which is so far either widely ignored or, at least, not described theoretically in a quantitative way in the analysis of crystallization processes in glass-forming melts and polymers. This circle of problems consists in the quantitative description of the dependence of the kinetics of crystallization on prehistory and not only on the actual values of pressure and temperature the system is brought to.

Again, related effects have been studied for a long time, starting already with Aristotle. According to [128], he wrote: "The fact that water has previously been warmed contributes to its freezing quickly; for so it cools sooner …" [129]. Similar considerations were advanced by F. Bacon and R. Descartes. In 1620, Bacon stated "Water slightly warm is more easily frozen than quite cold" [130], while somewhat later Descartes claimed "Experience shows that water that has been kept for a long time on the fire freezes sooner than other water" [131]. This effect is sometimes denoted today as "*Mpemba effect*" [132]. Mpemba arrived at his conclusion as he wrote "by misusing a refrigerator" in order to make ice-cream as quickly as possible.

Independently and somewhat earlier in the same year (in 1969), Kell [133] reported a thermodynamic analysis using Newton's law of cooling and his own experiment performed to test a widely held belief in Canada that hot water freezes faster than cold water. He showed that "a volume of water starting at 100 °C would finish freezing in 90 % of the time taken by an equal volume starting at room temperature" and more significantly that, "The mass lost when cooling by evaporation is not negligible. Water cooling from 100 °C has lost 16 % of its mass by 0 °C, and loses a further 12 % on freezing, for a total loss of 26 %. We also point out that hot liquids have less dissolved gases than cold liquids, and the effect of these dissolved gases on the rate of cooling and nucleation-crystallization is not known.".

Without going into a detailed discussion of the origin of this particular effect (i.e. whether it is inherent to the system or due – as mentioned already by Turnbull (1950) with respect to crystal nucleation in liquid metals [8] – for example, to the variation of the density of catalytic inclusions or the properties of the container walls by the heat treatment), we would like only to note that it represents a particular example of the dependence of the freezing (or crystallization) behavior on prehistory. Such kind of dependence of crystallization behavior on prehistory has to occur necessarily – to a more or less advanced degree – in crystallization processes in glass-forming systems and polymers. The origin of such dependence will be sketched below.

10.7.2 Kinetic Criteria for Glass-formation

As is well-known, liquids can be transferred into a glass by cooling the systems below the melting temperature with a certain by necessity finite rate. If crystallization can be avoided, then the – below the melting temperature meta-stable – liquid becomes frozen-in and is transformed into a glass. Thus, the glass transition is a kinetic phenomenon and depends on cooling or heating rates or more generally on the rates of change of the external control parameters.

Based on intensive discussions of the response of liquids to external disturbances as performed by a number of scientists, for example, by Bragg and Williams [134], by Frenkel [135], Kobeko [136], Jones [137] and others, as it seems for the first time Bartenev [138] formulated analytical kinetic convection for dynamic glass transitions, i.e., the qualitative change of the response of a liquid with respect to periodic external disturbances (periodic deformations with an angular frequency, ω). Considering a certain model system, he arrived at the relation

$$\omega \tau_R \big|_{T=T_g} \cong \text{constant} \tag{10.40}$$

for the determination of the value of T_g at which a dynamic glass transition is expected to occur, i.e., when the system cannot follow any more the changes of the external control parameters. Assuming, in addition, the relaxation time, τ_R, to be of Arrhenius type, i.e.,

$$\tau_R \cong \tau_0 \exp\left(\frac{U}{k_B T}\right), \tag{10.41}$$

he arrived at a relation connecting dynamic glass transition temperature, T_g, and angular frequency, ω, of external perturbations, being of the form

$$\frac{1}{T_g} = C_1 - C_2 \log \omega. \tag{10.42}$$

These considerations were extended by Bartenev (1951) [139], Bartenev and Lukyanov (1955) [140], and independently by Ritland (1954) [141] to glass transition at cooling

(cooling rate, q) with the results

$$\left|\frac{dT}{dt}\tau_R\right|_{T_g} = \text{constant},\tag{10.43}$$

$$\frac{1}{T_g} = C_1 - C_2 \log\left|\frac{dT}{dt}\right|.\tag{10.44}$$

Latter relation, determining the dependence of the glass transition temperature on cooling or heating rate, is commonly denoted as Bartenev-Ritland equation [1, 142].

In 1955/56 Volkenstein and Ptizyn [143] reconsidered these problems based on a chemical reaction model. They arrived at

$$\left|q(T_g)\right|\left|\frac{d\tau}{dT}\right|_{T_g} \cong 1, \qquad \left|\frac{d(q\tau)}{dT}\right|_{T_g} \cong 1.\tag{10.45}$$

and rederived also the result of Bartenev (they denoted to have been obtained by him by "qualitative arguments")

$$\left|\frac{dT}{dt}\tau\right|_{T_g} = \text{constant}.\tag{10.46}$$

They note already as well that in Bartenev's equation, the constant on the right hand side of Eq. (10.46) has to be a weak function of T_g.

Partly similar or related considerations can be found in the papers by Moynihan et al. (1974) [144, 145] and Mazurin (2007) [146]. A criterion identical to Eq. (10.45) was advanced as a reasonable assumption later by Cooper and Gupta [147] (Gupta and Cooper wrote in their paper: "The result emerges that the glass transition is predominantly affected by the value of the single dimensionless parameter, $-q(d\tau/dT)$" (in our notations). "This is not surprising when it is recognized that when $-q(d\tau/dT) = -1$, $d\tau = dt$; i.e. the relaxation time is increasing as fast as time is increasing. Thus, despite the simplifications and the approximations utilized above, it appears that 'frozen-in' transitions, whether the transition caused by temperature decrease, by pressure increase, or by the change in another external variable, will occur in the vicinity of $(dt/d\tau) = 1$."). Later it was proposed by Cooper to denote this number as Lillie number [148, 149] (for a detailed discussion cf. also [1, 142, 150, 151]).

There exists, however, also a widely employed alternative approach to the derivation of kinetic criteria for glass-formation connected with the notion of the Deborah number introduced in 1964 by Reiner [152]. According to Reiner: "Heraclitus' *everything flows* was not entirely satisfactory. Were we to disregard the solid and deal with fluids only? The way out of this difficulty had been shown by the Prophetess Deborah even before Heraclitus. In her famous song after the victory over the Philistines, she sang, *The mountains flowed before the Lord* ... Deborah knew two things. First, that the mountains flow, as everything flows. But, secondly, that they flowed before the Lord, and not before the man, for the simple reason that man in his short life-time cannot see them flowing, while the time of observation of God is infinite ... We may therefore well define as a non-dimensional number the Deborah number

$$Dh = \frac{\text{Time of relaxation}}{\text{Time of observation}}.$$

(10.47)

The difference between solids and fluids is then defined by the magnitude of Dh...In every problem of rheology make sure that you use the right Deborah number".

However, according to our point of view, the Deborah number concept cannot be employed as an appropriate starting point for the derivation of kinetic criteria of glass-formation for the following reasons: (i) The criterion is formulated and refers consequently to systems at time-independent boundary conditions. (ii) Reiner talks about relativity of liquids and solids and not about glasses. (iii) A subjective element enters the theory, the not well-defined "*time of observation*". (iv) In order to derive the kinetic criterion for glass-formation based on the Deborah number concept, one has to introduce *de facto* additional assumptions beyond the Deborah number concept [1, 151]. Actually, one employs in addition the criterion given by Eq. (10.45) (cf. [1, 147, 151]). The next question then is whether it is possible to derive a general model independent criterion of glass-formation covering above mentioned and further approaches (i.e. to find out *the right Deborah number* for the description of the glass transition to determine the relevant for this case parameter). This can be done as shown below repeating the derivations as outlined first in [1, 151].

In order to thermodynamically describe glass-forming melts and their transition to a glass, one has to employ in addition to the conventional thermodynamic state parameters, at least, one additional structural order parameter, which we denote here by ξ. The change in time of this additional structural order parameter can be expressed generally as

$$\frac{d\xi}{dt} = -\frac{1}{\tau_R(p, T, \xi)}(\xi - \xi_e), \qquad q = \frac{dT}{dt}.$$

(10.48)

Here ξ_e is the equilibrium value of the structural order parameter. In addition to the relaxation time, we can introduced a characteristic time of change of temperature, τ_T, via

$$\frac{dT}{dt} = -\frac{1}{\tau_T}T, \qquad \tau_T = \left\{\frac{1}{T}\frac{dT}{dt}\right\}^{-1}.$$

(10.49)

For thermodynamic equilibrium states and quasi-stationary processes in between them, always the relation $\tau_R \ll \tau_T$ has to be fulfilled. Alternatively, for frozen-in liquids, i.e., glasses, the relation $\tau_R \gg \tau_T$ holds. The condition for the transition from thermodynamic equilibrium to the vitreous (frozen-in) state has to be, consequently, by necessity of the form

$$\tau_R \cong \tau_T \quad \Longrightarrow \quad \left\{\frac{1}{T}\frac{dT}{dt}\tau_R\right\}\Big|_{T=T_g} \cong 1.$$

(10.50)

As easily can be verified, this general and primary criterion contains above mentioned criteria as special cases. For example, multiplying both sides with temperature and taking the derivative with respect to time, we arrive at Eq. (10.45), i.e., reestablish in a model-independent way and without introduction of additional assumptions and simplifications the results obtained by Volkenstein and Ptizyn [143], respectively, the identical relation as proposed by Cooper and Gupta [147]. Further, it can be easily applied to dynamic glass transitions.

Indeed, let us assume that the equilibrium state of the system, ξ_e, is changed with some characteristic angular frequency, ω

$$\xi_e \propto \exp(i\omega t). \tag{10.51}$$

The characteristic time of change of state, ξ_e, is given then by τ_D defined via the following relations

$$\frac{d\xi_e}{dt} \propto i\omega\xi_e, \qquad \frac{d\xi_e}{dt} = -\frac{1}{\tau_D}\xi_e, \qquad i\omega = -\frac{1}{\tau_D}. \tag{10.52}$$

Taking the absolute value, we arrive at

$$\tau_D\omega \cong 1. \tag{10.53}$$

The kinetic criterion for glass-formation is given consequently in this case via

$$\tau_R \cong \tau_D \qquad \Longrightarrow \qquad \omega\tau_R \cong 1, \tag{10.54}$$

reestablishing in this way the prediction of Bartenev [138].

In the case that several independent structural order parameters exist, above approach is applicable straightforwardly to any of them allowing the possibility of several glass transitions to occur in the system. Hereby the structural order parameters will depend as a rule not only on the conventional thermodynamic state parameters like pressure and temperature but also on the structural order parameters. In any case, the glass transition interval is reached when the characteristic times of temperature change (or, in general, change of the external control parameters) are comparable in magnitude with the relaxation times to the respective meta-stable equilibrium state of the liquid (e.g. [151]). This result, expressed mathematically via Eq. (10.50), underlines the importance of the detailed knowledge of the relaxation kinetics for the description of the glass transition as stressed, for example, in [37].

Completing this analysis, we would like to underline once again the following main conclusions: in the sketched here general approach, the glass transition is described in terms of only two characteristic time scales: (i) The time of change of external factors (temperature, pressure, ...) and (ii) the response time of the system under consideration. The concept of observation time and/or Deborah number does not enter the description. The concept of observation times comes into play, if one wants to study reactions of a system with respect to external variations of parameters. Here the values of τ_R and τ_D or similar become significant since the observation times required to observe changes must be larger than the respective relaxation times. However, the notion of observation time is not required in order to formulate kinetic criteria for glass-formation.

10.7.3 On the Dependence of the State of the Melt on Cooling and Heating Rates and Its Relevance for Crystal Nucleus Formation and Growth

As discussed in detail in the previous section, the state of the glass-forming melt and the glass can be described thermodynamically by a set of structural order parameters, $\{\xi_i\}$ (like fictive pressure or fictive temperature) in addition to the conventional thermodynamic control parameters like pressure and temperature. The actual values of the respective sets of structural order parameters depend on the cooling rate, or more generally, on the prehistory of the system. An example for such kind of behavior is shown in Fig. 10.14. So, for given fixed values of the conventional thermodynamic control parameters, the thermodynamic functions of the glasses are different in dependence on the values of the structural order parameters.

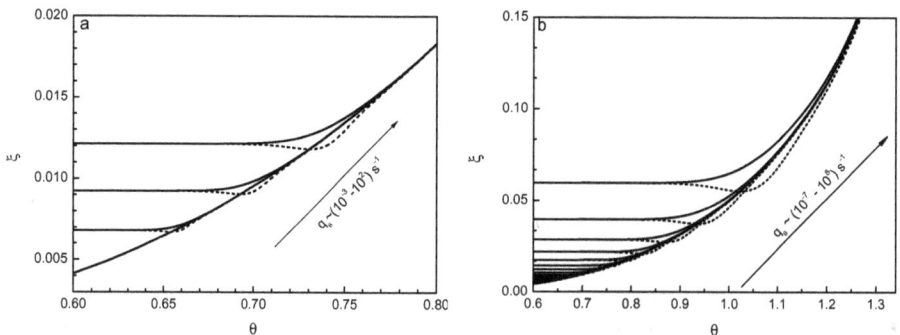

Fig. 10.14: Dependence of the curves $\xi = \xi(\theta)$ (with $\theta = (T/T_m)$) on cooling & heating rates: (a) in the range $10^{-3} \leq q_\theta \leq 10^2$ (more or less easily accessible experimentally); (b) in a larger range of q_θ-values ($10^{-7} \leq q_\theta \leq 10^8$) exceeding partly the ranges of cooling & heating rates accessible, at least, at present (q_θ is given here in s^{-1}, see also [150]).

The values of the structural order parameters depend on prehistory not only for glasses – which are kinetically frozen-in – but also for the glass-forming melts in the vitrification (glass transition) interval. On the other hand, due to the interplay between increase of viscosity (inhibiting crystallization) and increase of the thermodynamic driving force for crystallization (stimulating crystallization) with decreasing temperature, the maximum of the steady-state crystal nucleation rate is usually found near to the glass transition temperature [153], i.e., in a range of temperatures, where the dependence of the state parameters of the glass-forming melts on prehistory becomes already very significant. The findings in Tables 10.1 and 10.2 that both the enthalpy of crystallization and the parameter m vary with the temperature of isothermal crystallization and the heating rate may also partly indicate this role of thermal history.

Qualitatively, the dependence of crystal nucleation kinetics on prehistory is well-known and different attempts have been reported to study such dependence. For example, at the International Congress on Glass in September 2010 in Brazil (ICG 2010) there was organized a special symposium *"Structural basis of glass crystallization"* continuing at a modern level the mentioned earlier analysis and also the studies of Röntgen on the structure dependence of freezing of water and other subsequent similar investigations for a variety of glass-forming systems (for a detailed overview on different approaches to the description of the structure of glasses and its effects on the crystallization kinetics see also the excellent review given in [154]). In particular (as announced by the organizers), "the factors that govern the relationship between the structure of the initial glass and that of the final crystalline product and crystallization mechanism will be discussed, together with those that influence crystalline growth. Further insight to the nucleation and growth stages will be sought from the results of both atomistic and thermodynamic modeling". In terms of the described thermodynamic approach different structures refer to different values of the set of structural order parameters. Consequently, a not resolved so far problem in the description of crystal nucleation and growth consists in the development of a theory of nucleation-growth processes allowing one to account in a quantitative manner for the dependence of the structure (or the set of structural order parameters) on prehistory and its effect on crystallization kinetics. A first quantitative attempt in this direction has been advanced earlier in Schmelzer and Schick [25].

10.8 Conclusions

A critical analysis of the theories of nucleation and crystal growth in relation to the theories of glass formation indicates that there are a number of features that cannot be reconciled with the classical theories, particularly in the case of ultraviscous melts obtained by heating the glassy state. The difficulty arises from the views that the nanometer supercritical crystal nuclei size is about the same as the size of the co-operatively rearranging regions of the entropy theory and the size of dynamically heterogeneous domains in a theory that rationalizes the existence of the distribution of diffusion or relaxation times. It would seem unreasonable to assume that structural fluctuations that produce nuclei in an ultraviscous liquid are temporally and spatially different from the structural fluctuations that produce co-operatively rearranging regions (CRR) and the domains of dynamic heterogeneity (DDH). Since the CRR, DDH and an apparent decoupling of the rotational and translational diffusion coefficient are interrelated, an experimental test for the validity of one view would indicate the validity of all three views.

A study of a metal-alloy melt shows that while crystallization of an ultraviscous melt at a fixed temperature may show only one exothermic feature, the rate-heating crystallization may show two exothermic features over a shorter total time. The second exotherm is attributed to the melt's crystallization to a different phase with a much slower kinetics, to solid-solid transformation or to grain-growth. Therefore, were studies performed only by rate-heating, this would lead to misinterpretation of the crystallization kinetics.

Crystallization of ultraviscous melts at a fixed temperature may appear to be interpretable in terms of classical nucleation and growth kinetics and the KJMA equation with a temperature dependent exponent m whose magnitude may indicate that crystal growth occurring isothermally is predominantly interface-controlled with decreasing nucleation rate. It is also interface-controlled on rate heating, but the nucleation rate increases. The activation energy determined from isothermal crystallization kinetics is ~ 60 % of that determined from rate-heating kinetics.

As crystallization approaches completion, the fraction crystallized is less than that deduced from the KJMA equation, and the rate of crystallization becomes slower. This indicates persistence of inter-granular ultraviscous melt. The second exotherm that appears at a temperature of 66 K above the observed T_g has been attributed to the irreversible transformation of a metastable phase formed on crystallization. This formation may be relevant to the recent discussion of crystallization of melts and glasses of multi-component systems.

It is argued that crystallization kinetics would be influenced by the occurrence of mass transport by buoancy convection, which can be due to thermal gradients caused by the exothermic crystallization, and/or by the compositional gradients caused by incongruent crystallization. These may result from the density gradient, composition difference between the crystal and melt phases, and by surface tension gradients in the bulk or when the samples have a large surface area. Sedimentation of crystal heavier than the melt and upward motion of crystals lighter than the melt would show opposite effects. These effects are greatly reduced in microgravity environment, and greatly enhanced when crystallization occurs in a liquid subjected to a centrifugal force. It is suggested that these effect cause the activation energy for the crystallization kinetics to differ from the activation energy for viscous flow or molecular diffusion, and that the effects can be tested by performing experiments in different conditions. One simple consequence of this view is that a material that crystallizes with a large heat release, and the density and surface tension of the melt are strongly temperature dependent, would show these effects much more than a material that crystallizes with a relatively small heat release.

The classical approach to crystal nucleation and growth processes makes the assumption that bulk and surface properties of the aggregates of the newly evolving phases resemble widely the properties of the newly evolving macroscopic phases. These assumptions are shown to be, in general, not true. The crystalline clusters evolving in the course of nucleation and growth have as a rule different bulk and surface properties as compared with the respective macroscopic samples. These differences lead to a variety of consequences discussed in detail in the present chapter. Approaches are discussed as well allowing one to determine these cluster properties in dependence on the size of the aggregates. In particular, a generalization of the Skapski-Turnbull relation is formulated for the specific interfacial energy of crystallites not being in thermodynamic equilibrium with the ambient phase and valid also for incongruent melt crystallization. It is shown as well that in a number of cases the transition to the new phase does not proceed via an evolution path passing the saddle point (critical cluster) but via a ridge of the thermodynamic potential barrier. Such processes occur in the vicinity of the respective spinodal curves. They are expected by this reason to be of minor importance for congruent melt crystallization characterized by the absence of a spinodal. Finally, it is shown that new effects in the description of crystal formation in highly viscous glass-forming melts may be also due to prehistory effects, the dependence of the properties of the glass-forming melts on the way they are brought to the respective metastable states. Consequently, despite an extensive work performed over more than ten decades and existing impressive results, a broad spectrum of problems in the description of crystal nucleation and growth processes requires intensive further studies.

Bibliography

[1]   I. Gutzow and J.W.P. Schmelzer, *The Vitreous State: Thermodynamics, Structure, Rheology, and Crystallization* (First Edition, Springer, Berlin, 1995; Second enlarged edition, Springer, Heidelberg, 2013).

[2]   K.F. Kelton and A.L. Greer, *Nucleation in Condensed Matter: Applications in Materials and Biology* (Elsevier, Amsterdam, 2010).

[3]   J.W. Gibbs, *The Collected Works*, vol. 1, *Thermodynamics* (Longmans & Green, New York – London – Toronto, 1928).

[4]   G. Tammann, *Kristallisieren und Schmelzen* (Johann Ambrosius Barth, Leipzig, 1903); *Die Aggregatzustände* (Leopold Voss Verlag, Leipzig, 1922); *Der Glaszustand* (Leopold Voss Verlag, Leipzig, 1933).

[5]   W.E. Garner, *The Tammann Memorial Lecture*, J. Chem. Society (Resumed) 1961–1973 (1952).

[6]   W. Kossel, Nachrichten Gesellschaft der Wissenschaften Göttingen, Mathematisch – Physikalische Klasse 135 (1927); Naturwissenschaften **18**, 901 (1928); **42**, 296 (1955).

[7]   I.N. Stranski, Z. Phys. Chem. **136**, 259 (1928), **A 142**, 453 (1929); I.N. Stranski and R. Kaischew, Z. Krist. **78**, 373 (1931); I.N. Stranski and L. Krastanov, Sitzungsberichte Akademie der Wissenschaften Wien **146**, 797 (1938); Monatshefte für Chemie **71**, 351 (1938); I.N. Stranski, Disc. Faraday Society **5**, 66 (1949); I.N. Stranski, Naturwissenschaften **37**, 289 (1950).

[8]   D. Turnbull and J.C. Fisher, J. Chem. Phys. **17**, 71 (1949); D. Turnbull, J. Appl. Phys. **21**, 1022 (1950).

[9]   D.R. Uhlmann, J. Non-Crystalline Solids **7**, 337 (1972).

[10]   V.M. Fokin, E.D. Zanotto, N.S. Yuritsyn, and J.W.P. Schmelzer, J. Non-Crystalline Solids **352**, 2681 (2006).

[11]   J.W.P. Schmelzer, J. Non-Crystalline Solids **354**, 269 (2008).

[12]   J.W.P. Schmelzer, V.M. Fokin, A.S. Abyzov, E.D. Zanotto, and I. Gutzow, International J. Applied Glass Science **1**, 16 (2010).

[13]   V.V. Slezov, *Kinetics of First-Order Phase Transitions* (WILEY-VCH, Weinheim, 2009).

[14]   J. Möller, J. Schmelzer, and I. Gutzow, Z. Phys. Chemie **204**, 171 (1998).

[15]   V.M. Fokin, A.S. Abyzov, J.W.P. Schmelzer, and E.D. Zanotto, J. Non-Crystalline Solids **356**, 1679 (2010).

[16]   J.W.P. Schmelzer, J. Non-Crystalline Solids **356**, 2901 (2010).

[17]   J. Bartels, U. Lembke, R. Pascova, J.W.P. Schmelzer, and I. Gutzow, J. Non-Crystalline Solids **136**, 181 (1991).

[18]   J. Schmelzer Jr., U. Lembke, and R. Kranold, J. Chem. Phys. **113**, 1268 (2000).

[19]   A.S. Abyzov, J.W.P. Schmelzer, A.A. Kovalchuk, and V.V. Slezov, J. Non-Crystalline Solids **356**, 2915 (2010).

[20]   A.S. Abyzov and J.W.P. Schmelzer, J. Non-Crystalline Solids **384**, 8 (2014).

[21]   A.N. Kolmogorov, Izv. Acad. Nauk SSSR, Seriya Mathematica **1**, 355 (1937).

[22]   W.A. Johnson and R.F. Mehl, Trans. Metallurgical Society AIME **135**, 416 (1939).

[23]   M. Avrami, J. Chem. Phys. **7**, 1103 (1939); **8**, 212 (1940).

[24]   D.P.B. Aji and G.P. Johari, Thermochimica Acta **510**, 144 (2010).

[25]   J.W.P. Schmelzer and C. Schick, Phys. Chem. Glasses: European J. Glass Science Technology **B 53**, 99 (2012).

[26]   M. Hillert, Acta Metallurgica **1**, 764 (1953).

[27]   J.W. Cahn and J.E. Hilliard, J. Chem. Phys. **28**, 258 (1958); **31**, 688 (1959).

[28]   A.S. Abyzov and J.W.P. Schmelzer, J. Chem. Phys. **127**, 114504 (2007).

[29]   B. Chalmers, *Principles of Solidification* (WILEY, New York, 1964).

[30]   V.P. Skripov, *Metastable Liquids* (WILEY, New York, 1974); V.P. Skripov and V.P. Koverda, *Spontane Kristallisation unterkühlter Flüssigkeiten* (Nauka, Moskau, 1984 (in Russian)); V.P. Skripov and M.Z. Faizullin, *Solid-Liquid-Gas Phase Transitions and Thermodynamic Similarity* (WILEY-VCH, Berlin-Weinheim, 2006).

[31]   J.W. Christian, *The Theory of Transformations in Metals and Alloys* (Oxford University Press, Oxford, 1975).

[32]   J.W.P. Schmelzer, A.R. Gokhman, and V.M. Fokin, J. Colloid Interface Science **272**, 109 (2004).

[33]   J.W.P. Schmelzer, A.S. Abyzov, and J. Möller, J. Chem. Phys. **121**, 6900 (2004).

[34]   J.W.P. Schmelzer and A.S. Abyzov, J. Chem. Phys. **136**, 107101 (2012).

[35]   V.G. Baidakov, *Explosive Boiling of Superheated Cryogenic Liquids* (WILEY-VCH, Berlin-Weinheim, 2007).

[36]   A.M. Gusak, *Diffusion-Controlled Solid-State Reactions* (WILEY-VCH, Berlin-Weinheim, 2010).

[37]   K.L. Ngai, *Relaxation and Diffusion in Complex Systems* (Springer, Heidelberg, 2011).

[38]   G.P. Johari and J. Khouri, J. Chem. Phys. **138**, 12A511 (2013).

[39]   G. Adam and J.H. Gibbs, J. Chem. Phys. **43**, 139 (1965).

[40]   G.P. Johari and M. Goldstein, J. Chem. Phys. **53**, 2372 (1970).

[41]   J.K. Vij and G. Power, J. Non-Crystalline Solids **357**, 783 (2011).

[42]   T. Hikima, M. Hanaya, and M. Oguni, Bull. Chem. Society Japan **69**, 1863 (1996).

[43]   T. Hikima, M. Hanaya, and M. Oguni, J. Molecular Structure **479**, 245 (1999).

[44]   O. Norimaru and M. Oguni, Solid State Communications **99**, 53 (1996).

[45]   M. Hatase, M. Hanaya, T. Hikima, and M. Oguni, J. Non-Crystalline Solids **307–310**, 257 (2003).

[46]   F. Paladi and M. Oguni, J. Phys. Condens. Matter **15**, 3909 (2003).

[47]   F. Paladi and M. Oguni, Phys. Rev. **B 65**, 144202 (2002).

[48]   T. Ichitsubo, E. Matsubara, T. Yamamoto, H.S. Chen, N. Nishiyama, J. Saida, and K. Anazawa, Phys. Rev. Lett. **95**, 245501 (2005).

[49]   T. Ichitsubo, E. Matsubara, H.S. Chen, J. Saida, T. Yamamoto, and N. Nishiyama, J. Chem. Phys. **125**, 154502 (2006).

[50]   T. Ichitsubo, E. Matsubara, J. Saida, H.S. Chen, N. Nishiyama, and T. Yamamoto, Advances Materials Science **18**, 37 (2008).

[51]   G.P. Johari, Annals N.Y. Academy Sciences **279**, 117 (1976).

[52]   J.D. Bernal, Proc. Royal Society, London **A280**, 299 (1964).

[53]   J.M. Ziman, *Models of disorder, the theoretical physics of homogeneously disordered systems* (Cambridge Univ. Press, N.Y., 1982).

[54]   W.W. Graessley, J. Chem. Phys. **130**, 164502 (2009).

[55]   R. Böhmer, R.V. Chamberlin, G. Diezemann, B. Geil, A. Heuer, H. Hinze, S.C. Kuebler, R. Richert, B. Schiener, H. Sillescu, H.W. Spiess, U. Tracht, and W. Wilhelm, J. Non-Crystalline Solids **235**, 1 (1998).

[56]   U. Tracht, M. Wilhelm, A. Heuer, H. Feng, K. Schmidt-Rohr, and H.W. Spiess, Phys. Rev. Lett. **81**, 2727 (1998).

[57]   S.A. Reinsberg, X.H. Qiu, M. Wilhelm, H.W. Spiess, and M.D. Ediger, J. Chem. Phys. **114**, 7299 (2001).

[58]   S.A. Mackowiak, L.M. Leone, and L.J. Kaufman, Phys. Chem. Chem. Phys. **13**, 1786 (2011).

[59]   L. Berthier, G. Biroli, J.-P. Bouchaud, W. Kob, K. Miyazaki, and D.R. Reichman, J. Chem. Phys. **126**, 184503 (2007).

[60]   L. Berthier, Physics **4**, 42 (2011).

[61]   D. Coslovich and C.M. Roland, J. Chem. Phys. **131**, 151103 (2009).

[62]   C.M. Roland, D. Fragiadakis, D. Coslovich, S. Capaccioli, and K.L. Ngai, J. Chem. Phys. **133**, 124507 (2010).

[63]   C.T. Moynihan and J. Shroeder, J. Non-Crystalline Solids **160**, 52 (1993); **161**, 148 (1993).

[64]   R. Zangi, S.A. Mackowiak, and L.J. Kaufman, J. Chem. Phys.**126**, 104501 (2007).

[65]   S.A. Mackowiak, J.M. Noble, and L.J. Kaufman, J. Chem. Phys. **135**, 214503 (2011).

[66]   G.P. Johari and J. Khouri, J. Chem. Phys. **137**, 104502 (2012).

[67]   N.A. Bokov, J. Non-Crystalline Solids **17**, 74 (1994).

[68]   N.A. Bokov, J. Non-Crystalline Solids **354**, 1119 (2008).

[69]   D. Tatchev, A. Hoell, R. Kranold, and S. Armyanov, Physica **B 369**, 8 (2005).

[70]   J.W.P. Schmelzer and A.S. Abyzov, *Generalized Gibbs' Approach to the Thermodynamics of Heterogeneous Systems and the Kinetics of First-Order Phase Transitions*, in *Nucleation Theory and Applications*, Eds. J.W.P. Schmelzer, G. Röpke, and V.B. Priezzhev (Joint Institute for Nuclear Research Publishing Department, Dubna, Russia, 2005).

[71]   J.W.P. Schmelzer and A.S. Abyzov, J. Engineering Thermophysics **16**, 119 (2007).

[72]   W. Kauzmann, Chem. Rev. **43**, 219 (1948).

[73]   J.D. Stevenson and P.G. Wolynes, J. Phys. Chem. **A 115**, 3713 (2011).

[74]   M. Goldstein, J. Chem. Phys. **51**, 3728 (1969).

[75]   D.J. Wales, *Energy Landscapes: With Applications to Clusters, Biomolecules, and Glasses* (Cambridge University Press, New York, 2003).

[76]   A. Heuer, J. Phys.: Condens. Matter **20**, 373101 (2008).

[77]   F. Sciortino, J. Statistical Mechanics **2005**, P05015, May 2005, on line.

[78]   I. Gutzow, J. Schmelzer, and A. Dobreva, J. Non-Crystalline Solids **219**, 1 (1997).

[79]   Ya. B. Zeldovich, Sov. Phys. JETP **12**, 525 (1942); Acta Physicochim. USSR **18**, 1 (1943).

[80]   I. Gutzow, D. Kashchiev, and I. Avramov, J. Non-Crystalline Solids **274**, 208 (1985).

[81]   I. Gutzow, A. Dobreva, and J. Schmelzer, J. Materials Science **28**, 890, 901 (1993).

[82]   F. Fujara, B. Geil, H. Sillescu, and G. Fleischer, Z. Phys. **B 88**, 195 (1992).

[83]   M.S. Arzhakov, International Journal Polymer Materials, Biological Materials **39**, 289 (1998).

[84]   S.T. Reis, Cheol-Woon Kim, R.K. Brow, and C.S. Ray, J. Materials Science **39**, 6539 (2004).

[85]   D.S. Sanditov, Polymer Science **A 49**, 549 (2007); Original text published in Vysokomolekulyarnye Soedineniya, Ser. A, **49**, 832 (2007).

[86]   J.W.P. Schmelzer (Ed.), *Nucleation Theory and Applications* (WILEY-VCH, Berlin-Weinheim, 2005).

[87]   E. Donth, *The Glass Transition: Relaxation Dynamics in Liquids and Disordered Materials* (Springer, Berlin, 2001).

[88]   G.P. Johari, J. Chem. Phys. **58**, 1766 (1973).

[89]   K.L. Ngai and M. Paluch, J. Chem. Phys. **120**, 857 (2004).

[90]   A.A. Burbelko, E. Fra's, and W. Kapturkiewicz, Materials Science Engineering **A 429**, 413 (2005).

[91]   H.E. Kissinger, Analytical Chemistry **29**, 1702 (1957).

[92]   R.L. Blaine and H.E. Kissinger, Thermochimica Acta **540**, 1 (2012).

[93]   L.V. Meisel and P.J. Cote, Acta Metallurgica **31**, 1053 (1983).

[94]   H. Yinnon and D.R. Uhlmann, J. Non-Crystalline Solids **54**, 253 (1983).

[95]   A.L. Greer, Acta Metallurgica **30**, 171 (1982).

[96]   H. Lu and S. Nutt, J. Applied Polymer Science **89**, 3464 (2003).

[97]   H.S. Chen, J. Non-Crystalline Solids **27**, 257 (1978).

[98]   D. Turnbull and B.G. Bagley, *Treatise on Solid State Chemistry* (Plenum Press, New York, 1975, vol. 5, p. 513).

[99]   I. Gutzow, R. Pascova, A. Karamanov, and J. Schmelzer, J. Materials Science **33**, 5265 (1998).

[100] M.J. Davis, Glastechnische Berichte: Glass Science and Technology **73 C1**, 170 (2000); J. Amer. Ceramic Society **84**, 492 (2001).

[101] W. Ostwald, Z. Phys. Chemie **22**, 289 (1897).

[102] I.N. Stranski and D. Totomanov, Naturwissenschaften **20**, 905 (1932).

[103] I.M. Lifshitz and V.V. Slezov, J. Phys. Chem. Solids **19**, 35 (1961).

[104] G.P. Johari, J. Chem. Phys. **122**, 194504 (2005).

[105] P. Bérge and M. Dubois, Contemporary Physics **25**, 535 (1984).

[106] A.L. Tsekis, N.M. Kortsenstein, and J.W.P. Schmelzer, in *Nucleation Theory and Applications*, Eds. J.W.P. Schmelzer, G. Röpke, and V.B. Priezzhev (Joint Institute for Nuclear Research Publishing Department, Dubna, Russia, 2008).

[107] D.T.J. Hurle, G. Müller, and R. Nitsche, *Crystal Growth from the Melt*, Chapter 10, Fluid Science and Materials Science in Space, Ed. H.U. Walter (Springer-Verlag, Berlin, 1987).

[108] J. Leung, M.Sc.-thesis, 2009.

[109] S. Chandrasekhar, *Hydrodynamic and Hydromagnetic Stability* (Clarendon Press, Oxford, 1961).

[110] J.W.P. Schmelzer, G. Röpke, and V.B. Priezzhev, (Eds.) *Nucleation Theory and Applications*, 5 volumes (Joint Institute for Nuclear Research Publishing Department, Dubna, Russia, 1999, 2002, 2005, 2008, 2011); copies of the proceedings can be ordered via the author (Email: juern@theor.jinr.ru or via juern-w.schmelzer@uni-rostock.de).

[111] M. Rukeyser, *Willard Gibbs* (Doubleday, Doran & Company, New York, 1942).

[112] J.W.P. Schmelzer, J. Schmelzer Jr., and I. Gutzow, J. Chem. Phys. **112**, 3820 (2000).

[113] J.W.P. Schmelzer, G. Sh. Boltachev, and V.G. Baidakov, J. Chem. Phys. **124**, 194503 (2006).

[114] H. Ulbricht, J. Schmelzer, R. Mahnke, and F. Schweitzer, *Thermodynamics of Finite Systems and the Kinetics of First-Order Phase Transitions* (Teubner, Leipzig, 1988).

[115] K.F. Kelton, Solid State Physics **45**, 75 (1991).

[116] G. Sh. Boltachev and J.W.P. Schmelzer, J. Chem. Phys. **133**, 134509 (2010).

[117] J.W.P. Schmelzer, G. Sh. Boltachev, and A.S. Abyzov, J. Chem. Phys. **139**, 034702 (2013).

[118] A.S. Abyzov and J.W.P. Schmelzer, J. Chem. Phys. **138**, 164504 (2013).

[119] J.W.P. Schmelzer, A.S. Abyzov, J. Chem. Phys. **134**, 054511 (2011).

[120] J.W.P. Schmelzer and A.S. Abyzov, J. Non-Crystalline Solids **384**, 2 (2014).

[121] J.W.P. Schmelzer, *Dynamics of First-Order Phase Transitions in Multi-Component Systems: A New Theoretical Approach*, Proc. 20th International Conference on Glass (ICG), Kyoto, September 27 – October 1, 2004, Ref. O-08-002.

[122] V.M. Fokin, R.M.C.V. Reis, A.S. Abyzov, C.R. Chinagli, J.W.P. Schmelzer, and E.D. Zanotto, J. Non-Crystalline Solids **379**, 131 (2013).

[123] J.W.P. Schmelzer and V.G. Baidakov, J. Phys. Chem. **105**, 11595 (2001).

[124] J. Stefan, Wiedemanns Annalen der Physik und Chemie **29**, 655 (1886).

[125] M.Z. Faizullin, Fluid Phase Equilibria **211**, 75 (2003).

[126] V.P. Skripov and V.G. Baidakov, Teplofizika Vysokikh Temperatur **10**, 1226 (1972); English version is available in High Temperatures **10**, 1102 (1972).

[127] V.G. Baidakov, S.P. Protsenko, and A.O. Tipeev, J. Chemical Physics **139**, 224703 (2013).

[128] Monwhea Jeng, *Can hot water freeze faster than cold water?*,
http : //www.desy.de/user/projects/Physics/General/hot_water.html.

[129] Aristotle, in E.W. Webster, *Meteorologica I* (Oxford University Press, Oxford, 1923, pgs. 348b-349a); cited after [128].

[130] F. Bacon, 1620, Novum Organum, volume VIII, of *The Works of Francis Bacon*, ed. J. Spedding, R.L. Ellis and D.D. Heath (New York, 1869, pp 235, 337) quoted in T.S. Kuhn, *The Structure of Scientific Revolutions* (2nd edition, Chicago, University of Chicago Press, 1970, p. 16); cited after [128].

[131] R. Descartes, 1637, *Les Meteores* published with *Discours de la Methode* (Leyden, Ian Marie, 1637) quoted in *Oeuvres de Descartes*, volume VI, ed. Adam and Tannery (Paris, Leopold Cerf, 1902, p. 238, trans. F.C. Frank); cited after [128].

[132] E.B. Mpemba and D.G. Osborne, Physics Education **4**, 172 (1969).

[133] G.S. Kell, Amer. J. Physics **37**, 564 (1969).

[134] W.I. Bragg and E.J. Williams, Proc. Royal Society London A **145**, 699 (1934).

[135] Ya. I. Frenkel, *The Kinetic Theory of Liquids* (Oxford University Press, Oxford, 1946).

[136] P.P. Kobeko, *Amorphous Materials* (Academy of Sciences USSR, Moscow, Leningrad, 1933 (in Russian)).

[137] G.O. Jones, Rep. Prog. Phys. **12**, 133 (1949).

[138] G.M. Bartenev, Dokl. Akad. Nauk SSSR **69**, 373 (1949).

[139] G.M. Bartenev, Dokl. Akad. Nauk SSSR **76**, 227 (1951).

[140] G.M. Bartenev and I.A. Lukyanov, Zh. Fiz. Khimii **29**, 1486 (1955).

[141] H.N. Ritland, J. American Ceramic Society **37**, 370 (1954).

[142] J.W.P. Schmelzer and I.S. Gutzow, *Glasses and the Glass Transition* (WILEY-VCH, Berlin-Weinheim, 2011).

[143] M.V. Volkenstein and O.B. Ptizyn, Dokl. Akad. Nauk SSSR **103**, 795 (1955); Zh. Tekh. Fiziki **26**, 2204 (1956).

[144] C.T. Moynihan, A.J. Easteal, J. Wilder, and J. Tucker, J. Phys. Chem. **78**, 2673 (1974).

[145] C.T. Moynihan, A.J. Easteal, M.A. DeBolt, and J. Tucker, J. American Ceramic Society **59**, 12 (1976).

[146] O.V. Mazurin, Glass Phys. Chem. **33**, 22 (2007).

[147] A.R. Cooper Jr. and P.K. Gupta, Phys. Chem. Glasses **23**, 44 (1982).

[148] A.R. Cooper, Glastechnische Berichte **56**, 1160 (1983).

[149] A.R. Cooper, J. Non-Crystalline Solids **71**, 5 (1985).

[150] T.V. Tropin, J.W.P. Schmelzer, and C. Schick, J. Non-Crystalline Solids **357**, 1291, 1303 (2011).

[151] J.W.P. Schmelzer, J. Chem. Phys. **136**, 074512 (2012).

[152] M. Reiner, Phys. Today **17**, 62 (1964).

[153] V.M. Fokin, E.D. Zanotto, and J.W.P. Schmelzer, J. Non-Crystalline Solids **321**, 52 (2003).

[154] A.C. Wright, International Journal Applied Glass Science (2013), in press; *A Historical Introduction to the Constitution and Structure of Glass* (Society Glass Technology, Sheffield, in preparation).

Index

Amorphous modifications of silica 137–140, 148, 163, 164, 173, 189

Chemical structure of glass, 269, 277, 284, 291, 293, 296

Classical nucleation theory, 1, 75, 108, 116, 180, 183, 444, 455, 458, 466, 481, 482, 484, 518, 526, 560, 561, 565

Coesite 138, 140, 142–149, 159–169, 173, 176–178, 183, 185, 186, 190

Co-operatively rearranging regions, 521, 528–534, 579

Convection effects on crystallization, 10, 248, 257, 521, 524, 525, 534, 550–554, 580

Cristobalite, 137, 139, 140, 142–170, 172, 173, 176, 177, 179, 181, 183, 185, 187–191, 193

Critical clusters in nucleation, 61, 79, 180, 459, 481, 522, 526, 554–560, 562, 563, 565, 569

Crystallization, 1, 2, 5–11, 14, 16–54, 59–66, 68–75, 77–87, 89–92, 95–102, 104–124, 126, 137, 148, 153, 154, 157–159, 163, 179, 187, 188, 190, 191, 275, 276, 278, 280, 377–398, 400–439, 441–447, 449, 450, 453, 454, 457, 464, 466–468, 471–476, 478, 479, 481–484, 487, 491, 492, 497, 498, 503, 504, 506–511, 513, 515, 517, 518, 521, 522, 524–554, 560, 566, 567, 570–574, 578–581

Crystallization-induced porosity, 187, 335, 338–340, 344, 345, 467, 469–473, 475, 478

Deformation-induced nanocrystal formation, 126

Diffusion in glass-forming melts, 8, 77–79, 81, 84, 101, 104–109, 112, 115, 116, 121, 125, 126, 131, 224, 259, 260, 302, 309, 311, 312, 315, 317–328, 339, 346–349, 358, 360–373, 375, 384, 390, 392, 394, 399, 411, 422–425, 428–430, 438, 439, 442, 481, 486, 493, 504, 508, 510, 512, 513, 518, 521, 523, 525, 527–529, 531–535, 540, 541, 546, 550, 551, 562, 579, 580

Domains of heterogenous dynamics, 521, 530–534

Elastic stress effects, 79, 441–445, 451, 453–455, 466, 534

Enthalpy relaxation, 2, 51–62, 64, 66–71, 73–75, 77, 78, 81

Fast scanning calorimetry, 1, 2, 6, 9, 80, 84, 89

Glass transition, 1–3, 5–8, 11, 14, 16, 17, 19, 22–27, 36, 37, 39, 42, 45–47, 49–56, 58–60, 62, 65, 67, 69, 71, 73–77, 79–85, 88, 90, 95–97, 100, 102, 103, 105, 110–113, 116–119, 121–123, 127, 128, 131, 165, 184, 207, 266, 326, 327, 445, 462, 471, 532, 533, 537, 549, 574–578

Heterogeneous nucleation, 6, 20, 23, 24, 28–30, 38, 43, 74, 79, 84, 481, 482, 517, 528, 530, 541, 559, 561

Homogeneous nucleation, 1, 2, 6, 7, 13, 16, 20, 23, 25, 30, 32, 33, 35, 36, 40, 43, 45, 47, 49, 52, 62, 64, 74, 77–81, 84, 86, 87, 100, 441, 481, 482, 506, 508, 510, 512, 517, 527, 528, 530, 533, 540, 559, 560

Hydrothermal synthesis, 137, 139, 140, 155, 162, 177, 178, 183, 186, 189, 191, 197, 206, 222, 246–249, 251, 255–258

Isothermal ordering kinetics, 2, 3, 6, 7, 9, 11, 16, 17, 19–22, 28, 33, 35–37, 42, 43, 45–47, 49–51, 53, 56, 59, 61, 65, 66, 68, 69, 71, 72, 78, 84, 86, 88, 92, 99, 100, 102, 106, 108, 109, 116–118, 120, 121, 246, 247, 251, 310, 333, 335, 340, 343–345, 370, 373, 478, 485, 486, 517, 521, 534, 536, 537, 543–547, 549, 557, 578, 580

Johnson–Mehl–Avrami–Kolmogorov kinetics, 7, 39–42, 46, 91, 107, 478, 521, 525, 541, 542, 544, 582

Keatite, 138, 140, 142, 143, 148, 149, 159, 167–169, 173

Kinetic criteria for glass-formation, 574–577

Metallic alloys, 95–98, 102, 525

Modifications of silica, 137–146, 148–151, 153–172, 179, 183, 184, 186, 189, 190, 191, 202, 325, 405

Molecular dynamics studies, 6, 77, 80, 214, 219, 245, 481–484, 487, 495, 503, 508, 516–518, 540, 557

Non-isothermal ordering kinetics, 11, 17, 21, 35, 49, 59, 66, 84, 310, 335, 478, 544

Nucleation of pores, 441

Phase diagram of silica, 137–139, 141–148, 177, 192, 193

Phase selection, 441

Phases of silica, 146, 160, 176, 209, 255

Polyamorphism, 197, 224, 225, 228

Quartz, 137–140, 142–151, 154, 159–179, 183, 185–191, 193, 194

Quartz glass, 145, 149, 150, 153, 157, 158, 164, 182, 183, 185, 190, 191, 194

Relaxation, 1, 2, 4, 7, 16, 41, 42, 51–62, 64, 66–71, 73–75, 77–79, 81, 87, 88, 92, 93, 101, 103, 110, 111, 117, 118, 120, 127–129, 131, 348, 372, 442, 455, 528–532, 534, 540, 541, 546, 574–577, 579

Rigid amorphous fraction, 29, 39, 49, 53, 76, 79, 81–83

Sinter-crystallization, 441, 442, 444, 467, 472, 473, 475, 476

Sources of bubbles in glass, 304

Stishovite, 138, 139, 142–149, 159–169, 173, 175–178, 186, 188, 189, 194, 195

Stokes-Einstein relation, 78, 79, 533

Structural order-parameter, 1–5

Structure-property relationships, 269, 270, 273, 297

Surface nucleation, 377, 423

Thermal history, 1, 5, 17, 23, 37, 59–64, 67, 68, 70–72, 312, 313, 521, 545, 578

Thermodynamic properties of silica, 137, 139, 141, 142, 144, 146, 164, 167–169, 176, 178, 192, 193

Tridymite, 137, 138, 142, 143, 147, 149, 159, 165–170, 173, 175, 193

Ultraviscous melts, 521, 527–529, 532, 533, 535, 545, 552, 553, 579, 580

Viscosity, 16, 19, 78–81, 84, 96, 97, 99, 115, 121, 131, 302, 303, 312–314, 324, 326–328, 331, 334, 335, 338, 340–342, 344, 345, 347, 348, 370, 372, 376–379, 384, 399, 426–434, 436, 439, 442, 456, 457, 464, 467, 468, 472, 486, 507, 508, 513, 524, 525, 527, 528, 530–534, 540, 541, 546, 547, 551–553, 578